Nanoencapsulation Technologies for the Food and Nutraceutical Industries

Nanoencapsulation Technologies for the Food and Nutraceutical Industries

Edited by

Seid Mahdi Jafari

Gorgan University of Agricultural Sciences and Natural Resources, Gorgan, Iran

ELSEVIER

ACADEMIC PRESS
An imprint of Elsevier

Academic Press is an imprint of Elsevier
125 London Wall, London EC2Y 5AS, United Kingdom
525 B Street, Suite 1800, San Diego, CA 92101-4495, United States
50 Hampshire Street, 5th Floor, Cambridge, MA 02139, United States
The Boulevard, Langford Lane, Kidlington, Oxford OX5 1GB, United Kingdom

Notices

Knowledge and best practice in this field are constantly changing. As new research and experience broaden our understanding, changes in research methods, professional practices, or medical treatment may become necessary.

Practitioners and researchers must always rely on their own experience and knowledge in evaluating and using any information, methods, compounds, or experiments described herein. In using such information or methods they should be mindful of their own safety and the safety of others, including parties for whom they have a professional responsibility.

To the fullest extent of the law, neither the Publisher nor the authors, contributors, or editors, assume any liability for any injury and/or damage to persons or property as a matter of products liability, negligence or otherwise, or from any use or operation of any methods, products, instructions, or ideas contained in the material herein.

Library of Congress Cataloging-in-Publication Data
A catalog record for this book is available from the Library of Congress

British Library Cataloguing-in-Publication Data
A catalogue record for this book is available from the British Library

ISBN: 978-0-12-809436-5

Publisher: Nikki Levy
Acquisition Editor: Nina D. Bandeira
Editorial Project Manager: Mariana Kühl Leme
Production Project Manager: Julie-Ann Stansfield
Designer: Greg Harris

Typeset by Thomson Digital

To my beloved mother and father. Thanks for
your kindness and devotion.
To my love, Elham. Thanks for your endless support.
To my fabulous son, Amir Reza and my beautiful daughter, Elina.

Be happy for this moment. This moment is your life.

—Omar Khayyám, Persian poet and mathematician (1048–1131)

Khayyam monument on his grave, Neyshabour, North East of IRAN

Contents

Front cover figure description

The prepared illustration on the book cover exhibits a nanospray drier in which nanoparticles are achieved via spray drying by which particle sizes can be reduced to 300 nm and yields up to 90% are possible, moreover the sample amount can be as low as 1 mL. The bioactive compounds within the foods illustrated above this device include: oleurupein for olive, lycopene for tomato, ascorbic acid (vitamin C) for berries, quercetin for apple, and safranal for saffron, which is then entrapped in the capsule. The capsule next to the device represents the process of encapsulation in which our bioactive compound (safranal) is incorporated within the nanocapsule. The scientific name of the flower in the first place of the capsule is *Crocus sativus*, commonly known as the "saffron crocus." The name saffron comes from the Persian word "za'faran"; it is this famous pistil made of three stigmas (filaments) which when dried up gives the spice saffron. To date the undisputed center for saffron production is Iran, where the production process dates back to over 3000 years ago. Iran manufactures over 90% of the 250 tons produced globally annually, which has the potential to be used in food, cosmetics, and traditional medicine areas. In the second place the chemical structure of safranal is illustrated, next the spiral structure of amylose molecule is evident, which can be used as the host molecule to incorporate guest molecules within it. Lastly, the 3D form of safranal is represented and included in the capsule as the core material.

List of contributors

Sahar Akhavan Gorgan University of Agricultural Sciences and Natural Resources, Gorgan, Iran

Ali Alehosseini Research Institute of Food Science and Technology (RIFST), Mashhad, Iran

Cordin Arpagaus NTB University of Applied Sciences of Technology Buchs, Institute for Energy Systems, Buchs, Switzerland

Elham Assadpour Gorgan University of Agricultural Sciences and Natural Resources, Gorgan, Iran

Wolfgang Brysch Athenion GmbH, Berlin, Germany

Muhammed Y. Caglar Istanbul Sabahattin Zaim University, Istanbul, Turkey

Bilal Cakir Istanbul Sabahattin Zaim University, Istanbul, Turkey

Andreas Collenberg BÜCHI Labortechnik AG, Flawil, Switzerland

Mehmet Demirci Istanbul Sabahattin Zaim University, Istanbul, Turkey

Afshin F. Esfanjani Ferdowsi University of Mashhad, Mashhad, Iran

Mohammad Ganje Gorgan University of Agricultural Sciences and Natural Resources, Gorgan, Iran

Seyed M.T. Gharibzahedi Islamic Azad University, Tehran, Iran

Fatemeh Ghiasi Shiraz University, Shiraz, Iran

Behrouz Ghorani Research Institute of Food Science and Technology (RIFST), Mashhad, Iran

İbrahim Gülseren Istanbul Sabahattin Zaim University, Istanbul, Turkey

Anilda Guri Canadian Research Institute for Food Safety, Guelph, ON, Canada

Sanaz Haratifar Canadian Research Institute for Food Safety, Guelph, ON, Canada

Seyed M.H. Hosseini Shiraz University, Shiraz, Iran

Seid Mahdi Jafari Gorgan University of Agricultural Sciences and Natural Resources, Gorgan, Iran

Mastaneh Jahromi Shiraz University, Shiraz, Iran

Philipp John BÜCHI Labortechnik AG, Flawil, Switzerland

Mahsan Karimi Islamic Azad University, Kermanshah, Iran

Iman Katouzian Gorgan University of Agricultural Sciences and Natural Resources, Gorgan; Nano-encapsulation in the Food, Nutraceutical, and Pharmaceutical Industries Group (NFNPIG), Universal Scientific Education and Research Network (USERN), Tehran, Iran

Cornelia M. Keck Free University of Berlin, Berlin, Germany; PharmaSol GmbH, Berlin, Germany

Jozef Kokini Purdue University, West Lafayette, IN, United States

Ioanna Mandala Agricultural University of Athens, Athens, Greece

Mohammad A. Mehrnia Gorgan University of Agricultural Sciences and Natural Resources, Gorgan, Iran

Laleh Mehryar University of Urmia, Urmia, Iran

Rainer H. Müller Free University of Berlin, Berlin, Germany

Paraskevi Paximada Agricultural University of Athens, Athens, Greece

Sung-Min Pyo Free University of Berlin, Berlin, Germany

Hamid Rajabi Gorgan University of Agricultural Sciences and Natural Resources, Gorgan, Iran

Francisco Rodríguez-Félix University of Sonora, Sonora, Mexico

Gregori B. Romero Free University of Berlin, Berlin, Germany

David Rütti BÜCHI Labortechnik AG, Flawil, Switzerland

Rohollah Sadeghi Islamic Azad University, Kermanshah, Iran; Purdue University, West Lafayette, IN, United States

José A. Tapia-Hernández University of Sonora, Sonora, Mexico

Nick Tucker University of Lincoln, Lincoln, United Kingdom

Preface

Enthusiasm for the consumption of healthy and functional food products has dramatically expanded with the growth of industrial life and obesity among people. Therefore, many researches have focused on novel topics; such as, nanotechnology in preparing healthy cuisine. Nanotechnology is a new field of science combining different fields of technology in general and food in particular. Encapsulation can be defined as the technology of encasing nutrients in solid, liquid, or gaseous states in matrices, which can be released under particular circumstances with a controlled rate. Recently, according to perception of material properties and their reaction at the nanoscale research, the encapsulation area has moved to the nanoencapsulation field. The fabricated nanocapsules provide better opportunity for interaction, high bioavailability, solubility, and permeation due to their larger surface area. Also, nanoencapsulated ingredients enable targeting release plus high stability against harsh digestive processes and environment stresses. Therefore, selecting the best method for nanoencapsulation of distinct sensitive ingredients is the main step for designing an efficient delivery system in healthy food and functional products.

Considering the different techniques applied for fabricating nanoscale capsules (10–1000 nm), there is not a comprehensive and clear classification of nanoencapsulation techniques applicable to food bioactive ingredients and nutraceuticals. According to this issue, for the first time in this book, we have classified nanoencapsulation technologies into five groups based on the main mechanism/ingredient, which is being employed to produce nanocapsules. This classification can facilitate the study and research activities on nutraceutical-loaded nanocapsules and using them in food products and biological systems with the aim of fortification.

This book is divided into five parts: After giving an overview of nanoencapsulation techniques plus their classification (Chapter 1), Part I explains lipid formulation–based nanoencapsulation technologies including: nanoemulsions (Chapter 2), nanoliposomes (Chapter 3), and nanostructured lipid carriers (Chapter 4). Part II describes nanoencapsulation technologies based on natural nanocarriers like caseins (Chapter 5), nanocrystals (Chapter 6), and cyclodextrins (Chapter 7).

The fast and economical techniques for nanoencapsulation of bioactive compounds based on special equipment are described in Part III comprising electrospinning (Chapter 8), electrospraying (Chapter 9), and nanospray dryer (Chapter 10). Part IV underlines nanoencapsulation technologies based on biopolymer nanoparticles including singular biopolymer nanoparticles (Chapter 11) and biopolymer complexion (Chapter 12).

In the final section (Part V), bioavailability, characterization, and safety of nanoencapsulated ingredients is discussed; particularly the fate and release profile of bioactive compounds (Chapter 13), analytical and instrumental methods for characterizing nanocapsules (Chapter 14), plus safety and regulatory issues of nanocapsules (Chapter 15).

This book would be useful for a diverse group of readers including food technologists, food engineers, nanotechnologists, nutritionists, food colloid experts, pharmacists, physicists, chemists, microbiologists, biotechnologists, and those who are interested in novel technologies in the area of food. I believe that this book will clarify some concerns and present the latest techniques in this novel technology in a practical and pragmatic style.

Finally, I would like to appreciate all the contributors for sharing their vast knowledge to both the researchers and those who are interested in the area of nanoencapsulation. We also thank the staff of Elsevier for their fruitful cooperation and patience during the preparation and publication task.

Seid Mahdi Jafari
Iran

An overview of nanoencapsulation techniques and their classification

1

Seid Mahdi Jafari
Gorgan University of Agricultural Sciences and Natural Resources, Gorgan, Iran

1.1 Introduction

Today, it is possible to fabricate functional and palatable foods by engineering several types of encapsulation systems. In such formulations, the bioactive compounds are encased by a substance or a combination of several materials termed as the coating or wall material. In these systems, the stability of the nutrients are enhanced as well as influencing the physicochemical properties compared to nonencapsulated ingredients. Moreover, the food industry directly influences the public health and therefore it is necessary to produce healthier foodstuff and expedite healthy lifestyles. The world population is expected to exceed 1.5 billion; especially the older generation will constitute most of the world's population. It is notable that in 1950, this proportion was only 5% (Haub, 2011). In recent years there has been a strong interest to mass produce functional and healthy food products by implementing encapsulated nutraceuticals so as to provide advantages for the consumers.

Nutraceuticals are the link between nutrition and medicine. This term was first used by DeFelice in 1979 and defined as "food ingredients that have health benefits and inhibit the advancement of diseases" (DeFelice, 1995). Recently, the application of different nutraceutical compounds, such as essential fatty acids (omega 3), carotenoids (β-carotene and lycopene), vitamins (D, thiamin, riboflavin), antioxidants (tocopherols, flavonoids, polyphenolic compounds), phytosterols (stigmasterol and β-sitosterol), dietary fibers (inulin), minerals (Fe^{+2}, Mg^{+2}), and bioactive peptides (casein hydrolysates) has attracted the attention of food scientist and industries for developing enriched healthy foods (Jafari and McClements, 2017). Most of the nutraceuticals are sensible to decomposition during processing and storage after being incorporated into the food structures. Bioactive compounds are typically introduced into foods using different types of encapsulation (delivery) systems (Shimoni, 2009). Targeted delivery is one of the most important issues in the fields of encapsulation in pharmaceutical and food science. The goal of encapsulating nutraceuticals is to reduce the damage and undesirable changes through processing stages as well as digestion conditions. Hence, developing an appropriate carrier system is quite necessary (Fuchs et al., 2010; Katouzian and Jafari, 2016).

Before going through the nanoencapsulation techniques, it is worthwhile to have a quick look at encapsulation (micro) process and its development in the food industry. So, in the following sections, a brief review of encapsulation terms and definitions has been provided along with the properties of shell and bioactive compounds and typical encapsulation methods. Then, we will cover the nanoencapsulation techniques applicable to the food and nutraceutical industries plus their classification to make the foundation of following chapters.

1.2 Encapsulation in the food sector

Encapsulation is the technology of encasing substances in solid, liquid, or gaseous states in matrices, which can release their payload under certain conditions at a controlled rate (Nedovic et al., 2011; Zuidam and Shimoni, 2010). Encapsulation was first introduced in the biotechnology science to optimize the production processes and induce the decomposition of generative cells plus their intermediate products. This technology has many applications in the pharmaceutical industries; nevertheless, it is also employed in the food industry (Barbosa-Cánovas et al., 2005). According to Jizomoto et al. (1993), the encapsulation concept emanated from the cell model in which the genetic information included in the nucleus is protected from the external medium and a semipermeable membrane controls the transportation of several agents. As mentioned before, the aim of encapsulation is mainly protecting the labile and sensitive bioactive agents from the undesirable circumstances.

Considering the uncomplicated form, a capsule is composed of a tiny sphere surrounded by a monolithic shell. The ingredient inside the capsule is ascribed to the core material, internal phase, encapsulant, payload phase, or fill, while the wall is sometimes called the shell, coating, wall material, membrane shell, carrier material, encapsulating agent, external phase, or matrix (Jafari et al., 2008; Mahdavi et al., 2014). Micro/nanocapsules can be classified into two basic categories of single-core or multicore and each one in a form of single-wall or multiwall is shown in Fig. 1.1A. For nanocarriers in particular, two structures of nanosphere and nanocapsule have been mentioned which can be seen in Fig. 1.1B.

Encapsulation technology preserves the nutrients against the unpredicted conditions until the controlled release procedure is fulfilled (Zuidam and Nedovic, 2010). The growing interest in the field of microencapsulation is evident according to the numerous published articles annually. Microencapsulation has the potential to mask the undesirable taste of some bioactive compounds in food formulations and also to simply change liquid phases into solid ones. However, lately, more attention has been given to the issue of controlled release. The controlled release procedure increases the reaction of bioactive compounds in the living system and ensures the optimal uptake of nutrients (Ray et al., 2016; Risch and Reineccius, 1995). Generally, in the food industry, the microencapsulation process has several advantages which are listed as follows (Barbosa-Cánovas et al., 2005; Nedovic et al., 2011; Zuidam and Nedovic, 2010):

- Protection of sensitive food ingredients from the ambient and processing conditions with the aim of extending their shelf life.
- Controlled release of encapsulated ingredients.

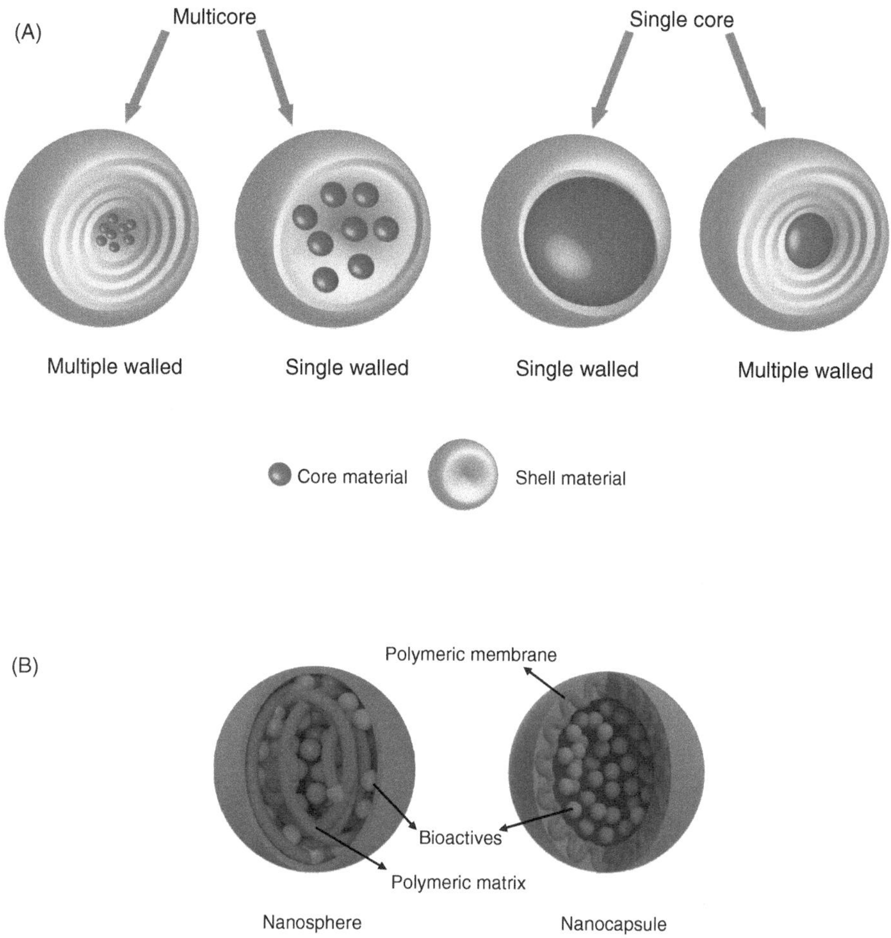

Figure 1.1 Schematic representation of (A) microcapsules morphology. (B) Nanosphere versus nanocapsule.

- Increasing the bioavailability of food ingredients.
- Modification of the physicochemical properties by converting solutions to fine powders and cutting down on the production costs as well as advancing easier handling of the bioactive component by improving flowability, preventing agglomeration and altering the particle density.
- Masking the unpleasant aromas and tastes, such as the undesirable taste of fish oil.
- Preventing the unnecessary interactions with other chemical structures, for example, microencapsulation of organic acids, such as citric acid and ascorbic acid to maintain the proper physicochemical properties and nutrient level of foods, along with encapsulation of choline chloride to limit the interactions with vitamin solutions.

Thereupon, nearly any food-grade component that should be gradually released or to be isolated from the ambient can be entrapped by encapsulation systems to enhance

Table 1.1 Overview of common food ingredients that can be encapsulated (core materials)

Food ingredients/ nutraceuticals	Purpose of encapsulation	Examples
Phenolic compounds and antioxidants	Stabilization, protection	Curcumin, oleorupein
Fish oil and essential fatty acids	Stabilization, controlled release	DHA, linolenic acid
Vitamins	Stabilization to oxidation	Fat soluble, water soluble
Antimicrobial agents	Controlled release	Nisin
Natural food colorants	Stabilization	Anthocyanins, carotenoids
Flavors and essential oils	Taste-masking, reduction of volatility, controlled release	Terpenes, lemon oil
Enzymes	Controlled release	Protease, glucose oxidase
Others	Stabilization, controlled release	Minerals, acids, alkalis, buffers, sweeteners, leavening agents, micronutrients

the quality of food networks. The types of food ingredients which can be encapsulated are presented in Table 1.1.

Human digestive tract has a linear form that might not be proper for the absorption of bioactive food components. Food ingredients can only be comfortably introduced by oral delivery and the processing of the food into a digestible form occurs in strict sequence thereafter. During chewing, the food network is influenced by the salivary amylase, which initiates the degradation of starches and yields simple sugars. Thereafter, the bolus is swallowed and transferred into the stomach, where digestion takes place in the acid solution (Jafari and McClements, 2017). The digested food then is absorbed gradually during its transfer to the small and large bowels. Through the passage along the digestive system the encapsulated nutrients are preserved from the undesirable conditions that cause their degradation in the body and liberate them in the determined zone to be absorbed (Katouzian and Jafari, 2016).

Several parameters should be considered before designing an encapsulation system, such as the capsule dimensions, final physical state and the expense. According to the function of the entrapped ingredients, the wall material should be formulated in a way that satisfies our needs. Furthermore, the physical conditions of the release environment are of utmost importance and should be incorporated in the simulation and design procedures (Jafari et al., 2015). The adequate level of the active ingredient, the type of release, the final particle size, density, and the resistance of the engineered microcapsules is also some of the crucial features that should be considered. Last but not least, the economic aspects are the most important items that determine the marketing of this novel method. Basically, according to the economic and technological viewpoints the shell of microcapsules must be resistant to withstand high shears and

Table 1.2 Some examples of biopolymers (wall materials) used in encapsulation of food ingredients

Polysaccharides	Fats and waxes	Proteins	Synthetic compounds
Gum Arabic Modified starches Maltodextrins Alginates Pectin Carrageenan Cellulose derivatives Chitosan Cyclodextrins Starch particles	Hydrogenated vegetable oils Bees wax Lecithin Medium chain triglycerides (MCT) Glyceryl behenate	Gelatin Whey proteins Sodium caseinate Soy proteins Gluten Caseins Zein Silk fibroin	Paclitaxel mPEG$_{5000}$-b-p Polyacrylonitrile Polycaprolactone Poly lactic acid

mechanical stresses along with the minimum thickness and high capacity for incorporating the core materials (Huang and Given, 2009; Lakkis, 2007).

Thus, an important step in formulating microcapsules is the fabrication of proper wall materials that meets the criteria mentioned previously (Jafari et al., 2008). Indeed, encapsulation efficiency and the durability of microcapsules rely on the wall material structure. This shell should act as a barrier, storing the core material as well as protecting it against oxygen, water, and light, or initiate a controlled diffusion in the specified spot of the body (Desai and Jin Park, 2005; Mahdavi et al., 2016). The characteristics of a proper shell in encapsulation systems include:

- stabilization of the payload;
- revealing no reactions with the bioactive compounds;
- starting a controlled release under the predetermined conditions;
- film-forming, flexible, stable;
- nonhygroscopic, low viscosity, cost-effective; and
- being able to be dissolved in an aqueous media or solvent to be molten.

The type of shells that form films is typically in the form of natural or synthetic polymers, concerning the structure of the coating material plus the desired conditions of the final microcapsules. A summary of the polymer wall materials, which are commonly used in the food sector is listed in Table 1.2.

1.3 Microencapsulation techniques

To develop an encapsulation system for a particular application, the most appropriate combination of elements and techniques must be selected. Jafari et al. (2015), and Gómez-Mascaraque et al. (2015) have suggested some of these key factors listed as follows.

1. Physical features like transparency or opacity, color, and so on.
2. Rheological properties and the flow behavior of the formulated structure.

3. Flavor and mouth feel of the product.
4. Stability: the chemical and physical stability of the encapsulated ingredients during the storage period.
5. Biological functionality: the release of bioactive components from the nanocarrier following ingestion at an apt site in the gastrointestinal tract.

Besides, the properties of the particles which are incorporated in the encapsulation systems are (Donsì et al., 2011; Khayata et al., 2012; McClements and Li, 2010; Sáiz-Abajo et al., 2013):

- digestibility,
- the chemical structure of particles,
- particle size distribution,
- charge density,
- particles' permeability, and
- resistance to unpleasant conditions, such as pH, temperature, and enzymes.

The degree resistance of the biopolymer components to the environment is another variable that significantly influences the extent and rate of the release (Chakraborty et al., 2009). Even though a number of different types of encapsulation systems have been developed and reported, still there is not a complete understanding of the main factors and thus an effective design of systems for specific applications is still somewhat of an innovation (Fathi et al., 2014). There are still concerns about the extent to which some very sensitive compounds, (proteins, enzymes, DNA, etc.) may be damaged by exposure to harsh conditions (Clarke and Jayasinghe, 2008).

Many encapsulation technologies are centered on making droplets from the core (in gas, liquid, or powder form). Next, the produced droplets are coated by the external agent via different physicochemical processes. Primarily, the encapsulation of bioactive agents comprises three stages:

1. development of the host molecule around the core material;
2. ensuring that undesired leakage does not occur; and
3. ensuring that undesired materials are not incorporated inside the capsule.

There are several traditional methods for microencapsulation of food ingredients but an individual method is not compatible with the whole core materials or product applications. As described before, the method of microencapsulation is determined according to the application and parameters, such as intended particle size, physicochemical characteristics of the external and internal phase, release mechanisms, total cost, etc. (Jafari et al., 2008; Sobel et al., 2014; Zuidam and Nedovic, 2010). As the encapsulated ingredients are normally in the liquid state, drying technologies are so common to obtain the stable powder form of the liquid phase. Different techniques, such as spray drying, spray chilling, spray cooling, extrusion, fluidized bed coating, centrifugal extrusion, freeze drying, coacervation, interfacial polymerization; molecular inclusion and so forth are commonly available to encapsulate active agents of food ingredients and nutraceuticals. Table 1.3 briefly outlines the conventional methods used to microencapsulate food ingredients.

Table 1.3 Overview of common food microencapsulation processes

Technology	Process steps	Particle size (μm)
Spray drying	1. Disperse or dissolve active in aqueous coating solution 2. Atomize 3. Dehydrate	10–400
Fluid bed coating	1. Fluidize active powder 2. Spray coating 3. Dehydrate or cool	20–200
Freeze or vacuum drying	1. Dissolve or disperse active agent and carrier material in water 2. Freeze the sample 3. Drying under low pressure 4. Grinding (option)	20–5000
Spray chilling/cooling	1. Disperse or dissolve active in heated lipid solution 2. Atomize 3. Cool	20–200
Melt injection	1. Melt the coating 2. Disperse or dissolve active in the coating 3. Extrude through filter 4. Cooling and dehydrating	200–2000
Melt extrusion	1. Melt the coating 2. Disperse or dissolve active in the coating 3. Extrude with twin-screw extruder 4. Cool	300–5000
Emulsification	1. Dissolve active and emulsifiers in water or oil phase 2. Mix oil and water phases under shear	0.2–5000
Coacervation	1. Prepare o/w emulsions with lipophilic active in oil phase 2. Mix under turbulent conditions 3. Induce three immiscible phases 4. Cool 5. Cross-link (optionally)	10–800
Encapsulation by rapid expansion of supercritical fluid (RESS)	1. Create a dispersion of active and dissolved or swollen shell material in supercritical fluid 2. Release the fluid to precipitate the shell onto the active	10–400
Liposome entrapment	1. Disperse lipid molecules in water, with active agent in lipid or water phase 2. Reduce size by high shear or extrusion 3. Remove free active (option)	10–1000
Inclusion complexation	1. Mix carrier, active, and water together 2. Incubate and dry if necessary	0.001–0.01
Preparation of microspheres via emulsification	1. Emulsify water with biopolymer in oil phase 2. Add gelling agent under shear	10–1000

1.4 Nanoencapsulation of food ingredients and nutraceuticals

Nanotechnology is able to revolutionize the food sector and human diet with its useful applications. The exclusive properties of nanoscale structures (10^{-9} m), being manipulated in the food network may modify the properties of food, such as texture, flavor, and other organoleptic features, processability and stability during storage.

The principle of nanotechnology stands on the fact that the physicochemical properties of materials are modified when they are reduced to the nanodimension, that is, when the particles are below approximately 1 μm in size (border length, compound-specific) (Müller and Keck, 2008). Within the pharmaceutical sciences, nanoparticles (NPs) are defined as particles with nanodimensions, that is, from a few nanometers to <1000 nm (<1 μm). In recent years, the regulatory authorities (e.g., FDA in the United States, EU commission in Europe) have introduced regulatory definitions for NPs and nanomaterials (EC, 2011; FDA, 2014). Unfortunately, these definitions are not uniformly enforced in all commercial areas. For example, in cosmetics, the EU regulation defines a population of particles as "nanoparticles" when 50% or more of the number distribution have a size of ≤100 nm (European Commission, 2009). In this case, the excipients need to be identified in a product by the addition "nano," for example, titanium dioxide (nano).

However, concerning the food applications, the European Food Safety Authority (EFSA) refers to NPs generally as engineered nanomaterial (ENM). EFSA adopts the nonspecific definition of an ENM as a material with the size of 1–100 nm in at least one dimension. If such a material is present in a product, it must go through a detailed risk assessment to prove that it is nonhazardous (EFSA, 2011).

In general, the encapsulated particles are classified to microcapsules (0.2–5000 μm), macrocapsules (larger than 5000 μm) and nanocapsules (smaller than 0.2 μm). There are numerous examples of naturally occurring nanoencapsulates like casein micelles in milk (<100 nm), mitochondria (500–10,000 nm), and viruses (10–300 nm).

Nanoencapsulation is a novel and practical branch of nanotechnology in the food industry. The term nanoencapsulation describes encapsulation on the nanometer scale using films, layers, or nanodispersions. The final capsule acts as a nanoscale shield for the food or nutraceutical molecules/ingredients. Often the active ingredient is in the molecule or nanoscale state. The major benefit is the induced homogeneity, leading to better encapsulation efficiency in addition to the improved physical and chemical properties (Khare and Vasisht, 2014; Sanguansri and Augustin, 2006). Nanoencapsulation is flourishing with a rapid pace in the food sector. In fact, it is possible to fabricate valued-added food products by employing the nanoencapsulation technology in the field of food and nutritional sciences (Matalanis et al., 2011). Several comprehensive articles have focused on the application of nanoencapsulation in different aspects of the food industry, such as enhancing the public health, supporting the safety of food products and designing the principles for the delivery of nutrients (Huang et al., 2010; Sozer and Kokini, 2009).

Considering the hydrophobic nature of some bioactive compounds, adding these materials to the liquid food and beverage systems are somehow challenging when it is compared to their addition in a solid-based system (Matalanis et al., 2011; Sagalowicz and Leser, 2010). Therefore, the payloads have to be stabilized in an aqueous network, unlike the procedure employed in the solid state. Spray-dried solid powders are not directly mixed with a liquid food product since the stability and integration of the wall material will be broken down and then the nutrients are liberated into the bulk phase (Sagalowicz and Leser, 2010). Therefore, delivering nutraceuticals in a liquid matrix demands planned and accurate encapsulation strategies. Since the utilization of food-grade delivery systems is under development and not very broad, finding suitable vehicles is more complicated for the food sector than other fields, such as pharmaceuticals, medical, or cosmetic applications (Sagalowicz and Leser, 2010).

A practical means to develop the fabrication of functional foods is the process of nanoencapsulating sensitive biological ingredients like antioxidants and vitamins (Jafari and McClements, 2017). Encapsulated ingredients can be formulated in a way to withstand the physical stresses in the digestive tract and deliver their payloads in a special site. Considering the semisolid and nonsolid food products, the reduction in the size of their network modifies their encapsulation potency without leaving any change in the sensory properties (López-Rubio and Lagaron, 2012). Besides, by reaching the nanoscale, highly controllable biochemical vehicles are obtained (López-Rubio and Lagaron, 2012). Meanwhile, the delivery rate scales directly with the particle size. Throughout the body, specific types of cells can absorb submicron NPs more efficiently (Ezhilarasi et al., 2013; Hughes, 2005). Larger particles tend to release their bioactive molecules more slowly and over longer time periods. In addition, the size reduction in the particles elevate the bioadhesive properties due to the increase in surface to volume ratio, which lastly heightens the bioavailability of the bioactive molecules by the prolonged transfer through the GI tract (Chen et al., 2006).

The food and beverage industry have invested greatly in the application of nanotechnology, which is a multibillion-dollar market expanding every year. Even though this approach is recently being employed in the food and beverage industries, it is assumed that it will expand globally (Blasco and Pico, 2011). It has been explained that the market of nanofood products encompasses the nanodelivery systems being applied to packaging materials and processed healthy foods (Chau et al., 2007). It is reported that over 400 companies are taking advantage of nanoscience to fabricate food products and packaging materials (Neethirajan and Jayas, 2011). Helmut Kaiser Consultancy has reported the growing trend of the nanofood market starting from a value of USD 2.6 billion in 2003 to the value of USD 20.4 billion in 2015. Accordingly, it can be predicted that this ascending trend will continue its growth in food and drink industry and companies should compete really hard to stay on top of this dynamic area (Titoria and Groves, 2015).

Many nanoencapsulation related papers have been published in the recent years (Cushen et al., 2012; Nazzaro et al., 2012; Neethirajan and Jayas, 2011). Nanocapsules have more advantages over micronsized capsules, which arise from their larger surface area and higher reaction rates resulting in higher solubility, enhanced bioavailability, and targeted liberation that differentiate them from the micro/macro counterparts (Anandharamakrishnan, 2014; Ezhilarasi et al., 2013). As an illustration, by implementing nanocarriers, a considerable amount of bioactive compounds are incorporated within the food network without changing its desirable sensory properties (McClements et al., 2009). In addition, the uptake rate is strongly related to the size of the carrier (Sabliov and Astete, 2008). In some cases, the absorption process only takes place if the carrier is under the submicron scale. Small particles can easily adhere to surfaces, also the charges and van der Waals forces significantly affect the release profile and can induce a sustained release and increase the transfer time within the digestive tract (Lafitte, 2008; Katouzian and Jafari, 2016; Khare and Vasisht, 2014; Weiss et al., 2009).

Generally, the major benefits of nanotechnology for food ingredient encapsulation are:

- As the available surface area increase, the bioavailability of food ingredients is also dramatically enhanced. This is of utmost importance for low-soluble flavor agents and/or low flavor and odor perception thresholds. Several bioactive compounds are slightly absorbed and metabolized rapidly following ingestion and therefore they will not reach the target organs to impart their beneficial effects. Nanocarriers are able to enhance the bioavailability of nutrients by manifolds by protecting them during digestion and imposing a programmed release during the digestion process.
- The nanodelivery systems are more reactive in relation to their micro- and macroscale counterparts due to their higher surface area.
- Inducing a targeted and programmed release and avoiding any unnecessary release in nontarget points.
- Improvement in solubility of poorly water soluble ingredients, for example, omega-3 fish oil solubilization by a micelle network.
- Optically transparent (important in beverage application): nano- and microemulsions that have oil droplets less than 100 nm are optically transparent.
- Higher bioactive retention in the processing time [volatile organic carbon (VOC) reduction] during spray drying.
- Closer to perfect molecular solution (homogeneity in structure, such as density), for example, molecular inclusion complexes based on amylose and cyclodextrins (CDs).
- Highly reactive encapsulated ingredients, for example, antimicrobials in nanoemulsion/microemulsion forms.
- Limiting the unessential reactions with other food components as well as inhibiting their degradation during and after consumption.

1.5 Nanoencapsulation techniques

Various nanoencapsulation techniques for food bioactive components and nutraceuticals have been studied in the last couple of years including nanoemulsions, nanostructured lipid carriers (NLCs), nanosuspensions, solid lipid nanoparticles (SLNs), nanosized liposomes, biopolymer NPs, and micelles made of proteins, polysaccharides, and their complexes or conjugates. These techniques yield nanoscale carriers (10–1000 nm).

From a classical viewpoint, nanoencapsulation technologies can be divided into two main approaches: "top-down" and "bottom-up." Regarding the "top-down" method, particle size is decreased during the encapsulation process; for example, by utilizing various mechanical forces, on the contrary in the "bottom-up" process, particle size is increased by methods; such as self-assembly. Furthermore, in some cases, a combination of both approaches has been used. This classification is not applicable nowadays and it is better to categorize nanoencapsulation techniques in more groups based on some other indices.

In this book for the first time, we have classified nanoencapsulation technologies into five groups based on the main mechanism/ingredient, which is being used to make nanocapsules. They include lipid formulations, natural nanocarriers, specialized equipment, functional biopolymeric NPs, and disparate techniques as shown in Table 1.4. In the following sections, we will discuss briefly different techniques used for the nanoencapsulation of food constituents. Much more details are provided in following chapters, each devoted to one nanoencapsulation technique. It should be mentioned that selection of a nanoencapsulation technology depends on several parameters, such as physicochemical features, required particle size, release type, delivery method, process cost, and so on.

1.5.1 Lipid-formulation nanoencapsulation technologies

Lipid-based nanoencapsulation systems are mostly used in the pharmaceutical and food industries and research programs (Assadpour et al., 2016a). Despite the major advantages of biopolymer nanocapsules, they cannot be mass produced due to the demand for complicated chemical and thermal processes that should be monitored permanently. On the contrary, water-insoluble nanocarriers have the possibility to be scaled up plus the potential of more encapsulation efficiency and low toxicity (Fathi et al., 2012).

Most of the bioactive compounds, such as hydrophobic vitamins, fatty acids, flavonoids, aromas, preservatives, etc. have hydrophobic natures. For encapsulating these lipophilic compounds, the addition of emulsifiers (typically, oil in water) are often required. In addition, the presence of digestible lipids simplifies the absorption of nutrients since it raises the content of mixed micelles needed to dissolve and transport hydrophobic materials (Jafari and McClements, 2017; Trevaskis et al., 2008). For this reason, numerous studies have been centered on the application of lipid-based nanocarriers (Tamjidi et al., 2013). Different categories of lipid-based nanocarriers have been reported and reviewed in the last decade. The main types which have been studied and developed in recent years include nanoemulsions, nanoliposomes, NLCs, and SLNs as shown in Fig. 1.2.

1.5.1.1 Encapsulation by nanoemulsions (Chapter 2)

Formulated nanoemulsions are either employed in the native form or dried via different techniques like spray or freeze drying to obtain the powder forms. The benefits of nanoemulsions concerning their high stability and bioavailability are extensively reported accordingly in the literature (Jafari et al., 2015). Nanoemulsions are considered as ideal vehicles for nanoencapsulation. They serve as nanocarriers in many different materials as they are capable of improving water solubility and preventing

Table 1.4 **Classification of nanoencapsulation technologies applicable to food ingredients and nutraceuticals**

Type of techniques	Nanocarriers	General description
Lipid formulations	Nanoemulsions	• Simple emulsions (O/W or W/O) or multiple emulsions. The dispersed phase droplet sizes are typically less than 500 nm. • The droplets are kinetically stable and need high energy input to obtain or low energy for microemulsions (<100 nm) • An appearance of milky to optically clear
	Nanoliposomes	• Spherical particles with a core containing water which is coated with a bilayer of lipids such as lecithin • Good carriers for both hydrophilic and lipophilic food bioactives
	NLCs	• Composed of solid lipids (SLNs) or binary mixture of solid lipid and spatially different liquid lipid as hybrid carrier • Average size of 10–500 nm
Natural nanocarriers	Caseins	• Self-assemblies of casein into nanostructures • Average size of 50–300 nm • Good for hydrophobic bioactives
	CDs	• Forming aggregates by self-assembly, a tendency that increases upon formation of inclusion complexes with lipophilic bioactives • Able to improve solubility of poorly water soluble compounds
	Nanocrystals	• Crystalline NPs with at least one diameter ≤100 nm • Able to provide special features including enhancement of saturation solubility, dissolution velocity, and adhesiveness to surface/cell membranes
Based on specialized equipment	Electrospun nanofibers	• Production possibility with an aqueous solution, at room temperature and without coagulation chemistry to produce matrices in nanorange • Production of nanofibers less than 1 μm • Vehicles for sustained and controlled release
	Electrosprayed NPs	• Very similar to electrospinning technique • Production of spheres from a polymer solution, by applying a high electric field and thereby encapsulating different nutraceuticals
	Nanospray dried particles	• Creating tiny droplets (before evaporation) in a size range of much smaller order of magnitude than in classical spray dryers • Capable of producing powder particles in the submicron size range with very narrow distributions and high formulation yields

Table 1.4 Classification of nanoencapsulation technologies applicable to food ingredients and nutraceuticals (*cont.*)

Type of techniques	Nanocarriers	General description
Biopolymer NPs	Single biopolymer NPs	• Colloidal structures which are assembled from different types of biopolymers • Production of protein NPs with desolvation technique • Preparation of carbohydrate (mainly starch) NPs via nanoprecipitation method
	Biopolymer nanocomplexes	• Nanostructures made with combination of biopolymers such as proteins and polysaccharides through various types of biopolymer–biopolymer associative interactions • Mainly are composed of protein anionic polysaccharide and protein cationic polysaccharide complexes
Others	Protein nanotubes	• Principally are produced from α-lactalbumin which is a fraction of whey proteins • Concentration of protein is critical and nanotubes are formed through enzymatic hydrolysis and calcium ion bridging
	Carbohydrate nanogels	• NPs composed of a hydrogel (for hydrophilic bioactives) or organogel (for hydrophobic compounds) • Most often are produced from synthetic polymers or biopolymers which are chemically or physically cross-linked such as alginate • The pores in nanogels can be filled with small molecules or macromolecules • Their properties, such as swelling, degradation and chemical functionality, can be controlled
	Amylose nanostructures	• Inexpensive, abundant, biocompatible, biodegradable, and nontoxic • Can be produced from high-amylose starches via different methods

CDs, Cyclodextrins; NPs, Nanoparticles; NLCs, nanostructured lipid carrier.

the degradation of bioactive food components. Oil-in-water (O/W) nanoemulsions are appropriate candidates for the incorporation of lipophilic bioactive agents. On the other hand, water-in-oil (W/O) emulsions can be also used to encapsulate hydrophilic compounds including most polyphenols (Mohammadi et al., 2016a) and vitamins (Assadpour and Jafari, 2017; Assadpour et al., 2017). Generally, the lipophilic compounds are protected in these emulsions since the oil phase entails nutritional oils. Recently, the application of double emulsions for nanoencapsulation of food bioactive components are being performed by food experts (Assadpour et al., 2016a,b; Esfanjani et al., 2015, 2017; Mehrnia et al., 2016, 2017; Mohammadi et al., 2016a,b).

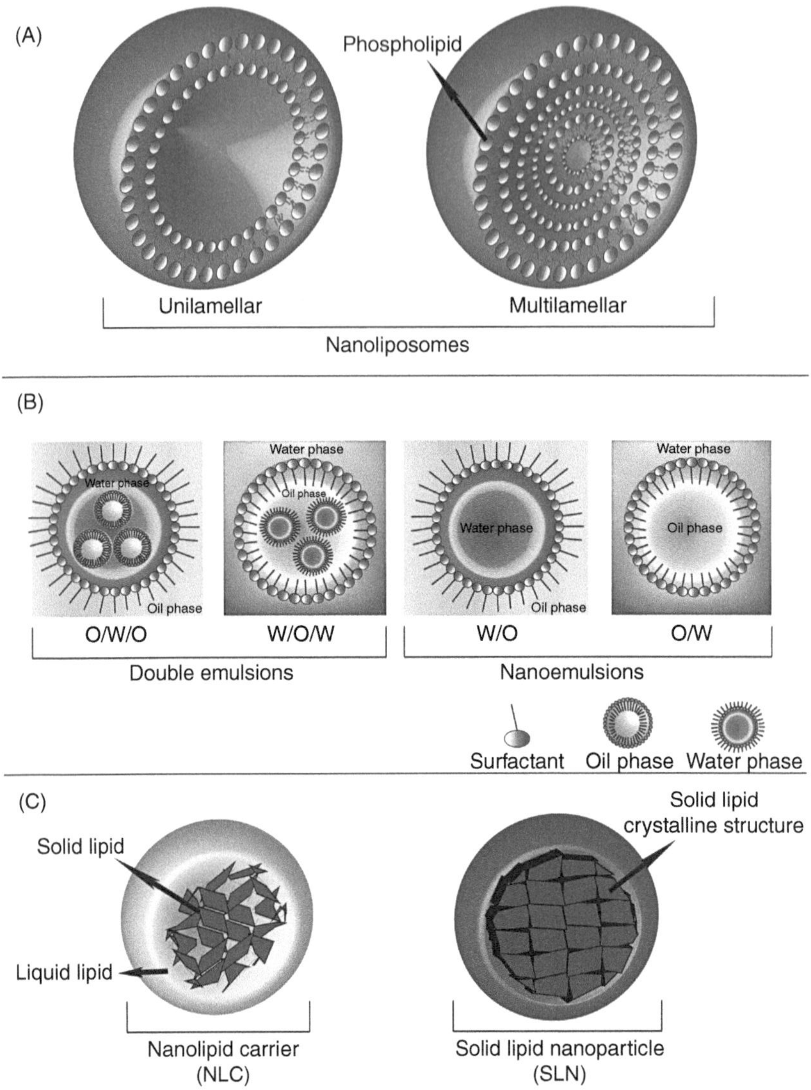

Figure 1.2 Schematic illustration of three important nanostructured lipid-based encapsulation formulations. (A) Nanoliposomes, (B) nanoemulsions, and (C) nanostructured lipid carriers *(NLCs).*

1.5.1.2 Encapsulation by nanoliposomes (Chapter 3)

Liposomes are vesicles composed of polar lipid bilayers, primarily of phospholipids, that surround an aqueous core. The formation of liposomes can be attributed to the self-assembly of amphiphilic lipids in aqueous solutions. Concentric packing of amphipathic lipids forms an interior aqueous phase, which acts as a reservoir for hydrophilic compound encapsulation (Briuglia et al. 2015). In addition, their lipid bilayers provide a relatively lipophilic environment compared to the liposomal core. Consequently, they are quite versatile in the sense that they have the potential to encapsulate and stabilize aqueous compounds in their aqueous core and hydrophobic compounds in the lipid bilayers, respectively. The nanoscale version of a liposome is named as nanoliposome. Although they have many physicochemical similarities to conventional liposomes, nanoliposomes bring along all the benefits of the nanocarriers mentioned earlier in this chapter (Ghorbanzade et al., 2017). The formation of a nanoliposome generates considerably more fresh surface areas which require an increased amount of energy input to generate nanoliposomes. In most of the recent applications, mean diameter range of nanoliposomes is referred to as approx. 50–150 nm.

1.5.1.3 Encapsulation by nanostructured lipid carriers (Chapter 4)

In comparison to other lipid-based nanocarriers; such as nanoliposomes and nanoemulsions, SLNs, and NLCs have recently drawn great attention from the food and pharmaceutical industries due to their advantages. SLNs are novel nanoparticulate vehicles developed to prolong the release period according to their solid state, protecting the active agents against unwanted chemical reactions like oxidation (Muller et al., 2000). SLNs contain fully crystallized and arranged lipid droplets incorporating the bioactive components in their lipid matrix (Weiss et al., 2008). SLN was designed as an alternative to the traditional lipid-based carriers by gathering the advantages of all the previous systems in its structure. They are composed of one solid lipid or mixture of diverse lipids (Müller et al., 2002). The solid matrix of the SLN is derived from O/W emulsions in which the liquid lipid (oil) is replaced by a rigid solid phase (i.e., maintains its solid structure in body temperature). Later, NLCs entered the nanoencapsulation field to cover the deficiencies found in SLNs. The main aim was to enhance the loading capacity and to inhibit the expulsion of bioactive compounds. They are produced by dispersion of solid and liquid lipids plus the active ingredients (as inner phase) in water containing emulsifiers (as outer phase). In contrast to SLNs, the presence of liquid lipid in the inner phase of NLCs causes the possibility of entrapping active ingredients which are better solubilized in liquid lipid. The encapsulated active ingredients are also protected from degradation by immobilization of them in solid lipid. On the other hand, the mixture of lipids in the NLCs has slower polymorphic transition and low crystallinity index. Thus, the composition of inner phase of the NLCs can provide higher loading volume, higher encapsulation efficiency, and higher bioavailability of nanoencapsulated ingredients in comparison to the SLNs (Müller et al., 2002).

1.5.2 Nanoencapsulation technologies based on natural nanocarriers

The idea of bioactives encapsulation using individual natural nanocarriers results from taking into account the nature-made functionalities of these NPs (Esfanjani and Jafari, 2016). As an example, casein micelles can naturally encapsulate and transport vital nutrients, amylose chains can bind a variety of flavor compounds developed during breadmaking, which can be released during subsequent heating, and starch derivatives (granules and CDs) can form inclusion complexes with many nutraceuticals. Nature-inspired nanocarriers have received great attention for the delivery of nutrients and drugs due to their safety and low cost. The functional properties of naturally occurring nanovehicles are characterized by their biological source, isolation methods, or physical and biochemical alterations. These tiny natural NPs such as caseins, nanocrystals, CDs, amylose nanostructures, and starch granules are employed in the nanoencapsulation of bioactive food ingredients and nutraceuticals (Fig. 1.3).

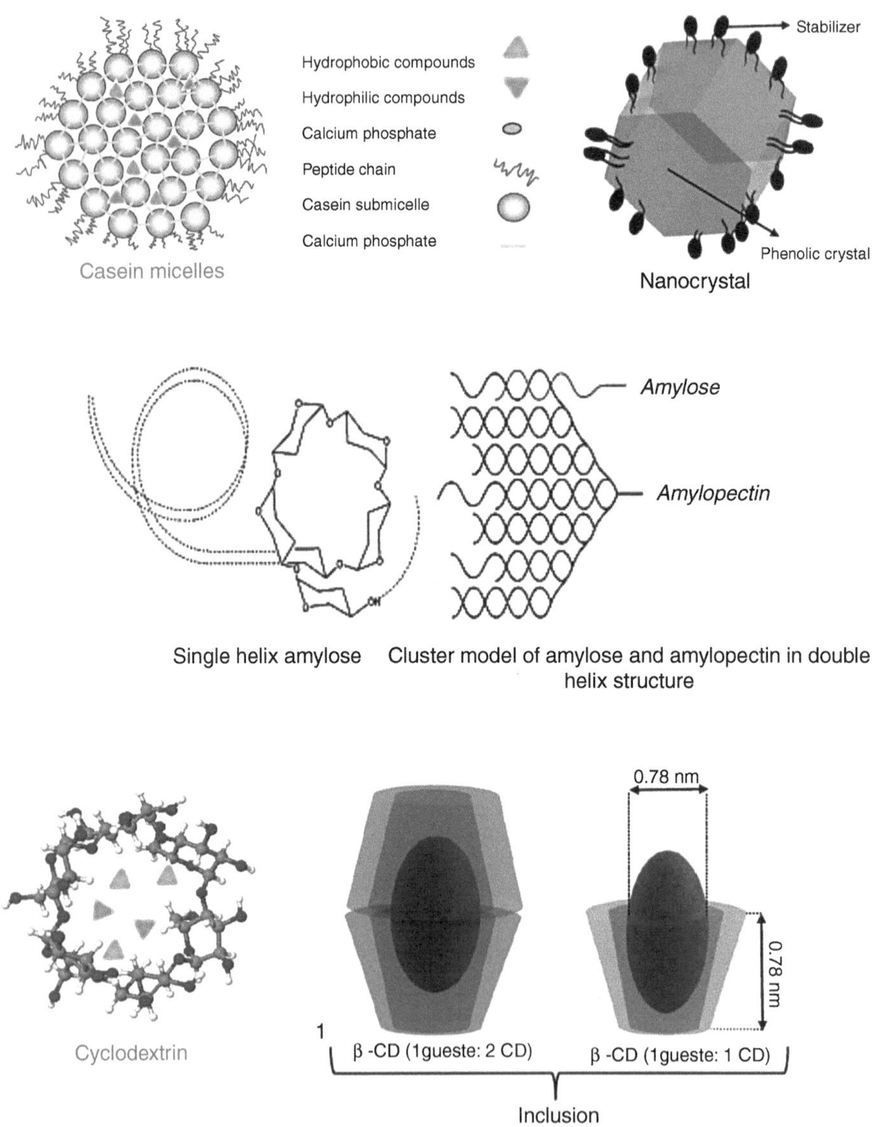

Figure 1.3 Schematic representation of natural nanocarriers.

1.5.2.1 Nanocapsule formation by caseins (Chapter 5)

Casein micelles are food-grade nanocarriers for encasing and delivering the unstable hydrophobic nutraceuticals that can be embedded in the foodstuff. The casein protein in bovine milk constitutes nearly 80% of the whole protein content (30–35 g/L) and is generally present as spherical micelles (5–500 nm). The naturally excellent digestibility of caseins and nanoscale micelles may develop the bioavailability of the encapsulated nutraceuticals as well as the stability of these micelles during processing. In fact, caseins have unique structural and functional properties which render them as suitable nanocarriers for bioactive components and nutraceuticals. Some of these properties are; binding with molecules and ions, binding with hydrophobic molecules, interactions with other proteins/polymers (e.g., covalent conjugations and noncovalent interactions), reassembly and self-assembly properties, gelation properties, and surface activity (Livney, 2015; Kimpel and Schmitt, 2015). The self-assembly of casein micelles due to the hydrophobic and electrostatic interactions renders them very suitable for incorporation of hydrophobic bioactive molecules. These highly nutritious naturally occurring protein structures are able to stabilize and transport essential nutrients. Considering the economical, nutritional, and safety issues, caseins and milk protein aggregates have been proved to be one of the most reliable nanovehicles (Tavares et al., 2014).

1.5.2.2 Nanocapsule formation by nanocrystals (Chapter 6)

Nanoencapsulation based on nanocrystals is presented for increasing the oral bioavailability since poorly soluble bioactives often have a low bioavailability, which may limit their effectiveness regarding the low doses used in formulated products. Nanocrystals have been described as any nanomaterial with at least one dimension $\leq$100 nm, which is single crystalline. As a matter of fact, the reduction in the size of formulations enhances the solubility rate and dissolution velocity. Nanocrystals could be made of pure bioactive ingredients. In the case of nanocrystals, the bioactives which are prone to degradation due to environmental/processing conditions, for example, pH, oxygen, light, temperature, can be coated with a protective polymer layer. This is also applicable to nanocrystals of bioactives, which are sensible to the different conditions in the GI tract after ingestion, for example, degradation in the stomach acid (Muller et al., 2013). Nanocrystals can be employed in functional drinks or nutraceutical products (tablets, capsules) and food formulations; examples are flavonoids of rutin and hesperidin, the antioxidant coenzyme Q10. Nowadays, nanocrystals are commercially available as water-based submicron (i.e., regulatorily no nanoproduct) particle suspensions (available preserved or nonpreserved), which are simply admixed during the production of the various products.

1.5.2.3 Nanocapsule formation by cyclodextrins (Chapter 7)

Molecular inclusion is generally achieved by using CDs as the encapsulating materials, which naturally have a nanoscale dimension. CDs are derived from starch molecules and comprise five or more α-D-glucopyranoside bodies, which form cyclic oligosaccharide structures. In general, CDs can be categorized into two main groups of natural and modified (Wang et al., 2014). Over 1500 CD subclasses have been recorded in different literatures. The typical types of CDs are α-, β-, and γ-CDs containing 6, 7, and 8 glucopyranose units, consecutively (Crini, 2014). The arrangement and stability of the CD-bioactive compound complex are related to the compatibility and shape fitness of the CD with the incorporated bioactive compound. Therefore, different types of CDs have different capacities to entrap the same bioactive compounds. The external structure of the CD molecules is hydrophilic and vice versa. These features make CDs a suitable medium for encasing molecules with less polarity (such as essential oils) within the nonpolar cavity through an interaction between water and hydrophobes (Astray et al., 2009). Many attempts have been made to enhance the functional properties of CDs, such as chemical modification of parent CDs which results in several derivatives with enhanced physiochemical characteristics and complexing potency (Loftsson and Duchêne, 2007). For instance, increased solubility level and optimized biocompatibility profile of hydroxyl propyl β-cyclodextrin (HP-b-CD) compared to typical CDs have been noted (López-Nicolás and García-Carmona, 2008).

1.5.2.4 Nanoencapsulation by amylose nanostructures

Different bioactive ingredients as ligands can be nanoencapsulated within amylose nanostructures (dos Santos et al., 2011; Kawai et al., 2012; Yang et al., 2009; Zabar et al., 2009, 2010). The presence of ligands induces a conformational change in amylose, namely the conversion of double helices into a unit strand. The V-shaped amylose formed is dense and has a central hydrophobic cavity, capable of binding lipophilic molecules via hydrophobic interactions (Putseys et al., 2010; Zabar et al., 2010). It seems that the ligands are present in specific sites inside the structure. As an example, the acyl chains and the carboxylic groups are included in crystalline and amorphous areas, respectively (Godet et al., 1995). Therefore, amylose molecular inclusion complexes are appropriate nanocarriers for fatty acids (Karkalas et al., 1995; Lalush et al., 2005; Zabar et al., 2009, 2010) and the other guest molecules (Cohen et al., 2008; Lorentz et al., 2012) by accommodating hydrophobic domains within the central cavity. Different types of complexes (polymorphs) can be formed under the effects of temperature and time of annealing (Karkalas et al., 1995). Regarding the long chain fatty acids, the high level of unsaturation hinders the formation of inclusion complexation structures (less organized structures) (Seo et al., 2015; Zabar et al., 2009) and results in the formation of larger and more dispersed particle populations (Lesmes et al., 2009). Increasing the amylose chain length [degree of polymerization (DP)] up to certain amounts, raises the complexation yields, relative crystallinities, and thermal stabilities (Gelders et al., 2004). Moreover, by the augmentation of the amylose DP, lipid chain length and processing temperature the stability of the structures increased against degradation by enzymes and acids (Gelders et al., 2005).

Self-assembly of conjugated linoleic acid (CLA) with either amylose or β-CD was reported by Yang et al. (2009). Amylose-CLA complex revealed more protection against oxidative degradation of CLA, empowering the reactions between CLA and amylose. By imitating the conditions of the small intestine, the release of ligand from amylose–CLA complex was higher than β-CD–CLA complex (Yang et al., 2009). The combinations of high amylose corn starch and amylose with genistein showed high encapsulation efficiency of genistein in simulated stomach ambient along with slow release during digestion by amylase in the pancreatin solution (Cohen et al., 2008). The complexation method affects the complex organization of V-amylose. Water/dimethyl sulfoxide (DMSO) complexation method resulted in amylose-CLA complexes of 150-nm spherical shape; while, KOH/HCl complexation technique led to the elongated structure with the width of 43–160 nm. The spherical structure exhibited superior protection to CLA against oxidation (Lalush et al., 2005). Recently, Marinopoulou et al. (2016) studied the dynamic interfacial tension of the host–guest networks containing amylose–fatty acid. By analyzing the systems, the dynamic interfacial tension of the complex dispersion was dramatically higher compared to the fatty acid solution, revealing that a portion of fatty acids are not adsorbed at the interface anymore and this point refers to the formation of the inclusion complex system. In addition, the dynamic interfacial tension of the combination was lower than the net amylose dispersion. Hence, portion of the fatty acid chains was not included in the complexes even at fatty acid/amylose contents underneath the amylose saturation level (Marinopoulou et al., 2016). Ryno et al. (2014) reported the application of microwave heating to reduce the annealing time required to prepare these amylose complexes from hours to minutes. Besides, Raphaelides et al. (2015) has reported a continuous method for the production of starch guest–host systems on an industrial scale.

1.5.3 Nanoencapsulation technologies based on specialized equipment

Although for the nanoencapsulation of food ingredients using different technologies, it is necessary to apply some general equipment including homogenizers, mills, mixing devices, etc., there are some nanoencapsulation techniques which are feasible to implement only by specialized developed equipment such as electrospinning, electrospraying, and nanospray dryer as illustrated in Fig. 1.4.

1.5.3.1 Nanocapsule formation by electrospinning (Chapter 8)

Electrospinning is a novel method for the continuous fabrication of nonwoven nanofibers with diameters of nearly a few nanometers that could be employed in the delivery of bioactive food components (Esfanjani and Jafari, 2016). Electrospinning is a straightforward process and robust method for the production of ultrathin eletrospuns, which involves injecting polymer/biopolymer solution from a spinneret toward a prepared collector. Finally, nonwoven ultrafine fibers are collected on the collector plate (Ghorani and Tucker, 2015). The fabricated nanoscale fiber mat exhibits exclusive properties, such as high surface area to volume ratio and considerable porosity which makes them suitable as the carriers of bioactive food ingredients and nutraceuticals so that the incorporated molecules have enhanced functionality due to the size of the ultrathin fibers (Katouzian and Jafari, 2016).

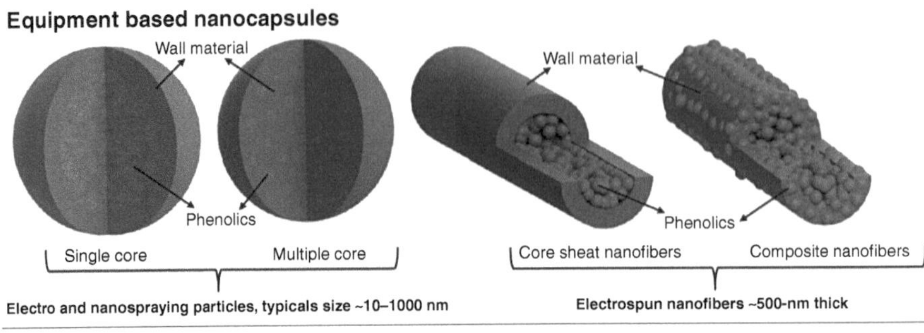

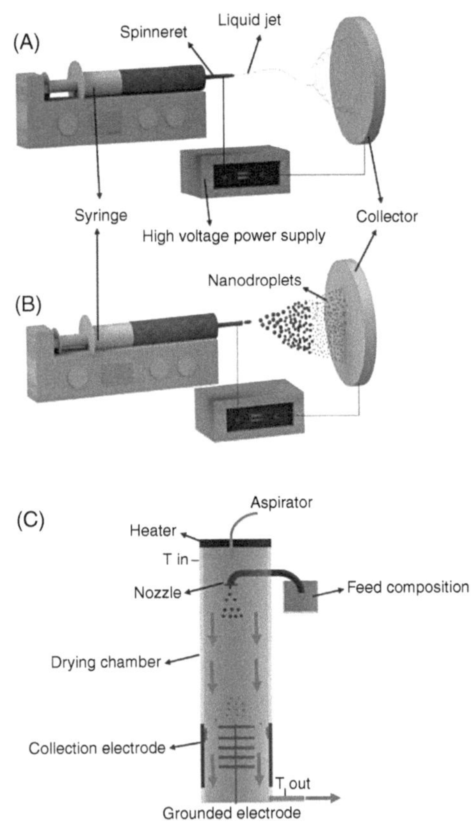

Figure 1.4 Equipment based nanoencapsulation processes. (A) Electrospinning, (B) electrospraying, and (C) nano Spray Dryer.
Source: Adapted from Faridi Esfanjani, A.F., Jafari, S.M., 2016. Biopolymer nano-particles and natural nano-carriers for nano-encapsulation of phenolic compounds. Colloids Surf. B 146, 532–543.

1.5.3.2 Nanocapsule formation by electrospraying (Chapter 9)

Electrospraying also known as electro-hydrodynamic atomization is a new nanoencapsulation technique akin to electrospinning with regard to its foundation; however

NPs are generated by this means instead of ultrathin fibers (Torres-Giner et al., 2010). In this method, high voltages are responsible for the atomization of liquids into fine droplets. As the droplets are stretched toward the ground electrode, the solvent is vaporized gradually. The noteworthy advantages of implementing this method include the single production process plus its significant encapsulation efficiency. The nanospheres and nanocapsules gained from the electrospraying technique may differ due to parameters, such as voltage, volume velocity, collector distance, and solution parameters like viscosity, density, and concentration. The ambient factors, such as temperature and pressure will also influence the characteristics of the resulting nanocapsules (Hao et al., 2013; Zhang and Kawakami, 2010).

1.5.3.3 Nanocapsule formation by nanospray dryer (Chapter 10)

A typical spray dryer is normally exerted to convert solutions into fine powders quickly and efficiently. The short time of the drying process is suited to even sensible nutrients without destructive effects (Jafari et al., 2008, 2015). Concerning the rapid progress of nanoencapsulation techniques in pharmaceutics, cosmetics, and functional food formulations, nanospray drying has been applied for encapsulation purposes (Esfanjani and Jafari, 2016). The nanospray dryer was introduced in 2009 to extend spray drying to the submicron scale. This is a milestone in spray drying technology, with the benefits of generating ultrafine powders with very narrow size distributions and high throughput. The Büchi nanospray dryer device is basically manufactured on the concept of spray drying technology, involving the production of submicron fragments from a solution, nanoemulsion nanosuspension (Arpagaus, 2011). Its technological novelty lies in the gentle flow of laminar drying gas, the vibrating mesh spray technology to form fine droplets and the highly efficient electrostatic precipitator to collect NPs.

1.5.4 Nanoencapsulation technologies based on biopolymer nanoparticles

Biopolymer NPs are formed by the assembly of one or more building blocks that constitute functional nanovehicles for the application in the food and nutraceutical industries. Biopolymer NPs are engineered with the goal of controlled and targeted delivery of nutri ents as well as fabricating value-added food products by improving their physicochemical properties, such as the texture and appearance to enhance the marketing chance for the final products. In recent years, biopolymer NPs have attracted significant attention in the food area and drug delivery research fields. The application of biopolymers and their complexes to develop nanodelivery systems to protect the bioactives and to enable subsequent fortification of food products is a highly desirable and promising technique. Considering the applications in other sectors (such as wound healing, tissue engineering, and drug delivery), biopolymers are considered as circa safe alternatives to synthetic polymers. Biopolymers are able to envelope both hydrophilic and hydrophobic bioactive ingredients and nutraceuticals. In addition, the usage of mild preparation process without involving destructive chemicals and organic solvents for the generation of nanovehicles are interesting options for green industrial applications (Mukhopadhyay and Kundu, 2015). Therefore, individual biopolymers and their mixtures introduce new possibilities for developing efficient food-grade nanobased systems for encapsulation, preservation, and targeted delivery of sensitive health-promoting substances using naturally GRAS (generally recognized as safe) ingredients (Arroyo-Maya and McClements, 2015; Hu and Huang, 2013).

1.5.4.1 Nanocapsule formation by individual biopolymer nanoparticles (Chapter 11)

1.5.4.1.1 Protein nanoparticles

There are many techniques for preparation of NPs from individual biopolymers for food and drug formulation and encapsulation, for example, desolvation, precipitation, thermal gelation, and self-assembly (Elzoghby et al., 2012). Among these techniques, desolvation is of great interest particularly for protein NPs, because of several advantages in terms of the procedure and the applied materials, including simplicity, wide range of used biopolymers, ability to use different solvent and desolvating agents, possibility to fabricate relatively uniform, stable, and spherical particles, etc. According to Sahoo et al. (2015), desolvation is defined as a thermodynamically oriented self-assembly process for biopolymers like proteins to prepare NPs, that is, based on a theory, by replacing some or all water molecules (solvent) around protein molecules with a desolvating agent (another solvent), leading to conformational changes in protein structure and a significant decrease in its solubility; therefore, resulting in phase separation and precipitation of proteins as NPs. Fabrication of protein NPs via desolvation method, such as gelatin, albumin, collagen, milk proteins, silk protein, elastin, zein, gliadin, soy protein, and so on has been reported in the literatures, which are explored in this book.

1.5.4.1.2 Carbohydrate nanoparticles

Polysaccharides are widely employed in the production of nanocarriers due to their unique structure and abundance plus their biodegradability. Mild methods for the production of carbohydrate nanocarriers are in priority to prevent degradation of the matrix, cut down on the additional costs, and enable scaling up and mass production by using these methods. Therefore, lipophilic and hydrophilic nutraceuticals may bind to the active sites of carbohydrate molecules by different types of reactions. Furthermore, polysaccharide molecules with ionic or nonpolar residues can adhere to lipids.

Among polysaccharides, there has been a growing interest on starch NPs as a proper material for formulating of nanodelivery systems. These NPs could be prepared from different starch sources which are found abundantly in nature. They facilitate the controlled release without undesirable liberation of bioactive compounds, reduce the present barriers due to overdose and enhance the performance of sensitive ingredients (Simi and Abraham, 2007). Starch is composed of a large number of glucose molecules and is a natural source of energy produced in green plants via photosynthesis. The structure and applications of this biodegradable compound are fascinating research topics, particularly in the field of delivery systems. It has a granular morphology (2–100 μm) including growth rings (120–500 nm) with nanoscale building blocks (20–50 nm) that mainly comprise amylopectin and amylose fractions (0.1–1 nm). The chief model for starch is a coaxial semicrystalline structure, which makes it a practical nanocarrier (1) starch nanocrystals are generated by the decomposition of the semi-crystalline framework of starch granules by means of hydrolyzing amorphous polymers and (2) starch NPs produced from gelatinized starch (Han et al., 2013; Le Corre and Angellier-Coussy, 2014).

1.5.4.2 Nanocapsule formation by complexation of biopolymers (Chapter 12)

The functional properties of individual biopolymers have been extensively explored over the last decades including molecular weight, conformation, concentration, polydispersity, charge density and distribution, pH, ionic strength, ion type, solvent type, and nature of inter/intramolecular forces (Goh et al., 2009). These properties are changeable according to the molecular structure of the biopolymers and the type of interactions between different types of biopolymers. Due to the synergistic combination effects between different biopolymers, the formation of biopolymer complexes yields enhanced functional properties that differentiate them from the maternal synergistic combination effects between different biopolymers (Schmitt et al., 2009). Tailor-made functionalities, such as nanoencapsulation and development of new functional nanostructures can often be obtained via noncovalent interactions or hydrogen bonding between a set of biopolymers. The new functionalities resulting from these self-assembling structures are of great applied significance for the improvement of many types of food formulations (van der Sman, 2012). The interactions between different biopolymers under the effect of distinct factors have been widely explored by different researchers in the previously published works. The expansive range of biopolymer mixtures (especially those of protein and polysaccharide) are available in the related review studies. Indeed, numerous classes of biopolymers are fit for forming molecular complexes through the assembly of lipophilic and hydrophilic molecules. The active molecules may then be nanoencapsulated and overprotected using protein–protein, polysaccharide–polysaccharide and protein–polysaccharide (nano) complexes.

1.5.5 Other nanoencapsulation technologies

There are some other nanoencapsulation technologies, which are employed in the protection and targeted delivery of food bioactive ingredients and nutraceuticals. As shown in Fig. 1.5, these include the formation of nanocarriers such as protein nanotubes, carbohydrate nanohydrogels, nanoorganogels, etc. Since there are limited published articles for these techniques in the field of food and nutraceutical studies, a brief description of each technology is provided here and hopefully in the next edition of the current book, separate chapters would be devoted to completely discuss them with more details.

1.5.5.1 Nanoencapsulation by protein nanotubes

The only food protein capable of producing nanotubes (as shown in Fig. 1.5), has been noted to be α-lactalbumin. Ipsen and Otte (2007) developed partially hydrolyzed self-assembled α-lactalbumin nanotubes from whey protein (20 nm in outer

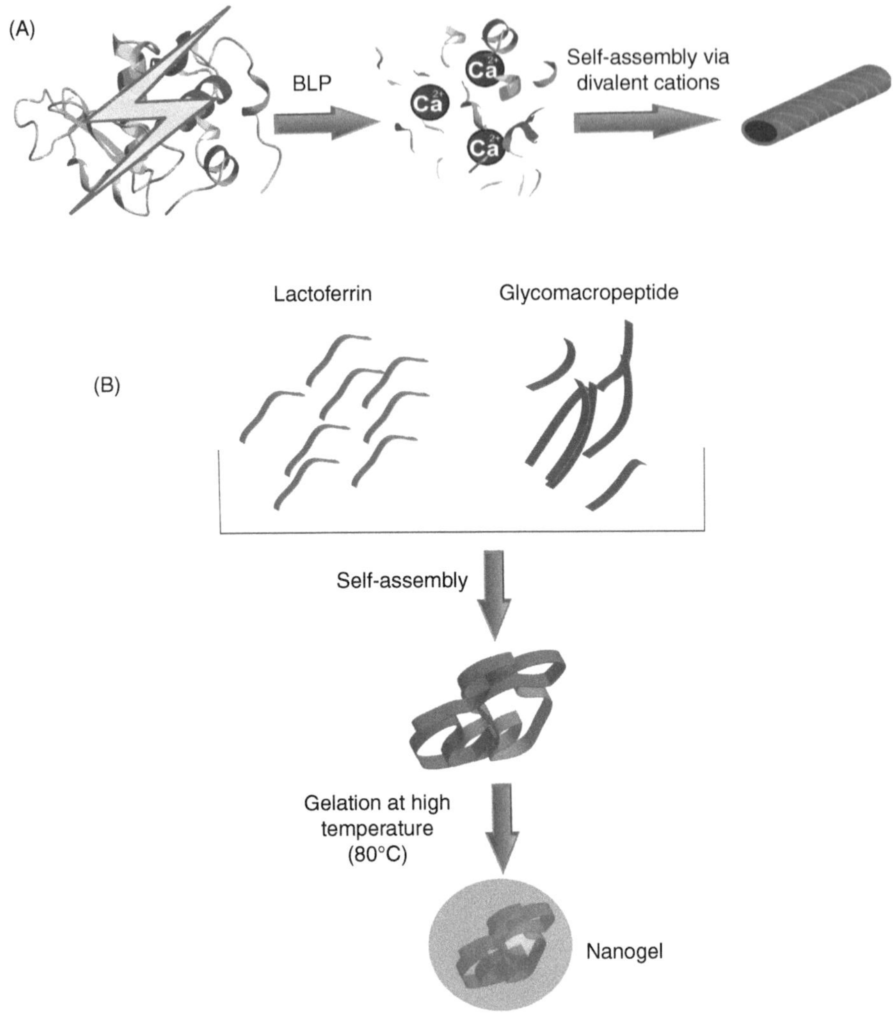

Figure 1.5 Miscellaneous nanocarriers and nanocapsules for food ingredients and nutraceuticals. (A) α-lactalbumin nanotubes: α-lactalbumin protein is partially hydrolyzed by serine protease secreted from *Bacillus licheniformis*, next nanotubes are formed by the process of self-assembly using divalent cations like Ca^{2+}. (B) Fabrication of nanohydrogels and nanoorganogels (here, nanohydrogels of glycomacropeptide-lactoferrin are formed via the interactions of two proteins along with the gelation at high temperature).

diameter). A concentration of 30 g/L protein was reported to be prerequisite for the tubular formation and it was concluded that these nanotubes are promising in future nanodelivery applications.

α-Lactalbumin nanotubular structures have some characteristics that make them unique to be used in nanotechnological systems. Some of these features include great aspect ratio, comparative stiffness, sustained release ability and their nanosized cavity.

These nanostructures can be employed as viscosifiers and thus it not only covers tubular shapes but also other lengthy and stiff protein-based aggregates would be practical as viscosifiers. α-Lactalbumin-derived gels are strong and reversible; moreover the yielded gel is transparent which makes it an interesting nanostructure. By lowering the pH, the gel structure will be degraded. Regarding the cavity that nanotubes possess they are able to encapsulate bioactives, such as vitamins, enzymes, and other sensitive compounds, which is very useful in the food and pharmaceutical industry (Graveland-Bikker and De Kruif, 2006).

Lately, liposome-microtubule compound has been developed including open or closed ends (Raviv et al., 2005). It is possible to obtain a controlled release through the shift between open and closed ends of nanotubes via lipid caps. Besides, by changing, for example, the charge of nanotubes, it may be possible to regulate the attachment of a specific molecule to the nanotube structure. Since α-lactalbumin is obtained from milk it is a food-grade nanocarrier and then can be utilized within food products as well as pharmaceutical sector. However, they cannot be considered as GRAS materials until the clinical tests approve it.

α-Lactalbumin also has important biological functions, for instance it is needed for the synthesis of lactose in the lactating mammary gland. Its amino acid array provides the requirement for an individual with its high digestibility. This protein is rich in tryptophan and has positive impacts on intellectual performance, mood, and feeling of repletion (Beulens et al., 2004). It has also been shown that the consumption of α-lactalbumin develops cognitive functionality in vulnerable topics (Markus et al., 2005), furthermore studies suggested that α-lactalbumin has antimicrobial and antitumor activities (Håkansson et al., 1999). HAMLET (human α-lactalbumin made lethal to tumor cells) and BAMLET (bovine α-lactalbumin made lethal to tumor cells) were implemented in in vitro models to induce apoptosis in tumor cells (Svensson et al., 2003). Lately, it has been depicted that oleic acid is not essential for the optimized reaction of α-lactalbumin with histone. In conclusion, α-lactalbumin has the potential to be utilized as antitumor agents via direct interaction between α-lactalbumin and alkaline proteins of histone (Permyakov et al., 2004).

1.5.5.2 Nanoencapsulation by carbohydrate nanogels

Nanogels are actually composed hydrophilic or amphiphilic biopolymers by physical or chemical cross-linking and are used for different purposes like drug/protein delivery systems, biocatalysis, biocidal agents, biosensors, and so on. Self-assembly is the classical physical approach based on a controlled aggregation of hydrophobic or electrostatic interactions and/or hydrogen bonding. These physical interactions usually take place under mild conditions, which is appealing for encapsulation of biologically sensitive proteins/drugs (Wu et al., 2015; Yan et al., 2006).

Nanogels are categorized into two different types; nanohydrogels and nanoorganogels (micelle nanogels). The former has a considerable capacity of water absorbance (high level of swelling in water), while the latter include water-insoluble nanogels with a high affinity for oily substances. In fact, nanoorganogels are aggregations of NPs in the shape of micelles in the presence of water in which the hydrophobic regions are navigated to the central part. The hydrophobic area contains the oily materials with functional units, such as carboxyl, amine, aldehyde, and so on attached to the polymeric skeleton (Khalili et al., 2015).

In recent years, hydrogel NPs are considered as the novel and practical delivery systems because of the combined characteristics of both hydrogels systems (hydrophilicity and retaining significant water levels in their matrix) and nanotechnology (very small size). Hydrogels are three-dimensional polymeric networks, with chemical or physical cross-links, capable of holding large amount of water (Hamidi et al., 2008). Single or mixed-type biopolymer clusters may be developed from hydrogel particles under certain conditions.

Among different biomacromolecules, which can be used for hydrogel formation, polysaccharides are extremely advantageous to physically or chemically form filled hydrogels. Polysaccharide-based hydrogels can be formed whenever an appropriate counter-ion is added to a polysaccharide dispersion of adequate concentration. As an example, sodium alginate dispersion transforms into alginate hydrogel after mixing (titration) with divalent cations (such as calcium ion). At low concentrations (diluted systems), NP formation is obtained instead of bulk phase gelation. Pistone et al. (2015) examined the chance of formulating stable alginate nanohydrogels through cross-linking with zinc cation in the presence of monovalent salt. A certain level of zinc was required to prepare NPs and in concentrations less than it, the formation of NPs are limited. An increase in the ionic strength of the solvent resulted in NPs of lower size as well as lower polydispersity. Rolland et al. (2014) studied the formation alginate hydrogel shells coating an aqueous core by the coextrusion process together with a sol–gel transition of the shell following submersion in a gelling medium. The hydrogel capsules have been suggested as useful tools for biomedical applications as well as designing versatile hydrogel-based cell cultures. To control the uptake and release of proteins, Li et al. (2009) fabricated a microencapsulation matrix on the basis of hydrogels composed of TEMPO [(2,2,6,6-tetramethylpiperidin-1-yl)oxyl]-oxidized potato starch polymers, held together by sodium trimetaphosphate as cross-linking agents. It has been shown that the enhancement swelling rate in microgels is related with the increasing pH and reduction in the salt concentration. Accordingly, the uptake of a globular enzyme (lysozyme) was altered by the degree of oxidation and the ratio of cross-linker to the polymer. Furthermore, highly charged microgels involving the average ratio of cross-linker to the macromolecule (0.15 and 0.2) were chosen as the optimal encapsulation factors (Li et al., 2009).

1.6 Conclusions and final remarks

Currently, the application of nanotechnology and nanostructured carriers has advanced in the food science and industry. Nanoencapsulation remains one of the practical and bright methods to improve the functional properties of the food formulations and to ensure the public health. In this way, various techniques for enveloping the bioactive ingredients are still introduced in this field having their advantages and drawbacks. The nanoencapsulation approach enables the food technologists to fabricate protective shields against unfavorable conditions, masking undesirable taste and aromas as well as providing enhanced uptake of nutrients, controlled release, targeted delivery and raising the solubility of lipophilic bioactive agents in aqueous solutions. Nanoencapsulation techniques are employed in a broad range of applications and the utilized are likely to overlap. Hence, a nanocarrier is chosen by considering the nature and characteristics of the nutrients plus the other factors, such as release patterns, safety and toxicity issues, or economic concerns. To achieve productive nanoencapsulation formulations the nanoencapsulation devices, they should be engineered and optimized concerning the physical, structural, and mass transport properties. Furthermore, the safety and toxicity of nanoencapsulation systems should be tested as they are directly consumed by human. This aspect is a hot topic in the research articles and even though the delivery systems might be food-grade they are not considered as GRAS since their physicochemical properties are thoroughly modified at the nanoscale and by running in vitro and in vivo tests deep insights on the safety of the nanodevices can be obtained. Any process or material not previously approved must be evaluated formally, whose approval requirements extend to methods of solvent removal, storage conditions, handling methods, and so on. Besides, the regulatory legislations on nanofoods are under advancement and all sectors in different continents should participate in this project to prepare a comprehensive set of principles for the fabrication, handling, and consumption of nanofoods. To sum up, there is an optimistic outlook for implementing nanoencapsulation and related technologies in the food industry, the number of companies that have initiated producing their nanofood commodities are few (like Tip-Top, Nutralease, Biodelivery sciences International, Aquanova, livethesource) but seems to expand in near future with the aim of reducing the symptoms of diseases and enhancing the general health of people and communities.

References

Anandharamakrishnan, C., 2014. Techniques for Nanoencapsulation of Food Ingredients. Springer-Verlag, New York, NY.

Arpagaus, C., 2011. Nano Spray Dryer B-90: Literature review and applications. Buchi Information Bulletin, Number 63/2011.

Arroyo-Maya, I.J., McClements, D.J., 2015. Biopolymer nanoparticles as potential delivery systems for anthocyanins: fabrication and properties. Food Res. Int. 69, 1–8.

Assadpour, E., Jafari, S.-M., 2017. Spray drying of folic acid within nano-emulsions; optimization by Taguchi approach. Drying Technol., doi: http://dx.doi.org/10.1080/07373937.2016.1242016.

Assadpour, E., Jafari, S.-M., Maghsoudlou, Y., 2017. Evaluation of folic acid release from spray dried powder particles of pectin-whey protein nano-capsules. Int. J. Biol. Macromol. 95, 238–247.

Assadpour, E., Maghsoudlou, Y., et al., 2016a. Optimization of folic acid nano-emulsification and encapsulation by maltodextrin-whey protein double emulsions. Int. J. Biol. Macromol. 86, 197–207.

Assadpour, E., Maghsoudlou, Y., Jafari, S.-M., Ghorbani, M., Aalami, M., 2016b. Evaluation of folic acid nano-encapsulation by double emulsions. Food Bioprocess Technol. 9, 2024–2032.

Astray, G., Gonzalez-Barreiro, C., Mejuto, J.C., Rial-Otero, R., Simal-Gándara, J., 2009. A review on the use of cyclodextrins in foods. Food Hydrocoll. 23, 1631–1640.
Barbosa-Cánovas, G.V., Ortega-Rivas, E., Juliano, P., Yan, H., 2005. Encapsulation processes. Springer, New York, US, 199–219.
Beulens, J.W., Bindels, J.G., de Graaf, C., Alles, M.S., Wouters-Wesseling, W., 2004. Alpha-lactalbumin combined with a regular diet increases plasma Trp–LNAA ratio. Physiol. Behav. 81 (4), 585–593.
Blasco, C., Pico, Y., 2011. Determining nanomaterials in food. Trends Anal. Chem. 30 (1), 84–99.
Briuglia, M.L., Rotella, C., McFarlane, A., Lamprou, D.A., 2015. Influence of cholesterol on liposome stability and on in vitro drug release. Drug Deliv. Transl. Res. 5 (3), 231–242.
Chakraborty, S., Liao, I.C., Adler, A., Leong, K.W., 2009. Electrohydrodynamics: a facile technique to fabricate drug delivery systems. Adv. Drug Deliv. Rev. 61 (12), 1043–1054.
Chau, C.-F., Wu, S.-H., Yen, G.-C., 2007. The development of regulations for food nanotechnology. Trends Food Sci. Technol. 18 (5), 269–280.
Chen, L., Remondetto, G.E., Subirade, M., 2006. Food protein-based materials as nutraceutical delivery systems. Trends Food Sci. Technol. 17 (5), 272–283.
Clarke, J.D., Jayasinghe, S.N., 2008. Bio-electrosprayed multicellular zebrafish embryos are viable and develop normally. Biomed. Mater. 3 (1), 011001.
Cohen, R., Orlova, Y., Kovalev, M., Ungar, Y., Shimoni, E., 2008. Structural and functional properties of amylose complexes with genistein. J. Agric. Food Chem. 56 (11), 4212–4218.
Crini, G., 2014. Review: a history of cyclodextrins. Chem. Rev. 114, 10940–10975.
Cushen, M., Kerry, J., Morris, M., Cruz-Romero, M., Cummins, E., 2012. Nanotechnologies in the food industry—recent developments, risks and regulation. Trends Food Sci. Technol. 24 (1), 30–46.
DeFelice, S., 1995. The nutraceutical revolution: its impact on food industry R&D. Trends Food Sci. Technol. 6, 59–61.
Desai, K.G.H., Jin Park, H., 2005. Recent developments in microencapsulation of food ingredients. Drying Technol. 23 (7), 1361–1394.
Donsì, F., Annunziata, M., Sessa, M., Ferrari, G., 2011. Nanoencapsulation of essential oils to enhance their antimicrobial activity in foods. LWT Food Sci. Technol. 44 (9), 1908–1914.
dos Santos, C., del Pilar Buera, M., Mazzobre, M.F., 2011. Phase solubility studies of terpineol with β-cyclodextrins and stability of the freeze-dried inclusion complex. Procedia Food Sci. 1, 355–362.
EC, 2001. European Commission recommendation of 18 October 2011 on the definition of nanomaterial (2011/696/EU), European Commission. Available from: http://eur-lex.europa.eu/legal-content/EN/TXT/?uri=CELEX:32011H0696
EFSA, 2011. Guidance on the risk assessment of the application of nanoscience and nanotechnologies in the food and feed chain. EFSA J. 9 (5), 2140, (36 p.).
Elzoghby, A.O., Samyand, W.M., Elgindy, N.A., 2012. Albumin-based nanoparticles as potential controlled release drug delivery systems. J. Control. Release 157 (No. 2), 168–182.
Esfanjani, A.F., Jafari, S.M., 2016. Biopolymer nano-particles and natural nano-carriers for nano-encapsulation of phenolic compounds. Colloids Surf. B 146, 532–543.
Esfanjani, A.F., Jafari, S.M., Assadpoor, E., Mohammadi, A., 2015. Nano-encapsulation of saffron extract through double-layered multiple emulsions of pectin and whey protein concentrate. J. Food Eng. 165, 149–155.
Esfanjani, A.F., Jafari, S.M., Assadpour, E., 2017. Preparation of a multiple emulsion based on pectin-whey protein complex for encapsulation of saffron extract nanodroplets. Food Chem. 221, 1962–1969.

European Commission, 2009. EU Regulation 1223/2009 (Cosmetics Regulation). Available from: http://eur-lex.europa.eu/

Ezhilarasi, P., Karthik, P., Chhanwal, N., Anandharamakrishnan, C., 2013. Nanoencapsulation techniques for food bioactive components: a review. Food Bioprocess Technol. 6 (3), 628–647.

Fathi, M., Martin, A., McClements, D.J., 2014. Nanoencapsulation of food ingredients using carbohydrate based delivery systems. Trends Food Sci. Technol. 39 (1), 18–39.

Fathi, M., Mozafari, M., Mohebbi, M., 2012. Nanoencapsulation of food ingredients using lipid based delivery systems. Trends Food Sci. Technol. 23 (1), 13–27.

FDA, 2014. Guidance for Industry Considering Whether an FDA-Regulated Product Involves the Application of Nanotechnology, Food and Drug Administration. Available from: http://www.fda.gov/RegulatoryInformation/Guidances/ucm257698.htm

Fuchs, S., Kutscher, M., Hertel, T., Winter, G., Pietzschand, M., Coester, C., 2010. Transglutaminase: new insights into gelatin nanoparticle cross-linking. J. Microencapsul. 27 (8), 747–754.

Gelders, G.G., Duyck, J.P., Goesaert, H., Delcour, J.A., 2005. Enzyme and acid resistance of amylose-lipid complexes differing in amylose chain length, lipid and complexation temperature. Carbohydr. Polym. 60 (3), 379–389.

Gelders, G.G., Vanderstukken, T.C., Goesaert, H., Delcour, J.A., 2004. Amylose–lipid complexation: a new fractionation method. Carbohydr. Polym. 56 (4), 447–458.

Ghorbanzade, T., Jafari, S.M., Akhavan, S., Hadavi, R., 2017. Nano-encapsulation of fish oil in nano-liposomes and its application in fortification of yogurt. Food Chem. 216, 146–152.

Ghorani, B., Tucker, N., 2015. Fundamentals of electrospinning as a novel delivery vehicle for bioactive compounds in food nanotechnology. Food Hydrocoll. 51, 227–240.

Godet, M.C., Tran, V., Colonna, P., Buleon, A., Pezolet, M., 1995. Inclusion/exclusion of fatty acids in amylose complexes as a function of the fatty acid chain length. Int. J. Biol. Macromol. 17 (6), 405–408.

Goh, K.K.T., Sarkar, A., Singh, H., 2009. Milk protein–polysaccharide interactions. In: Thompson, A., Boland, M., Singh, H. (Eds.), Milk Proteins: From Expression to Food. Elsevier, New York, NY, pp. 347–376.

Gómez-Mascaraque, L.G., Lagarón, J.M., López-Rubio, A., 2015. Electrosprayed gelatin submicroparticles as edible carriers for the encapsulation of polyphenols of interest in functional foods. Food Hydrocoll. 49, 42–52.

Graveland-Bikker, J.F., De Kruif, C.G., 2006. Unique milk protein based nanotubes: food and nanotechnology meet. Trends Food Sci. Technol. 17 (5), 196–203.

Håkansson, A., Andréasson, J., Zhivotovsky, B., Karpman, D., Orrenius, S., Svanborg, C., 1999. Multimeric α-lactalbumin from human milk induces apoptosis through a direct effect on cell nuclei. Exp. Cell Res. 246 (2), 451–460.

Hamidi, M., Azadi, A., Rafiei, P., 2008. Hydrogel nanoparticles in drug delivery. Adv. Drug Deliv. Rev. 60 (15), 1638–1649.

Han, F., Gao, C., Liu, M., 2013. Fabrication and characterization of size-controlled starch-based nanoparticles as hydrophobic drug carriers. J. Nanosci. Nanotechnol. 13 (10), 6996–7007.

Hao, S., Wang, Y., Wang, B., Deng, J., Liu, X., Liu, J., 2013. Rapid preparation of pH-sensitive polymeric nanoparticle with high loading capacity using electrospray for oral drug delivery. Mater. Sci. Eng. C 33, 4562–4567.

Haub, C., 2011. World Population Aging: Clocks Illustrate Growth in Population Under Age 5 and Over Age 65, Population Reference Bureau (PRB). Available from: http://www.prb.org/Articles/2011/agingpopulationclocks.aspx

Hu, B., Huang, Q.R., 2013. Biopolymer based nano-delivery systems for enhancing bioavailability of nutraceuticals. Chin. J. Polym. Sci. 31 (9), 1190–1203.

Huang, Q., Given, P., 2009. Micro/nano encapsulation of active food ingredients. ACS Symposium Series, vol. 1007. American Chemical Society, New York, NY.
Huang, Q., Yu, H., Ru, Q., 2010. Bioavailability and delivery of nutraceuticals using nanotechnology. J. Food Sci. 75 (1), R50–R57.
Hughes, G.A., 2005. Nanostructure-mediated drug delivery. Nanomedicine 1 (1), 22–30.
Ipsen, R., Otte, J., 2007. Self-assembly of partially hydrolysed α-lactalbumin. Biotechnol. Adv. 25 (6), 602–605.
Jafari, S.M., Assadpoor, E., He, Y., Bhandari, B., 2008. Encapsulation efficiency of food flavours and oils during spray drying. Drying Technol. 26 (7), 816–835.
Jafari, S.M., Fathi, M., Mandala, I., 2015. Emerging product formation. Elsevier Inc., Academic Press, San Diego, CA, 293–317.
Jafari, S.M., McClements, D.J., 2017. Nanotechnology approaches for increasing nutrient bioavailability. Advances in Food and Nutrition Research Elsevier, New York.
Jizomoto, H., Kanaoka, E., Sugita, K., Hirano, K., 1993. Gelatin-acacia microcapsules for trapping micro oil droplets containing lipophilic drugs and ready disintegration in the gastrointestinal tract. Pharm. Res. 10 (8), 1115–1122.
Karkalas, J., Ma, S., Morrison, W.R., Pethrick, R.A., 1995. Some factors determining the thermal properties of amylose inclusion complexes with fatty acids. Carbohydr. Res. 268 (2), 233–247.
Katouzian, I., Jafari, S.M., 2016. Nano-encapsulation as a promising approach for targeted delivery and controlled release of vitamins. Trends Food Sci. Technol. 53, 34–48.
Kawai, K., Takato, S., Sasaki, T., Kajiwara, K., 2012. Complex formation, thermal properties, and in-vitro digestibility of gelatinized potato starch–fatty acid mixtures. Food Hydrocoll. 27 (1), 228–234.
Khalili, S.T., Mohsenifar, A., Beyki, M., Zhaveh, S., Rahmani-Cherati, T., Abdollahi, A., Bayat, M., Tabatabaei, M., 2015. Encapsulation of Thyme essential oils in chitosan-benzoic acid nanogel with enhanced antimicrobial activity against *Aspergillus flavus*. LWT Food Sci. Technol. 60 (1), 502–508.
Khare, A.R., Vasisht, N., 2014. Nanoencapsulation in the food Industry: technology of the future. Academic Press, San Diego, CA, pp. 151–155 (Chapter 14).
Khayata, N., Abdelwahed, W., Chehna, M., Charcosset, C., Fessi, H., 2012. Preparation of vitamin E loaded nanocapsules by the nanoprecipitation method: from laboratory scale to large scale using a membrane contactor. Int. J. Pharm. 423 (2), 419–427.
Kimpel, F., Schmitt, J.J., 2015. Review: milk proteins as nanocarrier systems for hydrophobic nutraceuticals. J. Food Sci. 80 (11), 2361–2366.
Lafitte, G., 2008. Structure of the gastrointestinal mucus layer and implications for controlled release and delivery of functional food ingredients. In: Garti, N. (Ed.), Delivery and Controlled Release of Bioactives in Foods and Nutraceuticals. Woodhead Publishing, Cambridge, England, pp. 26–52, (Chapter 2).
Lakkis, J.M., 2007. Introduction. Blackwell Publishing, New York, NY, pp. 1–11.
Lalush, I., Bar, H., Zakaria, I., Eichler, S., Shimoni, E., 2005. Utilization of amylose-lipid complexes as molecular nanocapsules for conjugated linoleic acid. Biomacromolecules 6 (1), 121–130.
Le Corre, D., Angellier-Coussy, H., 2014. Preparation and application of starch nanoparticles for nanocomposites: a review. React. Funct. Polym. 85, 97–120.
Lesmes, U., Cohen, S.H., Shener, Y., Shimoni, E., 2009. Effects of long chain fatty acid unsaturation on the structure and controlled release properties of amylose complexes. Food Hydrocoll. 23 (3), 667–675.
Li, Y., de Vries, R., Slaghek, T., Timmermans, J., Cohen Stuart, M.A., Norde, W., 2009. Preparation and characterization of oxidized starch polymer microgels for encapsulation and controlled release of functional ingredients. Biomacromolecules 10 (7), 1931–1938.

Livney, Y.D., 2015. Nanostructured delivery systems in food: latest developments and potential future directions. Curr. Opin. Food Sci. 3, 125–135.
Loftsson, T., Duchêne, D., 2007. Cyclodextrins and their pharmaceutical applications. Int. J. Pharm. 329, 1–11.
López-Nicolás, J.M., García-Carmona, F., 2008. Rapid, simple and sensitive determination of the apparent formation constants of *trans*-resveratrol complexes with natural cyclodextrins in aqueous medium using HPLC. Food Chem. 109 (4), 868–875.
López-Rubio, A., Lagaron, J.M., 2012. Whey protein capsules obtained through electrospraying for the encapsulation of bioactives. Innov. Food Sci. Emerg. Technol. 13 (0), 200–206.
Lorentz, C., Pencreac'h, G., Soultani-Vigneron, S., Rondeau-Mouro, C., de Carvalho, M., Pontoire, B., Ergan, F., Le Bail, P., 2012. Coupling lipophilization and amylose complexation to encapsulate chlorogenic acid. Carbohydr. Polym. 90 (1), 152–158.
Mahdavi, S.A., Jafari, S.M., Assadpoor, E., Dehnad, D., 2016. Microencapsulation optimization of natural anthocyanins with maltodextrin, gum Arabic and gelatin. Int. J. Biol. Macromol. 85, 379–385.
Mahdavi, S.A., Jafari, S.M., Ghorbani, M., Assadpoor, E., 2014. Spray-drying microencapsulation of anthocyanins by natural biopolymers: a review. Drying Technol. 32 (5), 509–518.
Marinopoulou, A., Kalogianni, E.P., Raphaelides, S.N., 2016. Amylose-fatty acid inclusion complexes as examined by interfacial tension measurements. Colloids Surf. B 137, 133–137.
Markus, C.R., Jonkman, L.M., Lammers, J.H., Deutz, N.E., Messer, M.H., Rigtering, N., 2005. Evening intake of α-lactalbumin increases plasma tryptophan availability and improves morning alertness and brain measures of attention. Am. J. Clin. Nutr. 81 (5), 1026–1033.
Matalanis, A., Jones, O.G., McClements, D.J., 2011. Structured biopolymer-based delivery systems for encapsulation, protection and release of lipophilic compounds. Food Hydrocoll. 25 (8), 1865–1880.
McClements, D.J., Decker, E.A., Park, Y., Weiss, J., 2009. Structural design principles for delivery of bioactive components in nutraceuticals and functional foods. Crit. Rev. Food Sci. Nutr. 49 (6), 577–606.
McClements, D.J., Li, Y., 2010. Structured emulsion-based delivery systems: controlling the digestion and release of lipophilic food components. Adv. Colloid Interface Sci. 159 (2), 213–228.
Mehrnia, M.A., Jafari, S.M., et al., 2016. Crocin loaded nano-emulsions: factors affecting emulsion properties in spontaneous emulsification. Int. J. Biol. Macromol. 84, 261–267.
Mehrnia, M.-A., Jafari, S.-M., Makhmal-Zadeh, B.S., Maghsoudlou, Y., 2017. Rheological and release properties of double nano-emulsions containing crocin prepared with Angum gum, Arabic gum and whey protein. Food Hydrocolloids., doi: http://dx.doi.org/10.1016/j.foodhyd.2016.11.033.
Mohammadi, A., Jafari, S.M., Assadpour, E., Faridi Esfanjani, A., 2016a. Nano-encapsulation of olive leaf phenolic compounds through WPC-pectin complexes and evaluating their release rate. Int. J Biol Macromol. 82, 816–822.
Mohammadi, A., Jafari, S.M., Esfanjani, A.F., Akhavan, S., 2016b. Application of nano-encapsulated olive leaf extract in controlling the oxidative stability of soybean oil. Food Chem. 190, 513–519.
Mukhopadhyay, P., Kundu, P.P., 2015. Chitosan-graft-PAMAM–alginate core–shell nanoparticles: a safe and promising oral insulin carrier in an animal model. RSC Adv. 5 (114), 93995–94007.
Muller, R.H., Chen, R., Keck, C.M., 2013. smartCrystals for consumer care & cosmetics: enhanced dermal delivery of poorly soluble plant actives. H&PC Today 8, 18–24.
Müller, R.H., Keck, C.M., 2008. Second generation of drug nanocrystals for delivery of poorly soluble drugs: smartCrystal technology. Eur. J. Pharm. Sci. 34, 20–21.

Muller, R.H., MaÈder, K., Gohla, S., 2000. Solid lipid nanoparticles (SLN) for controlled drug delivery–a review of the state of the art. Eur. J. Pharm. Biopharm. 50 (1), 161–177.

Müller, R., Radtke, M., Wissing, S., 2002. Nanostructured lipid matrices for improved microencapsulation of drugs. Int. J. Pharm. 242 (1), 121–128.

Nazzaro, F., Fratianni, F., Coppola, R., 2012. Microtechnology and nanotechnology in food science. Springer, Springer-Verlag, New York, NY, pp. 471–494.

Nedovic, V., Kalusevic, A., Manojlovic, V., Levic, S., Bugarski, B., 2011. An overview of encapsulation technologies for food applications. Procedia Food Sci. 1, 1806–1815.

Neethirajan, S., Jayas, D.S., 2011. Nanotechnology for the food and bioprocessing industries. Food Bioprocess Technol. 4 (1), 39–47.

Permyakov, S.E., Pershikova, I.V., Khokhlova, T.I., Uversky, V.N., Permyakov, E.A., 2004. No need to be HAMLET or BAMLET to interact with histones: binding of monomeric α-lactalbumin to histones and basic poly-amino acids. Biochemistry 43 (19), 5575–5582.

Pistone, S., Qoragllu, D., Smistad, G., Hiorth, M., 2015. Formulation and preparation of stable cross-linked alginate-zinc nanoparticles in the presence of a monovalent salt. Soft Matter 11 (28), 5765–5774.

Putseys, J.A., Lamberts, L., Delcour, J.A., 2010. Amylose-inclusion complexes: formation, identity and physico-chemical properties. J. Cereal Sci. 51 (3), 238–247.

Raphaelides, S.N., Dimitreli, G., Exarhopoulos, S., Ilia, E., Koutsomihali, P., 2015. A process designed for the continuous production of starch inclusion complexes on an industrial scale. Food Bioproducts Process. 96, 245–255.

Raviv, U., Needleman, D.J., Li, Y., Miller, H.P., Wilson, L., Safinya, C.R., 2005. Cationic liposome–microtubule complexes: pathways to the formation of two-state lipid–protein nanotubes with open or closed ends. Proc. Natl Acad. Sci. USA 102 (32), 11167–11172.

Ray, S., Raychaudhuri, U., Chakraborty, R., 2016. An overview of encapsulation of active compounds used in food products by drying technology. Food Biosci. 13, 76–83.

Risch, S.J., Reineccius, G., 1995. Encapsulation and controlled release of food ingredients. ACS Symposium Series (USA). American Chemical Society, Washington, DC.

Rolland, L., Santanach-Carreras, E., Delmas, T., Bibette, J., Bremond, N., 2014. Physicochemical properties of aqueous core hydrogel capsules. Soft Matter 10 (48), 9668–9674.

Ryno, L.M., Levine, Y., Iovine, P.M., 2014. Synthesis, characterization and comparative analysis of amylose–guest complexes prepared by microwave irradiation. Carbohydr. Res. 383, 82–88.

Sabliov, C.M., Astete, C.E., 2008. Encapsulation and controlled release of antioxidants and vitamins. In: Garti, N. (Ed.), Delivery and Controlled Release of Bioactives in Foods and Nutraceuticals. Woodhead Publishing, Cambridge, England, pp. 297–330, (Chapter 12).

Sagalowicz, L., Leser, M.E., 2010. Delivery systems for liquid food products. Curr. Opin. Colloid Interface Sci. 15 (1), 61–72.

Sahoo, N., Sahoo, R.K., Biswas, N., Guhaand, A., Kuotsu, K., 2015. Recent advancement of gelatin nanoparticles in drug and vaccine delivery. Int. J. Biol. Macromol. 81, 317–331.

Sáiz-Abajo, M.-J., González-Ferrero, C., Moreno-Ruiz, A., Romo-Hualde, A., González-Navarro, C.J., 2013. Thermal protection of β-carotene in re-assembled casein micelles during different processing technologies applied in food industry. Food Chem. 138 (2), 1581–1587.

Sanguansri, P., Augustin, M.A., 2006. Nanoscale materials development–a food industry perspective. Trends Food Sci. Technol. 17 (10), 547–556.

Schmitt, C., Aberkane, L., Sanchez, C., 2009. Protein-polysaccharide complexes and coacervates. In: Phillips, G.O., Williams, P.A. (Eds.), Handbook of Hydrocolloids. second ed. CRC Press, Florida, pp. 420–476.

Seo, T.R., Kim, J.Y., Lim, S.T., 2015. Preparation and characterization of crystalline complexes between amylose and C18 fatty acids. LWT Food Sci. Technol. 64 (2), 889–897.

Shimoni, E., 2009. Nanotechnology for foods: delivery systems. In: Barbosa-Cánovas, G., Mortimer, A., Lineback, D., Spiess, W., Buckle, K., Colonna, P. (Eds.), Global Issues in Food Science and Technology. Elsevier, New York, NY, pp. 411–424.

Simi, C., Abraham, T.E., 2007. Hydrophobic grafted and cross-linked starch nanoparticles for drug delivery. Bioprocess Biosyst. Eng. 30, 173–180.

Sobel, R., Versic, R., Gaonkar, A.G., 2014. Introduction to microencapsulation and controlled delivery in foods. Academic Press, San Diego, CA, pp. 3–12 (Chapter 1).

Sozer, N., Kokini, J.L., 2009. Nanotechnology and its applications in the food sector. Trends Biotechnol. 27 (2), 82–89.

Svensson, M., Fast, J., Mossberg, A.K., Düringer, C., Gustafsson, L., Hallgren, O., Brooks, C.L., Berliner, L., Linse, S., Svanborg, C., 2003. α-Lactalbumin unfolding is not sufficient to cause apoptosis, but is required for the conversion to HAMLET (human α-lactalbumin made lethal to tumor cells). Protein Sci. 12 (12), 2794–2804.

Tamjidi, F., Shahedi, M., Varshosaz, J., Nasirpour, A., 2013. Nanostructured lipid carriers (NLC): a potential delivery system for bioactive food molecules. Innov. Food Sci. Emerg. Technol. 19, 29–43.

Tavares, G.M., Croguennec, T., Carvalho, A.F., Bouhallab, S., 2014. Milk proteins as encapsulation devices and delivery vehicles: applications and trends. Trends Food Sci. Technol. 37, 5–20.

Titoria, P., Groves, K., 2015. Nanotechnology for the Food Industry. Nano Magazine. Available from: http://www.nanomagazine.co.uk/index.php?option=com_content&view=article&id=56:nanotechnology-for-the-food-industry&Itemid=151

Torres-Giner, S., Martinez-Abad, A., Ocio, M.J., Lagaron, J.M., 2010. Stabilization of a nutraceutical omega-3 fatty acid by encapsulation in ultrathin electrosprayed zein prolamine. J. Food Sci. 75 (6), N69–N79.

Trevaskis, N.L., Charman, W.N., Porter, C.J., 2008. Lipid-based delivery systems and intestinal lymphatic drug transport: a mechanistic update. Adv. Drug Deliv. Rev. 60 (6), 702–716.

van der Sman, R.G.M., 2012. Soft matter approaches to food structuring. Adv. Colloid Interface Sci. 176, 18–30.

Wang, X., Luo, Z., Xiao, Z., 2014. Preparation, characterizationand thermal stability of β-cyclodextrin/soybean lecithin inclusion complex. Carbohydr. Polym. 101, 1027–1032.

Weiss, J., Decker, E.A., McClements, D.J., Kristbergsson, K., Helgason, T., Awad, T., 2008. Solid lipid nanoparticles as delivery systems for bioactive food components. Food Biophys. 3 (2), 146–154.

Weiss, J., Gaysinsky, S., Davidson, M., McClements, J., 2009. Nanostructured encapsulation systems: food antimicrobials. Global Issues Food Sci. Technol. 1, 425–480.

Wu, C., Böttcher, C., Haag, R., 2015. Enzymatically crosslinked dendritic polyglycerol nanogels for encapsulation of catalytically active proteins. Soft Matter 11 (5), 972-L980.

Yan, M., Ge, J., Liu, Z., Ouyang, P., 2006. Encapsulation of single enzyme in nanogel with enhanced biocatalytic activity and stability. J. Am. Chem. Soc. 128 (34), 11008–11009.

Yang, Y., Gu, Z., Zhang, G., 2009. Delivery of bioactive conjugated linoleic acid with self-assembled amylose−CLA complex. J. Agric. Food Chem. 57 (15), 7125–7130.

Zabar, S., Lesmes, U., Katz, I., Shimoni, E., Bianco-Peled, H., 2009. Studying different dimensions of amylose–long chain fatty acid complexes: molecular, nano and micro level characteristics. Food Hydrocoll. 23 (7), 1918–1925.

Zabar, S., Lesmes, U., Katz, I., Shimoni, E., Bianco-Peled, H., 2010. Structural characterization of amylose-long chain fatty acid complexes produced via the acidification method. Food Hydrocoll. 24 (4), 347–357.

Zhang, S., Kawakami, K., 2010. One-step preparation of chitosan solid nanoparticles by electrospray deposition. Int. J. Pharm. 397, 211–217.

Zuidam, N., Nedovic, V., 2010. Encapsulation Technologies for Active Food Ingredients and Food Processing. Springer Science LLC, Springer-Verlag, New York, NY.

Zuidam, N.J., Shimoni, E., 2010. Overview of microencapsulates for use in food products or processes and methods to make them. Springer, Springer-Verlag, New York, NY, pp. 3–29.

Part One

Lipid-Formulation Based Nanoencapsulation Technologies

2 Encapsulation by nanoemulsions

Seid Mahdi Jafari, Paraskevi Paximada**, Ioanna Mandala**, Elham Assadpour*, Mohammad A. Mehrnia**
*Gorgan University of Agricultural Sciences and Natural Resources, Gorgan, Iran;
**Agricultural University of Athens, Athens, Greece

2.1 Introduction

Many food ingredients and products exist either entirely or partially in the form of emulsions, for example, beverages, butter, desserts, margarines, cake batter, dressings, sauces, soups (Chung and McClements, 2014; Dolz et al., 2006; McClements, 2005; Protonotariou et al., 2013), in which the majority belongs to the category of O/W emulsions (Chung and McClements, 2014).

The droplet size of an emulsion, in the order of 10 to a few hundred nanometers, is indicative for its characterization as nanoemulsion. However, nanoemulsions should not be confused with microemulsions, which have smaller droplet size (i.e., 5–100 nm) and are thermodynamically stable, but are spontaneously created by shelf-assembly (Jafari et al., 2006; McClements, 2012).

According to Sivakumar et al. (2014), nanoemulsions are isotropic colloidal systems. The nanometer-sized droplets overcome the usual problem of creaming. On the contrary, they present a long-term physical stability, known as steric stabilization. An explanation to that is the high Brownian motion of the droplets compared to their gravitational separation force resulting in increased resistance to flocculation (Maswal and Dar, 2014). Therefore, these fine emulsions of nanodimensions are sometimes referred as "approaching thermodynamic stability" (Tadros et al., 2004). In fact, they are considered as kinetically stable taking into account that the creaming or sedimentation velocity depends proportionally on the square of the droplet size (Ezhilarasi et al., 2013). In the current chapter, both nano- and microemulsions, as a potential encapsulation matrix, will be discussed; however, for readability we will use the term nanoemulsions for both of them.

Food industry has been interested in fabrication of emulsions exploiting green or new ingredients of different sources. In particular, food-grade emulsifiers or stabilizers are a new demand in the food industry via employing food-waste products. Emulsion technology can be then applied for the encapsulation of bioactive compounds during food storage.

Nanoencapsulation is defined as a technology to pack substances in miniature, in other words encasing the bioactive compounds at the nanoscale (Quintanilla-Carvajal et al., 2010). Nanoemulsions are considered as ideal vehicles for nanoencaplulation. They serve as nanocarriers in many different materials as they are capable of protecting the bioactive components from degradation due to environmental stress expose and improve their physical attributes, such as solubility.

Oil in water emulsions serve as carriers of lipophilic active agents; such as plant sterols, carotenoids (like β-carotene), α-tocopherol, dietary fats, or a combination of these compounds via different methods. Generally, the lipophilic compounds are protected in these emulsions since the oil phase entails nutritional oils. On the other hand, water in oil emulsions (W/O) can be used to encapsulate hydrophilic compounds including most polyphenols (Zuidam and Shimoni, 2010).

Formulated matrices can be used in their liquid state or be often dried via different techniques like, spray or freeze-drying to obtain the powder forms. The benefits of nanoemulsions concerning their high stability and bioavailability are extensively reported accordingly in the literature (Jafari et al., 2015). In this chapter, recent approaches employed to produce nanoemulsions are discussed, as well as applying these nanostructures in nanoencapsulating food bioactive compounds and nutraceuticals.

2.2 Materials used for preparing nanoemulsions

O/W or W/O nanoemulsions have an oil or a water phase, dispersed in a continuous phase of water or oil accordingly (McClements, 2005). A brief review of the major components that can be used to formulate nanoemulsions is presented in the current section.

2.2.1 Oil

When an O/W or W/O nanoemulsion is to be prepared the oil phase is quite important as its attributes influence the stability of the emulsion, but the emulsion sensory characteristics upon the final application. Triacylglycerols, free fatty acids, essential oils (EO), mineral oils, fat substitutes, waxes, or combination of them can be used.

Stability factors that are influenced could be the ratio of polar (water soluble) to nonpolar (water insoluble) components of an oil, with a great amount of polar components leading to the Oswald ripening, the molecular weight, the interfacial tension, the chemical stability upon stress, the density and the viscosity of the oil phase influencing also the emulsification process (McClements, 2011).

Some oils that are often found in the food industry include corn, olive, soybean, sunflower, coconut, canola, peanut, cottonseed, fish, and flaxseed or algae oils. Oils containing long-chain triglycerides (LCT) are mostly used, nevertheless medium-chain triglycerides (MCT), and in certain cases short-chain triglycerides (SCT) oils (McClements, 2011).

MCT and LCT oils have high interfacial tension, viscosity and low polarity that make them less efficient for producing emulsions of low particle size, but are considered to be more stable under storage as their large molar volume makes them insoluble in water (McClements and Rao, 2011; Wooster et al., 2008). For instance, nanoemulsions production containing these oils and using the phase-inversion temperature method is rather difficult as they are highly hydrophobic (Witthayapanyanon et al., 2006). Moreover, according to Chang et al. (2013), it was not possible to produce nanoemulsions, when LCTs (i.e., corn and canola oil) were used, whereas MCT oils did produce stable nanoemulsions. On the whole, the possible mechanism for the results were unknown and should be further studied.

Similarly, when a high pressure homogenization technique is used the high viscosity of these oils limits droplet disruption (Do et al., 2009; Palamakula and Khan, 2004; Wooster et al., 2008; Yu et al., 2012). Nanoemulsions made with high viscosity oils, such as LCT were considerably larger than nanoemulsions prepared with low viscosity oils. Moreover, it is the oil volume fraction ϕ that significantly contributes to the final droplet size. Most O/W nanoemulsions have a very low oil volume fraction (e.g., ϕ 0.01–0.1), as higher ones lead to an increase in droplet size; practically hindering the formation of nanoemulsions (Wooster et al., 2008). On the other hand, EO, known as "volatile oils" as well, because they contain volatile aroma compounds, seem to be quite attractive in producing nanoemulsions, mostly in blends with other oils. They present the advantage to serve as natural preservatives as well.

2.2.2 *Emulsifiers*

A successful design of nanoemulsion involves the use of the appropriate emulsifier(s) that will act on the surface of the dispersed phase, been adsorbed on it, leading to further droplet disruption and hindering droplets recoalescence or aggregation (Kralova and Sjöblom, 2009; McClements, 2005).

There are several emulsifiers been exploited in the food industry, categorized as small molecule surfactants, phospholipids, proteins, or even polysaccharides (e.g., casein, β-lactoglobulin). Food-grade emulsifiers, such as polysaccharides are recently used in the food industry including hydrocolloids or modified starches. Emulsifiers have a specific interfacial behavior, they decrease the interfacial tension of the oil and water phases, but could balance electrostatic forces or result in steric repulsion, rheology change, and increased loading capability (Donsì and Ferrari, 2016; Henry et al., 2009; Kralova and Sjöblom, 2009). When formulating nanoemulsions, the surfactant concentration is quite important. As the particle surface increases, a greater amount of emulsifier is required to completely cover the oil droplets avoiding any destabilization.

Concerning the small molecule surfactants use in nanoemulsion formulation, there is a great variety. Some examples of nonionic surfactants are sorbitan esters (e.g., Span and Tween series, mostly 40, 60, or 80), polyoxyethylene ethers (e.g., Brij), monoglycerides, sugar esters, and polyglycerol esters of monolaurate (Cheong et al., 2016; Henry et al., 2009; Yin et al., 2009). Small molecule surfactants can form easily small droplets, compared to biopolymers. Due to their large molecules, proteins, or polysaccharide-based emulsifiers, adsorbed at the interfacial layer, lead to a significant droplet size increase. The interested reader is referred to the articles published by: Kaltsa et al. (2014) and Qian and McClements (2011).

However, proteins and polysaccharides impart a good emulsion stability either by electrostatic or by steric stabilization (Khouryieh et al., 2015; Paximada et al., 2016a; Wilde et al., 2004; Zinoviadou et al., 2012).

Another point is that the emulsifier type is determined by the homogenization technique applied. Small molecule surfactants are considered to effectively produce nanoemulsions regardless of the method used. In contrast, proteins and polysaccharides are not typically suitable for formulating small droplets using low-energy methods; meanwhile they are less effective at forming nanoemulsions via high-energy methods (Henry et al., 2009). However, proteins and polysaccharides, as biopolymers, have the advantage to be considered "label-friendly", which makes them interesting agents for producing nanoemulsions.

2.2.3 *Stabilizers*

Substances with thickening or gelling properties act as stabilizers. In emulsions, they are added into the aqueous phase and due to a viscosity increase oil droplets movement is then retarded. However, modifying the aqueous phase rheology, changes also the mouth feel and texture (e.g., creaminess, gel strength) and new structures are created (Dickinson, 1992, 2003). Most stabilizers used in emulsions are hydrocolloids, such as gum arabic, pectins, xanthan, modified starch, alginates, galactomannans, and chitosan. Some of them present emulsification properties due to either the nonpolar chemical groups or the presence of protein moieties that enable them to act at the oil–water interface. Some galactomannans, gum arabic, and bacterial cellulose belong to that category (Deshmukh et al., 2012; Kong and Singh, 2008; Kaltsa et al., 2016; Paximada et al., 2016b). Paximada et al. (2016b) dealt in detail with the stabilizing methods of bacterial cellulose and they observed that it is capable of stabilizing O/W emulsions by electrostatic and steric repulsion to prevent droplet flocculation and coalescence.

Finally, a blend of protein and polysaccharide is often used in an emulsion contributing in both its formation and stabilization. There are two ways in the formulation procedure, either to use a protein emulsifier followed by the addition of a polysaccharide that will increase the viscosity, thereby boosting the overall emulsion stability (Dickinson, 2003) or to form a complex between the protein and polysaccharide that will act at the interface. Paximada et al. (2016a) studied the effect of bacterial cellulose on the properties of protein-coated O/W emulsions. All in all, the addition of bacterial cellulose increased the emulsion viscosity and inhibited creaming during storage.

2.3 Preparation methods

Nanoemulsions are fabricated utilizing different or combined techniques. Generally, there are two main categories known as either high-energy or low-energy approaches depending on the energy input requirements (Tadros et al., 2004).

High-energy methods use a high mechanical energy. Their main advantage is that the size distribution and composition is controlled. Typically, the droplet size is controllable as it decreases with increasing energy input and duration (Walker et al., 2015), although there are some reports about droplet recoalescence during a high energy emulsification (Jafari et al., 2008a).

High energy methods provide a broad choice of ingredients, which in turn determine the stability, rheology, and organoleptic properties, such as color and mouth feel resulting in nanoemulsions with desired properties (Chime et al., 2014). Their main disadvantages include: chemical degradation due to the intense production process, as well as not being potential to be scaled up (Walker et al., 2015). However, according to other scholars they have industrial scalability but the equipment is more sophisticated and requires a careful approach (Chime et al., 2014). Energy consumption is still a treatable issue. The actual amount of energy per unit volume expended in fabrication is very high, ranging from around 2×10^7 J/m^3 to 3×10^8 J/m^3 depending on the volume fraction and the droplet size. The energy is so high as only a small amount is spent in droplet breakup. Around 99.8% of the energy, in high pressure homogenizer (HPH), is converted to heat (Schubert & Ax, 2003) or according to Tadros et al. (2004) a HPH has an efficiency of only 0.1%, where 99.9% is dissipated as heat during preparing emulsion of 600 nm in diameter.

For most of the common homogenizers theoretical or semiempirical equations are applied to estimate the energy density. In a high-pressure valve homogenizer the operating pressure is equal to the energy density that is $P_{\mathrm{H}} = E_{\mathrm{v}}$ (Stang and Schubert, 2001).

Finally, high-energy methods alone normally do not result in oil droplets of low size (<100 nm) (Chime et al., 2014). In particular, utilizing natural emulsifiers of large molecular weight hinders the formation of very small droplets.

On the other hand, low-energy emulsification success depends on the surfactants' properties and the oily phase. Their main advantages are low energy consumption and that their scale-up is more straightforward as the costs are minimized, as well as employing simple production methods (Date et al., 2010; McClements and Rao, 2011; Walker et al., 2015).

Low-energy emulsification drawbacks, regardless of the method, can be the requirement of large amounts of surfactants, typically needed in some cases. Natural surfactants have to be further investigated for food applications. Moreover, emulsifiers that are natural biopolymers, such as proteins or polysaccharides, can only be used to form nanoemulsions by high-pressure methods (Qian and McClements, 2011).

Advantages and disadvantages according to preparation methods are summarized in Table 2.1 and an overview of nanoemulsion preparation techniques is given in Table 2.2.

Prior to analyzing the specific nanoemulsion preparation methods, a schematic overview of them is depicted in Fig. 2.1.

2.3.1 High energy methods

An emulsion can be formulated in a single step but more frequently by following a number of consecutive steps. At the beginning separate oil and water dispersions are prepared containing the ingredients that are soluble in those phases respectively. Oil-soluble surfactants and other bioactive ingredients, like lipophilic vitamins that are to be included in the final emulsions, are added in the oil, whereas water-soluble surfactants and other hydrophilic ingredients, such as sugars, salts antioxidants are added in

Table 2.1 Advantages and disadvantages of high and low energy methods of nanoemulsion fabrication

Technique/points	High energy	Low energy
Techniques	• High speed homogenizer • High pressure valve homogenizer • Microfluidizer • Ultrasounds-based devices	• Phase inversion by phase mixing • Phase inversion by altering the conditions of the phase • Membrane emulsification
Advantages	• Control of composition • Control of stability • Production of emulsions with desired properties	• Low energy consumption • Low manufacturing costs • Simple production methods
Disadvantages	• Energy consumption • Chemicals degradation • Limits in droplet size • Recoalescence	• Surfactant choice-edible materials • No use of biopolymers • Stability is influenced by environmental parameter • A large concentration of surfactant may be necessary • Droplet polydispersity
Under consideration	• Scale-up: sophisticated careful approach • Core degradation upon storage	• Surfactants' use • Core degradation upon storage

the aqueous phase. The uniformity in the desired size of the droplets is a critical issue to produce a product of high quality and it is therefore important to take into consideration the major factors that influence droplets size and uniformity minimizing the energy requirements as well.

2.3.1.1 High speed and high shear homogenizers

The most commonly used method to homogenize an oil and aqueous phase mix is by using a high-speed colloid mill. In a batch process, all dispersions and single ingredients are placed in a vessel, which may have a small volume of a few cubic centimeter for laboratory use or a larger one of several cubic meter for industrial use. A rotational head at a speed of up to 30,000 rpm is served to agitate all ingredients. A particular characteristic of a high speed homogenizer is the high distribution of the specific energy density creating low and high shear regions. The rotational speed and the geometry of the homogenizer are important parameters. In particular, the design of the homogenizer could be significant for the efficiency of the homogenization and the final viscosity. Propellers and turbines are used for preparing low and medium-viscosity emulsions, whereas an agitator capable of scraping the walls of the container is required in high viscosity emulsions. Moreover, air incorporation leading to foam formation is not desirable (Becker et al., 1985; Schubert et al., 2003).

Table 2.2 **Overview of nanoemulsion preparation techniques**

Technology	Process steps	References
Hot homogenization technique	1. Melting the lipid and dissolving/dispersing the bioactive compound in the lipid 2. Dispersing the bioactive-loaded lipid in hot aqueous surfactant mixture 3. Premixing using a stirrer to form a coarse preemulsion 4. High pressure homogenization at temperatures above lipid melting point 5. Hot O/W nanoemulsion 6. Solidification of the nanoemulsion by cooling down to room temperature	Radtke and Müller (2001)
Cold homogenization technique	1. Melting the lipid and dissolving/dispersing the bioactive compound in the lipid 2. Solidification of the bioactive loaded lipid in liquid nitrogen or dry ice 3. Grinding in a powder mill (50–100 μm) 4. Dispersing the powder in an aqueous surfactant dispersion medium (premix) 5. High pressure homogenization at room temperature or below	Mehnert and Mäder (2001)
High pressure homogenization	1. The lipid is pushed with high pressure (100–2000 bars) through a very high shear stress 2. Disruption of particles down to the submicrometer or nanometer range	Liedtke et al. (2000)
Solvent emulsification–evaporation method	1. Lipids and bioactive compounds are dissolved in a water immiscible organic solvent with low boiling point 2. The solution is then emulsified in the aqueous emulsifier solution 3. Solvent removal in rotary evaporation at 50–60°C	Sjöström and Bergenståhl (1992); Wissing et al. (2004)
Solvent emulsification-diffusion technique	1. Both the solvent and water are mutually saturated in order to ensure the initial thermodynamic equilibrium of both liquids 2. Lipid and bioactive compound are dissolved in water saturated solvent and this organic phase is stirred using mechanical stirrer 3. After formulation of O/W emulsion, water in typical ratios from 1:5 to 1:10, is added to the system in order to allow solvent diffusion into the continuous phase, thus leading to the aggregation of the lipid in the nanoparticles	Trotta et al. (2003)

Table 2.2 **Overview of nanoemulsion preparation techniques *(cont.)***

Technology	Process steps	References
Microemulsion technique	1. Lipids are melted and bioactive compound is incorporated into molten lipid 2. Mixture of water, cosurfactant and surfactant is heated to the same temperature as lipids and added under mild stirring to the lipid melt 3. A transparent, thermodynamically stable system is formed when the compounds are mixed in correct ratios for microemulsion formation 4. This microemulsion is then dispersed in a cold aqueous medium under mild mechanical mixing of hot microemulsion with water in a ratio in the range 1:25–1:50	Pardeike et al. (2009)
Melting dispersion method	1. Bioactive compound and solid lipid are melted in an organic solvent regarded as oil phase. Simultaneously water phase is also heated to the same temperature as oil phase 2. The oil phase is added to a small volume of water phase and the resulting emulsion is stirred at high speed for few hours 3. The emulsion is cooled down to room temperature to yield nanoparticles	Reithmeier et al. (2001)
US technique	1. The core material is melted 2. Addition of phospholipids along with an aqueous medium 3. Dispersing the melted material at increased temperatures by ultrasonication	Puglia et al. (2008)
Solvent injection	1. Lipids are dissolved in a water–miscible solvent (e.g., acetone, isopropanol and methanol) or water–miscible solvent mixture 2. Quickly injected into an aqueous solution of surfactants through an injection needle	Schubert and Müller-Goymann (2003)
Double emulsion technique	1. Bioactive compound (mainly hydrophilic ones) is dissolved in aqueous solution emulsified in melted lipid 2. The primary emulsion is stabilized by adding stabilizer that is dispersed in aqueous phase containing hydrophilic emulsifier 3. Emulsion is stirred and filtered.	Esfanjani et al., (2015)

High shear homogenizers (such as Silverson or Ultra-Turrax) are often used as a necessary device for emulsion production. Rotor geometries have coaxial intermeshing rings with radial openings. The rotor results in the fluid acceleration as it enters the center of the system. Thereafter subsequent accelerations and decelerations take place and as a result high tangential/shear forces are produced in a turbulent flow. The

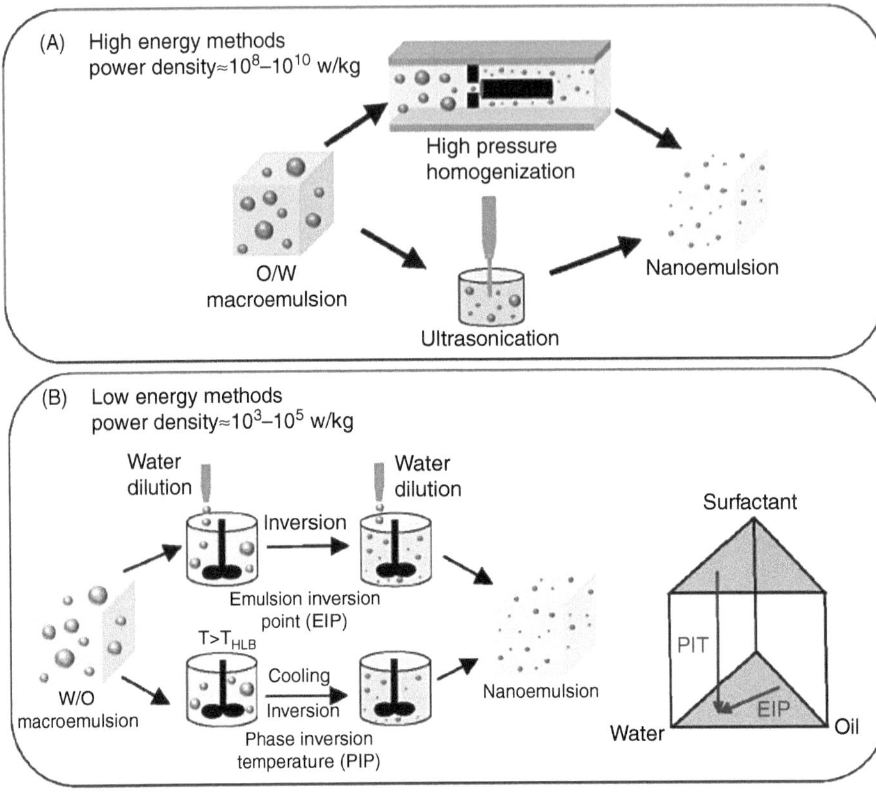

Figure 2.1 Overview of high and low energy methods for preparing nano-emulsions. (A) High energy, such as high pressure homogenization (HPH) and US. (B) Low energy methods, such as emulsion inversion point (EIP) and PIT.
Source: From Gupta, A., Eral, H.B., Hatton, T.A., Doyle, P., 2016. Nanoemulsions: formation, properties and applications. Soft Matter 12, 2826–2841.

droplets are then forced to break up into smaller segments. However, by using those homogenizers the reduction of the droplet size is still not sufficient nor the required monodispersity (Sanguansri et al., 2013; Schubert et al., 2003). Typically, the droplets diameter ranges between about 2 and 10 μm (McClements, 2011). It is of interest to point that some of the mechanical energy is converted into heat due to viscous dissipation leading to a slight temperature increase of an emulsion.

In nanoemulsion production, both approaches described, are used to prepare the primary emulsion. Thereafter a secondary homogenization takes place, where a reduction in droplet size of a coarse emulsion occurs. It is worth noticing that some homogenizers are more suitable to directly prepare a coarse emulsion, whereas others can only be used to reduce the size of the droplets in a primary emulsion. Microfluidizer and ultrasonic are rather suitable for primary homogenization, whereas high-pressure homogenizers normally use preemulsions (McClements, 2015). Different combinations among emulsification devices can be tried in order to achieve the desired droplet size.

However, some studies suggested that encapsulation of compounds in so called coarse emulsions could have beneficial outcomes (Table 2.2).

In particular, some studies have been conducted on the use of high speed homogenizers as a technique to formulate emulsions, with a final goal the encapsulation of different bioactives. Lante and Friso (2013) demonstrated that the production of W/O nanoemulsions as a vehicle to encapsulate epigallocatechin gallate (EGCG) increased the oxidative stability of the catechin and thus improving its bioavailability. Moreover, Di Mattia et al. (2009) dealt with the encapsulation of catechin, quercetin, and gallic acid in emulsions treated through high shear homogenizer. While gallic acid was found to delay lipid oxidation, catechin, and quercetin were not efficient in inhibition of the autooxidation. Their results confirmed that the efficiency of antioxidants is determined by their antioxidant activity apart from polarity. A high activity means that antioxidants are capable of rapidly donate a hydrogen atom in order to act against lipids autooxidation.

2.3.1.2 High pressure valve homogenization

High pressure valve homogenization is the most common high pressure device used to produce nanodroplets (Lee et al., 2014). High-pressure (valve) homogenizers (HPH) (Fig. 2.2A) have a piston pump to force under high pressure a premixed coarse emulsion through a valve of a specific design having a gap of 10–100 μm and developing velocities of hundreds of meters per second (Håkansson et al., 2011; Innings and Trägårdh, 2007; Lee et al., 2014). In addition, some homogenizers are equipped with one or two nozzles instead of valves. The operated pressure is usually in the range of 100–500 MPa.

Briefly, the droplet disruption occurs because of the laminar and turbulent flow at the entrance, therefore creating two flows fluid and cavitation bubbles by pressure decrease and the collapse of bubbles at the end by pressure increase.

There are some arguments about the type of the flow as following: the viscosity change does not influence the final size of the droplets indicating that the flow is laminar elongational rather than turbulent (Lee et al., 2014). However, it is believed that the flow is transformed into turbulent at the exit of the valve gap (Tcholakova et al., 2004). Moreover, tiny droplets can only be formed when viscosity ratio of oil to aqueous phase ranges between 0.05 and 5 in laminar flow conditions (Schubert et al., 2003).

High pressure processing always results in temperature increase of the sample depending on its specific physicochemical properties and the refrigeration system used in the homogenization valve case. At least a 15–20°C shear-induced increase in temperature is typically shown per 100 MPa increment in homogenization pressure.

Moreover, when several cycles are performed to decrease the particle droplet size, coalescence occurs due to the high kinetic energy of the particles (Floury et al., 2000).

Only when very high pressures and multiple passes are used the droplet size is reduced to nanoscale size. Even then droplets less than 100 nm in radius are produced only under certain circumstances (e.g., high emulsifier amounts, low interfacial tensions, and appropriate viscosity ratios).

(A)

(B)

(C)

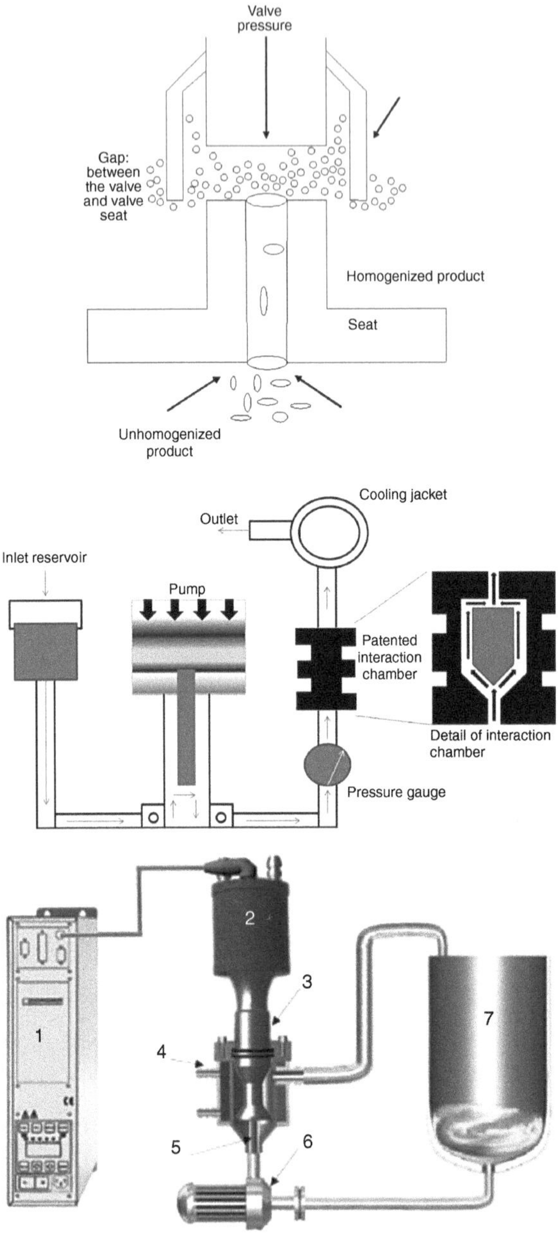

Figure 2.2 Schematic representation of various mechanical devices that can be used to produce food-grade nanoemulsions using a high-energy approach: high pressure valve homogenizer. (A) Ultrasonic probe homogenizer pilot-scale flow-through process schematic. (1) 1200 W ultrasonic generator, (2) piezoelectric transducer, (3) half-wave Barbell Horn, (4) reactor chamber with water-cooling jacket, (5) liquid delivery tube (used to bring the liquid close to the horn's lower output surface), (6) pump, (7) main storage/mixing beaker. (B); microfluidizer (C). *Source:* Part A and C: From Jafari, S.M., Fathi, M., Mandala, I., 2015. Emerging product formation. Food Waste Recovery: Processing Technologies and Industrial Techniques, Elsevier Inc.: pp. 293–317; part B: From Peshkovsky, A., Peshkovsky, S.L., Bystryak, S., 2013. Scalable high-power ultrasonic technology for the production of translucent nanoemulsions. Chem. Eng. Proc. 69, 77–82.

Small O/W nanoemulsion droplets can be prepared by using oils of high viscosities, in a method called evaporative ripening method mentioned by Fryd and Mason (2010). By this method the droplets produced by HPH are a mixture of oils of different molecular weights (low and high), that are volatile and nonvolatile respectively. Upon heating the volatile component in the droplets evaporates leading to droplet shrinkage. They remain then mostly containing the high molecular weight oil (Gupta et al., 2016).

Recently, nanoemulsions are investigated and further developed because there is a growing need to find out new delivery systems that can effectively encapsulate, protect, and release different bioactive ingredients. A number of recent researches that produced food-grade nanoemulsions by HPH are summarized in Table 2.3.

For instance, a lipophilic component is encapsulated for different reasons—to make easier its handling and utilization; to successfully incorporate it in a food; to protect it from environmental stresses; to improve its bioavailability by increasing its solubility and its efficient absorption through cells; to control the rate of its release, as well as localization. Lipophilic compounds are usually incorporated in the oil phase prior to nanoemulsion formation. When the emulsification process begins, they stay inside the oil droplets. Among others, characteristics, such as solubility and hydrophobicity, surface activity, molecular weight, and melting point determine their encapsulation efficacy.

Several investigations have exerted high pressure valve homogenization to encapsulate EO in nanoemulsions (Ghosh et al., 2013; Salvia-Trujillo et al., 2013, 2015).

Their direct incorporation into foods presents several technological limitations. They should be adequately dispersed in the food matrix, not interact with other ingredients, and maintain their activity for a prolonged time (Donsì et al., 2012, 2014). O/W nanoemulsions are ideal candidates as carriers for EOs, due to the affinity for EO components, their wide food compatibility, simple industrial scalability of their manufacturing process, as well as their ability to cross biological membranes (McClements et al., 2007).

2.3.1.3 Microfluidizer

A Microfluidizer is a patented device. It consists of an interaction chamber having microchannels of fixed geometry (Fig. 2.2B), where under high pressure (35–140 MPa) a coarse emulsion is added. The coarse emulsion flow is split into two streams that are redirected into the chamber allowing their impingement at approximately 180°C (Jafari et al., 2007a,b). The impact of the two high velocity and high pressure streams creates a region of high turbulence and shear capable of droplet disruption. At the exit of the chamber the fluid is subjected to elongational flow as in high-pressure homogenizers. Thanks to that elongation, the adsorption of the emulsifier at the newly formed interfaces is facilitated, resulting in lower coalescence (Henry et al., 2010). The process can be repeated until the desired particle size is obtained.

In general, microfluidizer and HPH have similar emulsification efficiency in W/O emulsions, forming small droplets of about 60 nm diameter in the same pressure of 50 MPa. Multiple passes are required to drop down the droplet size regardless of the device used (Lee et al., 2014). In order to reduce recoalescence of broken up droplets, the application of multiple axial flow nozzles (or orifices) in one microsystem has been proposed. A single passage is adequate to produce small droplets. It was

Table 2.3 An overview of recent studies on nanoencapsulation of food ingredients by nanoemulsion formation through high energy methods

Emulsification method	Oil phase	Emulsifier/surfactant	Applied core material	Droplet size (nm)	References
High-speed homogenizer (HS)	Olive oil	Tween 20	Gallic acid, catechin, quercetin	1716	Di Mattia et al. (2009)
HS	Soy, peanut, sunflower, corn oil	Lecithin	EGCG	20–50	Lante and Friso (2013)
HS	Canola oil	Maltodextrin–Whey protein concentrate	Folic acid	30–90	Assadpour et al. (2016a, b)
HS	Soybean oil	Whey protein concentrate, pectin, span	Olive leaf extract	6–1450	Mohammadi et al. (2016a,b)
High pressure homogenizer (HPH)	Stearin-rich milk fat	Sodium caseinate	α-Tocopherol	100	Relkin et al. (2008)
HPH	MCT	Soy lecithin	Ubiquinol-10	100–200	Zhao et al. (2013)
HPH	MCT	Tween 20, 40, 60, 80	β-Carotene	132–148	Yuan et al. (2008)
HPH	d-limonene	Angum gum, Arabic gum	D-limonene	1600–3200	Jafari et al. (2012)
High pressure valve homogenizer (HPH)	Palm kernel oil	Lecithin	Black tea extract, ascorbic acid, naringenin, tangeretin, β-carotene, tanshinone	100–220	Yang et al. (2011)
HPH	Peanut oil	Lecithin, Tween 30, Pea protein	Carvacrol	100–2000	Donsì et al. (2014)
Microfluidizer	MCT	Tween 20,80	β-Carotene	100	Jo and Kwon (2014)
Microfluidizer	MCT	Glycerol/soy lecithin	α-Tocopherol	85–381	Hatanaka et al. (2010)
Microfluidizer	Peanut oil	Tween 80	—	120	Wooster et al.

Microfluidizer	D-limonene or fish oil	Waxy starch/ maltodextrin	—	550	Jafari et al. (2007a,b)
Microfluidizer	Flax seed oil	Tween20	EGCG	200	Zhou and Elias (2013)
Microfluidizerand Ultrasounds	Aspirin	Cremophor EL	Aspirin	150–450	Tang et al. (2013)
Ultrasounds	Fish oil	Tween 80, Span 80	Fishoil	80–100	Dey et al. (2012)
Ultrasounds	Sunflower oil	Tween 80, Span 80	Canola oil	150–190	Leong et al. (2009)
Ultrasounds	Basil oil	Tween 80	Basil oil	30	Ghosh et al. (2013)
Ultrasounds	D-limonene	Waxy corn starch/ maltodextrin	—	650	Jafari et al. (2007a,b)
Ultrasounds	Paprika oleoresin	(Tween 40) and Span 20)	Carotenoidins, papika oleoresin	50–250	Pascual-Pineda et al. (2015)
Ultrasounds	D-limonene	Sorbitane, polyoxy ethylene (20)	D-limonene	50–350	Lu et al. (2014)
Ultrasounds	Sunflower oil	Tween 20, Bovine serum albumin	Caffeic extract		Almajano et al. (2007)

interesting to find that a double orifice operated more efficiently and to be virtually independent of the formulation properties (i.e., emulsifier adsorption kinetics, continuous phase viscosity).

With these properties, microfluidizer gains more and more interest as a process to produce nanoemulsions containing encapsulants. Some of the recent studies showing that have been summarized in Table 2.3.

It is of high interest the use of microfluidizer to encapsulate drugs for pharma applications. Pharmaceutical emulsion consists of various oils and/or fats dispersed throughout the aqueous phase in the presence of an emulsifier. Drugs; such as aspirin, and ubiquinol-10 has been successfully encapsulated with a microfluidizer device (Tang et al., 2013).

Polyphenols have been investigated in further applications in food, nutraceutical, and pharma (Scalbert et al., 2005).

Through encapsulation their stability is improved, as any unpleasant taste or flavor is masked. Nutritionally, their bioavailability and half-life in vivo and in vitro are improved. Microfluidizer exploitation in encapsulation technology is summarized via some interesting studies in Table 2.3.

A recent study indicates that β-carotene nanoemulsions, as potential active ingredients carrier in liquid form, can be prepared using microfluidizer (Jo and Kwon, 2014). They postulated that the pressure and number of cycles of an homogenizer had a positive effect on the emulsions' particle size and content of β-carotene. What is more, Zhou and Elias (2013) evaluated the effect of the pH and the phenolic content on the oxidation of nanoemulsions that contained EGCG. Specifically, low pH values showed a prooxidant effect on the nanoemulsions prepared with microfluidizer.

2.3.1.4 Ultrasound-based devices

Apart from high pressure devices, an ultrasound generator has been patented since 1944. Furthermore, it is considered an effective way of producing nanoemulsions. High intensity ultrasonication (US) with a frequency range between 16 and 100 kHz, and power density of 10 to 1000 W/cm^2 results in the formulation of oil droplets in nanoscale (up to 100 nm), with improved stability over time and the addition of smaller amounts of emulsifying agents (Chemat et al., 2011). When the Laplace pressure is dominant, microturbulent implosions of the produced cavitation bubbles provide huge forces that deform and finally break-off the droplets in nanometer dimensions (Sivakumar et al., 2014).

Commercial-scale ultrasonic emulsification requires the use of a sonochemical reactor in an industrial size in order to control the operating parameters. The ultrasonic amplitude, which is related to the intensity of cavitation-generated shear forces should be high enough in order that the emulsification process progresses efficiently. The problem is that in an industrial scale, when the size of ultrasonic horn-based processor is increased, the amplitude should be significantly decreased in order to be adjusted to a significant cavitation zone volume. Amplitude should be about 75–100 μm to produce high quality nanoemulsions, whereas the ultrasonic transducers generate vibration amplitudes of about 25 μm, thus a high-gain acoustic horn that augments the amplitude of vibration is required (Peshkovsky et al., 2013). When converging horns are used, the produced amplitudes are high, as long as their diameter remains very low at about 15–20 mm. Large volumes in process are then excluded.

The Barbell Horn Ultrasonic Technology (BHUT) was based on the discovery that the reduction in horn diameter is not necessary connected to the amplitude level. Thus, high-intensity ultrasound can be produced by high in diameter horn that have a specific shape and include a surface above the output section (Fig. 2.2C) (Peshkovsky et al., 2013).

US is then considered more appropriate in terms of production costs, as well as equipment contamination compared with microfluidization (Freitas et al., 2006; Jafari et al., 2008a; Karbstein and Schubert, 1995). Several studies have used ultrasonic homogenization to produce nanoemulsions with the possibility of being used in the food and pharma industry as shown in Table 2.3.

In a recent study, a successful oil-in-water nanoemulsion of paprika oleoresin was prepared using ultrasounds (Pascual-Pineda et al., 2015). The carotenoids stability was improved by particle size reduction. As far as the usage of ultrasounds in pharma application is concerned, several studies indicate the successful production of nanoemulsions for skin-care products. As it is known, a low droplet size may aid permeation through skin and transdermal delivery efficiency. Specifically, Lu et al. (2014) encapsulated D-limonene in nanoemulsions of a mean droplet size of 50–350 nm and the encapsulation efficiency varied from 70 to 92%. The droplet size determined the permeation rate; the emulsion with the lowest droplet size (54 nm) resulted in a high permeation rate. The epidermal region of permeated rat skin was compact with no obvious voids, so the D-limonene nanoemulsion may be an efficient matrix for transdermal drug delivery.

2.3.2 Low energy methods

Despite high energy approaches being quite straight forward for the formation of nanoemulsions, they are still not considered cost-effective methods. With this in mind, only a small percentage of the total energy available; approximately 0.1% is finally used for emulsification (Tadros et al., 2004), a route which is generally unfavorable for industrial applications.

Nanoemulsions may also be formed by employing/manipulating variables; such as the composition of the system (surfactants and cosurfactants), and mixing conditions (i.e., rate of addition, stirrer speed, temperature) generally referred to as low-energy emulsification methods. Actually, the energy of the system that is stored is being used to form a nanoemulsion and this procedure takes place when the change in entropy that favors dispersion is higher than the energy required to increase the surface area of the dispersion (Reiss, 1975).

Briefly, the low-energy methods can be distinguished in two main categories: phase inversion by simple phase mixing, and phase inversion by changing the hydrophilic–lipophilic balance through altering the conditions of the system.

The two commonly low-energy emulsification methods are the phase inversion and the spontaneous emulsification (SE). There are also two other methods, the phase inversion temperature (PIT) and the phase inversion composition (PIC) (Komaiko and McClements, 2014; McClements, 2011). Recent studies, which have used low energy methods to produce nanoemulsions are presented in Table 2.4 that could be potentially used in the food industry.

Table 2.4 An overview of recent studies on nano-emulsion formation using low energy methods

Emulsification method	Oil phase	Emulsifier/ surfactant	Applied encapsulant	Droplet size (nm)	References
Phase-inversiontemperature (PIT)	MCT	Span 80	NaCl	10–200	Anton et al. (2010)
Spontaneousemulsification	Palm oil	Tween 80	Vitamin E	90–140	Teo et al. (2010)
SE	MCT	Soy lecithin	Castor oil	150–200	Kelmann et al. (2007)
SE	Grape seed oil & orange oil	Tween 80	Reservatrol	140–180	Davidov-Pardo and McClements (2015)
SE	Soybean oil	Whey protein concentrate, pectin, span 80,	Olive leaf extract	6–1450	Mohammadi et al. (2016a,b)
SE	Olive oil	Span 80, PGPR	Crocin	10–40	Mehrnia et al. (2016)
SE	Caproyl MCM EP	Tween 60, PEG-400	Hydrochloride	120–170	Narkhede et al. (2014)
Phase inversion technique and self-emulsification	Avocado oil	Sucrose esters, glycerol	Avocado oil	106	Eid et al. (2015)
Membrane emulsification	MCT oil	Surfactant mixture Tween 80/Brij 35	Vitamin E	78	Laouini et al. (2012)
Membrane emulsification	Palm oil	Tween 20, whey protein isolate	Astaxanthin	2000	Ribeiro et al. (2005)

2.3.2.1 Phase inversion by phase mixing

Emulsion phase inversion or emulsion inversion point (EPI or EIP) is a method, in which an emulsion is formed by adding water to an oil-surfactant mixture. It relies on catastrophic phase inversion, meaning that the ratio between the oil and aqueous phase changes, whereas any other changes in surfactant's properties is not included. In this method, the water is added under stirring in the oily phase. Initially, a water in oil emulsion is formed, but when excess water is added a phase inversion occurs and the water becomes the continuous phase of dispersed oil droplets, forming an oil in water emulsion. Forgirani et al. (2001) observed that kinetically stable nanoemulsions having very small droplet size (~50 nm) can be then formulated. Meanwhile, there is another similar method called SE in which an emulsion is formed when an oil and surfactant mixture is added to water (Walker et al., 2015).

The main factor which differentiates the SE and EPI from all other preparation methods is that the oil and surfactant phases are intimately mixed together, prior to combining them with water. Furthermore, there must be miscibility between the two mixed components before production. The surfactant is also of great importance since it must be capable of moving from the oil into the water phase (Komaiko and McClements, 2014).

It is worth mentioning that the term emulsion phase inversion is used by many researchers when the nanoemulsions are formed either via progressive dilution with water or through dilution with oil (Anton et al., 2008; Sadtler et al., 2010).

2.3.2.2 Phase inversion by altering the conditions of the phase

A third method apart from SE and EPI is to rapidly change the temperature of a surfactant–oil–water mixture below the PIT under mixing, as this method employs the temperature-dependent solubility of nonionic surfactants that consequently changes their hydrophilic–lipophilic character (Chime et al., 2014). Typically, a mixture of appropriate composition of oil, water, and nonionic surfactants is prepared at ambient conditions, usually resulting in an O/W emulsion in excess oil. Then, it is heated above the PIT and rapidly cooled under continuous stirring. This method relies on the fact that temperature alterations modify the solubility or the molecular geometry of nonionic surfactants, which change the type of the emulsion as well. There is an intermediate temperature known as HLB temperature, where the surfactant affinity for aqueous and oil phase is very similar and the interfacial tension of the system becomes very low. At higher temperatures, the surfactant is completely solubilized in the oily phase and the initial O/W emulsion undergoes a phase inversion to W/O emulsion. Indeed, a suddenly broken-up nanoemulsion is maintained by a rapid cooling (Anton and Vandamme, 2009). A rapid cooling could begin either at HLB temperature, where a bicontinuous system is created, or at temperatures above, where phase inversion occurs (Fig. 2.3) (Chime et al., 2014; Shinoda and Saito, 1968; Walker et al., 2015).

The PIC method is similar to PIT. It is based again on a solubility change or optimum curvature of the surfactant. However, it is the composition, for example, salt concentration, that is changed and not the temperature (Walker et al., 2015). PIC method could be more suitable for large scale productions.

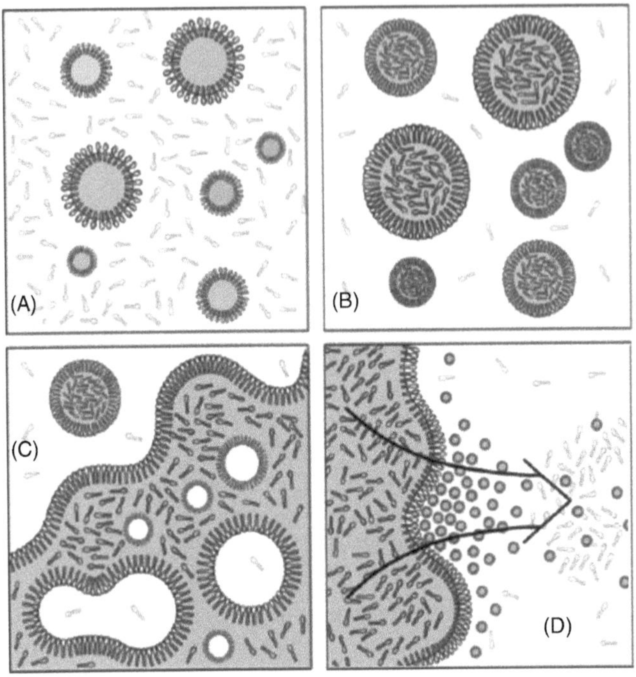

Figure 2.3 The mechanism of nanoemulsions generation by a PIT method. A water/nonionic surfactant/oil system undergoes phase inversion. (A) *T* is below the PIT, a macroemulsion where surfactants are mostly hydrophilic, (B) *T* is increased; the surfactants become lipophilic and are solubilized by the oily phase, (C) *T* is at the PIT, bicontinuous microemulsion form, (D) temperature is brought above the PIT, the emulsion is inverted and then water is dispersed into the mixture of oil and lipophilic surfactant. The system is then rapidly cooled using water dilution, making the surfactant suddenly hydrophilic, causing a spontaneous and rapid migration to the aqueous phase. As in spontaneous emulsification (SE), this turbulent displacement induces the generation of nanoemulsions.
Source: Adapted from Anton, N., Vandamme, T.F., 2009. The universality of low-energy nano-emulsification. Int. J. Pharm. 377 (1-2), 142–147.

As it is clear, all methods described are isothermal methods, whereas the PIT is an example of a thermal one (Komaiko and McClements, 2014). It includes the disruption of the microemulsion maintained at the phase inversion followed by a rapid cooling.

PIT technique has several limitations; such as the use of a large amount of surfactant, careful selection of both surfactant-cosurfactant and results in a high polydispersity index (PDI). After long-term storage stability problems can be observed and relatively difficult and tedious work load is required to identify the system inversion temperature (Sivakumar et al., 2014). However, stability can be controlled by choosing the appropriate storage temperature. The high amount of synthetic surfactants should be noted as they may be unsuitable for some food applications (Komaiko and McClements, 2014).

The PIT method has been used in many studies presents novel for encapsulating hydrophilic ingredients in lipid nanoemulsions (Anton et al., 2010). The nanoemulsions formed were stable for a long times, and exhibited droplet sizes were from 10 to 200 nm. Hydrophilic model molecules of fluorescein sodium salt were encapsulated in the oily core of these nanoemulsion droplets through their solubilization in the reverse micellar system. Multiscaled nanodroplets were produced with a "hydrophilic ingredient in a reverse-micelles-in-oil-in-water" structure. Once fluorescein had been encapsulated, it remained thermodynamically stable, and the encapsulation yields reached 90%.

2.3.2.3 Membrane emulsification

Another procedure to produce emulsions is membrane emulsification, which involves the formation of a phase dispersed through a membrane. It is a method that can be classified in low-energy processes, although, for some others, belongs to high energy approaches. According to Schubert (1997) a lower energy of about 100 times, less than that demanded by high pressure homogenization is used, whereas at the same energy, density membranes result in smaller droplets as less energy is dissipated as heat. Membrane emulsification allows the creation of emulsions by means of two methods:

1. Direct emulsification, where the dispersed phase is pressed and passed though the membrane, whereas the continuous phase flows on the other side;
2. Premix emulsification, which aims at reducing the particle size of an already formed coarse emulsion (Candéa, 2013; Sanguansri and Augustin, 2006; Suzuki et al., 1998);

It is then applied to both lipophilic and hydrophilic, double emulsions (Candéa, 2013; Vladisavljević et al., 2004)

Process parameters that should be controlled are: the membrane type and pore size, the crossflow velocity, the transmembrane pressure, and the emulsifier. Emulsions with very small droplets of narrow distribution can be produced having only to solve the problem of the low flux via the membrane for an industrial application (Joscelyne and Trägårdh, 2000).

It is notable that, in accordance with Darcy's law, by increasing the transmembrane pressure, the flux of dispersed phase through a membrane is also increased (Nazir et al., 2010). Moreover, the droplet size and the polydispersity will also increase as well. The same procedure occurs when a premix is used to pass through the membrane (Candéa, 2013). According to some references (Candéa, 2013; Hancocks et al., 2013, Joscelyne and Trägårdh, 2000; Vladisavljevic et al., 2000), membrane emulsification may be more suitable for generating individual ingredients with carefully selected structures, permitting also the use of shear sensitive products; starch or proteins.

Recent studies revealed the possibility of nanoemulsions produced with membrane emulsification to act as encapsulation matrices. Specifically, Laouini et al. (2012) developed a vitamin E-loaded nanoemulsion as a convenient drug carrier to be used for lungs targeting. Key parameters that influenced nanoemulsion characteristics were investigated. Small droplets and narrow size distribution were optimized at low transmembrane pressure, high continuous phase flow rate and high agitation speed. Under optimal conditions, nanoemulsion, with a span factor of 0.25 ± 0.01 and an average size of 78 ± 3 nm, was generated. Vitamin E was successfully loaded in the nanoemulsion with a high encapsulation efficiency.

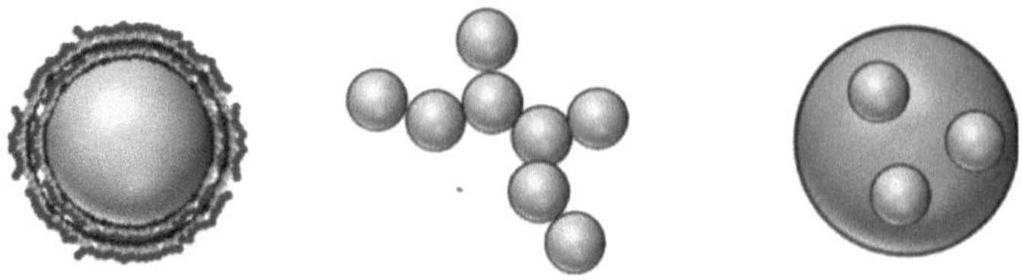

Figure 2.4 General approach for creating structural emulsions.
Source: Adapted from McClements, D.J., 2012. Advances in fabrication of emulsions with enhanced functionality using structural design principles. Curr. Opin. Colloid Interf. Sci. 17 (5) 235–245.

2.4 Structural emulsions

Recently, double emulsions, a more complex type of liquid–liquid dispersion, have been receiving increased interest. Double emulsions belong to a bigger group of multiple emulsions that themselves belong to structural emulsions. As shown in Fig. 2.4., methods of preparing structural emulsions can be divided into three broad categories:

1. *Layering*: dispersed droplets are coated with a layer of protecting substances. In this method droplets will be stabilized by applying an ionic emulsifier and afterward layers of charged particles will be coated on them. Using this method, emulsions with mono or multilayer coated droplets can be made (Chevalier and Bolzinger, 2013; Cui et al., 2009; McClements, 2012; Mun et al., 2005).
2. *Embedding*: an emulsion will be dispersed in another phase. Surrounding phase could be liquid as in multiple emulsions (Frasch-Melnik et al., 2010; Lutz et al., 2009; Márquez et al., 2007; Matos et al., 2014), semisolid as in filled hydrogels (Chung et al., 2013; Dickinson 2012; Komaiko and McClements 2015; McClements 2012) and solid as in micro and nanoencapsulated dried powders (Assadpour et al., 2016a, b; Esfanjani et al., 2015; Ezhilarasi et al., 2013; Jafari et al., 2008b).
3. *Clustering*: droplets of dispersed phase will be aggregated in a controlled manner by preventing any coalescence. Such process helps in producing emulsions with desired viscosity or gelling properties (Lombardo and Pozzo, 2015; Mao and McClements, 2011; McClements, 2012; Oliver et al., 2015).

2.5 Double emulsions

Double emulsions (also duplex or multiple emulsions) are emulsions of emulsions, such as an oil-in-water-in-oil (O/W/O) emulsion or a water-in-oil-in-water (W/O/W) emulsion. In the latter case, water droplets (W_1) are dispersed in a larger oil droplet (O) which in turn is dispersed in a continuous aqueous phase (W_2). Such structure has been called emulsion in emulsion or primary emulsion in a secondary emulsion (Assadpour et al., 2016a, b; Dickinson, 2011; Mehrnia et al., 2017).

Each double emulsion consists of three phases (Fig. 2.5): inner phase, middle phase, and outer phase; middle phase acts as a membrane that controls release of core

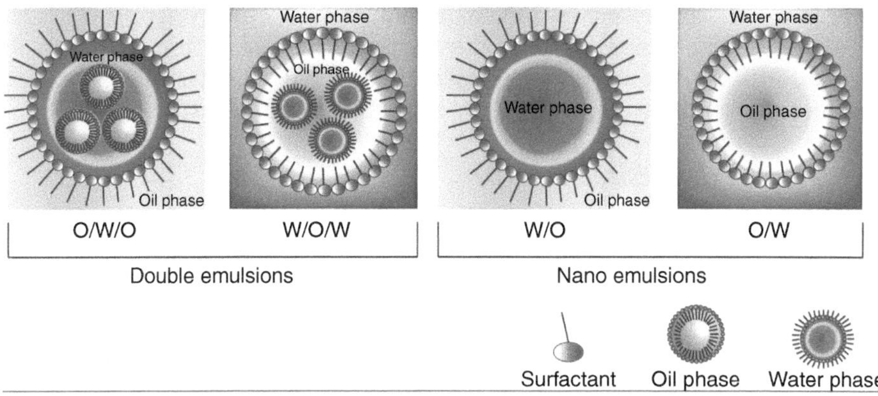

Figure 2.5 A schematic representation of double emulsions compared with single emulsions.

materials from inner phase into the outer phase. In other words, there are two different interfacial layers in this type of emulsion: For instance, in $W_1/O/W_2$ emulsions, the W_1–O interface surrounding the inner water droplets, and the O–W_2 interface surrounding the oil droplets. Consequently, two different types of emulsifier are usually needed to stabilize $W_1/O/W_2$ emulsions: an oil-soluble emulsifier for the inner water droplets and a water-soluble emulsifier for the oil droplets (Faridi Esfanjani et al., 2017; Mohammadi et al., 2016b). Due to their increased interfacial area, these double emulsions have a lower thermodynamic stability than single emulsions. In addition, the prepared double emulsions are susceptible to break-down (to become single emulsions) during processing, among other mechanisms by the migration of water between the inner and the outer aqueous phases (Assadpour and Jafari, 2017; Assadpour et al., 2017).

Although, double emulsions are less stable than single emulsions, they have some potential advantages. The inner water (W_1) phase can be used as a carrier of flavors, aromas, colors, preservatives, antimicrobials, vitamins, minerals, polyphenols, amino acids, bioactive proteins, enzymes, or peptides, keeping them apart from the continuous aqueous phase. Entrapment of these ingredients can protect them from environ mental effects; be used as delivery system; and mask undesired sensory attributes (Benichou et al., 2004; Dickinson, 2011; Hemar et al., 2010). Therefore, it has been recent investigations of double emulsion delivery systems to protect the bioactive components from detrimental effects.

2.5.1 Types of double emulsions

Based on structural definition of double emulsions, several types of them, with different primary and secondary emulsions could be formulated, such as water in oil in water (W/O/W), oil in water in oil (O/W/O), water in oil in oil (W/O/O) (Lee et al., 2000), and oil in water in water (O/W/W) (Matalanis and McClements, 2012). W/O/W and O/W/O are the most known types of double emulsions that many studies has been performed on their fabrication, stability, release of ingredients from inner phase and their applications in the food and pharmaceutical industries. These emulsions and their application in nanoencapsulation of food bioactive ingredients and nutraceuticals have been described briefly in the following sections.

2.5.1.1 W/O/W emulsions

The most common form of double emulsions is W/O/W emulsions. Some researchers have investigated the structural properties, such as double emulsion fractions, type of surfactants, release pattern of inner phase and their influence on stability of double W/O/W emulsions. To make primary W_1/O emulsions, usually a lipophilic emulsifier, such as Polyglycerol Polyricinoleate (PGPR) (Bonnet et al., 2009, 2010; Hattrem et al., 2014; Schuch et al., 2013) or Spans (Capdevila et al., 2010; Jiao and Burgess, 2003; Márquez et al., 2007) is utilized. It has been revealed that PGPR produces more stable and smaller droplets which may be due to its multianchoring nature (Le Révérend et al., 2011; Mehrnia et al., 2016). Also, it has shown that a mixture of lipophilic emulsifier of Span80 and hydrophilic emulsifier of Tween80 lowered interfacial tension more than single PGPR or Span80 (Le Révérend et al., 2011). For making secondary emulsion, that is, previous oil phase containing water (W_1/O) dispersed in outer water phase (O/W_2), nonionic hydrophilic surfactants like Tweens (Villa et al., 2003) and emulsifying biopolymers could be used. In food applications, using amphiphilic biopolymers, such as carbohydrates and proteins or mixture of them have some advantages over small molecule surfactants. Biopolymers have low diffusivity between different phases and stabilize emulsions by thickening and gelling properties at inner and outer aqueous phases (Dickinson, 2011).

2.5.1.2 O/W/O emulsions

Oil in water in oil double emulsions have been studied less than W/O/W emulsions, but they have some applications in food, cosmetic and drug delivery systems. These emulsions can be applied for encapsulation of lipophilic components since the inner phase is oil. For examples, highest stability of retinol has been reached in O/W/O emulsions compared with W/O and O/W emulsions (Yoshida et al., 1999). Also, they have resulted in better retention of flavors (Cho and Park, 2003).

Like other multiple emulsions, stability of O/W/O double emulsions could be improved by applying conjugate of protein/polysaccharide in aqueous phase (Benichou et al., 2007a,b), complex of hydrocolloids and blending lipophilic emulsifiers (Cho and Park, 2003), using different oils, especially oil containing solid fat (stearin) in outer oil which works like pickering emulsions (Jahaniaval et al., 2003) and fabricating double emulsion via one step process to avoid stresses happening in two step emulsification (Oh et al., 2004; Pradhan and Rousseau, 2012).

Besides food applications, O/W/O emulsions have been used for preparing microspheres of hydrophobic drugs using chitosan (Pavanetto et al., 1996), and for encapsulating and controlled release of effective skin care retinol using double emulsions stabilized by silica particles (Hwang et al., 2005; Lee et al., 2001).

2.5.1.3 O/W/W emulsions

Oil in water in water double emulsions are used for fabricating a group of structured emulsions called oil filled hydrogels (McClements, 2012). Water in water emulsions are prepared by mixing two biopolymers (usually a protein and a carbohydrate) which make separated solutions at natural conditions. Mixing phase separated solutions make W/W emulsion so that the phase occupying more volume is the continuous phase (Matalanis et al., 2010; Matalanis and McClements, 2011, 2012). Newly formed W/W emulsion is highly instable and tends to separate due to thermodynamic incompatibility. By dispersing oil phase within W/W emulsion, a O/W/W double emulsions will be made. To prevent phase separation of W/W or newly formed O/W/W emulsion, continuous phase will be gelled by adjusting pH (Chung et al., 2013) or chemically (Zeeb et al., 2015) and oil filled hydrogel will be formed.

2.5.2 Formulation of double emulsions

Double emulsions are usually made using a two-step emulsification technology. In the first step, primary emulsion will be formulated (Section 2.3) and in the second step, a further emulsification will be done by mild conditions to prevent coalescence or rupture of primary emulsion. Preparing primary emulsion is the key step for nanoencapsulation purposes, that is, smaller and more stable primary emulsions will provide smaller double emulsions, better encapsulation and higher stability during second step. In most of studies, primary emulsions have been made using high energy methods (Section 2.3.1) including US (Aditya et al., 2015), rotor-stator (Carrillo-Navas et al., 2012; Esfanjani et al., 2015; Frasch-Melnik et al., 2010; Hemar et al., 2010; Lutz et al., 2009), Couette's cell (Delample et al., 2014), high pressure valve homogenizer (O'Regan and Mulvihill, 2010) or microfluidizer (Bernewitz et al., 2014), and membrane emulsification (van der Graaf et al., 2005). Secondary step is usually performed by increasing the gap and decreasing the shear rate in rotor-stator apparatus or by decreasing pressure in HPHs.

An alternative method for preparing primary emulsions is low energy method, as described previously in Section 2.3.2. Low energy methods like SE, PIT and PIC have the ability to produce emulsions with high stability and small particle size within nanoscale range (Santana et al., 2013) which are key parameters for nanoencapsulation by double emulsions. Recently some effort has been made to fabricate double emulsions using one step emulsification process to prevent stresses caused in the secondary step of traditional method (Oh et al., 2004; Pradhan and Rousseau, 2012).

2.5.3 Stabilizing double emulsions

There are different mechanisms causing destabilization of double emulsion systems (Fig. 2.6); some of them are in common with normal emulsions. Destabilization mechanisms are coalescence of the outer droplets, coalescence of inner droplets, coalescence of inner droplets with outer phase due to thinning of middle (separating) phase causing release of core materials and shrinkage or swelling of inner phase due to diffusion of materials between inner and outer phases (Benichou et al., 2007;

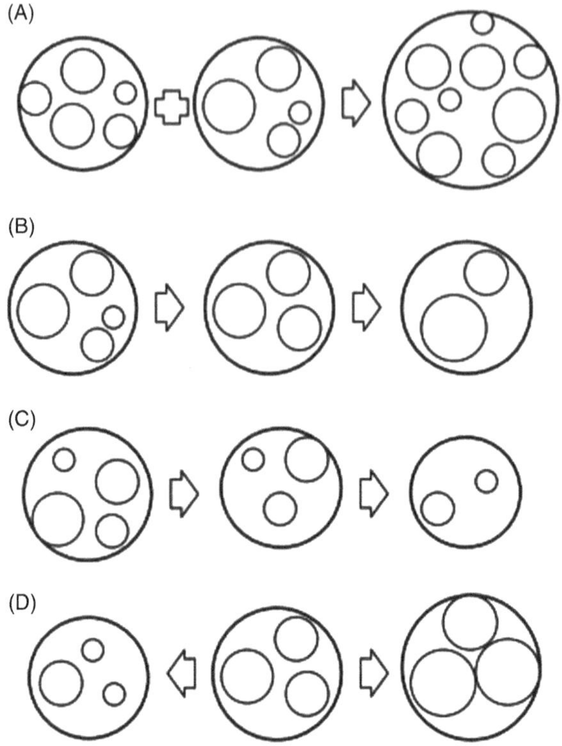

Figure 2.6 Schematic representation of the main breakdown mechanisms of a double emulsion. (A) Outer droplet coalescence, (B) inner droplet coalescence, (C) inner/outer droplet coalescence, and (D) inner droplet shrinkage or swelling.
Source: Adapted from Dickinson, E., 2011. Double emulsions stabilized by food biopolymers. Food Biophys. 6 (1), 1–11.

Dickinson, 2011). There are several ways for controlling destabilization mechanisms in double emulsions including adjustment of osmotic pressure between inner and outer phases, using biopolymers in outer phase for preventing coalescence of secondary emulsion, using biopolymers in inner phase and using a biopolymer in inner phase with ability to produce gel like structure at ambient temperatures (Aditya et al., 2015).

2.5.3.1 *Stabilizing primary emulsion*

Stabilization of primary emulsions could be improved in two ways: (1) improving structural parameters of primary emulsion to avoid coalescence of inner phase droplets and higher retention of core material; mostly could be done by using biopolymers within inner phase; (2) applying biopolymers in outer phase, which could have several benefits including higher yield of core material during second step of emulsification, higher resistance of primary emulsion droplets during second step of emulsification, and possibility of decreasing lipophilic surfactant usage in W/O/W emulsions (Oppermann et al., 2015; Perez-Moral et al., 2014).

2.5.3.2 Stabilizing secondary emulsion

Biopolymers with emulsifying properties could be used for preparing secondary emulsions. Moreover, stabilizing, thickening, and gelling biopolymers improve double emulsion stability (Hattrem et al., 2014; Matos et al., 2014; Schuch et al., 2015). To improve sensitivity of biopolymers to pH changes, decrease particle size, obtain narrow size distribution and increase steric protection, a complex (Carrillo-Navas et al., 2012; Murillo-Martínez et al., 2011; Rocha-Selmi et al., 2013) or conjugates (Benichou et al., 2007; Fechner et al., 2007; O'Regan and Mulvihill, 2010) of a protein as emulsifier and a carbohydrate as stabilizer can be applied.

2.5.3.3 Controlling osmotic pressure

Osmotic pressure gradient between inner and outer phases could be balanced to prevent migration of core materials between two phases (Lutz et al., 2009; Sapei et al., 2012). Although some research has shown that this kind of release is driven by the chemical potential difference between internal and external phases rather than osmotic pressure (Pawlik et al., 2010). Controlled migration of water from external phase into internal phase by adjusting osmotic pressure could be used for fabricating double emulsions with higher ratio of water and totally lower ratio of oil phase (Delample et al., 2014).

2.5.4 Application of double emulsions

Structural configuration of double emulsions make them suitable for several applications including reduced fat formulations, encapsulation of food bioactive compounds and nutraceuticals, and controlled release of core materials.

W/O/W emulsions have a promising future in reduced fat products (Muschiolik, 2007). Compared with O/W emulsions, in W/O/W emulsions, at the same surface level, a large volume of oil phase is composed of water which makes it possible to use real oil instead of fat replacers or use a combination of real oil and fat replacer. Recent studies have focused on making W/O/W emulsions with maximum amount of aqueous internal phase using food grade emulsifiers, which make double emulsions as a good ingredient for reduced fat products (Leal-Calderon et al., 2012). Moreover, nutraceuticals and food bioactive compounds with low bioavailability could be protected using double emulsions (Aditya et al., 2015; Assadpour et al., 2016a, b).

Double emulsions due to their special structure could be used for nanoencapsulation of hydrophilic compounds in W/O/W emulsions (Assadpour et al., 2016; Esfanjani et al., 2015; Hemar et al., 2010) or lipophilic compounds in O/W/O emulsions or both of them in a single double emulsion (Aditya et al., 2015). Except of rotor–stator devices with fixed gap, all equipment used for high energy emulsification, including colloid mills, HPHs and sonicators, as well as low energy methods can be used for nanoencapsulation.

In Table 2.5, a summary of W/O/W emulsion applications for encapsulation of different bioactive ingredients has been presented. It should be noted that for this type of emulsions, the core material should by hydrophilic since it will be located within the inner water phase.

Table 2.5 **Recent studies on applying W/O/W double emulsions for encapsulation purposes**

Primary/secondary surfactant	Core material	Fabrication method	W/O droplet size (μm)	W/O/W droplet size (μm)	References
PGPR/Tween 80	Curcumin/cathecin	Sonication		2.8–3	Aditya et al. (2015)
PGPR/WPI + Xanthan	Vitamin B1/Glucose	High pressure valve homogenizer		3–12	Benichou et al. (2007)
PGPR/NaCaseinate	Magnesium	Couette cell	1.1 ± 0.3	9.2 ± 2.3	Bonnet et al. (2010)
PGPR/mesquite gum + maltodexterin + wpc	Ascrobic Acid	Ultra-turrax		11.5–13.5	Carrillo-Navas et al. (2012)
PGPR/NaCaseinate + dextran	Vitamin B12	Rotor-stator	1.13–1.65	7.6–10	Fechner et al. (2007)
PGPR/NaCaseinate		Rotor-stator	3.5 ± 0.3	23–25.1	Frasch-Melnik et al. (2010)
PGPR/gelatin,polysorbate 80	Tartrazine	Rotor-stator		15–25	Hattrem et al. (2014)
PGPR/WPI,Gelatin	Resveratrol	HPV	0.5–0.7	7	Hemar et al. (2010)
PGPR/WPI + Pectin	Nacl, sodium ascorbate	Ultra-turrax		25–35	Lutz et al. (2009)
PGPR,Span80,Pecol/Tween 20, Tween 80, NaCaseinate	Resveratrol	High shear mixer, Membrane emulsification	0.74–10.34	Bimodal	Matos et al. (2014)
PGPR/WPI + Pectin, WPI + CMC		Ultra-turrax		2.47–10.68	Murillo-Martínez et al. (2011)
SLN/Tween 80	Raloxifene HCl	Solvent evaporation		0.197–0.445	Nabi-Meibodi et al. (2013)
PGPR/NaCN + Maltodextrin	Vitamin B12	Ultra-turrax		1.9–2.5	O'Regan and Mulvihill (2010)
Lecithin	Aspartame	Ultra-turrax		84–102	Rocha-Selmi et al. (2013)
PGPR/egg yolk powder		Colloid mill	0.05–1	50–150	Schuch et al. (2013)
Span80/WPC	Folic acid	Ultra-turrax	<100 nm	1–5	Assadpour et al. (2016a, b)

The main mechanisms of releasing core materials in double emulsions are thinning/rupturing the middle phase, and coalescence of inner and outer phases which results in diffusion of core material through middle phase (Pays et al., 2002). Diffusion could happen through direct solubilization and transfer of core material, permeation through thin middle film and by reverse micelles (Bonnet et al., 2009).

It was shown that release of magnesium in W/O/W emulsions was influenced by particle size (Bonnet et al., 2009). At the same particle size, release was dependent on oil type. Higher release happened in oils with lower viscosity and higher proportion of saturated fatty acids and percentage of magnesium released reduced by increasing globule fraction (Bonnet et al., 2010). Comparing release of inorganic sodium chloride and organic ascorbate in W/O/W double emulsions containing limonene and MCT oil as oil phase, Lutz et al. (2009) showed that electrolyte release was higher in MCT oils and NaCl released more slowly than ascorbate. Also, WPI/pectin soluble complex lowered the interfacial tension more than WPI alone (Lutz et al., 2009). In another research, controlled combination of gelatin and NaCl in inner aqueous phase provided promising mixture for controlling osmotic pressure and controlled gelation that influences stability and encapsulation of NaCl for a long period (Sapei et al., 2012). By balancing osmotic pressure of outer aqueous phase with internal aqueous phase, it has been shown there is no difference between salt and sugar on double emulsion stability (Frasch-Melnik et al., 2010). On the other hand, the use of WPI/xanthan gum complex in the middle phase of O/W/O emulsions highly increased drug entrapment during the second step of emulsification, made a barrier and prevented uncontrolled micellar diffusion of drug (Benichou et al., 2007).

2.6 Conclusions and further remarks

Nanoemulsion technology seems to be a promising method of encapsulating functional compounds. Nanoemulsions present some benefits compared to conventional emulsions. They can be added into clear or slightly turbid food without changing their appearance. Moreover, the larger their diameter the greater the Brownian motion of the droplets (Reddy and Folger, 1981), and suspensions of tiny droplets will be more stable avoiding any creaming or settling.

Nanoemulsions, as delivery systems, have a large surface area and a small volume and are then easily absorbed into the body increasing the bioavailability of the encapsulated bioactive compounds as well.

However, another issue, a bit neglected, is the diffusion rate of the encapsulated material. The particle size considerably affects the transport rate in both micro- and nanoparticles. According to an interesting, new research (Dan, 2016), the flux of compounds into or out of the particle takes place via diffusion and the characteristic time for transport is inversely related with the particle radius square. To conclude whether a diffusion rate is fast or slow, is strongly dependent on the particle size. Particle size should be taken into account, besides the difference in concentration between the core and the shell and besides the type of the compound, assuming that the greater the surface to volume ratio, the faster the transport rate becomes. This might be not always desirable, considering the controlled release of encapsulated ingredients.

Other drawbacks that should be taken into consideration are the amount and type of surfactants that might not be accepted by consumers. Finally, off-taste problems should also be considered.

Nanoemulsions have the potential to be used in their liquid form or further treated to produce nanoparticles via spray or freeze drying. Moreover, other methods, such as electrospraying can be used to produce particles or fibers having nanoemulsions as input sample. Recently, embedded nanoemulsions in a solid–liquid material, like gel are used to produce new structures combining a nanoemulsion and a hydrogel structure. It seems that more progresses will be made in the near future, introducing novel-structured nanoemulsions, consequently making them even more popular.

References

Aditya, N.P., Aditya, S., et al., 2015. Co-delivery of hydrophobic curcumin and hydrophilic catechin by a water-in-oil-in-water double emulsion. Food Chem. 173, 7–13.

Almajano, M.P., Carbo, R., Delgado, M.E., Gordon, M.H., 2007. Effect of pH on the antimicrobial activity and oxidative stability of oil-in-water emulsions containing caffeic acid. J. Food Sci. 72 (5), C258–263.

Anton, N., Vandamme, T.F., 2009. The universality of low-energy nano-emulsification. Int. J. Pharm. 377 (1–2), 142–147.

Anton, N., Benoit, J.P., Saulnier, P., 2008. Design and production of nanoparticles formulated from nanoemulsion templates—a review. J. Control Rel. 128, 185–199.

Anton, N., Mojzisova, H., Porcher, E., Benoit, J.P., Saulnier, P., 2010. Reverse micelle-loaded lipid nano-emulsions: new technology for nano-encapsulation of hydrophilic materials. Int. J. Pharm. 398 (1–2), 204–209.

Assadpour, E., Jafari, S.M., 2017. Spray drying of folic acid within nano-emulsions; optimization by Taguchi approach. Drying Technol. doi: 10.1080/07373937.2016.1242016.

Assadpour, E., Jafari, S.M., Maghsoudlou, Y., 2017. Evaluation of folic acid release from spray dried powder particles of pectin-whey protein nano-capsules. Int. J. Biol. Macromol. 95, 238–247.

Assadpour, E., Maghsoudlou, Y., Jafari, S.M., Ghorbani, M., Aalami, M., 2016a. Optimization of folic acid nano-emulsification and encapsulation by maltodextrin-whey protein double emulsions. Int. J. Biol. Macromol. 86, 197–207.

Assadpour, E., Maghsoudlou, Y., Jafari, S.M., Ghorbani, M., Aalami, M., 2016b. Evaluation of folic acid nano-encapsulation by double emulsions. Food Bioproc. Technol. 9, 2024–2032.

Becker, M., Klotzer, S., Moisar, E., 1985. US Patent No. 4,495,277. Washington, DC: US Patent and Trademark Office.

Benichou, A., Aserin, A., et al., 2004. Double emulsions stabilized with hybrids of natural polymers for entrapment and slow release of active matters. Adv. Colloid Interface Sci. 108–109 (0), 29–41.

Benichou, A., Aserin, A., et al., 2007a. O/W/O double emulsions stabilized with WPI–polysaccharide conjugates. Colloids Surf. A 297 (1–3), 211–220.

Benichou, A., Aserin, A., et al., 2007b. W/O/W double emulsions stabilized with WPI–polysaccharide complexes. Colloids Surf. A 294 (1–3), 20–32.

Bernewitz, R., Schmidt, U.S., Schuchmann, H.P., Guthausen, G., 2014. Structure of and diffusion in O/W/O double emulsions by CLSM and NMR—comparison with W/O/W. Colloids Surf. A 458, 10–18.

Bonnet, M., Cansell, M., et al., 2009. Release rate profiles of magnesium from multiple W/O/W emulsions. Food Hydrocoll. 23 (1), 92–101.

Bonnet, M., Cansell, M., et al., 2010. Influence of the oil globule fraction on the release rate profiles from multiple W/O/W emulsions. Colloids Surf. B 78 (1), 44–52.

Candéa T.V., 2013. Study of membrane emulsification process as a pre-step for the microencapsulation of lipid compounds by spray dryind. MSc Thesis. Universidade Nova de Lisboa.

Capdevila, M., Maestro, A., et al., 2010. Preparation of Span 80/oil/water highly concentrated emulsions: Influence of composition and formation variables and scale-up. J. Colloid Interface Sci. 345 (1), 27–33.

Carrillo-Navas, H., Cruz-Olivares, J., et al., 2012. Rheological properties of a double emulsion nutraceutical system incorporating chia essential oil and ascorbic acid stabilized by carbohydrate polymer–protein blends. Carbohydr. Polym. 87 (2), 1231–1235.

Chang, Y., McLandsborough, L., McClements, D.J., 2013. Physicochemical properties and antimicrobial efficacy of carvacrol nanoemulsions formed by spontaneous emulsification. J. Agric. Food Chem. 61 (37), 8906–8913.

Chemat, F., Zill e, H., Khan, M.K., 2011. Applications of ultrasound in food technology: processing, preservation and extraction. Ultrason. Sonochem. 18 (4), 813–835.

Cheong, A.M., Tan, K.W., Tan, C.P., Nyam, K.L., 2016. Kenaf (*Hibiscus cannabinus* L.) seed oil-in-water Pickering nanoemulsions stabilised by mixture of sodium caseinate, Tween 20 and β-cyclodextrin. Food Hydrocoll. 52, 934–941.

Chevalier, Y., Bolzinger, M.-A., 2013. Emulsions stabilized with solid nanoparticles: Pickering emulsions. Colloids Surf. A 439 (0), 23–34.

Chime, S.A., Kenechukwu, F.C., Attama, A.A., 2014. Nanoemulsions—advances in formulation, characterization and applications in drug delivery. In: Ali, D.S. (Ed.), Application of Nanotechnology in Drug Delivery. In Tech, Croatia, pp. 77–111.

Cho, Y.H., Park, J., 2003. Evaluation of process parameters in the O/W/O multiple emulsion method for flavor encapsulation. J. Food Sci. 68 (2), 534–538.

Chung, C., McClements, D.J., 2014. Structure–function relationships in food emulsions: improving food quality and sensory perception. Food Struc. 1 (2), 106–126.

Chung, C., Degner, B., et al., 2013. Oil-filled hydrogel particles for reduced-fat food applications: fabrication, characterization, and properties. Innov. Food Sci. Emerg. Technol. 20 (0), 324–334.

Cui, Z.G., Yang, L.L., et al., 2009. Effects of surfactant structure on the phase inversion of emulsions stabilized by mixtures of silica nanoparticles and cationic surfactant. Langmuir 26 (7), 4717–4724.

Dan, N., 2016. Transport and release in nano-carriers for food applications. J. Food Eng. 175, 136–144.

Date, A.A., Desai, N., Dixit, R., Nagarsenker, M., 2010. Self-nanoemulsifying drug delivery systems: formulation insights, applications and advances. Nanomedicine 5 (10), 1595–1616.

Davidov-Pardo, G., McClements, D.J., 2015. Nutraceutical delivery systems: resveratrol encapsulation in grape seed oil nanoemulsions formed by spontaneous emulsification. Food Chem. 167, 205–212.

Delample, M., Da Silva, F., et al., 2014. Osmotically driven gelation in double emulsions. Food Hydrocoll. 38 (0), 11–19.

Deshmukh, A.S., Setty, C.M., Badiger, A.M., Muralikrishna, K.S., 2012. Gum ghatti: a promising polysaccharide for pharmaceutical applications. Carbohyd. Polym. 87 (2), 980–986.
Dey, T.K., Ghosh, S., Ghosh, M., Koley, H., Dhar, P., 2012. Comparative study of gastrointestinal absorption of EPA & DHA rich fish oil from nano and conventional emulsion formulation in rats. Food Res. Int. 49 (1), 72–79.
Di Mattia, C.D., Sacchetti, G., Mastrocola, D., Pittia, P., 2009. Effect of phenolic antioxidants on the dispersion state and chemical stability of olive oil O/W emulsions. Food Res. Int. 42 (8), 1163–1170.
Dickinson, E., 1992. An Introduction to Food Colloids. Oxford University Press, Oxford.
Dickinson, E., 2003. Hydrocolloids at interfaces and the influence on the properties of dispersed systems. Food Hydrocoll. 17 (1), 25–39.
Dickinson, E., 2011. Double emulsions stabilized by food biopolymers. Food Biophys. 6 (1), 1–11.
Dickinson, E., 2012. Emulsion gels: the structuring of soft solids with protein-stabilized oil droplets. Food Hydrocoll. 28 (1), 224–241.
Do, L., Withayyapayanon, A., Harwell, J., Sabatini, D., 2009. Environmentally friendly vegetable oil microemulsions using extended surfactants and linkers. J. Surfactants Deterg. 12 (2), 91–99.
Dolz, M., Hernández, M.J., Delegido, J., 2006. Oscillatory measurements for salad dressings stabilized with modified starch, xanthan gum, and locust bean gum. J. Appl. Polym. Sci. 102 (1), 897–903.
Donsì, F., Ferrari, G., 2016. Essential oil nanoemulsions as antimicrobial agents in food. J. Biotechnol. 233, 106–120.
Donsì, F., Annunziata, M., Vincensi, M., Ferrari, G., 2012. Design of nanoemulsion-based delivery systems of natural antimicrobials: effect of the emulsifier. J. Biotechnol. 159 (4), 342–350.
Donsì, F., Cuomo, A., Marchese, E., Ferrari, G., 2014. Infusion of essential oils for food stabilization: unraveling the role of nanoemulsion-based delivery systems on mass transfer and antimicrobial activity. Innov. Food Sci. Emerg. Technol. 22, 212–220.
Eid, A.M., Elnattah, N., Elmahgoubi, A., Hamid, M.A., Hasham, R., Aziz, A., Ariffin, F.D., Salama, M.M., Elmarzugi, N.A., 2015. Usage of sugar ester in the preparation of avocado oil nanoemulsion. Asian J. Pharma. Clin. Res. 8 (4), 66–69.
Esfanjani, A.F., Jafari, S.M., et al., 2015. Nano-encapsulation of saffron extract through double-layered multiple emulsions of pectin and whey protein concentrate. J. Food Eng. 165, 149–155.
Ezhilarasi, P.N., Karthik, P., et al., 2013. Nanoencapsulation techniques for food bioactive components: a review. Food Bioproc. Technol. 6 (3), 628–647.
Faridi Esfanjani, A., Jafari, S.M., Assadpour, E., 2017. Preparation of a multiple emulsion based on pectin-whey protein complex for encapsulation of saffron extract nanodroplets. Food Chem. 221, 1962–1969.
Fechner, A., Knoth, A., et al., 2007. Stability and release properties of double-emulsions stabilised by caseinate–dextran conjugates. Food Hydrocoll. 21 (5–6), 943–952.
Floury, J., Desrumaux, A., Lardieres, J., 2000. Effect of high-pressure homogenization on droplet size distributions and rheological properties of model oil-in-water emulsions. Innov. Food Sci. Emerg. Technol. 1 (2), 127–134.
Forgiarini, A., Esquena, J., Gonzalez, C., Solans, C., 2001. Formation of nanoemulsions by low-energy emulsification methods at constant temperature. Langmuir 17, 2076–2083.
Frasch-Melnik, S., Spyropoulos, F., et al., 2010. W1/O/W2 double emulsions stabilised by fat crystals—formulation, stability and salt release. J. Colloid Interface Sci. 350 (1), 178–185.
Freitas, S., Hielscher, G., Merkle, H.P., Gander, B., 2006. Continuous contact- and contamination-free ultrasonic emulsification—a useful tool for pharmaceutical development and production. Ultrason. Sonochem. 13 (1), 76–85.

Fryd, M.M., Mason, T.G., 2010. Time-dependent nanoemulsion droplet size reduction by evaporative ripening. J. Phys. Chem. Lett. 1 (3), 3349–3353.

Ghosh, V., Mukherjee, A., Chandrasekaran, N., 2013. Ultrasonic emulsification of food-grade nanoemulsion formulation and evaluation of its bactericidal activity. Ultrason. Sonochem. 20 (1), 338–344.

Gupta, A., Eral, H.B., Hatton, T.A., Doyle, P., 2016. Nanoemulsions: formation, properties and applications. Soft Matter 12, 2826–2841.

Håkansson, A., Fuchs, L., Innings, F., Revstedt, J., Trägårdh, C., Bergenståhl, B., 2011. High resolution experimental measurement of turbulent flow field in a high pressure homogenizer model and its implications on turbulent drop fragmentation. Chem. Eng. Sci. 66 (8), 1790–1801.

Hancocks, R.D., Spyropoulos, F., Norton, I.T., 2013. Comparisons between membranes for use in cross flow membrane emulsification. J. Food Eng. 116, 382–389.

Hatanaka, J., Chikamori, H., Sato, H., Uchida, S., Debari, K., Onoue, S., Yamada, S., 2010. Physicochemical and pharmacological characterization of alpha-tocopherol-loaded nano-emulsion system. Int. J. Pharm. 396 (1–2), 188–193.

Hattrem, M.N., Dille, M.J., et al., 2014. Macro- vs. micromolecular stabilisation of W/O/W-emulsions. Food Hydrocoll. 37 (0), 77–85.

Hemar, Y., Cheng, L., et al., 2010. Encapsulation of resveratrol using water-in-Oil-in-water double emulsions. Food Biophys. 5 (2), 120–127.

Henry, J.V.L., Fryer, P.J., Frith, W.J., Norton, I.T., 2009. Emulsification mechanism and storage instabilities of hydrocarbon-in-water sub-micron emulsions stabilised with Tweens (20 and 80), Brij 96v and sucrose monoesters. J. Colloid Interface Sci. 338 (1), 201–206.

Henry, J.V.L., Fryer, P.J., Frith, W.J., Norton, I.T., 2010. The influence of phospholipids and food proteins on the size and stability of model sub-micron emulsions. Food Hydrocoll. 24 (1), 66–71.

Hwang, Y.-J., Oh, C., et al., 2005. Controlled release of retinol from silica particles prepared in O/W/O emulsion: the effects of surfactants and polymers. J. Control. Rel. 106 (3), 339–349.

Innings, F., Trägårdh, C., 2007. Analysis of the flow field in a high-pressure homogenizer. Exp. Therm. Fluid Sci. 32 (2), 345–354.

Jafari, S.M., He, Y., Bhandari, B., 2006. Nano-emulsion production by sonication and microfluidization—a comparison. Int. J. Food Prop. 9 (3), 475–485.

Jafari, S., He, Y., Bhandari, B., 2007a. Optimization of nano-emulsions production by microfluidization. Eur. Food Res. Technol. 225 (5–6), 733–741.

Jafari, S.M., He, Y., Bhandari, B., 2007b. Production of sub-micron emulsions by ultrasound and microfluidization techniques. J. Food Eng. 82 (4), 478–488.

Jafari, S.M., Assadpoor, E., He, Y., Bhandari, B., 2008a. Re-coalescence of emulsion droplets during high-energy emulsification. Food Hydrocoll. 22 (7), 1191–1202.

Jafari, S.M., Assadpoor, E., et al., 2008b. Encapsulation efficiency of food flavours and oils during spray drying. Dry. Technol. 26 (7), 816–835.

Jafari, S.M., Beheshti, P., Assadpoor, E., 2012. Rheological behavior and stability of D-limonene emulsions made by a novel hydrocolloid (Angum gum) compared with Arabic gum. J. Food Eng. 109 (1), 1–8.

Jafari, S.M., Fathi, M., Mandala, I., 2015. Emerging product formation. Food Waste Recovery: Processing Technologies and Industrial Techniques. Elsevier Inc., New York, pp. 293–317.

Jahaniaval, F., Kakuda, Y., et al., 2003. Characterization of a double emulsion system (oil-in-water-in-oil emulsion) with low solid fats: microstructure. J. Am. Oil Chem. Soc. 80 (1), 25–31.

Jiao, J., Burgess, D., 2003. Rheology and stability of water-in-oil-in-water multiple emulsions containing Span 83 and Tween 80. AAPS Pharm. Sci. 5 (1), 62–73.

Jo, Y.-J., Kwon, Y.-J., 2014. Characterization of β-carotene nanoemulsions prepared by microfluidization technique. Food Sci. Biotechnol. 23 (1), 107–113.

Joscelyne, S.M., Trägårdh, G., 2000. Membrane emulsification—a literature review. J. Membr. Sci. 169 (1), 107–117.

Kaltsa, O., Paximada, P., Mandala, I., Scholten, E., 2014. Physical characteristics of submicron emulsions upon partial displacement of whey protein by a small molecular weight surfactant and pectin addition. Food Res. Int. 66 (0), 401–408.

Kaltsa, O., Yanniotis, S., Mandala, I., 2016. Stability properties of different fenugreek galactomannans in emulsions prepared by high-shear and ultrasonic method. Food Hydrocoll. 52, 487–496.

Karbstein, H., Schubert, H., 1995. Developments in the continuous mechanical production of oil-in-water macro-emulsions. Chem. Eng. Proc. 34 (3), 205–211.

Kelmann, R.G., Kuminek, G., Teixeira, H.F., Koester, L.S., 2007. Carbamazepine parenteral nanoemulsions prepared by spontaneous emulsification process. Int. J. Pharm. 342 (1–2), 231–239.

Khouryieh, H., Puli, G., Williams, K., Aramouni, F., 2015. Effects of xanthan–locust bean gum mixtures on the physicochemical properties and oxidative stability of whey protein stabilised oil-in-water emulsions. Food Chem. 167 (0), 340–348.

Komaiko, J., McClements, D.J., 2014. Optimization of isothermal low-energy nanoemulsion formation: hydrocarbon oil, non-ionic surfactant, and water systems. J. Colloid Interface Sci. 425, 59–66.

Komaiko, J., McClements, D.J., 2015. Food-grade nanoemulsion filled hydrogels formed by spontaneous emulsification and gelation: optical properties, rheology, and stability. Food Hydrocoll. 46 (0), 67–75.

Kong, F., Singh, R.P., 2008. Disintegration of solid foods in human stomach. J. Food Sci. 73 (5), R67–80.

Kralova, I., Sjöblom, J., 2009. Surfactants used in food industry: a review. J. Disper. Sci. Technol. 30 (9), 1363–1383.

Lante, A., Friso, D., 2013. Oxidative stability and rheological properties of nanoemulsions with ultrasonic extracted green tea infusion. Food Res. Int. 54 (1), 269–276.

Laouini, A., Fessi, H., Charcosset, C., 2012. Membrane emulsification: a promising alternative for vitamin E encapsulation within nano-emulsion. J. Membr. Sci. 423–424, 85–96.

Le Révérend, B.J.D., Taylor, M.S., et al., 2011. Design and application of water-in-oil emulsions for use in lipstick formulations. Int. J. Cosmet. Sci. 33 (3), 263–268.

Leal-Calderon, F., Homer, S., et al., 2012. W/O/W emulsions with high internal droplet volume fraction. Food Hydrocoll. 27 (1), 30–41.

Lee, J.-H., Park, T.G., et al., 2000. Effect of formulation and processing variables on the characteristics of microspheres for water-soluble drugs prepared by w/o/o double emulsion solvent diffusion method. Int. J. Pharm. 196 (1), 75–83.

Lee, M.-H., Oh, S.-G., et al., 2001. Preparation of silica particles encapsulating retinol using O/W/O multiple emulsions. J. Colloid Interface Sci. 240 (1), 83–89.

Lee, L., Hancocks, R., Noble, I., Norton, I.T., 2014. Production of water-in-oil nanoemulsions using high pressure homogenisation: a study on droplet break-up. J. Food Eng. 131, 33–37.

Leong, T.S., Wooster, T.J., Kentish, S.E., Ashokkumar, M., 2009. Minimising oil droplet size using ultrasonic emulsification. Ultrason. Sonochem. 16 (6), 721–727.

Liedtke, S., Wissing, S., Müller, R.H., Mäder, K., 2000. Influence of high pressure homogenisation equipment on nanodispersions characteristics. Int. J. Pharm. 196 (2), 183–185.

Lombardo, M.T., Pozzo, L.D., 2015. Clusters and inverse emulsions from nanoparticle surfactants in organic solvents. Langmuir 31 (4), 1344–1352.

Lu, W.C., Chiang, B.H., Huang, D.W., Li, P.H., 2014. Skin permeation of D-limonene-based nanoemulsions as a transdermal carrier prepared by ultrasonic emulsification. Ultrason. Sonochem. 21 (2), 826–832.

Lutz, R., Aserin, A., et al., 2009. Release of electrolytes from W/O/W double emulsions stabilized by a soluble complex of modified pectin and whey protein isolate. Colloids Surf. B 74 (1), 178–185.

Mao, Y., McClements, D.J., 2011. Modulation of bulk physicochemical properties of emulsions by hetero-aggregation of oppositely charged protein-coated lipid droplets. Food Hydrocoll. 25 (5), 1201–1209.

Márquez, A., Palazolo, G., et al., 2007. Water in oil (w/o) and double (w/o/w) emulsions prepared with spans: microstructure, stability, and rheology. Colloid Polym. Sci. 285 (10), 1119–1128.

Maswal, M., Dar, A.A., 2014. Formulation challenges in encapsulation and delivery of citral for improved food quality. Food Hydrocoll. 37, 182–195.

Matalanis, A., McClements, D.J., 2011. Factors influencing the formation and stability of filled hydrogel particles fabricated by protein/polysaccharide phase separation and enzymatic cross-linking. Food Biophys. 7 (1), 72–83.

Matalanis, A., McClements, D.J., 2012. Impact of encapsulation within hydrogel microspheres on lipid digestion: an in vitro study. Food Biophys. 7 (2), 145–154.

Matalanis, A., Lesmes, U., et al., 2010. Fabrication and characterization of filled hydrogel particles based on sequential segregative and aggregative biopolymer phase separation. Food Hydrocoll. 24 (8), 689–701.

Matos, M., Gutiérrez, G., et al., 2014. Preparation of water-in-oil-in-water (W1/O/W2) double emulsions containing trans-resveratrol. Colloids Surf. A 442 (0), 69–79.

McClements, D.J., 2005. Food Emulsions: Principles, Practice and Techniques. CRC Press, Boca Raton.

McClements, D.J., 2011. Edible nanoemulsions: fabrication, properties, and functional performance. Soft Matter 7 (6), 2297–2316.

McClements, D.J., 2012. Advances in fabrication of emulsions with enhanced functionality using structural design principles. Curr. Opin. Colloid Interface Sci. 17 (5), 235–245.

McClements, D.J., 2015. Food emulsions Food Emulsions: Principles, Practices, and Techniques, third ed. CRC Press, Boca Raton.

McClements, D.J., Rao, J., 2011. Food-grade nanoemulsions.Formulation, fabrication, properties, performance, biological fate and potential toxicity. Crit. Rev. Food Sci. Nutr. 51, 285–330.

McClements, D.J., Decker, E.A., Weiss, J., 2007. Emulsion-based delivery systems for lipophilic bioactive components. J. Food. Sci. 72 (8), R109–124.

Mehnert, W., Mäder, K., 2001. Solid lipid nanoparticles: production, characterization and applications. Adv. Drug Deliv. Rev. 47 (2), 165–196.

Mehrnia, M.A., Jafari, S.M., Makhamal-Zadeh, B.S., Maghoudlou, Y., 2016. Crocin loaded nano-emulsions: factors affecting emulsion properties in spontaneous emulsification. Int. J. Biol. Macromol. 84, 261–267.

Mehrnia, M.A., Jafari, S.M., Makhmal-Zadeh, B.S., Maghsoudlou, Y., 2017. Rheological and release properties of double nano-emulsions containing crocin prepared with Angum gum. Arabic gum and whey protein. Food Hydrocoll. doi: 10.1016/j.foodhyd.2016.11.033.

Mohammadi, A., Jafari, S.M., Assadpour, E., FaridiEsfanjani, A., 2016a. Nano-encapsulation of olive leaf phenolic compounds through WPC-pectin complexes and evaluating their release rate. Int. J. Biol. Macromol. 82, 816–822.

Mohammadi, A., Jafari, S.M., Esfanjani, A.F., Akhavan, S., 2016b. Application of nano-encapsulated olive leaf extract in controlling the oxidative stability of soybean oil. Food Chem. 190, 513–519.

Mun, S., Decker, E.A., et al., 2005. Influence of droplet characteristics on the formation of oil-in-water emulsions stabilized by surfactant−chitosan layers. Langmuir 21 (14), 6228–6234.

Murillo-Martínez, M.M., Pedroza-Islas, R., et al., 2011. Designing W1/O/W2 double emulsions stabilized by protein–polysaccharide complexes for producing edible films: rheological, mechanical and water vapour properties. Food Hydrocoll. 25 (4), 577–585.

Muschiolik, G., 2007. Multiple emulsions for food use. Curr. Opin. Colloid Interface Sci. 12 (4–5), 213–220.

Nabi-Meibodi, M., Vatanara, A., et al., 2013. The effective encapsulation of a hydrophobic lipid-insoluble drug in solid lipid nanoparticles using a modified double emulsion solvent evaporation method. Colloids Surf. B 112, 408–414.

Narkhede, R.S., Gujar, K.N., Gambhire, V.M., 2014. Design and evaluation of self-nanoemulsifying drug delivery systems for nebivolol hydrochloride. Asian J. Pharm. 8 (3), 200–209.

Nazir, A., Schroën, K., Boom, R., 2010. Premix emulsification: a review. J. Membr. Sci. 362 (1–2), 1–11.

O'Regan, J., Mulvihill, D.M., 2010. Sodium caseinate–maltodextrin conjugate stabilized double emulsions: encapsulation and stability. Food Res. Int. 43 (1), 224–231.

Oh, C., Park, J.H., et al., 2004. O/W/O multiple emulsions via one-step emulsification process. J. Dispers. Sci. Technol. 25 (1), 53–62.

Oliver, L., Berndsen, L., et al., 2015. Influence of droplet clustering on the rheological properties of emulsion-filled gels. Food Hydrocoll. 50, 74–83.

Oppermann, A.K.L., Renssen, M., et al., 2015. Effect of gelation of inner dispersed phase on stability of (w1/o/w2) multiple emulsions. Food Hydrocoll. 48, 17–26.

Palamakula, A., Khan, M.A., 2004. Evaluation of cytotoxicity of oils used in coenzyme Q10 self-emulsifying drug delivery systems (SEDDS). Int. J. Pharm. 273 (1–2), 63–73.

Pardeike, J., Hommoss, A., Müller, R.H., 2009. Lipid nanoparticles (SLN, NLC) in cosmetic and pharmaceutical dermal products. Int. J. Pharm.V 366 (1), 170–184.

Pascual-Pineda, L.A., Flores-Andrade, E., Jiménez-Fernández, M., Beristain, C.I., 2015. Kinetic and thermodynamic stability of paprika nanoemulsions. Int. J. Food Sci. Technol. 50 (5), 1174–1181.

Pavanetto, F., Perugini, P., et al., 1996. Evaluation of process parameters involved in chitosan microsphere preparation by the o/w/o multiple emulsion method. J. Microencapsul. 13 (6), 679–688.

Pawlik, A., Cox, P.W., et al., 2010. Food grade duplex emulsions designed and stabilised with different osmotic pressures. J. Colloid Interface Sci. 352 (1), 59–67.

Paximada, P., Koutinas, A.A., Scholten, E., Mandala, I.G., 2016a. Effect of bacterial cellulose addition on physical properties of WPI emulsions. Comparison with common thickeners. Food Hydrocoll. 54 (Part B), 245–254.

Paximada, P., Tsouko, E., Kopsahelis, N., Koutinas, A.A., Mandala, I., 2016b. Bacterial cellulose as stabilizer of o/w emulsions. Food Hydrocoll. 53, 225–232.

Pays, K., Giermanska-Kahn, J., et al., 2002. Double emulsions: how does release occur? J. Control. Rel. 79 (1–3), 193–205.

Perez-Moral, N., Watt, S., et al., 2014. Comparative study of the stability of multiple emulsions containing a gelled or aqueous internal phase. Food Hydrocoll. 42 (Part 1), 215–222.

Peshkovsky, A., Peshkovsky, S.L., Bystryak, S., 2013. Scalable high-power ultrasonic technology for the production of translucent nanoemulsions. Chem. Eng. Proc. 69, 77–82.

Pradhan, M., Rousseau, D., 2012. A one-step process for oil-in-water-in-oil double emulsion formation using a single surfactant. J. Colloid Interface Sci. 386 (1), 398–404.
Protonotariou, S., Evageliou, V., Yanniotis, S., Mandala, I., 2013. The influence of different stabilizers and salt addition on the stability of model emulsions containing olive or sesame oil. J. Food Eng. 117 (1), 124–132.
Puglia, C., Blasi, P., Rizza, L., Schoubben, A., Bonina, F., Rossi, C., Ricci, M., 2008. Lipid nanoparticles for prolonged topical delivery: an in vitro and in vivo investigation. Int. J. Pharm. 357, 295–304.
Qian, C., McClements, D.J., 2011. Formation of nanoemulsions stabilized by model food-grade emulsifiers using high-pressure homogenization: factors affecting particle size. Food Hydrocoll. 25 (5), 1000–1008.
Quintanilla-Carvajal, M.X., Camacho-Díaz, B.H., Meraz-Torres, L.S., Chanona-Pérez, J.J., Alamilla-Beltrán, L., Jimenéz-Aparicio, A., Gutiérrez-López, G.F., 2010. Nanoencapsulation: a new trend in food engineering processing. Food Eng. Rev. 2 (1), 39–50.
Radtke, M., Müller, R.H., 2001. Nanostructured lipid drug carriers. New Drugs 2, 48–52.
Reddy, Folger, 1981. Emulsion stability—delineastion of different particle loss mechanisms. J. Colloid Interface Sci. 79 (1), 105–113.
Reiss, H., 1975. Entropy induced dispersion of bulk liquids. J. Colloid Interface Sci. 53, 61–70.
Reithmeier, H., Herrmann, J., Göpferich, A., 2001. Development and characterization of lipid microparticles as a drug carrier for somatostatin. Int. J. Pharm. 218 (1), 133–143.
Relkin, P., Yung, J.-M., Kalnin, D., Ollivon, M., 2008. Structural behaviour of lipid droplets in protein-stabilized nano-emulsions and stability of α-tocopherol. Food Biophys. 3 (2), 163–168.
Ribeiro, H.S., Lino, G.R., Badolato, G.G., Schubert, H., 2005. Production of o/w emulsions containing astaxanthin by repeated premix membrane emulsification. Food Sci. 70 (2), 117–123.
Rocha-Selmi, G.A., Bozza, F.T., et al., 2013. Microencapsulation of aspartame by double emulsion followed by complex coacervation to provide protection and prolong sweetness. Food Chem. 139 (1–4), 72–78.
Sadtler, V., Rondon-Gonzalez, M., Acrement, A., Choplin, L., Marie, E., 2010. PEO covered nanoparticles by emulsion inversion point (EIP) method. Macromol. Rapid Commun. 31 (11), 998–1002.
Salvia-Trujillo, L., Rojas-Graü, M.A., Soliva-Fortuny, R., Martín-Belloso, O., 2013. Effect of processing parameters on physicochemical characteristics of microfluidized lemongrass essential oil-alginate nanoemulsions. Food Hydrocoll. 30 (1), 401–407.
Salvia-Trujillo, L., Rojas-Graü, A., Soliva-Fortuny, R., Martín-Belloso, O., 2015. Physicochemical characterization and antimicrobial activity of food-grade emulsions and nanoemulsions incorporating essential oils. Food Hydrocoll. 43, 547–556.
Sanguansri, P., Augustin, M.A., 2006. Nanoscale materials development—a food industry perspective. Trends Food Sci. Technol. 17 (10), 547–556.
Sanguansri, L., Oliver, C.M., Leal-Calderon, F., 2013. Nanoemulsion technology for delivery of nutraceuticals and functional-food ingredients. In: Bagchi, D., Bagchi, M., Moriyama, H., Shahidi, F. (Eds.), Bio-Nanotechnology: A Revolution in Food, Biomedical and Health Sciences. Wiley-Blackwell, Washington, DC, pp. 667–672, (Chapter 38).
Santana, R.C., Perrechil, F.A., et al., 2013. High- and low-energy emulsifications for food applications: a focus on process parameters. Food Eng. Rev. 5 (2), 107–122.
Sapei, L., Naqvi, M.A., et al., 2012. Stability and release properties of double emulsions for food applications. Food Hydrocoll. 27 (2), 316–323.
Scalbert, A., Manach, C., Morand, C., Remesy, C., Jimenez, L., 2005. Dietary polyphenols and the prevention of diseases. Crit. Rev. Food Sci. Nutr. 45 (4), 287–306.

Schubert, H., 1997. Advances in the production of food emulsions. Proceedings of Engineering and Food at ICEF 7, ICEF-Conference, Academic Press, Brighton, pp. AA82–AA87.
Schubert, M.A., Müller-Goymann, C.C., 2003. Solvent injection as a new approach for manufacturing lipid nanoparticles–evaluation of the method and process parameters. Eur. J. Pharm. Biopharm. 55 (1), 125–131.
Schubert, H., Ax, K., Behrend, O., 2003. Product engineering of dispersed systems. Trends Food Sci. Technol. 14 (1–2), 9–16.
Schuch, A., Deiters, P., et al., 2013. Production of W/O/W (water-in-oil-in-water) multiple emulsions: droplet breakup and release of water. J. Colloid Interface Sci. 402 (0), 157–164.
Schuch, A., Helfenritter, C., et al., 2015. Observations on the influence of different biopolymers on coalescence of inner water droplets in W/O/W (water-in-oil-in-water) double emulsions. Colloids Surf. A 475, 2–8.
Shinoda, K., Saito, H., 1968. The effect of temperature on the phase equilibria and the types of dispersions of the ternary system composed of water, cyclohexane, and non- ionic surfactant. J. Colloid Interf. Sci. 26 (1), 70–74.
Sivakumar, M., Tang, S.Y., Tan, K.W., 2014. Cavitation technology - a greener processing technique for the generation of pharmaceutical nanoemulsions. Ultrason. Sonochem. 21 (6), 2069–2083.
Sjöström, B., Bergenståhl, B., 1992. Preparation of submicron drug particles in lecithin-stabilized o/w emulsions I. Model studies of the precipitation of cholesteryl acetate. Int. J. Pharm. 88 (1), 53–62.
Stang, M., Schubert, H., 2001. Emulsification in high-pressure homogenizers. Chem. Eng. Technol. 24 (10), 151–157.
Suzuki, K., Fujiki, I., Hagura, Y., 1998. Preparation of corn oil/water and water/corn oil emulsions using PTFE membranes. Food Sci. Technol. Int. 4, 164–167.
Tadros, T., Izquierdo, P., Esquena, J., Solans, C., 2004. Formation and stability of nano-emulsions. Adv. Colloid Interface Sci. 108–109, 303–318.
Tang, S.Y., Shridharan, P., Sivakumar, M., 2013. Impact of process parameters in the generation of novel aspirin nanoemulsions—comparative studies between ultrasound cavitation and microfluidizer. Ultrason. Sonochem. 20 (1), 485–497.
Tcholakova, S., Denkov, N.D., Danner, T., 2004. Role of surfactant type and concentration for the mean drop size during emulsification in turbulent flow. Langmuir 20 (18), 7444–7458.
Teo, B.S., Basri, M., Zakaria, M.R., Salleh, A.B., Rahman, R.N., Rahman, M.B., 2010. A potential tocopherol acetate loaded palm oil esters-in-water nanoemulsions for nanocosmeceuticals. J. Nanobiotechnol. 8, 4.
Trotta, M., Debernardi, F., Caputo, O., 2003. Preparation of solid lipid nanoparticles by a solvent emulsification–diffusion technique. Int. J. Pharm. 257 (1), 153–160.
van der Graaf, S., Schroën, C.G.P.H., et al., 2005. Preparation of double emulsions by membrane emulsification—a review. J. Membr. Sci. 251 (1–2), 7–15.
Villa, C.H., Lawson, L.B., et al., 2003. Internal coalescence as a mechanism of instability in water-in-oil-in-water double-emulsion globules. Langmuir 19 (2), 244–249.
Vladisavljevic, G.T., Brösel, S., Schubert, H., 2000. Preparation of water-in-oil emulsions using microporous polypropylene hollow fibres: conditions for producing small uniform droplets (PDF). Chem. Papers 54 (6a), 383–388.
Vladisavljević, G.T., Shimizu, M., Nakashima, T., 2004. Preparation of monodisperse multiple emulsions at high production rates by multi-stage premix membrane emulsification. J. Membr. Sci. 244 (1), 97–106.
Walker, R., Decker, E.A., McClements, D.J., 2015. Development of food-grade nanoemulsionsandemulsions for delivery of omega-3 fatty acids:opportunities and obstacles in the food industry. Food Funct. 6, 42–54.

Wilde, P., Mackie, A., Husband, F., Gunning, P., Morris, V., 2004. Proteins and emulsifiers at liquid interfaces. Adv. Colloid Interface Sci. 108–109, 63–71.

Wissing, S.A., Kayser, O., Müller, R.H., 2004. Solid lipid nanoparticles for parenteral drug delivery. Adv. Drug Deliv. Rev. 56 (9), 1257–1272.

Witthayapanyanon, A., Acosta, E.J., Harwell, J.H., Sabatini, D.A., 2006. Formulation of ultralow interfacial tension systems using extended surfactants. J. Surfact. Deterg. 9 (4), 331–339.

Wooster, T.J., Golding, M., Sanguansri, P., 2008. Impact of oil type on nanoemulsion formation and Ostwald ripening stability. Langmuir 24 (22), 12758–12765.

Yang, X., Tian, H., Ho, C.T., Huang, Q., 2011. Inhibition of citral degradation by oil-in-water nanoemulsions combined with antioxidants. J. Agric. Food Chem. 59 (11), 6113–6119.

Yin, L.-J., Chu, B.-S., Kobayashi, I., Nakajima, M., 2009. Performance of selected emulsifiers and their combinations in the preparation of β-carotene nanodispersions. Food Hydrocoll. 23 (6), 1617–1622.

Yoshida, K., Sekine, T., et al., 1999. Stability of vitamin A in oil-in-water-in-oil-type multiple emulsions. J. Am. Oil Chem. Soc. 76 (2), 1–6.

Yu, L., Li, C., Xu, J., Hao, J., Sun, D., 2012. Highly stable concentrated nanoemulsions by the phase inversion composition method at elevated temperature. Langmuir 28 (41), 14547–14552.

Yuan, Y., Gao, Y., Zhao, J., Mao, L., 2008. Characterization and stability evaluation of β-carotene nanoemulsions prepared by high pressure homogenization under various emulsifying conditions. Food Res. Int. 41 (1), 61–68.

Zeeb, B., Saberi, A.H., et al., 2015. Retention and release of oil-in-water emulsions from filled hydrogel beads composed of calcium alginate: impact of emulsifier type and pH. Soft Matter 11 (11), 2228–2236.

Zhao, Q., Ho, C.-T., Huang, Q., 2013. Effect of ubiquinol-10 on citral stability and off-flavor formation in oil-in-water (O/W) nanoemulsions. J. Agri. Food Chem. 61 (31), 7462–7469.

Zhou, L., Elias, R.J., 2013. Antioxidant and pro-oxidant activity of (−)-epigallocatechin-3-gallate in food emulsions: influence of pH and phenolic concentration. Food Chem. 138 (2–3), 1503–1509.

Zinoviadou, K.G., Scholten, E., Moschakis, T., Biliaderis, C.G., 2012. Properties of emulsions stabilised by sodium caseinate–chitosan complexes. Int. Dairy J. 26 (1), 94–101.

Zuidam, N.J., Shimoni, E., 2010. Overview of microencapsulates for use in food products or processes and methods to make them. Encapsulation Technologies for Active Food Ingredients and Food Processing. Springer, New York, pp. 3–29.

3 Encapsulation by nanoliposomes

Mehmet Demirci, Muhammed Y. Caglar, Bilal Cakir, İbrahim Gülseren
Istanbul Sabahattin Zaim University, Istanbul, Turkey

3.1 Introduction

Nanocarriers and their utilization in food and nutraceutical production have been increasingly popular over the recent years. It is well established that systems with nanoscale structures represent a variety of physical, chemical, and biological properties that have been repeatedly demonstrated to differ significantly from macroscale materials with similar compositions (Weiss et al., 2006). Since the majority of biological processes occur at the nanoscale (i.e., at least parts of digestion, absorption, and mucosal entrapment processes), it is realistic to anticipate nanocarriers to become successful tools in the delivery of bioactives. Therefore, nanocarriers including nanoliposomes do not only enhance the technical attributes of final products, but also improve their biological functionalities (Jafari et al., 2015).

Liposomes are vesicles composed of polar lipid bilayers, primarily of those phospholipids, that surround an aqueous core. In order to form liposomal structure, the investigators can rely on the self-assembly of amphiphilic lipids in aqueous solutions. Concentric packing of amphipathic lipids form an interior aqueous phase, which acts as a reservoir for hydrophilic compound encapsulation (Briuglia et al., 2015). In addition, their lipid bilayers provide a relatively lipophilic environment compared to the liposomal core. Consequently they are quite versatile in the sense that they have the potential to encapsulate and stabilize aqueous compounds in their aqueous core and hydrophobic compounds in the lipid bilayers, respectively (Katouzian and Jafari, 2016). This concept is detailed on Fig. 3.1, where the simultaneous encapsulation of vitamin C (a hydrophilic vitamin) and vitamin E (a lipophilic vitamin) in liposomes is illustrated.

Due to structural similarities between the liposomal bilayers and cellular membranes, liposomes have been regarded as easy-to-prepare model systems in the investigation of particle-cell interactions. A variety of natural compounds with low solubility can be incorporated into liposomal bilayers, which generated a myriad of pharmaceutical applications over the years (see e.g., Coimbra et al., 2011). Liposomal quercetin was administered intranasally to rats and the extent of delivery was found to be enhanced with the liposomal formulation (Priprem et al., 2008). Using targeted liposomal products, both therapeutic materials and imaging agents were efficiently delivered to the brain (Mourtas et al., 2014). Liposomal encapsulation of bioactives and drugs alter the spatial and temporal distribution of such molecules in the body. Especially in the case of biomedical applications, undesirable toxic side effects were limited while the efficacy of the treatment was improved (Keller, 2001). Liposomal encapsulation of compounds that demonstrate biological and technical functionality in food systems have been reviewed several times, especially over the last 15–20 years (Gibbs et al., 1999; Keller, 2001; Liu et al., 2015a;

Nanoencapsulation Technologies for the Food and Nutraceutical Industries

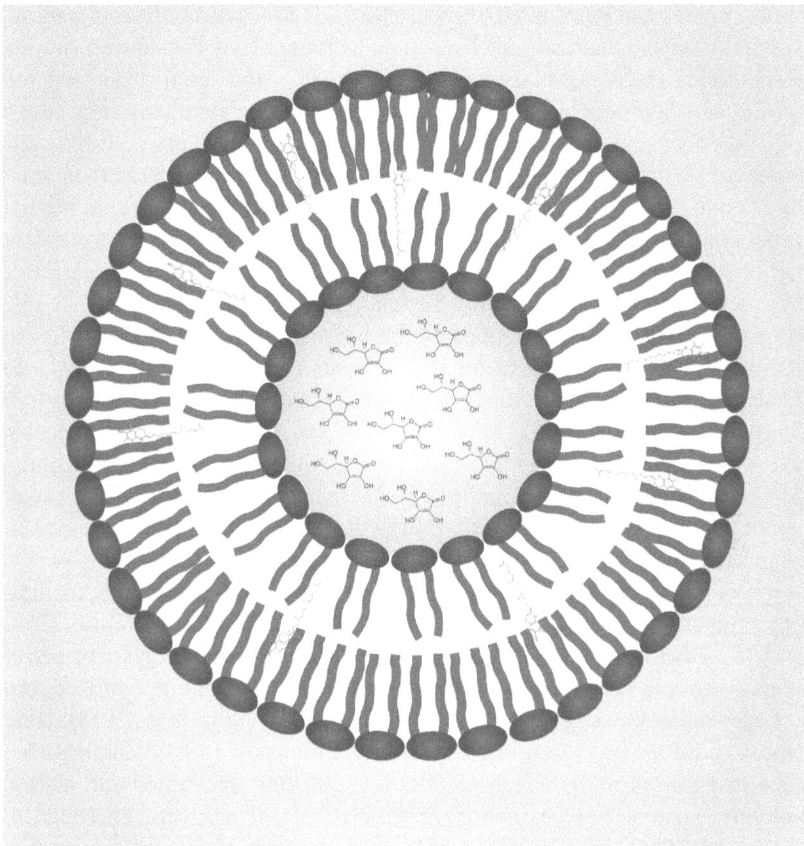

Figure 3.1 The general structure of liposomes. As model molecules, vitamin A and vitamin C were shown to be encapsulated in the liposomal bilayer and liposomal aqueous core, respectively (not to scale).

Maherani et al., 2011; Mozafari et al., 2008a,b; Taylor et al., 2005). In this chapter, the emphasis is mostly based on the more recent works and nanoscale liposomal dispersions.

The nanoscale version of liposomes is named as nanoliposomes. Although they have many physicochemical similarities to conventional liposomes, nanoliposomes bring along all the benefits of the nanocarriers mentioned earlier in this section. The formation of a nanoliposome generates considerably more fresh surface areas, which require an increased amount of energy input to generate nanoliposomes. In most of the recent applications, mean diameter range of nanoliposomes is referred to as approximately 50–150 nm. Although it must be noted that liposomes or nanoliposomes are not thermodynamically stable. Liposome manufacture practices need to address the destabilization of liposomes, degradation of encapsulated materials, the influence of environmental variables, such as composition, storage temperature, and pH, the nature of the encapsulated materials, ionic strength, exposure to light and oxygen (Liu et al., 2015a).

Nanoliposomes can be adapted to industrial scale production utilizing natural and relatively inexpensive materials, such as crude lecithins. They constitute a rich source of phospholipids including phosphatidylcholine (PC) and encapsulate and release hydrophilic and hydrophobic compounds, individually or simultaneously (de Assis et al., 2014; Mosquera et al., 2014). Compared to other colloidal delivery systems, liposomes are easily biodegraded and weakly immunogenic whereas their intrinsic toxicity is quite limited (Gülseren et al., 2012). The disadvantages of utilizing liposomal dispersions in nutraceuticals and functional foods include their relatively low encapsulation efficiency (especially for highly water soluble substances), performance variation between batches, difficulties in scaling up (Campardelli et al., 2016), the discontinous operation modes applied in the majority of the methods on liposomal preparation, and the utilization of organic solvents (Jackson and Lee, 1991). Especially solvent molecules that remain in the final formulations could impart toxicity to the products (Dwivedi, 2002; Vemuri and Rhodes, 1995). However, recent studies have shown that based on microfluidization, high pressure homogenization, or proliposomes technologies, the number of manufacturing steps can be decreased and food grade processing needs can be confirmed. Further details are presented in the following sections.

The predominant limitations in the usage of liposomal dispersions include the stability problem of the system (i.e., limited physicochemical stability) (Briuglia et al., 2015). Chemical destabilization could be due to the hydrolysis or peroxidation of phospholipid molecules. Liposomes might be physically destabilized through size enlargement due to aggregation and coalescence (Briuglia et al., 2015), which in turn would cause the reduction in encapsulation efficiency. In food applications, the materials that can be utilized in liposomal formulations are limited and quite critical problems, such as lipid oxidation, short shelf-life of products, microbial spoilage require attention, and take precedence over the other design objectives (Karimi et al., 2015). Utilization of liposomal delivery systems that help resolve some of these problems are highly sought after. Oxidation is among the major problems limiting the stability of liposomes. The effectiveness of natural antioxidants, for instance sesame ingredients, was proposed as strong inhibitors of lipid oxidation in liposomes (Lee and Jeon, 2015). Similar investigations on other novel natural ingredients can be expected to gain momentum in this area.

Liposomes can be classified based on their size, lamellarity, and vesicularity characteristics (Fig. 3.2). The chief characteristics of large unilamellar vesicles (LUV) include the presence of a single lipid bilayer, whereas a larger number of concentric lipid bilayers produce multilamellar vesicles (MLV). Meanwhile the presence of multiple vesicles that are coated with a shared bilayer are termed as multivesicular vesicles (MVV) and unilamellar vesicles with a double bilayer named as double bilayer vesicles (DBV) (Fig. 3.2) (Zawada et al., 2002). SUVs and LUVs are collectively named as ULVs (unilamellar vesicles).

Due to the complicated composition of industrial products, there are many ingredients in product formulations, such as proteins, polyphenolic compounds, sugars, salts, and other colloidal particles, such as casein micelles that could limit the stability of liposomes. In real products, SUVs could be more stable than MLVs due to the reduced

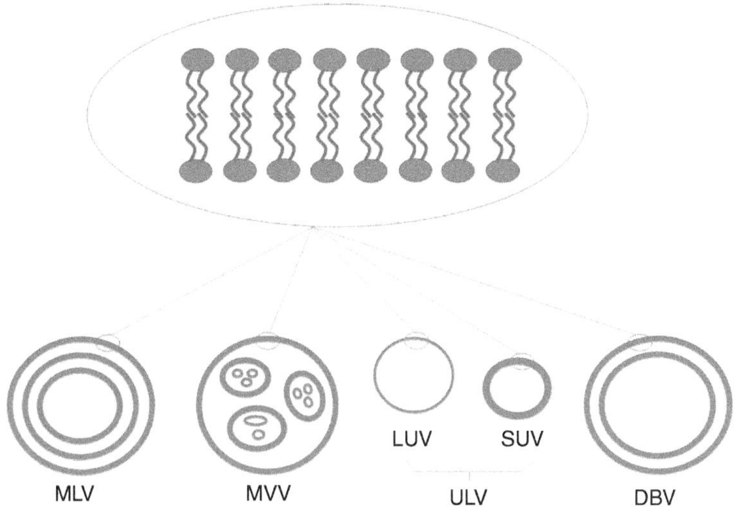

Figure 3.2 The major types of liposomes based on lamellarity, size and vesicularity.

particle diameters (Carvalho et al., 2015). However, MLVs benefit from the increased a number of layers (i.e., kinetic constraints) that prevents or at least delays the release of active molecules (Farhang et al., 2012). In most cases, the formation of a certain type of liposome may be directly linked to the method of liposome preparation.

The application of mechanical energy based on sonication, high-pressure homogenization or microfluidization makes it possible to generate LUVs from MLVs and eventually SUVs as the intensity of mechanical energy increases (da Silva Malheiros et al., 2011, 2012). It must be pointed out, however, that in any case the mechanical forces (Lagoueyte and Paquin, 1998) or ultrasonic waves (Gülseren et al., 2007) may alter the physicochemical and/or functional characteristics of natural biomolecules. Therefore, the liposome manufacture technologies have to be screened in order to overcome any negative consequences.

The major methods of liposome manufacture are represented on Fig. 3.3. Thin-film hydration (Fig. 3.3A) is based on the aqueous rehydration of thin phospholipid films which were initially dissolved in an appropriate organic solvent (e.g., chloroform) which was subjected to solvent evaporation. Afterward the dehydrated film is being hydrated in an appropriate buffer solution that contains the hydrophilic solutes, which is also instrumental for the encapsulation of the solutes. Generally, extrusion membranes or other homogenizers are utilized in order to generate ULVs, and preferably SLVs in this technique. During the application of solvent injection method (Fig. 3.3B), lipids are initially solubilized in an appropriate solvent, possibly ethanol, and ethanolic solutions are directly administered into aqueous systems where liposomes are formed. Detergent removal method (Fig. 3.3C) utilizes dialysis technique in the removal of surfactants from surfactant–phospholipid mixed micelles and leads to their transformation to liposomes. Finally, in the reverse phase method (Fig. 3.3D), polar lipids are solubilized in an appropriate organic solvent and a water-in-oil

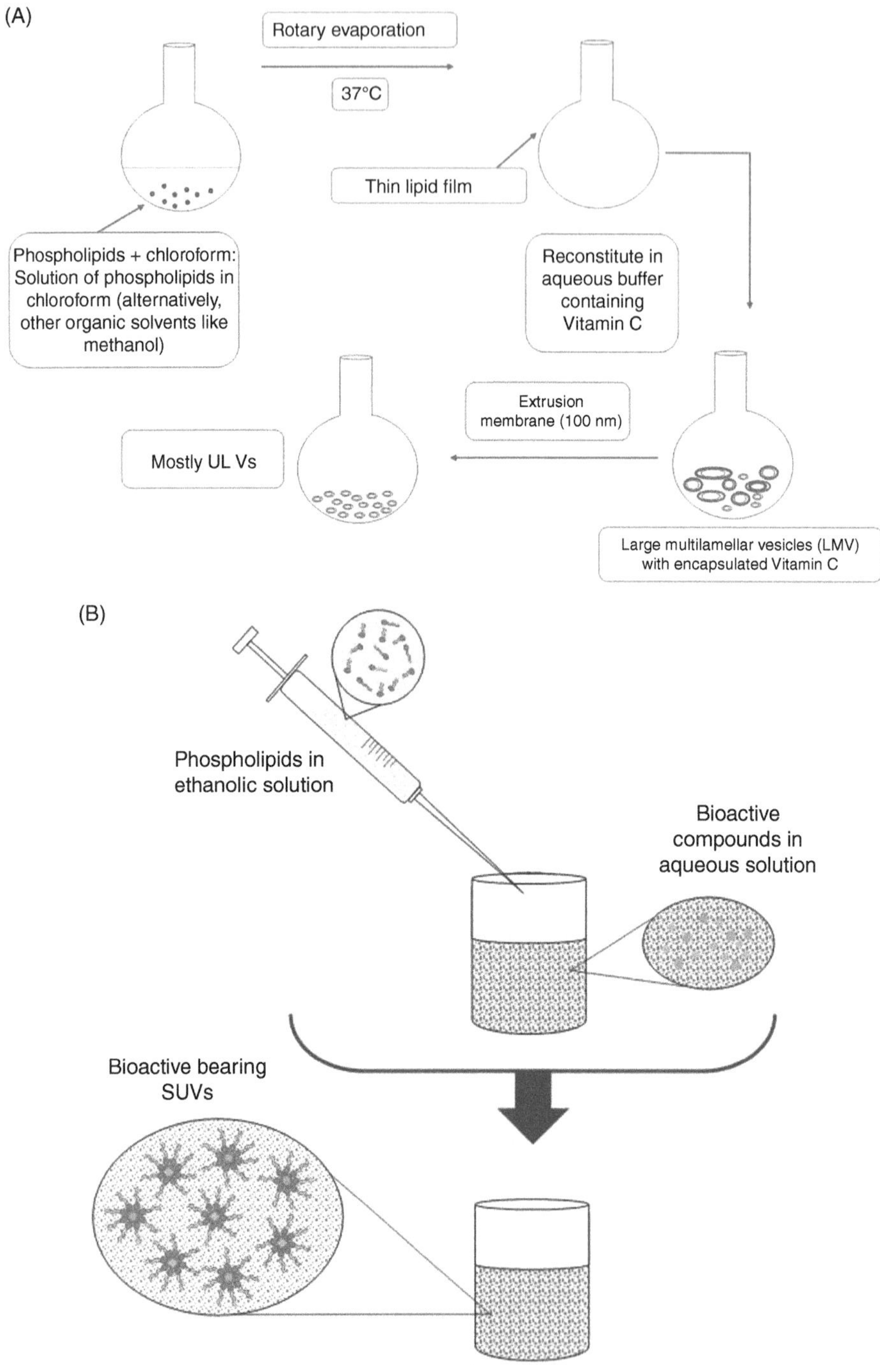

Figure 3.3 The major methods of liposome production. (A) Thin film hydration, (B) solvent injection, (C) detergent removal, and (D) reverse phase method.

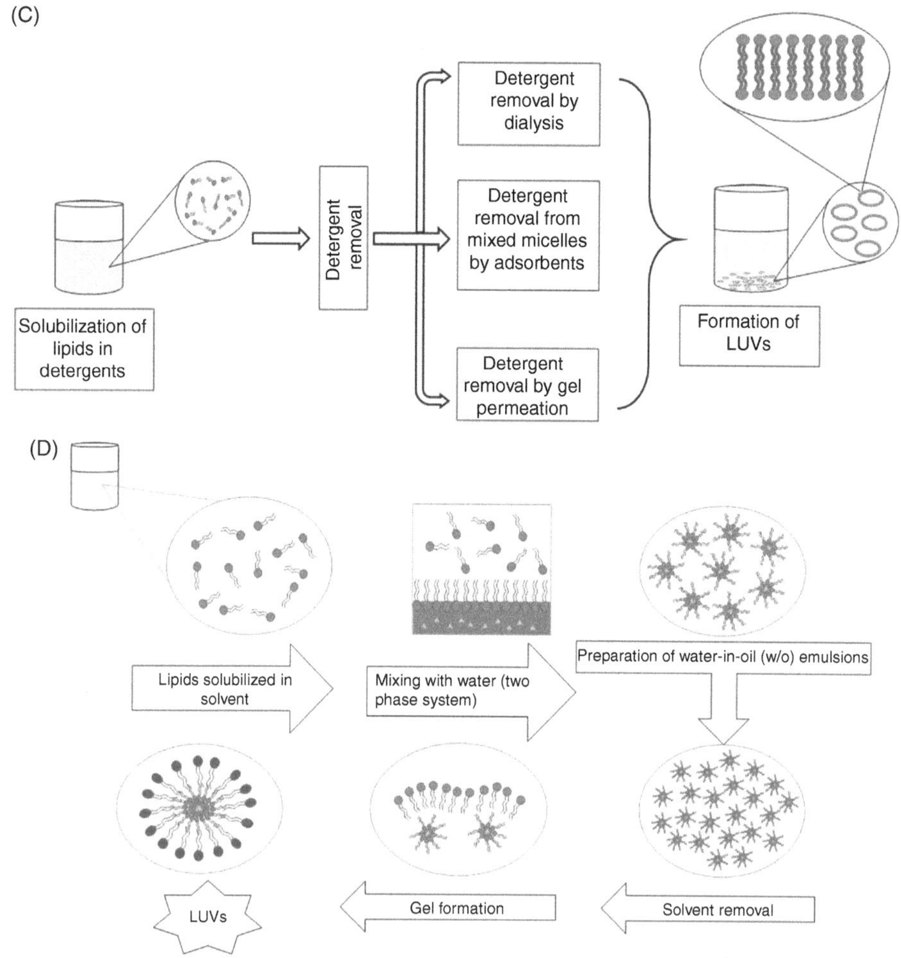

Figure 3.3 ***(cont.)***

emulsion is manufactured. Formation of liposomes will be preceded by the removal of the solvent, formation of a gel phase and aqueous dispersion.

3.2 Design of liposomes

Soy lecithin is composed primarily of PC, which is probably the most commonly employed phospholipid system in liposomes. Over the recent years, a number of alternative raw materials were proposed for the preparation of liposomes, such as egg lecithin and milk fat globule membrane (MFGM) phospholipids. Relatively new or completely novel materials from alternative sources are also being promoted (sunflower phospholipids, marine lipids, rapeseed lecithin etc.) in the literature. In various studies, the phospholipid composition was shown to determine the structural attributes of the liposomes. For example, egg yolk lecithin liposomes demonstrated a highly complex structure compared to soy liposomes, which were fairly uniform particles (Bryła et al., 2015). Furthermore, the lipid composition of the liposomal bilayers influence the particle size of liposomes, which in turn could alter the surface charge (i.e., ζ-potential) and consequently the colloidal stability of the system. Due to the crowding effects, phospholipids with large head groups were hypothesized to pack within the outer layer (i.e., instead of the inner layer) which could in turn increase the absolute valued surface charge of nanoliposomes (Liu et al., 2011). Consequently, it is critical to select the appropriate phospholipid source for the design of liposomes. Therefore, in this section, the influence of some of the liposomal ingredients on liposome structure and functionality are reviewed.

3.2.1 MFGM phospholipids

Compared to soy phospholipids, MFGM phospholipids had remarkably higher phase transition temperatures, the membranes they formed were thicker, and the permeability of the membranes were significantly lower, all of which could enhance their commercial viability in liposomal applications (Thompson et al., 2006a). The rigidity of the MFGM phospholipid bilayers was mainly determined by the presence of long-chain and very long-chain fatty acids that were available in the samples. The extent of saturation in MFGM phospholipids decreased the membrane fluidity characteristics of liposomes, which in turn limited the extent of bioactive compound release. A supportive argument can be borrowed from Maherani et al. (2011) who have effectively shown that the encapsulation efficiency of L-carnosine in nanoliposomes decreased with the presence of oleoyl groups in the phospholipids due to the increasing fluidity of liposomal membranes. Similarly, when MFGM and soy nanoliposomes were compared, liposomal stability (Thompson et al., 2006b) or the extent of polyphenol encapsulation (Gülseren et al., 2012) largely depended on the phospholipid composition. On the other hand, hydrogenated phospholipids from soy (e.g., Carvalho et al., 2015) or other sources could also alter the fluidity characteristics as well the biological functionality. In addition, processing techniques, such as centrifugation, spray-drying, homogenization, and churning altered the concentration of phospholipids in MFGM, as well as sphingolipids and other components (Gallier et al., 2010) which in turn affected the liposomal characteristics.

The entrapment efficiency of soya nanoliposomes was remarkably lower than the MFGM nanoliposomes for β-carotene (a model hydrophobic solute) and potassium chromate (a model hydrophilic solute). The differences in potassium chromate encapsulation were attributed to the larger mean diameter of the MFGM liposomes, whereas the differences in β-carotene were attributed to the variations in partitioning patterns between the bilayer and the liposomal core (Thompson et al., 2009). Similarly, Mozafari and Mortazavi (2005) have shown that α-tocopherol and glutathione, lipophilic, and hydrophilic compounds, respectively, could be encapsulated in nanoliposomes simultaneously. MFGM liposomes were also found to be superior to soy liposomes in the encapsulation and stabilization of curcumin in the presence of Fe^{3+}ions, during exposure to oxygen or light, or at elevated temperatures and relative humidity (Pezeshky et al., 2016).

As well as being a versatile ingredient in the formulation of nanoliposomes, MFGM is also known to demonstrate a variety of bioactivities on its own right. MFGM phospholipids were suggested as a functional food supplement for the prevention of cancer development, particularly human breast cancers, and possibly colon cancers due to the presence of minor amounts of β-glucuronidase inhibitor in its composition. Sphingomyelin, which is one of the minor MFGM lipids, was shown to inhibit intestinal cholesterol absorption in rats. In addition to polar lipids, a number of bioactive molecules are found in MFGM, such as proteins, glycoproteins, cholesterol, enzymes, and other minor components (Dewettinck et al., 2008; Lopez, 2011), which might alter the structural and performance related characteristics of MFGM. Bovine MFGM glycoproteins were also active in the inhibition of *Helicobacter pylori* infection in a BALB/cA mouse model (Spitsberg, 2005).

3.2.2 Marine lipids

Marine lipid based liposomes present an alternative strategy to deliver physiologically valuable compounds to our diets. Since the polyunsaturated fatty acid and esters of these acids (i.e., phospholipids) form a significant portion of marine lipid extracts, these nutritionally valuable compounds can be consumed through nanoliposomes (Nacka et al., 2001).

3.2.3 Cholesterol

Cholesterol has been widely utilized in nanoliposomes as a primary component. The inclusion of cholesterol in liposomes increases the membrane rigidity and limits the conformational alterations in the liposomal bilayer and eventually, decreases the rate of active compound release, especially that of hydrophilic compounds. Adjustment of cholesterol content and fatty acid composition is a mechanism utilized also by tumor cells in order to manipulate membrane fluidity (Socaciu et al., 2000). In certain cases, cholesterol inclusion can also affect the encapsulation efficiency (Briuglia et al., 2015). Increasing the liposomal cholesterol concentration lowered the extent of hydrophilic material encapsulation and enhanced the extent of release. On the other hand, cholesterol addition also lowered the encapsulation efficiency of hydrophobic materials but reduced their corresponding rates of release (Briuglia et al., 2015). Cholesterol incorporation improved the stability of unilamellar liposomes, but reduced the extent of vitamin A incorporation in the liposomal bilayer (Pezeshky et al., 2016). Cholesterol addition to lecithin bilayers was shown to increase the magnitude of ζ-potential values (i.e., from -29 to -43 mV) (Mohammadi et al., 2014), which also could affect the membrane rigidity. In order to limit cholesterol usage, the replacement of cholesterol with plant sterols was also considered in nanoliposome formulations. Plant sterols increased both the encapsulation of ascorbic acid (i.e., a hydrophilic solute) and the mean diameter of liposomes, therefore fine tuning of production methods would need to be carried out (Alexander et al., 2012).

3.2.4 Phenolic compounds

Similar to cholesterol, certain other compounds, such as phenolic compounds, have a bearing on the encapsulation, physical properties, and delivery characteristics of nanoliposomes. The variations in the structures of phenolic compounds can also be anticipated to alter encapsulation and stability characteristics of liposomal dispersions.

In the case of tea catechins, for example, the presence of gallic acid esters enhanced the affinity of EGCG molecules to the lipid/liposomal bilayers and induced the incorporation of significant amounts of polyphenols in the bilayers (Nakayama et al., 2000). That also affected the membrane fluidity and morphology (Ikigai et al., 1993). The presence of catechins in the membranes enhanced the rigidity of cellular membranes, as well (Tsuchiya et al., 2008). All these findings pointed out to liposomal encapsulation and stabilization of phenolic compounds.

Phenolic compounds are known to be stable at acidic pH and to some extent, around neutral pH, whereas rapid degradation can be observed at alkaline pH values. As anticipated, Zou et al. (2014a,b) have shown that oxidative and alkaline stability of green tea polyphenols were enhanced through liposomal encapsulation.

Based on layer-by-layer deposition technique, multiple polymer coatings (i.e., pectin and chitosan) both increased the physical and oxidative stability of liposomes and limited the interactions between the continuous phase and the encapsulated polyphenols (Gibis et al., 2012) while chitosan addition has not altered the polyphenol content of the liposomes (Gibis et al., 2013). Encapsulation of hibiscus extract also increased the oxidative stability of liposomal systems (Gibis et al., 2014a). Furthermore, unfavorable interactions between polyphenols with proteins, such as precipitation could be modulated through the careful design of polyphenol bearing nanoliposomes (Gibis et al., 2014b).

It is well known that oxidative deterioration of a certain food component in food systems will eventually spread oxidation reactions to other components. In a study by Viljanen et al. (2004), investigated how protein oxidation affected liposomal dispersions. Since proteins and lipids interact in liposomal dispersions, protein induced oxidation were inhibited by various bioactive components, such as anthocyanins, procyanidins, and ellagitannins included in the system, while the protein composition also had a bearing on the extent of lipid oxidation since casein, for example, acted as an inhibitor of protein oxidation. Therefore, oxidative stability is a function of both the inherent stability of the empty liposomes and their encapsulated contents. Using natural antioxidant agents, such as rosmarinate esters and chitosan coating, Panya et al. (2010) found that the lipid oxidation rate was significantly lowered. At appropriate chitosan concentrations, both physical and chemical stabilities of liposomal dispersions can be optimized (Laye et al., 2008).

The solvents utilized in liposome preparation (i.e., thin film hydration, solvent injection methods, etc.) also deeply affect the basic characteristics and liposomal performance. For example, methanolic extracts of polyphenols from cyanobacteria and algae were more concentrated in terms of polyphenol content compared to ethanolic extracts, whereas the encapsulation efficiency of the ethanolic extracts was found to be higher (de Assis et al., 2014). Especially in the context of food products and nutraceuticals, however, the extent of solvent and surfactant usage has to be clearly limited (i.e., toxicity, irritancy, etc.). Also, at high surfactant concentrations, vesicle size growth, and bilayer solubilization might take place (Alonso et al., 1981), which could rupture liposomal structures and render the encapsulation process unsuccessful.

3.3 Determination of encapsulation efficiency

The most critical property that successful nanoliposome formulations should encompass is the effective storage stability. This concept both comprises the physicochemical stability criteria, most importantly including oxidative, pH, and thermal stability, stability of encapsulated compounds within the liposomal nanoparticles, stability of the mean diameter of liposomal nanoparticles themselves and the prevention of the bioactive leakage from liposomes. The presence of surface-active compounds, such as surfactants (Sila et al., 1986), peptides (Gülseren and Corredig, 2013), or other ingredients may have a bearing on liposomal stability, all of which may positively or negatively affect the extent of encapsulation and stability of the encapsulated materials. In that sense, the quantification of encapsulation efficiency and stability of encapsulated compounds have to be carried out highly accurately.

In most protocols used for the determination of percent encapsulation efficiency (%EE), free bioactives are removed from the medium based on techniques including chromatography, dialysis, centrifugation, etc., prior to the %EE measurements. These procdures have to fully warrant that vesicle leakage is inhibited or separation has been fully completed. Otherwise, for most of the papers published in the literature the validity of the findings become debatable.

The molar extinction coefficient or molar fluorescence coefficients are highly dependent on the medium where the encapsulated compound is located (Guri et al., 2013; Tønnesen, 2002) and it is extremely difficult to prepare a reference sample for encapsulated bioactives since the partitioning coefficients of the bioactive compounds between the hydrophilic core, hydrophobic bilayer(s), and the continuous phase are unknown. Any HPLC method based on the usage of FLD or UV based detectors or any fluorescence spectroscopy based method would need to consider this difficulty. Maherani et al. (2011), for example, considered the utilization of ^{1}H NMR spectroscopy in order to overcome this problem. In the following paragraphs, a brief summary and critique of the relevant analysis methods (old and new) are presented.

One of the oldest methods used in the separation of free drugs from liposomes was the protamine aggregation method (Gunter et al., 1982). Protamine aggregation coupled with centrifugation enabled the separation of encapsulated and free materials from each other, thereby effectively allowing the measurement of encapsulation efficiency (Li and Chen, 2015). Alternatively, it was also possible to separate free and encapsulated fractions with the use of appropriate cartridges or columns (Farhang et al., 2012; Gülseren et al., 2012). Dialysis of liposomal dispersions enabled the separation of nonencapsulated bioactives primarily that of hydrophilic materials from the system. Meanwhile the release kinetics of the active materials could also be studied (Marsanasco et al., 2011).

In any of these protocols, it is truly necessary to ensure that no transfer of the active material should take place between the two phases, which could cause the under- or overestimation of the encapsulation efficiency. Although water-soluble bioactives are generally partitioned between the liposomal core and the surrounding continuous phase thus yield a low encapsulation efficiency, it is safe to assume that some hydrophobic solutes can fully locate themselves at the bilayer and yield much higher encapsulation efficiencies (i.e., ~100%) (Bryła et al., 2015).

Based on an innovative recent report, an alternative method to study encapsulation efficiency could be fluorescence quenching. Once the interactions between free butein and HSA were studied based on the strong quenching characteristics of butein, it became possible to quantitatively analyze in situ the extent of butein that was encapsulated (Toprak, 2016). Based on fluorescence spectroscopy or radioactivity, the stability of liposomal dispersions was monitored by the release of fluorescent (i.e., carboxyfluorescein) or radioactive markers [i.e., (^{14}C) sucrose] (Sila et al., 1986). In this particular case, if any other molecule was encapsulated along with the marker, the analysts would need to ensure whether the behavior of the markers are representative for their system.

3.4 Encapsulation of hydrophilic materials

Mozafari et al. (2008a) gave a possibly comprehensive list of components that have been encapsulated in liposomal dispersions which included "flavoring agents, pH modulation and control agents, redox agents and colorants, enzymes, microorganisms, essential oils, amino acids, vitamins, minerals, leavening agents, antioxidants, antimicrobials, preservatives, cross-linking agents, setting agents, and bioactives with undesired sensory attributes". Since a very wide variety of materials can be encapsulated in nanoliposomes, a brief review of recent studies is presented here with a specific emphasis on polyphenolic compounds and bioactive peptides. Some of the selected findings are also summarized on Table 3.1.

3.4.1 Phenolic compounds

A good deal of investigations on nanoliposomal encapsulation was based on the stabilization of phenolic compounds. For example, ellagic acid bearing soy lecithin nanoliposomes were prepared from concentrated soy lecithin dispersions and layer-by-layer (LbL) coating was carried out using chitosan and dextran sulfate. The authors reported the formation of spherical, monodisperse, and stable hybrid nanosized particles. Polymer coated liposomes exhibited better thermal and pH stability compared to the bare liposomes during cumulative release studies (15 days) (Madrigal-Carballo et al., 2010). Using a Franz diffusion cell, the diffusion characteristics of free and encapsulated polyphenols from thyme tea extract were studied (Pravilović et al., 2015). Tween 80 was utilized in the surface modification of liposomal dispersions which enhanced the diffusion resistance of liposomes approximately 4–5 times compared to that of free polyphenols, which is promising for the sustained release of polyphenols

Table 3.1 Selected recent examples on the encapsulation of active compounds in liposomal dispersions

Active compound	Liposome composition	Liposome preparation method	Area of investigation	References
Antihypertensive oligopeptides from tuna	Lecithin, cholesterol, and phytosterols, such as β-sitosterol or stigmasterol	Thin film hydration	Influence of replacing cholesterol with phytosterols on liposomal characteristics and efficacy of encapsulated antihypertensive peptides	Hwang et al. (2010)
Ascorbic acid	MFGM phospholipids	Microfluidization	Ascorbic acid encapsulation efficiency and stability of MFGM liposomes over extended storage	Farhang et al. (2012)
β-Carotene and potassium chromate	MFGM phospholipids and soya phospholipids	Microfluidization or thin film hydration	Comparison of hydrophilic and hydrophobic compound encapsulation characteristics of MFGM and soy liposomes	Thompson et al. (2009)
BSA	Soybean phospholipids, cholesterol, Tween-80, and vitamin E	Thin film hydration	BSA encapsulation and the study of in vitro digestion characteristics	Liu et al. (2015b)
Butein and HSA	L-α-phosphatidylcholine from egg lecithin	Thin film hydration	The study of interaction between butein and HAS via fluorescence spectroscopy in a liposomal dispersion	Toprak (2016)
Carotenoids (lycopene, β-carotene, lutein and canthaxanthin)	Egg yolk phospholipid (EYPC), Tween 80	Thin film hydration	The extent of liposomal encapsulation of carotenoids on membrane morphology, structure, and hydrophobicity of the liposomes as affected by carotenoid concentration	Xia et al. (2015)

(Continued)

Table 3.1 **Selected recent examples on the encapsulation of active compounds in liposomal dispersions** ***(cont.)***

Active compound	Liposome composition	Liposome preparation method	Area of investigation	References
Clove essential oil	Saturated (Phospholipon 80H, Phospholipon 90H) and unsaturated soybean (Lipoid S100) phospholipids, in combination with cholesterol	Ethanol injection method	Encapsulation of clove essential oil in liposomes, determination of morphological characteristics and resistance to UV radiation	Sebaaly et al. (2015)
Cucurbitacin E and D	DPPC, DPPE, DPPG, cholesterol, and/or Sulforhodamine B (SRB)	Thin film hydration	Influence of cucurbatin encapsulation on liposomal membrane characteristics	Habib et al. (2013)
Curcumin	PC, PE, PI, and PA based phospholipid systems	Thin film hydration, homogenizer or microfluidizer	Liposomal encapsulation of curcumin using commercial phospholipid systems	Takahashi et al. (2006)
Elderberry extract	Soybean, sunflower, or egg yolk lecithin	Thin film hydration followed by a freeze and thaw procedure	Elderberry extract encapsulation in liposomes manufactured from various lecithins and the comparison of their performances and structural attributes	Bryła et al. (2015)
Grape seed extract	Soybean phospholipids, glycerol, pectin, and chitosan	High-pressure homogenization and coating with chitosan	The investigation of polyphenol-protein interactions in polyphenol bearing liposomes coated with chitosan and pectin	Gibis et al. (2014b)
Green tea catechin and EGCg	Soy phospholipids	High shear homogenization	Incorporation of catechin bearing liposomes in cheese products	Rashidinejad et al. (2014)

Nisin and EDTA	DSPC and DSPG	Thin film hydration	Antimicrobial activities of liposome encapsulated nisin and EDTA	Taylor et al. (2008)
Phenolic acids (Caffeic acid, chlorogenic acid, ferrulic acid) and α-tocopherol	PC	Thin film hydration and membrane extrusion	Interactions of hydrophilic and hydrophobic antioxidants in liposomes	Neunert et al. (2015)
Purified nisin	Marine lecithin (ML) or soy lecithin (SL)	Proliposome hydration in a continuous cell disruptor	Antimicrobial encapsulation in marine and soy phospholipid liposomes and the comparison of their performances toward *L. monocytogenes*	Imran et al. (2015)
Quercetin	Highly unsaturated soy phospholipids	Thin film hydration and WPI coating	Stability of WPI coated liposomes under in vitro gastric digestion conditions and manufacture of a quercetin enriched whey permeate drink utilizing liposomes	Frenzel et al. (2015)
Tea polyphenol extract	MFGM (MFGM) phospholipids, soy phospholipids	High pressure homogenization	Encapsulation of tea polyphenols in MFGM liposomes and the investigation of the efficacy of these liposomes against cancer cells isolated from human colon carcinomas (Caco-2)	Gülseren et al. (2012)
Tea polyphenol extract	MFGM (MFGM) phospholipids and soy phospholipids	High pressure homogenization	Detailed investigation of storage stability of tea polyphenol bearing MFGM liposomes at varying pH, during ultrafiltration or interaction with surface active materials and their comparison to similarly prepared soy liposomes	Gülseren and Corredig (2013)

(*Continued*)

Table 3.1 Selected recent examples on the encapsulation of active compounds in liposomal dispersions *(cont.)*

Active compound	Liposome composition	Liposome preparation method	Area of investigation	References
Thyme tea extract (polyphenols)	Phospholigon 90G (min. 94% phosphatidylcholine), Tween 20, Tween 40	Proliposome method	The release rate of polyphenols from surface modified liposomes	Pravilović et al. (2015)
Vitamin C	L-α-PC from soybean, cholesterol, Tween 80, and vitamin E. Liposomes were coated with chitosan and sodium alginate.	Thin film hydration	Preparation of multilayer liposomes and the stability of encapsulated vitamin C under simulated digestion conditions	Liu et al. (2016)
Vitamin D3 (Cholecalciferol)	Soybean PC, cholesterol	Thin film hydration - sonication	Temperature and light sensitivity of vitamin D3 bearing liposomes	Mohammadi et al. (2014)
Vitamin E, cytochrome-C	DMPC, DMPG, cholesterol	Thin film preparation or ultrasonication	The influence of production method and cholesterol incorporation on the encapsulation of hydrophilic and hydrophobic compounds	Cagdas et al. (2011)

Liposome compositions and the manufacture methods utilized are listed along with the encapsulated active compounds and the major theme of investigation for each study.

(Pravilović et al., 2015). The absorption of tea catechins to lipid bilayers generally reflected their hydrophobicity and determined the extent of catechin-bilayer interactions. This could both affect polyphenol encapsulation characteristics and packing of other encapsulated materials within liposomes. Among green tea catechins, liposomal encapsulation efficiency for epicatechin gallate (ECg) was the highest. In the order of epigallocatechin gallate (EGCg), epicatechin (EC), and epigallocatechin (EGc), the efficiency steadily decreased. ECg and EGCg were found to distract the lipid membranes as considerable amounts of these compounds adsorbed to the lipid bilayers. This observation can possibly be attributed to the gallic acid esters found in catechins (Hashimoto et al., 1999) and the extent of membrane destruction is highly concentration dependent (Gülseren et al., 2012). In addition to their interactions with cellular or liposomal membranes, polyphenols are well-known for their interactions with proteins that could lead to protein precipitation (Siebert et al., 1996). Nanoliposomal polyphenols (namely, grape seed extract polyphenols) were shown to demonstrate a lesser extent of protein precipitation in comparison to freely soluble extract, whereas chitosan decoration of the nanoliposomes further decreased the extent of protein precipitation (Gibis et al., 2014a). However, the addition of the second biopolymer layer (pectin) decreased the surface charge which once again promoted the aggregation of the liposomes and the precipitation of the protein (Gibis et al., 2014b). These findings can be instrumental in tailoring of sensory attributes in polyphenol containing food formulations.

Along with phenolic acids, liposomal encapsulation characteristics of α-tocopherol were studied in PC nanoliposomes (Neunert et al., 2015). The interactions between the lipophilic and hydrophilic antioxidants in liposomal dispersions were investigated and their resistance against lipids oxidation was examined. Spectroscopic investigations showed that the most effective interaction was found to take place between membrane lipids and ferulic acid possibly due to its relatively low polarity. Weaker synergistic effects were observed for caffeic acid, but not for chlorogenic acid (Neunert et al., 2015). These findings are also of critical importance since they are relevant for the major phenolic acids of coffee brews and their antioxidant potential toward membranes.

3.4.2 Bioactive peptides

Encapsulation of bioactive peptides is of critical importance in the exploitation of industrial by-processing products, such as sea bream scales collagen. Based on their research on the manufacture of bioactive peptidic fractions, Mosquera et al. (2014) have studied the liposomal encapsulation of an antioxidative and ACE-inhibitory peptide (MW $<$ 3 kDa) and optimized the encapsulation performance using response surface methodology (RSM). Unlike Taylor et al. (2007), they have not reported an increase in the size of nanoliposomes due to bioactive peptide encapsulation. While antioxidative and ACE-inhibitory effects of the liposomal system were comparable to the free peptides, liposomal encapsulation extended the stability of the active component upon 8 days refrigerated storage (Mosquera et al., 2014).

3.5 Encapsulation of hydrophobic materials

The absorption of many bioactive hydrophobic molecules is limited due to their low aqueous solubility. However, majority of such molecules can be incorporated into the nanoliposomal bilayers, which renders liposomes as appropriate vehicles for the delivery of hydrophobic materials. Essential oils, ω-3 fatty acids, carotenoids, oil soluble vitamins, hydrophobic polyphenols were among the many hydrophobic compounds that have been encapsulated and stabilized in nanoliposomes in the recent literature.

3.5.1 *Hydrophobic polyphenols*

In addition to the encapsulation hydrophilic polyphenols, nanoliposomes are versatile systems that enable the stabilization of water-insoluble polyphenolic compounds as well. According to Paini et al. (2015), nanoliposomal encapsulation of apigenin, a water-insoluble polyphenolic antioxidant, was shown to be feasible in rapeseed lecithin liposomes prepared via ultrasonication. As expected, particle size decreased with sonication time and amplitude, which in turn could positively affect the cellular uptake of apigenin. The investigators suggested that apigenin bearing liposomes could be utilized in the production of functional foods or nutraceutical products.

A variety of carotenoids were encapsulated using liposomes based on egg yolk phospholipids and Tween 80 (Xia et al., 2015). Similar to the behavior of hydrophilic polyphenols, the extent of encapsulation strongly depended on how carotenoids affected membrane morphology, structure, and hydrophobicity of the liposomes and also on the carotenoid concentration. As also previously demonstrated with EGCG loading (Gülseren and Corredig, 2013), there seemed to be a concentration limit beyond which further increases in bioactive concentrations destabilized the liposomes (Xia et al., 2015). Other biochemical compounds may also affect the membrane dynamics in liposomes. An interesting set of results were obtained on cucurbitacin E encapsulation in liposomes which caused a reduction in liposomal size without altering the surface charge, and an increase in liposomal permeability (Habib et al., 2013).

3.5.2 *Bioactive lipids*

Starfish and sea cucumber phospholipids were utilized in the preparation of EPA-enriched liposomes (Du et al., 2015). These nanoliposomal dispersions inhibited the cellular proliferation of a number of cancer cell lines including cervical (HeLa), liver (HepG2), leukemia (P388), and murine (S180) cancer cells. Antitumor activities including apoptotic effects and cellular uptake characteristics were improved by liposomal nanoencapsulation. In another study, quercetin was utilized along with Omega-3 fatty acids, where fish oil and quercetin were encapsulated in soy PC liposomes and the liposomal stability was investigated for extended storage (Frenzel et al., 2015). Fish oil loading increased the membrane fluidity, which could explain the leakage of encapsulated oil, while the leakage of quercetin took place up to a lower extent. Nanoliposomal encapsulation of quercetin inhibited the production of secondary oxidation products (namely, hexanal) while the radical formation was only limited by the liposomal quercetin (Frenzel et al., 2015). Recently, Ghorbanzade et al. (2017) encapsulated fish oil within nanoliposomes and added them into yogurt samples. Their results revealed that nanoliposomes could successfully protect fish oil against deterioration and more importantly, panelists gave a sensory score similar to yogurt without fish oil.

Similarly, sensitive and valuable essential oils and their major active compounds, such as eugenol and clove essential oil were encapsulated in soy phospholipid nanoliposomes. Liposomal encapsulation reduced the UV exposure related damage and maintained the DPPH-scavenging activity of eugenol (Sebaaly et al., 2015). It is noteworthy that these compounds are also recognized for their well-established antimicrobial activities.

In addition, the stabilization of other bioactive lipids was examined in nanoliposomes. α-Galactosylceramide (α-GalCer) is a synthetic glycosphingolipid that potentially demonstrates immunoregulatory effects in liposomal dispersions. Kaneko et al. (2015) have effectively shown and monitored the encapsulation of α-GalCer and its stability in cholesterol containing DPPC liposomes. Using an appropriate LC-MS/MS technique, it was possible to selectively quantify the encapsulation efficiency over an extended storage duration (30 days, 4°C).

In addition to the bioactive compounds encapsulated in liposomes, the phospholipids that constitute the lipid bilayers demonstrated bioactivities in their own right (Liu et al., 2011; Gülseren and Corredig, 2013). Nanoliposomes prepared from middle chain fatty acids were utilized as a fast supply of energy upon oral administration in rats during a swimming trial. These findings could be relevant, for example, for the manufacture of sports drinks. In addition to their capabilities in bioactive material encapsulation, liposomes were also demonstrated to act as a reaction medium in the manufacture of a bioactive phospholipid, phosphatidylserine (PS) (Pinsolle et al., 2014), which could constitute an alternative for the usage of liposomes in synthetic chemistry.

3.5.3 Vitamins

Unilamellar liposomes were prepared using DMPC and DMPG based on sonication or extrusion methods in order to encapsulate hydrophilic cytochrome-C and lipophilic vitamin E simultaneously (Cagdas et al., 2011). Vitamin E encapsulation was not dependent whether cholesterol was employed or which manufacture technique was utilized. As also discussed elsewhere (Gülseren and Corredig, 2013), ultrafiltration was utilized in the recovery of the compounds that were not readily encapsulated. There are also published studies on the liposomal encapsulation of other lipophilic vitamins, for example, that of vitamin D3 (Mohammadi et al., 2014).

3.6 Novel techniques in the design of nanoliposomes

The techniques utilized in liposome manufacture have been recently reviewed (Huang et al., 2014). In addition to some of the more conventional techniques reviewed here, novel techniques, such as dual asymmetric centrifugation, membrane contactor technology, cross-flow filtration detergent depletion method, freeze drying double emulsion method approaches were presented along with some novel and specialized supercritical extraction techniques (Huang et al., 2014). A variety of novel techniques commonly studied in the literature are summarized in this section.

3.6.1 LbL deposition

As already mentioned in the earlier sections, LbL deposition has been increasingly used in the stabilization of nanoliposomes. For example, coating of soy phophatidylcholine (SPC) liposomes by cationic (i.e., chitosan) or anionic and cationic (i.e., chitosan and sodium alginate, respectively) polysaccharides were shown to increase the stability of encapsulated vitamin C against environmental stresses and digestive processes in vitro (Liu et al., 2016). In this particular case, chitosan molecules have been adsorbed on the negatively charged liposomal surfaces while on the second adsorption layer, sodium alginate was utilized. Particle size was shown to increase after decoration of the liposomes with chitosan and alginate layers, respectively (Liu et al., 2016). Polymer adsorption generally decreased the extent of bioactive compound release from liposomes during simulated gastrointestinal digestion (Liu et al., 2016). Similarly, chitosan was utilized as a rigidifying agent to increase the stability and encapsulation characteristics of egg yolk phospholipid liposomes against thermal and digestive stresses and centrifugal forces (Tan et al., 2016). Addition of chitosan enabled the liposomes to encapsulate and protect carotenoids successfully. However, it must be noted that chitosan could have a dual role on the stability of liposomes (Tan et al., 2013). Although electrostatic adsorption of unfolded chitosan chains enhanced nanoliposomal stability, chitosan coils negatively affected it as they potentially lowered membrane fluidity when permeated within the membrane (Tan et al., 2013). WPI coating of liposomes were shown to increase the resistance to saline environments and gastric digestion as the membrane surfaces became more polar and osmotic effects became less significant (Frenzel and Steffen-Heins, 2015a).

3.6.2 Supercritical CO_2 technologies

Supercritical CO_2 (SC–CO_2) has been proposed as a novel technology that had the potential to be utilized in liposome preparation (Zhao et al., 2015). In addition to the elimination of organic solvents, this method could be instrumental in the reduction of organic solvent usage and also offer additional benefits, such as the preparation of nanosized liposomes with a relatively low polydispersity and enhanced morphological stability. Using this method, phytosterols were shown to replace cholesterol in soy lecithin nanoliposomes (Zhao et al., 2015). Based on a novel depressurization protocol, the investigators enhanced the controllability of the particle characteristics (Zhao and Temelli, 2015). Using a novel supercritical fluid method, Campardelli et al. (2016) have shown that BSA could be encapsulated more efficiently than the thin film hydration methods.

3.6.3 Combinatorial methods

Microfluidization technology is critical in the manufacture of liposomes on a large scale in continuous mode with reasonably reproducible properties. This method also eliminates the need for the usage of detergents, solvents, and alcohols. Microfluidizer is a specialized high-pressure homogenizer where impact forces that cause the breakage of the particles are combined with high frequency vibrations and rapid build-up of pressure (Liu et al., 2015a). Based on a the combination of thin film hydration methods with microfluidization, a variety of low cost and commercially available lecithins were utilized in the preparation of nanoliposomes and the encapsulation of curcumin from turmeric. In most cases, the authors have reported an elevated extent of curcumin encapsulation (i.e., %EE > 85%) (Takahashi et al., 2006). The influence of the method of manufacture on MFGM liposomes were investigated (Thompson et al., 2007). The investigators have found that liposomes prepared based on the thin film hydration or heating methods had significantly larger mean sizes compared to liposomes manufactured by a microfluidizer.

3.6.4 Drying of liposomes

The majority of liposomal dispersions are leaky to some extent, so the inhibition or delay of leakiness is critical. In this context, freeze–drying of liposomes was suggested to enhance the protection the encapsulated materials (Li and Chen, 2015). Although the liposomes could become loose and porous during freeze–drying, during rehydration liposomes recovered their effectiveness, for example, in the encapsulation of CoQ_{10}. Consequently freeze-drying and powdered storage of liposomes were proposed as an appropriate technology for extended durations (Li and Chen, 2015). When proper cryoprotectants were added in liposomal dispersions, utilizing techniques, such as freezing or freeze-drying enhanced liposomal stability (Strauss et al., 1986). For example, glucose, lactose, or mannitol were used for freezing and sucrose for freeze–drying (Lasic, 1993). When needed, hydration of lipids may form large multilamellar vesicles (Pidgeon, 1993).

In order to avoid the usage of organic solvents in food processing, proliposomes was recommended as a viable option. Proliposomes are dried phospholipid particles, possibly freeze- or spray-dried, where bioactive molecules are entrapped in the matrix. Upon rehydration under controlled temperature and stirring conditions, liposomes have formed (Toniazzo et al., 2014). Since spray drying is widely utilized in food processing, it is theoretically possible to manufacture proliposome powders or hydrate spray-dried powders in many industrial sectors. Toniazzo et al. (2014) prepared β-carotene bearing liposomes based on a proliposome approach and used these dispersions as natural colorants in yogurt manufacture. In another study, microfluidization was exploited in order to hydrate proliposomes and encapsulate an enzyme extract. The extent of enzyme partitioning between the aqueous core and the liposmal bilayer was related to the surface charge of the liposome (Nongonierma et al., 2009). Preparation of spray dried liposomes were prepared in the presence of coating polymers in addition to conventional wall materials, such as maltodextrin. With the addition of proper coating materials, it was possible to redisperse the dried liposomes (Karadag et al., 2013).

3.6.5 Interactions with mucus

The efficacy of nanoliposomes in vivo could be affected by their entrapment in mucus. Especially in pharmaceutical formulations, surface modification of liposomes has been exploited in order to obtain mucoadhesive or mucus-penetrating liposomes (see e.g., Chen et al., 2013). Both the problems arising due to digestion, thus losing the protective shield for bioactive compounds, or entrapment in mucus should be carefully investigated for the design of successful liposomes.

3.7 Phytosomes: highly efficient delivery of phytochemicals

The development of phytosomes (Fig. 3.4) has been a major enhancement in the nanoliposomal delivery of phytochemicals. Phytosomes are innovative drug delivery systems that are manufactured from solid dispersions of plant extracts in a phospholipid matrix. They are utilized in the delivery of low solubility and low bioavailability compounds where the production methods, such as solvent evaporation/antisolvent

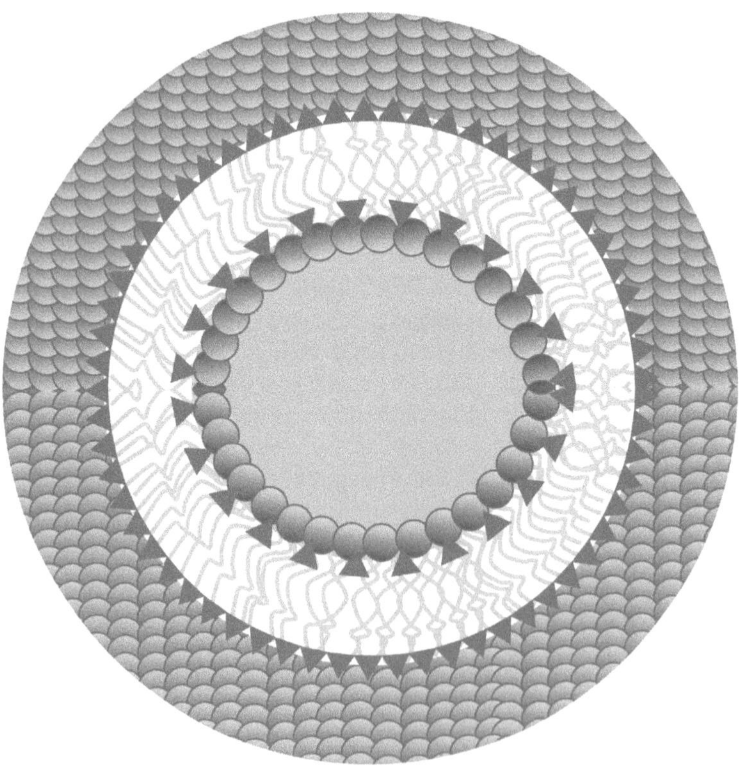

Figure 3.4 The general structure of phytosomes. Encapsulated bioactives were depicted as chemically bound to the polar head groups of phospholipids (not to scale).

precipitation and supercritical fluids may be utilized (Karataş and Turhan, 2015). Phytosomes are delivery systems that are related to liposomes in terms of their structural attributes. However, phytosomes differ from nanoliposomes based on an appropriate stoichiometric ratio that active compounds are chemically bound to the carrier structure, for example through H bonds (Bombardelli et al., 1989) which enhances the storage and digestive stability of the system. Since in liposome technology, bioactive compounds are not bound to the particle, this renders liposomes leaky and causes the loss of encapsulated bioactives over time. The chemical bonding ensures the stability of phytosomes, enhances the encapsulation efficiency and stability of bioactives, generally at a stoichiometric molar ratio of 1:1 or 1:2 (phospholipids: phytochemicals) (Amin and Bhat, 2012; Ting et al., 2014). In this patented process, PC is the most commonly utilized phospholipid. Phytosomes exhibit better pharmacokinetic and pharmacodynamic profiles than both liposomes and free plant extracts.

In a recent review, phytosomes were listed to enhance the bioavailability of phytochemicals from milk thistle, ginkgo biloba, grape seed, green tea, hawthorn, and ginseng (Jain et al., 2010) and further developments in this technology are anticipated. A brief summary of the recent findings is given in the following paragraphs. Cost-effective delivery of phytochemicals has also been successfully applied in cosmetics via phytosome technology (Amit et al., 2007). Current commercial applications of phytosomes are listed in Semalty et al. (2010).

Phytosomes were found to be an appropriate delivery system for curcumin both in vitro and in vivo in the delivery of hydrophobic molecules (Zhang et al., 2013). Another water-insoluble component (quercetin) was shown to be efficiently (approximately 98%) encapsulated in nanophytosomes with a mean diameter of 80 nm (Rasaie et al., 2014). Similarly, by the utilization of phytosome technology, the absorption of a variety of polyphenolic compounds (silymarin, curcumin, green tea, and grape seed extract) and their distribution in the intestinal tissue was enhanced compared to their liposomal counterparts (Kidd, 2009). Phytosomes were found to improve solubility, permeability rate and bioavailability of active compounds in various cases (Jain et al., 2010) and inhibit or delay physical and chemical degradation and could be implemented without generating any toxic effects (Karimi et al., 2015). Rutin incorporation in PC based phytosomes increased skin uptake rate of rutin, which could be a suitable dermal application against arthritis, rheumatism, athletic-aches (Das and Kalita, 2014). Green tea catechin, were found to be better absorbed from phytosomes than free cataehcins as determined from a clinical study. Human volunteers were provided equal amounts of free catechins and catechin bearing phytosomes and enhanced catechin uptake positively affected antioxidative parameters in the body (Pietta et al., 1998). In another clinical study, coated tablets of catechin phytosomes were found to be safe for human use in weight loss trials utilizing obese human subjects (Di Pierro et al., 2009).

Silybin–phospholipid complex phytosomes were found to be internalized by a human hepatoma cell line, whereas the location of the polyphenolic compound was found to be near the membrane layer by NMR spectroscopy. The active compound avoided any contact with the hydrophobic tails, which render phytosome technology more suitable for silybin encapsulation compared to liposomes (Angelico et al., 2014).

As an alternative delivery system, curcumin phytosomes loaded into chitosan microspheres were prepared, where microsphere gelation was based on an ionotropic mechanism (Zhang et al., 2013). Phytosome bearing microspheres slowed down the curcumin release in vitro and enhanced the curcumin stability in rat plasma compared to the other delivery systems (curcumin phytosomes, curcumin microspheres, etc.) (Zhang et al., 2013).

Phytosome formulations were shown to enhance tissue penetration of the major six boswellic acids in rats significantly. Tissue concentrations possibly suggested that boswellic acid phytosomes could be applied for their antiinflammatory activity (Hüsch et al., 2013).

3.8 Incorporation into food systems

Incorporation of nanoliposomes in food systems requires the utilization of inexpensive and natural ingredients, establishment of high performance systems that are capable of providing high encapsulation efficiency, extended storage stability, sustained and/or controlled release when necessary, ease of use, risk free implementation (absence of toxic or hazardous solvents, chemicals, detergents, cross-linkers) etc. Soy lecithin has possibly been the most commonly utilized polar lipid system in these investigations since it conforms to most of these features. Once again, some of the most recent investigations are summarized in this section with an emphasis on technological advantages and ingredient functionality. Bioactive properties and digestibility issues are discussed in the following sections.

3.8.1 Vitamins and minerals

Vitamin and mineral enrichment of foods is a critical issue globally. Soy phosphatidylcholine (SPC) based nanoliposomes were shown to be effective carriers of hydrophilic (vitamin C) and hydrophobic vitamins (vitamin E). Especially since vitamin C is a thermolabile compound, its stability in pasteurized fruit juices is important. Due to the positive effects of stearic acid on the oxidative and bilayer stability, SPC liposomes containing stearic acid enhanced the encapsulation efficiency of vitamin C. Addition of vitamin bearing nanoliposomes to orange juice formulations did not affect the sensory attributes negatively (Marsanasco et al., 2011). Wechtersbach et al. (2012) also demonstrated that liposomal encapsulation enhanced the stability of ascorbic acid in real food systems (apple juice and fermented milk product). At all cases, refrigeration and cholesterol addition enhanced the stability of ascorbic acid in the products. In addition, the stability of ascorbic acid in the dairy product was considerably lower compared to the apple juice. The authors attributed this finding to the interactions between dairy lipids and liposomal phospholipids. This finding is relevant for all lipid containing food systems. Similarly, utilization of liposomes in the vitamin D_3 fortification of beverages was regarded as a potentially successful product application (Mohammadi et al., 2014). The stability of folic acid in SPC, soy phosphatidylcholine:calcium stearate (SPC:CaS) and soy phosphatidylcholine: stearic acid (SPC:SA) liposomes and the inclusion of these liposomes in model systems and chocolate milk were investigated by the same group (Marsanasco et al., 2015). Although the liposomes based on SPC were observed to demonstrate a low extent of peroxidation, all three formulations bearing folic acid displayed remarkably higher TBARS values compared to the corresponding controls, possibly due to the prooxidant effects of folic acid. In most cases, thermal stability of folic acid, however, was enhanced and at least 50% of folic acid was kept stable after thermal processing. Once again, the inclusion of liposomal dispersions did not negatively affect the sensory characteristics (Marsanasco et al., 2015). A variety of encapsulation techniques were utilized in order to fortify milk with iron containing PC liposomes. Liposome fortified milk was found to offer an oily and unacceptable odor and taste due to the inherent flavor of PC and most of the sensory attributes were less acceptable compared to the control sample (Gupta et al., 2015).

3.8.2 Phenolic compounds

The inclusion of phenolic compound bearing nanoliposomes in food formulations is also of recent interest. Soy liposomes were coated with whey protein isolate (WPI) and spray dried, which was followed by the production of a quercetin enriched whey permeate drink from the spray dried powder (Frenzel et al., 2015). Liposomal encapsulation both improved the stability under in vitro gastric digestion conditions and the sensory attributes of the final product (Frenzel et al., 2015). WPI coating of liposomes permitted the hiding of oily aftertaste, whereas both uncoated and coated liposomes restricted bitterness of quercetin in whey drinks (Frenzel et al., 2015). Addition of catechin or EGCG to soy liposomes increased the mean diameter and phase transition temperature of the phospholipids. Although the antioxidant protection and release kinetics might need to be assessed during in vitro and in vivo digestion, green tea antioxidants were efficiently incorporated into the cheese matrix using liposomal dispersions (Rashidinejad et al., 2014).

3.8.3 Enzymes

Liposomal encapsulation of enzymes in cheese manufacture has been regarded for approximately 3 decades by now (Kirby et al., 1987). Added benefits of encapsulated enzymes in cheese manufacture might offer enhanced product texture, reduced time for flavor formation (down to about half the conventional storage time) and reduced enzyme concentrations (Mozafari, 2006).

3.9 Bioactivities of nanoliposomal encapsulation systems

Bioactive bearing nanoliposomes demonstrate a wide range of bioactivities in vitro and in vivo. A brief summary of relevant findings is presented in this section.

3.9.1 Antimicrobial activity

Assuring the safety of food products is possibly the most critical goal of all food processors. In a variety of recent studies, liposomal encapsulation of antimicrobial compounds has been investigated. Recently there has been increasing interest in the inclusion of natural antimicrobial compounds in food systems. Among such compounds, especially antimicrobial peptides including nisin, other bacteriocins or bacteriocin-like compounds (Laridi et al., 2003; Teixeira et al., 2008; Were et al., 2003) have been rigorously studied. The encapsulation of other antimicrobial compounds, such as plant oils and MFGM fractions are also summarized in this section.

Due to the increasing antimicrobial resistance of microorganisms, the concentration of antimicrobial agents in foods, nutraceuticals and related products has to be carefully adjusted, and if possible minimized. Therefore, the effectiveness of antimicrobial agents is of prime importance. Antimicrobial bearing nanoliposomes may offer an appropriate solution to this problem. For the utilization of antimicrobial bearing liposomes, some extent of leakiness might be necessary, which could lower the rate of antimicrobial action but render the system stable due to the sustained release of antimicrobials (Teixeira et al., 2008). Liposomal encapsulation and release characteristics of active compounds can be modulated based on molecular interactions including electrostatic and hydrophobic interactions (Were et al., 2003).

3.9.1.1 Antimicrobial peptides

Nisin or nisin and EDTA bearing PC and PC/PG liposomes demonstrated similar inhibition patterns as the unencapsulated mixture of nisin and EDTA, and in certain cases slightly more pronounced inhibition toward *Listeria monocytogenes*. PC liposomes showed the highest antimicrobial encapsulation efficiency, while coencapsulation of nisin and EDTA demonstrated maximum inhibition of *Escherichia coli* O157:H7. Triton-X-100 induced the lysis of nanoliposomes increasing the extent of pathogenic inhibition (Taylor et al., 2008), which could be interpreted as an additional mechanism for the controlled delivery of antimicrobials.

The stability and encapsulation efficiency characteristics of nisin bearing nanoliposomes (primarily made of PC and PG) were studied at a variety of testing conditions including elevated temperatures (25–75°C) and pH range (pH 5.5–11.0). At a particle size range of 100–240 nm, surface charge minimally depended on nisin incorporation or phospholipid composition, whereas encapsulation efficiency was found to be fairly high at elevated temperatures and varying pH values (i.e., acidic or alkaline) (Taylor et al., 2007). Liposomal antimicrobials were shown to be highly effective against *L. monocytogenes* and *E. coli* O157:H7 even at relatively low nisin concentrations (Taylor et al., 2008). Similarly, liposomal encapsulation was shown to enhance the antimicrobial potential of pediocin AcH against *L. monocytogenes* strains (Degnan et al., 1993). While shorter fatty acid chains of soy lecithin liposomes were more advantageous compared to marine phospholipid liposomes for the incorporation of nisin, 1:1 blend of free and liposome encapsulated nisin was highly effective in the inhibition of *L. monocytogenes* CIP 82110 (Imran et al., 2015).

Thermal and/or pressure induced denaturation or degradation of proteins and peptides could lower the stability of encapsulated antimicrobial agents, such as liposomal nisin, which could limit the extent of mechanical energy applied in liposome manufacture (da Silva Malheiros et al., 2011, 2012). Similarly, stresses that take place in the GI tract have to be considered for all encapsulated bioactive molecules including antimicrobials.

Both free and liposomal bacteriocins from *Lactobacillus sakei* subsp. *sakei 2a* showed inhibitory effect toward *L. monocytogenes* growth in microbial growth media and in experimentally contaminated goat milk (Malheiros et al., 2016). In this study, however, the authors have claimed that liposomal bacteriocins were not necessarily superior to free bacteriosins. In another study from the same group, Maillard reaction products were shown to limit the efficacy of bacteriocins, while liposomal encapsulation could protect bacteriosins from degradation induced by Maillard reaction products in thermally processed foods (Sant'Anna et al., 2011).

In another study where liposomal peptides proved less effective than free peptides, due to the complicated interactions between the liposomal membranes, peptides, and microorganisms, *L. monocytogenes* was inhibited more significantly by free bacteriocin than the encapsulated P34 peptide in short term experiments (da Silva Malheiros et al., 2011). When *L. monocytogenes* growth in Minas frescal cheese was studied, nisin and BLS P34 bearing liposomes demonstrated similar effects (da Silva Malheiros et al., 2012).

3.9.1.2 Natural oils

Microbial toxins from *S. aureus* and *Listeria* species have recently been utilized as a trigger in the release of natural antimicrobial oils from nanoliposomes, which can be defined as smart delivery systems. Toxin-triggered release of natural oil (salvia oil) from liposomes was found to be an effective method to inhibit the biofilm formation on the surfaces of milk tanks for which stainless steel chips were examined as a model system. As α-toxin was secreted from *Staphylococcus aureus*, salvia oil was released from the liposomes and it demonstrated antibacterial activity against *S. aureus* biofilms (Cui et al., 2016). Clove oil has been previously shown to demonstrate antimicrobial activity. However, its utilization in food systems is impractical due to its relative instability in food systems. Using a thin film hydration method, clove oil was encapsulated in soy liposomes containing cholesterol and polyvinylpyrrolidone (Cui et al., 2015). *S. aureus* toxins were utilized in the triggered release of clove oil from the liposomes which ensured the microbial safety of the medium. In addition, the authors have shown that these liposomal dispersions demonstrated anti-Staphylococcal activity also in tofu products (Cui et al., 2015). Similarly, pore-forming toxins (PFTs) from *Listeria* species triggered the release of nutmeg oil from liposomes. Consequently, liposomal encapsulation yielded sustained release of nutmeg oil and antibacterial effect was improved (Lin et al., 2016).

3.9.1.3 MFGM fractions

Shiga toxin produced by *E. coli* O157:H7 has the potential to cause damage to human colon blood vessels in the human colon and consequently cause bloody diarrhea.

MFGM fractions displayed inhibitory effect on the expression of Shiga toxin genes by *E. coli* O157:H7. This effect was more pronounced for heat-treated MFGM fractions (Tellez et al., 2012). These results indicated that along with other antimicrobials or separately, MFGM liposomes could be utilized in food safety and protection applications.

3.9.2 Anticarcinogenic activities

The liposomal nanoencapsulation of drugs and anticarcinogenic phytochemicals relevant to biomedical applications has been widely investigated. In these products, the exploitation of organic solvents, chemical cross-linkers, synthetic surfactants, etc., are much less restricted compared to foods and nutraceuticals. For example, according to Saengkrit et al. (2014), DDAB (didecyldimethylammonium bromide, a phase-transfer catalyst) containing liposome was demonstrated to enhance the sustained release characteristics of curcumin from liposomal formulations, due to which the liposomal formulation could effectively inhibit HeLa and SiHa cell growth.

The emphasis of food industry on functional foods is globally increasing and in that sense, chemoprevention based on the exploitation of anticarcinogenic food components in functional foods is a critical field. In the current literature, phenolic compounds, carotenoids, and bioactive polysaccharides are among the valuable food components that have been studied in cell culture environments or in animal trials.

3.9.2.1 Phenolic compounds and carotenoids

Curcumin is marginally soluble in water and nanoliposomal curcumin formulations have the potential to demonstrate anticarcinogenic and antioxidative activity. In addition to soy liposomes, curcumin bearing rapeseed, or salmon lecithin liposomes were also shown to reduce the proliferation of breast cancer cells (Hasan et al., 2014). To enhance its bioavailability and antitumor efficacy, hydrophobic flavonoid fisetin was encapsulated in liposomal dispersions. IV-administered liposomal fisetin was not only available at an elevated plasma concentration but in comparison to nonencapsulated fisetin, the liposomal formulation also decreased the tumor volume in Lewis lung carcinoma (Seguin et al., 2013). Beta-cryptoxanthin-bearing nanoliposomes showed a more pronounced antiproliferative activity, when compared to the antiproliferative activity of free beta-cryptoxanthin, for example, toward the inhibition of K562 cells, which was found to be dose and time dependent (Gharib et al., 2015).

3.9.2.2 Tea polyphenols

Gülseren et al. (2012) have studied the tea polyphenol encapsulation characteristics of MFGM and soy liposomes and tested their efficacy against HT-29 human colon carcinoma cells. They have effectively demonstrated that at all cases, the viability of cancer cells decreased with the extent of encapsulated polyphenols and during extended incubation with polyphenol bearing liposomes, the viability of cancer cells generally decreased over time. Under 500 nm, the influence of liposomal diameter on viability reduction was weak and although slightly larger, MFGM liposomes were more effective against the proliferation of HT-29 cells (Gülseren et al., 2012). Based on similarly prepared samples, detailed investigation of storage stability of tea polyphenol bearing MFGM and soy liposomes were carried out at varying pH, during ultrafiltration or interaction with surface active materials, such as tryptic peptides from casein that might be encountered during digestion (Gülseren and Corredig, 2013). These investigations have clearly shown that polyphenol bearing nanoliposomes had the potential to be utilized in food and nutraceutical production due to their anticarcinogenic properties, suitability to continuous mode production at high capacities and extended stability over storage and industrial processes. In a separate study, liposomal EGCG was found to be more effective than free EGCG against Caco-2 cells (Luo et al., 2014). While HT-29 cells present a heterogeneous multilayer of enterocytes and goblet cells, Caco-2 is characterized by a monolayer of enterocytes, which renders the results from both studies complementary to each other.

3.9.2.3 Bioactive polysaccharides

Fucoidans are a group of seaweed polysaccharides that demonstrate a myriad of bioactive properties including antioxidant and anticarcinogenic activities. Fucoidans were encapsulated in PC liposomes and immunomodulatory and anticancer activities were studied. Compared to free fucoidan, a small extent of enhancement was observed in the anticarcinogenic activity of liposomal fucoidan when multiple cell lines were tested (Qadir et al., 2008). The activity of fucoidan was further elaborated based on confocal microscopy (i.e., CLSM) by visualizing the cellular uptake and compartmentalization of fucoidan. In addition, immunomodulatory effects were investigated (for example, B and T cell growth and the level of secretory activities in the immune system) (Qadir et al., 2008).

Finally, it is noteworthy that coencapsulation of bioactive compounds and anticancer drugs in liposomes were previously studied in unilamellar liposomes (Cosco et al., 2015). Since resveratrol and 5-fluorouracil both demonstrate anticarcinogenic activity (antioxidative or antiproliferative), the authors have effectively shown that their liposomal coencapsulation favored their permeation from skin cancer medications with further promise toward the treatment of other cancer types (Cosco et al., 2015). When simultaneous nanoliposomal encapsulation of hydrophilic bioactives were carried out along with medical compounds, such as in the case of liposomal β-glucans and an antitumor compound commonly used in cancer chemotherapy (doxorubicin), the antitumor activity was significantly enhanced (Halwani et al., 2015).

3.9.3 Other bioactivities demonstrated by liposomal dispersions

Liposomal dispersions that bear bioactive compounds demonstrated various other bioactivities in addition to antimicrobial and anticarcinogenic activities. A brief compilation of these findings is summarized in this section.

The extent of release from liposomes effectively determines the extent of antioxidative activity in vitro. Especially in the case of antioxidants, moderate rates of release could be useful in food and pharmaceutical formulations in order to ensure extended oxidative stability of the system during storage. Ascorbic acid encapsulation in MFGM nanoliposomes was found to maintain its stability at least for 7 weeks (Farhang et al., 2012), which possibly could be a valuable finding in the enhancement of antioxidative activity in food applications.

Incorporation of phytosterols (namely β-sitosterol or stigmasterol) in liposome formulations generated a variety of liposomes with comparable encapsulation efficiency, storage, pH, oxidative, and enzymatic stability as those prepared by the incorporation of cholesterol. Phytosterol containing liposomes were successfully utilized in the encapsulation of antihypertensive oligopeptides from tuna cooking juice, which renders replacing cholesterol with phyosterols feasible (Hwang et al., 2010). In either case, the blood pressure of spontaneously hypertensive rats were significantly lowered.

Zinc-specific FluoZin-3 bearing liposomes were utilized for testing of zinc-ionophore activity in polyphenolic compounds.

This liposomal dispersion could be readily utilized in the analysis of polyphenols leading us to a better understanding on the structure-biological activity relation of phenolic compounds (Clergeaud et al., 2016). In addition, liposomal encapsulation was shown to enhance the antioxidant (Gortzi et al., 2007; Kong and Xiong, 2006), blood pressure lowering (Chen et al., 2003), antiviral (Sinico et al., 2005), and immune modulatory (Szebeni and Moghimi, 2009) activities of phytochemicals along with a myriad of other possible effects.

Finally, a very interesting liposomal dispersion with pronounced bioactivities was prepared based on the utilization of *Epimedium* polysaccharide and propolis flavone. The authors proposed that *Epimedium* polysaccharide—propolis flavone mixture carried immune system boasting activities and have shown that liposomal encapsulation strongly enhanced Kuppfer cell activity (Fan et al., 2015). Similarly, the antioxidative and immunoregulatory activities of Ophiopogon polysaccharide were enhanced after encapsulation in soy liposomes (Fan et al., 2016).

3.10 Digestion of bioactive bearing nanoliposomes

Nanoscale delivery systems need to be stable during the formation of bolus, swallowing, and transit through the GI tract (Liu et al., 2015b). The enzymatic and digestive stability of liposomes have been a matter of investigation for a long time. As early as 1978, the influence of bile salts on the leakiness of liposomes was discussed (Richards and Gardner, 1978). It is fairly difficult to ensure or monitor the stability of liposomes in biological systems. Due to the presence of bile salts and lipases, liposomal dispersions might lose their payload in the small intestine (Rowland and Woodley, 1980). Surface modification or coating can be utilized in order to improve the stability of nanoliposomes during digestion and also in circulation. For example, surface modification of phosphatidylethanolamine liposomes by covalent binding of lectins was shown to enhance the encapsulation efficiency of insulin and caused hypoglycemic effects in diabetic mice (Zhang et al., 2005). Although this effect seemed to be only slight, lectins have the drug and bioactive targeting potential through glycosylated membrane components, which deserves being further utilized along with other novel technologies (Zhang et al., 2005). WPI coating of liposomes were shown to decrease the permeability of liposomes and the osmotic effects that they need to endure during in vitro gastric digestion. The removal of the WPI coating in simulated intestinal fluids (SIF, neutral pH) indicated that the coating should not decrease the extent of bioactive delivery of encapsulated bioactives, although without the WPI coating, liposomal surfaces were more fissured and were more sensitive to enzymatic attack (Frenzel and Steffen-Heins, 2015b). In the rest of this section, studies on in vitro gastric and intestinal digestion of nanoliposomal dispersions are briefly reviewed.

3.10.1 In vitro digestion

During peptic hydrolysis under simulated gastric conditions, membrane structure of soy phospholipid liposomes was shown to remain intact and entrapped BSA was not released. However, in the second part of the digestion experiments configured using-SIF, damage to liposomal membrane and the release of free fatty acids (FFA) from the system clearly took place. This finding might be attributed to partial hydrolysis using pancreatic lipase. Both bile salt concentration and ionic strength increased the rate of FFA release (Liu et al., 2015b). When MFGM and soy liposomes were compared based on in vitro digestion assays, MFGM liposomes were shown to represent more pronounced membrane integrity (Liu et al., 2012).

EGCG bearing chitosan nanoliposomes were coated with dextran sulfate and digestive stability was monitored in vitro. Liposomal encapsulation reduced the extent of EGCG degradation in the SIF. In most cases, $<10\%$ EGCG was lost from the liposomes based on a 24 h storage. However, the presence of pancreatin in SIF digestion caused the precipitation of the liposomes (Zou et al., 2015). Liposomal encapsulation was also found to be promising for the protection of antioxidant activities of EGCG during in vitro digestion (Zou et al., 2014b). During in vitro gastric digestion of EGCG bearing nanoliposomes, about 21% of EGCG was released from the liposomes during a 4 h study, whereas during intestinal digestion, approximately 35% was released (Luo et al., 2014). Carotenoids were encapsulated in egg yolk phospholipid liposomes and their bioaccesibility was monitored immediately after in vitro digestion (Tan et al., 2014). The authors suggested that the relative polarity and concentration of the carotenoids determined their localization in the liposomal membrane, which consequently affected their micellization characteristics (Tan et al., 2014).

Milk protein allergy, primarily based on the indigestibility of β-lactoglobulin, is a common problem globally. Nanoliposomes might be an appropriate tool in the alleviation of allergy symptoms. Ma et al. (2014) made an attempt to eliminate the heat treatment in β-lactoglobulin dispersions, consequently heat induced allergenicity of β-lactoglobulin. β-lactoglobulin was encapsulated in PC nanoliposomes, which were stable under simulated gastric digestion conditions in vitro for 4 h (Ma et al., 2014). In another study, in vitro gastric digestion stability of thiamine was enhanced upon liposomal encapsulation in highly thermostable nanoliposomes (Fathima et al., 2016).

3.11 Conclusions and future perspectives

An increasing number of studies are being carried out on the nanoliposomal encapsulation of technically and biologically active compounds with the aim of their inclusion in nutraceuticals and food products. Nanoliposomes have the advantages of increasing the bioavailability of such compounds, increasing their solubility and dispersibility in food systems, stabilizing them against the stresses generated during food processing and digestion. Protection and controlled/sustained release of sensitive bioactives assist the manufacturers in preparing high quality, reproducible, safe product formulations, functional foods, and nutraceuticals.

Future food or nutraceutical formulations that bear nanoliposomal dispersions should focus on the generation of one- or reduced step operations in liposome manufacture, the possibility of large scale and continuous mode production processes that preferably utilize no solvents or detergents. The clean labels sought highly after by the consumers can be further reinforced with the emphasis on the ingredients themselves being bioactive (e.g., sphingolipids and phospholipids).

Microfluidization and especially high-pressure homogenization, the infrastructure for which is widely available in the food processing industries, could be instrumental in providing both large capacity and continuous mode production possibilities. In addition, novel and sustainable sources of liposomes are currently being proposed. In the future, a greater variety of liposome forming lipids may be expected to commercially exist at reasonable costs. In order to enhance the efficacy and reproducibility of liposome manufacture practices, process optimization methods based on RSM or similar techniques may be increasingly utilized. Spray-dried liposomal systems or proliposomes may be increasingly popular in functional food production. If this approach could lessen the number of steps in bioactive incorporation, reduce the burden on the manufacturer, possibly enhance the profitability of novel functional foods, it could be a step forward in this area.

Especially from the pharmaceutical sector, there is a wealth of information on the interactions of liposomes with viable cells or living organisms. We anticipate an increasing extent of investigations in the near future oriented toward the cellular uptake and interactions of bioactive bearing nanoliposomes in cell culture environments or in vivo investigations, whether such liposomes are formulated for analytical and R&D purposes or incorporated into model or real food systems. Based on the pharmaceutical progress, surface modification to enhance targeted delivery, manufacture of decorated liposomes, inhibition of entrapment in mucus and enhanced stability during storage and digestion have to be properly addressed in the areas of foods and nutraceuticals. Similarly, the importance of phytosomes can safely be expected to increase, while the bioavailability of active compounds from phytosomes has to be keenly investigated.

Acknowledgments

This study was partially funded by a grant from TÜBİTAK 2232 Program (Grant No. 114C049; The Scientific and Technological Research Council of Turkey). We would like to thank Mr. İbrahim Yılmaz (graphic designer, İZÜ) for his technical assistance in the preparation of Figs. 4.1 and 4.4.

Further Reading

Bezelgues, J.B., Morgan, F., Palomo, G., Crosset-Perrotin, L., Ducret, P., 2009. Short communication: milk fat globule membrane as a potential delivery system for liposoluble nutrients. J. Dairy Sci. 92 (6), 2524–2528.

Colas, J.C., Shi, W., Rao, V.M., Omri, A., Mozafari, M.R., Singh, H., 2007. Microscopical investigations of nisin-loaded nanoliposomes prepared by Mozafari method and their bacterial targeting. Micron 38 (8), 841–847.

Iavicoli, P., Urbán, P., Bella, A., Ryadnov, M.G., Rossi, F., Calzolai, L., 2015. Application of asymmetric flow field-flow fractionation hyphenations for liposome–antimicrobial peptide interaction. J. Chromatogr. A 1422, 260–269.

References

Alexander, M., Lopez, A.A., Fang, Y., Corredig, M., 2012. Incorporation of phytosterols in soy phospholipids nanoliposomes: encapsulation efficiency and stability. LWT Food Sci. Technol. 47 (2), 427–436.

Alonso, A., Villena, A., Goñi, F.M., 1981. Lysis and reassembly of sonicated lecithin vesicles in the presence of Triton X-100. FEBS Lett. 123 (2), 200–204.

Amin, T., Bhat, S.V., 2012. A review on phytosome technology as a novel approach to improve the bioavailability of nutraceuticals. Int. J. Adv. Res. Technol. 1 (3), 43–57.

Amit, G., Ashawat, M.S., Shailendra, S., Swarnlata, S., 2007. Phytosome: a novel approach towards functional cosmetics. Plant Sci. 2 (6), 644–649.

Angelico, R., Ceglie, A., Sacco, P., Colafemmina, G., Ripoli, M., Mangia, A., 2014. Phyto-liposomes as nanoshuttles for water-insoluble silybin–phospholipid complex. Int. J. Pharm. 471 (1), 173–181.

Bombardelli, E., Curri, S.B., Della, R.L., Del, N.P., Tubaro, A., Gariboldi, P., 1989. Complexes between phospholipids and vegetal derivatives of biological interest. Fitoterapia 60, 1–9.

Briuglia, M.L., Rotella, C., McFarlane, A., Lamprou, D.A., 2015. Influence of cholesterol on liposome stability and on in vitro drug release. Drug Deliv. Transl. Res. 5 (3), 231–242.

Bryła, A., Lewandowicz, G., Juzwa, W., 2015. Encapsulation of elderberry extract into phospholipid nanoparticles. J. Food Eng. 167, 189–195.

Cagdas, F.M., Ertugral, N., Bucak, S., Atay, N.Z., 2011. Effect of preparation method and cholesterol on drug encapsulation studies by phospholipid liposomes. Pharm. Dev. Technol. 16 (4), 408–414.

Campardelli, R., Santo, I.E., Albuquerque, E.C., de Melo, S.V., Della Porta, G., Reverchon, E., 2016. Efficient encapsulation of proteins in submicro liposomes using a supercritical fluid assisted continuous process. J. Supercrit. Fluids 107, 163–169.

Carvalho, J.M.P., Toniazzo, T., Cavalcanti, L.P., Moraes, C.L.P., Oliveira, C.L.P., Pinho, S.C., 2015. Physico-chemical stability and structural characterization of thickened multilamellar beta-carotene-loaded liposome dispersions produced using a proliposome method. Colloid Polym. Sci. 293 (8), 2171–2179.

Chen, T.L., Lo, Y.C., Hu, W.T., Wu, M.C., Chen, S.T., Chang, H.M., 2003. Microencapsulation and modification of synthetic peptides of food proteins reduces the blood pressure of spontaneously hypertensive rats. J. Agric. Food Chem. 51 (6), 1671–1675.

Chen, D., Xia, D., Li, X., Zhu, Q., Yu, H., Zhu, C., Gan, Y., 2013. Comparative study of Pluronic® F127-modified liposomes and chitosan-modified liposomes for mucus penetration and oral absorption of cyclosporine A in rats. Int. J. Pharm. 449 (1), 1–9.

Clergeaud, G., Dabbagh-Bazarbachi, H., Ortiz, M., Fernández-Larrea, J.B., O'Sullivan, C.K., 2016. A simple liposome assay for the screening of zinc ionophore activity of polyphenols. Food Chem. 197, 916–923.

Coimbra, M., Isacchi, B., Bloois, L.V., Torano, J.S., Ket, A., Wu, X., Broere, F., Metselaar, J.M., Rijcken, C.J.F., Storm, G., Bilia, R., Schiffelers, R.M., 2011. Improving solubility and chemical stability of natural compounds for medicinal use by incorporation into liposomes. Int. J. Pharm. 416 (2), 433–442.

Cosco, D., Paolino, D., Maiuolo, J., Di Marzio, L., Carafa, M., Ventura, C.A., Fresta, M., 2015. Ultradeformable liposomes as multidrug carrier of resveratrol and 5-fluorouracil for their topical delivery. Int. J. Pharm. 489 (1), 1–10.

Cui, H., Zhao, C., Lin, L., 2015. The specific antibacterial activity of liposome-encapsulated clove oil and its application in tofu. Food Control 56, 128–134.

Cui, H., Zhou, H., Lin, L., 2016. The specific antibacterial effect of the Salvia oil nanoliposomes against *Staphylococcus aureus* biofilms on milk container. Food Control 61, 92–98.

da Silva Malheiros, P., Sant'Anna, V., Micheletto, Y.M.S., da Silveira, N.P., Brandelli, A., 2011. Nanovesicle encapsulation of antimicrobial peptide P34: physicochemical characterization and mode of action on *Listeria monocytogenes*. J. Nanopart. Res. 13 (8), 3545–3552.

da Silva Malheiros, P., Sant'Anna, V., de Souza Barbosa, M., Brandelli, A., de Melo Franco, B.D.G., 2012. Effect of liposome-encapsulated nisin and bacteriocin-like substance P34 on *Listeria monocytogenes* growth in Minas frescal cheese. Int. J. Food Microbiol. 156 (3), 272–277.

Das, M.K., Kalita, B., 2014. Design and evaluation of phyto-phospholipid complexes (Phytosomes) of rutin for transdermal application. J. Appl. Pharm. Sci. 4 (10), 51–57.

de Assis, L.M., Machado, A.R., da Motta, A.S., Costa, J.A.V., de Souza-Soares, L.A., 2014. Development and characterization of nanovesicles containing phenolic compounds of microalgae *Spirulina* strain LEB-18 and *Chlorella pyrenoidosa*. Adv. Mater. Phy. Chem. 4, 6–12.

Degnan, A.J., Buyong, N., Luchansky, J.B., 1993. Antilisterial activity of pediocin AcH in model food systems in the presence of an emulsifier or encapsulated within liposomes. Int. J. Food Microbiol. 18 (2), 127–138.

Dewettinck, K., Rombaut, R., Thienpont, N., Le, T.T., Messens, K., Van Camp, J., 2008. Nutritional and technological aspects of milk fat globule membrane material. Int. Dairy J. 18 (5), 436–457.

Di Pierro, F., Menghi, A.B., Barreca, A., Lucarelli, M., Calandrelli, A., 2009. Greenselect Phytosome as an adjunct to a low-calorie diet for treatment of obesity: a clinical trial. Alt. Med. Rev. 14 (2), 154–160.

Du, L., Yang, Y.H., Wang, Y.M., Xue, C.H., Kurihara, H., Takahashi, K., 2015. Antitumour activity of EPA-enriched phospholipids liposomes against S180 ascitic tumour-bearing mice. J. Funct. Foods 19, 970–982.

Dwivedi, A.M., 2002. Residual solvent analysis in pharmaceuticals. Pharm. Technol. Eur. 14, 26–28.

Fan, Y., Ren, M., Hou, W., Guo, C., Tong, D., Ma, L., Zhang, W., He, M., Song, X., 2015. The activation of *Epimedium* polysaccharide-propolis flavone liposome on Kupffer cells. Carbohydr. Polym. 133, 613–623.

Fan, Y., Ma, X., Ma, L., Zhang, J., Zhang, W., Song, X., 2016. Antioxidative and immunological activities of ophiopogon polysaccharide liposome from the root of *Ophiopogon japonicus*. Carbohydr. Polym. 135, 110–120.

Farhang, B., Kakuda, Y., Corredig, M., 2012. Encapsulation of ascorbic acid in liposomes prepared with milk fat globule membrane-derived phospholipids. Dairy Sci. Technol. 92 (4), 353–366.

Fathima, S.J., Fathima, I., Abhishek, V., Khanum, F., 2016. Phosphatidylcholine, an edible carrier for nanoencapsulation of unstable thiamine. Food Chem. 197, 562–570.

Frenzel, M., Steffen-Heins, A., 2015a. Whey protein coating increases bilayer rigidity and stability of liposomes in food-like matrices. Food Chem. 173, 1090–1099.

Frenzel, M., Steffen-Heins, A., 2015b. Impact of quercetin and fish oil encapsulation on bilayer membrane and oxidation stability of liposomes. Food Chem. 185, 48–57.

Frenzel, M., Krolak, E., Wagner, A.E., Steffen-Heins, A., 2015. Physicochemical properties of WPI coated liposomes serving as stable transporters in a real food matrix. LWT Food Sci. Technol. 63 (1), 527–534.

Gallier, S., Gragson, D., Cabral, C., Jiménez-Flores, R., Everett, D.W., 2010. Composition and fatty acid distribution of bovine milk phospholipids from processed milk products. J. Agri. Food Chem. 58 (19), 10503–10511.

Gharib, A., Faezizadeh, Z., Godarzee, M., 2015. Preparation and characterization of nanoliposomal beta-cryptoxanthin and its effect on proliferation and apoptosis in human leukemia cell line K562. Trop. J. Pharm. Res. 14 (2), 187–194.

Ghorbanzade, T., Jafari, S.M., Akhavan, S., Hadavi, R., 2017. Nano-encapsulation of fish oil in nano-liposomes and its application in fortification of yogurt. Food Chem. 216, 146–152.

Gibbs, B.F., Kermasha, S., Alli, I., Mulligan, C.N., 1999. Encapsulation in the food industry: a review. Int. J. Food Sci. Nutr. 50 (3), 213–224.

Gibis, M., Vogt, E., Weiss, J., 2012. Encapsulation of polyphenolic grape seed extract in polymer-coated liposomes. Food Funct. 3 (3), 246–254.

Gibis, M., Rahn, N., Weiss, J., 2013. Physical and oxidative stability of uncoated and chitosan-coated liposomes containing grape seed extract. Pharmaceutics 5 (3), 421–433.

Gibis, M., Zeeb, B., Weiss, J., 2014a. Formation, characterization, and stability of encapsulated hibiscus extract in multilayered liposomes. Food Hydrocol. 38, 28–39.

Gibis, M., Thellmann, K., Thongkaew, C., Weiss, J., 2014b. Interaction of polyphenols and multilayered liposomal-encapsulated grape seed extract with native and heat-treated proteins. Food Hydrocol. 41, 119–131.

Gortzi, O., Lala, S., Chinou, I., Tsaknis, J., 2007. Evaluation of the antimicrobial and antioxidant activities of *Origanum dictamnus* extracts before and after encapsulation in liposomes. Molecules 12 (5), 932–945.

Gülseren, İ., Corredig, M., 2013. Storage stability and physical characteristics of tea-polyphenol-bearing nanoliposomes prepared with milk fat globule membrane phospholipids. J. Agri. Food Chem. 61 (13), 3242–3251.

Gülseren, İ., Güzey, D., Bruce, B.D., Weiss, J., 2007. Structural and functional changes in ultrasonicated bovine serum albumin solutions. Ultrason. Sonochem. 14 (2), 173–183.

Gülseren, İ., Guri, A., Corredig, M., 2012. Encapsulation of tea polyphenols in nanoliposomes prepared with milk phospholipids and their effect on the viability of HT-29 human carcinoma cells. Food Dig. 3 (1–3), 36–45.

Gunter, K.K., Gunter, T.E., Jarkowski, A., Rosier, R.N., 1982. A method of resuspending small vesicles separated from suspension by protamine aggregation and centrifugation. Anal. Biochem. 120 (1), 113–124.

Gupta, C., Chawla, P., Arora, S., 2015. Development and evaluation of iron microencapsules for milk fortification. CyTA-J. Food 13 (1), 116–123.

Guri, A., Gülseren, I., Corredig, M., 2013. Utilization of solid lipid nanoparticles for enhanced delivery of curcumin in cocultures of HT29-MTX and Caco-2 cells. Food Funct. 4 (9), 1410–1419.

Habib, L., Khreich, N., Jraij, A., Abbas, S., Magdalou, J., Charcosset, C., Greige-Gerges, H., 2013. Preparation and characterization of liposomes incorporating cucurbitacin E, a natural cytotoxic triterpene. Int. J. Pharm. 448 (1), 313–319.

Halwani, M., Hossain, Z., Khiyami, M.A., Omri, A., 2015. Liposomal β-glucan: preparation, characterization and anticancer activities. J. Nanomed. Nanotechnol. 6 (319), 2.
Hasan, M., Belhaj, N., Benachour, H., Barberi-Heyob, M., Kahn, C.J.F., Jabbari, E., Linder, M., Arab-Tehrany, E., 2014. Liposome encapsulation of curcumin: physico-chemical characterizations and effects on MCF7 cancer cell proliferation. Int. J. Pharm. 461 (1), 519–528.
Hashimoto, T., Kumazawa, S., Nanjo, F., Hara, Y., Nakayama, T., 1999. Interaction of tea catechins with lipid bilayers investigated with liposome systems. Biosci. Biotechnol. Biochem. 63 (12), 2252–2255.
Huang, Z., Li, X., Zhang, T., Song, Y., She, Z., Li, J., Deng, Y., 2014. Progress involving new techniques for liposome preparation. Asian J. Pharm. Sci. 9 (4), 176–182.
Hüsch, J., Bohnet, J., Fricker, G., Skarke, C., Artaria, C., Appendino, G., et al., 2013. Enhanced absorption of boswellic acids by a lecithin delivery form (Phytosome®) of Boswellia extract. Fitoterapia 84, 89–98.
Hwang, J.S., Tsai, Y.L., Hsu, K.C., 2010. The feasibility of antihypertensive oligopeptides encapsulated in liposomes prepared with phytosterols-β-sitosterol or stigmasterol. Food Res. Int. 43 (1), 133–139.
Ikigai, H., Nakae, T., Hara, Y., Shimamura, T., 1993. Bactericidal catechins damage the lipid bilayer. Biochim. Biophy. Acta (BBA)-Biomemb. 1147 (1), 132–136.
Imran, M., Revol-Junelles, A.M., Paris, C., Guedon, E., Linder, M., Desobry, S., 2015. Liposomal nanodelivery systems using soy and marine lecithin to encapsulate food biopreservative nisin. LWT Food Sci. Technol. 62 (1), 341–349.
Jackson, L.S., Lee, K., 1991. Microencapsulation and the food industry. Lebensmittel Wissenschaft Technologie 24 (4), 289–297.
Jafari, S.M., Fathi, M., Mandala, I., 2015. Emerging product formation. Food waste recovery: processing technologies andindustrial techniques. Elsevier, Academic Press, San Diego, CA, pp. 293–317.
Jain, N., Gupta, B.P., Thakur, N., Jain, R., Banweer, J., Jain, D.K., Jain, S., 2010. Phytosome: a novel drug delivery system for herbal medicine. Int. J. Pharm. Sci. Drug Res. 2 (4), 224–228.
Kaneko, K., McDowell, A., Ishii, Y., Hook, S., 2015. Selective quantitation of the incorporation of the immunomodulator α-galactosylceramide in liposomes using LC–MS/MS. Int. J. Mass Spectrom. 392, 96–101.
Karadag, A., Özçelik, B., Sramek, M., Gibis, M., Kohlus, R., Weiss, J., 2013. Presence of electrostatically adsorbed polysaccharides improves spray drying of liposomes. J. Food Sci. 78 (2), E206–E221.
Karataş, A., Turhan, F., 2015. Phyto-phospholipid complexes as drug delivery system for herbal extracts/molecules. Turkish J. Pharm. Sci. 12 (1), 93–102.
Karimi, N., Ghanbarzadeh, B., Hamishehkar, H., Keivani, F., Pezeshki, A., Gholian, M.M., 2015. Phytosome and liposome: the beneficial encapsulation systems in drug delivery and food application. Appl. Food Biotechnol. 2 (3), 17–27.
Katouzian, I., Jafari, S.M., 2016. Nano-encapsulation as a promising approach for targeted delivery and controlled release of vitamins. Trends Food Sci. Technol. 53, 34–48.
Keller, B.C., 2001. Liposomes in nutrition. Trends Food Sci. Technol. 12 (1), 25–31.
Kidd, P.M., 2009. Bioavailability and activity of phytosome complexes from botanical polyphenols: the silymarin, curcumin, green tea, and grape seed extracts. Alt. Med. Rev. 14 (3), 226–246.
Kirby, C.J., Brooker, B.E., Law, B.A., 1987. Accelerated ripening of cheese using liposome-encapsulated enzyme. Int. J. Food Sci. Technol. 22 (4), 355–375.
Kong, B., Xiong, Y.L., 2006. Antioxidant activity of zein hydrolysates in a liposome system and the possible mode of action. J. Agri. Food Chem. 54 (16), 6059–6068.

Lagoueyte, N., Paquin, P., 1998. Effects of microfluidization on the functional properties of xanthan gum. Food Hydrocol. 12 (3), 365–371.

Laridi, R., Kheadr, E.E., Benech, R.O., Vuillemard, J.C., Lacroix, C., Fliss, I., 2003. Liposome encapsulated nisin Z: optimization, stability and release during milk fermentation. Int. Dairy J. 13 (4), 325–336.

Lasic, D.D., 1993. Liposomes: From Physics to Applications. Elsevier, Amsterdam, The Netherlands.

Laye, C., McClements, D.J., Weiss, J., 2008. Formation of biopolymer-coated liposomes by electrostatic deposition of chitosan. J. Food Sci. 73 (5), N7–N15.

Lee, S.C., Jeon, Y.J., 2015. Effect of defatted sesame meal extract on phospholipid peroxidation in liposomes. Food Sci. Biotechnol. 24 (4), 1549–1554.

Li, H., Chen, F., 2015. Preparation and quality evaluation of coenzyme Q_{10} long-circulating liposomes. Saudi Journal of Biological Sciences, Available online.

Lin, L., Zhang, X., Zhao, C., Cui, H., 2016. Liposome containing nutmeg oil as the targeted preservative against *Listeria monocytogenes* in dumplings. RSC Advances 6 (2), 978–986.

Liu, W., Liu, W.L., Liu, M.C., Liu, J.H., Yang, S.B., Zheng, H.J., Lei, H.W., Ruan, R., Li, T., Tu, Z.C., Song, X.Y., 2011. Medium-chain fatty acid nanoliposomes for easy energy supply. Nutrition 27 (6), 700–706.

Liu, W., Ye, A., Liu, C., Liu, W., Singh, H., 2012. Structure and integrity of liposomes prepared from milk-or soybean-derived phospholipids during in vitro digestion. Food Res. Int. 48 (2), 499–506.

Liu, W., Ye, A., Liu, W., Liu, C., Han, J., Singh, H., 2015a. Behaviour of liposomes loaded with bovine serum albumin during in vitro digestion. Food Chem. 175, 16–24.

Liu, W., Aiqian, Ye., Singh, H., 2015b. Progress in applications of liposomes in food systems. In: Sagis, L.M.C. (Ed.), Microencapsulation and Microspheres for Food Applications. Academic Press, New York, NY, USA, pp. 151–170, (Chapter 8).

Liu, W., Liu, W., Ye, A., Peng, S., Wei, F., Liu, C., Han, J., 2016. Environmental stress stability of microencapsules based on liposomes decorated with chitosan and sodium alginate. Food Chem. 196, 396–404.

Lopez, C., 2011. Milk fat globules enveloped by their biological membrane: unique colloidal assemblies with a specific composition and structure. Curr. Opin. Colloid Interface Sci. 16 (5), 391–404.

Luo, X., Guan, R., Chen, X., Tao, M., Ma, J., Zhao, J., 2014. Optimization on condition of epigallocatechin-3-gallate (EGCG) nanoliposomes by response surface methodology and cellular uptake studies in Caco-2 cells. Nanoscale Res. Lett. 9 (1), 1–9.

Ma, J., Guan, R., Chen, X., Wang, Y., Hao, Y., Ye, X., Liu, M., 2014. Response surface methodology for the optimization of beta-lactoglobulin nano-liposomes. Food Funct. 5 (4), 748–754.

Madrigal-Carballo, S., Lim, S., Rodriguez, G., Vila, A.O., Krueger, C.G., Gunasekaran, S., Reed, J.D., 2010. Biopolymer coating of soybean lecithin liposomes via layer-by-layer self-assembly as novel delivery system for ellagic acid. J. Funct. Foods 2 (2), 99–106.

Maherani, B., Arab-Tehrany, E., Mozafari, M.R., Gaiani, C., Linder, M., 2011. Liposomes: a review of manufacturing techniques and targeting strategies. Curr. Nanosci. 7 (3), 436–452.

Malheiros, P.S., Cuccovia, I.M., Franco, B.D., 2016. Inhibition of *Listeria monocytogenes* in vitro and in goat milk by liposomal nanovesicles containing bacteriocins produced by *Lactobacillus sakei* subsp. *sakei* 2a. Food Control 63, 158–164.

Marsanasco, M., Márquez, A.L., Wagner, J.R., Alonso, S.D.V., Chiaramoni, N.S., 2011. Liposomes as vehicles for vitamins E and C: an alternative to fortify orange juice and offer vitamin C protection after heat treatment. Food Res. Int. 44 (9), 3039–3046.

Marsanasco, M., Márquez, A.L., Wagner, J.R., Chiaramoni, N.S., Alonso, S.D.V., 2015. Bioactive compounds as functional food ingredients: characterization in model system and sensory evaluation in chocolate milk. J. Food Eng. 166, 55–63.

Mohammadi, M., Ghanbarzadeh, B., Hamishehkar, H., 2014. Formulation of nanoliposomal vitamin D3 for potential application in beverage fortification. Adv. Pharm. Bull. 4 (Suppl 2), 569–575.

Mosquera, M., Giménez, B., da Silva, I.M., Boelter, J.F., Montero, P., Gómez-Guillén, M.C., Brandelli, A., 2014. Nanoencapsulation of an active peptidic fraction from sea bream scales collagen. Food Chem. 156, 144–150.

Mourtas, S., Lazar, A.N., Markoutsa, E., Duyckaerts, C., Antimisiaris, S.G., 2014. Multifunctional nanoliposomes with curcumin–lipid derivative and brain targeting functionality with potential applications for Alzheimer disease. Eur. J. Med. Chem. 80, 175–183.

Mozafari, M.R., 2006. Bioactive entrapment and targeting using nanocarrier technologies: an introduction. Nanocarrier technologies. Springer, Netherlands, pp. 1–16.

Mozafari, M.R., Mortazavi, S.M., 2005. Nanoliposomes: From Fundamentals to Recent Developments. Trafford Publishing, Australia.

Mozafari, M.R., Khosravi-Darani, K., Borazan, G.G., Cui, J., Pardakhty, A., Yurdugul, S., 2008a. Encapsulation of food ingredients using nanoliposome technology. Int. J. Food Prop. 11 (4), 833–844.

Mozafari, M.R., Johnson, C., Hatziantoniou, S., Demetzos, C., 2008b. Nanoliposomes and their applications in food nanotechnology. J. Liposome Res. 18 (4), 309–327.

Nacka, F., Cansell, M., Gouygou, J.P., Gerbeaud, C., Méléard, P., Entressangles, B., 2001. Physical and chemical stability of marine lipid-based liposomes under acid conditions. Colloids Surf. B 20 (3), 257–266.

Nakayama, T., Hashimoto, T., Kajiya, K., Kumazawa, S., 2000. Affinity of polyphenols for lipid bilayers. Biofactors 13 (1–4), 147–151.

Neunert, G., Górnaś, P., Dwiecki, K., Siger, A., Polewski, K., 2015. Synergistic and antagonistic effects between alpha-tocopherol and phenolic acids in liposome system: spectroscopic study. Eur. Food Res. Technol. 241 (6), 749–757.

Nongonierma, A.B., Abrlova, M., Fenelon, M.A., Kilcawley, K.N., 2009. Evaluation of two food grade proliposomes to encapsulate an extract of a commercial enzyme preparation by microfluidization. J. Agri. Food Chem. 57 (8), 3291–3297.

Paini, M., Dalyc, S.R., Aliakbarian, B., Fathi, A., Tehrany, E.A., Peregoa, B., Dehghani, F., Valtchev, P., 2015. An efficient liposome based method for antioxidants encapsulation. Colloids Surf. B 136, 1067–1072.

Panya, A., Laguerre, M., Lecomte, J., Villeneuve, P., Weiss, J., McClements, D.J., Decker, E.A., 2010. Effects of chitosan and rosmarinate esters on the physical and oxidative stability of liposomes. J. Agri. Food Chem. 58 (9), 5679–5684.

Pezeshky, A., Ghanbarzadeh, B., Hamishehkar, H., Moghadam, M., Babazadeh, A., 2016. Vitamin A palmitate-bearing nanoliposomes: preparation and characterization. Food Biosci. 13, 49–55.

Pidgeon, C., 1993. In: Gregordias, G. (Ed.), Liposome Technology. CRC Press, Boca Raton, FL, USA, pp. 99–110.

Pietta, P., Simonetti, P., Gardana, C., Brusamolino, A., Morazzoni, P., Bombardelli, E., 1998. Relationship between rate and extent of catechin absorption and plasma antioxidant status. Biochem. Mol. Biol. Int. 46 (5), 895–903.

Pinsolle, A., Roy, P., Cansell, M., 2014. Modulation of enzymatic PS synthesis by liposome membrane composition. Colloids Surf. B 115, 157–163.

Pravilović, R.N., Radunović, V.S., Bošković-Vragolović, N.M., Bugarski, B.M., Pjanović, R.V., 2015. The influence of membrane composition on the release of polyphenols from liposomes. Hemijska Industrija 69 (4), 347–353.
Priprem, A., Watanatorn, J., Sutthiparinyanont, S., Phachonpai, W., Muchimapura, S., 2008. Anxiety and cognitive effects of quercetin liposomes in rats. Nanomed. Nanotechnol. Biol. Med. 4 (1), 70–78.
Qadir, S.A., Kwon, M.C., Han, J.G., Ha, J.H., Jin, L., Jeong, H.S., Kim, J.C., You, S.G., Lee, H.Y., 2008. Enhancement of immunomodulatory and anticancer activity of fucoidan by nano encapsulation. Food Sci. Biotechnol. 17 (6), 1254–1260.
Rasaie, S., Ghanbarzadeh, S., Mohammadi, M., Hamishehkar, H., 2014. Nano phytosomes of quercetin: a promising formulation for fortification of food products with antioxidants. Pharm. Sci. 20, 96–101.
Rashidinejad, A., Birch, E.J., Sun-Waterhouse, D., Everett, D.W., 2014. Delivery of green tea catechin and epigallocatechin gallate in liposomes incorporated into low-fat hard cheese. Food Chem. 156, 176–183.
Richards, M.H., Gardner, C.R., 1978. Effects of bile salts on the structural integrity of liposomes. Biochim. Biophy. Acta (BBA) Gen. Subjects 543 (4), 508–522.
Rowland, R.N., Woodley, J.F., 1980. The stability of liposomes in vitro to pH, bile salts and pancreatic lipase. Biochim. Biophy. Acta (BBA)-Lipids Lipid Metabol. 620 (3), 400–409.
Saengkrit, N., Saesoo, S., Srinuanchai, W., Phunpee, S., Ruktanonchai, U.R., 2014. Influence of curcumin-loaded cationic liposome on anticancer activity for cervical cancer therapy. Colloids Surf. B 114, 349–356.
Sant'Anna, V., da Silva Malheiros, P., Brandelli, A., 2011. Liposome encapsulation protects bacteriocin-like substance P34 against inhibition by Maillard reaction products. Food Res. Int. 44 (1), 326–330.
Sebaaly, C., Jraij, A., Fessi, H., Charcosset, C., Greige-Gerges, H., 2015. Preparation and characterization of clove essential oil-loaded liposomes. Food Chem. 178, 52–62.
Seguin, J., Brullé, L., Boyer, R., Lu, Y.M., Romano, M.R., Touil, Y.S., Scherman, D., Bessodes, M., Mignet, N., Chabot, G.G., 2013. Liposomal encapsulation of the natural flavonoid fisetin improves bioavailability and antitumor efficacy. Int. J. Pharm. 444 (1), 146–154.
Semalty, A., Semalty, M., Rawat, M.S.M., Franceschi, F., 2010. Supramolecular phospholipids–polyphenolics interactions: the PHYTOSOME® strategy to improve the bioavailability of phytochemicals. Fitoterapia 81, 306–314.
Siebert, K.J., Troukhanova, N.V., Lynn, P.Y., 1996. Nature of polyphenol-protein interactions. J. Agri. Food Chem. 44 (1), 80–85.
Sila, M., Au, S., Weiner, N., 1986. Effects of Triton X-100 concentration and incubation temperature on carboxyfluorescein release from multilamellar liposomes. Biochim. Biophy. Acta (BBA)-Biomembr. 859 (2), 165–170.
Sinico, C., De Logu, A., Lai, F., Valenti, D., Manconi, M., Loy, G., et al., 2005. Liposomal incorporation of *Artemisia arborescens* L. essential oil and in vitro antiviral activity. Eur. J. Pharm. Biopharm. 59 (1), 161–168.
Socaciu, C., Jessel, R., Diehl, H.A., 2000. Competitive carotenoid and cholesterol incorporation into liposomes: effects on membrane phase transition, fluidity, polarity and anisotropy. Chem. Phy. Lipids 106 (1), 79–88.
Spitsberg, V.L., 2005. Invited review: bovine milk fat globule membrane as a potential nutraceutical. J. Dairy Sci. 88 (7), 2289–2294.
Strauss, G., Schurtenberger, P., Hauser, H., 1986. The interaction of saccharides with lipid bilayer vesicles: stabilization during freeze-thawing and freeze-drying. Biochim. Biophy. Acta (BBA)-Biomembr. 858 (1), 169–180.

Szebeni, J., Moghimi, S.M., 2009. Liposome triggering of innate immune responses: a perspective on benefits and adverse reactions: biological recognition and interactions of liposomes. J. Liposome Res. 19 (2), 85–90.
Takahashi, M., Inafuku, K.I., Miyagi, T., Oku, H., Wada, K., Imura, T., Kitamoto, D., 2006. Efficient preparation of liposomes encapsulating food materials using lecithins by a mechanochemical method. J. Oleo Sci. 56 (1), 35–42.
Tan, C., Xue, J., Eric, K., Feng, B., Zhang, X., Xia, S., 2013. Dual effects of chitosan decoration on the liposomal membrane physicochemical properties as affected by chitosan concentration and molecular conformation. J. Agri. Food Chem. 61 (28), 6901–6910.
Tan, C., Zhang, Y., Abbas, S., Feng, B., Zhang, X., Xia, S., 2014. Modulation of the carotenoid bioaccessibility through liposomal encapsulation. Colloids Surf. B 123, 692–700.
Tan, C., Feng, B., Zhang, X., Xia, W., Xia, S., 2016. Biopolymer-coated liposomes by electrostatic adsorption of chitosan (chitosomes) as novel delivery systems for carotenoids. Food Hydrocol. 52, 774–784.
Taylor, T.M., Weiss, J., Davidson, P.M., Bruce, B.D., 2005. Liposomal nanocapsules in food science and agriculture. Crit. Rev. Food Sci. Nutr. 45 (7–8), 587–605.
Taylor, T.M., Gaysinsky, S., Davidson, P.M., Bruce, B.D., Weiss, J., 2007. Characterization of antimicrobial-bearing liposomes by ζ-potential, vesicle size, and encapsulation efficiency. Food Biophy. 2 (1), 1–9.
Taylor, T., Bruce, B.D., Weiss, J., Davidson, P.M., 2008. *Listeria monocytogenes* and *Escherichia coli* O157:H7 inhibition in vitro by liposome-encapsulated nisin and ethylene diaminetetraacetic acid. J. Food Saf. 28 (2), 183–197.
Teixeira, M.L., dos Santos, J., Silveira, N.P., Brandelli, A., 2008. Phospholipid nanovesicles containing a bacteriocin-like substance for control of *Listeria monocytogenes*. Innov. Food Sci. Emerg. Technol. 9 (1), 49–53.
Tellez, A., Corredig, M., Guri, A., Zanabria, R., Griffiths, M.W., Delcenserie, V., 2012. Bovine milk fat globule membrane affects virulence expression in *Escherichia coli* O157: H7. J. Dairy Sci. 95 (11), 6313–6319.
Thompson, A.K., Haisman, D., Singh, H., 2006a. Physical stability of liposomes prepared from milk fat globule membrane and soya phospholipids. J. Agri. Food Chem. 54 (17), 6390–6397.
Thompson, A.K., Hindmarsh, J.P., Haisman, D., Rades, T., Singh, H., 2006b. Comparison of the structure and properties of liposomes prepared from milk fat globule membrane and soy phospholipids. J. Agri. Food Chem. 54 (10), 3704–3711.
Thompson, A.K., Mozafari, M.R., Singh, H., 2007. The properties of liposomes produced from milk fat globule membrane material using different techniques. Le Lait 87 (4–5), 349–360.
Thompson, A.K., Couchoud, A., Singh, H., 2009. Comparison of hydrophobic and hydrophilic encapsulation using liposomes prepared from milk fat globule-derived phospholipids and soya phospholipids. Dairy Sci. Technol. 89 (1), 99–113.
Ting, Y., Jiang, Y., Ho, C.T., Huang, Q., 2014. Common delivery systems for enhancing in vivo bioavailability and biological efficacy of nutraceuticals. J. Funct. Foods 7, 112–128.
Toniazzo, T., Berbel, I.F., Cho, S., Fávaro-Trindade, C.S., Moraes, I.C., Pinho, S.C., 2014. β-carotene-loaded liposome dispersions stabilized with xanthan and guar gums: physico-chemical stability and feasibility of application in yogurt. LWT Food Sci. Technol. 59 (2), 1265–1273.
Tønnesen, H.H., 2002. Solubility, chemical and photochemical stability of curcumin in surfactant solutions. Studies of curcumin and curcuminoids, XXVIII. Die Pharmazie 57 (12), 820–824.

Toprak, M., 2016. Fluorescence study on the interaction of human serum albumin with Butein in liposomes. Spectrochim. Acta Part A 154, 108–113.

Tsuchiya, H., Tanaka, T., Nagayama, M., 2008. Antiproliferative effects associated with membrane lipid interaction of green tea catechins. J. Health Sci. 54 (5), 576–580.

Vemuri, S., Rhodes, C.T., 1995. Preparation and characterization of liposomes as therapeutic delivery systems: a review. Pharmaceutica Acta Helvetiae 70 (2), 95–111.

Viljanen, K., Kivikari, R., Heinonen, M., 2004. Protein-lipid interactions during liposome oxidation with added anthocyanin and other phenolic compounds. J. Agri. Food Chem. 52 (5), 1104–1111.

Wechtersbach, L., Ulrih, N.P., Cigić, B., 2012. Liposomal stabilization of ascorbic acid in model systems and in food matrices. LWT Food Sci. Technol. 45 (1), 43–49.

Weiss, J., Takhistov, P., McClements, D.J., 2006. Functional materials in food nanotechnology. J. Food Sci. 71 (9), R107–R116.

Were, L.M., Bruce, B.D., Davidson, P.M., Weiss, J., 2003. Size, stability, and entrapment efficiency of phospholipid nanocapsules containing polypeptide antimicrobials. J. Agri. Food Chem. 51 (27), 8073–8079.

Xia, S., Tan, C., Zhang, Y., Abbas, S., Feng, B., Zhang, X., Qin, F., 2015. Modulating effect of lipid bilayer–carotenoid interactions on the property of liposome encapsulation. Colloids Surf. B 128, 172–180.

Zawada, Z., Michnik, A., Drzazga, Z., 2002. Liposomes with a double bilayer—a study on stability. Cell. Mol. Biol. Lett. 7 (2), 305.

Zhang, N., Ping, Q.N., Huang, G.H., Xu, W.F., 2005. Investigation of lectin-modified insulin liposomes as carriers for oral administration. Int. J. Pharm. 294 (1), 247–259.

Zhang, J., Tang, Q., Xu, X., Li, N., 2013. Development and evaluation of a novel phytosome-loaded chitosan microsphere system for curcumin delivery. Int. J. Pharm. 448, 168–174.

Zhao, L., Temelli, F., 2015. Preparation of liposomes using a modified supercritical process via depressurization of liquid phase. J. Supercrit. Fluids 100, 110–120.

Zhao, L., Temelli, F., Curtis, J.M., Chen, L., 2015. Preparation of liposomes using supercritical carbon dioxide technology: effects of phospholipids and sterols. Food Res. Int. 77, 63–72.

Zou, L.Q., Liu, W., Liu, W.L., Liang, R.H., Li, T., Liu, C.M., et al., 2014a. Characterization and bioavailability of tea polyphenol nanoliposome prepared by combining an ethanol injection method with dynamic high-pressure microfluidization. J. Agri. Food Chem. 62 (4), 934–941.

Zou, L.Q., Peng, S.F., Liu, W., Gan, L., Liu, W.L., Liang, R.H., Liu, C.M., Niu, J., Cao, Y.L., Liu, Z., Chen, X., 2014b. Improved in vitro digestion stability of (−)-epigallocatechin gallate through nanoliposome encapsulation. Food Res. Int. 64, 492–499.

Zou, L., Peng, S., Liu, W., Chen, X., Liu, C., 2015. A novel delivery system dextran sulfate coated amphiphilic chitosan derivatives-based nanoliposome: capacity to improve in vitro digestion stability of (−)-epigallocatechin gallate. Food Res. Int. 69, 114–120.

4 Encapsulation by nanostructured lipid carriers

Sung-Min Pyo, Rainer H. Müller, Cornelia M. Keck
Free University of Berlin, Berlin, Germany

4.1 Introduction

There is an increasing awareness by the consumers about the correlation between the quality of food and disease prevention. Consequently "health promoting substances" are introduced into food and beverages for the production of "functional foods" (Ganje et al., 2016; Ghorbanzade et al., 2017). Health promoting substances are bioactive molecules and bioactive living cells (probiotics) and can be added to "health drinks" or just "well-being drinks". The bioactive molecules can be essential molecules needed by the body (e.g., vitamins, fatty acids), molecules being protective to the body (e.g., antioxidants), or having a certain positive supporting activity to the body (e.g., phytosterols). The added value of bioactive molecules is also very important for advertising the products and their market differentiation in this heavily competitive market. The technologies applied in the product can be highlighted in marketing (e.g., superactivity of nano Q_{10} nutraceutical oral products, biosmarts technology in drinks).

Apart from marketing, there is an even more important reason for manufacturers, at least serious manufacturers, to focus on delivery technologies: the increase in bioavailability to make their products really active. Unfortunately many actives with high interest for food are poorly soluble and show a low oral bioavailability (e.g., antioxidant flavonoids). One approach is to use nanosized delivery systems, in which the actives are nanoencapsulated (Faridi Esfanjani and Jafari, 2016; Katouzian and Jafari, 2016). The nanosize itself has advantages for delivery, for example, the high adhesiveness to the membranes of the gastrointestinal tract (GIT). The active releases exactly at the site of its absorption, thus the bioavailability is increased. Encapsulation in the narrower sense is a process in which droplets or particles are surrounded by a coat (Fig. 4.1, left), for example, by a polymer or a lipid. As another kind of encapsulation can be seen the enclosure of actives (in molecular or nanocrystalline form) in a matrix material. The actives can be dispersed in the matrix, or being enriched in the center forming an active core surrounded by an encapsulating wall of matrix material (Fig. 4.1, middle and right). The latter is the case for lipid nanoparticles with a bioactive-enriched core (zur Mühlen et al., 1998). Both approaches—coat and use of matrix material—can protect chemically labile actives and provides the control of its release.

Nanoencapsulation Technologies for the Food and Nutraceutical Industries

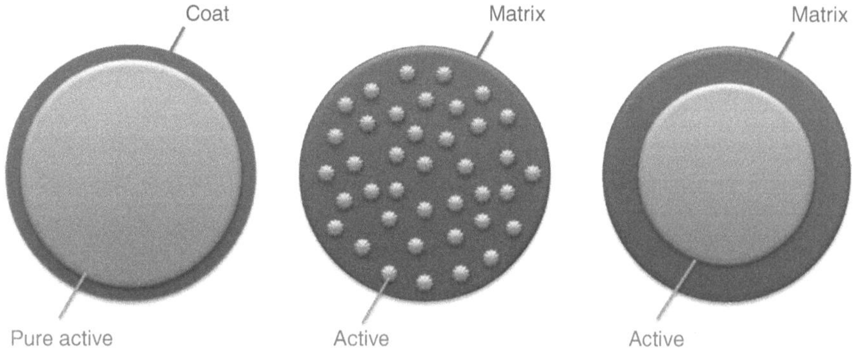

Figure 4.1 Encapsulation of actives via a coat (left), or use of a carrier matrix: enclosure of actives dispersed in the carrier matrix (middle) or an active core surrounded by a thick wall of carrier matrix material (right).

4.2 The logic behind the development of solid lipid nanoparticles

There is a broad variety on nanosized delivery systems, most of them are still in the academic labs, only a very few made it to the market. A nanosized system is scientifically defined in the pharmaceutical area as particles with a size of a few nanometer to just below 1,000 nm (=1 μm). It should not be mixed up with the legal definition of nanoparticles/nanomaterial for labeling consumer products, for example, labeling of cosmetic ingredients in the European Union with "nano" are necessary when more than 50% (by number distribution) of the particles have sizes below 100 nm (Mildau and Huber, 2010).

The most successful system is clearly the nanoemulsions (Chapter 2), easily accessible since invention of the high pressure homogenizer by Auguste Gaulin, presented at the World's Fair in Paris 1900. Apart from food industry, nanoemulsions are also used broadly in pharma products, for example, intravenous infusion for parenteral nutrition (Intralipid, Lipofundin, etc.) and for intravenous bioactive delivery (e.g., propofol). Very successful are also the liposomes (Chapter 3), introduced to the cosmetic market in 1986 by Dior (product Capture, still on the market) and in pharma products around 1990 (e.g., Aleveofact, Dr. Karl Thomae GmbH at this time). Advantages of these lipid-based systems are the ease of production and the unproblematic regulatory status of the excipients used (excipients often natural, accepted status, well tolerated). A disadvantage is the limited ability to stabilize chemically labile actives and the lack of controlled release. Due to the relatively low viscosity of the oils, the actives can easily diffuse into the surrounding water phase, where they can be degraded. After degradation, new nondegraded active diffuses into the water phase (according to partitioning coefficient) and the degradation continues (Fig. 4.2, left). After administration, the release into surrounding body fluids is very fast (small diffusional distance due to small size in combination with low viscosity). In addition, liposomes are of very limited use for peroral application due to their physical instability.

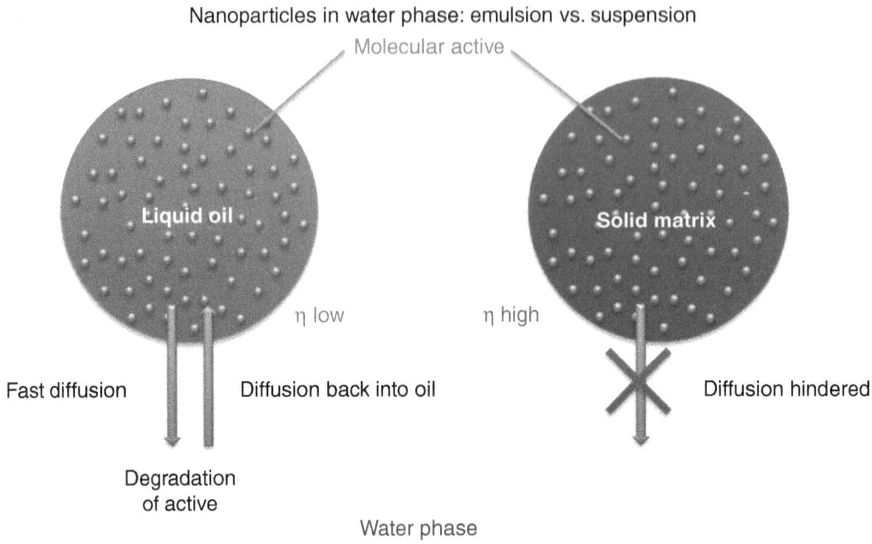

Figure 4.2 Degradation of lipophilic bioactives in oil droplets due to fast diffusional exchange with water phase (left) and slowing down of this process in particles with highly viscous solid matrix (right), for example, polymeric and solid lipid nanoparticles (SLNs).

These limitations can be overcome by using nanoparticles having a solid matrix with a high viscosity, since the diffusional coefficient D is then lowered according to the Einstein equation:

$$D = \frac{KT}{6\pi r\eta}$$

(K, Boltzmann constant; T, absolute temperature; r, particle radius; η, viscosity of matrix)

The slow diffusion in the viscous matrix minimizes the exchange of incorporated actives with the surrounding water phase of a nanoparticle suspension (Fig. 4.2, right). By modulating the size and type of matrix material, and also the internal structure of the solid particles, the release can be controlled from fast to very slow (c.f. Section 4.10). The classic SLNs are the pharmaceutical polymeric nanoparticles invented by P.P. Speiser and his PhD student G. Birrenbach (Birrenbach and Speiser, 1976) in the middle of the 1970s. Despite by now, about 40 years of research, the polymeric nanoparticles made it to, for example, cosmetic products, but not to a decent number of food and pharmaceutical products which have higher regulatory hurdles. Problems of the polymeric nanoparticles are the lack of large scale production methods or rather the costs of such methods, and on top the regulatory hurdles. Nicely functioning polymers do not have regulatory acceptance, and no company wants to invest the substantial amount of money for registration of such new excipients.

Thus, a solution was needed combining the desired solid state of the particles (=protection, controlled release) with accepted state of excipients, good tolerability, and ease of large scale production being also cost-effective. This was realized by replacing the polymers against solid lipids, and producing the particles by high pressure homogenization (HPH). The lipids used are solid at body temperature (i.e., they keep their properties also after the administration to the body), but for production by HPH they are processed in the melted state at elevated temperature.

4.3 First two generations of lipid nanoparticles: SLN vs. NLC

The first generation was developed 1991 in parallel by Prof. Maria Rosa Gasco in Turin (Gasco, 1993) and Prof. Rainer H. Müller in Berlin (Müller and Lucks, 1996) and submitted as patent applications. Some of the first experimental particles were made from lipids used for suppository production (e.g., Witepsol). The liquid oil of nanoemulsion was simply replaced by typically one single solid lipid. The particles differed in the production technology. The particles by Gasco were produced using a microemulsion technique, where the particles by Müller were produced using HPH. To clearly differentiate them from nanoemulsions, the term SLNs was consensually created by the inventors.

Of course the solubility of actives in solid lipids is lower than in liquid lipids, thus the SLN had a lower loading capacity than emulsions. The solubility of actives also depends on the polymorphic modification of the lipid matrix, showing less solubility in highly ordered β modification. Also the solubility was reduced using highly purified lipids, for example, tristearin (Bunjes et al., 1996).

The highly ordered matrix had little imperfections for accommodation of cosmetic active or bioactive, bioactive expulsion took sometimes place directly after production. Bioactive expulsion also occurred when less ordered α and β′ modifications were transformed into highly ordered β modification during storage.

To reduce these problems, the nanostructured lipid carriers (NLC) were developed in 1999 (Tamjidi, et al., 2013; Müller et al., 2014a). The philosophy of NLC was to combine spatially different lipid molecules which are not or less able to form perfectly ordered lipid structures in the particle matrix. Preferentially they should have as many imperfections as possible. For this, typically a solid lipid based on long chain fatty acids (e.g., behenic acid, C22) was mixed with an oil based on short chain fatty acids (e.g., Miglyol 812, containing C8 and C10). The differences in fatty acid chain length should distort perfect ordering. Typically one solid lipid was mixed with one oil. Thus, NLC are a blend of solid and liquid fat (oil). Blending with an oil had additionally the advantage that the bioactive loading increases due to the better solubility of actives in oils in general. In case of "ideal" mixtures, the solubility will be additive based on the molar ratios, and from this it can be roughly estimated from the solubility of the active in the single components. The differences between the two generations are shown in Fig. 4.3.

4.4 The third generation: smartLipids

By using the NLC, a clear increase in bioactive loading could be achieved, for example, for retinol from 1% in SLN to 5% in the new mixture (Jenning et al., 2000). The distinct increase in bioactive loading was paid dearly for, because addition of liquid lipid to a

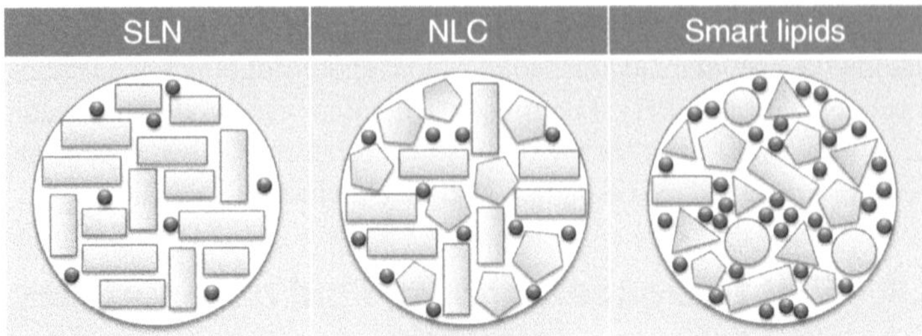

	SLN	NLC	Smart lipids
Composition	1 solid lipid	1 solid lipid+ 1 liquid lipid	Up to 10 lipids (solid or solid + liquid)
Structure	Ordered	Less ordered	Highly chaotic
Active loading	Low	Medium	High
Retinol	1 %	5 %	> 15%
Lidocaine	15 %	25 %	40%

Figure 4.3 Overview of the basic structures of SLN, NLC, and the latest generation smartLipids with their relevant properties and examples for increased loading capacities (with the bioactive retinol).

solid lipid can also accelerate transition to more ordered structures. There are two effects opposing each other: increase in loading due to better solubility in the oil fraction and reduction of loading by potential promotion of reordering. To avoid the reordering in smartLipids highly chaotic lipid mixtures were created, going up to 10 different blended lipids (Ding et al., 2015; Müller and Keck, 2015; Müller et al., 2014b,c). One option is to blend only solid lipids and creating many imperfections, or blending a majority of solid lipids with some liquid lipids to further enhance bioactive loading beyond the generated imperfections. Examples for compositions are presented in Table 4.1.

The stability of the crystalline state could be distinctly enhanced. Fig. 4.4 shows that in SLN made from Dynasan 118 transition to the β-modification occurred already

Table 4.1 **Examples for smartLipids compositions with solid lipids only or solid lipids combined with liquid lipid**

Mixture of solid lipids only			Mixture of solid and liquid lipids		
Lipid	**C-No.**	**%-age**	**Lipid**	**C-No.**	**%-age**
Dynasan 118	18	37.5	Dynasan 114	14	8.4
Compritol 888 ATO	22	12.5	Compritol 888 ATO	22	16.7
Geleol	16 + 18	12.5	Geleol	16 + 18	8.3
Precirol ATO 5	16 + 18	12.5	Precirol ATO 5	16 + 18	8.3
Imwitor 312	12	12.5	Imwitor 312	12	8.3
Imwitor 900 (F) P	16 + 18	12.5	Imwitor 900 (F) P	16 + 18	8.3
			Softisan 154	16 + 18	8.3
			Miglyol 812	8 + 10	33.4

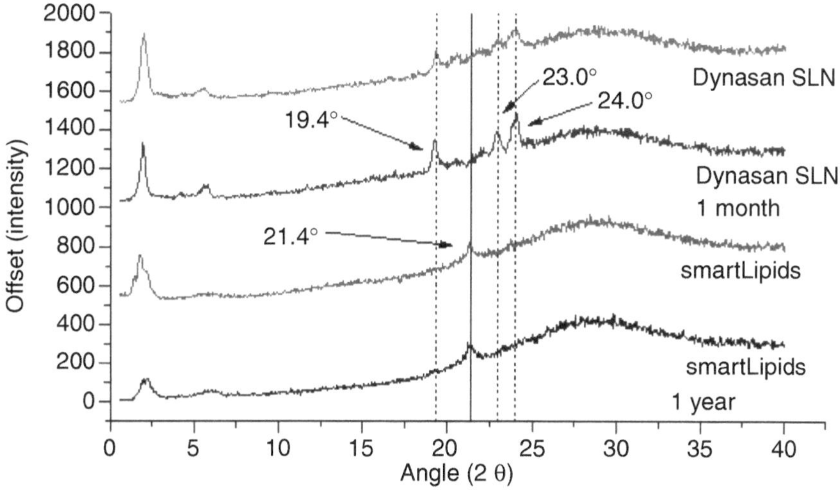

Figure 4.4 Examples of wide angle x-ray diffraction patterns of SLN made from Dynasan 118 and smartLipids formulation, both prepared with SDS. SLN showed a fast transition toward the β-modification during 1 month storage, where smartLipids show α-modification during 1 year of storage. Curves are detached from the *x*-axis for better visualization.
Source: With permission from Müller, R.H., Ruick, R., Keck, C.M., 2014. SmartLipids—the next generation of lipid nanoparticles by optimized design of particle matrix, PT.27. DPhG-Jahrestagung, Frankfurt, 24–26.

after 1 month of storage, where the smartLipids mixture remained primarily in α modification unchanged for 1 year of storage at room temperature.

In summary, the smartLipids are strictly speaking a specialized version of the NLC. Prerequisite for the particle matrix is that it contains a multiple number of different solid lipids. Optionally it contains one or more additional oils. This complex chaotic lipid mixture is also achieved when using one commercial solid lipid in NLC which is already a complex mixture of various lipids, for example, Cutina LM. Cutina LM is a mixture of fatty alcohols, vegetable waxes, and oils (cetearyl alcohol, cetearyl glucoside, carnauba wax, candelilla wax, beeswax, octyldodecanol). Thus, by definition, NLC are also smartLipids when they are made from one solid lipid, but being a complex commercial lipid mixture.

4.5 Selection of ingredients for SLN/NLC production—screening

Selection need to be made first under regulatory aspects (c.f. Section 4.8), and then subsequently under technical aspects. For commercial product development, only ingredients with regulatorily accepted status should be used for the product development process, which reduces the screening parameters. Most important is the selection of the solid and liquid lipids to provide maximum solubility for the active to be loaded.

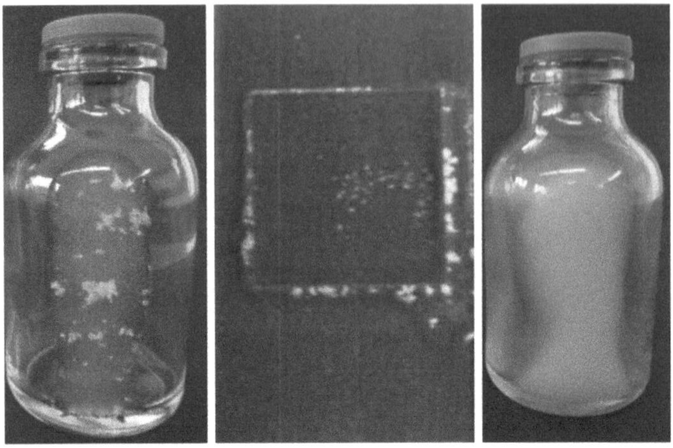

Figure 4.5 Picture of a lipid film in a test vial (left) and on a microscope slide (middle) showing phase separation and uniform appearance of a solid solution (right).

Often a lipid is just taken randomly, without assessing the solubility at all. A systematic solubility screening is recommended, determining the solubility in the melted state and the solid state of the solid lipid, and the solubility in oils at both production temperature and at room temperature.

Oil screening: In case the production temperature is 60°C, the oils are heated to this temperature (e.g., in 20 mL vials) and the active added successively in steps to generate increasing concentrations. After shaking the amount of active which can be completely dissolved is determined visually. In the next step, the oil solutions are cooled down to room temperature. It has to be investigated which amount of active remains completely in solution also after cooling down to room temperature and in addition after for example, 1 day of storage (in case of delayed crystallization).

Solid lipid screening: The lipids (e.g., 2 g) are melted in glass vials and active added in increasing amounts to investigate its maximum solubility in the melted lipid. In the next step, the lipid is cooled down and the solubility checked in the solid lipid (which is logically lower than in the melted and hot state). The investigation can be done macroscopically by spinning the vial during the solidification process by hand, that the lipid solidifies in a thin film on the vial wall. This film can then be inspected for homogeneity or separation in two phases (crystallized active and separate lipid phase with maximum dissolved active). Alternatively, or in parallel, a drop of the hot melted lipid solution can be placed on a microscope slide, covered with a cover glass and pressed to a thin film. After solidification, microscopic inspection for phase separation can be performed (Fig. 4.5, middle).

Result of this lipid solubility study is the knowledge which concentration leads to solid solution within the particle matrix, and which creates an active core with prolonged release (c.f. Section 4.10).

Stabilizer: The next important point is the selection of stabilizer. Two options are possible: electrostatic stabilizers (e.g., charged ionic surfactants) or steric stabilizers (e.g., poloxamer type). In case of oral or peroral delivery, the particles are generally exposed to an electrolyte-containing environment. The electrolytes reduce the zeta potential, thus the electrostatic stabilization leads to aggregation which impairs the functioning of the carriers (special nanoproperties are lost). Thus, it is generally recommended to go for steric stabilization via polymers, or a combined electrostatic and steric stabilization. Screening for optimal stabilizers can be performed by contact angle measurements (Fan et al., 2014). A lipid film of the final active-lipid mixture to be used for particle production is casted. Then the contact angles of different stabilizers in increasing concentrations are measured, selecting the stabilizer solution (both type and concentration) with lowest angles. As a rule of thumb, typical stabilizer concentrations are about 1.0%–1.5% for 10% lipid nanoparticle suspensions, and about 2% for 20% suspensions.

4.6 Industrial relevant production processes: high pressure homogenization on lab scale

Meanwhile many different production processes are described in literature (Mehnert and Mäder, 2001), which are typically used on lab scale when expensive equipment is not available. The most relevant processes are the microemulsion technology by Gasco and—most important—the HPH by Müller.

In the microemulsion process, a hot microemulsion is formed with the melted lipids and bioactive, then this mixture is poured into cold water. The dilution leads to breaking of the microemulsion, forming lipid droplets in the nanometer range (e.g., 100–200 nm), which immediately solidify due to the cold temperature. Regarding the microemulsion, it should be clarified, that it is not an ultrafine o/w emulsion, but a microemulsion in the sense of a thermodynamically stable, isotropic liquid mixture (water, oil, surfactant, and optionally cosurfactant). This process was scaled up in the 1990s by the Italian company Vectorpharma SpA in Trieste. However, a clear disadvantage of this process is the dilution of the particles with water. Often only a suspension with 1% particles or less can be achieved. This is unfavorable in case of the particles should be admixed to other products (e.g., creams) or the water should be removed to incorporate the particles into a tablet or capsule. Just too much water needs to be removed, which is costly and uneconomically and process challenging.

Much more feasible is the HPH, which allows the production of suspensions with particle contents up to about 40%. HPH is an industrially well-established process (Fig. 4.6). Homogenizers are available from very small volumes to safe expensive material (e.g., 3 mL, Avestin B3, Avestin/Canada), lab scale volumes from 40 mL (Micron LAB 40, APV Deutschland GmbH) (Fig. 4.6, left) to 100–200 mL (e.g., PANDA, Niro Soavi, Germany). Most suitable from our experiences is for broad formulation screening the Micron LAB 40, because it provides as yield about 3 × 12 mL vials (for 3 the standard storage temperatures), is fast to use, and avoids consumption of larger amounts of materials as with the 200 mL machines. Most of the produced suspension in screening with such machines will be discarded anyway.

The production process is very simple, typically the so called "hot homogenization technique" is used. The lipid blend is melted (typically 5–10°C above the melting

Figure 4.6 Homogenizers of different capacity: Micron LAB 40 with 40 mL (A) modified LAB 60 with batch size of 2–10 L (B) and Gaulin 5.5 with 150 L capacity per hour.

point), the active dissolved in this melt, and the active containing melt dispersed in a hot surfactant/stabilizer solution of equal temperature by high speed stirring. The obtained preemulsion is subsequently passed typically three times through a temperature controlled high pressure homogenizer at 600 bar pressure (=8702 psi). Alternatively, often 1 passage at 800 bar (=11,603 psi) is sufficient. The mean particle size typically obtained with most surfactants or polymeric stabilizers is about 200–300 nm.

Analytics of the lipid nanoparticles suspension should be made regarding particle size by photon correlation spectroscopy (PCS, determination of bulk population), laser diffraction (detection of potential larger remaining particles or formed aggregates) and light microscopy. Light microscopy should be performed using undiluted suspension to screen for larger particles/aggregates which are below the detection limit of the lazer diffraction. Magnification should be 1000-fold with oil immersion, allowing detection close to 500 nm. In case of smaller suspensions, the picture should appear as a grey shade with no detectable larger particles.

4.7 Medium and large scale industrial production

Medium scale production can be performed in homogenizers with a capacity of about 50 kg (liter) per hour, such as the LAB 60. Fig. 4.6 (middle) shows a modified version equipped with two 10-L containers (source container, product container). Both containers are double walled and temperature controlled by hot water. The source container (visible on the left) is filled with hot stabilizer solution, and the melted active-containing lipid blend added and dispersed by the build-in stirrer to generate the coarse raw emulsion. A batch size of 1 kg can be run in the continuous mode, i.e., the product circulates from the source container into the homogenizer and from there back to the source container. After about 20 min the production is finished, about at least 99.9% of the droplets have passed the homogenizer at least once. To guarantee the same passage probability for all droplets for a 10 kg batch cycling for hours is required leading to too long heat exposure of the product. Thus, this production has to be performed discontinuously. The suspension is passed from the source container via the homogenizer to the product container. After complete passage, the content of the product container is fed back to the source container for a second passage. Typically two passages at 600 bar are sufficient, because larger machines have two homogenization valves in series (one passage is practically equivalent to two passages on lab scale).

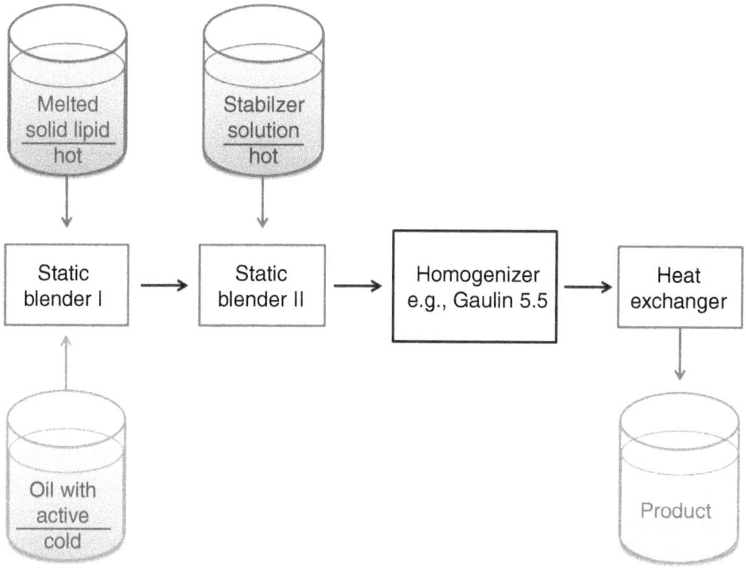

Figure 4.7 Large scale production line for producing lipid nanoparticle suspensions (NLC, smartLipids) loaded with heat sensitive actives (explanation cf. text).

Large scale production is performed on homogenizer of the Gaulin 5.5 type and larger. Typically one passage at 800 bar is sufficient, equivalent to 3 passages at 600 bar in a lab scale homogenizer. However, it needs about 1 h to process 150 kg, that means the respective active is exposed to heat for 1 h. This might be no problem for many actives, but not for labile compounds. In this case, the melted solid lipid should be admixed to cold oil containing the active by a static blender before subsequently being admixed to the aqueous stabilizer phase in a second static blender. The lipid melt should have a higher temperature to compensate the room temperature of the admixed active-containing oil, yielding finally the required temperature for homogenization (Fig. 4.7). To minimize heat exposure, the product should be cooled immediately by a heat exchanger.

4.8 Regulatory aspects—nanotechnology

As outlined previously, the regulatory acceptance of ingredients should be checked in the relevant legislation. Apart from the chemical nature of the ingredients, also the concentrations accepted need to be checked. When producing lipid nanoparticle concentrates for admixture, the concentration of excipients might be higher than specified, because the concentrate will be admixed to a final product. Thus, the resulting final concentrations in the product need to be considered.

There is an increasing concern of the consumers about nanoparticles, especially in food (analogous to absence of GMO). The European Union (EU) defined a nanomaterial as: "A natural, incidental or manufactured material containing particles, in an unbound state or as an aggregate or as an agglomerate and where, for 50% or more of the particles in the number size distribution, one or more external dimensions is in the size range 1–100 nm." (Potocnik, 2011). Based on this definition it is recommended to produce lipid nanoparticles above this size. Then the product can be clearly advertised as "nanofree". For better differentiation it would be smart to declare them as "submicron particles" (>100 nm to <1,000 nm).

It should be pointed out that this EU regulation intends to cover nanomatreials which are primarily insoluble, poorly soluble, or biopersistent, meaning they will never disappear (not biodegradable). Targets are not particles which can be degraded in biological systems (e.g., liposome). Based on this the lipid nanoparticles are a regulatorily highly acceptable delivery system. They can be produced well above 100 nm in size and on top they are perfectly biodegradable. This is reflected by their localization in the nanotoxicological classification system (NCS) (Keck and Müller, 2013), belonging to the "green" class out of three classes (yellow, red). More information about safety aspects and legislation of different countries for nanoparticles and nanoencapsulated ingredients is provided in Chapter 15.

4.9 Chemical stabilization of actives

A solid matrix is a priori suitable to stabilize chemically labile actives, because the exchange of the loaded active ingredients with a surrounding water phase is minimized (Fig. 4.2), in addition oxygen has limited access into the solid matrix, and also for example, light/UV cannot deeply penetrate and contribute to degradation. Stabilization of actives inside lipid nanoparticles has been reported for coenzyme Q_{10}, ascorbyl palmitate, all-*trans* retinol, α-tocopherol, β-carotene, chlorambucil, topotecan (Hentschel et al., 2008; Jenning, 1999; Junyaprasert et al., 2009; Sharma et al., 2009; Souza et al., 2011; Teeranachaideekul et al., 2007; Trombino et al., 2009), and so on.

Studies have been performed with retinol. Retinol is prone to very fast degradation. Following the literature, only about 40% remain stable in a nanoemulsion after 6 months of storage at room temperature (Jenning, 1999). Retinol was incorporated into 3 different smartLipids mixtures, and the stability was investigated over 6 months at 40°C to show the effectiveness of the protective solid particle matrix (Fig. 4.8, blue bars). In addition, Tinogard TT was added as antioxidant to further optimize the system. With this combination, stabilities of about 60% could be achieved even at the stress temperature of 40°C (Fig. 4.8, red bars). It can be summarized, that lipid nanoparticles are highly suitable for the protection of chemically labile actives. Main factors are the solid state of the particle matrix and the noninteraction of lipids used for particle production with the active. Thus, the stabilizing principle is valid for all three generations of lipid nanoparticles.

4.10 Controlled release—structures of particle matrix

Depending on the application purpose in consumer products, the desired release can vary from very fast (e.g., in local anesthetic creams) to extremely prolonged (e.g., practically no release of molecular sunscreens in dermal creams). The release can be controlled by the structure of the particle matrix. In case the active is enriched in

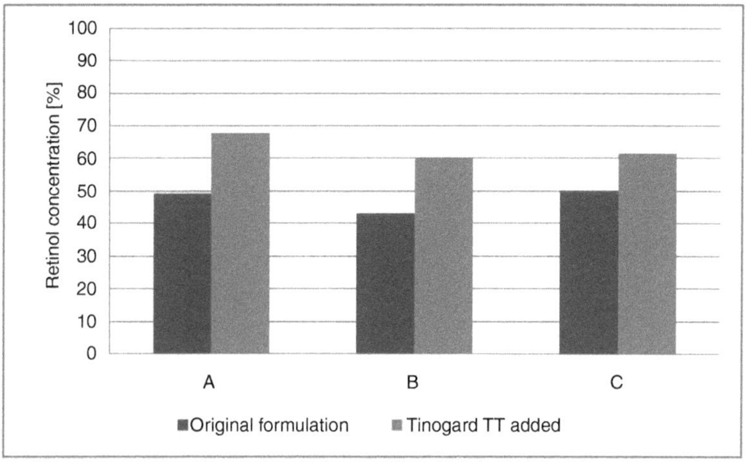

Figure 4.8 Retinol was incorporated into 3 different smart Lipids mixtures (A–C), and the stability was investigated over 6 months at 40°C to show the effectiveness of the protective solid particle matrix [*dark gray bars (blue bars in the web version)*]. In addition, Tinogard TT was added as antioxidant to further optimize the system. With this combination, stabilities of about 60% could be achieved even at the stress temperature of 40°C [*light gray bars (red bars in the web version)*].

the outer shell, release will be very fast [e.g., as shown for the bioactive cyclosporine (Runge, 1998)]. Releases will be slower in case the bioactive is dispersed in the particle in the form of a solid bioactive solution (molecular dispersion), and relatively slow if the active is located primarily in the core (Fig. 4.9).

The matrix structure can be controlled by the lipid composition and the solubility of the active in this lipid mixture. If the active concentration in the lipid mixture is close to the eutectic mixture, a solid solution will be formed preferentially. In case the concentration of active is soluble in the melted lipid but not soluble in the solidified lipid, the active will precipitate during cooling down process, forming an active core surrounded by a lipid coat with some dissolved active (Fig. 4.9). In case the lipid solidifies first, an active enriched shell will result (zur Mühlen et al., 1998). Fig. 4.10 shows the very fast release of cyclosporine from SLN (Runge, 1998) and Fig. 4.11 the extremely prolonged release of prednisolone from SLN. Of course, these are in vitro releases. In the GIT the particles will be degraded by the lipase–colipase complex and thus release within a few hours. This has to be considered when developing peroral formulations for food and nutraceuticals.

4.11 Oral delivery in mouth cavity—mechanisms

Oral delivery to the mouth cavity is used for quite a number of bioactives, but also consumer products, such as nicotine in nicotine replacement treatments (NRT). Advantages are the fast absorption leading to fast action, and/or the circumventing of the first pass effect of the liver. This might also make the oral absorption attractive for

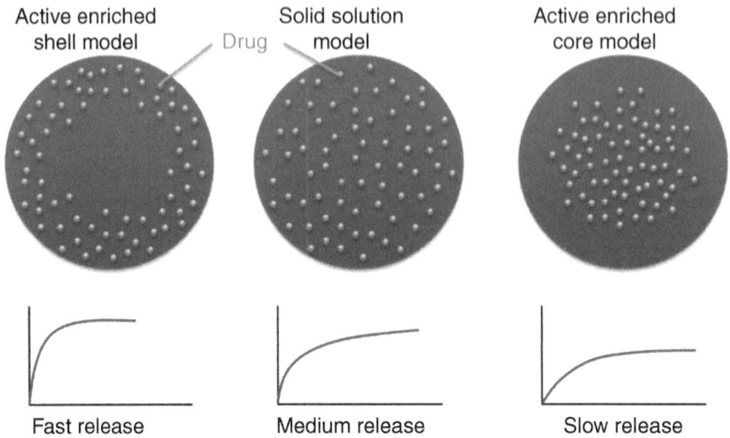

Figure 4.9 Release profiles of actives from lipid nanoparticles as a function of the particle matrix structure.

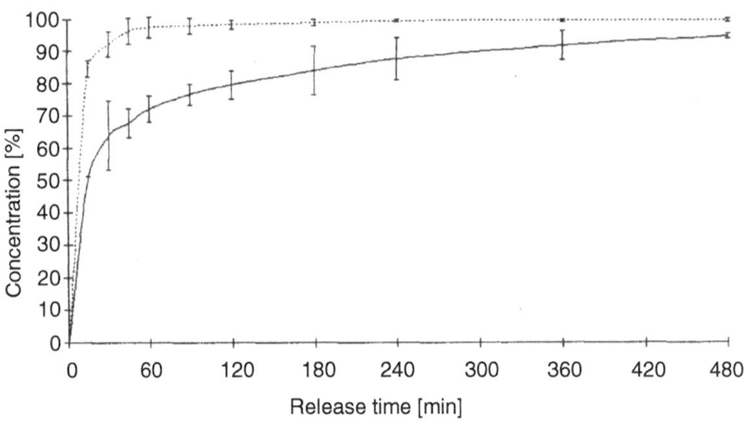

Figure 4.10 In vitro fast release of cyclosporine, attributed to enrichment in the particle shell (Runge, 1998).

nutraceuticals. In case of oral delivery, swallowing of administered active is in many cases undesired. For example, in case of nicotine swallowing leads to an undesired side effect, for example, hiccups. Thus, the aim is often to retain as much active in the mouth cavity as possible and make it available for oral absorption.

One approach to retain active in the moth cavity is the use of mucoadhesive carriers, for example, lipid nanoparticles. Due to their small size they show pronounced adhesion. In addition, the negatively charged lipid nanoparticles can be made positively charged, for example, by stabilizing them with cationic preservatives which are often anyway in the formulation (e.g., benzalkonium chloride or cetylpyridinium chloride). The positively charged particles adhere firmly to the negatively charged mucosa.

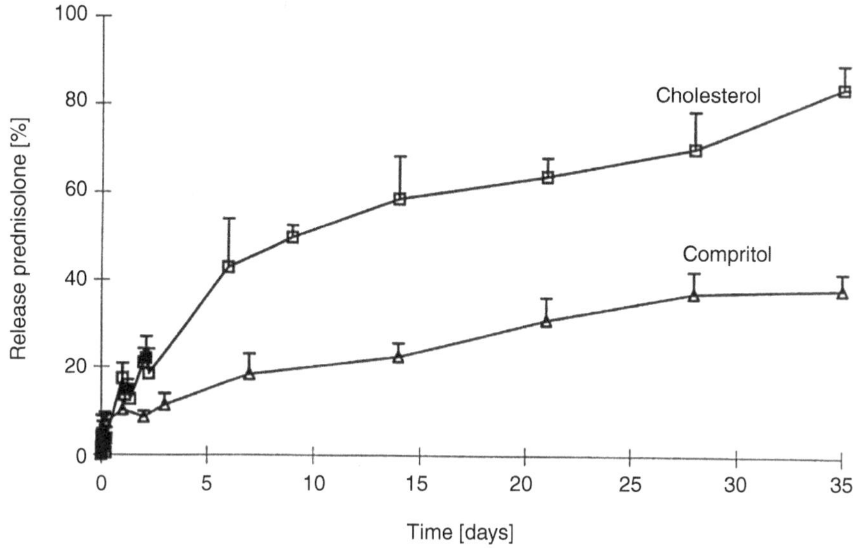

Figure 4.11 In vitro extremely slow release of prednisolone attributed to enrichment in the particle core (zur Mühlen et al., 1998).

4.12 Peroral bioavailability enhancement—mechanism and efficiency

It is well known that food taken simultaneously with a number of bioactives can enhance their absorption from the GIT. The absorption of lipophilic vitamins like vitamin A and E is for example enhanced in the presence of lipids/oils. Thus, when going for a diet, preparing the salad at least with some oil is beneficial for the vitamin household of the body.

A model bioactive for these effects is cyclosporin. The first oral formulation of cyclosporine Sandimmun by Sandoz (nowadays Novartis) was an oil based formulation. The bioavailability is highly dependent on the presence of bile salts to disperse the oil to fine droplets, thus the bioavailability was very variable between approximately 30%–60%. The subsequent formulation Sandimmun Neoral/Optoral was based on a microemulsion. Breaking of the microemulsion in the stomach lead to the formation of nanosized oil droplets having a size range of about 70–100 nm. Bile salts were not necessary anymore and therefore the bioavailability variation was distinctly reduced. In addition, the bioavailability was high due to the fineness of the cyclosporin loaded oil droplets. In summary, to have an optimum effect:

1. the bioactive should be closely associated with the oil/lipid, i.e., dissolved in it, and
2. the lipid droplets/particles should be as small as possible.

Both prerequisites are realized in the lipid nanoparticles. The active/bioactive is dissolved in the particles. In addition, with a size in preferentially low nanometer range they are much smaller than many oil droplets dispersed by the bile salts in the

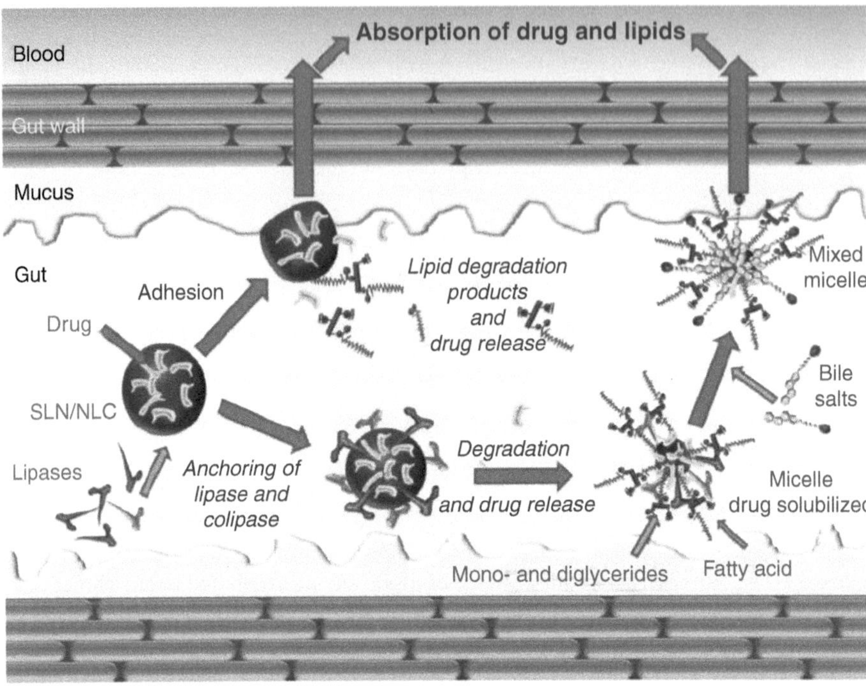

Figure 4.12 Mechanism of absorption enhancing effect of lipid nanoparticles in the intestine. The lipid is absorbed, and the bioactive "goes" for the body unnoticed with it (so called "Trojan horse effect" of lipid nanoparticles.
Source: With permission from Muller, R.H., Keck, C.M., 2004. Challenges and solutions for the delivery of biotech drugs–a review of drug nanocrystal technology and lipid nanoparticles. J. Biotechnol. 113 (1), 151–170.

motile GIT. Thus, lipid nanoparticles are an ideal delivery system to exploit the lipid enhancing effect on absorption from the GIT.

Fig. 4.12 explains the mechanism in more detail. The lipid nanoparticles, in this case SLN, enter the intestine. They are already small-sized and thus the dispersion by bile salts is not necessary anymore. That means the absorption process is independent on the presence of bile salts. In the next step, the lipase and colipase adsorb onto the particle surface. The lipid particles are surface-degraded. The triglycerides of the lipid nanoparticles are decomposed to di- and monoglycerides, which have surfactant properties. Both lipid and bioactive are solubilized in micelles. By the interaction with bile salts mixed micelles can be formed. These micelles undergo the lipid absorption process, ending up also in the lymph (=lymphatic absorption). To promote this micelle formation process, lipids can be used for the production of lipid nanoparticle having a higher content of monoglycerides.

The lipase–colipase degradation of lipid nanoparticles was studied in vitro by Olbrich and coworkers (Olbrich, 2002; Olbrich et al., 2002). SLN were used, stabilized with different stabilizers, resulting in different surface modifications. As stabilizers sodium

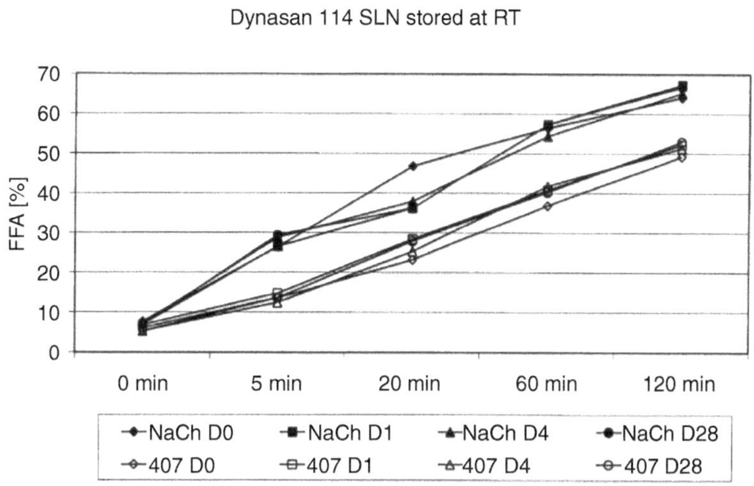

Figure 4.13 In vitro lipase–colipase degradation of Dynasan 114 SLN, stored at room temperature, as a function of time from day of production (D0) up to 28 days (D28). The SLN were stabilized either with cholic acid sodium salt (NaCh) or Poloxamer 407 (407). *Source:* With permission from Olbrich, C. (2002). Development of optimized lipid carriers and their investigation in biological systems. Free University of Berlin, Berlin.

cholate and the sterically stabilizing Poloxamer 407 (molecular weight approximately 12,500) were used. The particles were stored up to 28 days to assess if storage time (i.e., changes in matrix structure due to reordering) had any effect on the degradation velocity. Fig. 4.13 shows the free fatty acid (FFA) formed from the lipids as the function of degradation time in the lipase–colipase assay. Despite of identical lipid composition, the cholate-stabilized particles showed a faster degradation. The sterically stabilizing Poloxamer 407 seemed to sterically hinder the anchoring of the lipase–colipase complex, thus slowing down degradation velocity. The data show nicely, how the degradation velocity can be modulated by modifying the surface. In case a certain lipid is required for making the lipid nanoparticles (e.g., showing good solubility for active), but degrading too fast or to slow, degradation can be optimized via the stabilizer selection.

4.13 Examples of SLN and NLC formulations from food industry

Many polyphenols are of high interest for nutritional purposes, thus they are one of the primary groups of actives incorporated into lipid nanoparticles (Mohammadi, et al., 2016a,b; Faridi Esfanjani and Jafari, 2016). The flavonon hesperetin was incorporated into both SLN and NLC. NLC showed higher loading due to its less perfect crystal structure. General encapsulation was about 3%–6%. One of the aims was taste masking and general improving of sensory properties (e.g., homogeneity of appearance of milk fortified with hesperetin). Addition of hesperetin to milk led to bitterness, increased yellowness, decreased degree of homogeneity and general acceptance. The hesperetin loaded lipid carriers distributed evenly in the milk, and no sensory perception differences were noticed compared to blank milk (Fathi et al., 2013). In vitro release was, depending on the formulation between 50%–90% within 8 h. This will be accelerated in vivo due to the presence of lipase–colipase in the GIT. The lipid degradation will lead to the formation of mixed micelles with bile salts, and consequently increased oral bioavailability.

Quercetin, a very common dietary polyphenol, was also incorporated into NLC (Ni et al., 2015), having the same problems as all other polyphenols, such as poor solubility and instability. These problems might be overcome by incorporation in a protecting and bioavailability enhancing nanocarrier. The NLC were produced by HPH, having sizes about 130 nm with about 94% entrapment efficiency. The in vitro release model showed a sustained release/dissolution, being slower than from free quercetin. The distinctly slower release in the stomach might protect quercetin against degradation by the acidic environment, thus increasing its bioaccessibility. In addition, the physical stability of the NLC was studied in a simulated beverage. After an initial slight increase from about 130 to 150 nm, the size remained stable over 2 months.

Resveratrol is well known for its big health promoting potential. Formulation problems are its low water solubility, its low oral bioavailability and chemical instability (photosensitivity). Resveratrol was loaded onto SLN made from stearic acid and stabilized with Poloxamer 188, having particle sizes about 130 nm and loading efficiency of 88.9% (Pandita et al., 2014). In vivo in Wistar rats, the SLN formulation had an about eight times higher bioavailability compared to the resveratrol suspension. Apart from the absorption enhancing effect of the lipid matrix, this might also be attributed to the small particle size allowing partially lymphatic absorption.

The polyphenol rosmarinic acid has antioxidant, but also, for example, antibacterial and antiviral properties. It was loaded onto SLN made from carnauba wax and poylsorbate 80 (Madureira et al., 2015). A high loading efficiency was achieved (99%), which can be attributed to carnauba wax being a lipid mixture of rather smartLipids type. A close association between lipid and rosmarinic acid was shown by FTIR analysis. Storage stability was shown over 28 days with no rosmarinic acid release (=no expulsion of active during storage).

Alpha-lipoic acid (ALA) is a powerful antioxidant for the supplement and food area. It functions as cofactor in mitochondrial, multienzyme complexes but the use is limited to its instability (air, day light) and the unpleasant odor which is generated during degradation (due to sulfur content). ALA was incorporated into NLC (Wang et al., 2014). The NLC were effective in protecting ALA against natural day light. About 88.5% of ALA remained in NLC after 120 days of storage but only 0.7% of free ALA. This showed nicely the protective ability of solid lipid particle matrices.

Carotenoids are also a compound of high interest for functional foods (Hejri et al., 2013; Hentschel et al., 2008; Qian et al., 2013; Tamjidi et al., 2014). Astaxanthin, a type of β-carotenoid, was incorporated into NLC (Tamjidi et al., 2014). It has stronger antioxidant activity than vitamin E and β-carotene, but is low bioavailable, poor water soluble and shows limited stability—thus is an ideal compound for incorporation into lipid particles. Formulation optimization was performed by response surface methodology (RSM), and the variables identified with significant effects on formulation characteristics. This data might be a useful basis for controlled design of NLC for foods and beverages. β-carotene is sensitive to heat, light, and oxygen, i.e., has chemical stability problems. It was loaded onto SLN and compared with an o/w emulsion (Qian et al., 2013). Compared to the emulsion, the SLN showed lower physical stability and faster color change (less stabilization). The reduced stabilization effect was explained by the expulsion of β-carotene from the SLN particle matrix due to too high crystallinity of the matrix. This supports the use of NLC. β-carotene was also loaded onto NLC (Hejri et al., 2013). Formulation optimization was performed using again RSM in combination with central composite design (CCD). A reasonable agreement was found between predicted and experimental values, showing the usefulness of the approach. In optimized formulations, the NLC were able to protect β-carotene against degradation during the heat exposure of the production process (only 0%–3% degradation). This demonstrates the usefulness of the NLC system.

Lipophilic vitamins are also predestinated for incorporation into lipid particles(Katouzian and Jafari, 2016). There are reports about the incorporation of vitamin A and its derivatives, mainly for cosmetics, but in principle the same or slightly modified formulations can be used in foods or Nutraceuticals. The same applies to vitamin E. Optimized production of its isomer α-tocopherol has been reported (de Carvalho et al., 2013). RSM was again employed for formulation optimization. The reported physical stability was 21 days at 6°C (investigation period). However, for food products longer investigation periods are required, ideally at higher temperatures to meet the needs of such products. In addition, the particles need to be stable, in presence of electrolytes and electrolyte-type food preservatives (i.e., zeta potential reduction).

4.14 Examples of oral bioavailability enhancement

In the literature many examples are described for the increase of oral absorption. In following, cyclosporine and testosterone are discussed as examples.

As outlined in Section 4.12, cyclosporine has a high variability in the blood profiles, which was overcome by the microemulsion formulation. Disadvantage of the microemulsion formulation is the high plasma peak at the beginning of the absorption process (Fig. 4.14, blue line), which can potentially cause nephrotoxic problems. In a pig study, cyclosporine loaded SLN were applied (Müller et al., 2006). The SLN formulation could avoid the undesired plasma peak, but having a similar long plasma level in the therapeutic range compared to Sandimmun Neoral. As reference, cyclosporine nanocrystals were administered, which showed in this study a rather low absorption (Fig. 4.14, red line). Obviously the presence of lipids had a positive effect on absorption.

Another example is testosterone. Its oral bioavailability is relatively low (about 2%–4%) due to the high first pass metabolism of the liver. The testosterone reaching the blood originates mainly from lymphatic absorption, for example, from the product Andriol Testocaps. Particles can be taken up by lymphatic absorption if their size is below approximately 100 nm. Testosterone lipid nanoparticles were produced with different sizes, and the oral bioavailability were determined in rats and compared to the commercial product Andriol Testocaps (Muchow, 2009; Muchow et al., 2013). The area under the curve (AUC) in the blood for both lipid nanoparticles

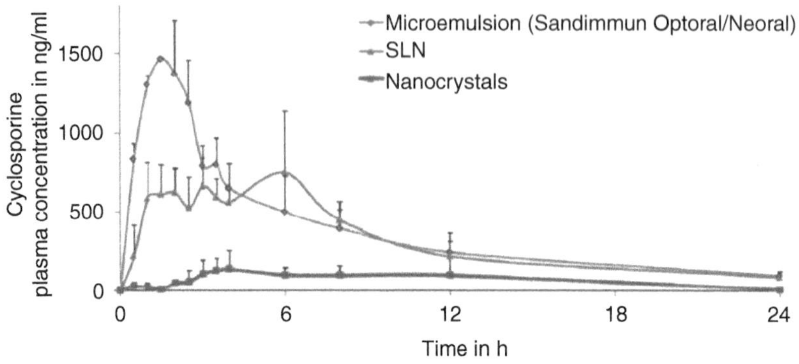

Figure 4.14 Blood profiles (pig study) of cyclosporin-loaded lipid nanoparticles (SLN) compared to microemulsion formulation Sandimmun Optoral and cyclosporin nanocrystals. *Source:* With permission from Müller, R.H., Runge, S., Ravelli, V., Mehnert, W., Thünemann, A.F., Souto, E.B., 2006. Oral bioavailability of cyclosporine: solid lipid nanoparticles (SLN®) versus bioactive nanocrystals. Int. J. Pharma. 317(1), 82–89.

formulations was higher compared to the commercial product, where the highest AUC can be found for the smaller sized (200 nm) lipid nanoparticle (Fig. 4.15). The lipid nanoparticles are surface degraded in the intestine, thus they will shrink in size and finally reaching the small size required for lymphatic uptake. From this data, lipid nanoparticles have high potential to exploit the lymphatic absorption, for both bioactives and nutraceuticals.

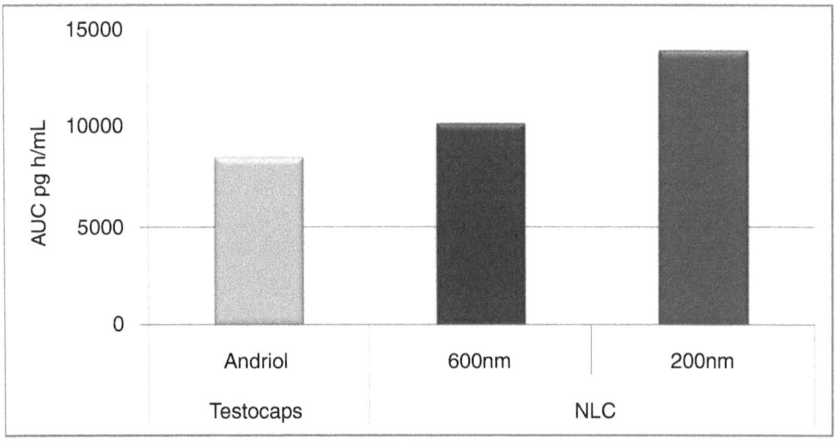

Figure 4.15 Bioavailability and effect of size of testosterone loaded NLC compared to the marketed product Andriol Testocaps in nonfed state.
Source: Modified after Muchow, M., Maincent, P., Müller, R.H., Keck, C.M., 2013. Testosterone undecanoate—increase of oral bioavailability by nanostructured lipid carriers (NLC). J. Pharma. Technol. Bioact. Res. 2(1), 4.

4.15 Lipid nanoparticle products on the market

In 2005 the first cosmetic products were introduced to the market, series Nanorepair by Dr. Rimpler GmbH Germany (http://www.rimpler.de). In the following year, AmorePacific, the leading cosmetic company in South Korea, introduced its first products in the line IOPE (Fig. 4.16). More product examples can be found in (Pardeike et al., 2009), meanwhile estimated a few hundred products are using lipid nanoparticles. All these products are based on NLC, exploiting their superior properties compared to SLN.

These cosmetic products exploit features, such as:

1. restoration of protective lipid barrier on the skin
2. formation of occlusive film which
   a. increases skin hydration and reduces wrinkles
   b. increases penetration of actives
3. stabilization of cosmetic actives.

Stabilization of incorporated actives is also of interest for the food and nutraceutical sector. Fig. 4.17 shows the stability of coenzyme Q_{10} in SLN compared to a supercooled melt suspension. After 1 year, even at the stress temperature of 40°C about 95% of Q_{10} remain stable in the SLN suspensions.

4.16 Commercial suppliers of lipid nanoparticle concentrates

A prerequisite for use of a delivery system in marketed products is the availability of the system, either by having the option of own production or alternatively—often preferred—having a commercial supplier. Presently the only worldwide distributor of

Figure 4.16 AmorePacific line IOPE SuperVital cream (left), SuperVital serum (middle) and SuperVital eye cream (right)—first three cosmetic products based on NLC on the South Korean cosmetic market.

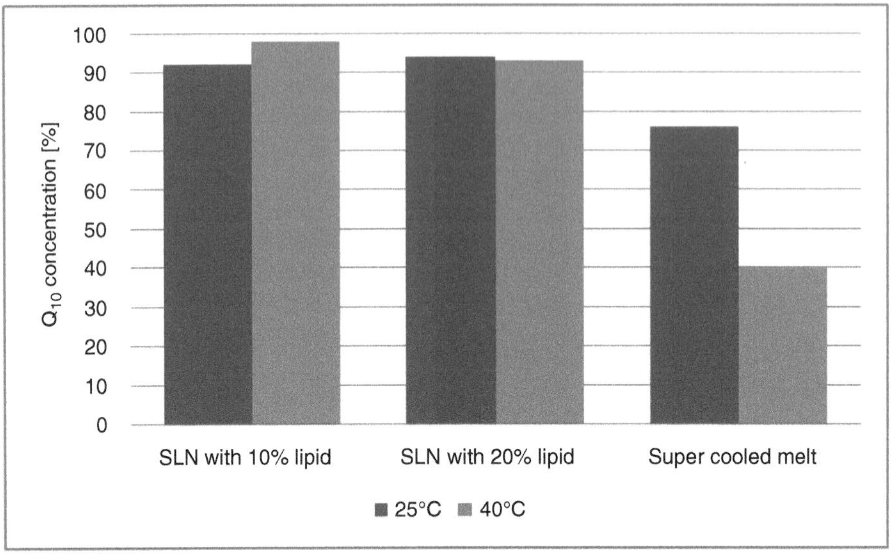

Figure 4.17 Stability of coenzyme Q_{10} incorporated into SLN suspension (with 10 and 20% lipid phase) versus a suspension of supercooled lipid melt particles (with identical lipid composition) after 1 year storage.

lipid nanoparticles is the company Berg + Schmidt in Hamburg, Germany (http://www.berg-schmidt.de/). They introduced their cosmetic lipid nanoparticles (trade name: BergaCare smartLipids) to the market at the in-cosmetics in Paris in April 2016.

The BergaCare smartLipids are presently available loaded with coenzyme Q_{10} "superactive" Q_{10} and retinol (with chemical stabilization). They are sold as concentrates, with a concentration of either 10 or 20%, depending on product type. The Q_{10} BergaCare smartLipids would also be suitable for oral nutraceutical products. In addition, customized products are offered. The customer can specify its desired actives, and also its own new chemical entities, and the preference of lipids and stabilizers, and a customized formulation will be developed.

4.17 Perspectives for food and nutraceutical products

Looking at the increasing market for "well-being" and awareness of the importance of a good health, the market with nutraceuticals, functional/well-being drinks and respective food additives will steadily increase. Important is, that the consumer is prepared to spend a decent amount of money for high quality products. This enables the producers to exploit more expensive, but more effective delivery technologies in their products. Claim and reality of product performance will be more in harmony.

At present, cf. the example of Q_{10} discussed previously, many products lack in in vivo performance/sufficient bioavailability with corresponding effect. In addition, many products have chemical stability problems. This can be overcome by using an appropriate delivery system, especially regarding the stability problems with encapsulation technologies.

Important is, that the delivery system is industrially feasible.

That means for example—besides others:

- using regulatorily accepted ingredients,
- having proven effects and properties,
- being producible on large industrial scale,
- ideally being commercially available by a certified supplier,
- being sufficiently cost-effective for the respective market segment, and
- ideally the make-ability of products is already shown by product examples on the market.

This is all fulfilled by the lipid nanoparticles from Berg + Schmidt. The worldwide commercial supplier exists now since April 2016. Based on this, all requirements are fulfilled for using this technology in nutraceuticals, drinks, and specialized food product.

References

Birrenbach, G., Speiser, P.P., 1976. Polymerized micelles and their use as adjuvants in immunology. J. Pharm. Sci. 65 (12), 1763–1766.

Bunjes, H., Westesen, K., Koch, M.H., 1996. Crystallization tendency and polymorphic transitions in triglyceride nanoparticles. Int. J. Pharma. 129 (1), 159–173.

de Carvalho, S.M., Noronha, C.M., Floriani, C.L., Lino, R.C., Rocha, G., Bellettini, I.C., et al., 2013. Optimization of α-tocopherol loaded solid lipid nanoparticles by central composite design. Ind. Crop. Prod. 49, 278–285.

Ding, Y., Staufenbiel, S., Keck, C.M., Müller, R.H., 2015. The smartLipids—new lipid nanoparticle generation for vitamin A derlivery, PT-03, 8th Polish–German Symposium on Pharmaceutical Sciences. Retrospects, Insights and Prospects, Kiel, 29–30.

Fan, H., Liu, G., Huang, Y., Li, Y., Xia, Q., 2014. Development of a nanostructured lipid carrier formulation for increasing photo-stability and water solubility of Phenylethyl Resorcinol. Appl. Surf. Sci. 288, 193–200.

Faridi Esfanjani, A.F., Jafari, S.M., 2016. Biopolymer nano-particles and natural nano-carriers for nano-encapsulation of phenolic compounds. Colloids Surf. B 146, 532–543.

Fathi, M., Varshosaz, J., Mohebbi, M., Shahidi, F., 2013. Hesperetin-loaded solid lipid nanoparticles and nanostructure lipid carriers for food fortification: preparation, characterization, and modeling. Food Bioproc. Technol. 6 (6), 1464–1475.

Ganje, M., Jafari, S.M., Dusti, A., Dehnad, D., Amanjani, M., Ghanbari, V., 2016. Modeling quality changes in tomato paste containing microencapsulated olive leaf extract by accelerated shelf life testing. [Article]. Food and Bioproducts Processing 97, 12–19.

Gasco, M.R., 1993. US Patent No. 5,250,236. Washington, DC: US Patent and Trademark Office.

Hejri, A., Khosravi, A., Gharanjig, K., Hejazi, M., 2013. Optimisation of the formulation of β-carotene loaded nanostructured lipid carriers prepared by solvent diffusion method. Food Chem. 141 (1), 117–123.

Ghorbanzade, T., Jafari, S.M., Akhavan, S., Hadavi, R., 2017. Nano-encapsulation of fish oil in nano-liposomes and its application in fortification of yogurt. Food Chemistry 216, 146–152.

Hentschel, A., Gramdorf, S., Müller, R.H., Kurz, T., 2008. β-Carotene-loaded nanostructured lipid carriers. J. Food Sci. 73 (2), N1–N6.

Jenning, V., 1999. Solid Lipid Nanoparticles (SLN) as a Carrier System for the Dermal Application of Retinol. Free University of Berlin, Berlin.
Jenning, V., Thünemann, A.F., Gohla, S.H., 2000. Characterisation of a novel solid lipid nanoparticle carrier system based on binary mixtures of liquid and solid lipids. Int. J. Pharma. 199 (2), 167–177.
Junyaprasert, V.B., Teeranachaideekul, V., Souto, E.B., Boonme, P., Müller, R.H., 2009. Q 10-loaded NLC versus nanoemulsions: stability, rheology and in vitro skin permeation. Int. J. Pharma. 377 (1), 207–214.
Katouzian, I., Jafari, S.M., 2016. Nano-encapsulation as a promising approach for targeted delivery and controlled release of vitamins. Trends Food Sci. Technol. 53, 34–48.
Keck, C.M., Müller, R.H., 2013. Nanotoxicological classification system (NCS)—a guide for the risk-benefit assessment of nanoparticulate bioactive delivery systems. European Journal of Pharmaceutics and Biopharmaceutics 84 (3), 445–448.
Madureira, A.R., Campos, D.A., Fonte, P., Nunes, S., Reis, F., Gomes, A.M., et al., 2015. Characterization of solid lipid nanoparticles produced with carnauba wax for rosmarinic acid oral delivery. RSC Adv. 5 (29), 22665–22673.
Mehnert, W., Mäder, K., 2001. Solid lipid nanoparticles: production, characterization and applications. Adv. Bioact. Deliv. Rev. 47 (2), 165–196.
Mildau, H., Huber, B., 2010. Die neue EG-Kosmetikverordnung 1223/2009–Inhalte und erste Erläuterungen. SOFW-J. (3), 40.
Muchow, M., 2009. Nanocarriers for Oral Bioavailability Enhancement. Free University of Berlin, Berlin.
Muchow, M., Maincent, P., Müller, R.H., Keck, C.M., 2013. Testosterone undecanoate–increase of oral bioavailability by nanostructured lipid carriers (NLC). J. Pharma. Technol. Bioact. Res. 2 (1), 4.
Müller, R.H., Keck, C.M., 2015. Next Generation After SLN and NLC—the "Chaotic" SmartLipids, P9. Jahrestagung der Gesellschaft für Dermopharmazie (GD), Berlin, March, 16–18.
Müller, R.H., Lucks, J.S., 1996. European Patent, 605497. Arzneistoffträger aus festen lipidteilchen, feste lipidnanosphären (sln).
Müller, R.H., Runge, S., Ravelli, V., Mehnert, W., Thünemann, A.F., Souto, E.B., 2006. Oral bioavailability of cyclosporine: solid lipid nanoparticles (SLN®) versus bioactive nanocrystals. Int. J. Pharma. 317 (1), 82–89.
Müller, R.H., Jenning, V., Mader, K., Lippacher, A., 2014a. US Patent No. 8,663,692. Washington, DC. US Patent and Trademark Office.
Müller, R.H., Ruick, R., Keck, C.M., 2014b. SmartLipids—the Next Generation of Lipid Nanoparticles by Optimized Design of Particle Matrix, PT.27. DPhG-Jahrestagung, Frankfurt, September, 24–26.
Müller, R.H., Ruick, R., Keck, C.M., 2014c. SmartLipids®—the new Generation of Lipid Nanoparticles after SLN and NLC, T3134, AAPS Annual Meeting, San Diego, November, 2–6.
Ni, S., Sun, R., Zhao, G., Xia, Q., 2015. Quercetin loaded nanostructured lipid carrier for food fortification: preparation, characterization and in vitro study. J. Food Proc. Eng. 38 (1), 93–106.
Olbrich, C., 2002. Development of Optimized Lipid Carriers and Their Investigation in Biological Systems. Free University of Berlin, Berlin.
Olbrich, C., Kayser, O., Müller, R.H., 2002. Lipase degradation of Dynasan 114 and 116 solid lipid nanoparticles (SLN)—effect of surfactants, storage time and crystallinity. Int. J. Pharma. 237 (1), 119–128.

Pandita, D., Kumar, S., Poonia, N., Lather, V., 2014. Solid lipid nanoparticles enhance oral bioavailability of resveratrol, a natural polyphenol. Food Res. Int. 62, 1165–1174.

Pardeike, J., Hommoss, A., Müller, R.H., 2009. Lipid nanoparticles (SLN, NLC) in cosmetic and pharmaceutical dermal products. Int. J. Pharma. 366 (1), 170–184.

Potocnik, J., 2011. Commission Recommendation of 18 October 2011 on the definition of nanomaterial (2011/696/EU). Off. J. Eur. Union 275, 38–40.

Qian, C., Decker, E.A., Xiao, H., McClements, D.J., 2013. Impact of lipid nanoparticle physical state on particle aggregation and β-carotene degradation: potential limitations of solid lipid nanoparticles. Food Res. Int. 52 (1), 342–349.

Runge, S.A., 1998. Solid lipid nanoparticles (SLN®) as colloidal bioactive carriers for the oral administration of ciclosporin A. Free University of Berlin, Berlin.

Sharma, P., Ganta, S., Denny, W.A., Garg, S., 2009. Formulation and pharmacokinetics of lipid nanoparticles of a chemically sensitive nitrogen mustard derivative: Chlorambucil. Int. J. Pharma. 367 (1), 187–194.

Souza, L.G., Silva, E.J., Martins, A.L.L., Mota, M.F., Braga, R.C., Lima, E.M., 2011. Development of topotecan loaded lipid nanoparticles for chemical stabilization and prolonged release. Eur. J. Pharma. Biopharma. 79 (1), 189–196.

Tamjidi, F., Shahedi, M., Varshosaz, J., Nasirpour, A., 2013. Nanostructured lipid carriers (NLC): a potential delivery system for bioactive food molecules. Innov. Food Sci. Emerg. Technol. 19, 29–43.

Tamjidi, F., Shahedi, M., Varshosaz, J., Nasirpour, A., 2014. Design and characterization of astaxanthin-loaded nanostructured lipid carriers. Inn. Food Sci. Emerg. Technol. 26, 366–374.

Teeranachaideekul, V., Müller, R.H., Junyaprasert, V.B., 2007. Encapsulation of ascorbyl palmitate in nanostructured lipid carriers (NLC)—effects of formulation parameters on physicochemical stability. Int. J. Pharma. 340 (1), 198–206.

Trombino, S., Cassano, R., Muzzalupo, R., Pingitore, A., Cione, E., Picci, N., 2009. Stearyl ferulate-based solid lipid nanoparticles for the encapsulation and stabilization of β-carotene and α-tocopherol. Colloids Surf. B 72 (2), 181–187.

Wang, J., Tang, J., Zhou, X., Xia, Q., 2014. Physicochemical characterization, identification and improved photo-stability of alpha-lipoic acid-loaded nanostructured lipid carrier. Bioact. Dev. Ind. Pharm. 40 (2), 201–210.

zur Mühlen, A., Schwarz, C., Mehnert, W., 1998. Solid lipid nanoparticles (SLN) for controlled bioactive delivery—bioactive release and release mechanism. Eur. J. Pharma. Biopharma. 45 (2), 149–155.

Part Two

Natural Nanocarrier-Based Nanoencapsulation Technologies

5 Nanocapsule formation by caseins

Sanaz Haratifar, Anilda Guri
Canadian Research Institute for Food Safety, Guelph, ON, Canada

5.1 Introduction

For enhancing the nutritional value of foods and fabricating healthy products, there is a need for nutritious materials, such as antioxidants, vitamins, fatty acids, probiotics, etc. These bioactive molecules are susceptible to changes in pH, ionic strength, and oxidation, thus degradation might take place due to the aforementioned medium alterations. Additionally, these compounds are mostly hydrophobic and their transportation through hydrophilic food matrices remains a challenge. The vehicles which undertake the transportation of bioactives should remain stable in the food matrix, so as to ensure a safe, functional, and targeted delivery (Jafari et al., 2015). Transport systems like common emulsion-based carriers have been successfully applied, furthermore novel nanostructured carriers have also been established (Gülseren et al., 2014; Guri et al., 2013; Haratifar and Corredig, 2014; Walker et al., 2015; Zhang et al., 2014). Incorporating the sensitive ingredients into the nanostructures will increase their stability and maintain their bioavailability against external degradation factors, ensuring a controlled and targeted release (Assadpour et al., 2016, 2017). Protein-based food materials are proved to be safe and are considered appropriate to deliver bioactive compounds as well as retaining their functionality in the body. In many nanoencapsulation studies, milk proteins, particularly caseins, have been the center of attention as agents of encapsulation of bioactives (Haratifar and Corredig, 2014; Livney, 2010). Milk proteins have unique structural and functional properties which makes them suitable nanocarriers for bioactive components and nutraceuticals. Some of these properties are binding with molecules and ions (e.g., ion binding), binding with hydrophobic molecules, interactions with other proteins/polymers (e.g., covalent conjugations and noncovalent interactions), reassembly and self-assembly properties, gelation properties, and surface activity (Kimpel and Schmitt, 2015; Livney, 2015).

The way how caseins self-assemble into micelles due to hydrophobic interactions and electrostatic interactions renders them very suitable for incorporation of hydrophobic active molecules. These protein assemblies are perfectly designed to transport all the essential nutrients from the mother to the offspring in the most precise way. Regarding the economical, nutritional, and safety issues, milk proteins and milk protein aggregates have been proven to be one of the most reliable nanovehicles (Tavares et al., 2014). Several reviews have been published on the functionality of the milk proteins more specifically, caseins (de Kruif and Holt, 2003; Elzoghby et al., 2011; Fox and Brodkorb, 2008; Guri and Corredig, 2014; Kimpel and Schmitt, 2015; Livney, 2010; Tunick and Van Hekken, 2015). Here we will summarize the use of caseins as nanocarriers for biologically active species and will highlight the strategies that have been developed to better explore the potential of these proteins as nanocarriers.

Nanoencapsulation Technologies for the Food and Nutraceutical Industries

5.1.1 *Milk caseins, structure and composition*

Milk is a unique colloidal suspension of proteins, fat fraction, lactose, and minerals that supplies nourishment and growth to the neonate (Fox et al., 2000). The protein assembly of milk is composed of insoluble proteins named as caseins (α_{s1}, α_{s2}, β-, κ-casein) and soluble proteins named as whey proteins (β-lactoglobulin, α-lactalbumin, and bovine serum albumin) which range from 80% to 20% in bovine milk, as to 60% to 40% in humans (Fox et al., 2000). The four casein subspecies account differences on the degree of phosphorylation, glycosylation (i.e., κ–casein), amino acid profile and their hydrophobicity. The dry matter in the micelle consists of 94% protein and the remaining 6% is composed of minerals, mainly calcium, phosphate magnesium, citrate, and few others in lower amounts. They interact with calcium phosphate nanoclusters, or so-called colloidal calcium phosphate (CCP) and self-assemble in stable, and polydispersed supramolecular structures called casein micelles (Fox et al., 2000; Holt et al., 2003; Horne, 2006). Caseins appear of spherical shape with molecular mass of 106–109 Da (Fox and Brodkorb, 2008; Fox and McSweeney, 2003) and vary in diameter from 80-400 nm (Dalgleish and Corredig, 2012; de Kruif, 1998). Casein micelles are highly hydrated structures, which according to Jeurnink and de Kruif (1993) may retain up to 3.5 kg of water/kg protein. With respect to their volume in milk they constitute almost 10% of the total volume and 2.5% of the total milk weight (Dalgleish and Corredig, 2012). The four individual types of casein molecules (α_{s1}, α_{s2}, β, κ-casein) found in a micelle are in relative proportions of 4:1:3.5:1.5, respectively. αs-Casein and β-casein are the main casein subspecies that interact with calcium phosphate cluster due to serine phosphate residues present within casein molecules, while κ-casein is positioned on the outside layer of the associated casein nanoclusters. It is believed that since there is an insufficient number of κ-casein molecules to stabilize the entire surface of the micelles, β-casein may play a role in stabilizing the protein rich regions in the inner core of the micelle (Dalgleish, 2011).

The structural organization of caseins in micelles is still an ongoing debate for the past 50 years and several competing theories have been introduced. A more comprehensive understanding can be found on the following publications and reviews (Dalgleish, 2011; Dalgleish and Corredig, 2012; de Kruif and Holt, 2003; Farrell et al., 2006; Fox et al., 2000; Horne, 2006, 2009). Generally, it is believed that submicelles are held together by an equilibrium between electrostatic and hydrophobic interactions as well as van der Waals forces (Dalgleish, 2011; Horne, 2006, 2009). The presence of calcium phosphate clusters in their structure contributes to the elevated stability. Scattering techniques like small-angle neutron (SANS) and X-ray (SAXS) and the most advanced imaging techniques like cryo-transmission electron microscopy (cryo-TEM) and atomic force microscopy (AFM) have been used to reveal more information regarding the internal structure of the casein micelles and CPP nanocluster organization (Anema and de Kruif, 2012; Dalgleish, 2011; Ouanezar et al., 2012; Silva et al., 2015; Trejo et al., 2011). The structural organization of caseins molecules in a submicelle and their arrangement in colloidal micelles is illustrated in Fig. 5.1.

The primary structure of caseins is characterized by a high amount of proline, which impedes the formation of significant ordered structures. For this reason, caseins

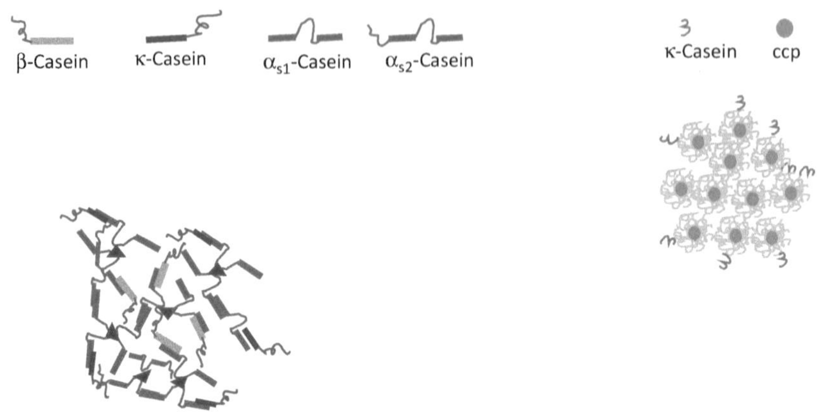

Figure 5.1 Casein structure indicating the main subspecies; α_{s1}-casein, α_{s2}-casein, k-casein, β-casein.

are very flexible and are defined rheomorphic, as they can adapt their conformation to changing environmental conditions (Holt and Sawyer, 1993). Casein micelles are negatively charged, representing a ζ-potential of about −20 mV and an isoelectric point of 4.6, thus precipitation occurs under acidic conditions (pH ~4.6). However, when pH is increased above the isoelectric point closer to the neutral conditions, caseins have the ability to resolubilize (Fox and McSweeney, 2003; Walstra et al., 1999).

It has been proposed that the composition of casein micelles may be somehow correlated with their size. Therefore, smaller micelles may contain higher percentage of κ-casein, and relatively low β-casein, while such correlation has not been observed for α_s-caseins (Marchin et al., 2007). Thereby, it is assumed that κ-caseins are mostly distributed on the surface of the micelles, β-caseins mainly in the inner core, and α_s-caseins can be found anywhere in the structure (Dalgleish and Corredig, 2012).

Certain factors could influence the stability of casein micelles, such as specific proteolitic enzymes, acids and addition of excessive amounts of calcium ions or ethanol. All these procedures are essential in many dairy products (de Kruif, 1998; de Kruif and Zhulina, 1996). Casein micelles are relatively stable structures that can withstand moderate temperature changes (heating or cooling) without significant changes in their basic structures (e.g., aggregation or disruption). At temperatures above 60°C, caseins still remain relatively stable to heating, while whey proteins are denatured via self-aggregation and thus aggregate with casein micelles ruled by disulfide bonding and hydrophobic interactions. To sum up, heat stability of caseins makes them ideal nanocarriers for heat-sensitive bioactive compounds.

5.1.2 Chemistry of caseins

The optimal functionality of casein proteins is strongly correlated to its chemistry and has been well studied in the last decades. The primary structures of casein subunits and their genetic variations have been extensively reviewed (Fox and McSweeney, 2003; Swaisgood, 1993).

Caseins contain an abundance of cyclic amino acids, such as proline residues, specifically in the case of β-caseins (35 out of 209 amino acid residues). κ-Caseins have 20 proline residues in their structure, followed by, α_{s1} and α_{s2}-caseins containing 17 and 10 residues, respectively. Sequences of relatively polar, nonpolar, and charged residues do not follow any particular order in the casein structure, but in overall they count for the strong amphipathic nature of caseins itself. The four types of caseins also differ in the number of hydrophobic/hydrophilic patches and phosphoserine clusters they contain in their structures. For more details refer to following publications (Dalgleish, 2011; Farrell et al., 2006; Fox and McSweeney, 2003; Horne, 2006).

Apart from proline, tryptophan residues are found in the casein molecular structure, but they differ with respect to the number of the residues present in each type and their position in the sequence. Two tryptophan residues are found in α_{s1} and α_{s2}-casein, positioned at 164 and 199 for α_{s1} and 109 and 193 for α_{s2}, and only one residue for β- and κ-casein structures at 143 and 76, respectively. It has been accepted that the hydrophobic portion of the casein micelles locks internally α_s-caseins and β-casein where most of the tryptophan residues are found, while κ-casein is occupying mainly the outer shell of the structure (Dalgleish, 2011).

Unlike whey proteins, caseins have phosphoserine residues in their structure which are organized in clusters. These groups are essential not only for their nutritional value but they also serve as binding sites for polyvalent cations. Therefore, apart from κ-casein, all of the other three caseins are susceptible to calcium ions causing precipitation (Fox and McSweeney, 2003). κ-Casein is the only domain of casein structure that is highly glycosylated containing sugar residues (galactose, galactosamine, and *N*-acetylneuraminic (sialic) acid). They are present in tri or tetrasaccharide form linked to threonine residues and referred as casein macropeptide (CMP). Its hydrophilic, glycosylated C-terminal region protrudes into solution providing steric and electrostatic stabilization to the protein assembly (Horne, 2006). The N-terminal so-called *para*-κ-casein, refers to hydrophobic domain, usually accessed by chymosin and other related enzymes during milk coagulation process (de Kruif and Holt, 2003; Fox and McSweeney, 2003).

5.1.3 Why choosing caseins as nanodelivery vehicles?

Recently, investigations have contributed to the elucidation of casein structures, which has built knowledge on better understanding of their binding capacities as well as has opened new horizons for the encapsulation of hydrophobic molecules. Casein is increasingly being used as a natural nanocarrier for biologically active compounds especially for hydrophobic substances to be applied in either food or pharmaceutical products (Faridi Esfanjani and Jafari, 2016; Katouzian and Jafari, 2016). Formerly, casein micelles were utilized as nanocarriers within pharmaceutical research, afterward, they became established in the field of food and nutrition, especially the delivery systems prepared with natural biopolymers (Baracat et al., 2012; Elzoghby et al., 2011; Jarunglumlert et al., 2015; Livney et al., 2009). The presence of acid-soluble calcium phosphate bridging, makes caseins readily available to proteolysis during gastric digestion. The assembly of nutrients provided from casein micelles is a unique example of a successful delivery in nanoscale. Although the principal mechanisms are quite similar in both fields, the limitation to food grade materials requires considerable attention and can be often challenging in food processing and product development (Jarunglumlert et al., 2015).

Caseins existing in their natural environment are extremely stable displaying good solubility, heat resistance, and highly surface active. However, caseins are also available commercially, in the form of sodium caseinate and calcium caseinate. They are prepared industrially by separating native caseins from skimmed milk fraction in the presence of acids, then resolubilized under alkaline conditions and finally processed through a drying process to yield caseinates. Likewise, native structures, caseinate powders have high solubility in water, additionally they are able to rearrange and form aggregated casein clusters in ionic environments. Studies being carried out on caseins reassembly prepared by various techniques, have proven that the organization of casein micelles in micellar substructures when calcium, citrate, or phosphate ions were added to aqueous casein solutions, have been very similar to that of natural casein assemblies (Knoop et al., 1979). The aggregation dynamics of caseins after solubilization can be achieved by simply controlling the pH of the solution during synthesis. Hence, studies have shown that during the aggregation process of caseinate solutions considering the pH changes, hydrophobic substances can be easily entrapped in the clusters (Jarunglumlert and Nakagawa, 2013). Studies suggest that the mechanism of micelle formation and casein aggregation during the encapsulation process should not be neglected, as it is a crucial factor that directly influences the encapsulation efficiency and core materials stability (Jarunglumlert et al., 2015).

Additionally, caseinate solutions have been introduced as stability and solubility amplifiers of lipophilic substances, such as curcumin (Pan et al., 2013), β-carotene (Sáiz-Abajo et al., 2013) vitamin D (Haham et al., 2012), omega-3 fatty acids (Zimet and Livney, 2011), and essential oils (Arranz et al., 2015). Other forms of purified casein proteins were also used successfully in the encapsulation procedure to enhance the bioavailability of hydrophobic substances like pure αs1-casein or β-casein (Sahu et al., 2008; Sneharani et al., 2010). To sum up, the aforementioned studies suggest that the entrapment of hydrophobic compounds in micellar fractions protects them against degradation induced by oxidation and processing.

5.2 Nanoencapsulation of food bioactive components and nutraceuticals by caseins

Encapsulation of the bioactives to increase their performance and maintain their functionality during industrial processing or gastrointestinal tract requires development of certain techniques. Pharmaceutical research has been the lead in this field considering that less manipulation of the matrix is required compared to the food industry. These requirements make the encapsulation of the bioactives more challenging for the food sector, however there has been great improvement in the engineered nanostructure and respective technological aspects. In general, nanoencapsulation technologies can be divided in two main approaches: the so-called "top-down" and "bottom-up." The first technique is more known as traditional "colloidal science," where the particle size is reduced during the encapsulation process, for example, by different mechanical forces, whereas in the "bottom-up" process, particle size is increased by methods, such as self-assembly or reassembly of caseins (Semenova and Dickinson, 2010). The concept of structure formation known as "bottom-up approach" is based on the ability to fabricate structures using a precise knowledge of the chemistry and physics underlying the individual building blocks by controlling the structure and behavior on the nanoscale level (Semenova and Dickinson, 2010). In some cases, a combination of the two approaches have been employed as their operational length scale overlaps considerably (McClements, 2015).

Many of the techniques used for encapsulation of bioactives and nutraceuticals that involve milk proteins are similar. However, since whey proteins and caseins have different structural and functional properties, different approaches based on their characteristics are applied to increase their encapsulation potential (Tavares et al., 2014). Selecting the most suitable system to encapsulate a bioactive compound, mainly depends on the type of bioactive, its solubility, functionality, the range of particle size, its rheological behavior, digestive release, and the degree of bioavailability and in overall what is the net profit over the cost on investment (Dordevic et al., 2014). The main encapsulation techniques based on casein functional properties are: (1) binding with ions, (2) hydrophobic interactions, (3) self-assembly and reassembly of caseins, (4) surface activity, and (5) interactions with other biopolymers. For some bioactives (e.g., polyphenols, vitamins, etc.) different casein nanosystems and accordingly different encapsulation techniques may be used. Table 5.1 represents a summary of the latest studies with regard to nanoencapsulation of biological active structures like polyphenols, essential oils, carotenoids, vitamins, and minerals, and examples of hydrophobic drugs in casein nanocarriers, and it is by no means a comprehensive list.

A more detailed discussion on the main nanoencapsulation techniques based on the casein functional properties which have been applied in recent studies for different bioactive components and nutraceuticals are discussed as follows.

5.2.1 Encapsulation by caseins through binding with ions (Fig. 5.2A)

It was aforementioned that all three types of caseins (α_{s1}, α_{s2}, and β-casein) have the ability to bind calcium, and calcium phosphate in their structure due to the presence of phosphate residues. Knowing that researchers have practically looked at likelihood of utilizing caseins as carriers of other metal ions, such as iron (Fe^{+2}) and magnesium (Mg^{+2}). A reversible acidification process by carbonation was used to enhance iron content in the caseins of skimmed milk, which resulted in the recovery of a significant high proportion of iron (i.e., 95%) in casein fractions (Raouche et al., 2009). In addition, the binding of iron to sodium caseinate and whey proteins has also been studied showing that sodium caseinate can bind more iron ions compared to whey proteins, and owing that to the presence phosphoserine clusters in casein micelles (Sugiarto and Singh, 2009).

Bonnet et al. (2009) employed sodium caseinate as an emulsifier and encapsulated magnesium within the inner aqueous droplets. In summary, magnesium leakage was reduced due to apparent ion binding to caseins.

Table 5.1 Summary of the main applications of caseins (based on their functional properties) as bioactive nanocarriers along with the encapsulation techniques employed and highlighted results

Bioactive compounds	Examples	Casein nanocarrier system	Encapsulation technique (strategy used)	Highlighted results
Polyphenols	Curcumin	• Caseins (Rahimi Yazdi and Corredig, 2012; Sahu et al., 2008) • Sodium caseinate (casein nanoparticles) (Pan et al., 2013, 2014) • β-Casein (Esmaili et al., 2011; Mehranfar et al., 2013, 2015) • Phosphocasein micelles (Benzaria et al., 2013)	• Hydrophobic interactions (Rahimi Yazdi and Corredig, 2012; Sahu et al., 2008) • Self-assembly of sodium caseinate (Pan et al., 2013, 2014) • Self-assembly of β-casein (Esmaili et al., 2011; Mehranfar et al., 2013, 2015) • Hydrophobic interactions (Benzaria et al., 2013)	• Enhanced solubility and dispersibility (Esmaili et al., 2011; Mehranfar et al., 2013, 2015; Pan et al., 2013, 2014; Rahimi Yazdi and Corredig, 2012; Sahu et al., 2008) • Antioxidant activity and cytotoxicity against cancer cell lines (Pan et al., 2013, 2014; Rahimi Yazdi and Corredig, 2012; Sahu et al., 2008) • Increased encapsulation efficiency (Benzaria et al., 2013; Pan et al., 2014) • Controlled release and protection of bound curcumin (Benzaria et al., 2013)
	EGCG–tea catechins	• Casein micelles (Guri and Corredig, 2014; Haratifar and Corredig, 2014; Haratifar et al., 2014a,b; Hu et al., 2012) • β-Casein nanoparticles (Bohin et al., 2012)	• Noncovalent forces and hydrophobic interactions (Guri and Corredig, 2014; Haratifar and Corredig, 2014; Haratifar et al., 2014a,b; Hu et al., 2012) • Self-assembly of β-casein (Bohin et al., 2012)	• Increased bioavailability (Guri and Corredig, 2014; Haratifar and Corredig, 2014; Haratifar et al., 2014a,b; Hu et al., 2012) • Stable nanoparticles (Guri and Corredig, 2014; Haratifar and Corredig, 2014; Haratifar et al., 2014a,b; Hu et al., 2012) • Great carrier for high doses of catechins (Guri and Corredig, 2014; Haratifar and Corredig, 2014; Haratifar et al., 2014a,b; Hu et al., 2012) • Increased encapsulation efficiency (Bohin et al., 2012)
	Flavonoids isolated from *Phaleria macrocarpa* leaf	Casein micelles (Sahlan and Pramadewi, 2012)	• Noncovalent forces and hydrophobic interactions (Sahlan and Pramadewi, 2012)	• Novel drug-delivery system (Sahlan and Pramadewi, 2012) • High encapsulation efficiency (Sahlan and Pramadewi, 2012)

	Resveratrol	• α- and β-Caseins (Bourassa et al., 2013) • Sodium caseinate (Acharya et al., 2013; Davidov-Pardo et al., 2015; Joye et al., 2015)	• Hydrophilic and hydrophobic interactions (Bourassa et al., 2013) • Biopolymer nanoparticles (Acharya et al., 2013; Davidov-Pardo et al., 2015; Joye et al., 2015)	• Potential carriers for in vitro transportation (Acharya et al., 2013; Bourassa et al., 2013; Davidov-Pardo et al., 2015; Joye et al., 2015) • Protection against UV-light (Acharya et al., 2013; Davidov-Pardo et al., 2015; Joye et al., 2015)
Vitamins	Beta-carotene	• Sodium caseinate (casein nanoparticles) (Jarunglumlert et al., 2015) • Sodium caseinate (Jarunglumlert et al., 2015; Sáiz-Abajo et al., 2013; Yi et al., 2014, 2015) • Caseins (Pan et al., 2007)	• Self-assembly of sodium caseinate (Jarunglumlert et al., 2015) • Reassembly of caseins (addition of salts to the sodium caseinate solutions) (Jarunglumlert et al., 2015; Sáiz-Abajo et al., 2013) • Surface activity, emulsifier (Yi et al., 2014, 2015) • Maillard reaction based casein–dextran conjugates (Pan et al., 2007)	• Protection against degradation during heating, high pressure processing, and baking (Jarunglumlert et al., 2015; Sáiz-Abajo et al., 2013) • High encapsulation efficiency (Jarunglumlert et al., 2015) • Stable carrier systems and high bioavailability (Jarunglumlert et al., 2015) • Decreased oxidation (Jarunglumlert et al., 2015) • Improved solubilization and protection against oxidation, and enabled long-time storage (Yi et al., 2014, 2015) • Higher cellular uptake and higher cellular antioxidant activity compared to free beta carotene (Yi et al., 2014, 2015) • Stable nanoparticles against dilution, pH change, ionic strength variations, and long-time storage (Pan et al., 2007)
	Vitamin A	Casein micelles (Mohan et al., 2013)	Hydrophobic interaction	Fortification
	Vitamin D	• Sodium caseinate (Haham et al., 2012; Semo et al., 2007) • β-Casein (Forrest et al., 2005)	• Reassembly of caseins (added salts) (Haham et al., 2012; Pan et al., 2007) • Self-assembly of β-casein (Forrest et al., 2005)	Protect against heat and storage-induced degradation, improved bioavailability (Forrest et al., 2005; Haham et al., 2012; Semo et al., 2007)
	Folic acid	Sodium caseinate (Penalva et al., 2015)	Reassembly of caseins (added salts)	• Higher oral bioavailability (Penalva et al., 2015) • Gastro-resistant carrier (Penalva et al., 2015)

(Continued)

Table 5.1 Summary of the main applications of caseins (based on their functional properties) as bioactive nanocarriers along with the encapsulation techniques employed and highlighted results (*cont.*)

Bioactive compounds	Examples	Casein nanocarrier system	Encapsulation technique (strategy used)	Highlighted results
Fatty acids	Omega-3 docosahexaenoic acid (DHA)	Sodium caseinate (Zimet and Livney, 2011)	Self-assembly of sodium caseinate (casein nanoparticles without salts) and reassembled caseins micelles (addition of salts)	• Protective effect against DHA oxidation • Great colloidal stability Formation of nanoparticles
	Omega-3	Casein micelles (Ghasemi and Abbassi, 2014)	pH increase–hydrophobic interactions	Increased protection against UV and enhanced oxidative stability
Drugs	Chemotherapeutic drug (paclitaxel)	β- Casein (Shapira et al., 2010, 2012; Zeev et al., 2016)	Self-assembly of β-casein	Stable nanovehicles Target-activated oral delivery of hydrophobic chemotherapeutics
	Chemotherapeutic drug (mitoxantrone)	β-Casein (Sripriyalakshmi et al., 2014)	Self-assembly of β-casein	Stable nanovehicles of hydrophobic anticancer drugs
	Celecoxib (Cx)	β-Casein (Perlstein et al., 2014)	Self-assembly of β-casein	Improved bioavailability
	Metformin hydrochloride (MET)	Sodium caseinate (casein nanoparticles) (Raj and Uppuluri, 2015)	Self-assembly of sodium caseinate	Stable nanoparticles Improvement of drug release
	Triclosan (hydrophobic drug)	Casein micelles (Roach et al., 2009)	Dissociation and reassociaiton of casein micelle	Potential delivery system, bioaccessibility

Minerals	Fe^{+2}, Mg^{+2}	• Caseins (Raouche et al., 2009) • Sodium caseinate (Sugiarto and Singh, 2009)	Metal ion binding	• Fortification of dairy products (Raouche et al., 2009; Sugiarto and Singh, 2009) • Stable delivery system (Raouche et al., 2009; Sugiarto and Singh, 2009)
Essential oils	Rosemary Extract (Arranz et al., 2015)	Casein micelles (Arranz et al., 2015)	Hydrophobic interactions	Improved bioavailability
Other proteins	Lactoferrin (Anema and de Kruif, 2012)	Casein micelle	Hydrophobic interactions	Potential nanocarrier

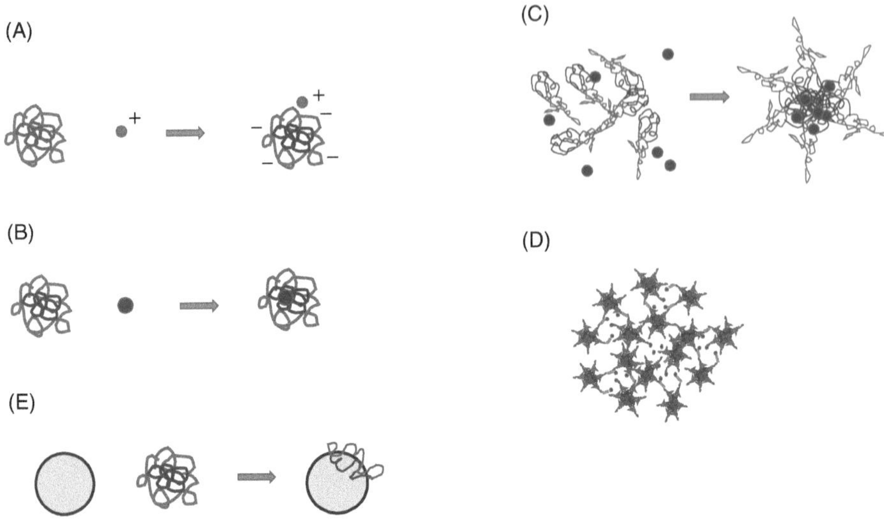

Figure 5.2 Encapsulation techniques based on casein functional properties. (A) Ion binding, (B) hydrophobic binding, (C) self-assembly, (D) reassembly (i.e., addition of salts), (E) surface activity.

5.2.2 *Encapsulation of hydrophobic molecules and other molecules by caseins*

5.2.2.1 *Encapsulation by caseins through hydrophobic interactions (Fig. 5.2B)*

Based on the chemistry of casein micellar organization, it is believed that the core of casein micelles provides an environment where nonpolar compounds could be accommodated for a safe delivery. It is well known that milk proteins mainly bind with hydrophobic components via hydrophobic and hydrogen bonding combined with van der Waals forces (Livney, 2010). The reversible binding of hydrophobic structures (e.g., vitamins, polyphenols, and fatty acids) can be through noncovalent interactions between the protein and ligand. To ensure the efficient binding and delivery of hydrophobic molecules generally, irreversible mechanisms are used through strong covalent binding with amino and thiol groups which can either be situated in the inner core or the surface of the milk proteins (Kimpel and Schmitt, 2015).

In an attempt to study the interactions of curcumin as hydrophobic active molecule with casein micelles, it was observed that curcumin interacts hydrophobically with casein molecules under physiological buffer conditions (Rahimi Yazdi and Corredig, 2012), where the nanoencapsulated curcumin in casein micelles showed the same cytotoxic activity on cancer cells as to free curcumin (Sahu et al., 2008). Curcumin has been also encapsulated in native-like phosphocasein micelles by means of ultrahigh pressure homogenization in which an increase in the encapsulation efficiency was reported. Similarly in vitro simulated digestion studies showed a controlled release of the encapsulated curcumin overtime (Benzaria et al., 2013).

The potential of milk proteins to transport and protect hydrophobic bioactives; such as vitamins and fatty acids has been another interesting area of casein-binding studies. Mohan et al. (2013) showed that casein structures in fortified commercial skim milk have strong binding affinity for vitamin A, where they carry up to 40% of the total vitamin A added to milk. These studies reconfirm that unmodified casein micelles could protect and safely deliver vitamin A. However, not only native caseins but also individual casein proteins have the ability to safely bind vitamins and deliver them. Forrest et al. (2005) suggested that vitamin D_3 and β-casein have binding affinities through hydrophobic interactions that can improve vitamins stability, where binding strength varies based on pH and respective ionic strength.

Ghasemi and Abbassi (2014) studied the encapsulation of omega-3 fatty acids from fish and vegetable oils in native casein micelles during changes in pH and processing. The alkaline pH, favored an overall increase in the hydrophobic regions in the core of casein micelles causing an increase in particle size and consequently increasing the capacity for encapsulating more bioactive in their structure. Therefore, while maintaining the native functional structure, an increase of 92%–97% was achieved in their encapsulation efficiency, according to the different techniques applied and oil content. In addition, a higher oxidative stability of the encapsulated oils was observed, resulting in better protection against UV oxidation.

Another recent application has been introduced from researchers where casein micelles were efficiently used as nanocarriers for essential oils, like rosemary extracts. Carnosic acid is one of the principal diterpenes of rosemary leaves which exhibits anti-inflammatory and anticancer properties. The encapsulation of the rosemary oil significantly protected the extract from degradation during digestion and a recovery of about 81% was noticed and respectively higher bioavailability of carnosic acid in human colon cancer cells. This opens up new horizons in healthy food formulations for a successful utilization of casein micelles for bioactive enrichment of foods (Arranz et al., 2015).

There is considerable research that claims strong interactions of milk caseins and polyphenol molecules, achieved from mainly hydrophobic and hydrogen binding (Yan et al., 2009). Tea catechins have strong binding affinity to proteins containing proline residues, such as the saliva proteins and bovine caseins. Studies have demonstrated that casein micelles were saturated at an approximate concentration of 0.08 mg epigallocatechin gallate (EGCG) per milligram of casein, where the nanoencapsulation of tea catechins specifically EGCG in casein micelles, did not diminish their antiproliferative activity on colon cancer cells compared with free tea catechins both before and after digestion (Guri et al., 2014; Haratifar and Corredig, 2014; Haratifar et al., 2014a,b). Similarly, an increase in the bioavailability of EGCG using bioactive peptides/chitosan nanoparticles was obtained by the release of tryptic digestion of milk caseins (Hu et al., 2012).

Casein nanoparticles with similar diameter of native casein micelles were obtained by slight reduction of pH and rennet hydrolysis of bovine caseins. These particles ensured an encapsulation efficiency of about 42%, corresponding to an encapsulation of almost 1.0 mg flavonoid per gram of casein (Sahlan and Pramadewi, 2012).

Apart from casein micelles, sodium caseinate and isolated individual caseins (i.e., α- and β-caseins) have been employed as nanovehicles for other classes of polyphenols, such as flavonoids (e.g., naringenin and resveratrol) (Table 5.1). Naringenin is found in grapefruit and exhibits biological activity, but high hydrophobicity reduces its solubility and hence bioavailability. However, a water-solubility increase was shown when naringenin was transported through β-casein nanocomplexes. The possibility of delivering naringenin through β-caseins was studied by means of fluorescence spectroscopy. It was found that naringenin binds spontaneously to caseins through hydrophobic, hydrogen boding, and van der Waals forces (Moeiniafshari et al., 2015).

It has been already confirmed through several studies that encapsulation of resveratrol using caseins and sodium caseinate can protect it from UV-light degradation and have great potential as carriers for in vitro transportation of resveratrol (Acharya et al., 2013; Bourassa et al., 2013; Davidov-Pardo et al., 2015; Joye et al., 2015). Spectroscopic methods and molecular modeling results showed that resveratrol binds through both hydrophobic and hydrophilic bonding to isolated caseins (i.e., α- and β-caseins). However, β-caseins forms stable complexes compared to α-caseins (Bourassa et al., 2013).

The ability of caseins to form cobiopolymers has also been exploited in the nanoencapsulation of resveratrol. Biopolymers nanoparticles of zein or gliadin with sodium caseinate or pectin encapsulation resveratrol by antisolvent precipitation method has been investigated. The results showed that when ionic strength was increased at certain experimental conditions (i.e., temperatures up to 90°C for 30 min), the sodium caseinate coated nanoparticles showed higher stability when compared to pectin-coated particles. Moreover those particles were highly dispersed in water after freeze drying. This successful incorporation of reservatrol in sodium caseinate–coated zein nanoparticles ensured a better protection for resveratrol against UV-light (Joye et al., 2015).

Apart from the aforementioned bioactives encapsulated in casein micelles the ability to deliver proteins through their assembled structure has also been explored. Anema and de Kruif (2012) studied the delivery of lactoferrin, through casein micelles. Lactoferrin could successfully bind to skim milk, however, a slow process of casein micelle disintegration (i.e., with a kinetic about 10-fold slower than the binding kinetic) was observed increasing milk transparency. The disintegration of the casein micelle was prevented by simulation of cross-linking the casein molecules by transglutaminase which was added right before the addition of lactoferrin to milk. It was suggested that interactions between casein micelles and lactoferrin are merely electrostatic, whereas the accessibility of some other hidden binding sites within the internal hydrophobic region of the casein can be controlled by manipulating the physicochemical conditions of the medium.

5.2.2.2 *Encapsulation by caseins through self-assembly and reassembly mechanisms combined with hydrophobic interactions (Fig. 5.2C–D)*

In many studies, casein micelles and their isolated individual forms (i.e., α- and β-caseins) have not required any specific modification to encapsulate and/or interact with hydrophobic compounds, which makes them even more appealing as bioactive nanocarriers. However, in some cases, modifications are applied to enhance the encapsulation properties of caseins. Different techniques have been applied to take advantage of caseins reassembling functional properties and self-assembly for β-casein (Tavares et al., 2014). The self-assembly of casein in micelles have been classified as a "bottom-up" approach (Livney, 2015).

In the case of β-casein, the presence of two domains, one negatively charged highly polar (N-terminal) and a highly nonpolar (C-terminal), justify its highly amphiphilic character displaying a self-association behavior. At low temperatures, this protein is monomeric and at concentrations 0.5–2 mg/mL, neutral pH at 25°C which is considered as critical micellization concentration (CMC), itself assembles into nanosized micelles in water medium enclosing a hydrophobic domain and surrounded by hydrophilic patches (Moeiniafshari et al., 2015; Portnaya et al., 2006).

Due to the aforementioned structural and physicochemical characteristics and their ability for hydrophobic interactions, caseinates and β-casein have been used to encapsulate different hydrophobic active molecules, such as hydrophobic vitamins, curcumin, fatty acids, etc.

Pan et al. (2014) showed that a high amount of curcumin was encapsulated in casein nanoparticles using spray-drying technique. Aqueous ethanol (i.e., containing 35% ethanol) was used as a solvent to dissolve curcumin. The casein micelles were dissociated at temperatures higher than 60°C in the mentioned aqueous ethanol. After spray drying, nanoparticles were formed entrapping curcumin due to the self-assembly of sodium caseinate as well as through hydrophobic interactions. The dissociation of sodium caseinate at higher temperatures facilitated higher curcumin binding as more hydrophobic regions of caseins were exposed. After rehydrating the resulted powder, the entrapment of curcumin was increased up to fourfold compared to simple mixing techniques. More information on the binding site of curcumin to caseins was given by fluorescence spectroscopy, where the results showed that through hydrophobic interactions with caseins, curcumin was mostly seen in the core of the nanoparticles. Other studies have looked at the possibility of sodium caseinate to dissociate and to self-assemble in casein nanoparticles by modifying the pH of the medium (pH 12 @ 21°C for 30 min, following neutralization). The dissociation of sodium caseinate and its reassociation after neutralization were confirmed using dynamic light scattering and analytical ultracentrifugation. This nanoencapulation ensured a higher biological activity compared to pristine curcumin when assessed by antioxidant and cell proliferation assays (Pan et al., 2014).

Since curcumin is known for its high functionality (e.g., antioxidant, antiinflammatory, anticancer effects), attempts to encapsulate pure compounds from curcumin have been reported. Thus, bisdemethoxycurcumin (BDMC), one of the main active components of turmeric and curcumin, was encapsulated in the hydrophobic core of β-casein micelle nanoparticles and its biological activity was measured in cancer cells (Mehranfar et al., 2013). Higher cytotoxic effects of encapsulated BDMC and curcumin on human breast cancer cell line MCF7 was observed compared to equal dose of free BDMC or free curcumin. The β-casein micelle solutions were prepared by filtering each protein solution to avoid large protein aggregates. Afterward, the solutions were dialyzed against buffer for 24 h at 4°C, using 1 mM ethylene diamine tetraacetic acid (EDTA). Computational docking studies showed that curcumin binds at the hydrophobic core of β-casein and a combination of van der Waals interactions and hydrogen bonds are also involved in the nanoformulation process. As β-casein is edible, the complex formed has the potential to consume curcumin orally (Mehranfar et al., 2013, 2015). Furthermore, the bioefficiency of curcumin in self-assembled β-casein increased at least 2500 times, likewise, its antioxidant activity and cytotoxicity against human leukemia cells was significantly higher compared to its free form, as reported by Esmaili et al. (2011).

Bohin et al. (2012) produced β-casein micelles at 25°C and pH 7.0 for the encapsulation of epigallocatechin-3-gallate (EGCG). Nanosized β-casein micelles with a mean diameter of about 13 nm were formed with a high encapsulation efficiency of EGCG (i.e., 38 g EGCG per 100 g of protein).

Jarunglumlert et al. (2015) evaluated the encapsulation properties of casein nanoparticles (i.e., aggregated casein formed by adjusting the pH of the sodium caseinate solutions) and reassembled micelles (i.e., prepared by the addition of salts to the sodium caseinate solutions) loaded with β-carotene via hydrophobic interactions. The results showed that both encapsulated forms improved storage stability of β-carotene; however, the aggregated casein nanoparticles prepared at pH 6.0 had higher encapsulation efficiency than that of the reassembled casein micelles. Another study showed that encapsulated β-carotene by reassembled casein micelles using sodium caseinate, was significantly protected from degradation during heating, high pressure processing, baking, and long-time storage compared to its free form (Sáiz-Abajo et al., 2013).

Other similar studies have reported casein nanoencapuslation of vitamin D_2 (Semo et al., 2007) and vitamin D_3 (Haham et al., 2012). Vitamin D_2 was nanoencapsulated by the reassembly of caseinate, where the obtained casein-micelle-like structures with a diameter of about 156 nm encapsulated an efficient amount of the vitamin and protected it against photochemical degradation. Casein micelles were reformed, after ligand (i.e., vitamin D_2) and soluble caseinate binding and reconstituting the mineral composition of milk caseins (Semo et al., 2007). Another study reported the reassembly process of sodium caseinate and the obtained casein micelles including vitamin D_3. All in all, the encapsulated vitamin was more stable and had less degradation compared to free nonencapsulated vitamin (i.e., degradation of 40% and >70%, respectively). Furthermore, clinical trials showed great bioavailability of the vitamin encapsulated in the casein matrix (Haham et al., 2012).

Folic acid encapsulated in casein nanoparticles showed 52% higher oral bioavailability compared to the traditional aqueous solution providing increased bioavailability. In vitro release studies, indicated that folic acid was released when simulated intentional conditions were applied therefore the casein nanoparticles acted as gastroresistant carriers. The nanoparticles were prepared by simple coacervation procedure followed by ultrafiltration and spray drying. Briefly, the nanoparticles were formed by incorporating a calcium chloride solution to the mixture of folic acid in aqueous sodium caseinate suspension. The suspension was then purified and spray dried (Penalva et al., 2015; Zhang et al., 2014). A study has shown that folic acid binds to β-casein via hydrophobic interactions without causing any change in the structure of the protein (Zhang et al., 2014).

Other attempts to deliver functionality through casein micelles have included compounds like polyunsaturated fatty acids (PUFAs) of the omega-3 series (ω-3), which have positive impacts on human health. Zimet and Livney (2011) showed that docosahexaenoic acid (DHA) which is an important omega-3 PUFA, has high affinity to bind with caseins. An ethanolic solution of DHA was added to sodium caseinate solution at room temperature to a final pH of 6.7. The casein nanoparticles were formed in two different ways, with and without the addition of salts at 4°C (i.e., calcium and phosphate) to the DHA-sodium caseinate solution and DHA loaded reassembled casein nanoparticles were formed. Particle size experiments showed 4 times decrease when method of salt addition was performed, 50–60 and 288.9 nm, respectively. However, both systems protected DHA against oxidation, displaying good colloidal stability throughout shelf life studies at 4°C.

Different studies have attempted to introduce friendly biodegradable nanoparticles to solve some of the main challenges imposed from pharmaceutical agents. Caseins due to their known functionalities have been studied and used as such nanovehicles for drug delivery. For example, to facilitate the oral delivery of a chemotherapeutic drug called mitoxantrone, β-casein was proposed as a nanocarrier based on self-assembly of this protein which enhanced entrapping and solubilization of mitoxantrone. The encapsulation of the drug was enabled by the addition of a dimethyl-sulfoxide solution to a phosphate-buffered saline containing β-casein, offering good colloidal stability (Shapira et al., 2010; Sripriyalakshmi et al., 2014). Similarly, paclitaxel, a hydrophobic anticancer drug used in the treatment of gastric carcinoma, was entrapped within β-casein nanoparticles with an entrapment efficiency of nearly 100% forming stable colloidal nanovehicles. Afterward, simulated digestion was applied and the results showed that the cytotoxic activity of the encapsulated drug on human gastric cancer cells were similar to the nonencapsulated drug (Shapira et al., 2010, 2012).

Zeev et al. (2016) used a novel nanosystem based on β-casein to orally deliver a synergistic combination of a chemotherapeutic drug (paclitaxel) and a chemosensitizer (tariquidar) to assist in overcoming multidrug resistance (MDR) in gastric cancer cells. Their studies demonstrated high loading capacity and good encapsulation efficiency of the drugs to β-casein. In vitro cytotoxicity assays also showed a significant synergistic efficacy against human MDR gastric carcinoma cells. It was concluded that this novel nanosystem has the potential for target-activated release of chemotherapeutic drugs in the treatment of gastric cancer cells.

Perlstein et al. (2014) were able to load a high amount of celecoxib (Cx) (a drug used for osteoarthritis, rheumatoid arthritis, and acute pain) within small, uniform, and stable β-casein micelles, in the absence of any additional polymers or cryoprotectants. An improved oral bioavailability of Cx was seen compared to its commercial Cx formulation, which could be due to improvement of its solubility in intestinal fluids facilitated by β-casein micelles (Perlstein et al., 2014).

Other drug studies on metformin hydrochloride (MET) (Raj and Uppuluri, 2015) or triclosan (Roach et al., 2009) delivery through casein micelle have been reported. Met is a widely used biguanide oral antidiabetic agent against type II diabetes possessing poor oral bioavailability. Stable and uniform casein-MET micelles were formed using sodium caseinate. In vitro drug release analysis showed that approximately half of the drug is released within the 3 h after administration and 85% of the drug is released after 15 h, which is a significant improvement compared to its current release behavior (Raj and Uppuluri, 2015).

In the case of triclosan encapsulation, high pressure homogenization and solvent-mediated pressure-induced dissociation were applied. Triclosan binds naturally to casein micelles in skim milk (up to 70%) however the association was 30% increased through homogenization at pressures up to 300 MPa. The release of triclosan in a simulated intestinal solution, was governed by the disruption of micelle integrity, therefore caseins prevented its release in simulated gastric solutions, demonstrating the ability of caseins to ensure its bioaccessibility (Roach et al., 2009).

5.2.3 Surface activity of caseins (Fig. 5.2E)

Surface activity of caseins has been used to create emulsion-based carrier systems for hydrophobic compounds. Nanoemulsion-based carriers have been used as vehicles for hydrophobic bioactives that are highly insoluble in aqueous systems. Due to their amphiphilic structure and flexibility, milk proteins have great surface activity, as a result they are absorbed at oil–water interfaces to stabilize the emulsions. Caseins have highly flexible conformations, thus they can act as emulsion stabilizers. The surface activity of the individual milk caseins is as follows: β-casein mono dispersed casein micelles > αs-caseins = κ-casein. It has been noted that caseinates have higher surface activity than casein micelles (Kimpel and Schmitt, 2015; Livney, 2010; Sabouri et al., 2015; Singh and Ye, 2009). Courthaudon et al. (1999) showed that caseinates have higher surface activity than casein micelles by using interfacial tension results. Their results showed that the ultrastructural state of caseins impacts both their surface activity and their adsorption at the oil–water interface, hence, causing differences in the emulsion microstructure. The emulsions made with casein micelles were coarser, had higher protein surface coverage as well as higher extent of droplet aggregation compared to emulsions made with caseinate.

A combination of homogenization and evaporation method was used to obtain β-carotene loaded nanoparticles using different ratios of whey protein isolate, sodium caseinate, or soy protein isolate. The results showed that stable carrier systems were developed with sodium caseinate as emulsifier, and cellular antioxidant activity was significantly increased through nanoparticle delivery systems compared to free β-carotene (Yi et al., 2014, 2015).

5.2.4 Interaction of caseins with other biopolymers

Milk proteins mainly interact with other types of polymers through covalent and noncovalent interactions. Covalent conjugation can take place during the Maillard process between the hydrophobic parts of proteins and polysaccharides, resulting in copolymers. Whereas, noncovalent interactions take place when electrostatic polymers interact with their opposite charge (Kimpel and Schmitt, 2015).

Pan et al. (2007) produced a casein–graft–dextran copolymer via Maillard reaction, afterward nanoparticles were synthesized by casein and β-carotene as core materials and dextran for the shell material by means of dialysis or evaporation and dispersion in water. They are not only used in microencapsulation techniques, but also in nanoencapsulation approaches as well (Livney, 2010). As a result, the nanoparticles were stable in aqueous solution against pH change, dilution, ionic strength variations, and extended storage times. These nanoparticles were prepared in a green process (Maillard reactions are considered as green process since their products are natural) in which the only chemicals utilized were alkali and ethanol.

5.3 Advantages and disadvantages

Nanoencapsulation has been rapidly growing in the last recent years in both food and pharmaceutical industries, while numerous potential benefits count, there are also some concerns alongside the great promises. Some of the main concerns are over its safety (e.g., toxic above a certain dose, possible toxicity due to dramatically higher permeability of nanoparticles through biological barriers), the impact on the sensory properties as well as the physicochemical properties of the food product, regulations and acceptance by the industry and consumers (Handford et al., 2014; Raynes et al., 2014).

In many cases, a significant increase in the bioavailability of the compound is seen when encapsulated by caseins, which can be a safety issue if the bioactive compound becomes toxic at a certain dose. Therefore, thorough safety evaluations, regulatory approvals and even perhaps reevaluation of the Recommended Daily Allowance (RDA) and Tolerable Upper Intake Level (UL) of the bioactive, may be required. Broader and more realistic studies must take place using in in vitro/in vivo screening methods validations and clinical trials, to determine safe application of the materials as well as providing scientific proof-evidence based information to the consumers before commercial exploitation (Livney, 2015; Raynes et al., 2014).

The advantages of using milk proteins, especially caseins as nanodelivery systems for bioactives is not limited to the fact that caseins are generally recognized as safe (GRAS) but that they are also inexpensive, and easily available to use. Other advantages with regard to casein nanoencapsulation are; protection and stabilization of the bioactive compound, increased bioavailability, as well as controlled release and targeted delivery along the digestive tract (Joye et al., 2014; Kimpel and Schmitt, 2015).

As caseins and other milk proteins have shown to be ideal nanocarriers, they have the potential to be used in the food industry in two main forms; delivery systems for food additives and delivery systems for bioactive compounds. For example, caseins can be used to fortify processed dairy and food products with nanoencapsulation of nutrients to influence texture, flavor, functionality, and nutritional quality. They also have shown to be great delivery vehicles of bioactives, such as proteins, lipids, vitamins, and polyphenols. Hence, using caseins is highly efficient in manufacturing of functional foods (Radha et al., 2014). Casein nanoencapsulation have caused great savings for formulators, since using caseins decreases the amount of other active ingredients needed and therefore, the respective costs.

In the case of using caseins as nanovehicles for drugs, the encapsulation has shown to facilitate the bioavailability problems which many hydrophobic drugs encounter. However, much more research is required to develop nanocasein drug that have minimal or no adverse effects (Pathak and Raghuvanshi, 2015).

5.4 Insight for future work

Caseins exhibit a wide diversity of functional properties which are strongly correlated with their physicochemical characteristics. The unique properties that they possess like self-assembly in supramolecular structures, high binding capacity of minerals, coassembly with other polymers, relative stability to changes in pH, and temperature makes them very attractive for not only nutritional values but also as nanocarriers in delivering functionality of other bioactive structures. The most recent studies have been carried out trying to probe the interactions of biologically active molecules and casein micelles in the food area and pharmaceutical applications and there are already products in the market (Livney, 2010; Esmaili et al., 2011). In addition, there is grown evidence of the maintained functionality of the structures during processing and digestion (Guri et al., 2014). With regard to food more attention has been given to structures which are considered GRAS (generally recognized as safe) and also the matrix interference is minimized to ensure and maintain the functionality. Although drug treatments are effective in treating the diseases, the diet based therapies are also promoted to preserve and enhance health. This becomes more prevalent when naturally occurring compounds with GRAS status are used, with known functionality which assert consumers' choice and demand in the functional market (Handford et al., 2014; Livney, 2015). The chemistry of the proteins in general and its behavior under different mechanical and analytical conditions is the tapping into the key when encapsulation of different bioactive structures is considered. To exhibit the right biological activity, the chemical bonding of bioactive and casein proteins should be well aligned with their final performance (Raynes et al., 2014). Depending on what we want to achieve through nanoencapsulation, that is, to mask the flavor, protect form degradation and deterioration, or to be released on specific time and tissue proper techniques and caution should be taken since a nonproper bonding of caseins with the molecule of interest may not be able to assure the expected outcome. Thus, precise applications and evaluations need to be considered in each case. Since most of research on casein encapsulation done so far is lab scale, upscaling production of nanoencapsulated casein–based materials needs further attention. Furthermore, more fundamental research is required to fully understand and control the assembly and disassembly mechanisms of the structures; stability of the encapsulated molecules and the stability of the carrier itself during processing and storage. Finally, applications of caseins as an ideal source of not only nutrition, but also drug and bioactive carrier for oral-delivery of functional constituents, will undoubtedly become more strongly established in the functional market through the upcoming developments in food and pharmaceutical research.

References

Acharya, D.P., Sanguansri, L., Augustin, M., 2013. Binding of resveratrol with sodium caseinate in aqueous solutions. Food Chem. 141 (2), 1050–1054.

Anema, S.G., de Kruif, C.G., 2012. Lactoferrin binding to transglutaminase cross-linked casein micelles. Int. Dairy J. 26, 83–87.

Arranz, E., Santoyo, S., Jaime, L., Fornari, T., Reglero, G., Guri, A., Corredig, M., 2015. Improved bioavailability of supercritical rosemary extract through encapsulation in different delivery systems after in vitro digestion. Food Dig. Res. Curr. Opin. 6 (1–3), 30–37.

Assadpour, E., Maghsoudlou, Y., Jafari, S.M., Ghorbani, M., Aalami, M., 2016. Evaluation of folic acid nano-encapsulation by double emulsions. Food Bioproc. Technol. 9, 2024–2032.

Assadpour, E., Jafari, S.M., Maghsoudlou, Y., 2017. Evaluation of folic acid release from spray dried powder particles of pectin-whey protein nano-capsules. Int. J. Biol. Macromol. 95, 238–247.

Baracat, M.M., Nakagawa, A.M., Casagrande, R., Georgetti, S.R., Verri, Jr., W.A., de Freitas, O., 2012. Preparation and characterization of microcapsules based on biodegradable polymers: pectin/casein complex for controlled drug release systems. AAPS Pharm. Sci. Tech. 13 (2), 364–372.

Benzaria, A., Maresca, M., Taieb, N., Dumay, E., 2013. Interaction of curcumin with phosphocasein micelles processed or not by dynamic high-pressure. Food Chem. 138 (4), 2327–2337.

Bohin, M.C., Vincken, J.P., van der Hijden, H.T., Gruppen, H., 2012. Efficacy of food proteins as carriers for flavonoids. J. Agric. Food Chem. 60, 4136–4143.

Bonnet, M., Cansell, M., Berkaoui, A., Ropers, M.H., Anton, M., Leal-Calderon, F., 2009. Release rate profiles of magnesium from multiple W/O/W emulsions. Food Hydrocoll. 23, 92–101.

Bourassa, P., Bariyanga, J., Tajmir-Riahi, H.A., 2013. Binding sites of resveratrol, genistein, and curcumin with milk α- and β-caseins. J. Phys. Chem. B 117 (5), 1287–1295.

Courthaudon, J.L., Girardet, J.M., Campagne, S., Rouhier, L.M., Campagna, S., Linden, G., Lorient, L., 1999. Surface active and emulsifying properties of casein micelles compared to those of sodium caseinate. Int. Dairy J. 9, 411–412.

Dalgleish, D.G., 2011. On the structural models of bovine casein micelles—review and possible improvements. Soft Matter 7 (7), 2265–2272.

Dalgleish, D.G., Corredig, M., 2012. The structure of the casein micelle of milk and its changes during processing. Annu. Rev. Food. Sci. Technol. 3, 449–467.

Davidov-Pardo, G., Perez-Ciordia, S., Marín-Arroyo, M.R., McClements, D.J., 2015. Improving resveratrol bioaccessibility using biopolymer nanoparticles and complexes: impact of protein–carbohydrate Maillard conjugation. J. Agric. Food Chem. 63 (15), 3915–3923.

de Kruif, C.G., 1998. Supra-aggregates of casein micelles as a prelude to coagulation. J. Dairy Sci. 81, 3019–3028.

de Kruif, C.G., Holt, C., 2003. Casein micelle structure, functions and interactions. In: Fox, P.F., McSweeney, P.L.H. (Eds.), Advanced Dairy Chemistry-1 Proteins. Springer, US, pp. 233–276.

de Kruif, C.G., Zhulina, E.B., 1996. κ-Casein as a polyelectrolyte brush on the surface of casein micelles. Colloids Surf. A 117, 151–159.

Dordevic, V., Balanc, B., Belscak-Cvitanovic, A., Levic, S., Trifkovic, K., Kalusevic, A., Kostic, I., 2014. Review article: trends in encapsulation technologies for delivery of food bioactive compounds. Food. Eng. Rev. 7, 452–490.

Elzoghby, A.O., El-Fotoh, W.S., Elgindy, N.A., 2011. Casein based formulations as promising controlled release drug delivery systems. J. Control. Release 153, 206–216.

Esmaili, M., Ghaffari, S.M., Moosavi-Movahedi, Z., Atri, M.S., Sharifizadeh, A., Farhadi, M., 2011. Beta casein-micelle as a nano vehicle for solubility enhancement of curcumin; food industry application. LWT Food Sci. Technol. 44 (10), 2166–2172.

Faridi, Esfanjani A., Jafari, S.M., 2016. Biopolymer nano-particles and natural nano-carriers for nano-encapsulation of phenolic compounds. Colloids Surf. B. 146, 532–543.

Farrell, H.M., Malin, E.L., Brown, E.M., Qi, P.X., 2006. Casein micelle structure: what can be learned from milk synthesis and structural biology? Curr. Opin. Colloid Interface Sci. 11, 135–147.

Forrest, S.A., Yada, R.Y., Rousseau, D., 2005. Interactions of vitamin D_3 with bovine β-lactoglobulin A and β-casein. J. Agric. Food Chem. 53, 8003–8009.

Fox, P.F., Brodkorb, A., 2008. The casein micelle: historical aspects, current concepts and significance. Int. Dairy J. 18, 677–684.

Fox, P.F., Guinee, T.P., Cogan, T.M., McSweeney, P.L.H., 2000. Fundamentals of Cheese Science. Aspen Pub, Gaithersburg, MD, (Chapter 1).

Fox, P.F., McSweeney, P.L.H., 2003. Milk proteins: general and historical aspects. Advanced Dairy Chemistry. Vol. 1: Proteins. third ed. Kluwer Academic/Plenum Publisher, New York, NY, USA.

Ghasemi, S., Abbassi, S., 2014. Formation of natural casein micelle nanocapsule by means of pH changes and ultrasound. Food Hydrocoll. 42, 42–47.

Gülseren, İ., Guri, A., Corredig, M., 2014. Effect of interfacial composition on uptake of curcumin–piperine mixtures in oil in water emulsions by Caco-2 cells. Food Funct. 5 (6), 1218–1223.

Guri, A., Corredig, M., 2014. Dairy materials as delivery tools for bioactive components in dairy platforms. In: Boland, M., Golding, M., Singh, H. (Eds.), Food Structure, Digestion and Health. Riddet Institute, Massey University, Palmerston North, New Zealand, pp. 465–488, (Chapter 17).

Guri, A., Gülseren, I., Corredig, M., 2013. Utilization of solid lipid nanoparticles for enhanced delivery of curcumin in cocultures of HT29-MTX and Caco-2 cells. Food Funct. 4 (9), 1410–1419.

Guri, A., Haratifar, S., Corredig, M., 2014. Bioefficacy of tea catechins associated with milk caseins tested using different in vitro digestion models. Food Dig. 5 (1–3), 8–18.

Haham, M., Ish-Shalom, S., Nodelman, M., Duek, I., Segal, E., Kustanovich, M., Livney, Y.D., 2012. Stability and bioavailability of vitamin D nanoencapsulated in casein micelles. Food Funct. 3 (7), 737–744.

Handford, C.E., Dean, M., Henchion, M., Spence, M., Elliott, C.T., Campbell, K., 2014. Implications of nanotechnology for the agri-food industry: Opportunities, benefits and risks. Trends Food Sci. Technol. 40, 226–241.

Haratifar, S., Corredig, M., 2014. Interactions between tea catechins and casein micelles and their impact on renneting functionality. Food Chem. 143, 27–32.

Haratifar, S., Meckling, K.A., Corredig, M., 2014a. Antiproliferative activity of tea catechins associated with casein micelles, using HT29 colon cancer cells. J. Dairy Sci. 97, 672–678.

Haratifar, S., Meckling, K.A., Corredig, M., Haratifar, S., Meckling, K., Corredig, M., 2014b. Bioefficacy of tea catechins encapsulated in casein micelles tested on normal mouse cell line (4D/WT) and its cancerous counterpart (D/v-src) before and after in vitro digestion. Food Funct. 5 (6), 1160–1166.

Holt, C., de Kruif, C.G., Tuinier, R., Timmins, P.A., 2003. Substructure of bovine casein micelles by small-angle X-ray and neutron scattering. Colloids Surf. A 213, 275–284.

Holt, C., Sawyer, L., 1993. Caseins as rheomorphic proteins—interpretation of primary and secondary structures of the αs1-caseins, β-caseins and κ-caseins. J. Chem. Soc. Faraday Trans. 8, 2683–2692.

Horne, D.S., 2009. Casein micelle structure and stability. In: Thompson, A., Boland, M., Singh, H. (Eds.), Milk proteins-from expression to food. Academic Press, New York, NY, pp. 442–473.

Horne, D.S., 2006. Casein micelle structure: models and muddles. Curr. Opin. Colloid Interface Sci. 11 (12), 148–153.
Hu, B., Yang, X., Tang, W., Zeng, X., Huang, Q., 2012. Nanochemo prevention by encapsulation of (−)-epigallocatechin-3-gallate with bioactive peptides/chitosan nanoparticles for enhancement of its bioavailability. Chem. Commun. 48 (18), 2421–2423.
Jafari, S.M., Fathi, M., Mandala, I., 2015. Emerging product formation. Food waste recovery: processing technologies and industrial techniques. Elsevier, Academic Press, San Diego, CA, 293-317.
Jarunglumlert, T., Nakagawa, K., 2013. Spray drying of casein aggregates loaded with β-carotene: influences of acidic conditions and storage time on surface structure and encapsulation efficiencies. Drying Technol. 31 (13–14), 1459–1465.
Jarunglumlert, T., Nakagawa, K., Adachi, S., 2015. Influence of aggregate structure of casein on the encapsulation efficiency of β-carotene entrapped via hydrophobic interaction. Food Struct. 5, 42–50.
Jeurnink, T.J.M., de Kruif, K.G., 1993. Changes in milk on heating: viscosity measurements. J. Dairy Res. 60 (2), 139–150.
Joye, I.J., Davidov-Pardo, G., McClements, D.J., 2014. Nanotechnology for increased micronutrient bioavailability. Trends Food Sci. Technol. 40, 168–182.
Joye, I.J., Davidov-Pardo, G., McClements, D.J., 2015. Encapsulation of resveratrol in biopolymer particles produced using liquid antisolvent precipitation. Part 2: stability and functionality. Food Hydrocoll. 49, 127–134.
Katouzian, I., Jafari, S.M., 2016. Nano-encapsulation as a promising approach for targeted delivery and controlled release of vitamins. Trends Food Sci. Technol. 53, 34–48.
Kimpel, F., Schmitt, J.J., 2015. Review: milk proteins as nanocarrier systems for hydrophobic nutraceuticals. J. Food Sci. 80 (11), 2361–2366.
Knoop, A.M., Knoop, E., Wiechen, A., 1979. Sub-structure of synthetic casein micelles. J. Dairy Res. 46 (2), 347–350.
Livney, Y.D., 2010. Milk proteins as vehicles for bioactives. Curr. Opin. Colloid Interface Sci. 15 (1), 73–83.
Livney, Y.D., 2015. Nanostructured delivery systems in food: latest developments and potential future directions. Curr. Opin. Food Sci. 3, 125–135.
Livney, Y.D., Assaraf, Y.G., Shapira, A., 2009. U.S. Patent Application No. 12/867,219c.
Marchin, S., Putaux, J.L., Pignon, F., Léonil, J., 2007. Effects of the environmental factors on the casein micelle structure studied by cryo transmission electron microscopy and small-angle x-ray scattering/ultra small-angle x-ray scattering. J. Chem. Phys. 126 (4), 045101.
McClements, D.J., 2015. Nanoparticle- and microparticle-based delivery systems. CRC Press, Taylor & Francis Group, LLC, New York.
Mehranfar, F., Bordbar, A.-K., Amiri, R., 2015. In vitro cytotoxic activity and binding properties of curcumin in the presence of β-casein micelle nanoparticles. Biomacromol. J. Food Sci. 1, 69–79.
Mehranfar, F., Bordbar, A.K., Fani, N., Keyhanfar, M., 2013. Binding analysis for interaction of diacetylcurcumin with b-casein nanoparticles by using fluorescence spectroscopy and molecular docking calculations. Spectrochim. Acta A 115, 629–635.
Moeiniafshari, A.A., Zarrabi, A., Abdol-Khalegh, B., 2015. Exploring the interaction of naringenin with bovine beta-casein nanoparticles using spectroscopy. Food Hydrocoll. 51, 1–6.
Mohan, M.S., Jurat-Fuentes, J.L., Harte, F., 2013. Binding of vitamin A by casein micelles in commercial skim milk. J. Dairy Sci. 96, 790–798.
Ouanezar, M., Guyomarc'h, F., Bouchoux, A., 2012. AFM imaging of milk casein micelles: evidence for structural rearrangement upon acidification. Langmuir 28 (11), 4915–4919.

Pan, K., Luo, Y., Gan, Y., Baek, S.J., Zhong, Q., 2014. pH-driven encapsulation of curcumin in self assembled casein nanoparticles for enhanced dispersibility and bioactivity. Soft Matter 10, 6820–6830.

Pan, X., Yao, P., Jiang, M., 2007. Simultaneous nanoparticle formation and encapsulation driven by hydrophobic interaction of casein-graft-dextran and beta-carotene. J. Colloid Interface Sci. 315, 456–463.

Pan, K., Zhong, Q., Baek, S.J., 2013. Enhanced dispersibility and bioactivity of curcumin by encapsulation in casein nanocapsules. J. Agric. Food Chem. 61 (25), 6036–6043.

Pathak, K., Raghuvanshi, S., 2015. Review article: oral bioavailability: issues and solutions via nanoformulations. Clin. Pharmacokinet. 54 (4), 325–357.

Penalva, R., Esparza, I., Agüeros, M., Gonzalez-Navarro, C.J., Gonzalez-Ferrero, C., Irache, J.M., 2015. Casein nanoparticles as carriers for the oral delivery of folic acid. Food Hydrocoll. 44, 399–406.

Perlstein, H., Yaelle, B., Turovsky, T., Rubinstein, A., Danino, D., Stepensky, D., Barenholz, Y., 2014. Beta-casein nanocarriers of celecoxib for improved oral bioavailability. Eur. J. Nanomed. 6 (4), 217–226.

Portnaya, I., Cogan, U., Livney, Y.D., Ramon, O., Shimoni, K., Rosenberg, M., Danino, D., 2006. Micellization of bovine β-casein studied by isothermal titration microcalorimetry and cryogenic transmission electron microscopy. J. Agric. Food Chem. 54 (15), 5555–5561.

Radha, K., Thomas, A., Sathian, C.T., 2014. Application of nano technology in dairy industry: prospects and challenges—a review. Indian J. Dairy Sci. 67 (5), 367–374.

Rahimi Yazdi, S., Corredig, M., 2012. Heating of milk alters the binding of curcumin to casein micelles. A fluorescence spectroscopy study. Food Chem. 132, 1143–1149.

Raj, J., Uppuluri, K.B., 2015. Metformin loaded casein micelles for sustained delivery: formulation, characterization and in-vitro evaluation. Biomed. Pharmacol. 8 (1), 83–89.

Raouche, S., Dobenesque, M., Bot, A., Lagaude, A., Marchesseau, S., 2009. Casein micelles as a vehicle for iron fortification of foods. Eur. Food Res. Technol. 229, 929–935.

Raynes, J.K., Carver, J.A., Gras, S.L., Gerrard, J.A., 2014. Protein nano structures in food—should we be worried? Trends Food Sci. Technol. (37), 42–50.

Roach, A., Dunlap, J., Harte, F., 2009. Association of triclosan to casein proteins through solvent-mediated high-pressure homogenization. J. Food Sci. 74, 23–29.

Sabouri, S., Geng, J., Corredig, M., 2015. Tea polyphenols association to caseinate-stabilized oil–water interfaces. Food Hydrocoll. 51, 95–100.

Sahlan, M., Pramadewi, I., 2012. Nanoencapsulation of the flavonoids isolated from *Phaleria macrocarpa* leaf by casein micelle. Int. J. Pharm. Bio. Sci. 3 (3), 472–478.

Sahu, A., Kasoju, N., Bora, U., 2008. Fluorescence study of the curcumin-casein micelle complexation and its application as a drug nanocarrier to cancer cells. Biomacromolecules 9, 2905–2912.

Sáiz-Abajo, M.J., González-Ferrero, C., Moreno-Ruiz, A., Romo-Hualde, A., González-Navarro, C.J., 2013. Thermal protection of b-carotene in re-assembled casein micelles during different processing technologies applied in food industry. Food Chem. 138, 1581–1587.

Semenova, M.G., Dickinson, E., 2010. Biopolymers in Food Colloids: Thermodynamics and Molecular Interactions. CRC Press, Leiden, Boston.

Semo, E., Kesselman, E., Dganit, D., Livney, Y.D., 2007. Casein micelle as a natural nanocapsular vehicle for nutraceuticals. Food Hydrocoll. 21, 936–942.

Shapira, A., Assaraf, Y.G., Epstein, D., Livney, Y.D., 2010. Beta-casein nanoparticles as an oral delivery system for chemotherapeutic drugs: impact of drug structure and properties on co-assembly. Pharm Res. 27, 2175–2186.

Shapira, A., Davidson, I., Avni, N., Assaraf, Y.G., Livney, Y.D., 2012. β-Casein nanoparticle-based oral drug delivery system for potential treatment of gastric carcinoma: stability, target-activated release and cytotoxicity. Eur. J. Pharm. Biopharm. 80, 298–305.
Silva, N.N., Bahri, A., Guyomarc'h, F., Beaucher, E., Gaucheron, F., 2015. AFM study of casein micelles cross-linked by genipin: effects of acid pH and citrate. Dairy Sci. Technol. 95 (1), 75–86.
Singh, H., Ye, A., 2009. Interactions and functionality of milk proteins in food emulsions. In: Thompson, A., Boland, M.J., Singh, H. (Eds.), Milk Proteins, From Expression to Food, first ed. Elsevier, Academic Press, Amsterdam. Taylor, S.L. (Series Editor), Food Sci. Tech. Int. Series, pp. 321–345.
Sneharani, A.H., Karakkat, V.J., Singh, S.A., Appu Rao, A.G., 2010. Interaction of curcumin with β-lactoglobulin—stability, spectroscopic analysis, and molecular modeling of the complex. J. Agric. Food Chem. 58 (20), 11130–11139.
Sripriyalakshmi, S., Jose, P., Ravindran, A., Anjali, C.H., 2014. Recent trends in drug delivery system using protein nanoparticles. Cell Biochem. Biophys. 70, 17–26.
Sugiarto, M., Singh, H., 2009. Characterisation of binding of iron to sodium caseinate and whey protein isolate. Food Chem. 114, 1007–1013.
Swaisgood, H.E., 1993. Review and update of casein chemistry 1, 2. J. Dairy Sci. 76 (10), 3054–3061.
Tavares, G.M., Croguennec, T., Carvalho, A.F., Bouhallab, S., 2014. Milk proteins as encapsulation devices and delivery vehicles: applications and trends. Trends Food Sci. Technol. 37, 5–20.
Trejo, R., Dokland, T., Jurat-Fuentes, J., Harte, F., 2011. Cryo-transmission electron tomography of native casein micelles from bovine milk. J. Dairy Sci. 94 (12), 5770–5775.
Tunick, M.H., Van Hekken, D.L., 2015. Dairy products and health: recent insights. J. Agric. Food Chem. 63 (43), 9381–9388.
Walker, R., Decker, E.A., McClements, D.J., 2015. Development of food-grade nanoemulsions and emulsions for delivery of omega-3 fatty acids: opportunities and obstacles in the food industry. Food Funct. 6 (1), 42–54.
Walstra, P., Geurts, T.J., Noomen, A., Jellema, A., Van Boekel, M.A.J.S., 1999. Dairy pechnology: principles of milk properties and processes. Food Sci. Technol., pp. 471–483.
Yan, Y., Hu, J., Yao, P., 2009. Effects of casein, ovalbumin, and dextran on the astringency of tea polyphenols determined by quartz crystal microbalance with dissipation. Langmuir 25, 397–402.
Yi, J., Lam, T.I., Yokoyama, W., Cheng, L.W., Zhong, F., 2015. Beta-carotene encapsulated in food protein nanoparticles reduces peroxyl radical oxidation in Caco-2 cells. Food Hydrocoll. 43, 31–40.
Yi, J., Lam, T.I., Yokoyama, W., Cheng, L.W., Zhong, F., 2014. Cellular uptake of beta-carotene from protein stabilized solid lipid nanoparticles prepared by homogenization–evaporation method. J. Agric. Food Chem. 62 (5), 1096–1104.
Zeev, M.B., Assaraf, Y.G., Livney, Y.D., 2016. β-Casein nanovehicles for oral delivery of chemotherapeutic drug combinations overcoming P-glycoprotein-mediated multidrug resistance in human gastric cancer cells. Oncotarget 7, 1–13.
Zhang, J., Liu, Y., Liu, X., Li, Y., Yin, X., Subirade, M., Zhou, P., Liang, L., 2014. The folic acid/β-casein complex: characteristics and physicochemical implications. Food Res. Int. 57, 162–167.
Zimet, P.D.R., Livney, Y.D., 2011. Re-assembled casein micelles and casein nanoparticles as nano-vehicles for omega-3 polyunsaturated fatty acids. Food Hydrocoll. 25 (5), 1270–1276.

Further Reading

Cerqueira, M.A., Pinheiro, A.C., Silva, H.D., Ramos, P.E., Azevedo, M.A., Flores-Lopez, M.L., Rivera, M.C., Bourbon, A.I., Ramos, O.L., Vicente, A.A., 2014. Design of bio-nanosystems for oral delivery of functional compounds. Food Eng. Rev. 6 (1–2), 1–19.

Dalgleish, D.G., 1998. Casein micelles as colloids: surface structures and stabilities. J. Dairy Sci. 81, 3013–3018.

Sood, S.M., Erickson, G., Slattery, C.W., 2003. κ-Casein interactions in the suspension of the two major calcium-sensitive human β-caseins. J. Dairy Sci. 86 (7), 2269–2275.

Yao, M., McClements, D.J., Xiao, H., 2015. Improving oral bioavailability of nutraceuticals by engineered nanoparticle-based delivery systems. Curr. Opin. Food Sci. 2, 14–19.

Yao, M., Xiao, H., McClements, D.J., 2014. Delivery of lipophilic bioactives: assembly, disassembly, and reassembly of lipid nanoparticles. Annu. Rev. Food Sci. Technol. 5, 53–81.

Nanocapsule formation by nanocrystals

6

Gregori B. Romero, Wolfgang Brysch**, Cornelia M. Keck†, Rainer H. Müller**
*Free University of Berlin, Berlin, Germany; **Athenion GmbH, Berlin, Germany;
†PharmaSol GmbH, Berlin, Germany

6.1 Introduction

To produce high quality, biologically effective food additives, nutraceuticals and functional/well-being drinks, the incorporated bioactives need to have a sufficiently high bioavailability (Jafari and McClements, 2017). "Sufficiently high bioavailability" means that a sufficient percentage of the incorporated bioactives that are absorbed into the blood stream from the gastrointestinal tract (GIT), yielding an improved blood profile that is indicative of health and well-being. However, many poorly soluble bioactives show a very low bioavailability (Faridi Esfanjani and Jafari, 2016; Katouzian and Jafari, 2016). A classical example is the broadly marketed coenzyme Q_{10}, which is contained in many oral products (tablets, capsules) and may also be incorporated into creams and cosmetics and topically applied. The oral bioavailability is typically very poor. Human studies showed that the fraction absorbed (percentage of drug in the blood out of the administered dose) in most oral formulations was about 1%–12% (Barakat et al., 2013). Considering the fact that the dose in nutraceuticals is rather low (e.g., compared to pharma products), a biological effect is questionable. The same applies to the various flavonoids, which are becoming increasingly popular as antioxidants, for example, rutin, apigenin, hesperidin, hesperetin, quercetin (Mohammadi, et al., 2016a,b).

In general, molecules are best absorbed when they are dissolved in the GIT. Bioactives can be poorly soluble in water, but can still be oil soluble (e.g., coenzyme Q_{10}). In some instances, actives are poorly soluble in both aqueous media and in oils/organic media (e.g., many flavonoids). Oil soluble bioactives can be dissolved in oils or solid lipids, and administered orally as an emulsion or in the form of a lipid based tablet. However, the surrounding water phase of the GIT fluids still represents an absorption barrier if the molecules have low water solubility. This partially explains the low bioavailability of Q_{10}. In the case of molecules with poor solubility in all media, one approach is the use of delivery technologies to increase the water solubility (saturation solubility, C_s), and optionally enhance absorption by various other mechanisms.

Technologies developed in pharma to deliver drugs more efficiently have been followed with keen interest by other sectors, such as food industry for many years. A particular focus over the past 25 years has been on nanotechnologies. The classical example for such very successful technology transfer are liposomes, first described in the mid-1960ies by Bangham (Bangham and Haydon, 1968; Bangham and Horne, 1964). Pharma developments take more time, thus liposomes first appeared on the cosmetic market in 1986 with the product Capture, launched by the company Dior (Diederichs

Nanoencapsulation Technologies for the Food and Nutraceutical Industries and Müller, 1994). At around 1990, liposomes entered the pharma market in the form of the pulmonary Alveofact (Schubert, 1998) and intravenous Doxil (Schubert, 1998), chiefly developed by Barenholz (Barenholz, 2012; Gabizon et al., 1994). Nanocrystals are a highly successful pharma nanotechnology that increases the oral bioavailability of poorly soluble compounds (Merisko-Liversidge, 2002; Müller et al., 2000; Rabinow, 2004). The first oral product (Rapamune) entered the market in 2000 (Müller and Junghanns, 2006). The first six products to enter the market have reached very high sales volumes. For example Tricor from Abbott is a blockbuster product with more than 1 billion USD annual sales in the US (Downing et al., 2012). About 20 products are estimated to be currently undergoing clinical trial (Rabinow, 2004).

Of course the nanocrystal technology can also be applied to other delivery routes, as for example dermal (Al Shaal et al., 2011; Müller et al., 2016), as well as to other markets, such as nutraceuticals. In 2007, the nanocrystal technology was successfully transferred to the cosmetic market (trade name: smartCrystals). The smartCrystals appeared first in the product line JUVEDICAL by the Juvena of Switzerland (Müller et al., 2013). The third-most expensive cosmetic on the market, Platinum Rare by La Prairie (Switzerland) also contains nanocrystals (hesperidin crystals, 50 g cream for about 1000 €) (weblink). In human studies, it could be shown that the nanocrystal technology could increase the biological effect by a factor of 1000 (Müller et al., 2013).

As a next step, the nanocrystal technology was transferred to the food/nutraceutical sector with the BioSmart Technology. Within this process, it was necessary to adapt the nanocrystals for use in these products, for example, regarding regulatory requirements (e.g., excipients) and their physical stability (e.g., preservation). These food-adapted nanocrystals are available as BioSmarts, to be admixed with various products. This chapter covers the regulatory aspects of these nanocrystals (i.e., the prerequisites for their use in marketed products), their properties, mechanism of action, and production.

6.2 Definitions of nanocrystals

Based on the regulatory legal definitions, nanocrystals can be "nano" or "not nano", depending whether their size is below or above 100 nm. This is very important for the marketing of products. In pharma, nanotechnology is broadly accepted by patients, especially when it is the only life-saving technology. In cosmetics, the advertisement of "nano" products has clear market advantages in many countries. In contrast to this, in the food sector there is an increasing awareness/concern about nano in products, similar to absence of GMOs (genetically modified organisms) (Baker and Burnham, 2001). Taking this into account, BioSmarts are sized above the legal nano definition. Special desired nanoproperties, such as increased saturation solubility are present when the particles are clearly below 1 μm, for example, 400 nm. They do not need to be below 100 nm in size. Thus, crystals in this size range are legally not nanoproducts, yet they still benefit from nanotechnology properties. For definition purposes it would be more appropriate to differentiate within "nanocrystals" between "submicron-crystals" and "real legal" nanocrystals, the former measuring ≤100 nm (Fig. 6.1) and the latter falling between 100 and 1,000 nm (<1 μm). In this chapter the pharmaceutical definition of nanocrystals is used, that is, a few nm to <1000 nm.

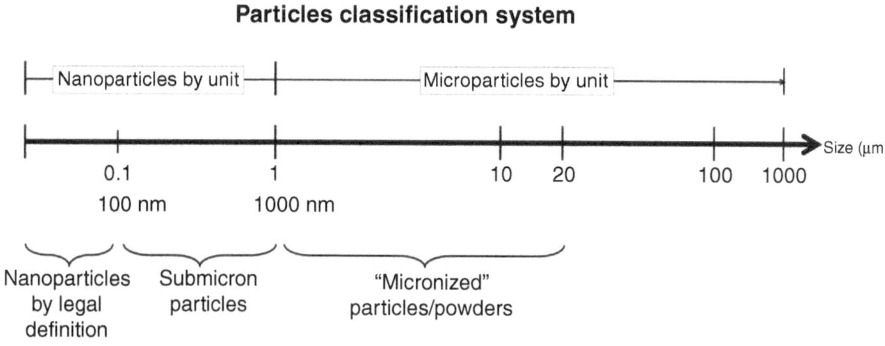

Figure 6.1 Proposed new classification scheme of nanoparticles to account for the legal definitions, and the change in physico–chemical properties below approximately 1 μm.

6.3 Special properties of nanocrystals

Transfer of materials into the nanodimension changes many properties. The relevant properties of nanocrystals to increase oral bioavailability are (Keck and Muller, 2008):

1. increase in saturation solubility C_s
2. increase in dissolution velocity *dc/dt* (*c*, concentration; *t*, time)
3. adhesiveness to surfaces (e.g., biological membranes)

The increase in saturation solubility is due to an increased dissolution pressure of the nanocrystals. In the equilibrium of saturation solubility, the number of molecules dissolving from crystals is equal to the number that recrystallizes. In lower nanodimension, the increased dissolution pressure shifts the equilibrium to the dissolved molecules. Physically, this can be explained by the Kelvin equation (Müller and Keck, 2012), for example, applied in spray-drying. The Kelvin equation describes the increase in vapor pressure of liquid droplets in a gas phase (curved surface) with decreasing droplet size (increasing curvature). When below about 1 μm, the vapor pressure increases with decreasing droplets size, which means more molecules transfer from the liquid phase to the gas phase (Fig. 6.2, left). The same situation applies to dissolution of crystals. In the transfer from solid to liquid phase, vapor pressure is replaced by dissolution pressure (Fig. 6.2, right).

The dissolution velocity of particles is described in the well-known Noyes–Whitney equation (Noyes and Whitney, 1897) (Eq. 6.1):

$$\frac{dc}{dt} = A\frac{D}{d}(C_s - C_b) \tag{6.1}$$

dc/dt, dissolution velocity (mg/s)
A, surface area (cm^2)
D, diffusion coefficient (cm^2/s)
d, thickness of diffusion layer (cm)
C_s, saturation solubility (mg/cm^3)
C_b, bulk concentration (mg/cm^3)

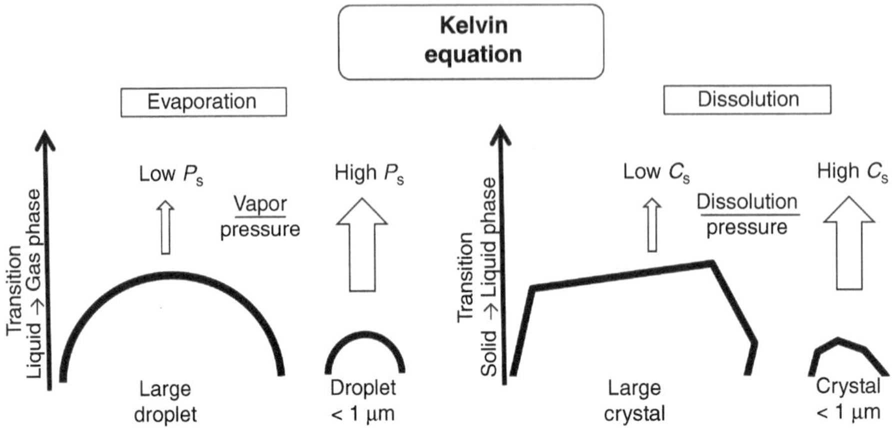

Figure 6.2 Based on the Kelvin equation, the vapor pressure of liquid droplets increases with increasing curvature in a gas phase, that is, decreasing size (left), corresponding to the transition of molecules from solid crystals to a surrounding liquid phase (=dissolution) (right), the vapor pressure is replaced by the dissolution pressure. (P_s, saturation vapor pressure; C_s, saturation solubility).

In Eq. (6.1) the dissolution velocity is proportional to the surface area A, the surface area increases by an order of magnitude when moving, for example, from a 50 μm crystal to a 5 μm crystal (micronized powder) and finally to a 500 nm nanocrystal (Fig. 6.3). The equation also shows that dc/dt is proportional to the difference C_s–C_b, that is, the velocity increases also by the increase in the saturation solubility C_s of the nanocrystals. Thus, two factors increase dc/dt increased A and increased C_s.

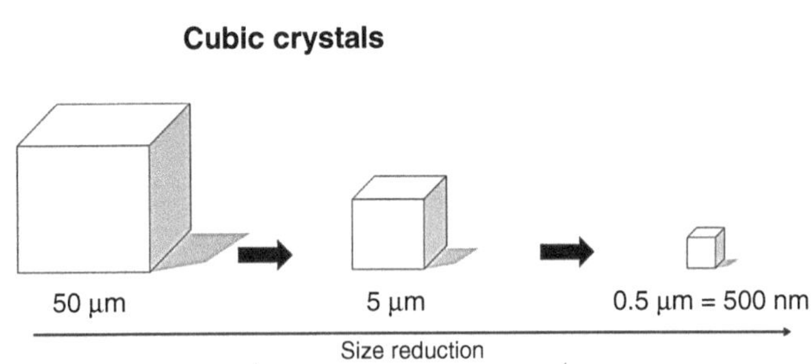

	50 μm	5 μm	0.5 μm = 500 nm
Factor reduction	—	× 10	× 100
Number of cubes	1	1,000	1,000,000
Surface area	15,000 μm^2	150,000 μm^2	1,500,000 μm^2
Factor increase	—	10	100

Figure 6.3 Increase in surface area when moving from a 50 to a 5 μm microcrystal, and finally to a 500 nm nanocrystal.

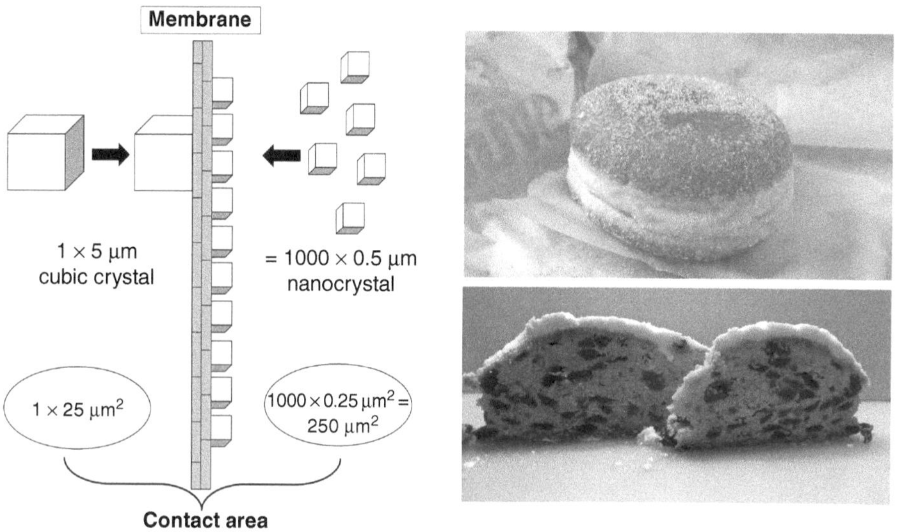

Figure 6.4 Contact surface between one large particle compared to many small particles of same total mass (left), and a bakery example showing less well sticking crystalline sugar on "Berliner Pfannkuchen" (upper) versus iced sugar on Christmas cake from Dresden/East Germany (so called "Dresdner Christmas Stollen") (lower).

The adhesiveness of particles to surfaces increases with decreasing particle size. This can be explained by the increase in contact area between the particles and the respective surface (Fig. 6.4, left) (Stieß, 2013). From food technology it is well known that crystalline sugar sticks less well to a bakery product than iced sugar does, as shown in traditional German baked goods (Fig. 6.4, right).

6.4 Mechanisms of absorption enhancement

In pharmaceuticals, the bioavailability of a solution is of great importance, and other formulations (e.g., tablets) are compared to a solution formulation for determining their "relative bioavailability". Based on this, it is desirable to dissolve, as much as possible from the poorly soluble active. By increasing the saturation solubility C_s using nanocrystals, as opposed to a micrometer-sized crystal, the concentration gradient increases, and consequently the passive diffusional flux does as well (Fig. 6.5). This principle is exploited for class II drugs of the biopharmaceutical classification system (BCS) (Amidon et al., 1995) (low solubility, good permeability through membranes). The principle can also be employed for drugs of the BCS class IV (low solubility, poor permeability), whereby the membrane is "flooded" with a surplus of drug to increase absorption. The same principle can be applied to bioactives in food, functional drinks, and nutraceuticals that show the same absorption obstacles.

In case that a molecule is very well absorbed, the dissolution velocity from the undissolved crystals in the GIT can be the limiting step. In such cases, an increase in dissolution velocity of the crystals provides a solution. This can be achieved simply

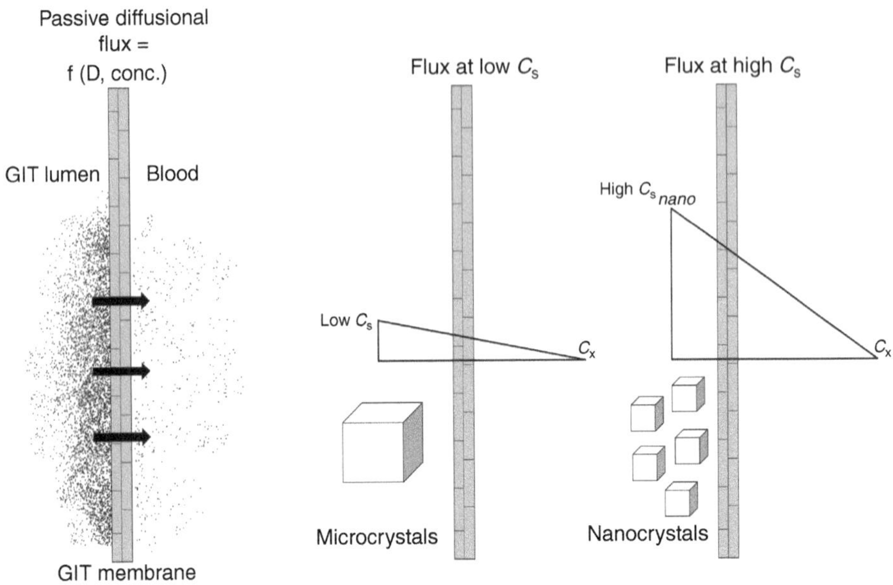

Figure 6.5 Increased concentration gradient generated by nanocrystals increase the passive diffusion through the gastrointestinal tract (GIT) membrane. (*D*, diffusion coefficient; *conc.*, concentration; C_s, saturation solubility; C_{snano}, saturation solubility of nanocrystals; C_X, concentration in the blood)

by enlarging the surface area—as done in the nanocrystal technology (Fig. 6.6). Of course, in the case of highly hydrophobic actives (high contact angle) that have only been partially exposed to GIT fluids, the submersion, and subsequent dissolution of the actives can be further enhanced by additional fluid agents. In general, it is preferable to wetten the actives with GIT fluids than with pure water, as the bile salts present reduce the surface tension. The stabilizers used in the production of nanocrystals are surface active, and thus reduce further interfacial tension between crystal surface and GIT fluids.

The adhesion of nanocrystalsto membrane surfaces increases the drug permeation through these membranes. As nanocrystals move randomly around in the gastrointestinal content, there is a statistically probability that they will come in contact with the GIT wall and adhere to it. This effect increases with the decrease in particle size, mucoadhesion of smaller particles is higher (Ponchel et al., 1997). In the next step the crystals continue to dissolve exactly at the desired site of absorption, being surrounded by a saturated layer of dissolved molecules directly in contact with the absorbing membrane (Fig. 6.7).

This effect is also very reproducible. Many drugs are known for variable oral bioavailability, meaning that their bioavailability varies between administrations, even when conditions remain constant (e.g., in nonfed state). Some drugs show even stronger variations when administered in nonfed and fed state. For danazol nanocrystals it could be shown that the variation in bioavailability was reduced (Liversidge and

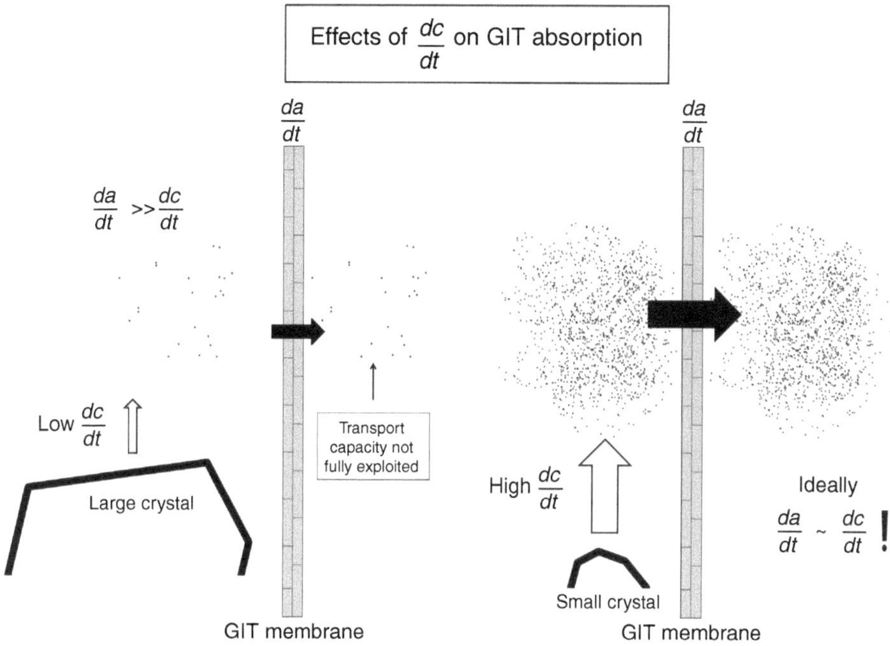

Figure 6.6 Effect of increased dissolution velocity on absorption from the GIT. Left: low dissolution velocity of large microcrystals; Right: high dissolution velocity of small crystals due to large total surface area (*dc/dt*, dissolution velocity; *da/dt*, absorption velocity across GIT membrane).

Cundy, 1995). For the drug fenofibrate, formulation as nanocrystal (product: Tricor) showed a clear reduction in bioavailability as function of food intake (Junghanns and Müller, 2008). The movement of crystals in the GIT content is much faster when ingested in a viscous form as opposed to a liquid form. The low viscosity enables faster movement, better attachment and subsequent higher absorption. A liquid pharmaceutical suspension product currently on the pharma market is Megace ES, which also shows good physical stability over its shelf life. This proves the make-ability of long-term stable nanocrystals in liquids, for example, in functional drinks.

6.5 Encapsulated (coated) nanocrystals

In case of nanocrystals of bioactives which are prone to degradation due to environmental/processing conditions, for example, pH, oxygen, light, temperature, they can be coated by a protective polymer layer. This is also applicable for nanocrystals of bioactives which are sensible to the different conditions in the GI tract after ingestion, for example, degradation in the acidic environment of the stomach.

Nanocrystals start dissolving immediately when brought in contact with unsaturated fluids. This can happen after ingestion or even during the manufacturing of the

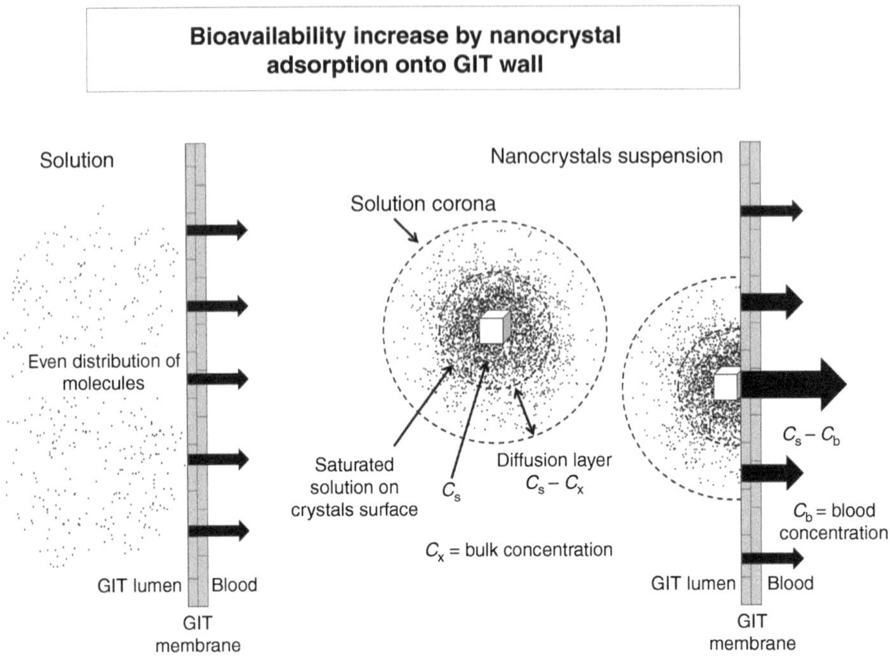

Figure 6.7 Mechanism of bioavailability increase by adsorption of nanocrystals onto GIT membrane: nanocrystals with surrounding highly saturated solution corona adsorb exactly at the place of absorption (=membrane), creating locally a high concentration gradient ($C_s - C_b$).

final food/nutraceutic product. However, it is not always desirable that this happens. There are two main reasons for that:

1. The place of dissolution leads to decomposition of the dissolved active, for example, acid pH of a beverage or stomach.
2. The nanocrystal has not yet reached its desired site of absorption or action on the GIT.

For example, for nanocrystals for food/nutraceutics products, the respective site of absorption in the GIT has not yet been reached. Dissolved nanocrystals cannot adhere any more to the mucosa of the absorption site, creating a high concentration gradient. Too early dissolution will lead to a distribution of dissolved molecules across the content of the GIT, the reduction in the concentration gradient will lead subsequently to a reduced bioavailability. Also if the bioactive is absorbed in the intestine, degradation in the stomach will reduce bioavailability. In case of nanocrystals in food products, dissolving already before ingestion, for example, functional drinks, should be avoided, as much as possible and delayed until the nanocrystals have reached the GIT.

The solution for this is coating the nanocrystals with polymers. A full range of various coating polymers is available, for example, well known from pharma polymethacrylate copolymers (Eudragit series, Evonik). Polylactides or poly(lactic-*co*-glycolic) acid copolymers are also an option. It appears also feasible to use water swellable polymers (hydrocolloids) which generate a viscous coat after getting in contact with

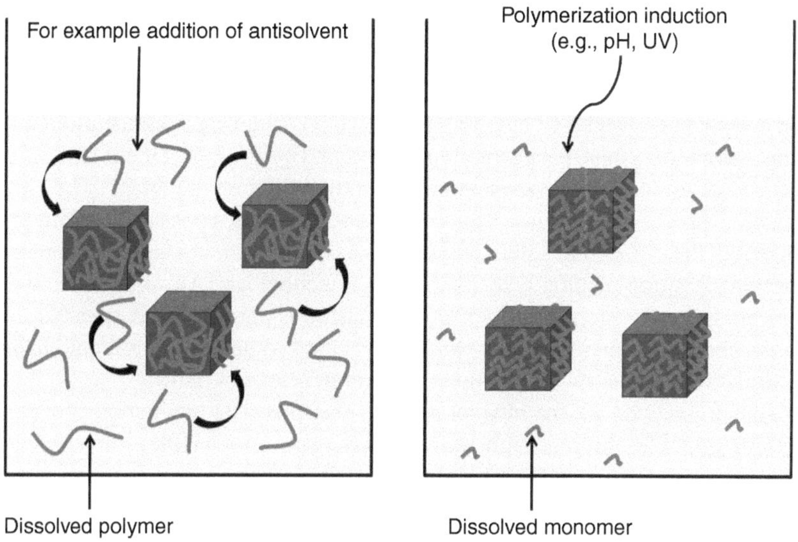

Figure 6.8 Coating of nanocrystals performed by polymers via solubility reduction (e.g., coacervation), or in a monomer solution forming the coating after initiation of the polymerization process.

water, slowing down the nanocrystal dissolution process. Normally these are all preformed, but it is also possible to generate coatings from the monomers (analogous to producing latex particles). The monomers are dissolved in the water phase of the nanosuspension, and coating is performed after initiation of the polymerization process. Fig. 6.8 shows the two possible principles. Polymers for classical coating starting from monomers are in pharma, for example, the cyanoacrylates (Damge et al., 1997). Coating with polymers can also change their surface properties, from hydrophobic to hydrophilic, when using amphiphilic coatings (Pellegrino et al., 2004).

The coating of single nanocrystal is often a more tedious and costly process. A smarter and more cost effective solution is the encapsulation in larger units, more or less "containers" for nanocrystals—the "container approach". Such containers are typically pellets. These are useful for example to protect the nanocrystals until they are incorporated into the final composition of the product being produced or, in case of nutraceuticals, they can be loaded into, for example, hard gelatin capsules. The nanocrystals are admixed to an aqueous extrusion mass and the pellets are coated with a polymer dissolving in specific conditions, for example, acid pH of the stomach or more alkaline pH in the intestine. Differently coated pellets can also be mixed to generate continuous release over a longer part of the GIT. Eudragits are again polymers of first choice in many cases. In case of nutraceutics, the largest "container" is a tablet, which shows the appropriate coating, for example, enteric coating. A more simple solution is using multilayered tablets, with layers dissolving in different parts or velocities, for example, fast dissolution (equal to burst release for, fast onset of action) and then followed by slow release.

6.6 Lab scale and industrial scale production of nanocrystals

Nanocrystals can be produced by bottom–up methods (starting from the molecule and growing it to nanocrystals through precipitation), and by top–down methods (by reducing the size of μm crystals to the nanodimension) (Müller et al., 2011). The classical precipitation process, also known as via humida paratum, consists of addition of a nonsolvent to the compound solution to obtain amorphous or crystalline nanoparticles. Another precipitation method is the NanoMorph technology (Auweter et al., 2002) to produce amorphous nanopaticles. An O/W emulsion is produced and after lyophilization, amorphous nanoparticles are obtained. It is used to produce carotenoid nanoparticles for the food industry, for example, Locarotin, Lucantin (BASF).

Top–down methods are the mostrelevant for the industrial production of nanocrystals, as through bead milling, high pressure homogenization (HPH), the combination of these processes, the smartCrystal technology. All of these are wet milling processes.

The bead milling process was used by the company Nanosystems (later ėlan, now the technology is owned by Alkermes) for their NanoCrystal technology in 1991 (Liversidge et al., 1991). The powder is dispersed in a stabilizer solution, and a macrosuspension produced. This suspension is passed several times through a bead mill. The milling chamber is filled with fine milling pearls, typical sizes are between 0.2 and 0.5 mm of diameter. The crystals are ground between the moving beads, and a nanosuspension is obtained (=suspension of nanocrystals in a liquid phase). The bead milling technology is an established process in many different industrial sectors (e.g., paint industry). Mills are available on lab scale (e.g., 200 mL milling chamber volume), but also on medium scale (e.g., 1 L volume), and also in form of large industrial towers. The mills can be run continuously in a loop process, that is, even by using a 1 L milling chamber, batches of 50 kg and more can be produced, for example, with the PML 2 from Bühler (Switzerland) (Fig. 6.9).

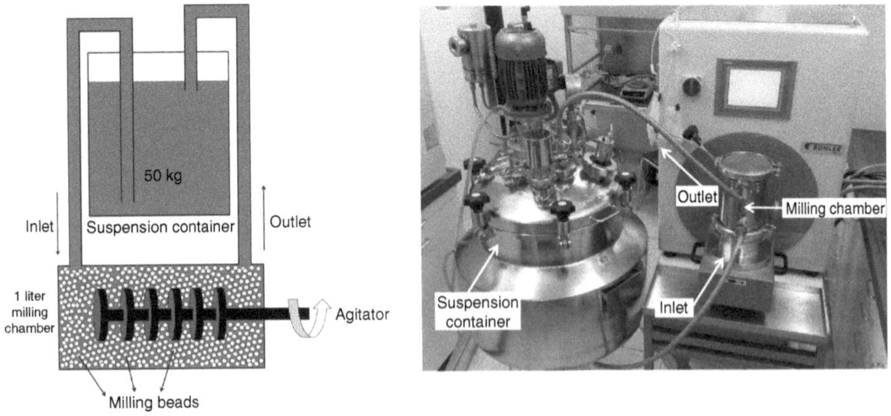

Figure 6.9 Production principle of loop process in bead milling (left) and picture of PML 2 Bühler bead mill with product container (right).

HPH was invented by Auguste Gaulin, who presented the first homogenizer at the World Exhibition in Paris in 1900. HPH is a widely used technique, employed in the food industry and in the production of emulsions (Chapter 2) and creams. The process is based on passing a coarse preemulsion (μm-sized) at a high velocity through a very tiny gap (e.g., 5 μm) at high pressures of typically 200–600 bar (2×10^4–6×10^4 kPa). According to the law of Bernoulli, the dynamic pressure on the liquid falls below its vapor pressure, and as a result the water starts boiling and gas bubbles are formed. This generates shock waves which disrupt oil droplets and lead to droplet diminution. When leaving the gap, the pressure on the fluid increases to normal atmospheric pressure (101 kPa), the water stops boiling and the gas bubbles collapse, again generating shock waves which contribute to the diminution (=cavitation). This principle was transferred from emulsions to suspensions (Müller et al., 1996). It may also be shown that a suspension can be effectively diminished to the nanosize if it is passed 5–20 times through a high pressure homogenizer, typical pressure of 1500 bar (15×10^4 kPa) (i.e., clearly higher than typically used for emulsions). The technology was introduced to the market in the 1990s as DissoCubes by PharmaSol GmbH Germany (Müller et al., 1999, 2002).

Subsequently several "combination technologies" have been developed, which typically include two subsequent steps, for details see (Salazar et al., 2014). As example of this is the combination of precipitation and high energy input (e.g., HPH), as in the NANOEDGE technology by Baxter Healthcare US (Kipp et al., 2003). PharmSol GmbH Germany developed a family of technologies to produce nanocrystals for various special applications (smartCrystals family, now owned by Abbvie, US). The technologies H42 [spray-drying and subsequent HPH (Möschwitzer, 2005)], H69 [precipitation and HPH (Müller and Möschwitzer, 2015)], and H96 [lyophilisation and HPH (Möschwitzer and Lemke, 2006)] allow the production of nanocrystals below 100 nm, which provide important models for intravenous injection of pharma products. The smartCrystal CT technology which combines bead milling and subsequent HPH (Fig. 6.10) leads to nanocrystals that have a higher physical stability (Petersen, 2006).

This is valuable for products containing destabilizing excipients, for example, food products and functional drinks with electrolytes and/or preservatives. This CT technology is used for the production of crystals in the BioSmarts Technology. A summary of nanocrystals production techniques is shown on Fig. 6.11.

The availability of contract manufacturers and suppliers is highly important when placing the final product on the market. BioSmarts technology nanocrystal suspensions can be obtained from Athenion GmbH (Berlin, Germany). They are sold as concentrates (5 or 10%) to be admixed in the production process of nutraceuticals and drinks.

6.7 Nanocrystals in functional drinks

There are a number of drinks currently on the market containing a variety of nutrients and bioactive compounds, especially antioxidants and plant extracts (Corbo et al., 2014). Of interest for functional drinks, poorly soluble bioactives, such as curcumin, rutin, and quercetin could be cited. However, the critical question remains: what percent of the compounds present in nutritional drinks are ultimately traced in the blood?

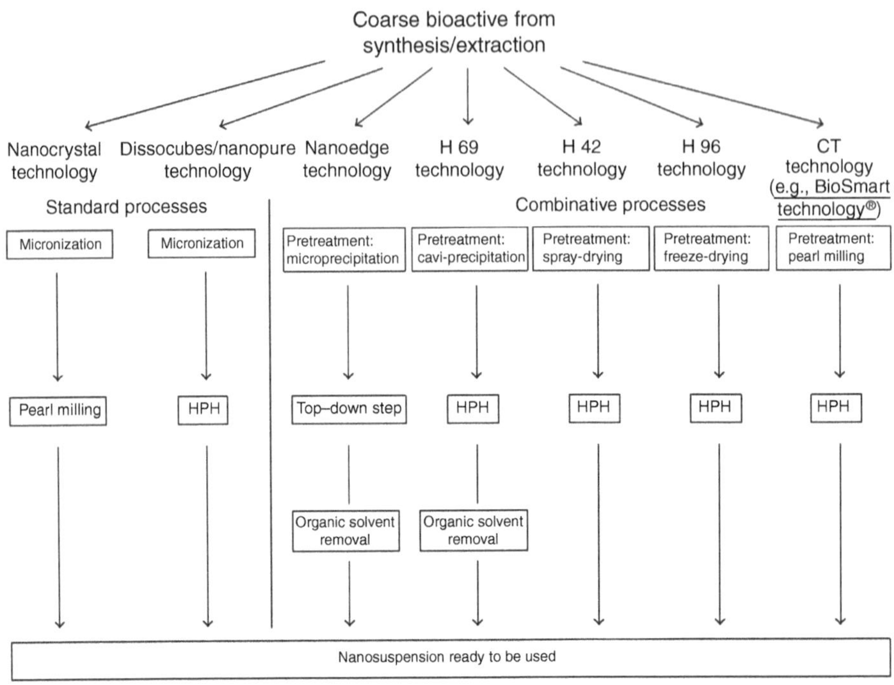

Figure 6.10 Production scheme of BioSmarts technology: in the first step, the suspension is passed 2–5 times through a bead mill (left), yielding a nanosuspension with still large crystals/aggregates present (upper right, shoulder 1–10 μm), which is then processed by HPH yielding a more uniform nanosuspension (lower right).

This is even more critical when considering the small quantities typically used in such drinks. Higher quantities might be affordable from the cost side, but lead to turbidity in drinks. In the case these compounds are poorly soluble; they must be suspended in the drinks, that is, creating a suspension. In certain Asian countries, however, an optically clear appearance might be preferred. This explains why, for example, coenzyme Q_{10} is solubilized in certain Asian products to obtain a clear solution.

In case of nanocrystals, a low concentration of a bioactive can still be used because the bioavailability will be distinctly higher. The claimed effects are more likely when considering the bioactivity increase of up to a factor 1000 as observed in dermal application in cosmetics. Due to the small size of the nanocrystals it is possible to increase the concentration of a bioactive with minimal effects on turbidity. Also undesired sedimentation effects—as they occur with micrometer-sized particles are less pronounced or can be avoided in case of very small nanocrystals. Fig. 6.12 summarizes the effects of nanocrystals in functional drinks.

To maintain their bioavailability enhancing effect, the nanocrystals need to be physically stable in the drinks, that is, they should not aggregate to micrometer-sized aggregates. Nanocrystals in functional drinks are from the technical point a "suspension", and as such, they are subjected to destabilizing effects typical within a suspension. The nanocrystals are stabilized by high zeta potential (mV) absolute values,

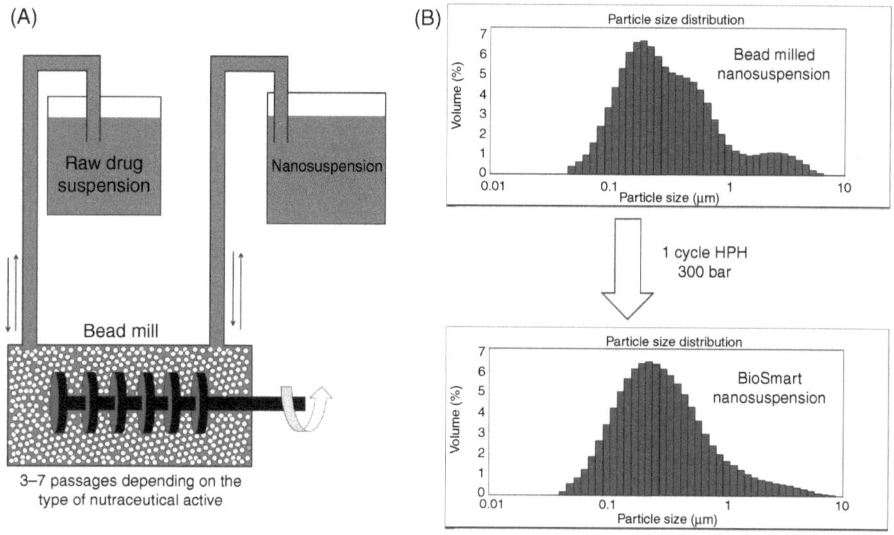

Figure 6.11 Schematic description of different nanocrystal production techniques. The CT technology (first on the right) is employed for the production of nanocrystals for the food/nutraceutic industries (BioSmart technology).
Source: Modified after Salazar, J., Müller, R.H., Möschwitzer, J.P., 2014. Combinative particle size reduction technologies for the production of drug nanocrystals. J. Pharma. 2014, 14.

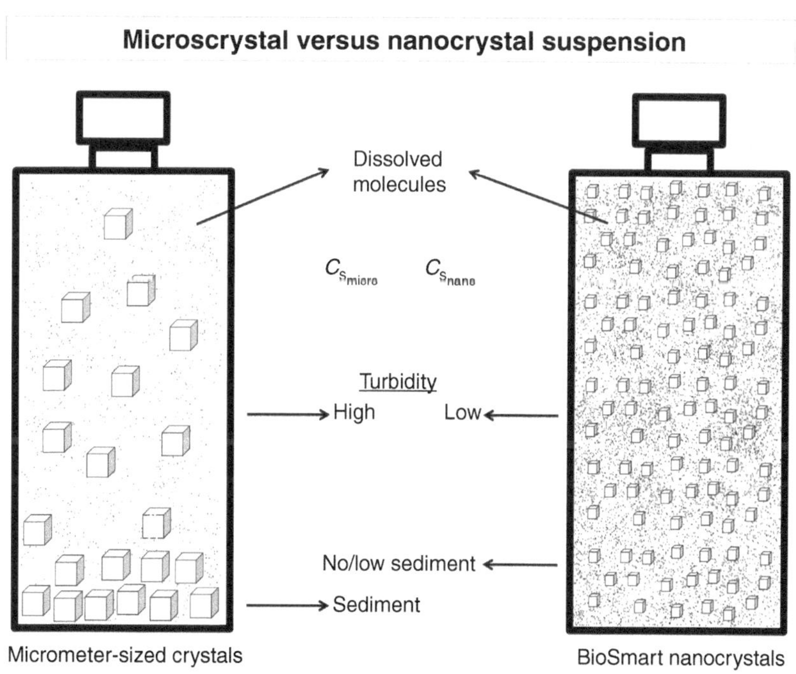

Figure 6.12 Effects of nanocrystals on the performance of functional drinks.

whereby equally charged particles repel each other and stay separated. In addition, they are stabilized by the surrounding stabilizer layer. Destabilization is mainly caused by electrolytes, preservatives (in case they are electrolytes) and osmotic/dehydrating effects of additives. Electrolytes reduce the zeta potential, thus the electrostatic repulsion between crystals, and promote aggregation. Dehydrating effects are caused, for example, by the addition of ethanol. The ethanol molecules are hydrated by the bulk water, and thus dehydrate the stabilizer layer making it less effective. Therefore, it is important to use nanocrystals with sufficient physical stability—as provided by the special production processes and selected stabilizers of the BioSmarts technology—and preferably steric stabilizers of low electrolyte susceptibility, for example, cellulose derivatives, polysorbates, arabic gum, xantan gum. In addition, the excipients used in the final products need to be optimized.

To produce functional drinks with nanocrystals, nanocrystal concentrates (5%–10%) are simply admixed in the final production step. For reasons of microbiological safety, the concentrates should be preserved using food approved preservatives or compounds with preservative action, but not listed in the list of preservatives. In this case, the nanocrystal concentrates are legally "preservative-free". The agents with preservative action should be selected in a screening process to make sure that they are not affecting the physical stability. Analysis was performed by photon correlation spectroscopy (PCS) and light microscopy of the concentrates (without dilution). Fig. 6.13 (upper) shows the PCS size distribution of unpreserved nanosuspension, and after addition of 20% glycerol for preservation, no change in the size distribution. Fig. 6.13 (lower) shows light microscopy pictures of unpreserved nanosuspension

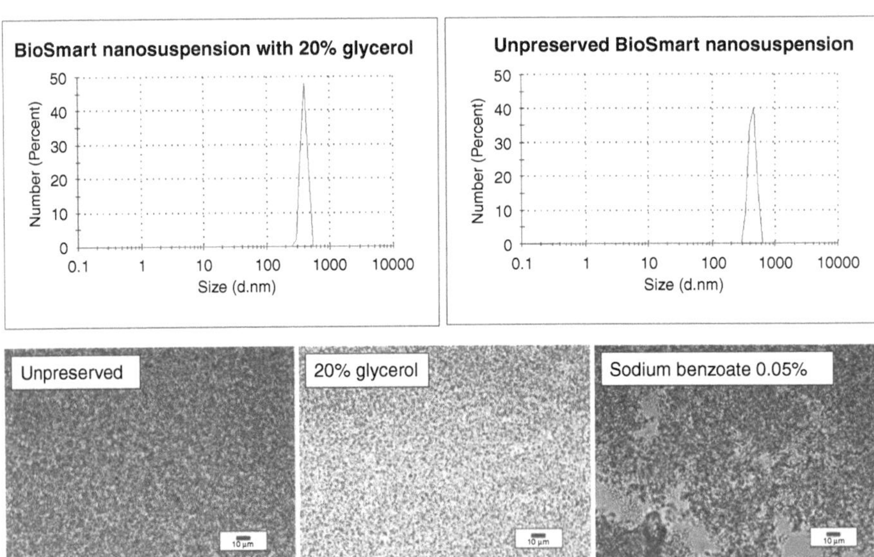

Figure 6.13 Upper: PCS size distributions of unpreserved rutin BioSmart nanosuspension (right), and after addition of 20% glycerol (=stable suspension) (left). Lower: Light microscopy pictures of unpreserved rutin BioSmart nanosuspension, after addition of 20% glycerol (=stable) and destabilizing preservative sodium benzoate 0.5% (from left to right).

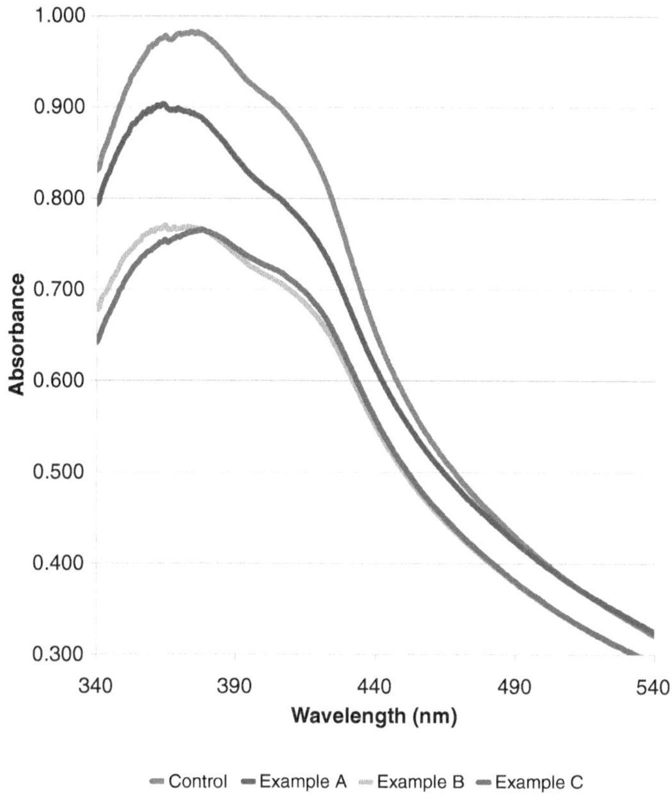

Figure 6.14 UV/VIS scan of three examples of nanosuspensions, example A being stable, examples B and C of equal instability.

concentrate, concentrate preserved with glycerol, and exemplarily preserved with the destabilizing preservative sodium benzoate 0.5%.

The stability of nanocrystals in functional drinks should be assessed before placing a product in the market. This can be done by PCS measurements, or alternatively by simple UV/VIS measurements. In the case that the particles initially aggregate, they will have increased absorption potential. Continuing further aggregation leads to a reduction in absorbance—a few large particles let the light pass better than many small particles, less turbidity in case of large aggregates (Schumann, 1995). That means that in the case of no or little change in absorbance, the drink products are stable. Fig. 6.14 shows a UV/VIS scan of nanosuspensions examples where destabilization can be identified.

6.8 Nanocrystal technology in oral nutraceutical products

The transfer of nanocrystals to nutraceutical products in form of tablets and capsules is very straight forward. All the principles from the pharma sector can be applied, and it makes no difference whether this is done by processing a nutraceutical active or a

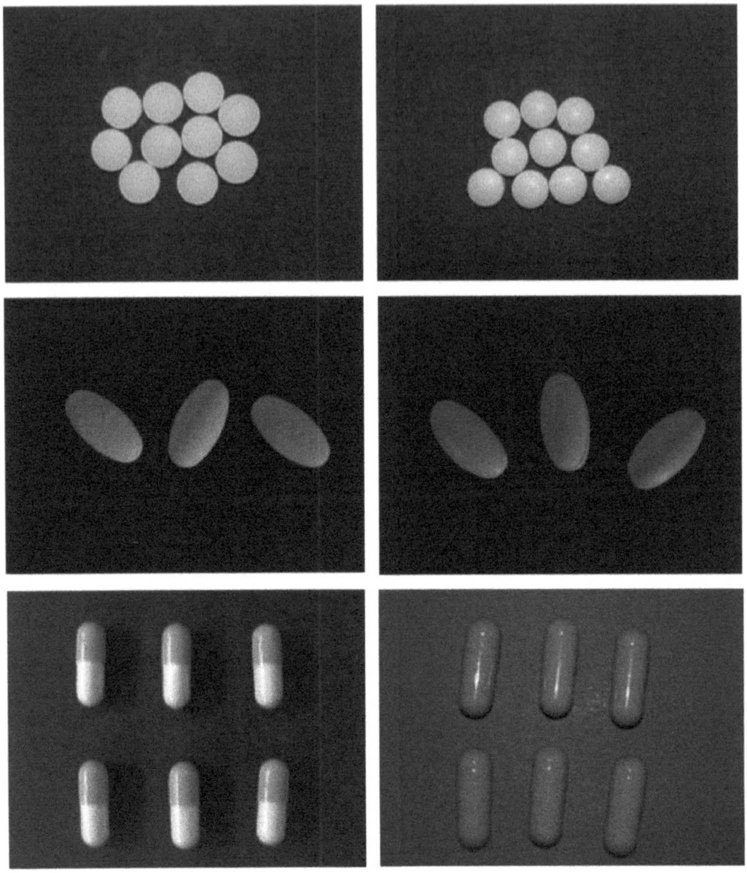

Figure 6.15 Upper: rutinnanocrystal-loaded tablets (left) and marketed tablets (right) middle: hesperidin nanocrystal-loaded tablets (left) and marketed tablets (right). Lower: coenzyme Q_{10} nanocrystal loaded capsules (left) and marketed capsules (right).
Source: With permission from Mauludin, R., 2009. Nanosuspensions of poorly soluble drugs for oral administration. PhD thesis, Freie Universität Berlin, Germany.

drug as a nanocrystal tablet or capsule. The importance of an optimized formulation can be nicely demonstrated when determining the in vitro release profile and the saturation solubility with the paddle method according to the US pharmacopeia (USP). Investigated were rutin and hesperidin tablets, and coenzyme Q_{10} capsules (Fig. 6.15).

A clear superiority was found for rutin nanocrystal tablets versus a marketed product (Fig. 6.16). About 100% rutin dissolved from the nanocrystal tablets with 10–15 min, but only about 50% of the marketed tablet after 30 min. This data shows the potential of the nanocrystal technology to improve the bioavailability of nutraceutical products. For production, the nanocrystal concentrates are added to the granulation fluid, and tablets are produced via granulation technology. Alternatively, the concentrates can be mixed with the extrusion mass for producing pellets, or added to the spraying solution in case of using the Pelletier process, or spraying onto nonpareils.

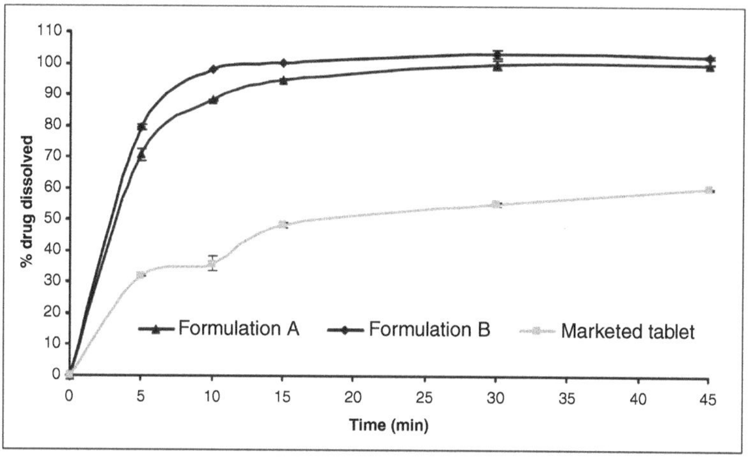

Figure 6.16 Dissolution profiles of rutin nanocrystal tablet formulations A and B versus marketed tablet in water, USP paddle method, 25°C.
Source: With permission from Mauludin, R., 2009. Nanosuspensions of poorly soluble drugs for oral administration. PhD thesis, Freie Universität Berlin, Germany.

Fig. 6.17-upper shows approximately 3 times higher saturation solubility for hesperidin nanocrystals compared to the raw drug powder. Very important for the oral bioavailability is the very fast increase in solubility within the first 0.5–1 h of dissolution. The active is rapidly available for absorption, and active absorbed will be immediately replaced from the fast dissolving active of the nanocrystals.

Further, two hesperidin nanocrystal tablet formulations *X* and *Y* were compared to a marketed tablet regarding the dissolution velocity. Even after half an hour, only about 1%–2% is dissolved from the marketed tablet, whereas about 20% dissolved from the nanocrystals formulations after only 10 min (Fig. 6.17-lower). The same picture was obtained when comparing a capsule filled with coenzyme Q_{10} nanocrystals versus Q_{10} microcrystals (Fig. 6.18). This might explain the poor oral bioavailability of many nutraceutical coenzyme Q_{10} products.

6.9 Nanocrystal technology in food products

To our knowledge, nanocrystals have not yet been directly admixed in food products. This will be a completely new commercial sector, opening a lot of opportunities. Of course, food naturally contains many different compounds that might impair the physical stability of the nanocrystals. This ranges from electrolytes to polymers/natural macromolecules, which are often in concentrations that cause the bridging of crystals. On the other hand, most food products possess a relatively high viscosity compared to fluids (e.g., functional drinks). The viscosity slows down the diffusional velocity of the nanocrystals (lower diffusion coefficient *D*), thus the nanocrystals have less kinetic energy to overcome the repulsive electrostatic barrier due to the particle charge (zeta potential). The viscosity of food products can rather be compared to the viscosity

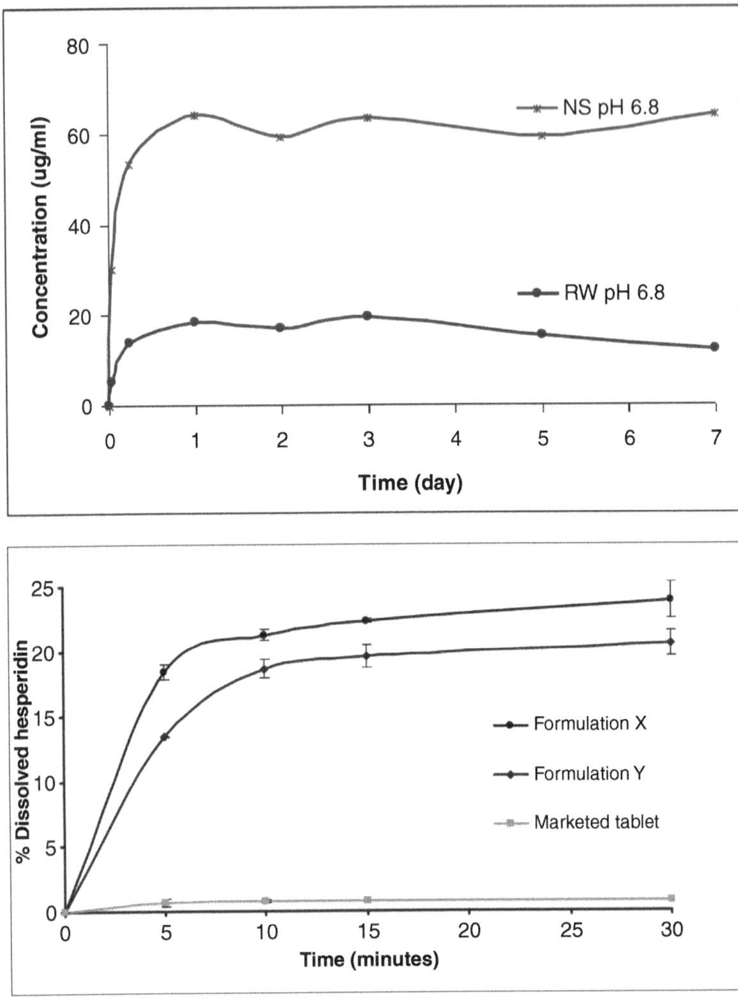

Figure 6.17 Upper: saturation solubility of hesperidin as a function of time, nanosuspension (NS) versus raw drug powder (RW) at pH 6.8 in water, at 25°C. Lower: Dissolution profiles of hesperidin nanocrystal tablets X and Y in comparison to a marketed tablet in water, USP paddle method.
Source: With permission from Mauludin, R., 2009. Nanosuspensions of poorly soluble drugs for oral administration. PhD thesis, Freie Universität Berlin, Germany.; With permission from Mauludin, R., Müller, R.H., 2013. Physicochemical properties of heperidin nanocrystals. Int. J. Pharm. Pharm. Sci. 5, 954–960.

of cosmetic or pharma dermal gels. It is known that nanoparticles can be stabilized through their incorporation with more viscous gels, as reported for solid lipid nanoparticles (SLN) (Lippacher, 2000). Ideal are products like yoghurts and smoothies.

Of course for quality assurance, the stability of the nanocrystals in food products should be monitored, which represents a challenge in this complex environment. The same analytics can be applied as in gels, that is, labeling the crystals fluorescently

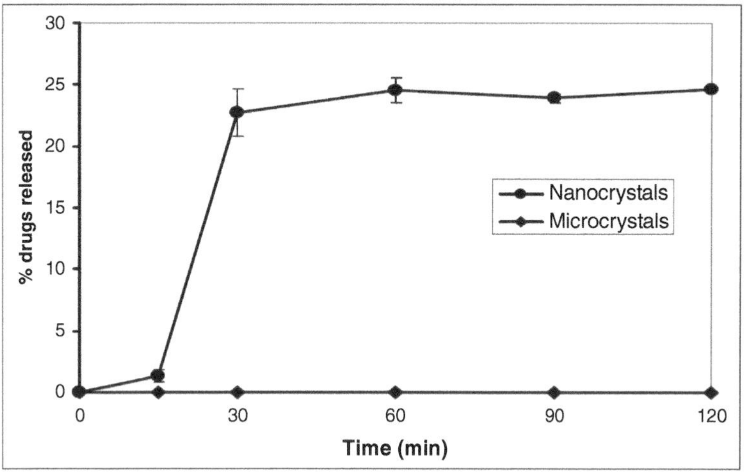

Figure 6.18 Dissolution profiles of capsules filled with coenzyme Q_{10} nanocrystal versus microcrystals in water, USP paddle method, PCS diameter of nanocrystal was 263 nm, laser diffraction diameter 50% of microcrystals was 111 μm.
Source: With permission from Mauludin, R., 2009. Nanosuspensions of poorly soluble drugs for oral administration. PhD thesis, Freie Universität Berlin, Germany.

and monitoring their even distribution in the food matrix by fluorescence microscopy. Through 1000-fold magnification and oil immersion, it is possible to screen for crystals around >500 nm. It is particularly easy to detect potential undesired aggregates of 1 μm and larger. The samples are screened undiluted. Fig. 6.19 shows the even distribution of fluorescent curcumin nanocrystals in a hydroxypropyl cellulose (HPC) gel as model system. The same analytics can be applied to food to assure a high quality (i.e., stable, nonaggregated nanocrystals).

6.10 Conclusions and perspectives

Nanocrystals have been successfully introduced to the pharmaceutical market, and considering sales per product, they belong to the most successful pharmaceutical nanotechnologies. Their approval as pharma products, a strictly regulated process overseen by national authorities, guarantees that they are (1) a functioning, effective technology, and (2) a safe technology.

Nanocrystal technology was shown to distinctly increase the oral bioavailability of poorly soluble compounds after oral administration, and also to make bioavailability food-independent and very reproducible. These findings are of great interest for developers of top quality nutraceutical products, functional drinks, and future "high performance" designer food.

A prerequisite for the use of nanocrystal technology in this new sector is the availability of physically stable nanocrystals, because in case of aggregation they would lose their special, outstanding performance. Such stable nanocrystals were realized by BioSmart technology, which provides nanocrystal concentrates, also available

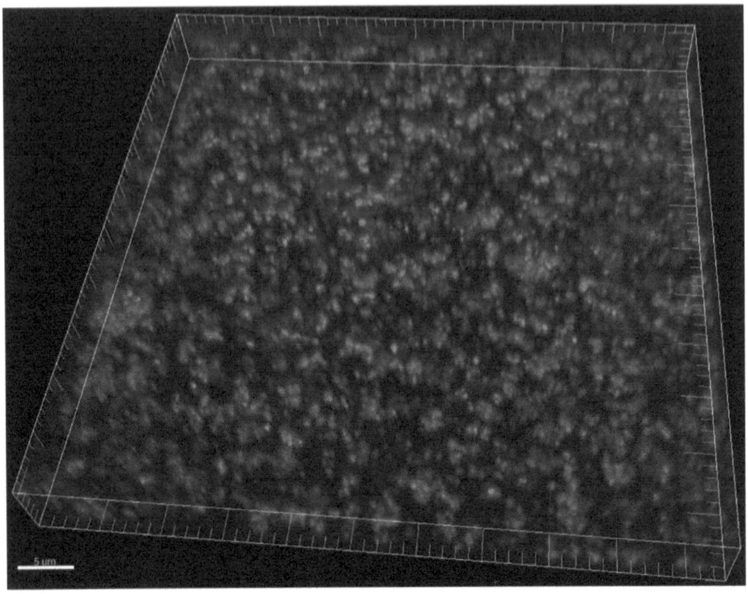

Figure 6.19 Even distribution of fluorescent, physically stable curcumin nanocrystals in a viscous gel matrix, similar viscosity to many food products.

preserved. Following the successful introduction of nanocrystal technology to the cosmetic/consumer care market in the creation of top price class products (e.g., platinum rare), the sector of nutraceuticals, functional drinks and designer food products will be the next target market. Let us have our "biosmart yoghurt" or "biosmart health drink"!

References

Al Shaal, L., Shegokar, R., Muller, R.H., 2011. Production and characterization of antioxidant apigenin nanocrystals as a novel UV skin protective formulation. Int. J. Pharma. 420, 133–140.

Amidon, G.L., Lennernas, H., Shah, V.P., Crison, J.R., 1995. A theoretical basis for a biopharmaceutic drug classification: the correlation of in vitro drug product dissolution and in vivo bioavailability. Pharma. Res. 12, 413–420.

Auweter, H., Bohn, H., Heger, R., Horn, D., Siegel, B., Siemensmeyer, K., 2002. US Patent 6,494,924. Precipitated water-insoluble colorants in colloid disperse form.

Baker, G.A., Burnham, T.A., 2001. The market for genetically modified foods: consumer characteristics and policy implications. Int. Food Agribus. Man. Rev. 4, 351–360.

Bangham, A.D., Haydon, D.A., 1968. Ultrastructure of membranes: biomolecular organization. Brit. Med. Bull. 24, 124–126.

Bangham, A.D., Horne, R., 1964. Negative staining of phospholipids and their structural modification by surface-active agents as observed in the electron microscope. J. Mol. Biol. 8, 660IN2–668IN10.

Barakat, A., Shegokar, R., Dittgen, M., Muller, R.H., 2013. Coenzyme Q_{10} oral bioavailability: effect of formulation type. J. Pharma. Investig. 43, 431–451.

Barenholz, Y.C., 2012. Doxil®—The First FDA-Approved Nano-Drug: From an Idea to a Product. Handbook of Harnessing Biomaterials in Nanomedicine. Pan Standford Publishing Pte. Ltd, Singapure, pp. 335-398.

Corbo, M.R., Bevilacqua, A., Petruzzi, L., Casanova, F.P., Sinigaglia, M., 2014. Functional beverages: the emerging side of functional foods. Compr. Rev. Food Sci. Food Saf. 13, 1192–1206.

Damge, C., Vranckx, H., Balschmidt, P., Couvreur, P., 1997. Poly (alkyl cyanoacrylate) nanospheres for oral administration of insulin. J. Pharma. Sci. 86, 1403–1409.

Diederichs, J.E., Muller, R.H., 1994. Liposome in Kosmetika und Arzeimitteln. Die Pharmazeutische Industrie 56, 267–275.

Downing, N.S., Ross, J.S., Jackevicius, C.A., Krumholz, H.M., 2012. How Abbott's fenofibrate franchise avoided generic competition. Arch. Intern. Med. 172, 724–730.

Faridi Esfanjani, A.F., Jafari, S.M., 2016. Biopolymer nano-particles and natural nano-carriers for nano-encapsulation of phenolic compounds. Colloids Surf. B 146, 532–543.

Gabizon, A., Catane, R., Uziely, B., Kaufman, B., Safra, T., Cohen, R., Martin, F., Huang, A., Barenholz, Y., 1994. Prolonged circulation time and enhanced accumulation in malignant exudates of doxorubicin encapsulated in polyethylene-glycol coated liposomes. Cancer Res. 54, 987–992.

Jafari, S.M., McClements, D.J., 2017. Nanotechnology approaches for increasing nutrient bioavailability. Advances in Food and Nutrition Researchvol. 81Elsevier, New York.

Junghanns, J.-U., Muller, R.H., 2008. Nanocrystal technology, drug delivery and clinical applications. Int. J. Nanomed. 3, 295–310.

Katouzian, I., Jafari, S.M., 2016. Nano-encapsulation as a promising approach for targeted delivery and controlled release of vitamins. Trends Food Sci. Technol. 53, 34–48.

Keck, C.M., Muller, R.H., 2008. Smart crystals—review of the second generation of drug nanocrystals. In: Torchilin, V.P. (Ed.), Nanoparticlulates as drug carriers. Imperial College Press, London.

Kipp, J.E., Wong, J.C.T., Doty, M.J., Rebbeck, C.L., 2003. USA Patent Application. Microprecipitation method for preparing submicron suspensions.

Lippacher, A., 2000. Pharmaceutical characterisation of liquid and semi-solid SLN dispersions for topical application. PhD-Thesis, Freie Universität Berlin, Berlin.

Liversidge, G.G., Cundy, K.C., 1995. Particle size reduction for improvement of oral bioavailability of hydrophobic drugs: I. absolute oral bioavailability of nanocrystalline danazol in beagle dogs. Int. J. Pharma. 125, 91–97.

Liversidge, G.G., Cundy, K.C., Bishop, J., Czekai, D., 1991. US Patent 5,145,684. Surface modified drug nanoparticles.

Merisko-Liversidge, E., 2002. Nanocrystals: resolving pharmaceutical formulation issues associated with poorly water-soluble compounds. In: Marty, J.J. (Ed.), Particles. Orlando, Marcel Dekker.

Mohammadi, A., Jafari, S.M., Assadpour, E., Faridi Esfanjani, A., 2016a. Nano-encapsulation of olive leaf phenolic compounds through WPC-pectin complexes and evaluating their release rate. Int. J. Biol. Macromolec. 82, 816–822.

Mohammadi, A., Jafari, S.M., Esfanjani, A.F., Akhavan, S., 2016b. Application of nano-encapsulated olive leaf extract in controlling the oxidative stability of soybean oil. Food Chem. 190, 513–519.

Möschwitzer, J.P., 2005. Method for the production of ultra-fine submicron suspensions. DE1020050117864.

Möschwitzer, J., Lemke, A., 2006. Method for carefully producing ultrafine particle suspensions and ultrafine particles and use thereof.

Müller, R.H., Junghanns, J.-U., 2006. Drug nanocrystals/nanosuspensions for the delivery of poorly soluble drugs. In: Torchilin, V.P. (Ed.), Nanoparticulates as Drug Carriers. first ed. Imperial College Press, London.
Müller, R.H., Keck, C.M., 2012. Twenty years of drug nanocrystals: where are we, and where do we go? Eur. J. Pharma. Biopharma. 80, 1–3.
Müller, R.H., Möschwitzer, J., 2015. Method and device for producing very fine particles and coating such particles. U.S. Patent No. 9,168,498.
Müller, R.H., Becker, R., Kruss, B., Peters, K., 1996. US patent 5,858,410. Pharmaceutical nanosuspensions for medicament administration as systems with increased saturation solubility and rate of solution.
Müller, R.H., Becker, R., Kruss, B., Peters, K., 1999. Pharmaceutical nanosuspensions for medicament administration as systems with increased saturation solubility and rate of solution. U.S. Patent No. 5,858,410.
Müller, R.H., Jacobs, C., Kayser, O., 2000. Nanosuspensions for the formulation of poorly soluble drugs. In: Nielloud, F., Marti-Mestres, G. (Eds.), Pharmaceutical Emulsions and Suspensions. Marcel Dekker, New York, pp. 383–407.
Müller, R.H., Jacobs, C., Kayser, O., 2002. DissoCubes—a novel formulation for poorly soluble and poorly bioavailable drugs. In: Rathbone, M.J., Hadgraft, J., Roberts, M.S. (Eds.), Modified-release drug delivery systems. Marcel Dekker, New York, pp. 135–149.
Müller, R.H., Gohla, S., Keck, C.M., 2011. State of the art of nanocrystals—special features, production, nanotoxicology aspects and intracellular delivery. Eur. J. Pharma. Biopharma. 78, 1–9.
Müller, R.H., Chen, R., Keck, C.M., 2013. Smartcrystals for consumer care & cosmetics: enhanced dermal delivery of poorly soluble plant actives. H&PC Today 8, 18–24.
Müller, R.H., Zhai, X., Romero, G.B., Keck, C.M., 2016. Nanocrystals for passive dermal penetration enhancement. In: Dragicevic, N., Maibach, H.I. (Eds.), Percutaneous Penetration Enhancers Chemical Methods in Penetration Enhancement. Springer, Berlin, Heidelberg, pp. 283–295.
Noyes, A., Whitney, W., 1897. The rate of solution of solid substances in their own solutions. J. Am. Chem. Society 19, 930–934.
Pellegrino, T., Manna, L., Kudera, S., Liedl, T., Koktysh, D., Rogach, A.L., Keller, S., Rädler, J., Natile, G., Parak, W.J., 2004. Hydrophobic nanocrystals coated with an amphiphilic polymer shell: a general route to water soluble nanocrystals. Nano Lett. 4, 703–707.
Petersen, R. 2006. Patent WO2008058755. Nanocrystals for use in topical cosmetic formulations and method of production thereof.
Ponchel, G., Montisci, M.-J., Dembri, A., Durrer, C., Duchene, D., 1997. Mucoadhesion of colloidal particulate systems in the gastro-intestinal tract. Eur. J. Pharma. Biopharma. 44, 25–31.
Rabinow, B.E., 2004. Nanosuspensions in drug delivery. Nat. Rev. Drug Discov. 3, 785–796.
Salazar, J., Müller, R.H., Möschwitzer, J.P., 2014. Combinative particle size reduction technologies for the production of drug nanocrystals. J. Pharma. 2014, 14.
Schubert, R., 1998. Liposomen in arzneimitteln. In: Muller, R.H., Hildebrand, G.E. (Eds.), Pharmazeutische Technologie: Moderne Arzneiformen. Wissenschaftliche Verlagsgesellschaft mbH, Stuttgart.
Schumann, R., 1995. Physical stability of parenteral fat emulsions: development of a characterisation procedure with special focus on analytical techniques. PhD-Thesis, Freie Universität Berlin, Berlin.
Stieß, M., 2013. Mechanische Verfahrenstechnik 1. Springer-Verlag, Weblink. Available from: http://www.laprairie.com/default/platinum-collection/cellular-cream-platinum-rare/95790-00153-14.html#start=1&cgid=platinum-collection.

Nanocapsule formation by cyclodextrins

7

*Seyed M.T. Gharibzahedi**, Seid Mahdi Jafari****
*Islamic Azad University, Tehran, Iran; **Gorgan University of Agricultural Sciences and Natural Resources, Gorgan, Iran

7.1 Historical background of cyclodextrins

Since the discovery of cyclodextrins (CDs) in 1891 by Villiers, these components have been broadly scrutinized in various aspects. Villiers conducted pioneering studies on composition and chemical features of CDs. In the first decade of the 20th century, Schardinger isolated the types of bacteria responsible for CD synthesis and his efforts lead to the discrimination between various CDs in 1903. For 25 years, between 1911 and 1935, Schardinger and his workers determined CDs as a family of macrocyclic oligosaccharides.

The macrocycle carbohydrates with D-glucopyranose units (α1→4) in a ring formation constitute the CD molecules. In 1935–36, after discovering γ-CD, Freudenberg determined the cyclic structure of these oligosaccharides. In the early 1940s, Cramer suggested the idea of inclusion complex formation by CDs. Freudenberg and Borchert determined polymeric matrix of parent CDs (α-, β-, and γ-) during two separate studies via analytical methods, especially X-ray diffraction. After studying inclusion complexation characteristics of CDs by Cramer in the early 1950s, French initially published a basic review article on β-CD toxicity. However, the first patent concerning applications in pharmaceutical formulations was published in 1953.

In the late 1950s, researchers conducted investigations on the usage of different methods and involvement of hydrophobic forces to form these inclusion complexes. In 1965–75, Higuchi and Connors developed a mathematical model describing the inclusion complexation mechanism. At this time, with first publication on CD polymers by Furue, the release of prostarmon E as the first medicine complexed by CD was studied. In the early 1980s, CDs were industrially applied in food and cosmetic products and after organizing the "first international cyclodextrin symposium" in the field of CDs, the first cyclodextrin book was published by Szejtli in 1982. In the USA and Europe, production of modified forms of CD based on patenting hydroxypropylated (HP)-β-CD was founded and continued with sulfobutylether (SBE)-β-CD in 1991. During years between 1990 and 2000, many scientists carried out numerous activities on catenanes, rotaxanes and aggregation of CDs (Buschmann and Schollmeyer, 2002; Crini, 2014; Del Valle, 2004; French et al., 1950; Furue et al., 1975; Loftsson and Duchêne, 2007; Messner et al., 2010; Szejtli, 1998, 2004; Villiers, 1891).

There are over 1500 derivatives for CDs which were generally categorized into two main groups of natural and modified. The parent CDs containing 6, 7, and 8 glucopyranose units are α-, β-, and γ-CDs (Crini, 2014). Chemical modification of parent cyclodextrins resulted in numerous safe derivatives (e.g., 2-HP-, maltosyl-, and SBE-β-CDs) with improved physiochemical properties and complexing abilities (Loftsson and Duchêne, 2007). Starch as the easily available carbohydrate polymer is the raw material for the industrial-scale production of CDs. These natural structures are formeds from the starch transformation by certain bacteria, such as *Bacillus macerans* and *B. circulans* through the CD-glycosyltransferase (CGTase) enzyme, which partially hydrolyzes starch into α- ($C_{36}H_{60}O_{30}$), β- ($C_{42}H_{70}O_{35}$), and γ- ($C_{45}H_{80}O_{40}$) CDs (Crini, 2014; Singh et al., 2010; Zhou and Ritter, 2010). Table 7.1 depicts the physicochemical properties of the parent CDs. Among these CDs, β- and γ-CDs significantly have the highest stability and water solubility, respectively (Del Valle, 2004). However, β-CD has the largest contribution (97%) in the international markets of CD sales due to its lower purification cost compared with that of α- and γ-CDs (Astray et al., 2009).

Table 7.1 Physicochemical properties of three common cyclodextrins

Characteristics	α-CD	β-CD	γ-CD
Number of glucopyranose units	6	7	8
Molecular weight (Da)	972	1135	1297
Solubility in water (g/100 mL, 25°C)	14.5	1.85	23.2
[α]D at 25°C (°C)	150	162.5	177.4
Internal diameter (nm)	0.47–0.53	0.60–0.65	0.75–0.83
Height of torus (nm)	0.79	0.79	0.79
External diameter (nm)	1.46	1.54	1.75
Approximate volume of cavity (nm^3)	0.174	0.262	0.427
Approximate cavity volume in 1 mol of CD (mL)	104	1257	256
Number of water molecules taken by cavity	6	11	17
Crystal form (from water)	Hexagonal plates	Monoclinic parallelogram	Quadratic prisms
Hydrophobic interaction	CD-cavity	CD-cavity	CD-cavity
Hydrogen bond	Glucose-OH	Glucose-OH	Glucose-OH
Surface tension (mN/m)	71	71	71
Melting range (°C)	255–260	255–265	240–245
Water of hydration (wt.%)	10.2	13.2–14.5	8.13–17.7
Water molecules in cavity	6	11	17
Diffusion constant at 40°C	3.443	3.224	3.000
p_K (by potentiometry) at 25°C	12.332	12.202	12.081
Hydrolysis by *A. oryzae* α-amylase	Negligible	Slow	Rapid

7.2 Regulatory issues of cyclodextrins

An annual increase of about 20%–30% in the use of CDs has been reported. Meanwhile, application of these components in food products had the highest proportion (80%–90%) among other industrial manufactured goods including pharmaceutical, cosmetic, and chemical products (Astray et al., 2009). Over 3 recent decades, the regulatory aspects of these functional constituents have been developed. The JECFA (Joint FAO/WHO Expert Committee on Food Additives), US-EPA (US Environmental Protection Agency), and US-FDA (US Food and Drug Administration) established different regulatory conditions for CDs; such as, safety (GRAS, Generally Recognized as Safe), limit of addition, acceptable daily intake (ADI), etc. (Asp et al., 2006). Nonetheless, different countries have various regulatory rules for the application of CDs in foods. For instance, the parent CDs as GRAS food additives in the USA are approved to be in the FDA list. Purity is only limiting factor to use them in natural food products in Japan. α-CD in 2003 and γ-CD in 2004 were considered as "Novel Foods" in Australia and New Zealand (Cravotto et al., 2006). The Codex Alimentarius of JECFA (JECFA, 2013) also recorded three main CDs with various INS numbers of 457 (α-CD), 459 (β-CD), and 458 (γ-CD). A maximum allowable level for CD residues in foods was determined by the US-EPA in 2005 (US Federal Register, 2005). Fenyvesi et al. (2015) reported registration of β-CD as a safe additive (E-459) in EU. The highest amount of β-CDs (0–5 mg/kg body weight) used in food formulations was also recommended by the JECFA. β-CD in Germany as an additive in producing foods were presented in 2000 because this compound based on the OECD's experimental protocols has shown no allergic impact (Astray et al., 2009; Cravotto et al., 2006). Oral quantities of β-CD-LD_{50} for dog, mice, and rat were 5, >12.5, and 18.8 kg/kg, respectively (Szejtli, 1988). Irie and Uekama (1997) by feeding rats with a diet containing 10% β-CD for 3 months found that there were no substantial toxic effects in the animal models. In regard to the rigorous clinical and biological investigations, no ADI "not specified" was defined for α- and γ-CDs. Moreover, there was not any reported toxicological result for α- and γ-CDs in the FDA database (Waalkens-Berendsen et al., 1998). Table 7.2 shows the results of a global safety assessment of parent CDs for food purposes.

Table 7.2 Global safety evaluation of the parent CDs for food applications

Safety assessment	Cyclodextrin type[a]			
	α-CD	β-CD	γ-CD	Branched-CD
United States (US)	GRAS approved in 2004 (wide application)	GRAS approved in 2001 (carrier for food and cosmetics)	GRAS approved in 2000 (wide application)	Not approved
European Union (EU)	Novel food	Available (1 g/1 kg food)	Novel food	Not approved
JECFE (WHO/FAO)	Without ADI	ADI (0–5 mg/kg per day)	Without ADI	Not evaluated
Japan	Available	Available	Available	Available

[a]ADI and GRAS are "acceptable daily intake" and "generally recognized as safe," respectively.

7.3 Principles of encapsulation by cyclodextrins

Molecular inclusion of the host–guest complex can be considered as a unique "molecular nanoencapsulation" for many bioactive components (Faridi Esfanjani and Jafari, 2016). In other words, CDs are unfilled molecular nanocapsules with certain sizes to encapsulate many different ingredients through the insertion of appropriate "guest" structure into CDs cavity(Katouzian and Jafari, 2016). This apolar cavity is less polar than water (Augustin and Hemar, 2009); because the CD cyclic oligosaccharides with glucopyranose units in a cylindrical configuration have a water-insoluble inner and a water-soluble outer linked to the hydrogen and glycosidic oxygen bands. The internal diameter of the water-insoluble cavity with a certain depth (0.78 nm) depends on the number of glucose units in α-, β-, and γ-CDs, which is 0.57, 0.78, and 0.95 nm, respectively (Fig. 7.1) (Crini, 2014; Song et al., 2009).

Fig. 7.1 schematically depicts the complex formation between substrate and CDs by the designing CD (host) and substrate (guest) in different stoichiometric ratios. Generally, the inclusion is formed by complexing one CD molecule and one guest molecule in a stoichiometry ratio of 1:1. However, it is possible in the design of CD complex to include over one guest with low and/or high-molecular-weights into the cavity (Del Valle, 2004). Noncovalent interactions; such as Van der Waals and hydrophobic forces compared with the steric effects and hydrogen bonding have a key role in forming the stable inclusion complexes or encapsulation processes (Rekharsky and Inoue, 1998). This unique structure can potentially protect guest molecule(s) against negative effects of many reactions; such as photodegradation, enzymatic decomposition, steric rearrangement, racemization, oxidation, and even hydrolysis (Gharibzahedi et al., 2015; Mielcarek, 1996). CDs have a relatively rigid structural behavior with two freedom degrees so that there are two rotations at the glucosidic (C1—O4 and C4—O4) links and hydroxyl (—OH; C5—C6) groups (Fig. 7.1). Therefore, all the primary (at C6) and secondary (at C2 and C3) —OH groups in this semiflexible structure are respectively placed on the narrower and larger sides of the torus-like ring. High water-solubility of CDs can be ensured through conformation of the —OH groups at C2 and C3 positions because these groups are toward and outward of the cavity, respectively (Fig. 7.1).

Apolar nature of the CD cavity is due to the presence of three rings including C—H groups in C3 and C5 and glycosidic oxygen atoms are located in. Lewis base character and considerable electron mass confirm this intrinsic feature (Crini, 2014; Kfoury et al., 2014). This exclusive structure is responsible for their physicochemical characteristics, which can provide full or fractional inclusion of extensive numbers of functional ingredients (Del Valle, 2004).

Change of guest's solubility and stabilization against effects of heat, light, oxidation and reduction, inhibition of its volatility, and guest's physiological responses can be formed with inclusion complexes of the host molecules (Maswal and Dar, 2014). Meanwhile, the encapsulation efficiency (EE, 30%–99%) of CDs inclusion to form stable complexes is affected by the core molecules as a greater affinity for the CDs can be obtained for the higher hydrophobic and smaller guest molecules (Fang and Bhandari, 2010). The EE of CDs is not only affected by the hydrophilicity/hydrophobicity of guest compounds, but also by the operating parameters including process

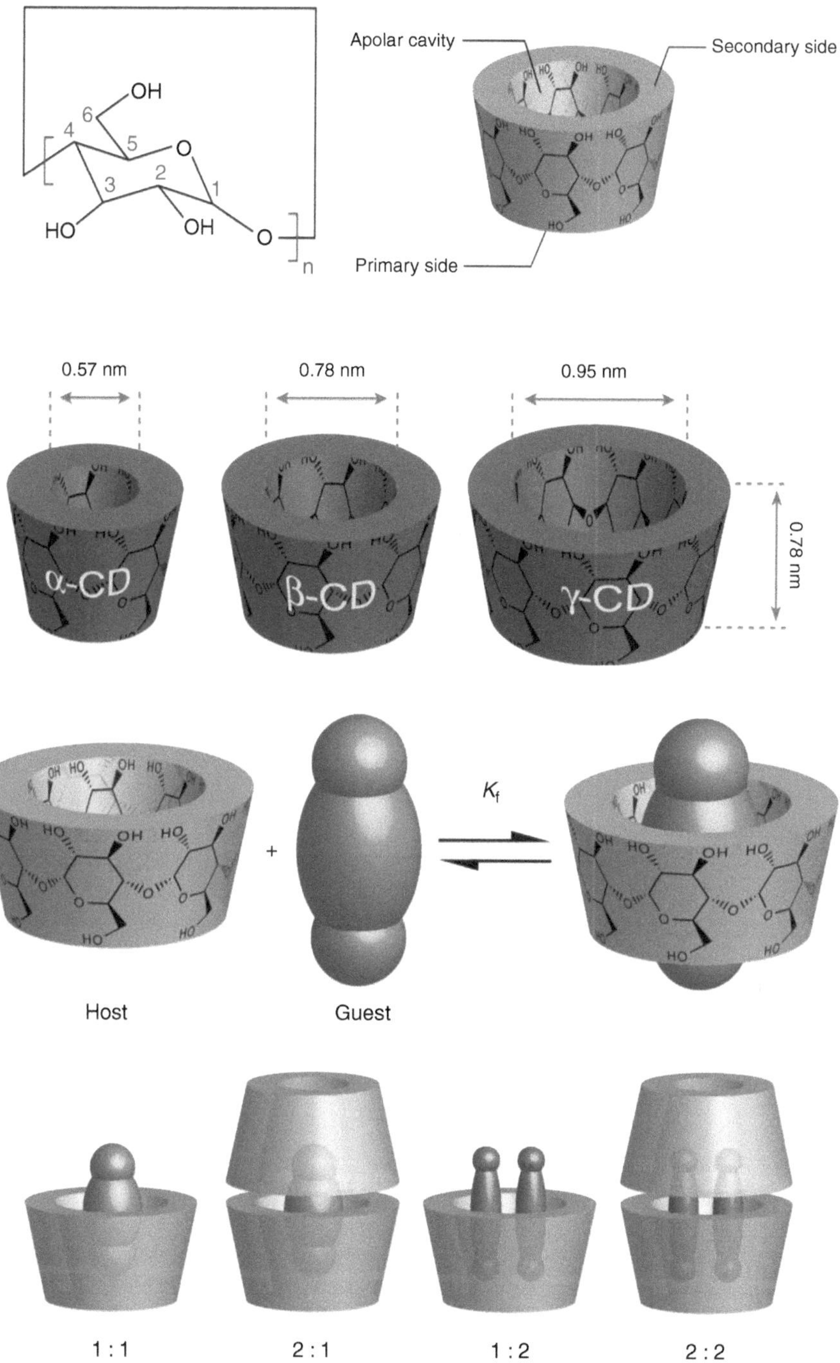

Figure 7.1 Functional structural scheme of α- (n = 6), β- (n = 7), and γ-CD (n = 8); geometric dimensions of CD types, and formation of substrate-CD complexes with the association of free-CD (host) and substrate (guest) in the various stoichiometric ratios. *Source:* Reprinted with permission from Crini, G., 2014. Review: a history of cyclodextrins. Chem. Rev. 114, 10940–10975, Copyright © 2016, ACS.

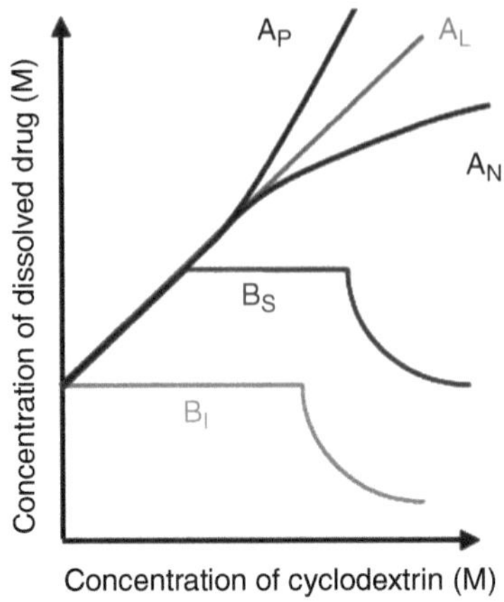

Figure 7.2 Phase solubility profiles and classification of the host–guest complexes according to Higuchi and Connors (1965).
Source: Reprinted with permission from Cabral Marques, H.M., 2010. A review on cyclodextrin encapsulation of essential oils and volatiles. Flavour Fragr. J. 25, 313-326, Copyright © 2016, John Wiley & Sons.

temperature, agitation speed, mixing time, and separation conditions of complex from uncomplexed forms, such as centrifugation rate, which has a noticeable impact on the guest entrapment (Ciobanu et al., 2013).

A critical classification on the host–guest complexes is based on the phase-solubility profiles in a solution (Higuchi and Connors, 1965). Fig. 7.2 briefly illustrates A- and B-kind curvatures showing the inclusion development of strong- and poorly soluble complexes, consecutively. Moreover, B_I- and B_S-type curves represent the formation of insoluble- and limited solubility-complexes, respectively. In addition, A_L-, A_P-, and A_N-subtypes can be obtained by dividing A-type curvatures. A_L-subtype shows a linear relationship between CD concentration and solubility value. A_P- and A_N-subtypes also represent positively- and negatively-deviating isotherms, respectively (Fig. 7.2).

7.4 Encapsulation technologies with cyclodextrins

Generally, there is no single methodology for forming the complexes as a process must be developed for each guest to be encapsulated by the CD. Based on this fact, the use of existing available equipment can be the first selection for developing the solid complexes. Several parameters including the amount of water, temperature, time, and drying conditions must be optimized to establish the formation process of CD–guest complexes (Hedges, 2009). Table 7.3 represents some encapsulation methods with the increased complexation efficiency and solubility rate. Table 7.4 also summarizes various techniques to characterize CD–guest inclusion complexes in both the solid and solution states. Furthermore, Table 7.5 summarizes some host–guest complexes containing different CDs and food bioactive ingredients, which have been formed by

Table 7.3 Methods for enhancing the complexation efficiency by CDs

Method	Consequences
Salt formation	Enhancement of apparent intrinsic solubility of a drug/food via salt formation
Ionization	Unionized drugs/foods do commonly form more stable complexes than their ionic counterparts. Enhancement of the apparent intrinsic solubility by this process resulting in improved complexation.
Complex-in-complex	Improvement of apparent intrinsic solubility of a drug/food via formation of metal complexes.
The acid/base ternary complexes	Enhancing the complexation efficiency using certain organic hydroxy acids (such as citric acid) and certain organic bases by formation of ternary drug (or food)-CD-acid (or base) complexes
Polymer complexes	Increasing the observed stability constant and complexation efficiency of the drug(or food)-CD complex through formation of a ternary complex with drug (or food)-CD-water soluble polymers
Solubilization of CD aggregates	Solubilizing the uncharged complexes of drug (or food)-CD and enhancing their complexation efficiency with organic cations and anions.
Combination of two/ more methods	Enhancing the complexation efficiency by combining two or more of the polymer method. For example, drug (or food) ionization and the polymer method or solubilization of the CD aggregates by adding both polymers and cations or anions to the aqueous complexation medium.

Table 7.4 Techniques for characterization of inclusion complexes formed between the guest and CD molecules

Inclusion complexation in the solid state	Inclusion complexation in the solution state
1. Thermoanalytical methods	1. Electrochemistry
2. Scanning electron microscopy (SEM)	(1-1) Polarography
3. X-ray diffractiometry	(1-2) Conductivity
4. Single crystal X-ray structure analysis	(1-3) Polarimetry
5. Wettability and dissolution tests	2. Solubility studies
6. Infra-red (IR) spectroscopy	3. Spectroscopy methods
7. Thin layer chromatography (TLC)	(3-1) Nuclear Magnetic Resonance (NMR) spectroscopy
	(3-2) Electron spin resonance (ESR)
	(3-3) Ultraviolet/visible (UV/VIS) spectroscopy
	(3-4) Fluorescence spectroscopy
	(3-5) Circular dichroism spectroscopy
	4. pH-potentiometric titration
	5. Microcalorimetry

Table 7.5 The CD-guest inclusion complexes with suitable molar ratios formed by the different preparation methods

	Main character		CD-guest inclusion complexes				
				Special guest ingredient			
Preparation methods	Advantages	Limitations	CD type	Food/ nutraceutical	Drug	Molar ratio	References
Dry mixing	No adding water	Caking risk on scale-up	β	Lemon oil	—	1:1	Nakai et al. (1987)
		Prolonged mixing time	β	Azo-dyes	—	2:1	Parlati et al. (2007)
			β	Quercetin	—	1:1	Borghetti et al. (2009)
Milling/ cogrinding	Fast dissolution and high solubility	—	β	—	Epichlorohydrin	1:1	Cirri et al. (2012)
	Eco-friendly process		β	—	Pranlukast	2:1	Wongmekiat et al. (2007)
	No use any toxic solvents		HP-β	—	Rofecoxib	1:1	Abou-Taleb et al. (2006)
	Short operating time						
Kneading	The most common and simple method		β	—	Gliclazide	2:1	Sapkal et al. (2007)
	Low-cost and short-time process						
	Good yield of inclusion formation for poorly water-soluble guests						
Coprecipitation	The most used method in the laboratory	Unsuitable for large-scale preparations	β	—	Gliclazide	2:1	Sapkal et al. (2007)

		Use of organic solvents as precipitants Removal of the mother liquor	β	Curcumin	—	2:1	Marcolino et al. (2011)
			β	Bixin	—	1:1	Marcolino et al. (2011)
			β	Vanillin	—	1:1	Karathanos et al. (2007)
			β	Flavonoid	—	1:1	Kalogeropoulos et al. (2010)
			α,β,HP-β, Me-β	—	*Trans*-anethole	1:1	Kfoury et al. (2014)
			β	α-Tocopherol	—	1:1.7	Koontz et al. (2009)
			γ	Quercetin	—	1:1.7	Koontz et al. (2009)
Slurry-complexation	Decrease of required water Reduction of the reactor size Good yield of the complex without heating during large-scale production	Severe necessity to control the process parameters	β	Essential oil	—	1:1	Menezes et al. (2015)
Paste complexation	Suitable for heat-labile and reactive guests Suitable for large-scale production		β	Essential oil	—	1:1	Menezes et al. (2015)

(*Continued*)

Table 7.5 The CD-guest inclusion complexes with suitable molar ratios formed by the different preparation methods *(cont.)*

	Main character		CD-guest inclusion complexes				
				Special guest ingredient			
Preparation methods	**Advantages**	**Limitations**	**CD type**	**Food/ nutraceutical**	**Drug**	**Molar ratio**	**References**
Solvent evaporation	Easy setup in both laboratory and large-scale systems	Unsuitable in enhancing the dissolution of lipid complexes	β, HP-β, hydroxy-ethyl-β	—	Indomethacin	1:1	Jambhekar et al. (2004)
	Low-cost and quite simple process	Multistage processing (Long process time)	β	—	Racecadotril	1:1	Semalty et al. (2014b)
Damp mixing and heating	Little or no addition of water	Necessary optimization for the process parameters	β	—	Nimesulide	1:1	Madhulita et al. (2010)
Neutralization precipitation	An industrialized method for most food, drug and nutraceutical ingredients	Unsuitable for acid and alkaline susceptible guests	β	—	Terfenadine	2:1	Choi et al. (2001)
			β	—	Piroxicam	1:1	Doijad et al. (2007)
High pressure homogenization	A more industrialized method	Unsuitable for acid and alkaline susceptible guests	HP-β	Osthole	—	1:1	Liu et al. (2010)
Freeze drying	Providing the fast-dissolving products	More expensive	β	—	Scutellarin	1:1	Cao et al. (2005)

		Time-consuming process	β	—	Racecadotril	1:1	Semalty et al. (2014b)
			HP-β	Canthaxanthin	—	1:1	Gharibzahedi et al. (2014)
Spray drying	A one-step procedure	More expensive	β	—	Omeprazole	1:1	Figueiras et al. (2007)
	Decrease of preparation steps		Me-β	—	Miconazole	1:1	Ribeiro et al. (2008)
	Ideal process control		β, SBEβ	—	Vinpocetine	1:1	Ribeiro et al. (2003)
	Save-consuming time		di-Me-β	—	Manidipine	1:1	Csabai et al. (1998)
			β	d-Limonene	—	1:1	Yuliani et al. (2006)
			α	Shiitake flavors	—	1:1	Yoshii et al. (2005)
			β, di-Me-β, HP-β	—	Budesonide	1:1	Vozone et al. (2003)
Microwave irradiation	An industrial-scale preparation	Absence or low value of true-inclusion	β, di-Me-β	—	Ketoprofen	1:1	Cirri et al. (2009)
	Short reaction time and high product yield		β,HP-β	—	Ziprasidone	1:1	Deshmukh et al. (2007)
Supercritical antisolvent	One-step process	An experimental method	β	—	Benzocaine	1:1 or 2:1	Al-Marzouqi et al. (2007)
	High complexation efficiency		β	Lycopene	—	1:1	Nerome et al. (2013)
	High dissolution rate of the guest molecule		HP-β	Resveratrol	—	1:1	Zhou et al. (2012)

(Continued)

Table 7.5 **The CD-guest inclusion complexes with suitable molar ratios formed by the different preparation methods *(cont.)***

Preparation methods	Main character		CD-guest inclusion complexes				References
				Special guest ingredient			
	Advantages	Limitations	CD type	Food/ nutraceutical	Drug	Molar ratio	
	Minimum contact level between the CD with guest during the process		β	Carvacrol, Thymol, Eugenol	—	1:2.5	Locci et al. (2004)
Extrusion	A continuous industrialized process	Only for heat-stable components	HP-β	—	Indomethacin	1:1	Yano and Kleinebudde (2010)
	Reduction of water consumption						
Gas–liquid		Nonindustrialized method Low yield of the product	β	Volatile compounds Essential oils		1:1	Pagington (1987)
Melting	Cheap and easy	Limited application Only for some sublimable guests	α,β,γ	—	Indomethacin, Griseofulvin	1:1	Wulff and Aldén (1999)
Sealing	Suitable for thermo-stable volatiles	Nonindustrialized method	α,γ	—	Benzoic acid	1:1	Nakai et al. (1987)
		Low yield of the product	γ	—	Salicylic acid	1:2	Higashi et al. (2010)

the various preparation techniques. The most important methods involved in the complexes formation with CDs have been briefly explained below.

7.4.1 Dry mixing (physical blending) method

Many guest molecules can be encapsulated by easily adding these substrates to CDs and their mixing without dissolving in water. In a laboratory scale, the CD and guest molecules using trituration are carefully blended and passed via suitable sieves to obtain the fine particle size. Production of the physical blends in the industry using a fast mass granulator is carried out according to the severe mixing of the guest and CD for 30 min. The powdered mixtures are stored under controlled temperatures and humidity conditions. The best guests to form complexes in this technique are oils or liquid guests (Del Valle, 2004).

Dry mixing is a variation of the paste method, which is done at ambient temperature. The duration of mixing required for the complete reaction is variable and depending on the guest can range from hours to days (Hedges, 2009). A β-CD-disperse/reactive dye complex using a modified dry mixing method (DMM) by milling the reactants in the molar ratio 2:1 for 3 days was prepared by Parlati et al. (2007). However, complexation of lemon oil into CD using DMM is an exception, because the complexation process completes in few minutes. The guest molecule in these cases might also be serving as a solvent for the CD (Nakai et al., 1987).

Overall, this method is not very efficient because it can be rather slow compared with other techniques. Dry mixing usually needs approximately 60 min to form the CD–guest complexes. The rate and efficiency of complexation can be increased with adding a slight quantity of water and drying/warming in a closed container (Giordano et al., 2001). Removing the water as quickly as possible is very desirable, particularly after complexing a volatile guest. The key benefit of the present technique is the complexation development without adding water or using a washing stage (Hedges, 1998). The main drawbacks in large scale are the caking problem, leading to an incomplete blending and complexation, therefore the mixing time is prolonged (Del Valle, 2004).

7.4.2 Milling/cogrinding method

There is a main difference between techniques of cogrinding and physical mixing. Although plain mixing is adequate for physical blending method, extensive combined attrition in cogrinding technique is required to affect the powder blend (Friedrich et al., 2005; Mulay, 2011; Patel et al., 2010). Mechanical devices in this method are used to grind/mill the guest and CDs for preparing solid binary inclusion compounds. The guest and CD molecules are well mixed and the resulted blend, in order to grind, is transferred into a swinging grinder during an appropriate time. Also, binary structure of the guest-CD can be alternatively prepared using a ball milling system. Spherical-shaped balls used in the conventional systems have different mass which is worked at an encoded speed and time and finally filtered with a 60 mesh sieve (Mulay, 2011; Patel et al., 2010). This method provides a nearly amorphous powder with the improved solubility rate. Cogrinding method led to an enhancement in dissolution rate of rofecoxib (Abou-Taleb et al., 2006) and benzocaine (Al-Marzouqi et al., 2007) in comparison to the other preparation processes. From commercial and ecofriendly viewpoints, cogrinding technique because of devoid of any toxic solvents is outstanding compared with other complexation methods (Friedrich et al., 2005).

7.4.3 Kneading method

Kneading is a typical processing step in the preparation of host–guest complexes containing CDs. In this technique, the CD is initially saturated using a minimum amount of aqueous and hydroalcoholic solutions for obtaining slurry with a suitable consistency. The liquid or dissolved compound of guest is then added to the prepared slurry in a mortar to achieve a uniform paste by carefully kneading using a mortar and pestle for a specific time. The kneading process is similar to the wet granulation procedure and requires conventional kneaders, such as low and high shear mixers (Das et al., 2013). This process under an industrial scale can be achieved by using extruders and other machines (Bhati et al., 2012; Semalty et al., 2014b). A slight quantity of solvent (e.g., ethanol) is used to wash the obtained solid with the aim of removing the noncomplexed guest molecules from the formed complexes. A vacuum-drying step is then done to reduce the moisture content of the developed product at room temperature (Cabral Marques, 2010), passed through an appropriate sieve and finally placed into a desiccator to carry out additional evaluations.

Kneading considered as the cheapest method to develop the inclusion complexes of guest-CD (Patil et al., 2010). Use of this complexation approach can significantly improve the inclusion efficiency for guests with poor solubility (e.g., flavoring agents and essential oils), due to their slow dissolving into the CD cavity to form inclusion complexes (Madene et al., 2006; Waleczek et al., 2003). For example, kneading method led to an enhancement in nimesulide dissolution (Parikh et al., 2005). Encapsulation of D-limonene with CDs and maltodextrins (MDs) using kneading technique has been performed by Furuta and Yoshii (1994). Also, Bhandari et al. (1999) by studying the complex-encapsulated profiles of selected volatiles, following the process of kneading found that there were no significant differences between the compositional properties of the resulting powdered complexes, which were obtained using a vacuum oven and various spray-drying techniques. Ascenso et al. (2011) investigated the complexation of tretinoin and di-Me-β-CD using different inclusion methods including coevaporation, freeze drying, spray drying, and kneading. Koeda et al. (2014) also used kneading method to form complexes of retinyl palmitate using CDs and MD with different values of dextrose equivalent (DE; 17.5, 28, and 47) at low moisture content.

7.4.4 Coprecipitation method

In coprecipitation method (CPM), the guest molecule is dissolved in minimum quantity of an organic solvent (diethyl ether, acetone, chloroform, benzene, etc) and is then added dropwise to the CD solubilized in low volume of water under stirring/sonication and/or heating (Karathanos et al., 2007; Kalogeropoulos et al., 2010; Kfoury et al., 2014). If a guest molecule is thermo-stable, the solubility of CD–guest complex by heating will increase up to 20%. The solubility rate at high amounts of the guest component can be gradually increased by the prolonging complexation reaction or using a cooling step (Del Valle, 2004). Before forming precipitates, cooling the CD–guest mixture to room temperature is gradually done under agitation and then collection of crystals or precipitates is carried out using vacuum-filtration, decanting, or centrifugation processes (Pitha and Hoshino, 1992). A washing step might sometime be performed on the obtained precipitates with a little water and/or hydrophobic solvents (e.g., methanol, ethanol, diethyl ether, or acetone). Finally, the washed precipitates are dried (about 50°C) to yield a powdered sample, afterward they are stored at ambient temperature and relative humidity of 40%–50% (Sapkal et al., 2007). Formation capability of CD complexation in this method can be significantly affected by adding various chemical agents. Earlier investigations have demonstrated the inclusion of diazepam and different steroids with CD could be respectively reduced by adding nonionic surfactants and stabilizers (Loftsson et al., 1992), while the use of ethanol can considerably improve the formation of solid/semisolid complexes (Furuta et al., 1993). In pharmacy, it is possible to form inclusion complexes of CD-unionized drugs with more stability than ones of CD-ionized drugs. For this reason, the researchers found the ammonia addition to the reactive aqueous environment can considerably improve the complexation efficacy of essential drugs (Del Valle, 2004). The solubilization of HP-β-CD complex with pancratistatin was also optimized by adding NH_4OH to the mixture by Torres-Labandeira et al. (1990).

The CPM is the most common technique applied in laboratory scale because the coprecipitation is a principle for all the complexation methods. This axiom not only is fairly plain but also complexation parameters for each individual guest molecule are specific (Koontz et al., 2009). Evaluating the main conditions of CPM in the laboratory revealed the production of guest-CD complex in an industrial scale through slurry (SCM)—and paste (PCM)—complexation methods were much more suitable than CPM because these techniques require removing lower water/humidity in the subsequent drying (Qi and Hedges, 1995).

Since a low solubility for the applied CD and complex to precipitate without any cosolvent is necessary, there is a limitation for this technique in relation to β-CD type (Del Valle, 2004). Therefore, a system with phase-solubility behavior of A-type cannot be appropriate to form a soluble inclusion complex using CPM (Higuchi and Connors, 1965). This method for high-volume production of the complexes between CD and bioactive components is inappropriate because of low yield, use of high amounts of water and time-consuming character (Cabral Marques, 2010). Precipitants in this method are organic solvents which can inhibit the successful formation of inclusion complexes. This important disadvantage limits the application of CPM compared to the kneading approach. Therefore, this major restriction should be considered before the launching (Del Valle, 2004). Tank valence, process time, and required energy for heating and cooling steps are assumed to be the main costs. The main defect in this process is collection of high volume of mother liquor after forming the inclusion complex. However, recovering the mother liquor can efficiently remove this problem (Loftsson et al., 1992; Pitha and Hoshino, 1992).

7.4.5 Slurry-complexation method

In slurry-complexation method (SCM), the whole dissolution of CD is not essential to produce stable conclusion complexes. CD is thus suspended in slurry up to a 50–60 wt% concentration and stirred in a reactor. The guest molecule with CD in solution usually forms complexes at ambient temperatures. Forming the inclusion complex as crystalline or precipitate is happened by saturating the water phase (Shimpi et al., 2005). In SCM similar to the CPM (filtration or centrifugation), the complex usually collected into a stirred reservoir. However, a roller/ball mill is used for agitating high-concentrated slurries (Del Valle, 2004). Sometimes it is essential to use lower levels of guest molecules because some these compounds intrinsically are sticky which can provide a difficulty in mixing step. Although the slurry is more flowable by mixing, a stoppage in this process can produce a firm mixture, needing more energy to resume mixing (Del Valle, 2004; Shimpi et al., 2005). Since there are no visual indications for the formation of inclusion complex, it is vital to empirically assess the amount of stirring time in the CPM. The complexation is completed at variable time, being different from guest to guest and depends on the particular characteristics of guest plus the stirring speed (Wang et al., 2007). Reduction in the level of required water and the reactor size are the main benefits of SCM. Some guests are sensitive to heat, and thus limiting the usage of thermal processing can lead to a considerable yield in the complex formation (Hedges, 2009). Nonetheless, complexation rate of the many guest molecules can be effectively increased by applying some heat. Thus, a proper precision is necessary during the heating process because the complex destabilization and incomplete complexation reaction can occur under this condition (Loftsson et al., 1993).

7.4.6 Paste complexation method

Paste complexation method (PCM) is a variation of the SCM as a paste is produced by mixing a little volume of water (20% maximum) with the guest by means of a mortar and pestle, or a kneader system. After implementing drying or water washing steps on the resultant complex, it is then retained via filtration and centrifugation processes (Del Valle, 2004; Shimpi et al., 2005). Similar to the SCM, the guest type determines the time required to form the complex. Sometimes, a rigid bulk rather than a very small powder is formed by pastes drying. This fact relies on the volume of added water and the amount of entrapped guest during the paste preparation. Under this circumstance, a higher temperature is employed to decrease viscosity and associated mixing problems. After removing the complex from the blender and its cooling, a glass is formed. Milling the glass and its careful drying can attain a powdered form of the complex (Del Valle, 2004; Hedges, 2009). Bhandari et al. (1999) produced fine powders by complexing lemon oil and β-CD using PCM. Apart from the length of mixing period in the kneading process, the present process leads to a similar profile in the encapsulated flavor into the complex.

7.4.7 Solvent evaporation (coevaporation/solid dispersion) method

In this technique, after dissolving the guest and CD molecules in a common organic solvent, such as methanol and acetone, stirring for 24 h at ambient temperature is carried out to prepare inclusion complex by noncovalent bonds. The obtained molecular dispersion in solvent evaporation method (SEM) is then evaporated off under suitable processing conditions (particularly at temperature of 45°C) in a vacuum oven or a rotary evaporator to eliminate the remaining solvent (Al-Marzouqi et al., 2007, 2009; Jambhekar et al., 2004; Ribeiro et al., 2003, 2008; Semalty et al., 2014a). The solid powdered inclusion complex is formed when the consumed solvent is completely evaporated. The mixture after concentrating is filtered and vacuum dried for 1 day at 25°C to obtain the final complex (Semalty, 2014). With regard to the components nature and type of used drying procedure, crystallinity amount of the obtained complex can be variable. The dissolution rate and bioavailability of target bioactive compounds can be influenced by the factors involved in the shape of solid particles, such as crystal size and pattern (Perrut et al., 2005). To achieve a constant weight in the produced solid mass, it is placed under vacuum in a desiccator. The dried solid particles are triturated and separated via a 60-mesh sieve and finally kept in an air-tight container.

In bio- and pharma-industries, the ratio between bioactive component and CD depends on the geometrical fitness of the biomolecule which is usually taken in 1:1 or 1:2 molar ratios (Semalty, 2014). Semalty et al. (2014b) studied the complexation between racecadotril and β-CD in (1:1 and 1:2) using two methods of solvent evaporation and kneading. Racecadotril as an antisecretory and antidiarrheal agent has a week dissolution/solubility degree in water along with an incomplete uptake. They found that the complexes prepared in a molar ratio of 1:1 could improve solubility/dissolution profile compared to those made in a molar ratio of 1:2. Moreover, the complex prepared by SEM compared with kneading technique showed better solubility and the dissolution due to better amorphization of racecadotril (Semalty et al., 2014b).

7.4.8 Damp mixing and heating method

In damp mixing method (DMM) a small amount of water is utilized only for hydrating the guest molecule (equal to 20%–25%) in the CD matrix. Usually, the exerted water quantity is equal to the amount of water present in filter-cake obtained from CPM and SCM (Shimpi et al., 2005). Sometimes, no water in DMM is added to the mixture with regard to the guest nature. Having fully mixed molecules of guest and CD in a closed vessel, heating at ~100°C is done and subsequently, the dried particles are collected (Del Valle, 2004; Shimpi et al., 2005). In the next step, the dried solids are crushed, triturated, and forced through sieve No. 100. For each guest, it is necessary to optimize the added quantity of water, the mixing speed, and the required time for heating process in DMM (Del Valle, 2004).

7.4.9 Neutralization precipitation method

Neutralization precipitation method (NPM) is performed according to the precipitation of inclusion compounds by neutralization technique. The ionizable guests after dissolving in high-pH solutions (0.1 N NH_4OH/NaOH) mixed with CD for about half an hour resulting in a clear solution. This solution is then neutralized under agitation by adding drop wise hydrochloric acid solution (0.1 N) till reaching the equivalent point (about pH = 7.5). Proper pH adjustments in this technique provide the reduction of guest solubility and the formation of complex outside the mixture. A white precipitate under this condition is being formed parallel with the inclusion development of guest-CD. Filtration, washing (chlorine removal), drying (at 250°C for 1 day) and storage (into a desiccator) are next steps in order to achieve the final product (Patil et al., 2010; Saravana et al., 2013). Generally, the main drawback of NPM is the degradation of acid and alkaline-susceptible bioactive components in this procedure. An enhancement in solubility of terfenadine and piroxicam by forming their complexes with the utilization of β-CD NPM was reported (Choi et al., 2001; Doijad et al., 2007).

7.4.10 High-pressure homogenization method

In high-pressure homogenization method (HPHM), the guest and CD molecules into a suitable solvent are sufficiently blended using a thermostatic stirring valve in a water-bath for 3 days at 25°C. The dispersion homogenization to the breakdown of particles/all over the product is done via a high-pressure homogenizer (Nasir et al., 2012). Providing the cavitation and shear breakdowns induces the particle disintegration. Gas bubbles are formed by boiling the solution under stationary pressure. These bubbles gradually depart from the gap and can be burst by air pressure and shocking waves (Chaudhary et al., 2012). The complex formation is achieved after filtering and evaporation processes (Nasir et al., 2012). Formation of inclusion complexes osthole/HP-β-CD with NPM and HPHM was analyzed by (Liu et al., 2010). Particle size, thermal, and structural examinations showed that solid complexes formed by HPHM had the lowest mean particle size and led to a complete complexation than those obtained by NPM. The results also revealed that the solid complexes obtained from HPHM dissolved faster than those developed by NPM and dry mixing method (Liu et al., 2010).

7.4.11 Freeze drying (lyophilization) method

Freeze-drying method (FDM) is also termed lyophilization, which can be considered as a suitable technique for obtaining an irregular, spongy powder of the complex guest-CDs. Freeze-drying is an ideal method for forming CD inclusion complexes due to their favorable stability at low temperatures (Cao et al., 2005). Generally, FDM is taken into account as the best common technique to dry heat-labile components for forming a stable complex (Patil et al., 2010). For the preparation, the guest component and CD is respectively solubilized in water using a shaking device. In this approach, required concentration of ammonia has been used to produce the stable complexes between slightly soluble, poorly acidic ingredients and CD in watery medium (Cabral Marques, 2010). A preliminary freezing (at about −30 to −40°C) leads to the solvent removal from the solution. The solvent-free solution is then dried with two molecules of guest and CD at a decreased pressure (about 30–40 N/m^2) (Patil et al., 2010). Having frozen the material, the vapor sublimes from the surface and this fact leads to the receding of the ice front into the solid in a similar manner to drying (stage of the second falling rate). Traces of moisture are removed by secondary drying, which involves exposure to a higher temperature (50–60°C) in a short time (Bhati et al., 2012; Semalty, 2014).

FDM is a replacement method for SEM so that it involves blending the guest/CD in an organic solvent (Tsinontides et al., 2004). An inclusion complex powder with an excellent output developed using FDM. It is also possible to scale up the process (Cabral Marques, 2010). This technique has a high potential to entrap the volatile compounds inside a matrix with hollow orifices because of sublimation of ice crystals (Madene et al., 2006). Choi et al. (2008) concluded the damage to really volatile compounds can be minimized at low temperatures particularly in soluble complexes like HP-β-CD complexes. Furthermore, powders obtained by freeze drying are able to keep their primary color so that can be rapidly reconstructed by adding water. However, extended time and formation of a powdered product with low flowing rate are the limitations present in this method (Semalty, 2014).

7.4.12 Spray drying (atomization) method

Spray-drying method (SDM) as a derivative of the SEM is a common and widely used technique to produce fine dry particles in pharmaceutical and food industries. SDM signifies as an ideal technique for developing the CD complex beginning from a solution. CDs (especially β-CD) are applied to improve aroma retention/encapsulation (about 1%) during SDM to protect unsteady aroma compounds in soft drinks (Figueiras et al., 2007; Shiga et al., 2004; Yoshii et al., 2005) or production of extruded snacks (Yuliani et al., 2006). In this technique, CD is dissolved in a solution (200 mL) previously alkalinized with 25% aqueous ammonia (final pH = 9.5). The guest is liquefied in 96% ethanol (100 mL). The two prepared solutions are blended at room temperature for 24–48 h and sonicated to attain equilibrium and provide a high-quality mixture suitable for drying operation. The eventual monophasic solution is atomized and then the yield is dried with the tangentially entering hot air leading to the spray-dried CD complex to form the complexes (Csabai et al., 1998). A control system for precipitation process should be employed to prevent the formation of bulk and aggregated particles and occlusion of the atomizer nozzle. The product obtained by SDM yields the porous particles in a more uniform size and shape, which consecutively enhances the solubility amount of the complexed guest. Guest and CD molecules have an adequate and effective contact to form an ideal complex is an extra benefit for SDM. However, major drawbacks of this complexation method are the formation of low-yield product plus being sensitive to high temperatures (Cabral Marques, 2010). Previous results have shown that the shape of particles obtained from SDM is fine spheres, while that of those developed based on FDM is amorphous with a small number of crystalline parts (Ribeiro et al., 2003; Figueiras et al., 2007; Ribeiro et al., 2008). Although dissolution of both products is very rapid because of the amorphization, FDM lead to the fastest dissolution (Figueiras et al., 2007). SDM is not considered as an achievable technique to generate complexes between CDs and extremely-volatile and thermo-sensitive molecules. Thus, it is essential to optimize its operating parameters, such as flow rates of air and feed, inlet and outlet temperatures, aspirator speed, and so on (Devil et al., 2010). An optimization study on the formation of CD-budesonide complex using SDM and evaluation of aerodynamic properties of the produced particles for dry powder aspiration were conducted by Vozone and Marques (2003).

7.4.13 Microwave irradiation method

Microwave irradiation method (MIM) is mainly developed to prepare rapid organic synthesis in an industrial-scale because this process forms a high-yield product in a short time (Deshmukh et al., 2007; Shin-ichi et al., 2004; Patil et al., 2010). In this present novel technique, a microwave oven is applied to perform the microwave irradiation between the reactive molecules of guest and CD. At first, the two reactants in a certain molar proportion are well solubilized in a blend with definite amounts of water and organic solvent. The resulted mixture was heated at 60°C for 60–120 s in a microwave oven (150 W) to carry out the complexation reaction. An alternative treatment for the physical mixture is using microwave at 500–750 W for 5–10 min (Chaudhary et al., 2012). The remaining and uncomplexed guest and CD were eliminated at the end of the reaction by adding an adequate amount of solvent mixture to the reaction medium. A filter was used to separate the formed complex as the precipitate. In the next step, an oven system was applied to vacuum-dry it at 40°C for 2 days (Saharan et al., 2009; Chaudhary et al., 2012). The complexes were produced via MIM. This formulation not only has a high stability at room temperature and relative humidity but also show a practically unmodified solid-state (Cirri et al., 2009). Deshmukh et al. (2007) developed inclusion complexes between ziprasidone hydrochloride and β-CD or HP-β-CD to design a fast dissolving formulation by employing different superdisintegrants. MIM regarding the dissolution rate can be considered as a higher effective method compared to kneading and cogrinding techniques (Cirri et al., 2009).

7.4.14 Supercritical antisolvent method

In the late 1980s, supercritical antisolvent method (SASM) was presented for the first time and soon a number of methods to develop solid particles using supercritical fluid were patented (Muhrer et al., 2006; Mulay, 2011). Supercritical antisolvent (CO_2) due to having a low critical temperature (31°C) and pressure (73.8 bar) is an attractive candidate for processing heat-labile active compounds and improving their bioavailability (Fages et al., 2004; Muhrer et al., 2006). Also, CO_2 have the benefits of being nonhazardous, anticorrosion, antifulminant, antieruptive, abundant and inexpensive. Desirable physical properties of CO_2; such as nonsignificant consistency and considerable mass transfer and surface activity provide far better conditions for its elimination from the polymer matrix after the process is completed. However, CO_2 nontoxic residuals in slight quantities are ensnared into the polymer structure (Tirucherai and Mitra, 2003; Fages et al., 2004). Although SASM requires a quite-high initial cost, this fast process has favorable outcomes without the use of organic solvent and high storage costs. Since CO_2 is nonpolar, this solvent has restrictions in its ability to dissolve many guests. The power of CO_2 may be dramatically increased with the addition of low cosolvent quantities (such as 2.5% ethanol) into a stainless steel cylinder (Tirucherai and Mitra, 2003). An interesting alternative method to raise interaction in the CD–guest complex is SASM combined with molecular inclusion because of the improved diffusion coefficient and solvating efficiency. SASM is an innovative approach to form inclusion complexes of guest-CD with less operating stages, avoiding the application of hazardous solvents like ammonium hydroxide (Charoenchaitrakool et al., 2009; Tirucherai and Mitra, 2003; Van Hees et al., 1999). In the SASM, after dissolving the guest and CD molecules in a proper solvent, the solute through a nozzle is sprayed into a container under supercritical pressure conditions. In this step, CO_2 quickly transfers into the carrier liquid solvent which diffuses into the antisolvent. The mixture supersaturation owing to the large expandability of CO_2 precipitates the solute and also flow of supercritical fluid moves the used liquid solvent (Nasir et al., 2012; Patil et al., 2010).

Nerome et al. (2013) have lately produced lycopene/β-CD inclusion complexes using SASM. They demonstrated that spherical/agglomerated nanoparticles (NPs) obtained from inclusion complex of lycopene/β-CD can be produced according to the operating factors; such as temperature, pressure, solution flow rate, and CO_2 flow rate (Nerome et al., 2013). Using this technique, Zhou et al. (2012) produced unstructured and uniform powders of inclusion complexes of resveratrol and HP-β-CD with a great variation in the structural shape of raw materials. These researchers reported a considerable enhancement in the dissolution profile of resveratrol using the antisolvent method combined with inclusion in comparison to the complexes formed by physical mixture (Zhou et al., 2012). β-CD inclusion complexes with carvacrol, thymol, eugenol, and citral flavors have also been prepared by the SASM (Locci et al., 2004; Marongui et al., 2002).

7.4.15 Extrusion method

In recent years, great progress in extrusion technologies have been fulfilled to develop encapsulation processes. Application of multiple nozzles, spinning disc atomizer, and jet-cutter systems can lead to the high number of droplets to produce micro- and nanocapsules. A smaller internal diameter for the nozzle can produce the finer capsules. In general, extrusion as a continuous system is a modification of thermal and blending processes. CD, guest, and very little water in this technique are premixed or mixed into the extruder. An accurate monitoring into extruder container can be performed on the required mixing and heating degrees and process time (Del Valle, 2004). The yield can be dried by heating in an oven or by cooling with regard to the content of added water. High lateral temperature induced by considerable shear forces into the extruder system can lead to a significant decomposition in heat-sensitive guests (Del Valle, 2004; Shimpi et al., 2005). The existence of a continuous process and use of very little water are advantages present in extrusion method (Shimpi et al., 2005).

7.4.16 Gas–liquid method

Gas–liquid method (GLM) is a traditional procedure to encapsulate essential oils and volatile compounds. Solvents and other chemical materials form complexes by transferring the vapor via a warm/cool CD solution. The formed inclusion in GLM not only can isolate through filtration, but also a recovery process for volatile compounds can be obtained via a steam distillation system (Pagington, 1987).

7.4.17 Melting method

Preparing the complexes in this method can be carried out by the simple melting process of guest molecule and following mixing with CD. A melting point (MP) of < 180°C can be ideal because this temperature must not cause a thermal degradation for the CD. However, Chaudhary et al. (2012) reported that MP of a binary system is affected by its forming compounds, CD type and weight fraction of the guest molecule. The guest-CD mixture is cooled slowly to room temperature. A careful washing or a vacuum-sublimation is applied to eliminate the additional quantity of guest causing a poor inclusion complex. The solidified product is placed in desiccators, transferred to a clean mortar, triturated to prepare very fine powders, and finally passed through sieves with various pore sizes (Nasir et al., 2012).

7.4.18 Sealing method

Grinding the physical mixtures of the guest and CD molecules at a favorite molar ratio, sealing the mixture in a crystal vessel and keeping it in a 75–90°C oven for 10–60 min (in most cases for 180 min) can provide a possibility for the formation of solid complexes of guest-CD. Although the generated complex has quantitative crystallinity, a significant enhancement can be obtained in the inclusion solubility (Al-Marzouqi et al., 2007, 2009; Cirri et al., 2009). A physical mixture between the active compound of benzoic acid and α- or γ-CD was formed into a sealed vessel with adsorption of certain quantities of water vapor, and subsequent heating at 43–142°C. Increasing the container temperature to over 70°C provided a crystalline inclusion compound. Whenever the sealing technique under N_2 pressure was performed, the pressurized materials showed a higher combination ratio compared with nonpressurized ones (Nakai et al., 1987). A simple and quick "sealed-control temperature method" was designed to form paeonol-β-CD complexes (Wang et al., 2007). The creation of inclusion complex between guest and CD is confirmed by using analysis of infrared spectrum and powder X-ray diffraction (XRD). Many researchers demonstrated that the heating temperature, heating time, and crystallinity of β-CD can significantly affect the complex formation by sealing technique (Wang et al., 2007). Since the guest ingredient is successfully incorporated into the CD inside under the gaseous condition, the sealing method can be successfully used for thermostable volatiles because of guest components (Nakai et al., 1987).

7.5 Selecting an encapsulation technology with cyclodextrins

Determination of the best complexation technique for various guest molecules needs exact and comprehensive assessment. Apart from consideration of the efficiency of the final mass solids (such as solubility rate and stability), other parameters including simplicity, high yield, low cost, swiftness and scale-up ease have a specific importance in the choice of a perfect complexation method (Semalty, 2014). Table 7.5 shows some advantages and limitations for different processes to form guest-CD complexes.

Studies revealed that the incomplete complexation along with the formation of caking risk on scale-up in dry mixing leads to the fact that this very time-consuming method is not an appropriate technique to encapsulate the bioactive constituents (Del Valle, 2004; Nakai et al., 1987). Mura et al. (1999) found that both methods of kneading and cogrinding can be industrially applied to form the stable inclusion complexes between guest and CD molecules. These methods require only a short operating time to obtain an inadequate improvement in solubility/dissolution rate without the need to "true-inclusion" formation. Qi and Hedges (1995) earlier claimed that PCM or SCM was more suitable for large-scale production in comparison to laboratory method of CPM because less water must be subsequently removed during drying. Overall, the HPHM is a more industrialized method than the NPM because NPM is not appropriate to acid and alkaline susceptible guest molecules (Nasir et al., 2012). SEM is nor done in short time and includes multistage processing, neither very useful in improving the dissolution of lipid complexes (Li et al., 2007). However, SEM can be considered as an alternative to the SDM because this production process is fully easy and affordable under laboratory and industrial conditions (Patil et al., 2010). Although SDM is more expensive, this process offers a one-step procedure with the advantage of reducing preparation steps, saving time and better process control (Yoshii et al., 2005). Although FDM is a well-organized process to gain true-inclusion and amorphous complexes with quick dissolving ability, it is an expensive and time-consuming process (Mura et al., 1999).

A liquid medium can provide a possible means to obtain the true-inclusions in MIM (Nacsa et al., 2008); nonetheless a true-inclusion in this extremely fast technique cannot be formed in the presence of dried materials. Meanwhile, a small proportion of a true-inclusion in the obtained solid mass is showing a partial protection of the guest component against the surrounding (Cirri et al., 2009). Altogether, MIM, because of the short time of the process and great yield of the production as a new technique can be applied in an industrial scale (Deshmukh et al., 2007). Extrusion method as a continuous industrialized process with reduced water consumption is only able to form inclusion complexes of CD with heat-stable components because heat-labile components can deteriorate through high temperatures provided by shear forces in an extruder system (Shimpi et al., 2005). Although SASM is still an experimental process, it is considered as an innovative method for the forming solid state complexes between CD molecule and guest component (Patil et al., 2010). Advantages of SASM over other techniques are one-step preparation of the CD–guest complexes, achieving high complexation efficiency, minimizing the contact level of guest with CD molecules during the process, and enhancing the solubility amount of the guest molecule. These features are similar to the dissolution profile of micronized complexes of guest-CD (Patil et al., 2010). From an industrial viewpoint, the melting method is inexpensive and simple; nevertheless this technique is limited to sublimable guests like menthol (Devil et al., 2010). In contrast, sealing and GLM are not industrialized methods because only slight quantities of product during these processes can be obtained. The main limitation of sealing technique and GLM is small number of guest compounds encapsulated with CD (Al-Marzouqi et al., 2009).

7.6 Cyclodextrin modification

CD modification not only can provide its safety, but also can keep the formation ability of inclusion complexes with the different substrates. The modification usually converts natural CDs to their amorphous derivatives (Duchêne, 2011). Two primary chemical parameters including nucleophilicity of —OH groups and the production potential of complexes between CDs and applied reagents should be considered for their modification process (Wenz and Thomas, 1999).

Many chemical modifications on CDs were conducted through the grafting replacements to various sites namely primary and secondary rims or both faces (Fig. 7.3) in order to provide nonionic and cationic derivatives capable of producing nanoscale molecular cloud aggregates used in bio- and pharma-industries (Gèze et al., 2007). Two neutral CD derivatives of methylated (Me) and HP and also negatively-charged sulfobutylated CD derivative are frequently used (Duchêne, 2011). The most fundamental molecular units are —OH nucleophilic groups on the C6 site. The most acidic —OH groups are placed at the C2 site. Moreover, the —OH groups at the C3 position are the most unreachable (Saenger et al., 1976).

Methylation usually happens on the OH groups of C2 secondary and C6 primary (di-Me-CDs) and/or on all the —OH groups at C2, C3, and C6 positions (tri-Me-CDs). In contrast with the hydroxypropylation, use of the Me-CDs is restricted by sterilizing injectable solutions because dissolution rate of these derivatives decreases with increasing temperature. A more binding ability and better hemolytic properties can be obtained during fractional methylation of the OH groups present in the C2, C3, or C6 places of β-CD (Bonnett, 1995; Dougherty et al., 1998).

Hydroxypropyl -CDs are a mixture of products with different substitution degrees. This fact causes the formation of amorphous compounds if many types of HP-CDs in

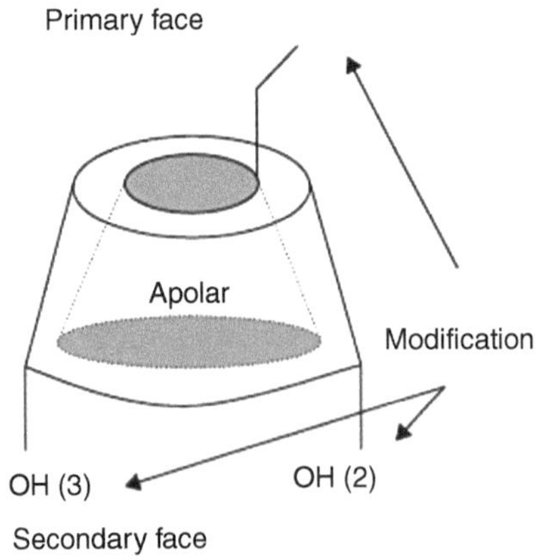

Figure 7.3 A schematic representation on the modification sites of CD.
Source: Reprinted with permission from Bilensoy, E., Hincal, A.A., 2009. Recent advances and future directions in amphiphilic cyclodextrin nanoparticles. Expert Opin. Drug Deliv. 6, 1161–1173, Copyright © 2016, Informa Healthcare.

the reaction medium would have existed (Uekama et al., 1998). β-CD is usually applied for some reasons, such as low-cost, high accessibility plus the cavity size, and denotes at least 95% of all the produced and consumed CDs (Fathi et al., 2014). On the basis of a hundred or so publications, HP-β-CD among the different derivatives compared to natural (parent) CD significantly showed a higher water-solubility and better biocompatibility profile (Fathi et al., 2014). There is an endothermic dissolution for HP-β-CD because no reduction in its dissolution rate is observed by raising the temperature of the reaction medium. An increase in the degree of HP substitution due to steric hindrance effects led to a decrease in the drug binding. It was demonstrated that the low toxicity and high solubility of HP derivatives of CD can be considered as a potential to formulate them in parenteral drugs (Müller and Brauns, 1986).

SBE-β-CDs similar to HP-β-CDs is suggested in osmotic drugs and oral/parental routes due to their low toxicity and high solubility (Loftsson and Duchêne, 2007). Higher replacement of sulfobutyl group than —OH group in β-CD structure can often lead to the higher drug complexing characteristics (Zia et al., 1997). The presence of SBE-β-CDs was found to be more suitable for promoting the peptide encapsulation and minimizing its presence at the surface of NPs (Trapani et al., 2008).

In addition to the main modifications, fatty acids with various chain lengths can be chemically/enzymatically grafted on the —OH groups of the secondary face in order to form new amphiphilic CDs (Perdersen et al., 2005; Zhang et al., 1991). These CDs have been mostly derived from β- and γ-CDs to reduce their hydrophilic properties which may increase the contact of CDs with biological membranes (Lemos-Senna et al., 1998a). These CD derivatives offer an extensive diversity in particle shapes, such as reservoir, rather-lamellar, and matrix structures. Nanospheres and nanocapsules have matrix and reservoir arrangements, respectively (Choisnard et al., 2006; Gèze et al., 2002; Memisoglu-Bilensoy et al., 2005). The amphiphilic CDs are individually discussed in the following sections.

7.7 Amphiphilic cyclodextrins

Chemical derivatives produced from parent CDs (α-, β-, and γ-CDs) are called amphiphilic CDs. These CDs are obtained by modifying the primary and/or secondary structures through linking aliphatic linear or branched chains having different lengths from 2 to 18 carbons with a dissimilar chemical bond (e.g., ester, ether, thiol, or amide) types (Memisoglu-Bilensoy et al., 2006c). In the past decades, amphiphilic CDs have been widely applied to prepare various carrier types especially in pharma industries (Memişoğlu et al., 2002a,b). These CD derivatives possess very promising self-assembling characteristics which can be adequate for spontaneously generating nanoparticulate systems (such as nanospheres and nanocapsules) by nanoprecipitation of amphiphilic. Furthermore, the preformed inclusion complexes between drug ingredients and amphiphilic CDs without the presence of a surfactant can be a direct route to prepare nanospheres and nanocapsules. Under this condition, achieving a high loading rate along with a delay in burst effect is significantly possible (Skiba et al., 1995, 1996a; Memişoğlu et al., 2002a,b; Péroche et al., 2005).

The linked amount of bioactive component to a polymer mass unit is named loading capacity (LC). Since NPs of amphiphilic CDs show high LC with favorable release characteristics, these nanomaterials are of great interest to be applied in dissimilar pharmaceutical and food applicaitons. An enhanced LC is indicating the high content of the guest molecule in polymerization environment on the guest-CD inclusion complex, and/or a significant quantity of water-insoluble positions in nanospheres structure for linking abundant amounts of CD aggregates to the NPs (Memişoğlu et al., 2003b; Silveira et al., 1998). Boudad et al. (2001) showed HP-β-CD using a soluble drug carrier providing the NP-formation mechanism in polymerization environment can increase saquinavir loading into poly (alkylcyanoacrylate)-NPs. Under this condition, there was a stable balance among the complex, disconnected molecular units, and formed polymeric NPs (PNPs).

Moreover, the in vivo studies showed amphiphilic CDs irrespective of their dissimilar molecular configurations are nonhemolytic and noncytotoxic (Memişoğlu et al., 2003a; Memisoglu-Bilensoy et al., 2006a). In conclusion, the aim of CD derivatives synthesis was to overcome some of the concerns of parent CDs which restrict industrial utilizations including: (1) interaction enhancement between CDs and biological membranes with the help of a relative external hydrophobicity, (2) permitting self-assembly of CDs as a result the spontaneous development of nanoscale materials (nanospheres/nanocapsules), and (3) a binding increase between CDs and water-insoluble guests due to having high quantities of various aliphatic long-chains and hydrophobic sites to establish potential molecular links with water-insoluble components (Bilensoy and Hincal, 2008).

Surface charge (nonionic and ionic) is the most important factor to classify amphiphilic CDs. Based on this factor; three categories of amphiphilic CDs can be prepared:

7.7.1 Nonionic amphiphilic cyclodextrins (NIA-CDs)

A grafting mechanism of CD aliphatic chains with various lengths on the primary and secondary rims of α-, β-, or γ-CD can develop NIA-CDs. Different derivatives of NIA-CDs depending on their structure are illustrated in Table 7.6. Among these nonionic amphiphilic CDs, medusa-like and skirt-shaped CDs due to the specific surfactant characteristics have revealed their ability to form nanocapsules or nanospheres. Thus, other parameters; such as length, structure of grafted chains, and bond type of the aliphatic chain can be effective in the surface activity of these amphiphilic CD-NPs (Duchêne et al., 1999).

7.7.2 Cationic amphiphilic cyclodextrins (CA-CDs)

Cationic amphiphilic-cyclodextrins (CA-CD)s can be synthesized by substituting an ionic amine group. Cryan et al. (2004b) by fully replacing amine groups at the 6-site produced a collection of polyamino-β-CDs. It was reported CA-CDs can be used to deliver gene (Cryan et al., 2004a; Méndez-Ardoy et al., 2011; Ortiz Mellet et al., 2011). Structural features of CA-CDs also can increase equilibrium between water-soluble (thioalkyl chains) and water-insoluble (ethylene glycol oligomers)

Table 7.6 Some NIA-CDs and their synthesis path

Nonionic amphiphilic CDs	Synthesis path	Schematic representation	References
Cup-and-ball	Adding a big group (i.e., tertiary butyl), which is linked to the end of the aliphatic chain to prevent self-inclusion of the pendant group		Dodziuk et al. (2000)
Lollipop	Grafting only one aliphatic chain to a 6-amino-β-CD		Bellanger and Perly (1992)
Medusa-like	Grafting hydrophobic aliphatic chains with length between C4 and C16 to all primary -OH of the CD molecule, that is, per (2,6-di-*O*-alkyl) CDs		Kawabata et al. (1983) Liu et al. (1992)
Bouquet-shaped	Grafting of 14 polymethylene chains to 3-monomethylated β-CD implying the existence of 7 chains on each side of the CD ring molecule		Wenz (1991) Canceill et al. (1992)
Skirt-shaped	Adding aliphatic esters or amides (C2-C14) into the secondary face of β- and γ-CDs		Zhang et al. (1991) Zhang et al. (1992)
Cholesteryl (a more complicated neutral derivatives)	Grafting a cholesterol group onto a CD Assumption: CD and cholesterol are the hydrophilic head group and the hydrophobic part, respectively.	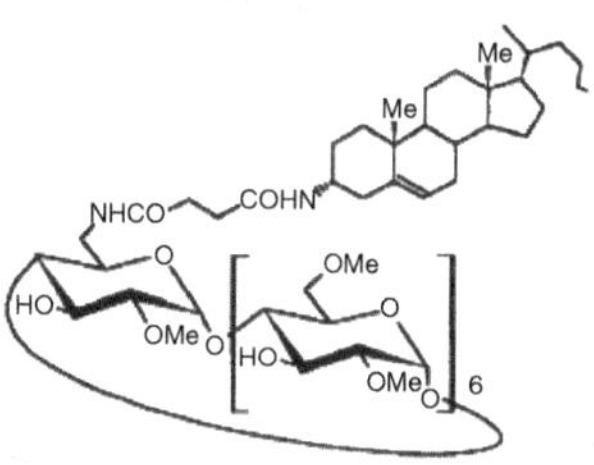	Roux et al. (2002)

groups. The colloidal stability of nanoaggregates of CA-CDs is mainly increased in the presence of ethylene glycol chains. Matsumoto et al. (2004) synthesized and characterized CA-α- and β-CDs and investigated their film-forming characteristics using hydrophilic compounds. Although in vivo transfection efficacy can be improved by interacting neutral CDs and nucleic acids/nucleotides, CA-CDs possess better binding potential with nucleotides for increasing the delivery by virus vectors. Therefore, the improved capability in more interacting with nucleic acids and self-assembly characteristics are main benefits of these amphiphilic CD-types (Cryan et al., 2004b).

7.7.3 Anionic amphiphilic cyclodextrins (AA-CDs)

Anionic nature in a CD molecule can arise with the incorporation of a sulfate group in its structure. Dubes et al. (2001) reported an effective regiospecific generation method for achieving acyl-sulfated β-CDs with formation ability of molecular aggregates in water medium so that the upper and lower rims are respectively functionalized with sulfates and fatty acid esters. Dubes et al. (2003) using ultraviolet-visible spectrophotometry and electrospray ionization-mass spectrometry confirmed that the molecular complexes between antivirus drug acyclovir and sulfated amphiphilic CDs in a molar ratio of 1:1 can be formed by noncovalent bonds. Conversely, interactions for sulfated AA-CDs are entirely taken place in the water-insoluble section of alkanoyl chains. Sukegawa et al. (2002) prepared sulfated amphiphiles using a similar procedure and showed that both sulfated and nonsulfated compounds form stable monolayers. Schwinte et al. (2003) synthesized sulfonated derivatives from acylated 6-per-bromo-β-CD via reaction with sulfonated thiols salts. Abdelwahed et al. (2008) investigated NPs preparation from amphiphilic β-CDs created with various acylation degrees at the secondary —OH face followed by varying the sulfatation degrees at the primary hydroxyl face and found a fascinating relationship between the structure of amphiphilic-β-CD and its potential to produce NPs. Péroche et al. (2005) after synthesizing novel amphiphilic perfluorohexyl- and perfluorooctyl-thio-β-CDs with their alkyl analog, nonanethio-β-CD, studied their capacity for producing NPs using photon autocorrelation function and imaging analyses.

7.8 Nanoencapsulation with amphiphilic cyclodextrins

Amphiphilic CDs can spontaneously form NPs in the shapes of nanospheres or nanocapsules[a] with regard to their preparation method. Several techniques to prepare NPs from amphiphilic CDs have recently been published, which will be discussed in the next sections.

[a]Nanoscale colloidal systems can be divided into two subgroups including nanospheres and nanocapsules. Nanospheres and nanocapsules are matrix and core-shell structures, respectively. Structure of a nanocapsule composes of an external polymeric wall and an internal liquid core.

7.8.1 Nanoprecipitation technique

Solvent displacement or nanoprecipitation method (NPM) is preferred among other techniques applied to produce both nanospheres and nanocapsules because no extra surfactant in NPM is used. Moreover, it considered as an easy method to achieve homogeneous size distribution, low polydispersity index, and rapid preparation with good reproducibility, involving mild solvents like ethanol or acetone (Duchêne et al., 1999; Fessi et al., 1997; Lemos-Senna et al., 1998d; Skiba et al., 1992; Wouessidjewe et al., 1996a,b). Nevertheless, a quick release rate for guest molecule can be observed due to its adsorption on the NP surface instead of inclusion into the CD cavity or the water-insoluble core of NP (Wouessidjewe et al., 1996a,b). Generation of the complex is happened into the used organic solvent or/and the period of nanoprecipitation development. Stability of NPs developed using NPM with respect to the number and length of water-insoluble chains of amphiphilic CDs has been proved (Fessi et al., 1988; Péroche et al., 2005). The design of a suitable NPM is dependent on selecting kind of the organic solvent and content of amphiphilic CDs because these components provide a proper medium to form a stable aqueous suspension of NPs. For example, fluorinated amphiphilic molecules into water and organic solvents usually have a trace solubility, which is a great problem to design the process (Bilensoy and Hincal, 2009).

Fig. 7.4a depicts a diagram to prepare NPs using NPM. In this method, the organic phase containing the amphiphilic CD solubilized in an organic solvent with high polarity, such as ethanol, dimethyl sulfoxide, or tetrahydrofuran is first prepared. The produced phase is then dropwise transferred into deionized water at ambient temperature at a certain shaking rate (about 30 min). Having diffused the organic solvent in water droplets, a supersaturation condition is accorded to leave the amphiphilic CD as NP precipitates in the solution (Memisoglu-Bilensoy et al., 2006b). A vacuum evaporator is used to remove the organic solvent and concentrate the nanosphere dispersion to a favorite amount. Some researchers reported that the emulsification process can fail when there is too much concentration of amphiphilic CD in water because the precipitation under this condition occurs from an extremely supersaturated solid solution (Ganachaud and Katz, 2005; Vitale and Katz, 2003).

NPM fundamentally can be an appropriate approach for nanoencapsulating lipid-soluble compounds due to the miscibility of two phases of solvent and water. This method is considered as a successful way to nanoencapsulate hydrophilic compounds. To produce nanocapsules using the NPM, oil phase containing benzyl benzoate or Miglyol 8121 as a definitive marketable mixture of caprylic and capric glycerides along with the amphiphilic CD in the presence or absent of guest component are added to the solvent phase composed of ethanol or acetone (Loukas et al., 1994; Tsorteki et al., 2004). Thus, nanocapsules involve an oil phase that is not present in nanospheres. The nanosphere size can be changed by modifying the position of CD primary or secondary face because it has been demonstrated that a greater surface area with reduced particle size can be obtained through modification functional groups situated in the secondary structure. Although the size of NPs is not influenced by the surfactant (e.g., Pluronic F68) attendance and its content, the nanocapsule size distribution can

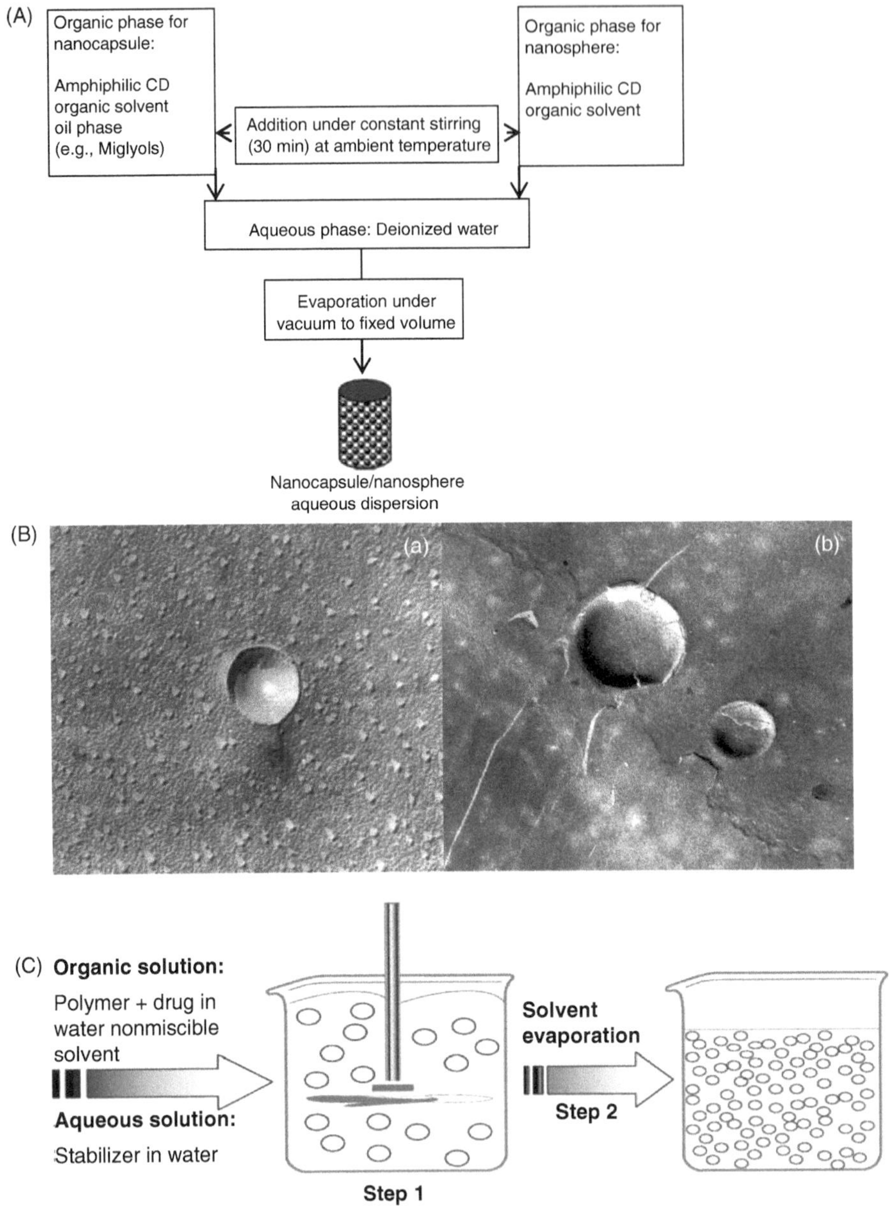

Figure 7.4 (A) A schematic diagram on the preparation of nanocapsules/nanospheres from the amphiphilic CD using nanoprecipitation method, (B) cryo-TEM photo-micrographs of 6-*O*-CAPRO-β-CD nanocapsules (a, Memişoğlu et al., 2002b) and nanospheres (b, Memişoğlu et al., 2003b) developed using nanoprecipitation method, prior to their loading (mag, 5000 ×), Copyright © 2016, Elsevier Ltd., and John Wiley & Sons, and (C) basic principles of the ESE technique for a guest (drug) molecule.
Source: Reprinted with permission from Pinto Reis, C., Neufeld, R.J., Ribeiro, A.J., Veiga, F., 2006. Nanoencapsulation I. Methods for preparation of drug-loaded polymeric nanoparticles. Nanomed. Nanotechnol. Biol. Med. 2, 8–21, Copyright © 2016, Elsevier Ltd.

be considerably changed by different parameters; such as size of oil droplets generated in the production steps, molarity, and amphiphilic CD character. However, nanospheres can be only formed in the presence of excessive contents of amphiphilic CDs. This fact causes the nanocapsules to be usually 2–3 times larger than nanospheres (Bilensoy and Hincal, 2009; Memisoglu-Bilensoy et al., 2006a; Skiba et al., 1996a,b). In Fig. 7.4b, a cryo-transmission electron microscopy (Cryo-TEM) images of nanospheres/nanocapsules of a β-CD derivative (6-*O*-CAPRO-β-CD) followed by freeze-fracture preparation is shown (Memişoğlu et al., 2002b, 2003b).

Amphiphilic CD-NPs using nanoprecipitation are loaded with subsequent procedures modified to attain higher encapsulation for poorly-soluble guests that can be complexed to amphiphilic CDs before NPs preparation (Memişoğlu et al., 2002a, 2003a,b). Skiba et al. (1992) pointed out that preparation of loaded-NPs is highly dependent on attendance of a surfactant; such as poloxamer 188.

1. Conventional loading: a solution containing guest molecule in production steps is poured into an organic phase based on the NPM. Basically, main mechanism in this process is adsorption on the NP surface. Nevertheless, there is a trace feasibility to include the guest inside the CD cavity.
2. Preloading: NPs are directly prepared from preformed guest-amphiphilic CD complexes. Application of this method cannot significantly enhance the entrapment efficiency or association's percentage in comparison to the conventional loading for drugs, such as progesterone, bifonazole, and clotrimazole.
3. High-loading: NPs are directly prepared through preformed complexes of guest-amphiphilic CD and are more loaded with the addition of the guest solution to the organic phase. In other words, this technique is a combination of the conventional and preloading methods. The amount of guest encapsulated within the NPs by this method can be significantly increased for drugs of progesterone, bifonazole, or clotrimazole.

Solvent displacement technique is used to prepare highly-loaded NP systems containing amphiphilic β-CD to promote the nonoral administration of slightly water-soluble antifungal drugs, such as bifonazole and clotrimazole (Memişoğlu et al., 2002a; Pinto Reis et al., 2006). Bilensoy et al. (2008) and Çirpanli et al. (2009) respectively prepared amphiphilic β-CD NPs containing anticancer drugs of camptothecin and paclitaxel by the NPM. The developed NPs based on amphiphilic CD also showed more LCs in comparison with the polymeric ones, formed with poly(lactic-*co*-glycolic acid) (PLGA) or poly-ε-caprolactone using the same NPM (Çirpanli et al., 2009).

7.8.2 Emulsion/solvent evaporation technique

Vanderhoff in 1979 for the first time applied emulsion/solvent evaporation (ESE) approach to prepare polymeric NPs (PNPs) (El-Aasser et al., 1979). ESE because of having an easy operating protocol is considered as the most usual technique to prepare NPs (Seo et al., 2003). This technique can control the stirring rates and conditions; moreover it can provide a high efficiency in the incorporation of lipophillic drugs and food ingredients (Freitas and Marchetti, 2005). ESE usually involves two major steps (Fig. 7.4c): (1) emulsification of the two oil and aqueous phases using a high-energy homogenizer and/or an ultrasound device. The water phase composed of a stabilizing agent, while the polymer and functional compounds are presence in the organic phase, and (2) evaporating the organic solvent during a constant mild shaking at ambient temperature or under high temperature and low pressure conditions for polymer precipitation and particulate formation (Miladi et al., 2013). Sah (1997) found that the implementing a gentle process in evaporating solvent can lead to the formation of polymer droplets with high concentration and followed by thicker particles with slower release rate. An efficient ultracentrifuge or filter can be applied to collect the solidified NPs which are washed by deionized water for removing additives (e.g., surfactants). Having washed, the resulted particles are lastly freeze-dried (Miladi et al., 2013). Although dichloromethane and chloroform (Hans and Lowman, 2002) were earlier used to develop the emulsion, ethyl acetate has recently exerted because of its superior toxicological character (Sani et al., 2009). A controlling mechanism on the formulation and operating parameters, such as stirring and homogenizer speed, process temperature, kind and quantity of dispersing and stabilizer agents, and consistency of both the phases of organic and aqueous can adjust the size of produced particles. Stabilizers (e.g., gelatin, polyvinyl alcohol (PVA), polysorbate-80 and poloxamer-188) either as film-forming and barrier agents can release drug/food ingredients on the interior boundary layer, or as a strong steric component can stabilize the exterior aqueous phase (Khoee and Yaghoobian, 2009; Mora-Huertas et al., 2010).

Two main strategies in the conventional processes are being used for the formation of emulsions (Chapter 2), which include the production of simple [e.g., oil/water (O/W)] or double [e.g., water/oil/water (W/O/W)] emulsions. In simple emulsification process, the selected polymer with desired characteristics into an oil/organic solvent is first dissolved and afterward the resulted polymeric solution in a water-continuous phase containing bioactive ingredients is emulsified by means of mechanical and ultrasonic homogenizing for preventing coalescence of droplets. This ESE type is usually applied to encapsulate water-insoluble drug/food ingredients (Miladi et al., 2013; O'Donnell and McGinity, 1997). Since water is used as a nonsolvent in these systems, formation of O/W emulsions is very interesting. This fact can simplify and improve process economics, because of no use of toxic solvents, application of an easy recovery and washing process, and diminishing the chances of aggregation of NPs (Aftabrouchard and Dorlker, 1992). Double-emulsion evaporation and multiple emulsion methods are modified forms of single ESE, which can favorably be applied to encapsulate any hydrophilic compounds (Hans and Lowman, 2002). In this method, the preliminary coarse emulsion produced through simple emulsification for the second time is dispersed by homogenization into an outer watery phase containing a stabilizer (Giri et al., 2013; Vila et al., 2002). A concern in using this technique is the toxicity of remaining organic solvents and surfactants into particles. However, double emulsions render good homogeneity and ideal dispersal characteristics (Murakami et al., 1999).

The ESE technique can be only applied to form nanospheres from amphiphilic CDs. Compared with nanospheres produced using NPM; nanospheres prepared by ESE not only have an ideal sphericity, but also cannot well form monodispersed solutions (Lemos-Senna et al., 1998b). Lemos-Senna and her coworkers used this method to encapsulate some important drugs, such as progesterone, hydrocortisone, and testosterone inside nanospheres developed based on amphiphilic γ-CD substituted at the position of C6 chains of the secondary face (Lemos-Senna et al., 1998a, c). The initial amounts of drug present in the organic phase, the used surfactant and amphiphilic CD had key roles in the drug encapsulation using the ESE technique (Lemos-Senna et al., 1998b). A medium loading rate of progesterone (1.0 mg), a high concentration of initial γ-CDC6, and a low content of surfactant could favor an ideal condition to retain progesterone in the lipid phase or in formed spheres (Skiba et al., 1992). Mura et al. (2010) showed the complex formation of oxaprozin-CD using milling/cogrinding method and its loading into PLGA-NPs by double ESE technique can significantly enhance the therapeutic efficacy of this nonsteroidal antiinflammatory drug. They added PLGA solution in dichloromethane comprising oxaprozin-CD complexes to the aqueous solution of PVA. After emulsifying the obtained mixture using a high-shear homogenizer, it was again transferred to a water solution containing PVA for producing W/O/W emulsion. Oxaprozin release rate from NPs collected by ultracentrifugation was highly influenced by the CD presence so that this rate during 24 h for the simple drug-loaded and drug-Me-βCD-loaded NPs was 16 and 50%, respectively (Mura et al., 2010).

7.8.3 Detergent removal technique

A surfactant subcategory with a high potential to dissolve lipophilic membranes is detergent. The reorganization of lipid bilayers for creating finer, soluble detergent–lipid aggregates of different morphologies can be obtained by enough quantity of detergents. In this technique, incorporating the lipophilic substances into the CDs is conducted by dissolving them with the detergent in an organic solvent (e.g., methanol, ethanol) or solvent mixture to attain a clear solution. A rotary vacuum-evaporator system at a decreased pressure and an enough heat are usually applied to remove the consumed solvent. The residual solvent can be also removed under vacuum for at least 1 h. Applying the bile salts and nonionic detergent can lead to the clear and turbid dry films. High EE for hydrophilic substances in this technique can be provided by adding a suitable buffer to the substrate and adjusting the process temperature (Schubert, 2003). Using dilution or dialysis of nonionic *n*-octylglucoside detergent can reconstruct NPs from the mixed micelles of detergent-amphiphilic CD. The present method in a moderate condition was operated without using any organic solvent and exposure to extreme heat (Lemos-Senna, 1998).

7.9 Effective factors on the characteristics of amphiphilic cyclodextrin nanoparticles

Previous investigations have revealed that higher concentrations of amphiphilic β-CD (0.05–1.0 mM) can lead to larger particle sizes (Memişoğlu et al., 2003a; Memisoglu-Bilensoy and Hincal, 2006). A high instability in the NPs at concentrations superior to 0.6 mM was observed according to the span measured using dynamic light scattering analysis (Memişoğlu et al., 2003b). Conversely, similar NPs depending on the chain length were produced at different concentrations of amphiphilic CD (0.3–0.6 mM)

(Choisnard et al., 2005, 2007), which can be correlated to the critical aggregation level of each component. It has been also found that the chain length has a substantial influence on 3D structural morphologies (spherical, barrel-like, or irregular) of NPs (Choisnard et al., 2005, 2007). Skiba et al. (1996a,b) by investigating NPs formation based on CDs grafted with alkyl chains of C6, C12, and C14 successfully produced NPs with 200-nm diameter by nanoprecipitation technique without surfactants. Although the size of supramolecular objects for a given chain length is comparable, their polydispersity is higher than those developed in the absent of surfactant (Gèze et al., 2009).

The structure of grafted chain and substitution on the primary/secondary face can play key roles in the correct alignment of CD in the boundary layer of O/W and subsequently the preparation possibility of nanocapsules. It was shown that the NPs can be simply formed based on alcohol as solvent when there was a linear chain CD. However, CDs having a ramified chain only generate NPs at very low concentrations (Memişoğlu et al., 2002b). Studies have demonstrated that CDs modified on the primary face in comparison to those modified on the secondary face show a far more surface activity depending upon their density and thus provide stable NPs (particularly nanocapsules) at lower concentrations by better stabilizing the O/W interface (Memişoğlu et al., 2002b; Ringard-Lefebvre et al., 2002).

Substitution degree is another factor affecting the characteristics of NPs. This index provides an opportunity to assess the molecular weight of the ionized products. Memişoğlu et al. (2002b) produced stable nanocapsules for 12 months based on the selectively of substituted derivatives grafted on the secondary/primary face, with linear or ramified alkyl chains, using ester or amide bonds with variable chain lengths (C6-C14). The NPs production in the attendance or absence of Pluronic F68 surfactant can be possible by using fully peracylated amphiphilic β-CDs (14 alkyl chains) (Szejtli et al., 1986). However, 200-nm particles can be formed by the mixture of amphiphilic CDs having 11 or 13 alkyl chains (underacylated) with peracylated CDs even without stabilizers could remain stable for 7 months at 6°C (Gèze et al., 2002).

7.10 Formation techniques of the cyclodextrin-based polymeric nanoparticles

7.10.1 Physicochemical processes

7.10.1.1 Dialysis method

In dialysis technique, having solubilized the selected polymer into an organic solvent, the obtained solution is poured into a dialysis tube with an appropriate molecular weight cut-off. This semipermeable membrane tube is a physical barrier which allows the passive transport of solvents (Chronopoulou et al., 2009). Mixing the solvent and nonsolvent inner side the membrane slows down the ability of the mixture to dissolve the polymer gradually. Displacing the solvent to internal sides of the membrane causes to avant-garde polymer aggregation because of solvability loss and formation of uniform solutions of NPs (Fig. 7.5a). The last product for attaining a steady-state

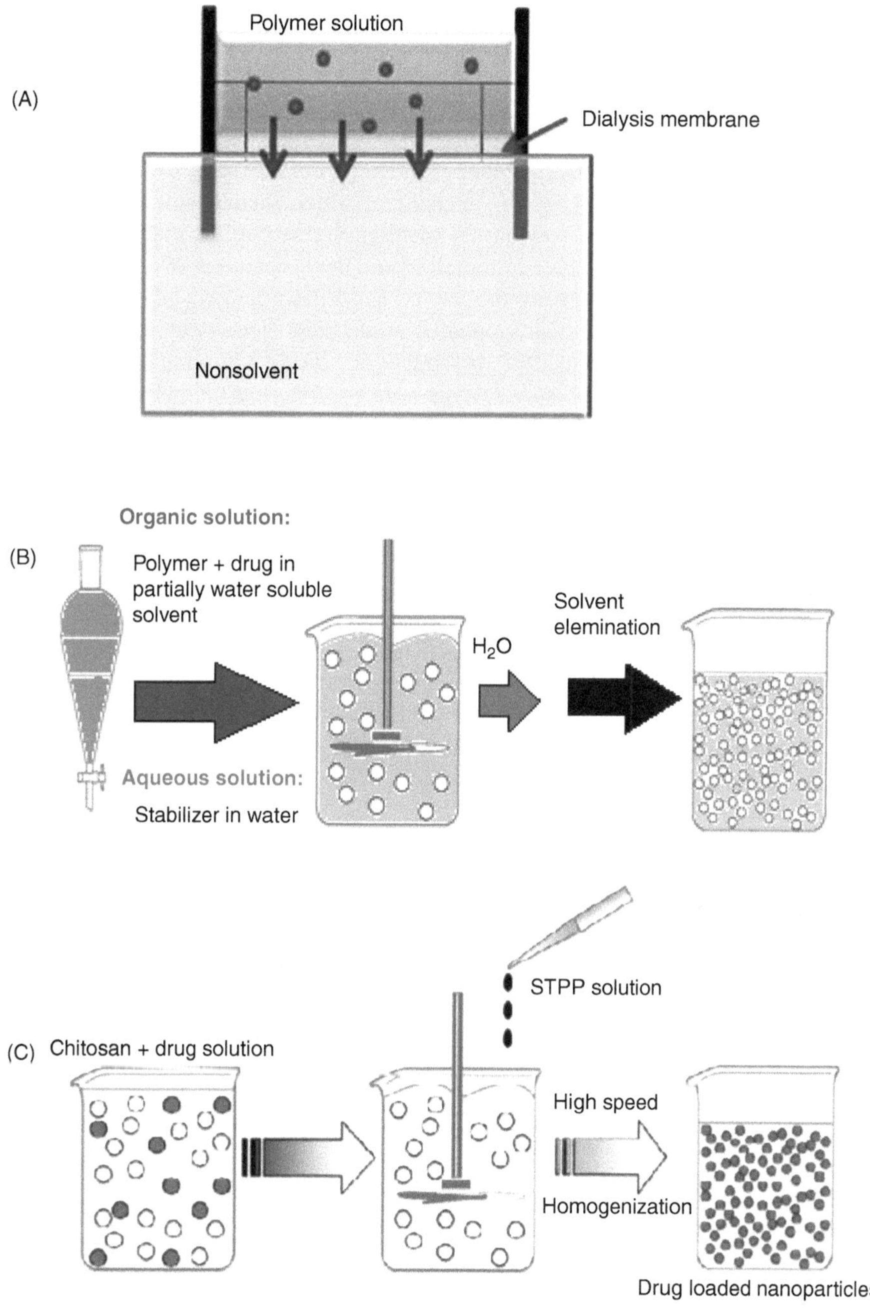

Figure 7.5 Schematic representation for PNPs synthesis from a guest (drug) molecule by dialysis (A) ESD (B), and ionic gelation technique (C) techniques.
Source: Reprinted with permission from Nasir, A., Kausar, A., Younus, A., 2015. A review on preparation, properties and applications of polymeric nanoparticle-based materials. Polym. Plast. Technol. Eng. 54, 325–341; Nagavarma, B.V.N., Yadav, H.K.S., Ayaz, A., Vasudha, L.S., Shivakumar, H.G., 2012. Different techniques for preparation of polymeric Nanoparticles-a review. Asian J. Pharm. Clin. Res. 5, 16–23.

in formed capsules is exposed to dehydration (Xie and Wang, 2005; Hornig and Heinze, 2007). At present, the production mechanism of PNPs using dialysis technique is not well identified. However, this procedure might be based on a protocol comparable to the principle of NPM recommended by Fessi et al. (1989).

There are many advantages of this technique, which include low cost, general applicability, easy recovery of solvents by distillation process, the pure NPs formation along with bioactive molecule immobilization, the preparation of small and narrow-distributed PNPs, and averting the use of emulsifiers and costabilizers due to the employed mild conditions (Chronopoulou et al., 2009; Jeon et al., 2000; Kostog et al., 2010; Lemoine and Preat, 1998; Masotti et al., 2008). Morphology and size of NPs are highly affected by the solvent type employed during the polymer solution preparation (Nasir et al., 2015).

The assemblies may serve as novel nanocarriers for the delivery of hydrophobic compounds. Zhang and Ma (2011) prepared β-CD based core-shell structured nanoassemblies through a dialysis technique using double hydrophilic copolymers of polyethylene glycol-*b*-water insoluble cyclodextrin polymers (PEG-*b*-PCDs) in the presence of a hydrophobic homopolymer poly (β-benzyl L-aspartate) (PBLA). Briefly, dissolution of PEG-*b*-PCD and PBLA in DMSO were distinctly done and afterward mixed with an appropriate weight ratio and finally dialyzed against deionized water for the whole day (Zhang and Ma, 2011). These researchers earlier prepared coumarin 102 containing assemblies by dissolving them in DMSO followed by addition of PBLA and PEG-*b*-PCD-3. Finally, the same dialysis process was done (Zhang and Ma, 2009).

7.10.1.2 Emulsion solvent diffusion (spontaneous emulsification) method

Although Leroux et al. (1995) for the first time designed emulsion solvent diffusion (ESD) technique; the development of ESD is actually based on the studies of Quintanar-Guerrero et al. (2012). Formation of PNPs using ESD is based on two processes of emulsification and diffusion. An aqueous phase containing stabilizer dispersed, an organic phase composing of both polymeric matrix and hydrophobic drug and a dilution phase are main liquid phases in ESD technique (Miladi et al., 2013). The polymer and the guest molecules in ESD are solubilized into a partially or water-soluble solvent (e.g., acetone, methanol, or propylene carbonate). To prevent the formation of aggregates in the emulsion system, the resulted mixture is then emulsified into a watery phase formulated with a stabilizer agent. In the next stage, adding an excess water amount to the system under shaking operation gives a possibility for the solvent to diffuse inside water phase for the formation of particles precipitation (Fig. 7.5b) with EE of more than 70% (Hans and Lowman, 2002; Kwon et al., 2001). Removing the consumed solvent is based on its boiling point either by filtration or evaporation. Water-logging and polymer concentration can highly affect PNPs size in this technique (Choi et al., 2010; Nasir et al., 2015). ESD as an attractive method is applied for encapsulating hydrophobic ingredients, which has many advantages, such as simplicity, easy scaling-up, not any requirement to homogenize, great batch-to-batch reproducibility, and narrow-size distribution for PNPs with good yields. Nevertheless, exerting large volume of water during the process is an important limitation for ESD (Vauthier and Bouchemal, 2008).

Makhlof et al. (2008) used various CDs as stabilizers without adding any surfactants to prepare nanocrystals of a slightly hydrophilic drug "indomethacin" using the ESD technique. In this process, indomethacin into ethanol solvent was solubilized to form an organic phase, then gradually added to an aqueous CD phase with mild stirring and ultimately centrifuged. By diffusing ethanol into bulk water phase during initial emulsion production, indomethacin nanocrystals (~300–500 nm) could be directly formed. Forming a network-like structure by CD in the aqueous medium through intermolecular bindings of CDs can inhibit large crystal aggregations of indomethacin NPs initially produced by the ESD (Makhlof et al., 2008). Similar stabilization mechanism was previously found in another research where budesonide microsuspension was stabilized with HP-β-CD in hydrofluoroalkane (HFA) medium (Steckel and Wehle, 2004). Lopedota et al. (2009) successfully developed NPs with average diameter of 99–156 nm containing CD (types of α, β, γ, HP-β, or Me-β) and Eudragit RS 100 to encapsulate and deliver a model peptide drug "glutathione" (EE of 14.8%–24%) using the ESD method, which was intended to be absorbed via the transmucosal route. In vitro release amount of glutathione was highly influenced by the applied CD type because NPs comprising of γ-CD had the highest release rate of this drug at pH 6.8 (pH of simulated intestinal fluid) over 6 h (Lopedota et al., 2009). The carvedilol-β-/HP-β CD agglomerates using strong (acetone) and poor (water) solvents and a bridging liquid (dichloromethane) were also prepared by the ESD. These structural forms can significantly improve micromeritic properties, solubility, and dissolution behaviors in comparison with pure carvedilol crystals (Tapas et al., 2011).

7.10.1.3 Ionic gelation (coacervation) method

In ionic gelation technique, gradual addition of charged polymer solution to an oppositely charged polyelectrolyte under constant stirring and subsequently homogenization leads to the NPs formation via electrostatic interactions and cross linkages between the two materials (Fan et al., 2012). Using ionic gelation method, Maestrellia et al. (2006) successfully formed NPs based on the 2-HP-β-CD and chitosan (CS) mixtures. They incubated water mixtures having CS and various contents of 2-HP-β-CD under shaking for 1-day and subsequently passed them through paper filters. Negatively charged tripolyphosphate (TPP) as a cross-linking material was consequently added to the prepared solution to have an adjusted gelation mechanism for CS and thus NPs formation. In brief, negatively charged TPP electrostatically interacts with a positively charged amino group of CS to form the nanoscale coacervates (Fig. 7.5c). These nanospheres using centrifuge process were finally separated and water-resuspended. Results demonstrated that there was a possibility to increase drug content in the developed CS/CD NPs using 2-HP-β-CD because of its significant dissolving characteristics (Maestrellia et al., 2006). Trapani et al. (2008) obtained a platform for hydrophilic CDs; such as, carboxymethyl-β-CD and SBE-β-CD and their NPs according to the CS ionic gelation by TPP to deliver both hydrophilic and hydrophobic drug agents. Ieva et al. (2009) encapsulated glutathione into CS/SBE-β-CD NPs (160 ± 30 nm) with an EE (25 ± 3%) based on the ionic gelation process. The NPs made with a weight ratio of 4:3:3 for CS/SBE-β-CD/glutathione. In this process, water-soluble mixtures comprising SBE-β-CD and glutathione were prepared, kept under stirring in nitrogen atmosphere for 1 day at 25°C and then the resulted solutions were added to an acidic solution of CS, next the NPs were formed spontaneously. The TPP presence as a cross-linking agent was not necessary because of the use of the anionic SBE-β-CD (Ieva et al., 2009).

Mahmoud et al. (2011) implemented the present technique to formulate CS/SBE-β-CD NPs (60-673 nm) with a positive surface charge density (22–33 mV) and a low EE (13%–45%) in order to deliver econazole. These researchers concluded that evaluating the ratio of CS and SBE-β-CD was very important to adjust the size of nanoparticles (Mahmoud et al., 2011). Yuan et al. (2013) after grafting a multifeatured polymeric nanocarrier "mono-6-deoxy-6-(*p*-toluenesulfonyl)-β-CD" on CS, cross-linked it with negatively-charged TPP through coacervation method. The aim of this research was to reach ketoprofen-NPs with the ability to be released for the mucus adsorption under a controlled manner with an improved solubility. The delivery NPs of albendazole functional drug based on thiolated carboxymethyl-CS-graft-CD using coacervation method were designed to be applied in mucosal practices (Alamdarnejad et al., 2013). Sajeesh and Sharma (2006) improved the delivery system by encapsulating an insulin/CD complex within NPs developed of poly(methacrylic acid), CS and poly(ethylene oxide)-*co*-poly(propylene oxide) block copolymers. CS, TPP, and carboxymethyl-β-CD were used to develop a novel NP for delivering insulin and heparin drugs. Results showed that the relatively suitable values for EE of insulin (~59%) and heparin (~45%). Although the release behavior revealed a very-fast rate for insulin from 84% to 97% for 15 min, heparin was retained in NPs structure for hours (Krauland and Alonso, 2007).

Nahaei et al. (2013) exerted two aqueous solutions in term of ionic gelation process to develop CS/β-CD NPs encapsulating plasmid DNA (pDNA; pUMVC3-hIL12) encoding forinterleukin-12, which contained CS solution and CD having TPP as a cross-linking agent. Spherical NPs (171.3 ± 2.1 nm) with a high EE (83%) were obtained by mixing the two aqueous phases prepared under vortexing for 10 min (Nahaei et al., 2013). Teijeiro-Osorio et al. (2009) also demonstrated that CS/β-CD NPs made using ionic gelation technique can be employed for the gene delivery into airway epithelial cells. The 100–200 nm NPs with a positive zeta potential (+22 to +35 mV) had considerable efficacy (more than 90%) of DNA association.

7.10.1.4 Spray drying method

Spray drying due to the simple operation, good yield, and easy scale up is one of the most frequently used techniques to prepare a dry powder from a liquid phase (Rajabi et al., 2015; Dehnad et al., 2016). It as a rustic and single-step production technique can provide a suitable physical and chemical stability for NPs (Jafari et al., 2007, 2008). NPs are usually obtained from the powders containing polymeric materials. Spray drying of a transparent solution prepared with fully solubilizing all ingredients into appropriate solvents can develop solid NPs. Evaporating the solvent and precipitating the polymer are done when the droplets are transferred into a high-temperature zone. Since NPs formed in this method have an irregular shape and/or surface, it is named "nanocorpuscle" (Peltonen et al., 2010).

Yousaf et al. (2015) have lately studied oral bioavailability of slightly hydrophilic fenofibrate using PNPs formed by spray drying method. They using a spray-dryer system developed NPs based on the various components under the constant operating parameters of feed flow rate (3 mL/min), air pressure (4 kg/cm^2), aspiration air flow (100%), and inlet (100°C) and outlet (60–65°C) temperatures. Each clear mixture formulated with wholly solubilized ingredients into 95% ethanol in an ideal proportion was sprayed into the dryer chamber. Nanospheres of polyvinylpyrrolidone (PVP) and nanocorpuscles of HP-β-CD were produced with two optimal combinations of fenofibrate-PVP-sodium lauryl sulfate (2.5:4.5:1.0 w/w/w) and HP-β-CD-fenofibrate (4:1 w/w), respectively. However, a weight ratio of 1:8 for fenofibrate/gelatin to form gelatin nanocorpuscles was considered. For this aim, having solubilized the constituents into ethanol solvent (40%) at 50°C, the resulted transparent mixture was spray-dried at the optimal inlet (120°C) and outlet (65°C) temperatures (Yousaf et al., 2015). Findings demonstrated that all formulations considerably had higher water-solubility, dissolution rate, and in vivo bioavailability of fenofibrate than the powdered one. Although there was an amorphous form of fenofibrate into HP-β-CD nanocorpuscles, this drug in other preparations was crystalline with a decreased stability. Therefore, an interesting oral medicinal product to administer slightly-hydrophilic fenofibrate with the increased bioaccessibility rate can be the fenofibrate-loaded gelatin nanocapsules prepared by spray-drying method (Yousaf et al., 2015).

7.10.2 Chemistry based processes

7.10.2.1 Emulsion polymerization method

In general, the polymerization techniques can be categorized into two groups; nanospheres and nanocapsules are commonly formed by emulsion polymerization (EP) and interfacial polycondensation in emulsions or microemulsions, respectively (Pinto Reis et al., 2006). These techniques have substantial limitations, particularly for NPs formation which are as follows (De Jaeghere et al., 1999):

1. Molecular weight prediction of the resultant substance is hard and somewhat impossible. Since this factor has a significant effect on the biorepartition and liberation of the polymer carrier, difficulty of the molecular weight determination can be considered as the main disadvantage.
2. The guest activity can be inhibited in the attendance of free-radicals or many ions of H^+ produced through anionic polymerization procedure.
3. Removal of poisonous unreacted monomer, initiator, and surfactant often is carried out during an inefficient and long-time process.

EP as one of the quickest preparation techniques for NPs is easily practicable (Pinto Reis et al., 2006). There are two subgroups for EP method according to the application of a water- or organic-continuous phase (conventional (O/W) and inverse (W/O) emulsions). Two dissimilar ways are applied to produce NPs using polymerization processes which include (Shapiro, 2004): (1) EP: polymerization of a monomer via its emulsification into a nonsolvent phase, and (2) dispersion polymerization: monomer polymerization by its dissolving into a solvent which is considered as a nonsolvent for the subsequent polymer.

EP process thus is a heterogeneous polymerization method which is applied water as a dispersion phase for polymerizing hydrophobic monomers in terms of suspended particles. The reagents involved in this process consist of distilled water, hydrophilic initiator (e.g., potassium persulfate), and monomers (e.g., styrene, methyl methacrylate, butyl acrylate, etc) (Rao and Geckeler, 2011). Solubilizing the emulsifiable monomer into water-continuous phase is done without the addition of surfactants or emulsifiers. This approach is called as surfactant free, emulsifier free, or soapless EP technique, if no surfactant is used in the system. However, the surfactants (cationic, anionic, or nonionic) are commonly applied to stabilize the colloidal system and produced micelles into the system wherein the polymerization process takes place (Fig. 7.6a).

Various mechanisms are usually involved in initiating the polymerization process. Generally, it happens by dissolving a monomer component into an oil/water-continuous phase (W/O or O/W emulsion) with an ion or a free-radical as the initiator. On the other hand, use of high-energy UV and visible radiations can transform the monomer molecule into a starter radical. The collision of initiated monomer ions/radicals with other monomers through an anionic polymerization process leads to a beginning in the chain growth (Vauthier et al., 2003). The phase differentiation and creation of solid particle are occurred before or after the end of the polymerization process. The difficulties present in controlling temperature and viscosity related with bulk polymerization can be generally solved using EP. An increase in the polymer molecular weight without decreasing the polymerization rate can be obtained by changing the surfactant amount, which is not possible in other modes of polymerization (Kreuter, 1982; Pinto Reis et al., 2006).

Monza da Silveira et al. (1999) prepared the parent/modified CDs-based NPs under acidic medium using anionic polymerization of alkyl cyanoacrylates. Then, they entrapped functional ingredients (e.g., taxol and doxorubicin, 0.01–300 mg/g) in the NPs prepared with CD (Monza da Silveira et al., 1999). Monza da Silveira et al. (1998) earlier combined 2HP-β-CD with polyisobutylcyanoacrylate (PIBCA) and prepared PIHCA nanospheres by EP of isobutyl- and isohexyl cyanoacrylate monomers to deliver progesterone with a significant increase in LC (50-fold). The size of NPs can be also reduced by rising quantities of CD. Monza da Silveira et al. (1998) introduced this mixture as a good system for delivering water-soluble ingredients. Boudad et al. (2001) utilized FDM formed complex HP-β-CD with saquinavir drug to enhance its solubility before carrying out polymerization. As a result, a 400-fold increase in saquinavir solubility using HP-β-CDs in neutral pH was observed. They then could easily prepare saquinavir-loaded NPs in the presence of a drug-CD complex. The particle diameter is variable (250–350 nm) according to the HP-β-CD amount. Applying the CD molecule led to a 20-times enhancement in LC of saquinavir into the NPs in comparison with the NPs produced without the presence of CD (Boudad et al., 2001). Rimmer (2000) investigated β-CD effect on the polymerization of various vinyl monomers, such as

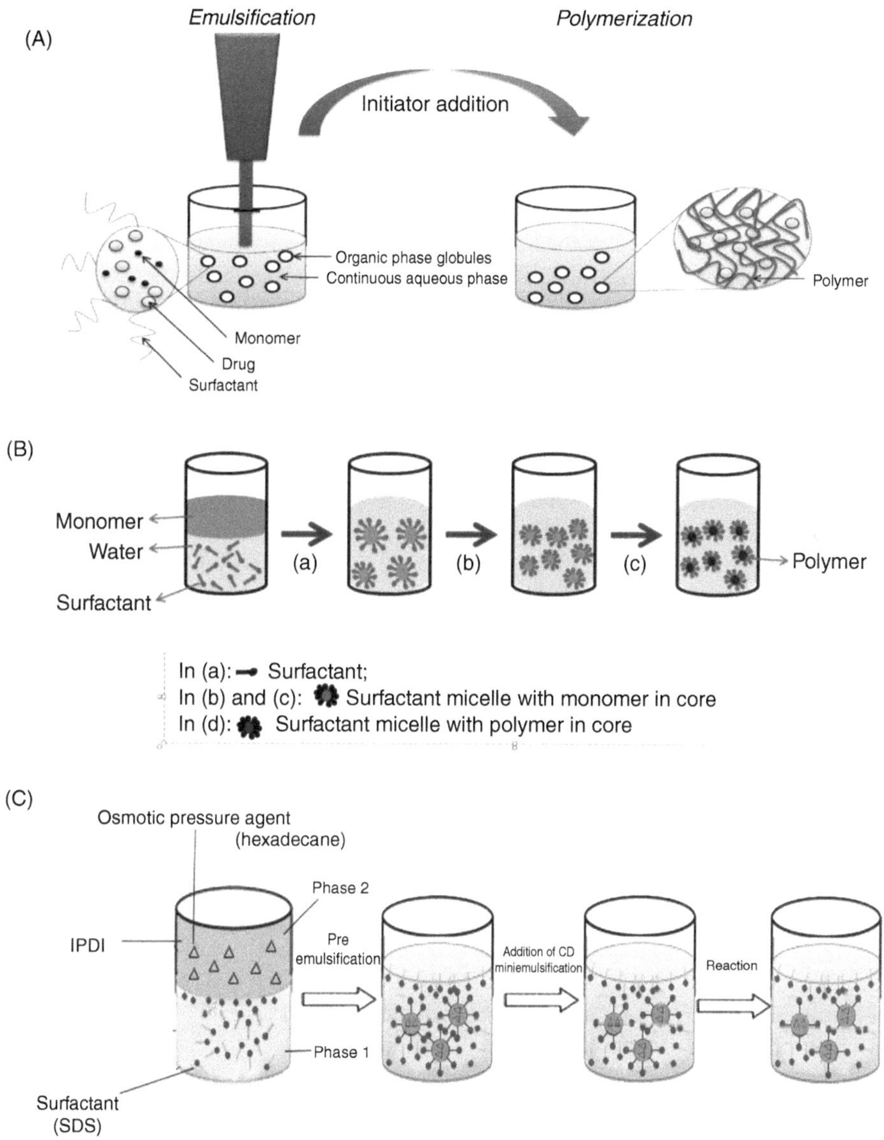

Figure 7.6 Schematic representation of (A) EP, (B) MEP [a, preemulsification; b, high-shear emulsification; and c, polymerization], (C) MIP processes for the CD-nanoencapsulation of guest (drug) molecules.
Source: Reprinted with permission from Zafar, N., Fessi, H., Elaissari, A., 2014. Cyclodextrin containing biodegradable particles: From preparation to drug delivery applications. Int. J. Pharm. 461, 351–366 and Rao, J.P., Geckeler, K.E., 2011. Polymer nanoparticles: preparation techniques and size-control parameters. Prog. Polym. Sci. 36, 887–913, Copyright © 2016, Elsevier Ltd.

butyl-methacrylate, styrene, dodecyl-methacrylate, and octadecyl-methacrylate. He used EP method to prepare particles of poly(butyl-methacrylate) (PBMA) using β-CD in place of surfactant (Rimmer, 2000). The inclusion of PBMA chains within β-CD led to a high colloidal stability in the polymerization of BMA and styrene monomers. The solubility of highly-insoluble dodecyl and octadecyl methacrylate monomers was also increased by β-CD and subsequently augmented emulsion homopolymerization of these monomers (Rimmer, 2000). Fresta et al. (2001) prepared nanospheres of acyclovir-loaded polyethyl-2-cyanoacrylate (PECA) using an EP method in the micellar phase. They then scrutinized the effect of nonionic surfactant and other formulation components, such as 2HP-β-CD and PEG on NPs formation and LC. The PEG-coated PECA nanospheres loaded with acyclovir revealed a considerable enhancement of 25-times in drug concentration into aqueous humor in comparison to the free-drug mixture. Using EP process and various forms of CDs (e.g., HP-β-CD, β-CD, Me-β-CD), Yang et al. (2011) formed semifluorinated acrylic NPs possessing nonfluorinated core and wall material comprising of copolymers of water-insoluble and soluble FMA (heptadecafluorodecyl methacrylate). Sajeesh and Sharma (2006) using EP technique prepared NPs based on polymethacrylic acid encapsulating CD–insulin complex to deliver insulin (EE of about 80%). An improvement in therapy by stabilization the protein counter to aggregation, as a result, thermal denaturation was obtained by insulin complexation with CD. Complexation with CD showed no change in release characteristics of insulin from the NPs (Sajeesh and Sharma, 2006). Finally, Van Herk (2012) also deliberated very valuable data about the latest advancement in EP and the main factors in the modeling kinetics of emulsion copolymerization.

7.10.2.2 Microemulsion polymerization method

Novel and efficient microemulsion polymerization (MEP) method has been recently used to prepare PNPs using an extensive number of suitable polymers. MEP provides versatile PNPs by employing different combinations of costabilizers and initiators (Bardajee et al., 2007; Baruch-Sharon and Margel, 2010; Crespy and Landfester, 2009; Ethirajan et al., 2009; Jiang et al., 2010; Rao and Geckeler, 2011; Rotureau et al., 2008; Wu et al., 2009b; Wang et al., 2007; Yildiz and Landfester, 2008). Water, monomer mixture, costabilizer, surfactant, and initiator are consisted of main ingredients of a classic formulation applied in MEP (Ham et al., 2006; Ziegler et al., 2009). Some of the key parameters affecting the MEP kinetics and the PNPs characteristics are the process temperature plus the type and concentration of the initiator, monomer, and surfactant (Fessi et al., 1989; Puig, 1996). Both EP and MEP techniques appear similar in the way that they are able to prepare colloidal PNPs with a considerable molar weight. However, EP and MEP are completely different when compared kinetically (Nagavarma et al., 2012). Rao and Geckeler (2011) pointed out the application of low-molecular weight costabilizer and high-shear devices like ultrasound are the main variance between these two methods. In general, the MEP has a smaller particle size and a lower mean quantity of polymer chains than those of the EP (Nagavarma et al., 2012).

MEP includes a proportionate emulsion or a steady dispersion containing monomer droplets within the size of 50–500 nm. In the MEP process (Fig. 7.6b); the organic and water phases at a certain shaking rate are mixed to attain a thermodynamically stable microemulsion. The aqueous phase is composed of deionized water and surfactant, whereas water-soluble initiator and solvent are derived from the organic phase. The application of a mechanical source (e.g., ultrasound) generates high energy for breaking down microemulsion droplets and their inversion into nanodroplets. The size of monomer droplets quickly vary during the mechanical shaking operation and lastly lead to a balance condition with the lowest possible particle size in nanoscale (Udagama et al., 2011; Yildiz and Landfester, 2008). The polymerization initiates under such stable conditions based on significant amounts of surfactant components to reduce interfacial tension in the O/W boundary layer approximately to zero. Owing to the use of high surfactant quantities, the surface active agent can completely cover the particles. Since the initiation cannot be concurrently achieved in all microdroplets, formation of polymer chains firstly took place in certain droplets. In the next step, the delicate microemulsions are destabilized by establishing osmotic and elastic forces in the chains. This fact usually causes an enhancement in the particle size, the development of unfilled micelles, and subordinate nucleation. Really fine latexes (5–50 nm) and a considerable amount of unfilled micelles could be together in the last sample (Rao and Geckeler, 2011).

Baruch-Teblum et al. (2010) utilized MEP to synthesize cross-linked CD-PNPs with different sizes (30–500 nm) as components for water purification systems. A cross-linking agent named isophorone diisocyanate (IPDI) modified β-CD. An emulsifier (sodium dodecyl) and a costabilizer (hexadecane) were used in the formulation. The applied costabilizer had a key role inside the formed droplets by inducing osmotic pressure (Fig. 7.6c). Adsorbing the organic chemicals particularly having aromatic rings (e.g., toluene and phenol) is an interesting practice for these CD-NPs. A considerable reduction in the organic pollutants of wastewater was occurred using the developed nanosphere particles. The advantage of these polymeric nanospheres to inorganic materials is their higher flexibility in processing which can provide a means to convert them into granulated solid materials/powders, edible nanocomposite films, and probably fantastic membranes. The high pliability provides a potential to utilize these nanomaterials in multiple applications and formats, thus accommodating various routs and needs in water-treatment. However, CD nanospheres in comparison to activated carbon have several limitations; such as their effectiveness loss due to no moisture absorption from the air (Baruch-Teblum et al., 2010).

7.10.2.3 Interfacial polycondensation method

Interfacial polymerization is a suitable technique to prepare PNPs. In this gradual-growth polymerization, two monomer components into two immiscible continuous- and dispersed-phases are polymerized to generate a polymer film layer in a small area next to the boundary layer of the phases (Charcosset and Fessi, 2006; Gaudin and Zydowicz, 2008; Karode et al., 1998; Lambert et al., 2001; Landfester et al., 2010; Wu et al., 2009a). On the contrary, in order to obtain oil-containing nanocapsules through interfacial polycondensation, a reaction is initiated in the boundary layer of oil droplets added into the water phase of a very fine O/W microemulsion fabricated using spontaneous emulsification method. There are water-soluble surfactant and monomer into the water phase, whereas the organic phase is composed of a water-insoluble solvent, monomer, and oil. Since monomers are present in both of the emulsion phases, the polycondensation can easily form at the interface of the oil droplets of O/W emulsion via spontaneous emulsification (Bouchemal et al., 2006a; Montasser et al., 2007). On the other hand, the interfacial polymerization reaction between the composed monomer units in W/O emulsions can supply water-containing nanocapsules. The polymer in such structures is locally precipitated in the boundary layer of W/O to form the shell of nanocapsules (Gasco and Trotta, 1986; Watnasirichaikul et al., 2000).

Nanoscale hollow polymer particles can be synthesized through interfacial cross-linking processes including poly-addition and polycondensation (Crespy et al., 2007; Danicher et al., 2000; Torini et al., 2005) and radical polymerization (Sarkar et al., 2005; Scott et al., 2005). The cover of nanocapsule wall can be formed with a polymer that can be defined by the nature of the monomers used. The polymeric wall of the nanocapsules usually involve some synthetic polymers, such as polyurethanes, poly(ether urethanes), polyamides, and polyurea (Bouchemal et al., 2004, 2006a; Vauthier and Bouchemal, 2008). The monomers concentration and their molecular weight can lead to a significant change in the wall thickness. These parameters along with porosity of nanocapsules are considered as the main influential factors on loading, preservation and release rate of foods or nutraceuticals (Bouchemal et al., 2006b). Aprotic solvents (e.g., acetone, acetonitrile, ethanol, *n*-butanol, and isopropanol) were proposed to promote the nanocapsule formation (Nagavarma et al., 2012). Puglisi et al. (1995) found that some of these solvents can provide an ideal opportunity to form NPs.

Chelli et al. (2009) prepared polyrotaxanes based on the polycondensation so that β-CD was threaded onto a polyether chain by polycondensation of β-CD/bisphenol A (PBA) inclusion complex with aromatic amines. The inclusion of polyether chains in the β-CD cavity was also demonstrated by the NMR spectrum. Skiba et al. (2005) encapsulated amikacin into 5 μm β-CD based microspheres with EE higher than 50%. For the polycondensation of β-CD by terephtaloyle chloride (TPC), TPC was dissolved in a mixture of chloroform/cyclohexane to constitute organic phase and then β-CD dissolved in NaOH was emulsified in cyclohexane solution containing sorbitan85 trioleate under constant stirring. In the next phase, microparticles were constituted by adding the organic phase to emulsion under stirring. Cyclohexane was added to finish the cross-linking reaction. Formulation of the desired microparticles was finally obtained by centrifugation and freeze-drying processes (Skiba et al., 2005).

7.11 Cyclodextrin-based magnetic nanoparticles

Progress in the fabrication of nano-structures and nanoobjects have recently led to the design of magnetic nanoparticles (MNPs). MNPs are very valuable for biomedical, biological, food, and environmental applications because of the exceptional size, biocompatibility, low toxicity, and superparamagnetic characteristics (Medeiros et al., 2011).

They have been highly employed in magnetic bioseparation, tumor thermotherapy, magnetic resonance imaging (MRI), diagnostic agents, encapsulation of bioactive ingredients, targeted drug delivery, biomolecule immobilization, and designing the iron oxide-based novel bio thin films (Ahmd et al., 2012; Ahsan et al., 2013; Laurent et al., 2010; Medeiros et al., 2013; Rahman and Elaissari, 2012; Shubayev et al., 2009; Yallapu et al., 2012).

Researchers studied a number of iron oxides to synthesize magnetic particles with a size of 5–20 nm (e.g., maghemite, γ-Fe_2O_3, or magnetite, Fe_3O_4). Meanwhile, Fe_3O_4 is taken into account as an attractive and popular option because of its biocompatibility character (Gupta and Gupta, 2005). Many synthetic techniques including coprecipitation (Jiang et al., 2004), formation of sol–gel, and microemulsion (da Costa et al., 1994; Deng et al., 2003), solution-based sonochemical or hydrothermal reactions (Abu Mukh-Qasem and Gedanken, 2005; Chen and Xu, 1998), thermal decomposition (Sun and Zeng, 2002), and laser pyrolysis (Veintemillas-Verdaguer et al., 1998) were used to fabricate iron oxide MNPs. Coprecipitation technique owing to the simplicity and efficiency is applied more than the other methods in order to form Fe_3O_4 particles. In this technique, Fe^{2+} and Fe^{3+} (in a 1:2 molar proportion) are mixed into an alkaline solution at ambient or higher temperatures. A modification of formed particles using CDs can provide the combined benefits of both CDs and magnetic particles (Wu et al., 2008).

There is no toxicological concern in relation to the iron oxide applied in MNPs fabrication due to its biodegradability and storage into body iron reservoirs. Nevertheless, a considerable accumulation of uncoated MNPs is the main problem not only for in vitro application but also for their in vivo biomedical usage. Two efficient solutions for this difficulty are inducing the steric stability in the involved biomolecules via their chemical grafting (Yallapu et al., 2012), and use of surfactants and polymers as stabilizing coatings through coating process in term of surface engineering of MNPs (Shubayev et al., 2009; Wu et al., 2008). Many research works for grafting β-CD to form a novel group of pharma carriers on the MNPs surface for biohealth applications have been recently conducted. CD-based MNPs are thus developed and fabricated by conjugation or grafting CDs with the drug through a linker moiety to achieve targeted delivery of hydrophobic anticancer drugs.

Banerjee and Chen (2007) developed a magnetic nanocarrier ($\sim$15 nm) for ketoprofen via grafting citrate-modified CD on MNPs with a unique LC and modified with arabic gum (AG). Attractive features of this system were the magnetic properties of Fe_3O_4, CD inclusion characteristics and bioadhesive nature of AG. The adsorption behavior of the exothermic complex between CD-citrate-AG-modified MNPs and ketoprofen followed the Langmuir isotherm. They demonstrated that the formed complex system can be an ideal carrier for oral administration of water-insoluble drugs due to the successful evaluations during the loading and release tests (Banerjee and Chen, 2007). Xia et al. (2007) prepared water-soluble MNPs with a surrounded layer by using PEG nonylphenyl ether (N5) and CD in an aqueous medium. Lee et al. (2013) functionalized some MNPs by adamantane units to incorporate into supramolecular aggregates of adamantanamine grafted polyamidoamine dendrimers, β-CD-grafted branched polyethyleneimine, and adamantane-PEG. The formed nanoexcipients had a specific capability to magnetothermal release of doxorubicin in a colorectal adenocarcinoma cell line.

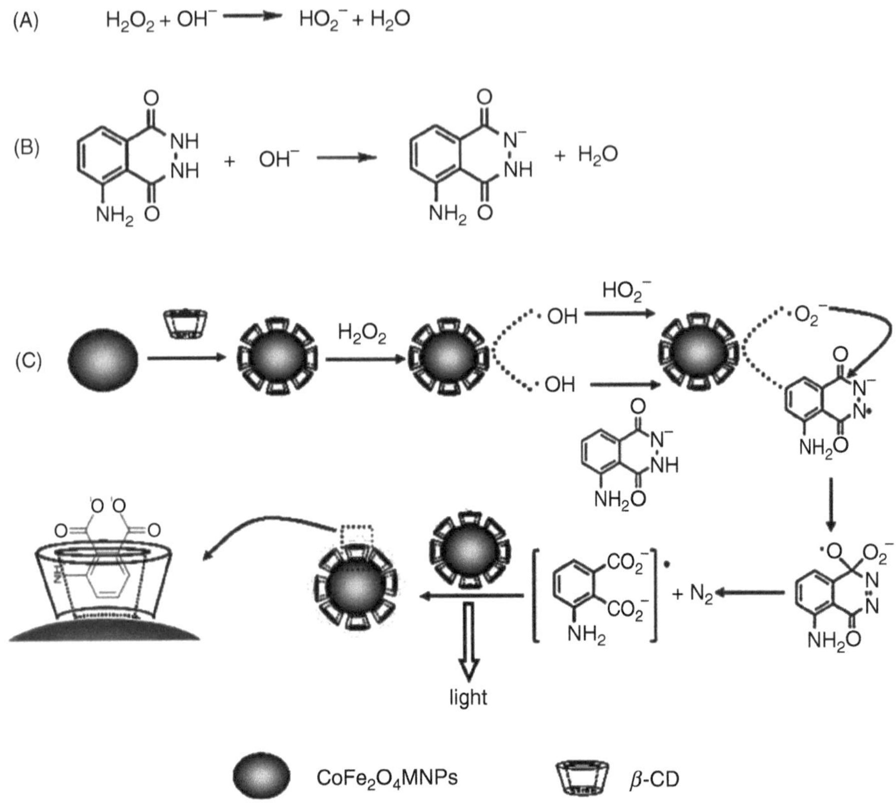

Figure 7.7 Possible CL process of luminol-H_2O_2 in the presence of inclusion complexes between β-CD and CoFe2O4 MNPs.

Source: Reprinted with permission from He, S., Shia, W., Zhang, X., Li, J., Huang, Y., 2010. β-cyclodextrins-based inclusion complexes of CoFe2O4 magnetic nanoparticles as catalyst for the luminol chemiluminescence system and their applications in hydrogen peroxide detection. Talanta 82, 377–383, Copyright © 2016, Elsevier Ltd.

The inclusion complex based on β-CD of $CoFe_2O_4$-MNPs was formed to catalyze chemiluminescence (CL) model system by the luminol-hydrogen peroxide CL reaction (Fig. 7.7). Results showed that the inclusion complexes between β-CD and $CoFe_2O_4$ MNPs could significantly increase the CL of the luminol-H_2O_2 (He et al., 2010). Nowicki et al. (2006) had previously proved that β-CD can stabilize the catalytically active NPs in water via both mechanisms of hydrophobic/hydrophobic interactions of metal NPs-CDs and hydrogen bonds between metal NPs and hydroxyl groups of native CDs. An efficient peroxidase-like catalyst (Prussian-blue-modified iron oxide MNPs) is also used to break down methylene blue by H_2O_2 (Wang and Huang, 2011). Guo et al. (2010) also reported CDs functionalized graphane nanosheets with considerable molecular-cloud diagnosis ability. They pointed out that the modification MNPs surface using CDs not only can enhance the MNPs stability and spreading, but can also increase the detection precision for several essential bioactive compounds via the production of molecular-cloud complexes of CD–guest (Guo et al., 2010).

Sahu and Mohapatra (2013) have recently prepared the modified fluorescent MNPs (~71 nm) with β-CD to magnetically deliver 5-fluorouracil. Results revealed that the complexation can lead to an increase in aqueous solubility and also fluorescent carrier can cause an enhancement in stability and magnetic delivery of a drug. The potential of β-CD-conjugated 70 nm MNPs with ideal monodispersity character as stripping agents for protein refolding was explored by Badruddoza et al. (2013). Magnetic Fe_3O_4 particles were encapsulated within the shell of silicon-dioxide using folic acid as the cell targeting ligand and β-CD as the vehicle. These theranostic nanocomposite magnetic particles had multi-functional characteristics, such as cell targeting, fluorescence imaging, and delayed release so that these NPs during the fluorescence imaging of cellular absorption had a more priority for cancer cells (Hela) than healthy cells (Badruddoza et al., 2013). CS-cross-linked β-CD-grafted silylated MNP was fabricated for delivering target indomethacin drug and for preventing unpleasant complications and gastrointestinal (GI) bleeding. Moreover, a pH-dependent behavior for the in vitro drug release was monitored (Anirudhan et al., 2013).

Ozyilmaz et al. (2014) immobilized a silica-based β-CD on MNPs for achieving a macrocyclic component with magnetic character. Next, the β-CD-grafted MNPs with *Candida rugosa* lipase using alkoxysilane precursors were encapsulated in sol–gel matrices. They found that the encapsulated lipase mainly showed higher conversion and enantioselectivity than those of sol–gel free lipase. Kang et al. (2011) synthesized β-CD-modified hybrid MNPs (Fe_3O_4@SiO_2-PGMACD) through both mechanisms of ring-opening reaction of epoxy groups and atom transfer radical polymerization (ATRP) on Fe_3O_4-NP surfaces coated with silica (Fe_3O_4@SiO_2; ≈ 60 nm). β-CD-modified hybrid MNPs (Fe_3O_4@SiO_2-PGMACD) prepared in this study had a potential to be applied as a catalyst in substrate-selective oxidation of alcohols. A similar catalytic effectiveness for these MNPs with the sole β-CD system was found. Moreover, the obtained NPs had significantly higher adsorption abilities than poly(glycidyl methacrylate) grafted-MNPs (Fe_3O_4@SiO_2-PGMA) in order to eliminate bisphenol A in water mixtures (Kang et al., 2011). TEM images of Fe_3O_4, Fe_3O_4@SiO_2 and Fe_3O_4@SiO_2-PGMACD are illustrated in Fig. 7.8.

7.12 Layer by layer (LBL): an ideal process to form nanoparticles

The layer-by-layer (LBL) assembly technique was first designed for producing colloidal particles. LBL provides ideal conditions for achieving typical spongy particles "polyelectrolyte capsules" with distinct structural and biochemical characteristics (Munin and Edwards-Lévy, 2011). This technique is based on the alternate adsorption of components with opposite charge especially linear polyelectrolytes on mineral or organic chemicals which can form the particle core through electrostatic interactions. In other words, mechanism of the nanocapsule production in LBL can be a self-assembly method according to irreversible electrostatic interaction of polymers

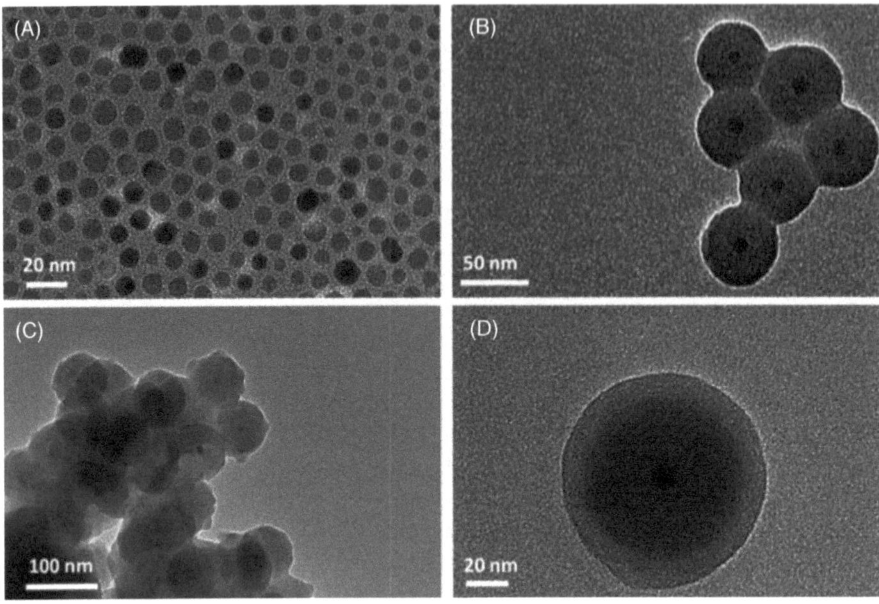

Figure 7.8 TEM images of (A) Fe_3O_4, (B) $Fe_3O_4@SiO_2$, (C) and (D) $Fe_3O_4@SiO_2$-PGMACD.
Source: Reprinted with permission from Kang, Y., Zhou, L., Lib, X., Yuan, J., 2011. β-Cyclodextrin-modified hybrid magnetic nanoparticles for catalysis and adsorption. J. Mater. Chem. 21, 3704–3710, Copyright © 2016, RSC.

with dissimilar charges; causing the generation of membranes with uniform and controlled thickness based on the count of accumulated layers (De Cock et al., 2010; Lvov et al., 1995).

The nanoencapsulation process by LBL involves following steps: (1) incubating a positively-charged solid support; such as, drug or food into a liquid mixture comprising of polyanion for 0.5 h for adsorption of a polyanion layer, (2) several times water-washing of the solid supports in order to eliminate extra free polyelectrolyte, (3) immersing the solid support into a liquid mixture containing cationic polyelectrolytes and adsorption of a layer. This process can lead to a 4–20 layers wall for capsules with 8–50 nm thickness (Bertrand et al., 2000; Mora-Huertas et al., 2010). Thin-film deposition method on a flat surface by LBL self-assembly and also bioactive molecules encapsulation using LBL self-assembled polyelectrolyte polymers are schematically shown in Fig. 7.9.

7.12.1 β-Cyclodextrin-modified Fe_3O_4 magnetic nanoparticles by LBL method

As mentioned earlier, LBL approach can be used to synthesize magnetic latex through the spontaneous electrostatic potential of polymer chains. LBL technique

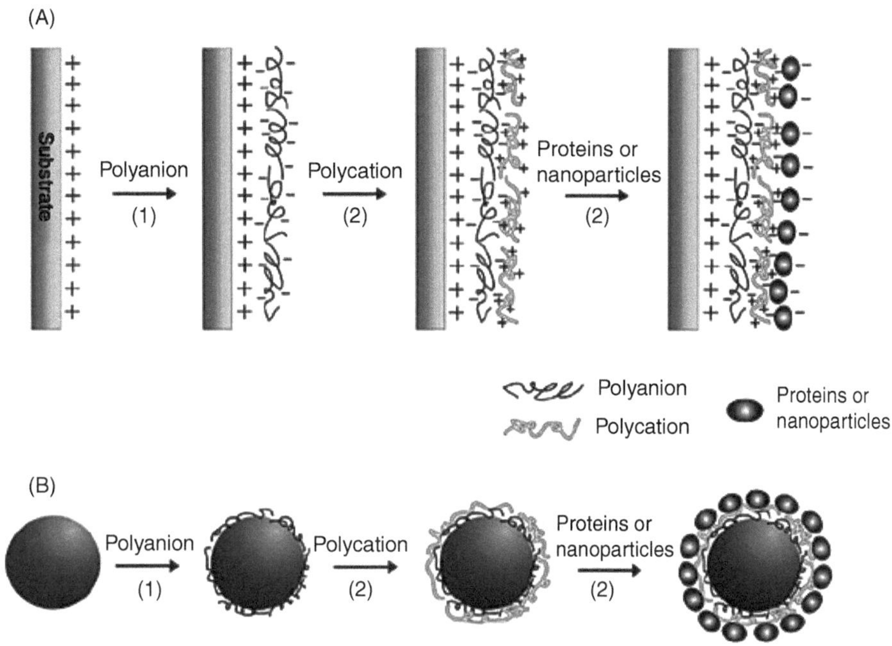

Figure 7.9 The procedure of electrostatic LBL self-assembly on 2-D substrates (A) and 3-D micro/nano-templates (B).
Source: Adapted from Ai, H., Jones, S.A., Lvov, Y.M., 2003. Biomedical applications of electrostatic layer-by-layer nano-assembly of polymers, enzymes, and nanoparticles. Cell Biochem. Biophys. 39, 23–43.

without an expensive device can provide a possibility to develop the coated latex having a homogeneous and controllable thickness in various morphologies and sizes (Rahman and Elaissari, 2010). Multifunctional magnetic latexes can be provided by introducing the different functional groups onto the particles surface (Kim et al., 2002).

By employing mono-6-deoxy-6-(*p*-tolysulfonyl)-CD (6-TsO-β-CD), Zhu et al. (2011) fabricated a novel surface-functionalized MNPs by the Fe_3O_4-NPs. They first prepared monotosyl-CD particles by modification of β-CD and then NaOH aqueous solution was added to β-CD suspended in water. The *p*-toluenesulfonylchloride (TsCl; in acetonitrile) was dropwise added to this mixture for the immediate formation of white precipitates of 6-TsO-CD. The γ-aminpropyltriethoxysilane (APTES) was used to modify the magnetic particles surface. Then, covalent interactions of monotosyl-CD onto the modified iron NPs surface by LBL process was formed (Fig. 7.10). The TEM photographs given in Fig. 7.11 illustrate the size and morphology of fabricated MNPs and CD-MNPs. The NPs (~16 nm) showed a good dispersion and a uniformity in their size and structure. Nevertheless, there were some agglomerated clusters without the presence of surfactants. Therefore, use of

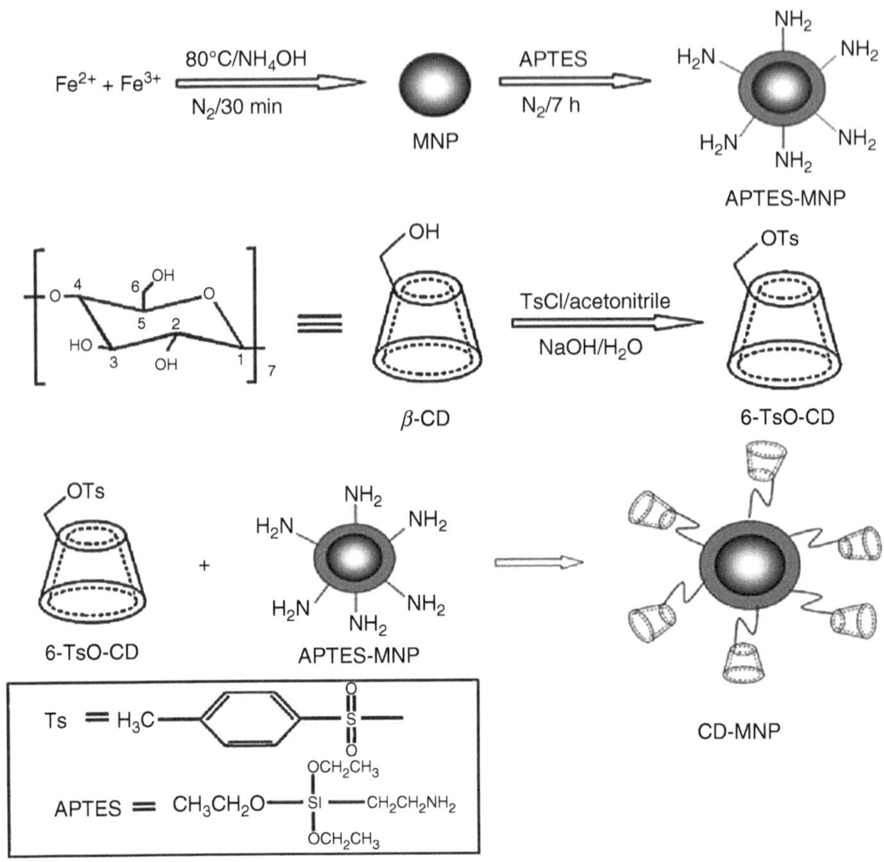

Figure 7.10 The preparing procedure of CD-APTES-MNPs.
Source: Reprinted with permission from Zhu, J., He, J., Du, X., Lu, R., Huang, L., Ge, X., 2011. A facile and flexible process of β-cyclodextrin grafted on Fe_3O_4 magnetic nanoparticles and host–guest inclusion studies. Appl. Surf. Sci. 257, 9056–9062, Copyright © 2016, Elsevier Ltd.

TEM can directly provide a chance to monitor the successful development of CD functionalized MNPs. Uric acid could produce the CDs-guest complexes with high stability. Furthermore, detection of NPs by using electrochemical techniques reveals excellent functional characteristics of CD-MNPs and molecular diagnosis characteristic of CDs. Regarding the obtained data, NPs can be applied as pharma carriers, bioseparators, and biosensors (Zhu et al., 2011).

Cao et al. (2009) effectively fabricated multifunctionalized Fe_3O_4 MNPs with diameters less than 20 nm using LBL method in order to form the complex with uric acid. They combined APTES-coated magnetic Fe_3O_4 NPs (APTES-MNPs) with 6-*O*-toluenesulfonyl (TsO)-β-CD (6-TsO-β-CD) via substituting the —NH_2 group on the

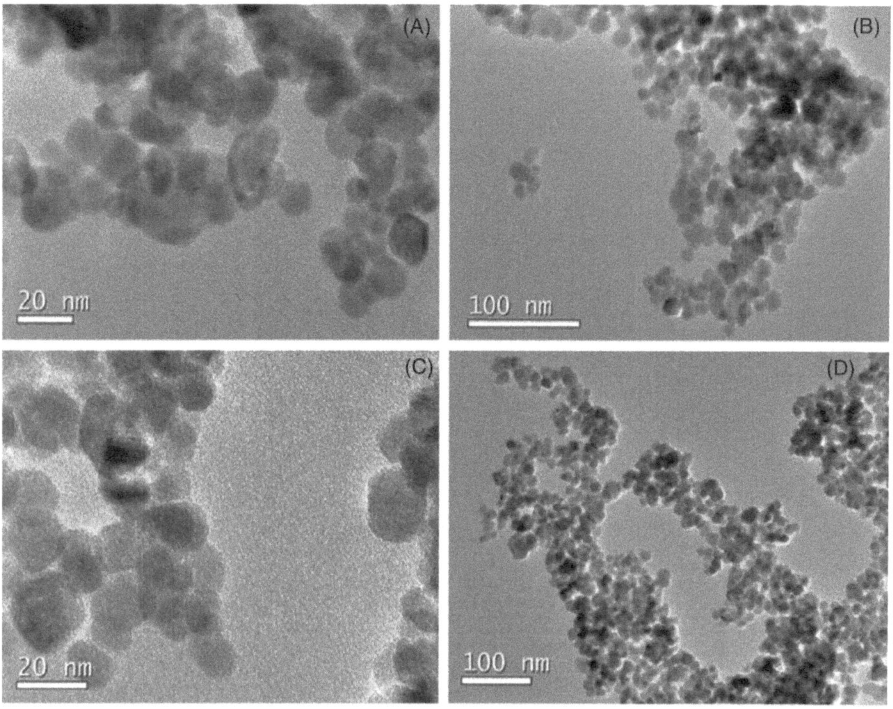

Figure 7.11 TEM images of Fe_3O_4-MNPs (A and B) and CD-MNPs (C and D).
Source: Reprinted with permission from Zhu, J., He, J., Du, X., Lu, R., Huang, L., Ge, X., 2011. A facile and flexible process of β-cyclodextrin grafted on Fe_3O_4 magnetic nanoparticles and host–guest inclusion studies. Appl. Surf. Sci. 257, 9056–9062, Copyright © 2016, Elsevier Ltd.

C-6 of 6-TsO-β-CD under moderate reaction conditions. To achieve this, monotosyl-CDs can covalently bond to the surface modified MNPs using LBL technique (Fig. 7.12).

7.12.2 Cyclodextrin-based biodegradable polyelectrolyte nanocapsules by LBL method

Since hydrolysis of peptides and proteins is simply done under the stomach acidic medium, proteolytic enzymes present in the GI tract and also enterohepatic circulation metabolism, these macromolecule drugs are commonly administered intravenously. There is a limited number of therapeutic peptide agents in the international market for clinical applications. The development of oral administration systems of macromolecules can be thus an important and challenging opportunity for scientists and industrialists (Shu et al., 2009).

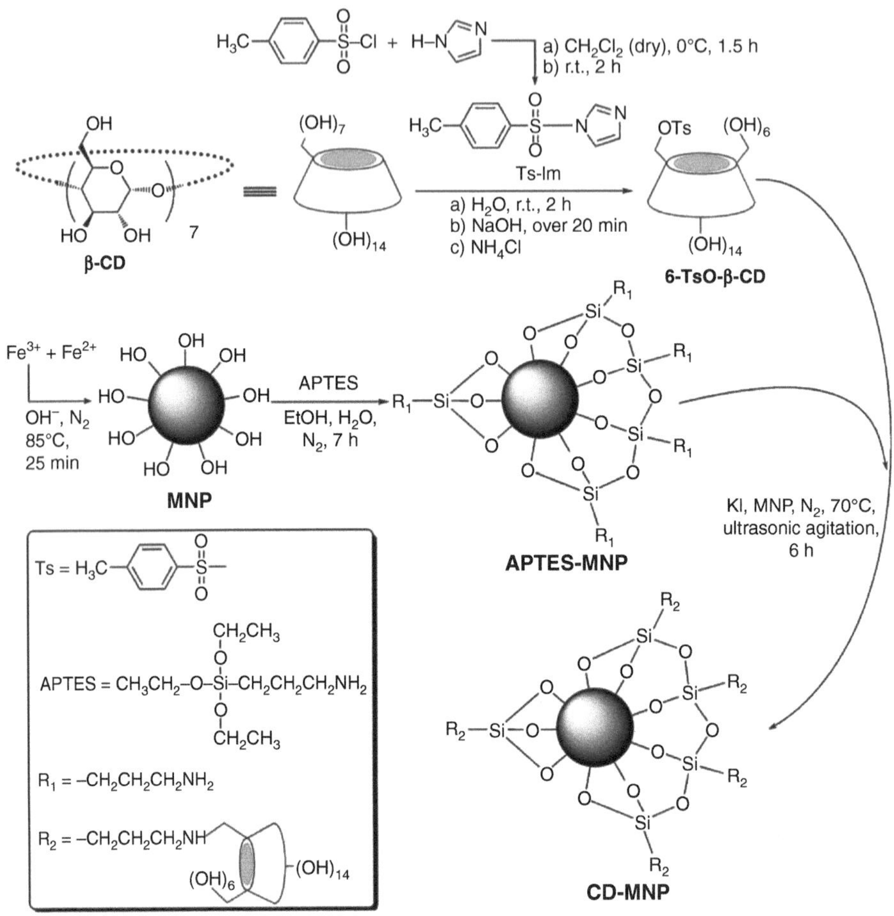

Figure 7.12 Preparation steps of CD-functionalized Fe_3O_4/SiO_2 core–shell nanocomposite (CD-MNP) fabrication.
Source: Reprinted with permission from Cao, H.N., He, J., Deng, L., Gao, X.Q., 2009. Fabrication of cyclodextrin-functionalized superparamagnetic Fe_3O_4/amino-silane core–shell nanoparticles via layer-by-layer method. Appl. Surf. Sci. 255, 7974–7980, Copyright © 2016, Elsevier Ltd.

Shu et al. (2010) used the LBL adsorption on β-CD functionalized silica spheres after that cross-linking thiols and exclusion of silica core prepared gradient shell cross-linked hollow polyelectrolyte nanocapsules comprising cysteamine conjugated CS and dextran sulfate. Synthesis of β-CD conjugated silica NPs and generation of shell gradient cross-linked hollow nanocapsules are schematically depicted in Figs. 7.13 and 7.14. The model protein drug in this study was bovine serum albumin (BSA), which is loaded within the spherical nanocapsules developed with an average size of 126.7 ± 8.2 nm. The BSA loaded polymeric outer shell consists dextran sulfate and CS decorated with

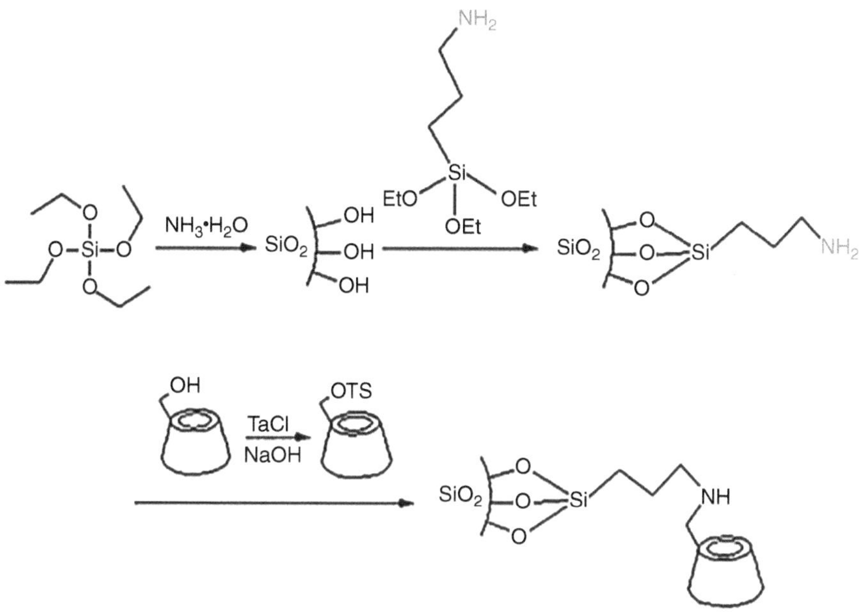

Figure 7.13 Synthesis route of β-CD conjugated silica NPs.
Source: Reprinted with permission from Heo, D.N., Yang, D.H., Moon, H.J., Lee, J.B., Bae, M.S., Lee, S.C., Lee, W.J., Sun, I.C., Kwon, I.K., 2012. Gold nanoparticles surface-functionalized with paclitaxel drug and biotin receptor as theranostic agents for cancer therapy. Biomaterials 33, 856–866, Copyright © 2016, Elsevier Ltd.

cysteamine was fabricated in four steps: (I) loading target protein to the β-CD functionalized silica spheres through physical adsorption; (II) LBL assembling polymers on silica spheres; (III) cross-linking thiols to generate polymeric matrix and (IV) removal of silica core-based templates. They used disulfide cross-linked nanocapsules in comparison with the uncross-linked ones for improving the stability under acidic conditions. Therefore, applying these nanocapsules not only could protect protein drugs against the gastric conditions, but also could lead to a release in the intracellular medium after glutathione reduction. Although the EE and LC were respectively decreased and enhanced by increasing the primary level of BSA, these nanocapsules hold promising applications in drug delivery and micro-reactors. Therefore, a noteworthy result of this research was a precise technique to not only reduce the protein loss in acidic environments, but also keep the sustained release of protein in the cytosol (Shu et al., 2010).

7.13 Cyclodextrins in gold nanoparticles

Gold (Au)-nanostructures have recently attracted many scientific interests due to their potential in enhancing both the diagnosis and treatment of cancer by their exclusive visual, catalytic, electric, and molecular-cloud features. Meanwhile, the considerable

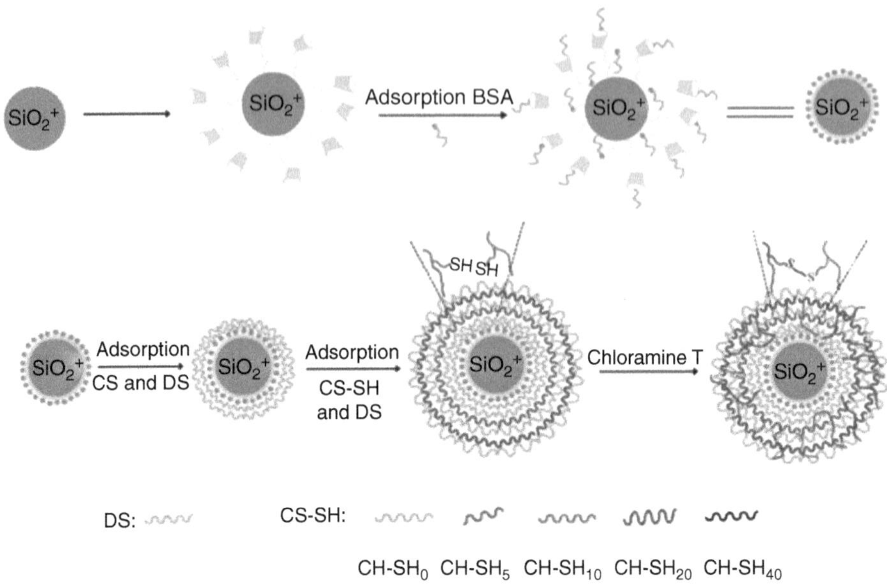

Figure 7.14 A schematic representation for loading BSA of gradient cross-linked NPs. The degree of thiol modification increased with the color become darker.
Source: Reprinted with permission from Shu, S., Zhang, X., Wu, Z., Wang, Z., Li, C., 2010. Gradient cross-linked biodegradable polyelectrolyte nanocapsules for intracellular protein drug delivery. Biomaterials 31, 6039–6049, Copyright © 2016, Elsevier Ltd.

chemical stability, biocompatibility, and easy interaction with many biocomponents of Au-nanostructures make them suitable to be applied in biomedical applications for a variety of purposes, such as carriers for the delivery of functional components, vectors for cell targeted gene delivery and molecular imaging of cancerous cells (Heo et al., 2012; Nalage et al., 2013; Xia et al., 2011). Incorporating Au-NPs into part of the assembled structures can be provided by the host–guest mechanism of CDs.

Mazzaglia (2011) pointed out that the Au-NPs capped with amphiphilic CDs can encapsulate drugs and therefore can be the starting point to develop effective systems for combined therapies (i.e., photothermal and photodynamic therapies). Heo et al. (2012) prepared Au-NPs (Fig. 7.15) containing PEG on the surface, biotin, model drug "paclitaxel," and rhodamine linked β-CD. An improvement in water-solubility of the model drug was found by forming a stable complex between β-CD and paclitaxel (Agüeros et al., 2010). The fabricated Au-NPs had a noticeable anticancer potential on Hela cells with low toxic side effects against NH3T3 cells (Heo et al., 2012). Liu et al. (1998) had earlier reported a novel technique to modify the Au-colloidal particles surface according to the water dissolution of aliphatic thiols using α-CD. The CD successfully interacts with the aliphatic chains and carries the water-insoluble thiol groups to the Au-NPs surface. Au-colloidal NPs (>10 nm), modified with long chain alkanethiolates could be prepared by this method.

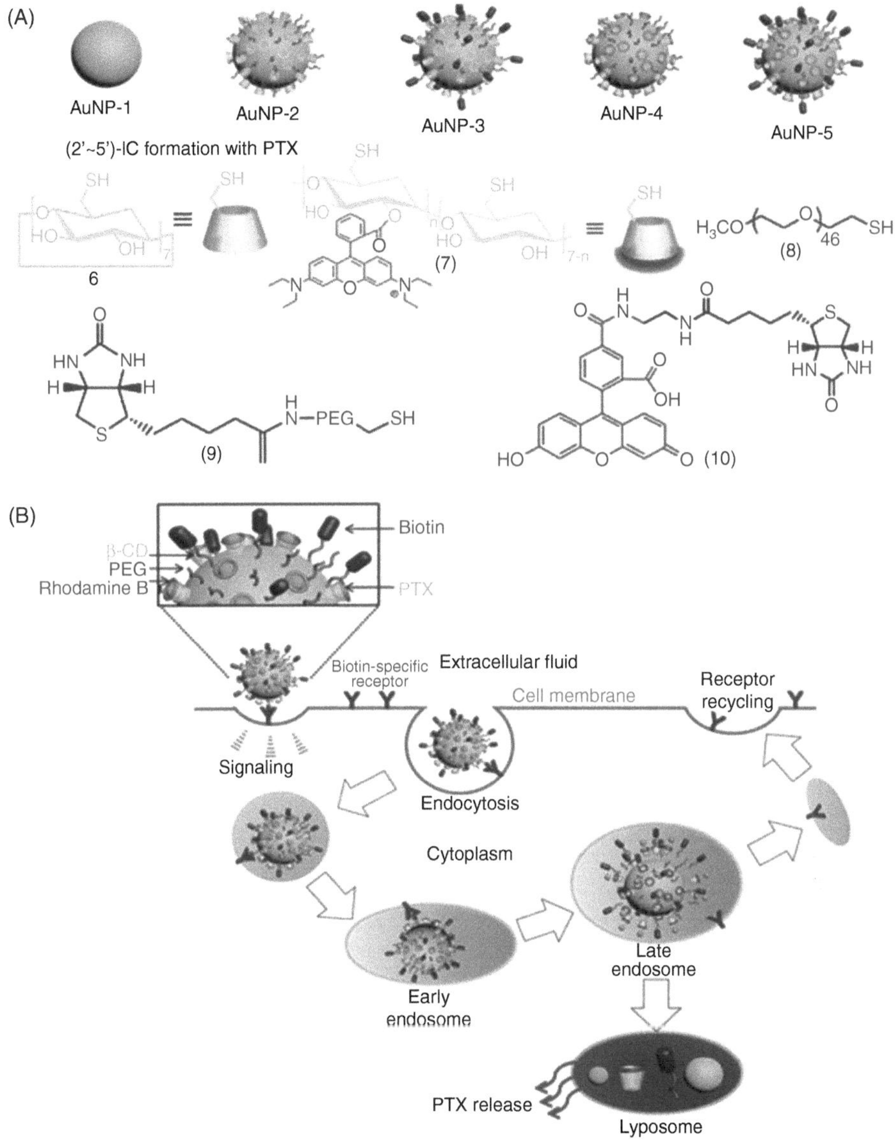

Figure 7.15 (A) Au-NPs (nos. 1–5) and compounds (nos. 6–10). (B) Schematic illustration of cancer-targeting (PTX) delivery mechanism of Au-NP as a theranostic agent.
Source: Reprinted with permission from Heo, D.N., Yang, D.H., Moon, H.J., Lee, J.B., Bae, M.S., Lee, S.C., Lee, W.J., Sun, I.C., Kwon, I.K., 2012. Gold nanoparticles surface-functionalized with paclitaxel drug and biotin receptor as theranostic agents for cancer therapy. Biomaterials 33, 856–866, Copyright © 2016, Elsevier Ltd.

7.14 Concluding remarks and future trends

CDs are useful functional carriers which are used in over forty marketed products worldwide. These complex agents can significantly improve bioavailability/delivery, physicochemical stability, and water solubility of slightly-soluble bioactive ingredients. A unique and novel class of encapsulation method is the formation of the strong inclusion complex between CDs and a number of guest functional components. The formed inclusion complex by CDs can considerably increase the EE of the guest molecules and can successfully regulate the release rate of bioactive compounds. The bioadaptability and multifunctional features of CDs along with their relatively low cost and easy acquisition have attracted great attention in the field of nanobiopharmaceutical technology to apply them in improving the bioavailability/bioaccessibility of bioactive ingredients. However, obtaining the information about the influencing parameters on the complex formation is mandatory for preparing CD–guest complexes with appropriate functionalities. Although CDs can enhance the delivery rate of functional components based on the data obtained from both human and animal studies, there is inadequate information in relation to the physicochemical stability of molecular-cloud molecules after using a systemic path and evaluation of release efficiency of the target bioactive component. Therefore, there is a research gap to optimize bioactive carriers in the form of CDs. The application of these CDs (especially β-CD) is restricted to oral and topical medication because of the renal toxicity of parent CDs by the accumulation of their crystals or CD–cholesterol complexes. Hence, three proprietary modified β-CDs (methoxyl-, hydroxypropyl-, and sulfobutylether-β-CD) were developed to dramatically increase water-solubility. Nevertheless, novel amphiphilic CD-based nanospheres or nanocapsules can potentially form drug delivery applications without the presence of surfactant through NPM. But, this technique is not suitable for the nanoencapsulation of hydrophilic drugs and is basically used for lipophilic moieties owing to the miscibility of organic and water phases. In recent years, the formation of the inclusion complex between CDs and other polymeric components (e.g., PNPs) has opened a new horizon towards practical uses of bioactive ingredients for pharma, bio- and food-applications, especially for lipophilic or unstable bioactive compounds. Superior qualities of the final products can be provided by combining CDs with biodegradable substances. These outstanding functionalities are nontoxicity, low cost, tolerability, greater specificity, protection from degradation in biological systems, inhibition of the interaction between formulation components, and reduction in drug irritancy by encapsulation. Selection of the best technique to fabricate CD-based biodegradable NPs with an ideal size and EE depends on physicochemical characteristics of the intended guest. Meanwhile, fewer requirements to dangerous reagents, the technique simplification to establish easy scale-up, and process optimization/modeling in order to enhance yield and EE are three important aspects in the development of NP preparation methods. Some emulsification and polymerization techniques (ESE, ESD, EP, and MEP) are the simple, safe, and reproducible methods. These desirable features can provide a valuable chance to develop the production process of nanoparticulate drug delivery systems from batch to continuous. Magnetic

NPs modified with CD can provide particles having dual functionalities of being both therapeutic and diagnostic in nature. Nonetheless, there are still some challenges which should be discussed. The process for all bioactive components is not appropriate. Moreover, further studies should be performed on some postpreparative stages in the preparation of nanocapsules; such as, purification and preservation especially from the perspective of the remaining amount of the used solvent. Formation of a partial or irregular film, and insufficient stability and nonreproducible release properties of some functional ingredients are other main difficulties/restrictions during the industrial production process. In spite of all these problems, NPs are promising materials to develop the drug administration. Therefore, investigations on the synthesis of NPs and PNPs based on CD should continue to find the best performance in food, and biomedicinal products. The present achievements on the application of parent CDs and their modified derivatives, thus is an interesting sign to have a shining future for CD-containing biopharmaceuticals and NPs.

References

Abdelwahed, W., Degobert, G., Dubes, A., Parrot-Lopez, H., Fessi, H., 2008. Sulfated and non-sulfated amphiphilic-beta-cyclodextrins: impact of their structural properties on the physicochemical properties of nanoparticles. Int. J. Pharm. 351, 289–295.

Abou-Taleb, A.E., Abdel-Rhman, A.A., Samy, E.M., Tawfeek, H.M., 2006. Interaction of rofecoxib with β-cyclodextrin and HP-β-cyclodextrin in aquoues solution and in solid state. Bull. Pharm. Sci. Assiut Univ. 29, 236–252.

Abu Mukh-Qasem, R., Gedanken, A., 2005. Sonochemical synthesis of stable hydrosol of Fe3O4 nanoparticles. J. Colloid Interface Sci. 284, 489–494.

Aftabrouchard, D., Dorlker, E., 1992. Preparation methods for biodegradable microparticles loaded with water-soluble drugs. STP Pharma. Sci. 2, 365–380.

Agüeros, M., Zabaleta, V., Espuelas, S., Campanero, M.A., Irache, J.M., 2010. Increased oral bioavailability of paclitaxel by its encapsulation through complex formation with cyclodextrins in poly(anhydride) nanoparticles. J. Control. Release 145, 2–8.

Ahmd, M., Rashid, K., Nadeem, M., Masood, K., Ali, S., Nafees, M., Gull, N., Mumtaz-ul-Haq, Ibrahim, N., Saeed, A., Qureshy, A., Aleem, F., Naseer, H., Mehmood, S., Hyder, S.W., 2012. A simple method to prepare aqueous dispersion of iron oxide nanoparticles and their biodistribution study. J. Colloid Sci. Biotechnol. 1, 201–209.

Ahsan, A., Aziz, A., Arshad, M.A., Ali, O., Nauman, M., Ahmad, N.M., Elaissari, A., 2013. Smart magnetically engineering colloids and biothin films for diagnostics applications. J. Colloid Sci. Biotechnol. 2, 19–26.

Alamdarnejad, G., Sharif, A., Taranejoo, S., Janmaleki, M., Kalaee, M.R., Dadgar, M., Khakpour, M., 2013. Synthesis and characterization of thiolated carboxymethyl chitosan–graft–cyclodextrin nanoparticles as a drug delivery vehicle for albendazole. J. Mater. Sci. Mater. Med. 24, 1939–1949.

Al-Marzouqi, A.H., Jobe, B., Dowaidar, A., Maestrelli, F., Mura, P., 2007. Evaluation of supercritical fluid technology as preparative technique of benzocaine-cyclodextrin complexes; comparison with conventional methods. J. Pharm. Biomed. Anal. 43, 566–574.

Al-Marzouqi, A.H., Elwy, H.M., Shehadi, I., Adem, A., 2009. Physicochemical properties of antifungal drug-cyclodextrin complexes prepared by supercritical carbon dioxide and by conventional techniques. J. Pharm. Biomed. Anal. 49, 227–233.

Anirudhan, T., Dilu, D., Sandeep, S., 2013. Synthesis and characterisation of chitosan crosslinked-β-cyclodextrin grafted silylated magnetic nanoparticles for controlled release of indomethacin. J. Magn. Magn. Mater. 343, 149–156.

Ascenso, A., Guedes, R., Bernardino, R., Diogo, H., Carvalho, F.A., Santos, N.C., Silva, A.M., Marques, H.C., 2011. Complexation and full characterization of the tretinoin and dimethyl-β-cyclodextrin complex. AAPS Pharm. Sci. Tech. 12, 553–563.

Asp, M.L., Hertzler, S.R., Chow, J., Wolf, B.W., 2006. Gamma-cyclodextrin lowers post-prandial glycemia and insulinemia without carbohydrate malabsorption in healthy adults. J. Am. Coll. Nutr. 25, 49–55.

Astray, G., Gonzalez-Barreiro, C., Mejuto, J.C., Rial-Otero, R., Simal-Gándara, J., 2009. A review on the use of cyclodextrins in foods. Food Hydrocolloid. 23, 1631–1640.

Augustin, M.A., Hemar, Y., 2009. Nano- and micro-structured assemblies for encapsulation of food ingredients. Chem. Soc. Rev. 38, 902–912.

Badruddoza, A.Z.M., Rahman, M.T., Ghosh, S., Hossain, M.Z., Shi, J., Hidajat, K., Uddin, M.S., 2013. β-Cyclodextrin conjugated magnetic, fluorescent silica core-shell nanoparticles for biomedical applications. Carbohydr. Polym. 95, 449–457.

Banerjee, S.S., Chen, D.H., 2007. Magnetic nanoparticles grafted with cyclodextrin for hydrophobic drug delivery. Chem. Mater. 19, 6345–6349.

Bardajee, G.R., Vancaeyzeele, C., Haley, J.C., Li, A.Y., Winnik, M.A., 2007. Synthesis, characterization, and energy transfer studies of dye-labeled poly(butylmethacrylate) latex particles prepared by miniemulsion polymerization. Polymer 48, 5839–5849.

Baruch-Sharon, S., Margel, S., 2010. Synthesis and characterization of poly-chloromethylstyrene nanoparticles of narrow size distribution by emulsion and miniemulsion polymerization processes. Colloid. Polym. Sci. 288, 869–877.

Baruch-Teblum, E., Mastai, Y., Landfester, K., 2010. Miniemulsion polymerization ofcyclodextrin nanospheres for water purification from organic pollutants. Eur. Polym. J. 46, 1671–1678.

Bellanger, N., Perly, B., 1992. NMR investigations of the conformation of new cyclodextrins-based amphiphilic transporters for hydrophobic drugs: Molecular lollipops. J. Mol. Struct. 15, 215–226.

Bertrand, P., Jonas, A., Laschevsky, A., Legras, R., 2000. Ultrathin polymer coating by complexation of polyelectrolytes. Macromol. Rapid Commun. 21, 319–348.

Bhandari, B.R., D'Arcy, B.R., Padukka, I., 1999. Encapsulation of lemon oil by paste method using β-cyclodextrin. J. Agric. Food Chem. 47, 5194–5197.

Bhati, L.K., Tiwari, G., Tiwari, R., Kumar, V., 2012. Enhancement of complexation efficiency of meloxicam using binary and ternary solid systems: formulation considerations. Am. J. Drug Discov. Dev. 2, 17–31.

Bilensoy, E., Hincal, A.A., 2008. Cyclodextrin-based nanomaterials in pharmaceutical field. In: Gad, S.C. (Ed.), Pharmaceutical Manufacturing Handbook: Production and Processes. John Wiley & Sons, NJ, USA, pp. 1225–1247.

Bilensoy, E., Hincal, A.A., 2009. Recent advances and future directions in amphiphilic cyclodextrin nanoparticles. Expert Opin. Drug Deliv. 6, 1161–1173.

Bilensoy, E., Gürkaynak, O., Dog˘an, A.L., Hıncal, A.A., 2008. Safety and efficacy of amphiphilic β-cyclodextrin nanoparticles for paclitaxel delivery. Int. J. Pharm. 347, 163–170.

Bonnett, R., 1995. Photosensitizers of the porphyrin and phthalocyanine series for photodynamic therapy. Chem. Soc. Rev. 24, 19–33.

Borghetti, G.S., Lula, I.S., Sinisterra, R.D., Bassani, V.L., 2009. Quercetin/β-cyclodextrin solid complexes prepared in aqueous solution followed by spray-drying or by physical mixture. AAPS Pharm. Sci. Tech. 10, 235–242.

Bouchemal, K., Briançon, S., Perrier, E., Fessi, H., Bonnet, I., Zydowicz, N., 2004. Synthesis and characterization of polyurethane and poly (ether urethane) nanocapsules using a new technique of interfacial polycondensation combined to spontaneous emulsification. Int. J. Pharm. 269, 89–100.

Bouchemal, K., Briançon, S., Fessi, H., Chevalier, Y., Bonnet, I., Perrier, E., 2006a. Simultaneous emulsification and interfacial polycondensation for the preparation of colloidal suspension of nanocapsules. Mater. Sci. Eng. C 26, 472–480.

Bouchemal, K., Couenne, F., Briançon, S., Fessi, H., Tayakout, M., 2006b. Polyamides nanocapsules: modelling and wall thickness estimation. AIChE J. 52, 1–10.

Boudad, H., Legrand, P., Lebas, G., Cheron, M., Duchêne, D., Ponchel, G., 2001. Combined hydroxypropyl-β-cyclodextrin and poly(alkylcyanoacrylate)nanoparticles intended for oral administration of saquinavir. Int. J. Pharm. 218, 113–124.

Buschmann, H.J., Schollmeyer, E., 2002. Applications of cyclodextrins in cosmetic products: a review. J. Cosmet. Sci. 53, 185–191.

Cabral Marques, H.M., 2010. A review on cyclodextrin encapsulation of essential oils and volatiles. Flavour Fragr. J. 25, 313–326.

Canceill, J., Jullien, L., Lacombe, L., Lehn, J.M., 1992. Channel type molecular structures, Part 2, synthesis of bouquet-shaped molecules based on a β-cyclodextrin core. Helv. Chim. Acta 75, 791–812.

Cao, F.T., Guo, J., Ping, Q., 2005. The physicochemical characteristics of freeze-dried scutellarin-cyclodextrin tetracomponent complexes. Drug Dev. Ind. Pharm. 31, 747–756.

Cao, H.N., He, J., Deng, L., Gao, X.Q., 2009. Fabrication of cyclodextrin-functionalized superparamagnetic Fe3O4/amino-silane core–shell nanoparticles via layer-by-layer method. Appl. Surf. Sci. 255, 7974–7980.

Charcosset, C., Fessi, H., 2006. Amembrane contactor for the preparation of nanoparticles. Desalination 200, 568–569.

Charoenchaitrakool, M., Dehghani, F., Foster, N.R., 2009. Utilization of supercritical carbon dioxide for complex formation of ibuprofen and methyl- betacyclodextrin. Int. J. Pharm. 239, 103–112.

Chaudhary, A., Nagaich, U., Gulati, N., Sharma, V.K., Khosa, R.L., 2012. Enhancement of solubilization and bioavailability of poorly soluble drugs by physical and chemical modifications: A recent review. J. Adv. Pharm. Edu. Res. 2, 32–67.

Chelli, S., Majdoub, M., Aeiyach, S., Jouini, M., 2009. Polyrotaxanes basedon polyethers and β-cyclodextrin. J. Polym. Sci. Polym. Chem. 47, 4391–4399.

Chen, D., Xu, R., 1998. Hydrothermal synthesis and characterization of nanocrystalline Fe3O4 powders. Mater. Res. Bull. 33, 1015–1021.

Choi, H.G., Lee, B.J., Han, J.H., Lee, M.K., Park, K.M., Yong, C.S., Rhee, J.D., Kim, Y.B., Kim, C.K., 2001. Terfenadine-β-cyclodextrin inclusion complex with antihistaminic activity enhancement. Drug Dev. Ind. Pharm. 27, 857–862.

Choi, M.J., Hong, G.P., Briancon, S., Fessi, H., Lee, M.Y., Min, S.G., 2008. Effect of a high-pressure-induced freezing process on the stability of freeze-dried nanocapsules. Drying Technol. 26, 1199–1207.

Choi, M.J., Ruktanonchai, U., Min, S.G., Chun, J.Y., Soottitantawat, A., 2010. Physical characteristics of fish oil encapsulated by β-cyclodextrin using an aggregation method or polycaprolactone using an emulsion–diffusion method. Food Chem. 119, 1694–1703.

Choisnard, L., Géze, A., Bigan, M., Putaux, J.L., Wouessidjewe, D., 2005. Efficient size control of amphiphilic cyclodextrin nanoparticles through a statistical mixture design methodology. J. Pharm. Pharm. Sci. 8, 593–600.

Choisnard, L., Gèze, A., Putaux, J.L., Wong, Y.S., Wouessidjewe, D., 2006. Nanoparticles of β-cyclodextrin esters obtained by self-assembling of biotransesterified β-cyclodextrins. Biomacromol. 7, 515–520.

Choisnard, L., Gèze, A., Yaméogo, B.G., Putaux, J.L., Wouessidjewe, D., 2007. Miscellaneous nanoaggregates made of β-CD esters synthesised by an enzymatic pathway. Int. J. Pharm. 344, 26–32.

Chronopoulou, L., Fratoddi, I., Palocci, C., Venditti, I., Russo, M.V., 2009. Osmosis based method drives the self-assembly of polymeric chains into micro and nanostructures. Langmuir 25, 119406.

Ciobanu, A., Landy, D., Fourmentin, S., 2013. Complexation efficiency of cyclodextrins for volatile flavor compounds. Food Res. Int. 53, 110–114.

Çirpanli, Y., Bilensoy, E., Lale Dog˘an, A., Çalis¸, S., 2009. Comparative evaluation of polymeric and amphiphilic cyclodextrin nanoparticles for effective camptothecin delivery. Eur. J. Pharm. Biopharm. 73, 82–89.

Cirri, M., Maestrelli, F., Mennini, N., Mura, P., 2009. Influence of the preparation method on the physical-chemical properties of ketoprofen-cyclodextrin-phosphatidylcholine ternary systems. J. Pharm. Biomed. Anal. 50, 690–694.

Cirri, M., Bragagni, M., Mennini, N., Mura, P., 2012. Development of a new delivery system consisting in 'drug-in cyclodextrin-in nanostructured lipid carriers' for ketoprofen topical delivery. Eur. J. Pharm. Biopharm. 80, 46–55.

Cravotto, G., Binello, A., Baranelli, E., Carraro, P., Trotta, F., 2006. Cyclodextrins as food additives and in food processing. Curr. Nutr. Food Sci. 2, 343–350.

Crespy, D., Landfester, K., 2009. Synthesis of polyvinylpyrrolidone/silver nanoparticles hybrid latex in non-aqueous miniemulsion at high temperature. Polymer 50, 1616–1620.

Crespy, D., Stark, M., Hoffmann-Richter, C., Ziener, U., Landfester, K., 2007. Polymeric nanoreactors for hydrophilic reagents synthesized by interfacial polycondensation on miniemulsion droplets. Macromolecules 40, 3122–3135.

Crini, G., 2014. Review: a history of cyclodextrins. Chem. Rev. 114, 10940–10975.

Cryan, S.A., Donohue, R., Ravo, B.J., Darcy, R., O'Driscoll, C.M., 2004a. Cationic cyclodextrin amphiphiles as gene delivery vectors. J. Drug Del. Sci. Technol. 14, 57–62.

Cryan, S.A., Holohan, A., Donohue, R., Darcy, R., O'Driscoll, C.M., 2004b. Cell transfection with polycationic cyclodextrin vectors. Eur. J. Pharm. Sci. 21, 625–633.

Csabai, K., Vikmon, M., Szejtli, J., Redenti, E., Poli, G., Ventura, P., 1998. Complexation of manidipine with cyclodextrins and their derivatives. J. Incl. Phenom. Mol. Recognit. Chem. 31, 169–178.

da Costa, G.M., Grave, E.D., de Bakker, P.M.A., Vandenberghe, R.E., 1994. Synthesis and characterization of some iron oxides by sol-gel method. J. Solid State Chem. 113, 405–412.

Danicher, L., Frere, Y., Calve, A.L., 2000. Synthesis by interfacial polycondensation of polyamide capsules with various sizes. Charecteristics and properties. Macromol. Symp. 151, 387–392.

Das, S.K., Rajabalaya, R., David, S., Gani, N., Khanam, J., Nanda, A., 2013. Cyclodextrins-the molecular container Res. J. Pharmaceut. Biol. Chem. Sci. 4, 1694–1720.

De Cock, L.J., De Koker, S., De Geest, B.G., Grooten, J., Vervaet, C., Remon, J.P., Sukhorukov, G.B., Antipina, M.N., 2010. Polymeric multilayer capsules in drug delivery. Angew. Chem. Int. Ed. 49, 6954–6973.

De Jaeghere, F., Doelker, E., Gurny, R., 1999. Nanoparticles. In: Mathiowitz, E. (Ed.), Encyclopedia of Controlled Drug Delivery. John Wiley & Sons, Inc, New York, pp. 641–664.

Dehnad, D., Jafari, S.M., Afrasiabi, M., 2016. Influence of drying on functional properties of food biopolymers: From traditional to novel dehydration techniques. Trends Food Sci. Tech. 57 (Part A), 116–131.
Del Valle, E.M.M., 2004. Cyclodextrins and their uses: a review. Process Biochem. 39, 1033–1046.
Deng, Y., Wang, L., Yang, W., Fu, S., Elaıssari, A., 2003. Preparation of magnetic polymeric particles via inverse microemulsion polymerization process. J. Magn. Magn. Mater. 257, 69–78.
Deshmukh, S.S., Potnis, V.V., Shelar, D.B., Mahaparale, P.R., 2007. Studies on inclusion complexes of ziprasidone hydrochloride with beta-cyclodextrin and hydroxypropyl beta-cyclodextrin. Indian Drug. 44, 677–682.
Devil, N.K.D., Rani, A.P., Muneer, J.M., Kumar, K.S., Kaushik, J., Sowjanya, V., 2010. Cyclodextrins in pharmacy-an overview. Pharmacophore 1, 155–165.
Dodziuk, H., Chmurski, K., Jurczak, J., Koźmiński, W., Lukin, O., Sitkowski, J., Stefaniak, L., 2000. A dynamic NMR study of self-inclusion of a pendant group in amphiphilic 6-thiophenyl-6-deoxycyclodextrins. J. Mol. Struct. 519, 33–36.
Doijad, R.C., Kanakal, M.M., Manvi, I.V., 2007. Studies on piroxicam-beta-cyclodextrin inclusion complexes. Indian Pharma. VI, 94–98.
Dougherty, T.J., Gomer, C.J., Henderson, B.W., Jori, G., Kessel, D., Korbelik, M., Moan, J., Peng, Q., 1998. Photodynamic therapy. J. Natl. Cancer Inst. 90, 889–905.
Dubes, A., Bouchu, D., Lamartine, R., Parrot-Lopez, H., 2001. An efficient regiospecific synthetic route to multiply substituted acyl-sulphated β-cyclodextrins. Tetrahedron Lett. 42, 9147–9151.
Dubes, A., Degobert, G., Fessi, H., Parrot-Lopez, H., 2003. Synthesis and characterisation of sulfated amphiphilic alpha-, beta- and gamma-cyclodextrins: application to the complexation of acyclovir. Carbohydr. Res. 338, 2185–2193.
Duchêne, D., 2011. Cyclodextrins and their inclusion complexes. In: Bilensoy, E. (Ed.), Cyclodextrins in Pharmaceutics, Cosmetics, and Biomedicine: Current and Future Industrial Applications. Wiley, NJ, USA.
Duchêne, D., Ponchel, G., Wouessidjewe, D., 1999. Cyclodextrins in targeting. Application to nanoparticles. Adv. Drug Deliv. Rev. 36, 29–40.
El-Aasser, M.S., Ugelstad, J., Vanderhoff, J.W., 1979. Polymer Emulsification Process. US4177177 A.
Ethirajan, A., Ziener, U., Landfester, K., 2009. Surface-functionalized polymeric nanoparticles as templates for biomimetic mineralization of hydroxyapatite. Chem. Mater. 21, 2218–2225.
Fages, J., Lochard, H., Letourneau, J.J., Sauceau, M., Rodier, E., 2004. Phase behavioral effects on particle formation processes using supercritical fluids. Powder Technol. 141, 219–226.
Fan, W., Yan, W., Xu, Z., Ni, H., 2012. Formation mechanism of monodisperse, lowmolecular weight chitosan nanoparticles by ionic gelation technique. Colloid. Surf. B Biointerfaces 90, 21–27.
Fang, Z., Bhandari, B., 2010. Encapsulation of polyphenols-a review. Trend. Food Sci. Technol. 21, 510–523.
Faridi Esfanjani, A., Jafari, S.M., 2016. Biopolymer nano-particles and natural nano-carriers for nano-encapsulation of phenolic compounds. Colloids Surf. B. 146, 532–543.
Fathi, M., Martín, A., McClements, D.J., 2014. Nanoencapsulation of food ingredients using carbohydrate based delivery systems. Trend. Food Sci. Technol. 39, 18–39.
Fenyvesi, É., Vikmon, M.A., Szente, L., 2015. Cyclodextrins in food technology and human nutrition: benefits and limitations in 2012. Crit. Rev. Food Sci. Nutr. 56 (12), 1981–2004.
Fessi, H., Puissieux, F., Devissaguet, J.P., Thies, C., 1988. Process for the preparation of dispersible colloidal systems of a substance in the form of nanospheres. US Patent No. 5,118,528.

Fessi, H., Puisieux, F., Devissaguet, J.P., Ammoury, N., Benita, S., 1989. Nanocapsule formation by interfacial deposition following solvent displacement. Int. J. Pharm. 55, R1–R4.

Fessi, H.C., Devissaguet, J.P., Puisieux, F., Thies, C., 1997. Process for the preparation of dispersible colloidal systems of a substance in the form of nanoparticles. US Patent 5,118,528.

Figueiras, A., Carvalho, R.A., Ribeiro, L., Torres-Labandeira, J.J., Veiga, F.J.B., 2007. Solid-state characterization and dissolution profiles of the inclusion complexes of omeprazole with native and chemically modified β-cyclodextrin. Eur. J. Pharm. Biopharm. 67, 531–539.

Freitas, M.N., Marchetti, J.M., 2005. Nimesulide PLA microspheres as a potential sustained release system for the treatment of inammatory diseases. Int. J. Pharm. 295, 201–211.

French, D., Knapp, D.W., Pazur, H., 1950. Studies of the Schardinger dextrins. VI. The molecular size and structure of γ-dextrin. J. Am. Chem. Soc. 72, 5120–5152.

Fresta, M., Fontana, G., Bucolo, C., Cavallaro, G., Giammona, G., Puglisi, G., 2001. Ocular tolerability and in vivo bioavailability of poly(ethylene glycol) (PEG)-coated polyethyl-2-cyanoacrylate nanosphere-encapsulated acyclovir. J. Pharm. Sci. 90, 288–297.

Friedrich, H., Nada, A., Bodmier, R., 2005. Solid state and dissolution rate characterization of co-ground mixtures of nifedipine and hydrophillic carriers. Drug Dev. Ind. Pharm. 31, 719–728.

Furue, M., Harada, A., Nozakura, S., 1975. Preparation of cyclodextrin-containing polymers and their catalysis in ester-hydrolysis. J. Polym. Sci., Polym. Lett. Ed. 13, 357–360.

Furuta, T., Yoshii, H., 1994. Powdery encapsulation of d-limonene by kneading with mixed powders of β-cyclodextrin and maltodextrin at low water content. Biosci. Biotechol. Biochem. 58, 847–850.

Furuta, T., Yoshii, H., Miyamoto, A., Yasunishi, A., Hirano, H., 1993. Effects of water of inclusion complexes of d-limonene and cyclodextrins. Supramol. Chem. 1, 321–325.

Ganachaud, F., Katz, J.L., 2005. Nanoparticles and nanocapsules created using the Ouzo effect: spontaneous emulsification as an alternative to ultrasonic and highshear devices. Chem. Phys. Chem. 6, 209–216.

Gasco, M., Trotta, M., 1986. Nanoparticles from microemulsions. Int. J. Pharm. 29, 267–268.

Gaudin, F., Zydowicz, N.S., 2008. Core-shell biocompatible polyurethane nanocapsules obtained by interfacial step polymerisation in miniemulsion. Colloid. Surf A 331, 133–142.

Gèze, A., Aous, S., Baussanne, I., Putaux, J.-L., Defaye, J., Wouessidjewe, D., 2002. Influence of chemical structure of amphiphilic β-cyclodextrins on their ability to form stable nanoparticles. Int. J. Pharm. 242, 301–305.

Gèze, A., Tieu Chau, L., Choisnard, L., Mathieu, J.P., Marti-Batlle, D., Riou, L., Putaux, J.L., Wouessidjewe, D., 2007. Biodistribution of intravenously administered amphiphilic β-cyclodextrin nanospheres. Int. J. Pharm. 344, 135–142.

Gèze, A., Choisnard, L., Putaux, J.L., Wouessidjewe, D., 2009. Colloidal systems made of biotransesterified alpha, beta and gamma cyclodextrins grafted with C10 alkyl chains. Mater. Sci. Eng. C 29, 458–462.

Gharibzahedi, S.M.T., Razavi, S.H., Mousavi, S.M., 2014. Characterizing the natural canthaxanthin/2-hydroxypropyl-β-cyclodextrin inclusion complex. Carbohydr. Polym. 101, 1147–1153.

Gharibzahedi, S.M.T., Razavi, S.H., Mousavi, S.M., 2015. Optimal development of a new stable nutraceutical nanoemulsion based on the inclusion complex of 2-hydroxypropyl-β-cyclodextrin with canthaxanthin accumulated by Dietzia natronolimnaea HS-1 using ultrasound-assisted emulsification. J. Disp. Sci. Technol. 36, 614–625.

Giordano, F., Novak, C., Moyano, J.R., 2001. Thermal analysis of cyclodextrins and their inclusion compounds. Thermochim. Acta 380, 123–151.

Giri, T.K., Choudhary, C., Ajazuddin, Alexander, A., Badwaik, H., Tripathi, D.K., 2013. Prospects of pharmaceuticals and biopharmaceuticals loaded microparticles prepared by double emulsion technique for controlled delivery. Saudi Pharm. J. 21, 125–141.
Guo, Y., Guo, S., Ren, J., Zhai, Y., Dong, S., Wang, E., 2010. Cyclodextrin functionalized graphene nanosheets with high supramolecular recognition capability: synthesis and host-guest inclusion for enhanced electrochemical performance. ACS Nano 4, 4001–4010.
Gupta, A.K., Gupta, M., 2005. Synthesis and surface engineering of iron oxide nanoparticles for biomedical applications. Biomaterials 26, 3995–4021.
Ham, H.T., Choi, Y.S., Chee, M.G., Chung, I.J., 2006. Single wall carbon nanotubes covered with polystyrene nanoparticles by in-situ miniemulsion polymerization. J. Polym. Sci. Part A Polym. Chem. 44, 573–584.
Hans, M.L., Lowman, A.M., 2002. Biodegradable nanoparticles for drug delivery and targeting. Curr. Opin. Solid State Mater. Sci. 6, 319–327.
He, S., Shia, W., Zhang, X., Li, J., Huang, Y., 2010. β-cyclodextrins-based inclusion complexes of CoFe2O4 magnetic nanoparticles as catalyst for the luminol chemiluminescence system and their applications in hydrogen peroxide detection. Talanta 82, 377–383.
Hedges, A.R., 1998. Industrial applications of cyclodextrins. Chem. Rev. 98, 2035–2044.
Hedges, A., 2009. Cyclodextrins: properties and applications. In: BeMiller, J., Whistler, R. (Eds.), Starch Chemistry and Technology. Academic Press, New York, USA, pp. 833–852.
Heo, D.N., Yang, D.H., Moon, H.J., Lee, J.B., Bae, M.S., Lee, S.C., Lee, W.J., Sun, I.C., Kwon, I.K., 2012. Gold nanoparticles surface-functionalized with paclitaxel drug and biotin receptor as theranostic agents for cancer therapy. Biomaterials 33, 856–866.
Higashi, K., Tozuka, Y., Moribe, K., Yamamoto, K., 2010. Salicylic acid/γ-cyclodextrin 2:1 and 4:1 complex formation by sealed-heating method. J. Pharm. Sci. 99, 4192–4200.
Higuchi, T., Connors, K.A., 1965. Phase-solubility techniques. Adv. Anal. Chem. Instrum. 4, 117–122.
Hornig, S., Heinze, T., 2007. Nanoscale structures of dextran esters. Carbohydr. Polym. 68, 280–286.
Ieva, E., Trapani, A., Cioffi, N., Ditaranto, N., Monopoli, A., Sabbatini, L., 2009. Analytical characterization of chitosan nanoparticles for peptide drug delivery applications. Anal. Bioanal. Chem. 393, 207–215.
Irie, T., Uekama, K., 1997. Pharmaceutical applications of cyclodextrins: III. Toxicological issues and safety evaluation. J. Pharm. Sci. 86, 147–162.
Jafari, S.M., He, Y., Bhandari, B., 2007. Encapsulation of nanoparticles of d-limonene by spray drying: role of emulsifiers and emulsifying techniques. Dry. Technol. 25, 1069–1079.
Jafari, S.M., Assadpoor, E., Bhandari, B., He, Y., 2008. Nano-particle encapsulation of fish oil by spray drying. Food Res. Int. 41, 172–183.
Jambhekar, S., Casella, R., Maher, T., 2004. The physicochemical characteristics and bioavailability of indomethacin from β-cyclodextrin, hydroxyethyl-β-cyclodextrin, and hydroxypropyl-β-cyclodextrin complexes. Int. J. Pharm. 270, 149–166.
JECFA, 2013. Codex Alimentarius gamma-cyclodextrin. Available from: http://www.fao.org/fao-who-codexalimentarius/sh-proxy/ar/?lnk=1&url=https%253A%252F%252Fworkspace.fao.org%252Fsites%252Fcodex%252FStandards%252FCAC%2BMISC%2B6-2015%252FCXA_006efs.pdf
Jeon, H.J., Jeong, Y.I., Jang, M.K., Park, Y.H., Nah, J.W., 2000. Effect of solvent on the preparation of surfactant free poly(DL-lactide-*co*-glycolide) nanoparticles and norfloxacin release characteristics. Int. J. Pharm. 207, 99–108.
Jiang, W., Yang, H.C., Yang, S.Y., Horng, H.E., Hung, J.C., Chen, Y.C., Hong, C.Y., 2004. Preparation and properties of superparamagnetic nanoparticles with narrow size distribution and biocompatible. J. Magn. Magn. Mater. 283, 210–214.

Jiang, X., Dausend, J., Hafner, M., Musyanovych, A., Röcker, C., Landfester, K., Mailan-der, V., Nienhaus, G.U., 2010. Specific effects of surface amines on polystyrenenanoparticles in their interactions with mesenchymal stem cells. Biomacromolecules 11, 748–753.
Kalogeropoulos, N., Yannakopoulou, K., Gioxari, A., Chiou, A., Makris, P.D., 2010. Polyphenol characterization and encapsulation in β-cyclodextrin of a flavonoid-rich Hypericum perforatum (St John's wort) extract. LWT-Food Sci. Technol. 43, 882–889.
Kang, Y., Zhou, L., Lib, X., Yuan, J., 2011. β-Cyclodextrin-modified hybrid magnetic nanoparticles for catalysis and adsorption. J. Mater. Chem. 21, 3704–3710.
Karathanos, V.T., Mourtzinos, I., Yannakopoulou, K., Andrikopoulos, N.K., 2007. Study of the solubility, antioxidant activity, and structure of inclusion complex of vanillin with beta-cyclodextrin. Food Chem. 101, 652–658.
Karode, S.K., Kulkarni, S.S., Suresh, A.K., Mashelkar, R.A., 1998. New insights into kinetics and thermodynamics of interfacial polymerization. Chem. Eng. Sci. 53, 2649–6263.
Katouzian, I., Jafari, S.M., 2016. Nano-encapsulation as a promising approach for targeted delivery and controlled release of vitamins. Trends Food Sci. Tech. 53, 34–48.
Kawabata, Y., Matsumoto, M., Tanaka, M., Takahashi, H., Irinatsu, Y., Tagaki, W., Nakahara, H., Fukuda, K., 1983. Formation and deposition of monolayers of amphiphilic β-cyclodextrin derivatives. Chem. Lett. 15, 1933–1934.
Kfoury, M., Auezova, L., Greige-Gerges, H., Ruellan, S., Fourmentin, S., 2014. Cyclodextrin, an efficient tool for trans-anethole encapsulation: chromatographic, spectroscopic, thermal and structural studies. Food Chem. 164, 454–461.
Khoee, S., Yaghoobian, M., 2009. An investigation into the role of surfactants in con-trolling particle size of polymeric nanocapsules containing penicillin-G in doubleemulsion. Eur. J. Med. Chem. 44, 2392–2399.
Kim, H.S., Sohn, B.H., Lee, W., Lee, J.K., Choi, S.J., Kwon, S.J., 2002. Multifunctional layer-by-layer self-assembly of conducting polymers and magnetic nanoparticles. Thin Solid Films 419, 173–177.
Koeda, T., Wada, Y., Neoh, T.L., Wada, T., Furuta, T., Yoshii, H., 2014. Encapsulation of retinyl palmitate with a mixture of cyclodextrins and maltodextrins by the kneading method. Food Sci. Technol. Res. 20, 529–535.
Koontz, J.L., Marcy, J.E., O'Keefe, S.F., Duncan, S.E., 2009. Cyclodextrin inclusion complex formation and solid-state characterization of the natural antioxidants alpha-tocopherol and quercetin. J. Agric. Food Chem. 57, 1162–1171.
Kostog, M., Kohler, S., Liebert, T., Heinze, T., 2010. Pure cellulose nanoparticles from trimethylsilyl cellulose. Macromol. Symp. 294, 96–106.
Krauland, A.H., Alonso, M.J., 2007. Chitosan/cyclodextrin nanoparticles as macromolecular drug delivery system. Int. J. Pharm. 340, 134–142.
Kreuter, J., 1982. The mechanism of termination in heterogeneous polymerization. J. Polym. Sci. 20, 543–545.
Kwon, H.Y., Lee, J.Y., Choi, S.W., Jang, Y., Kim, J.H., 2001. Preparation of PLGA nanoparticles containing estrogen by emulsification–diffusion method. Colloid Surf. A 182, 123–130.
Lambert, G., Fattal, E., Alphandary, H.P., Gulik, A., Couvreur, P., 2001. Poly isobutyl cyano acrylate nanocapsules containing an aqueous core for the delivery of oligonucleotides. Int. J. Pharm. 214, 13–16.
Landfester, K., Musyanovych, A., Mailander, V., 2010. From polymeric particles to multifunctional nanocapsules for biomedical applications using the miniemulsion process. J. Polym. Sci. Part A Polym. Chem. 48, 493–515.
Laurent, S., Bridot, J.L., Elst, L.V., Muller, R.N., 2010. Magnetic iron oxide nanoparticles for biomedical applications. Future Med. Chem. 2, 427–449.

Lee, J.H., Chen, K.J., Noh, S.H., Garcia, M.A., Wang, H., Lin, W.Y., Jeong, H., Kong, B.J., Stout, D.B., Cheon, J., Tseng, H.R., 2013. On-demand drug release system for in vivo cancer treatment via self assembled magnetic nanoparticles. Angew. Chem. Int. Edit. 52, 4384–4388.

Lemoine, D., Preat, V., 1998. Polymeric nanoparticles as delivery system for influenza virus glycoproteins. J. Control. Release 54, 15–27.

Lemos-Senna E., 1998. Contribution a l'etude pharmacotechnique et physico-chimique de nanospheres de cyclodextrins amphiphiles comme transporteurs de principes actifs, PhD thesis, Universite Paris-Sud.

Lemos-Senna, E., Wouessidjewe, D., Duchêne, D., Lesieur, S., 1998a. Amphiphilic cyclodextrin nanospheres: particle solubilization and reconstitution by the action of a non-ionic detergent. Col. Surf. B: Biointerfaces 10, 291–301.

Lemos-Senna, E., Wouessidjewe, D., Lesieur, S., Duchêne, D., 1998b. Preparation of amphiphilic cyclodextrin nanospheres using the emulsification solvent evaporation method. Influence of the surfactant on preparation and hydrophobic drug loading. Int. J. Pharm. 170, 119–128.

Lemos-Senna, E., Wouessidjewe, D., Lesieur, S., Puisieux, F., Couarraze, G., Duchêne, D., 1998c. Evaluation of the hydrophobic drug loading characteristics in nanoprecipitated amphiphilic nanospheres. Pharm. Dev. Technol. 3, 85–94.

Lemos-Senna, E., Wouessidjewe, D., Lesieur, S., Puisieux, F., Couarraze, G., Duchêne, D., 1998d. Evaluation of the hydrophobic drug loading characteristics in nanoprecipitated amphiphilic cyclodextrins nanospheres. Pharm. Dev. Technol. 3, 1–10.

Leroux, J.C., Allemann, E., Doelker, E., Gurny, R., 1995. New approach for the prepa-ration of nanoparticles by an emulsification–diffusion method. Eur. J. Pharm. Biopharm. 41, 14–18.

Li, Y., Yang, D.J., Chen, S.L., Chen, S.B., Chan, A.S., 2007. Comparative physicochemical characterization of phospholipids complex of puerarin formulated by conventional and supercritical methods. Pharm. Res. 25, 563–577.

Liu, F.Y., Kildsig, D.O., Mitra, A.K., 1992. Complexation of 6-acyl-O-β-cyclodextrin derivatives with steroids, effects of chain length and substitution degree. Drug Dev. Ind. Pharm. 18, 1599–1612.

Liu, J., Xu, R., Kaifer, A.E., 1998. In situ modification of the surface of gold colloidal particles. Preparation of cyclodextrin-based rotaxanes supported on gold nanospheres. Langmuir 14, 7337–7339.

Liu, H.K., Lo, Y.K., Tsai, T.R., Cham, T.M., 2010. Physicochemical characterizations of osthole hydroxypropyl-β-cyclodextrin inclusion complexes with high-pressure homogenization method. J. Food Drug Anal. 18, 391–397.

Locci, E., Lai, S., Piras, A., Marongiu, B., Lai, A., 2004. 13C-CPMAS and 1H-NMR study of the inclusion complexes of β-cyclodextrin with carvacrol, thymol, and eugenol prepared in supercritical carbon dioxide. Chem. Biodivers. 1, 1354–1366.

Loftsson, T., Duchêne, D., 2007. Cyclodextrins and their pharmaceutical applications. Int. J. Pharm. 329, 1–11.

Loftsson, T., Stefansdottir, Ó., Fridriksdottir, H., Gudmundsson, Ö., 1992. Interaction between preservatives and 2-hydroxypropyl-β-cyclodextrin. Drug Devel. Ind. Pharm. 18, 1477–1484.

Loftsson, T., Ólafsdóttir, B.J., Fridriksdóttir, H., Jónsdóttir, S., 1993. Cyclodextrin complexation of NSAIDs: physicochemical characteristics. Eur. J. Pharm. Sci. 1, 95–101.

Lopedota, A., Trapani, A., Cutrignelli, A., Chiarantini, L., Pantucci, E., Curci, R., Manuali, E., Trapani, G., 2009. The use of Eudragit®RS 100/cyclodextrin nanoparticles for the transmucosal administration of glutathione. Eur. J. Pharm. Biopharm. 72, 509–520.

Loukas, Y., Antoniadou-Vyza, E., Papadaki-Valiraki, A., Machera, K., 1994. Gamma-cyclodextrin inclusion complex of a new organophosphorus insecticide. Determination of stability constant with HPLC. J. Agric. Food Chem. 42, 944–948.

Lvov, Y., Ariga, K., Ichinose, I., Kunitake, T., 1995. Assembly of multicomponent protein films by means of electrostatic layer-by-layer adsorption. J. Am. Chem. Soc. 117, 6117–6123.

Madene, A., Jacquot, M., Scher, J., Desobry, S., 2006. Flavour encapsulation and controlled release-a review. Int. J. Food Sci. Technol. 41, 1–21.

Madhulita, P., Ganesh, P., Amol, M., Sunil, M., Bhanoji, R.M.E., 2010. Effect of β-cyclodextrin on physical stability of nimesulide suspension. Int. J. Drug Develop. Res. 2, 669–675.

Maestrellia, F., Garcia-Fuentes, M., Mura, P., Alonso, M.J., 2006. A new drug nanocarrier consisting of chitosan and hydroxypropyl cyclodextrin. Eur. J. Pharm. Biopharm. 63, 79–86.

Mahmoud, A.A., El-Feky, G.S., Kamel, R., Awad, G.E.A., 2011. Chitosan/sulfobuty-lether-β-cyclodextrin nanoparticles as a potential approach for ocular drug delivery. Int. J. Pharm. 413, 229–236.

Makhlof, A., Miyazaki, Y., Tozuka, Y., Takeuchi, H., 2008. Cyclodextrins as stabilizers for the preparation of drug nanocrystals by the emulsion solvent diffusion method. Int. J. Pharm. 357, 280–285.

Marcolino, V.A., Zanin, G.M., Durrant, L.R., Benassi, Mde.T., Matioli, G., 2011. Interaction of curcumin and bixin with β-cyclodextrin: complexation methods, stability, and applications in food. J. Agric. Food Chem. 59, 3348–3357.

Marongui, B., Pira, A., Porcedda, S., Delogu, G., Fabbri, D., Dettori, M.A., 2002. In: Besnard, M., Cansell, F. (Eds.), Proceedings of the Eighth Meeting on Supercritical Fluids, Bordeaux, France, p. 327.

Masotti, A., Bordi, F., Ortaggi, G., Marino, F., Palocci, C., 2008. A novel method to obtain chitosan/DNA nanospheres and a study of their release properties. Nanotechnology 19, 055302–055307.

Maswal, M., Dar, A.A., 2014. Formulation challenges in encapsulation and delivery of citral for improved food quality. Food Hydrocolloid. 37, 182–195.

Matsumoto, M., Matsuzawa, Y., Noguchi, S., Sakai, H., Abe, M., 2004. Structure of Langmuir-Blodgett films of amphiphilic cyclodextrin and water-soluble benzophenone. Mol. Cryst. Liq. Cryst. 425, 197–204.

Mazzaglia, A., 2011. Photodynamic tumor therapy with cyclodextrin nanoassemblies. In: Bilensoy, E. (Ed.), Cyclodextrins in Pharmaceutics, Cosmetics, and Biomedicine. John Wiley & Sons, Inc, Hoboken, New Jersey, pp. 343–361.

Medeiros, S.F., Santos, A.M., Fessi, H., Elaissari, A., 2011. Stimuli-responsive magnetic particles for biomedical applications. Int. J. Pharm. 403, 139–161.

Medeiros, S.F., Lara, B.R., Oliveira, P.F.M., Moraes, R.M., Alves, G.M., Elaissari, A., Santos, A.M., 2013. Stimuli-responsive and biocompatible poly(*N*-vinylcapro-lactam-*co*-acrylic acid)-coated iron oxide nanoparticles by nanoprecipitation technique. J. Colloid Sci. Biotechnol. 2, 180–194.

Memişoğlu, E., Bochot, A., Özalp, M., Sen, M., Duchêne, D., Hincal, A.A., 2002a. Direct formation of nanospheres from amphiphilic β-cyclodextrin inclusion complexes. Pharm. Res. 20, 117–125.

Memişoğlu, E., Bochot, A., Sen, M., Charon, D., Duchêne, D., Hıncal, A.A., 2002b. Amphiphilic β-cyclodextrins modified on the primary face: synthesis, characterization and evaluation of their potential as novel excipients in the preparation of nanocapsules. J. Pharm. Sci. 91, 1214–1224.

Memişoğlu, E., Bochot, A., Sen, M., Duchêne, D., Hıncal, A.A., 2003a. Direct formation of nanospheres from amphiphilic β-cyclodextrin inclusion complexes. Pharm. Res. 20, 117–125.

Memişoğlu, E., Bochot, A., Sen, M., Duchêne, D., Hıncal, A.A., 2003b. Non-surfactant nanospheres of progesterone inclusion complexes with amphiphilic β-cyclodextrins. Int. J. Pharm. 251, 143–153.
Memisoglu-Bilensoy, E., Hincal, A.A., 2006. Sterile, injectable cyclodextrin nanoparticles: effects of gamma irradiation and autoclaving. Int. J. Pharm. 311, 203–208.
Memisoglu-Bilensoy, E., Vural, I., Bochot, A., Rendir, J.-M., Duchêne, D., Hincal, A.A., 2005. Tamoxifen citrate loaded amphiphilic cyclodextrin nanoaprticles: In Vitro characterization and cytotoxicity. J. Control. Release 104, 489–496.
Memisoglu-Bilensoy, E., Dŏgan, A.L., Hıncal, A.A., 2006a. Cytotoxic evaluation of injectable cyclodextrin nanoparticles: surfactant effect. J Pharm. Pharmacol. 58, 585–589.
Memisoglu-Bilensoy, E., Sen, M., Hincal, A.A., 2006b. Effect of drug physicochemical properties on in vitro characteristics of amphiphilic cyclodextrin nanospheres and nanocapsules. J. Microencapsul. 23, 59–68.
Memisoglu-Bilensoy, E., Trichard, L., Bochot, A., Duchêne, D., Hıncal, A.A., 2006c. Amphiphilic cyclodextrins and microencapsulation. In: Benita, S. (Ed.), Microencapsulation-Methods and Industrial Applications. Taylor & Francis Ltd, New York, pp. 269–296.
Méndez-Ardoy, A., Urbiola, K., Aranda, C., Ortiz-Mellet, C., García-Fernández, J.M., de Ilarduya, C.T., 2011. Polycationic amphiphilic cyclodextrin-based nanoparticles for therapeutic gene delivery. Nanomedicine (Lond.) 6, 1697–1707.
Menezes, P.P., Araújo, A.A.S., Dória, G.A.A., Quintans-Júnior, L.J., Oliveira, M.G.B., Santos, M.R.V., Oliveira, J.F., Matos, J.R., Carvalho, F.M.S., Alves, P.B., Matos, I.L., Santos, D.A., Marreto, R.N., Silva, G.F., Serafini, M.R., 2015. Physicochemical characterization and analgesic effect of inclusion complexes of essential oil from *Hyptis pectinata* L. poit leaves with β-cyclodextrin. Curr. Pharm. Biotechnol. 16, 1–11.
Messner, M., Kurkov, S.V., Jansook, P., Loftsson, T., 2010. Self-assembled cyclodextrin aggregates and nanoparticles. Int. J. Pharmaceut. 387, 199–208.
Mielcarek, J., 1996. Photochemical stability of the inclusion complexes of nicardipine with α-, γ-cyclodextrin, methyl-β-cyclodextrin and hydroxypropyl-β-cyclodextrin in solid state and in solution. Pharmazie 51, 477–479.
Miladi, K., Sfar, S., Fessi, H., Elaissari, A., 2013. Drug carriers in osteoporosis: preparation, drug encapsulation and applications. Int. J. Pharm. 445, 181–195.
Montasser, I., Briançon, S.I., Fessi, H., 2007. The effect of monomers on the formulation of polymeric nanocapsules based on polyureas and polyamides. Int. J. Pharm. 335, 176–179.
Monza da Silveira, A., Ponchel, G., Puisieux, F., Duchêne, D., 1998. Combined poly(isobutylcyanoacrylate) and cyclodextrins nanoparticles for enhancing the encapsulation of lipophilic drugs. Pharm. Res. 15, 1051–1055.
Monza da Silveira, A., Ponchel, G., Duchene, D., Couvreur, P., Puisieux, F., 1999. Nanoparticles comprising polyisobutylcyanoacrylate and cuclodextrins. France, WO 99/43359.
Mora-Huertas, C.E., Fessi, H., Elaissari, A., 2010. Polymer-based nanocapsules for drug delivery. Int. J. Pharm. 385, 113–142.
Muhrer, G., Meier, V., Fusaro, F., Albano, S., Mazzotti, M., 2006. Use of compressed gas precipitation to enhance the dissolution behaviour of a poorly water-soluble drug: generation of drug microparticles and drug polymer solid dispersions. Int. J. Pharm. 308, 69–83.
Mulay, M.S., 2011. Taste masking by inclusion complexation: a review. Int. J. Pharm. Res. Dev. 3, 35–47.
Müller, B.W., Brauns, U., 1986. Hydroxypropyl-β-cyclodextrin derivatives: influence of average degree of substitution on complexing ability and surface activity. J. Pharm. Sci. 75, 571–572.
Munin, A., Edwards-Lévy, F., 2011. Encapsulation of natural polyphenolic compounds: a review. Pharmaceutics 3, 793–829.

Mura, P., Fauci, M.T., Parrini, P.L., Furlanetto, S., Pinzauti, S., 1999. Influence of the preparation method on the physicochemical properties of ketoprofen–cyclodextrin binary systems. Int. J. Pharm. 179, 117–128.
Mura, P., Maestrelli, F., Cecchi, M., Bragagni, M., Almeida, A., 2010. Development of anew delivery system consisting in "drug-in cyclodextrin-in PLGA nanoparticles". J. Microencapsul. 27, 479–486.
Murakami, H., Kobayashi, M., Takeuchi, H., kawashima, Y., 1999. Preparation of poly(DL-lactide-co-glycolide) nanoparticles by modified spontaneous emulsification solvent diffusion method. Int. J. Pharm. 178, 143–152.
Nacsa, Á., Ambrus, R., Berkesi, O., Szabó-Révész, P., Aigner, Z., 2008. Water-soluble loratadine inclusion complex: analytical control of the preparation by microwave irradiation. J. Pharm. Biomed. Anal. 48, 1020–1023.
Nagavarma, B.V.N., Yadav, H.K.S., Ayaz, A., Vasudha, L.S., Shivakumar, H.G., 2012. Different techniques for preparation of polymeric nanoparticles—a review. Asian J. Pharm. Clin. Res. 5, 16–23.
Nahaei, M., Valizadeh, H., Baradaran, B., Nahaei, M.R., Asgari, D., Hallaj-Nezhadi, S., Dastmalchi, S., Lotfipour, F., 2013. Preparation and characterization of chitosan/β-cyclodextrin nanoparticles containing plasmid DNA encodinginterleukin-12. Drug Res. 63, 7–12.
Nakai, Y., Yamamoto, K., Terada, K., Watanabe, D., 1987. New methods for preparing cyclodextrin inclusion compounds. I. Heating in a sealed container. Chem. Pharm. Bull. 35, 4609–4615.
Nalage, S.V., Bhosale, Sidhanath, V., Sheshanath, V., Bhosale, R.S., 2013. Kanamycin A-gold nanoparticles: a receptor for cation, anion and neutral species. J. Colloid Sci. Biotechnol. 2, 195–199.
Nasir, A., Harikumar, S.L., Kaur, A., 2012. Cyclodextrins: an excipients tool in drug delivery. Int. Res. J. Pharma. 3, 44–50.
Nasir, A., Kausar, A., Younus, A., 2015. A review on preparation, properties and applications of polymeric nanoparticle-based materials. Polym. Plast. Technol. Eng. 54, 325–341.
Nerome, H., Machmudah, S., Wahyudiono, Fukuzato, R., Higashiura, T., Youn, Y.S., Lee, Y.W., Goto, M., 2013. Nanoparticle formation of lycopene/β-cyclodextrin inclusion complex using supercritical antisolvent precipitation. J. Supercrit. Fluid. 83, 97–103.
Nowicki, A., Zhang, Y., Léger, B., Rolland, J.P., Bricout, H., Monflier, E., Roucoux, A., 2006. Supramolecular shuttle and protective agent: a multiple role of methylated cyclodextrins in the chemoselective hydrogenation of benzenederivatives with ruthenium nanoparticles. Chem. Commun. 45, 296–298.
O'Donnell, P.B., McGinity, J.W., 1997. Preparation of microspheres by the solvent evaporation technique. Adv. Drug Deliv. Rev. 28, 25–42.
Ortiz Mellet, C., García-Fernández, J.M., Benito, J.M., 2011. Cyclodextrin-based gene delivery systems. Chem. Soc. Rev. 40, 1586–1608.
Ozyilmaz, E., Sayin, S., Arslan, M., Yilmaz, M., 2014. Improving catalytic hydrolysis reaction efficiency of sol–gel-encapsulated *Candida rugosa* lipase with magnetic β-cyclodextrin nanoparticles. Colloid. Surf. B: Biointerfaces 113, 182–189.
Pagington, J.S., 1987. β-cyclodextrin: the success of molecular inclusion. Chem. Br. 23, 455.
Parikh, R.K., Mansuri, N.S., Gohel, M.C., Sonlwala, M.M., 2005. Dissolution enhancement of nimesulide using complexation and salt formation techniques. Indian Drug. 42, 149–153.
Parlati, S., Gobetto, R., Barolo, C., Arrais, A., Buscaino, R., Medana, C., Savarino, P., 2007. Preparation and application of a β-cyclodextrin-disperse/reactive dye complex. J. Incl. Phenom. Macrocycl. Chem. 57, 463–470.
Patel, R., Patel, M., Shah, D., 2010. Applications of cyclodextrin in drug delivery. Int. J. Pharma. World. Res. 1, 1–22.

Patil, J.S., Kadam, D.V., Marapur, S.C., Kamalapur, M.V., 2010. Inclusion complex system; a novel technique to improve the solubility and bioavailability of poorly soluble drugs. Int. J. Pharm. Sci. 2, 29–32.

Peltonen, L., Valo, H., Kolakovic, R., Laaksonen, T., Hirvonen, J., 2010. Electrospraying, spray drying and related techniques for production and formulation of drug nanoparticles. Expert Opin. Drug Deliv. 7, 705–719.

Perdersen, N.R., Kristensen, J.B., Bauw, G., Ravoo, B.J., Darcy, R., Larsen, K.L., Pedersen, L.H., 2005. Thermolysin catalyses the synthesis of cyclodextrin esters in DMSO. Tetrahedron Asymm. 16, 615–622.

Péroche, S., Degobert, G., Putaux, J.L., Blanchin, M.G., Fessi, H., Parrot-Lopez, H., 2005. Synthesis and characterization of novel nanospheres made from amphiphilic perfluoroalkylthio-β-cyclodextrins. Eur. J. Pharm. Biopharm. 60, 123–131.

Perrut, M., Jung, J., Leboeuf, F., 2005. Enhancement of dissolution rate of polysoluble active ingredients by supercritical fluid processes. Part I: micronization of neat particles. Int. J. Pharm. 288, 3–10.

Pinto Reis, C., Neufeld, R.J., Ribeiro, A.J., Veiga, F., 2006. Nanoencapsulation I. Methods for preparation of drug-loaded polymeric nanoparticles. Nanomed: Nanotechnol. Biol. Med. 2, 8–21.

Pitha, J., Hoshino, T., 1992. Effects of ethanol on formation of inclusion complexes of hydroxypropyl cyclodextrins with testosterone or with methyl organe. Int. J. Pharm. 80, 243–251.

Puglisi, G., Fresta, M., Giammona, G., Ventura, C.A., 1995. Influence of the preparation conditions in poly(ethylcyanoacrylate) nanocapsule formation. Int. J. Pharm. 125, 283–287.

Puig, J.E., 1996. Microemulsion polymerization (oil-in-water). In: Salamone, J.C. (Ed.), Polymeric Materials Encyclopedia. CRC Press, Boca Raton, FL, pp. 4333–4341.

Qi, Z.H., Hedges, A.R., 1995. Use of cyclodextrins for flavors. In: Ho, C.T., Tong, C.H. (Eds.), Flavor Technology: Physical Chemistry, Modification and Process. Washington DC, American Chemical Society, USA, pp. 231–243.

Quintanar-Guerrero, D., Zambrano-Zaragoza, M.L., Gutierrez-Cortez, E., Mendoza-Munoz, N., 2012. Impact of the emulsification-diffusion method on the development of pharmaceutical nanoparticles. Recent Pat. Drug Deliv. Formul. 6, 184–194.

Rahman, M.M., Elaissari, A., 2010. Organic–inorganic hybrid magnetic latex. In: Herk, A.M., van Landfester, K. (Eds.), Hybrid Latex Particles Advances in Polymer Science. Springer, Berlin, Heidelberg, pp. 237–281.

Rahman, M.M., Elaissari, A., 2012. Multi-stimuli responsive magnetic core–shell particles: synthesis, characterization and specific RNA recognition. J. Colloid Sci. Biotechnol. 1, 3–15.

Rajabi, H., Ghorbani, M., Jafari, S.M., Sadeghi Mahoonak, A., Rajabzadeh, G., 2015. Retention of saffron bioactive components by spray drying encapsulation using maltodextrin, gum Arabic and gelatin as wall materials. Food Hydrocoll. 51, 327–337.

Rao, J.P., Geckeler, K.E., 2011. Polymer nanoparticles: preparation techniques and size-control parameters. Prog. Polym. Sci. 36, 887–913.

Rekharsky, M.V., Inoue, Y., 1998. Complexation thermodynamics of cyclodextrins. Chem. Rev. 98, 1875–1918.

Ribeiro, L.S.S., Ferreira, D.C., Veiga, F.J.B., 2003. Physicochemical investigation of the effects of water-soluble polymers on vinpocetine complexation with β-cyclodextrin and its sulfobutyl ether derivative in solution and solid state. Eur. J. Pharm. Sci. 20, 253–266.

Ribeiro, A., Figueiras, A., Santos, D., Veiga, F., 2008. Preparation and solid state characterization on inclusion complexes formed between miconazole and methyl-β-cyclodextrin. AAPS Pharm. Sci. Technol. 9, 1102–1109.

Rimmer, S., 2000. Cyclodextrins in the emulsion polymerization of vinyl monomers. Macromol. Symp. 150, 149–154.

Ringard-Lefebvre, C., Bochot, A., Memişoğlu, E., Charon, D., Duchêne, D., Baszkin, A., 2002. Effect of spread amphiphilic β-cyclodextrins on interfacial properties of the oil/water system. Coll. Surf. B Biointerf. 25, 109–117.

Rotureau, E., Raynaud, J., Choquenet, B., Marie, E., Nouvel, C., Six, J.L., Dellacherie, E., Durand, A., 2008. Application of amphiphilic polysaccharides as stabilizers in direct and inverse free-radical miniemulsion polymerization. Colloid. Surf. Physicochem. Eng. Asp. 331, 84–90.

Roux, M., Auzély-Velty, R., Djedaini-Pilard, F., Perly, B., 2002. Cyclodextrin-induced lipid lateral separation in DMPC membranes: (2)H nuclear magnetic resonance study. Biophys. J. 82, 813–822.

Saenger, W., Noltemeyer, M., Manor, P.C., Hingerty, B., Klar, B., 1976. Induced-fit type complex formation of the model enzyme α-cyclodextrin. Bioorg. Chem. 5, 187–195.

Sah, H., 1997. Microencapsulation techniques using ethyl acetate as a dispersed solvent: effects of its extraction rate on the characteristics of PLGA microspheres. J. Control. Release 47, 233–245.

Saharan, V.A., Kukkar, V., Kataria, M., Gera, M., Choudhary, P.K., 2009. Dissolution enhancement of drugs. Part I: technologies and effect of carriers. Int. J. Health. Res. 2, 107–124.

Sahu, S., Mohapatra, S., 2013. Multifunctional magnetic fluorescent hybrid nanoparticles as carriers for the hydrophobic anticancer drug 5-fluorouracil. Dalton Trans. 42, 2224–2231.

Sajeesh, S., Sharma, C.P., 2006. Cyclodextrin–insulin complex encapsulated polymethacrylic acid based nanoparticles for oral insulin delivery. Int. J. Pharm. 325, 147–154.

Sani, S.N., Das, N.G., Das, S.K., 2009. Effect of microfluidization parameters on the physical properties of PEG-PLGA nanoparticles prepared using high pressure microfluidization. J. Microencapsul. 26, 556–561.

Sapkal, N.P., Kilor, V.A., Bhusari, K.P., Daud, A.S., 2007. Evaluation of some methods for preparing gliclazide beta-cyclodextrin inclusion complexes. Trop. J. Pharm. Res. 6, 833–840.

Saravana, K.K., Sushma, M., Prasanna, R.Y., 2013. Dissolution enhancement of poorly soluble drugs by using complexation technique—a review. J. Pharm. Sci. Res. 5, 120–124.

Sarkar, D., El Khoury, J., Lopina, S.T., Hu, J., 2005. An effective method for preparing polymer nanocapsules with hydrophobic acrylic shell and hydrophilic interior by inverse emulsion radical polymerization. Macromolecul. 38, 8603–8605.

Schubert, R., 2003. Liposome preparation by detergent removal. Method Enzymol. 367, 46–70.

Schwinte, P., Ramphul, M., Darcy, R., O'Sullivan, J.F., 2003. Amphiphilic cyclodextrin complexation of clofazimine. J. Incl. Phenom. Macrocyclic Chem. 47, 109–112.

Scott, C., Wu, D., Ho, C.C., Co, C.C., 2005. Liquid-core capsules via interfacial polymerization: a free-radical analogy of the nylon rope trick. J. Am. Chem. Soc. 127, 4160–4161.

Semalty, A., 2014. Cyclodextrin and phospholipid complexation in solubility and dissolution enhancement: a critical and meta-analysis. Exp. Opin. Drug Deliv. 11, 1255–1272.

Semalty, A., Tanwar, Y.S., Semalty, M., 2014a. Preparation and characterization of cyclodextrin inclusion complex of naringenin and critical comparison with phospholipid complexation for improving solubility and dissolution. J. Thermal Anal. Calorim. 115, 2471–2478.

Semalty, M., Panchpuri, M., Singh, D., Ajay, R., 2014b. Cyclodextrin inclusion complex of racecadotril: effect of drug-beta-cyclodextrin ratio and the method of complexation. Curr. Drug Discov. Tech. 11, 154–161.

Seo, S., Khang, G., Rhee, J., Kim, J., Lee, H., 2003. Study on in vitro release patterns of fentanyl-loaded PLGA microspheres. J. Microencapsul. 20, 569–579.

Shapiro, Y.E., 2004. Nanoencapsulation of bioactive substances. In: Schwarz, J.A., Contescu, C.I., Putyera, K. (Eds.), Dekker Encyclopedia of Nanoscience and Nanotechnology. Marcel Dekker, New York, USA, pp. 2339–2352.

Shiga, H., Yoshii, H., Ohe, H., Yasuda, M., Furuta, T., Kuwahara, H., Ohkawara, M., Linko, P., 2004. Encapsulation of Shiitake (Lenthinus edodes) flavors by spray drying. Biosci. Biotechnol. Biochem. 68, 66–71.

Shimpi, S., Chauhan, B., Shimpi, P., 2005. Cyclodextrins: application in different routes of drug administration. Acta Pharm. 55, 139–156.

Shin-ichi, Y., Katsuhiro, I., Keiichi, M., Hideo, T., Akira, O., 2004. Evaluation of ophthalmic suspensions using surface tension. Eur. J. Pharm. Biopharm. 57, 377–382.

Shu, S., Wang, X., Zhang, X., Zhang, X., Wang, Z., Li, C., 2009. Disulfide cross-linked biodegradable polyelectrolyte nanoparticles for the oral delivery of protein drugs. New J. Chem. 33, 1882–1887.

Shu, S., Zhang, X., Wu, Z., Wang, Z., Li, C., 2010. Gradient cross-linked biodegradable polyelectrolyte nanocapsules for intracellular protein drug delivery. Biomaterials 31, 6039–6049.

Shubayev, V.I., Pisanic, II., T.R., Jin, S., 2009. Magnetic nanoparticles for theragnostics. Adv. Drug Deliv. Rev. 61, 467–477.

Silveira, M.A., Ponchel, G., Puisieux, F., Duchêne, D., 1998. Combined poly (isobutylcyanoacrylate) and cyclodextrins nanoparticles for enhancing the encapsulation of lipophilic drugs. Pharm. Res. 15, 1051–1055.

Singh, R., Bharti, N., Madan, J., Hiremath, S.N., 2010. Characterization of cyclodextrin inclusion complexes-a review. J. Pharm. Sci. Technol. 2, 171–183.

Skiba, M., Wouessidjewe, D., Fessi, H., Devissaguet, J.P., Duchêne, D., Puisieux, F., 1992. Preparation et utilizations des nouveau systemes colloidaux dispersibles a base de cyclodextrines sous forme de nanocapsules. French Patent, 9207285.

Skiba, M., Morvan, C., Duchêne, D., Puisieux, F., 1995. Evaluation of the gastrointestinal behaviour in the rat of amphiphilic β-cyclodextrin nanocapsules loaded with indomethacin. Int. J. Pharm. 126, 275–279.

Skiba, M., Duchêne, D., Puisieux, F., Wouessidjewe, D., 1996a. Development of a new colloidal drug carrier from chemically modified cyclodextrins: nanospheres and influence of physicochemical and technological factors on particle size. Int. J. Pharm. 129, 113–121.

Skiba, M., Nemati, F., Puisieux, F., Duchêne, D., Wouessidjewe, D., 1996b. Spontaneous formation of drug-containing amphiphilic cyclodextrin nanocapsules. Int. J. Pharm. 145, 241–245.

Skiba, M., Bounoure, F., Barbot, C., Arnaud, P., Skiba, M., 2005. Development ofcyclodextrin microspheres for pulmonary drug delivery. J. Pharm. Pharmaceut. Sci. 8, 409–418.

Song, L.X., Bai, L., Xu, X.M., He, J., Pan, S.Z., 2009. Inclusion complexation, encapsulation interaction and inclusion number in cyclodextrin chemistry. Coordin. Chem. Rev. 253, 1276–1284.

Steckel, H., Wehle, S., 2004. A novel formulation technique for metered dose inhaler (MDI) suspensions. Int. J. Pharm. 284, 75–82.

Sukegawa, T., Furuike, T., Niikura, K., Yamagishi, A., Nishimura, S.I., 2002. Erythrocyte-like liposomes prepared by means of amphiphilic cyclodextrin sulfates. Chem. Commun. 5, 430–431.

Sun, S., Zeng, H., 2002. Size-controlled synthesis of magnetite nanoparticles. J. Am. Chem. Soc. 124, 8204–8205.

Szejtli, J., 1988. Topics in inclusion science-cyclodextrin technology. In: Davies, J.E.D. (Ed.), Cyclodextrin Technology. Kluwer Academic Publisher, Dordrecht, Netherlands.

Szejtli, J., 1998. Introduction and general overview of cyclodextrin chemistry. Chem. Rev. 98, 1743–1753.

Szejtli, J., 2004. Past, present, and future of cyclodextrin research. Pure Appl. Chem. 76, 1825–1845.

Szejtli, J., Cserhati, T., Szögyi, M., 1986. Interactions between cyclodextrins and cell-membrane phospholipids. Carbohydr. Polym. 6, 35–49.

Tapas, A.R., Kawtikwar, P.S., Sakarkar, D.M., 2011. Evaluation of physico-mechanical properties of carvedilol-cyclodextrin agglomerates obtained by emulsion solvent diffusion method. Lat. Am. J. Pharm. 30, 1461–1467.

Teijeiro-Osorio, D., Remunán-López, C., Alonso, M.J., 2009. Chitosan/cyclodextrin nanoparticles can efficiently transfect the airway epithelium in vitro. Eur. J. Pharm. Biopharm. 71, 257–263.

Tirucherai, G.S., Mitra, A.K., 2003. Effect of hydroxypropyl beta cyclodextrin complexation on aqueous solubility, stability, and corneal permeation of acyl ester prodrugs of ganciclovier. AAPS Pharm. Sci. Technol. 4, E45.

Torini, L., Argillier, J.F., Zydowicz, N., 2005. Interfacial polycondensation encapsulation in miniemulsion. Macromolecul. 38, 3225–3236.

Torres-Labandeira, J.J., Davignon, P., Pitha, J., 1990. Oversaturated solutions of drug in hydroxypropylcyclodextrins: parenteral preparation of pancratistatin. J. Pharm. Sci. 80, 384–386.

Trapani, A., Garcia-Fuentes, M., Alonso, M.J., 2008. Novel drug nanocarriers combining hydrophilic cyclodextrins and chitosan. Nanotechnology 19, 185101.

Tsinontides, S.C., Rajnaik, P., Pham, D., Hunke, W.A., Placek, J., Reynolds, S.D., 2004. Freeze drying-principles and practice for successful scale up to manufacturing. Int. J. Pharm. 28, 1–16.

Tsorteki, F., Bethanis, K., Pinotsis, N., Giastas, P., Mentzafos, D., 2004. Structure of the inclusion complexes of heptakis(2,3,6-tri-O-methyl)-beta-cyclodextrin withindole-3-butyric acid and 2,4-dichlorophenoxyacetic acid. Carbohydr. Res. 339, 233–240.

Udagama, R., Degrandi-Contraires, E., Creton, C., Graillat, C., McKenna, T.F.L., Bourgeat-Lami, E., 2011. Synthesis of polymer-polymer hybrids by miniemulsion polymerization and characterization of hybrid latex. Macromolecul. 44, 2632–2642.

Uekama, K., Hirayama, F., Irie, T., 1998. Cyclodextrin drug carrier systems. Chem. Rev. 98, 2045–2076.

US Federal Register, 2005. Alpha-cyclodextrin, beta-cyclodextrin, and gamma-cyclodextrin: exemption from the requirement of a tolerance. US Federal Register 70, 38780-38785.

Van Hees, T., Piel, G., Evrard, B., Otte, X., Thunus, L., Delattre, L., 1999. Application of supercritical carbon dioxide for the preparation of piroxicam-β-cyclodextrin inclusion compound. Pharm. Res. 16, 1864–1870.

Van Herk, A.M., 2012. Modeling of emulsion copolymerizations will it ever be possible? Part III conclusion. J. Colloid Sci. Biotechnol. 1, 169–174.

Vauthier, C., Bouchemal, K., 2008. Methods for the preparation and manufacture of polymeric nanoparticles. Pharmaceut. Res. 26, 1025–1058.

Vauthier, C., Dubernet, C., Fattal, E., Pinto-Alphandary, H., Couvreur, P., 2003. Poly(alkylcyanoacrylates) as biodegradable materials for biomedical applications. Adv. Drug Deliv. Rev. 55, 519–548.

Veintemillas-Verdaguer, S., Morales, M., Serna, C., 1998. Continuous production of γ-Fe_2O_3 ultrafine powders by laser pyrolysis. Mater. Lett. 35, 227–231.

Vila, A., Sanchez, A., Tobio, M., Calvo, P., Alonso, M.J., 2002. Design of biodegradable particles for protein delivery. J. Control. Release 78, 15–24.

Villiers, A., 1891. Sur la fermentation de la fe cule par l'action du ferment butyrique. C. R. Acad. Sci. 112, 536–538.

Vitale, S.A., Katz, J.L., 2003. Liquid droplet dispersions formed by homogeneous liquid-liquid nucleation: the ouzo effect. Langmuir 19, 4105–4110.

Vozone, C.M., Marques, H.M.C., 2003. Complexation of budesonide in cyclodextrins and particle aerodynamic characterization of the complex solid form for dry powder inhalation. J. Incl. Phenom. Macroc. Chem. 44, 111–115.

Waalkens-Berendsen, D.H., Smits van Prooije, A.E., Bar, A., 1998. Embryotoxicity and teratogenicity study with γ-cyclodextrin in rabbits. Regul. Toxicol. Pharmacol. 27, 172–177.

Waleczek, K.J., Cabral Marques, H.M., Hempel, B., Schmidt, P.C., 2003. Phase solubility studies of pure (-)-α-bisabolol and camomile essential oil with β-cyclodextrin. Eur. J. Pharm. Biopharm. 55, 247–251.

Wang, H., Huang, Y., 2011. Prussian-blue-modified iron oxide magnetic nanoparticles as effective peroxidase-like catalysts to degrade methylene blue with H_2O_2. J. Haz. Mat. 191, 163–169.

Wang, Q.F., Fan, X.W., Xu, L., Yao, Y., 2007. Preparation of inclusion complex of paeonol and beta-cyclodextrin by sealed-control temperature method. Chin. J. Chin. Mater. Med. 32, 218–221.

Watnasirichaikul, S., Davies, N.M., Rades, T., Tucker, I.G., 2000. Preparation of biodegradable insulin nanocapsules from biocompatible microemulsions. Pharm. Res. 17, 684–689.

Wenz, G., 1991. Synthesis and characterization of some lipophilic per (2,6-di-O-alkyl) cyclomaltooligosaccharides. Carbohydr. Res. 214, 257–265.

Wenz, G., Thomas, H., 1999. Synthesis of highly water-soluble cyclodextrin sulfonates by addition of hydrogen sulfite to cyclodextrin allyl ethers. Carbohydr Res. 322, 153–165.

Wongmekiat, A., Tozuka, Y., Moribe, K., Oguchi, T., Yamamoto, K., 2007. Preparation of drug nanoparticles by co-grinding with cyclodextrin: formation mechanism and factors affecting nanoparticle formation. Chem. Pharm. Bull. 55, 359–363.

Wouessidjewe, D., Skiba, M., Leroy-Lechat, F., Lemos-Senna, E., Puisieux, F., Duchêne, D., 1996a. A new concept in drug delivery based on 'skirt-shaped cyclodextrins aggregates' present state and future prospects. STP Pharma. Sci. 6, 21–26.

Wouessidjewe, D., Skiba, M., Leroy-Lechat, F., Lemos-Senna, E., Puisieux, F., Duchêne, D., 1996b. A new concept in drug delivery based on skirt-shaped cyclodextrin aggregates, present state and future prospects. STP Pharma. Sci. 6, 21–28.

Wu, W., He, Q., Jiang, C., 2008. Magnetic iron oxide nanoparticles: synthesis and surface functionalization strategies. Nanoscale Res. Lett. 3, 397–415.

Wu, M., Dellacherie, E., Durand, A., Marie, E., 2009a. Poly(*n*-butyl cyanoacrylate)nanoparticles via miniemulsion polymerization (1): dextran-based surfactants. Colloid. Surf. B Biointerfaces 69, 141–146.

Wu, M., Dellacheriea, E., Durand, A., Marie, E., 2009b. Poly(*n*-butyl cyanoacrylate) nanoparticles via miniemulsion polymerization. PEG-based surfactants. Colloid. Surf. B 69, 147–151.

Wulff, M., Aldén, M., 1999. Solid state studies of drug-cyclodextrin inclusion complexes in PEG 6000 prepared by a new method. Eur. J. Pharm. Sci. 8, 269–281.

Xia, H.B., Yi, J., Foo, P.-S., Liu, B., 2007. Facile fabrication of water-soluble magnetic nanoparticles and their spherical aggregates. Chem. Mater. 19, 4087–4091.
Xia, Y., Li, W., Cobley, C.M., Chen, J., Xia, X., Zhang, Q., Yang, M., Cho, E.C., Brown, P.K., 2011. Gold nanocages: from synthesis to theranostic applications. Acc. Chem. Res. 44, 914–924.
Xie, J., Wang, C., 2005. Self-assembled biodegradable nanoparticles developed by direct dialysis for the delivery of paclitaxel. Pharm. Res. 22, 2079–2090.
Yallapu, M.M., Othman, S.F., Curtis, E.T., Bauer, N.A., Chauhan, N., Kumar, D., Jaggi, M., Chauhan, S.C., 2012. Curcumin-loaded magnetic nanoparticles for breast cancer therapeutics and imaging applications. Int. J. Nanomed. 7, 1761–1779.
Yang, C., Castelvetro, V., Scalarone, D., Bianchi, S., Zhang, Y., 2011. Three different β-cyclodextrins direct the emulsion copolymerization of a highly fluorinated methacrylate toward distinctive nanostructured particle morphologies. J. Polym. Sci. Polym. Chem. 49, 4518–4530.
Yano, H., Kleinebudde, P., 2010. Improvement of dissolution behavior for poorly water-soluble drug by application of cyclodextrin in extrusion process: comparison between melt extrusion and wet extrusion. AAPS. Pharm. Sci. Tech. 11, 885–893.
Yildiz, U., Landfester, K., 2008. Miniemulsion polymerization of styrene in the presence of macromonomeric initiators. Polymer 49, 4930–4934.
Yoshii, H., Yasuda, M., Furuta, T., Kuwahara, H., Ohkawara, M., Linko, P., 2005. Retention of cyclodextrin complexed shiitake (Lentinus edodes) flavors with spray drying. Drying Technol. 23, 1205–1215.
Yousaf, A.M., Kim, D.W., Oh, Y.K., Yong, C.S., Kim, J.O., Choi, H.G., 2015. Enhanced oral bioavailability of fenofibrate using polymeric nanoparticulated systems: physicochemical characterization and in vivo investigation. Int. J. Nanomed. 10, 1819–1830.
Yuan, Z., Ye, Y., Gao, F., Yuan, H., Lan, M., Lou, K., Wang, W., 2013. Chitosan-graft-β-cyclodextrin nanoparticles as a carrier for controlled drug release. Int. J. Pharm. 446, 191–198.
Yuliani, S., Torley, P.J., D'Arcy, B., Nicholson, T., Bhandari, B., 2006. Extrusion of mixtures of starch and d-limonene encapsulated with β-cyclodextrin: flavor retention and physical properties. Food Res. Int. 39, 318–331.
Zhang, J., Ma, P.X., 2009. Host-guest interaction mediated polymeric core–shell assemblies: versatile nanocarriers for drug delivery. Angew. Chem. Int. Ed. Engl. 48, 964–968.
Zhang, J., Ma, P.X., 2011. Core–shell structured nanoassemblies based on β-cyclodextrin containing block copolymer and poly(β-benzyl L-aspartate) viahost–guest complexation. Polymer (Guildf.) 52, 4928–4937.
Zhang, P., Ling, C.C., Coleman, A.W., Parrot-Lopez, H., Galons, H., 1991. Formation of amphiphilic cyclodextrin via hydrophobic esterification at the secondary hydroxyl face. Tetrahedron Lett. 32, 2769–2770.
Zhang, P., Parrot-Lopez, H., Tchoreloff, P., Basakin, A., Ling, C.C., De Rango, C., Coleman, A.W., 1992. Self-organizing systems based on amphiphilic cyclodextrins diesters. J. Phys. Org. Chem. 5, 518–528.
Zhou, J., Ritter, H., 2010. Cyclodextrin functionalized polymers as drug delivery systems. Polym. Chem. 1, 1552–1559.
Zhou, R., Wang, F., Guo, Z., Zhao, Y.L., 2012. Preparation and characterization of resveratrol/hydroxypropyl-beta-cyclodextrin inclusion complex using supercritical antisolvent technology. J. Food Process Eng. 35, 677–686.

Zhu, J., He, J., Du, X., Lu, R., Huang, L., Ge, X., 2011. A facile and flexible process of β-cyclodextrin grafted on Fe3O4 magnetic nanoparticles and host–guest inclusion studies. Appl. Surf. Sci. 257, 9056–9062.

Zia, V., Rajewski, R., Bornancini, E.R., Luna, E.A., Stella, V.J., 1997. Effect of alkyl chain length and substitution degree on the complexation of sulfoalkyl ether β-cyclodextrins with steroids. J. Pharm. Sci. 86, 220–224.

Ziegler, A., Landfester, K., Musyanovych, A., 2009. Synthesis of phosphonate functionalized polystyrene and poly(methyl methacrylate) particles and their kinetic behavior in miniemulsion polymerization. Colloid. Polym. Sci. 287, 1261–1271.

Part Three

Nanoencapsulation Technologies Based on Special Equipment

8 Nanocapsule formation by electrospinning

*Behrouz Ghorani**, *Ali Alehosseini**, *Nick Tucker***
*Research Institute of Food Science and Technology (RIFST), Mashhad, Iran;
**University of Lincoln, Lincoln, United Kingdom

8.1 Introduction

The efficacy of micronutrients can be improved by controlling and targeting the site of release, thus broadening their range of application as high added value food ingredients by ensuring the delivery of optimal dosages in the right place in the gastrointestinal tract (Katouzian and Jafari, 2016; Faridi Esfanjani and Jafari, 2016). Electrospinning is an established method for manufacturing nanoscale products, the process produces long fibers that typically have diameters below 1 μm. The high specific surface area of such nanofibers attracts particular attention by imparting the ability to significantly affect the macroscale material properties, even when used in 1% fractions (Ghorani et al., 2013). Electrospun fibers have potential as vehicles for drug delivery, nanoencapsulators for bioactive molecules, as biomolecular sensors and ultrafiltration media (Anu Bhushani and Anandharamakrishnan, 2014). The earliest description of electrospinning dates from the middle of the 19th century with commercialization attempts being recorded throughout the last century. In the late 1930s Soviet scientists made a great breakthrough in the application of electrospinning to make ultrafine nonwoven filtration media; filtration is still the most popular commercial application of electrospinning (Tucker et al., 2012). The food industry has exhibited an increasing level of interest in the use of food industry applications of the technology as demonstrated by a rising number of technical publications (Ghorani and Tucker, 2015), concerning the use of electrospinning for food ingredient encapsulation, using bioactive compound and catalysts (Rezaei et al., 2015; Faridi Esfanjani and Jafari, 2016).

Recently, significant progress has been made in understanding the mechanisms of electrospinning and electrospun fiber morphologies have been developed to deliver supplementary substances along with the fiber forming polymer; these supplementary substances significantly influence the mechanical and properties and functionality of the fiber product (Ghorani and Tucker, 2015). These supplements may result in fiber products containing an array of soluble additives, insoluble nanoparticles, or nanotubes. In addition, fibers and particulates can be made with multilayer structures, typically with a resistant outer coating around a chemically active interior. As required, these materials can be immobilized on a permanent basis within the carrier substance (usually a polymer), as for the purpose of delivery pharmaceutical materials or later removed if molecular templating is the desired process (Ghorani and Tucker, 2015). For food ingredient manufacture, the process is highly appropriate. Both the formation of the preliminary

Nanoencapsulation Technologies for the Food and Nutraceutical Industries

droplet and the consequent fiber occur at ambient temperature—the bottom temperature processing limit is the feedstock freezing point. As a polar solvent, water can be used for a wide variety of both natural origin and synthetic materials. Solidification into a fiber is a purely physical process resulting from solvent evaporation or the freezing of a molten strand, so no chemical alteration of the material occurs. Thus, the method is an appropriate production route for environmentally sensitive ingredients because any biochemical degradation of the ingredient is minimized (Ghorani and Tucker, 2015).

A large number of electrospun fiber, different shaped cross sections, and fiber diameters result from using different polymers and solvents (Fig. 8.1). Electrospun micro- and nanofibers are monofilaments with a cylindrical cross-sections; the latter being a good indication that the fiber strand was solid at the point of impact. However under some specific conditions, other forms are observed (Spasova et al., 2006).

Branched fibers form because highly charged strands bifurcate in flight to reduce charge density, flat ribbons result when the strand does not have time to solidify in

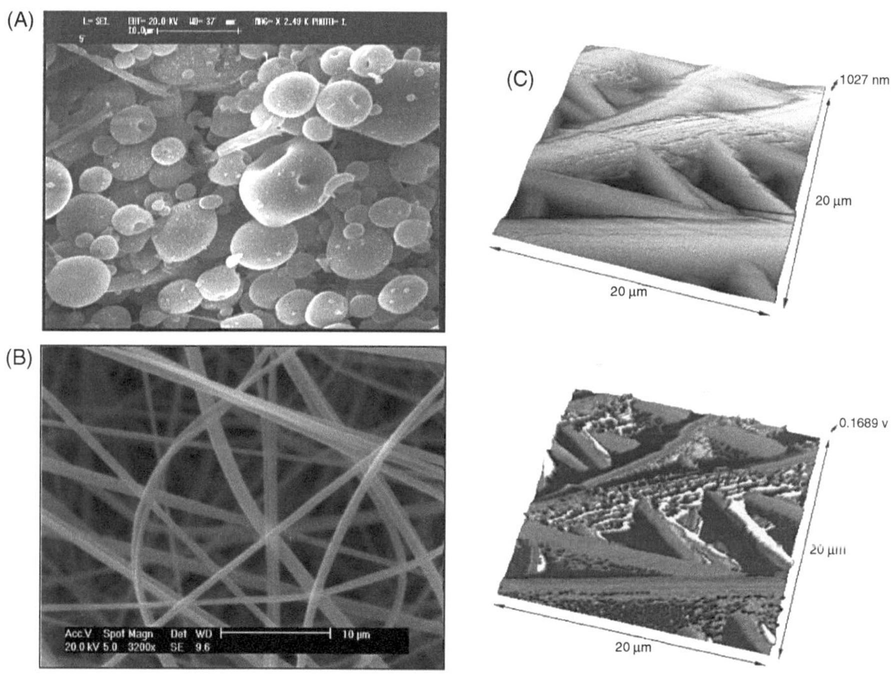

Figure 8.1 (A) Whey protein capsules formed by electro-hydrodynamic atomization technique. These capsules have been developed for encapsulation of probiotics as a joint venture project between Research Institute of Food Science & Technology (RIFST), Iran and King Mongkut's University of Technology Thonburi, Thailand. (B) SEM micrograph of ribbon-like structure in an electrospun CA fiber produced from the binary solvent system Acetone: DMAc at a polymer concentration of 25% (w/v). (C) AFM 3D-topograph (left) and amplitude (right) images of zein electrospun fibers with circular cross sections at a constant concentration of 26 w/v%, applied voltage of 20 kV, flow rate of 8 mL/h, and distance between spinneret and collector of 15 cm. Electrospun proteins have been developed for various applications in the Research Institute of Food Science & Technology (RIFST), Mashhad, Iran.

flight. For bent ribbon morphologies which are sometimes described, the method of formation is not well understood (Frenot and Chronakis, 2003; Koombhongse et al., 2001), similarly hollow cylindrical configurations and hollow capsules have been described (Koombhongse et al., 2001). Helical coils at the microscale (Kessick and Tepper, 2004) can be made by varying polymer solution properties (Koski et al., 2004). Similar helical morphology structures have been electrospun on to a conductive target, using a two-component solution (Kessick and Tepper, 2004). The average diameter of electrospun fibers depends on interactions between process conditions but typically varies between ≈5 nm and ≈10 μm, this range of variation in diameter being a characteristic. There is usually a wide range of diameters as compared to conventional fiber spinning processes, where fiber diameter variations are relatively small (Ghorani and Tucker, 2015). Other structural morphologies are observed, namely colloidal beads or capsules, "beads-on-a-string" or spindle morphologies either as infrequent inclusions in the fiber web or as the main product, dependent again on the interplay between process parameters and raw material characteristics. This chapter is in the form of a discussion on electrospinning as a production method for food-safe nanoscale fibers particularly as nanocarriers and for encapsulation.

8.2 Principles of electrospinning

Electrostatic force is used in electrospinning to draw a fine fiber out of a sessile polymer solution droplet, which flies toward the nearest lower electrical potential target (Frenot and Chronakis, 2003). Fundamentally, an electrospinning assembly (Fig. 8.2) comprises of three main components which are—supply of high voltage, a spinneret

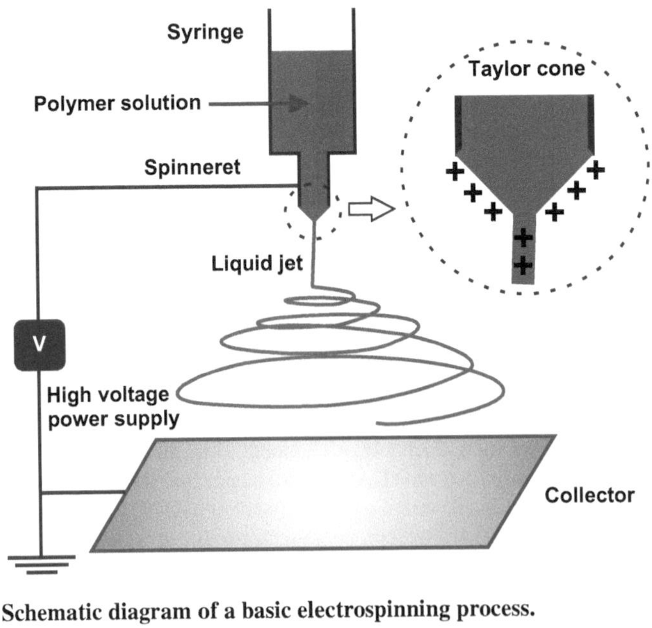

Figure 8.2 Schematic diagram of a basic electrospinning process.

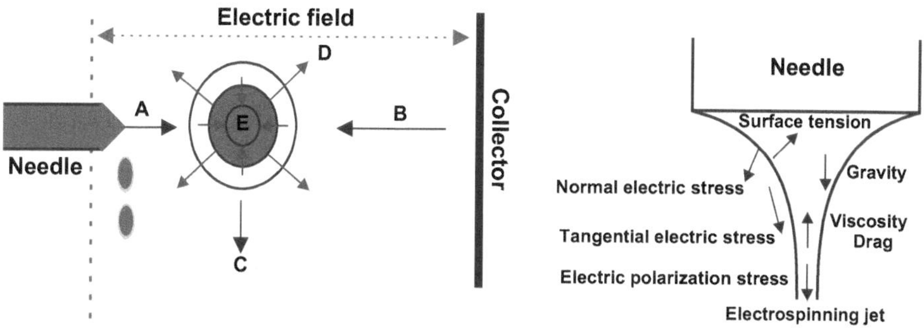

Figure 8.3 Forces acting on the charged droplet during the electrospinning. A, Electrostatic force; B, drag force; C, gravity; D, coulombic repulsion force; E, surface tension and viscoelastic force.

(typically made by repurposing a blunt hypodermic syringe), and a target-collector plate connected to an electrical earth (a metal screen or film). A kilovolt DC supply is used to induce a surplus electric charge in the liquid polymer (either a melt, or more commonly a solution). A fiber strand is pulled out of the droplet and accelerated toward the grounded collector. The liquid polymer is either pumped from the spinneret at a constant volume flow using a syringe pump or more traditionally fed from a reservoir at constant head (Fig. 8.2).

When a droplet of liquid polymer is charged with a high potential, the droplet deforms, being attracted to a site of low potential. This droplet distortion is a characteristic structure dubbed as the "Taylor cone" (Fig. 8.2) (Reneker et al., 2000; Yarin et al., 2001). When the electric charge rises to a certain value, internal electrostatic repulsion overcomes the intrinsic molecular tension forces present at the surface of the droplet at the tip of the Taylor cone, and a liquid polymer stream, still carrying a high electric charge is thrown out toward the nearest lower charged point (the collector) that is usually held at earth potential to encourage fiber capture (Doshi and Reneker, 1995). In flight between the spinneret and the collector the stream of polymer is elongated as an expanding helix (often incorrectly described as the "whipping instability"). Forces from the electric field, viscous drag effects, gravitational effects, repulsive forces of coulombic origin, surface tension, and viscoelastic properties all affect the charged polymer strand (Fig. 8.3) (Ding et al., 2006).

The polymer jet continues to be stretched as the solvent evaporates and a continuous fiber deposits onto the earthed target plate (Dabirian et al., 2007; Doshi and Reneker, 1995; Shin et al., 2001). When the fiber strand is thrown out by the repulsive force of electrical charge from the spinneret and takes flight in the direction of the lower potential collector, an increasing force resulting from viscous friction with the atmosphere is experienced by the accelerating jet. The repulsive columbic forces acting on the droplet oppose the forces of contraction originating in the solution viscoelasticity and from surface tension. The existence of the electric field between spinneret and collector overtly affects these electrical origin forces of repulsion. The forces resulting from the presence of the electric field initiates the change in shape of the droplet from spheroid into an onion or spindle-form and also whether electrospraying (droplets) or electrospinning (fibers) occurs (Almería et al., 2010; Grimm and Beauchamp, 2005).

8.3 Electrospinning versus electrospraying

The difference between electrospinning and electrospraying (Chapter 9) is controlled by how much molecular cohesion occurs—this is difficult to measure directly, but is readily inferred from the polymer solution viscosity and less directly from polymer concentration (Chakraborty et al., 2009; Enayati et al., 2011). Electrospraying or electro-hydrodynamic spraying as it is sometime called is proposed as method for nanoencapsulation, where the addition of a high level of electrostatic force reduces the size of the formation of aerosol droplets to the nanoscale. In other words, the electrostatic force resulting from the application of high voltages increases the degree of atomization, resulting in finer droplets than from the effect of surface tension alone. The removal of solvent from the strand occurs at an increased rate during the flight of the droplets because the surface to volume ratio is higher (Hao et al., 2013; Zhang and Kawakami, 2010). Understanding the electro-hydrodynamics of the fluids is a fundamental requirement for understanding and controlling both electrospinning and electrospraying (Chakraborty et al., 2009). Electrospinning and electrospraying are both based on the same physics governing the ejection of a continuous stream, however if the degree of molecular cohesion is below a critical level droplets are formed from the ejecta, and not a continuous fiber (Bock et al., 2011). Electrospraying of viscous polymer solutions tends to occur in three basic modes: a dripping mode, an intermittent ejection mode, and a single cone stream characterized by the presence of a stable Taylor cone (Jain et al., 2015). The diagnostic characteristic of electrospraying is the presence of fine charged droplets of about half the size of the Rayleigh limit where repulsive electrostatic force is said to overcome the attractive force of surface tension (Jaworek, 2007). Barrero and Loscertales (2007) quantified the minimum flow rate at where a steady state of the cone-jet mode could be achieved (Barrero and Loscertales, 2007). This flow rate ($Q_{\min}$) is calculated as follows,

$$Q_{\min} \approx \frac{\sigma_1 \varepsilon_0 \varepsilon_r}{\rho_1 \gamma_1} \tag{8.1}$$

where $Q_{\min}$ is minimum liquid volume flow rate, σ_1 is surface tension of the liquid, ε_0 is permittivity of the free space, ρ_1 is mass density of the liquid, and γ_1 is liquid bulk conductivity.

The theory is confirmed by experiment showing the droplet diameter to be proportional to the liquid flow rate and inversely proportional to the power of the magnitude conductivity of the liquid (De La Mora and Loscertales, 1994; Gañán-Calvo, 1999; Hartman et al., 2000; Ogata et al., 1978; Tomita et al., 1986).

Clearly, electrospraying is fundamentally similar to electrospinning. The main process factor that differentiates between electrospraying and electrospinning is the polymer solution concentration (Chakraborty et al., 2009; Shenoy et al., 2005a). The

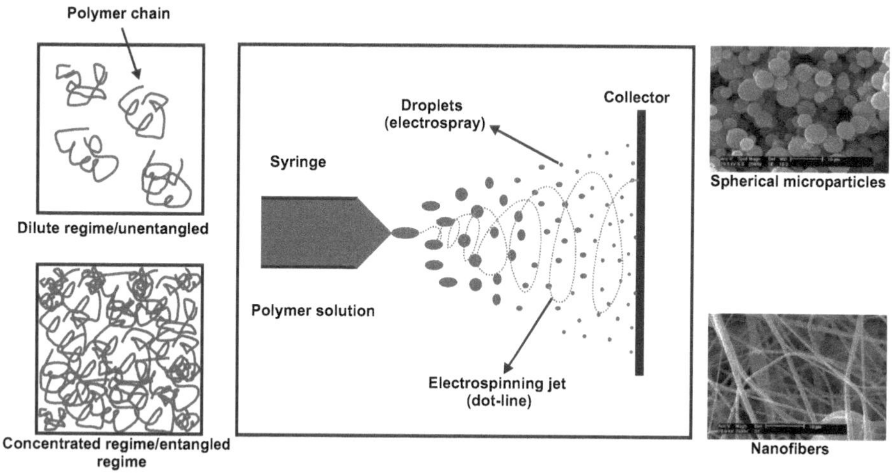

Figure 8.4 (Left) Schematic of physical representation at the molecular level of entanglement regimes for dilute and concentrated polymer concentration. (Middle) Schematic diagram of a basic electrospinning (jet formation) and electrospraying (liquid-droplet atomization) process. (Right) Examples of SEM image of microspheres (electrospraying) and nanofibers (electrospinning).

influence of the degree of chain entanglements in the polymer solution on the formation bead or fiber morphology from the electrospinning or spraying process is critical. When the %solids dissolved in the solution and thus the degree of entanglement among the molecules is high, the jet formed is stable, and elongation happens in flight in the initial straight flight path and continues at an increased rate after the onset of the so-called "whipping instability" (actually a consistent expanding helix) (Bock et al., 2012) so that the jet does not form a stream of discrete droplets but rather a micro- or nanoscale fiber (Fig. 8.4) (Jaworek and Sobczyk, 2008).

The properties of solution surface tension and viscosity are strongly influenced by the size of the polymer molecule and the concentration of the polymer in the liquid; it also has significance for the nature of the particulate product. The particle size distribution can be influenced by altering the amount of material entrained or in solution in the prespun liquid (Jaworek, 2007). The degree of chain entanglement in the liquid containing the polymer has a notable effect on product morphology, which can range from beading, spindling, or fiber formation (Shenoy et al., 2005a; Zhang et al., 2012). A low level of chain entanglement results in the formation of particulates rather than fibers, although there will be a cross-over level of entanglement where both morphologies are observed (Almería et al., 2010). In summary, limited molecular entanglement densities result in electrospraying (Bock et al., 2011; Shenoy et al., 2005a). Hence the polymer concentration should be optimized for the desired particulate or continuous product. It should also be noted that in the entanglement range for fiber production increasing the solution concentration will reduce the occurrence of spindle-like morphologies until uniform fibers with a tendency toward greater diameters are observed (Fathi et al., 2014).

Whenever biopolymers are utilized, whether nanofibers or droplets are made is influenced by variation in molecular mass, quantity of molecular chain entanglement, and the evaporation rate of solvent from the biopolymer solution. A nanoscale fiber is created when the product of biopolymer concentration (c) and solution intrinsic viscosity (η), known as the Berry number, $Be = [\eta]c$, is above some threshold value (Be_{cr}), which is intrinsic to an individual biopolymer (Jaworek et al., 2009). If the Berry number of a solution is below this value, then instabilities on the surface resulting from the motion of the jet and charge interactions can result in nanobeads instead of a fiber.

Electrosprayed droplets result from the interaction surface and bulk electro-hydrodynamic forces cause the fracture of the polymer strand into drops (Chakraborty et al., 2009). If the concentration of the polymer, perceived as viscosity, is below the threshold for Rayleigh shattering then the jet produced will be a continuous fiber, but will consist of spindle-shaped particles joined by thin threads (Shin et al., 2001). Summarizing, high viscosity polymeric liquids produce continuous fibers, but a dilute polymer jet will form droplets. Electrospraying and electrospinning are two techniques that can use identical equipment and are based on the same physical rules with solutions of higher molecular entanglement (perceived as being more viscous) being required for electrospinning (Bohr et al., 2015). As the electric field strength is increased, the process moves from dripping to the formation of a stable cone-jet. However, perhaps in an attempt to increase production rate the electric field strength is raised too far, jet ejection destabilizes into a multijet spray with a consequent reduction in the degree of particle uniformity. Stable emission is a requirement for the production of low variation droplets. The acceptable range of processing voltages becomes wider as the solution viscosity is increased. It has been reported that the mode of spray production is significantly affected by the strength of the electric field (Grigoriev and Edirisinghe, 2002; Jaworek and Krupa, 1999; Smith et al., 2006). Jaworek et al. (2009) noted that a slow rate of solvent evaporation or a very low concentration polymer solution, may result in the appearance of the beads-on-a-string fiber morphology or the jet disrupting into droplets (Jaworek et al., 2009). The electrical conductivity and permittivity values are constant for a particular polymer solution, consequently the flow rate is the principle mechanism to influence the size distribution of droplet product. The use of electrospray process to make systems for the delivery of drugs (Bock et al., 2012; Chakraborty et al., 2009), film coatings for foodstuffs (Khan et al., 2012), making chocolate (Luo et al., 2012), making solid phase lipid nanoparticles to entrain active compounds (Eltayeb et al., 2013), stabilized nutraceuticals (Torres-Giner et al., 2010), and target encapsulated bioactives and probiotics (López-Rubio and Lagaron, 2012; López-Rubio et al., 2012) have all been demonstrated (see Chapter 9).

Generally, electrospun fibers are fabricated as nonwoven mats characterized by a structure of intertwined fibers and require some form of physical modification to be suitable for use as food ingredients. Bioactives encapsulated in electrospun fibers combine stability and functionality and are thus ideal for use in functional foodstuffs (Pérez-Masiá et al., 2014a). The benign processing conditions (ambient temperature, aqueous solution, and only physical processes to cause a change in state) make electrospraying an attractive method when biochemically sensitive chemicals are to be used; however, there are certain difficulties to be overcome to make manufacturing in bulk entirely profitable.

8.4 The electrospinning process

Electrospinning is a tested and practical route for the production of submicron or nanometer scale diameter polymer fibers. A range of cross-section shapes and fiber diameters can be made by varying polymer solution properties and manufacturing parameters. Electrospun fibers usually possess circular cross sections, however, other forms have been reported (Spasova et al., 2006). Branched shapes, ribbons, and bent ribbon structures have all been observed, if not entirely explained (Frenot and Chronakis, 2003; Koombhongse et al., 2001), in addition hollow cylinders and capsules have been reported (Koombhongse et al., 2001; Koski et al., 2004). Branched fibers are associated with very high levels of charge present in the spinning liquid and resulting from high driving voltages (Filatov et al., 2007), the bifurcation in flight is caused by the need to increase the surface area of the fibers to normalize the surface charge. Regular circular cross sections are evidence that the fiber was solid before it hit the target. Ribbon morphologies are formed when a gel or liquid state jet collapses on hitting the target. Other noncircular cross sections are caused by shrinkage when the strand solidifies after hitting the target; and the fiber is no longer being stretched. Helical coils at the microscale (Kessick and Tepper, 2004) have been recorded (Koski et al., 2004) these helices can be made by electrospinning from a bicomponent solution consisting of a nonconductive polymer such as polyethylene oxide (PEO) mixed with a conductive polymer such as poly(aniline sulfonic acid) (PASA) (Kessick and Tepper, 2004). The average diameter of electrospun fibers varies with process conditions between ≈ 5 nm to ≈ 10 μm and is typically possessed of a distribution of fiber diameters greater than the distribution of sizes observed in conventional fiber spinning. Processing a polymer melt rather than a solution will result in diameter increase, because loss of heat happens faster than solvent evaporation. The so-called whipping instability is not observed when spinning from melt either. Other structural morphologies such as capsules or beads, "beads-on-a-string," or spindle structures maybe present in varying degrees, depending on the choice of operating parameters and selection of the parent polymer (Fig. 8.5) (Spasova et al., 2006).

The formation of spindle-type fiber geometries is due to uneven drawing of the fiber, perhaps as a result of minor in homogeneities of concentration in the parent solution, and does not occur because of stress relaxation happening after fiber deposition. Bead morphologies are well known in conventionally drawn fibers (e.g., glass fibers) (Mohr and Rowe, 1978).

The principle factors influencing bead formation are said to be the polymer solution surface tension and viscosity, coupled with a high charge density within and on the fiber strand (Fong et al., 1999; Spasova et al., 2006), however observing a trend is some distance away from being able to control the phenomenon. The strength of

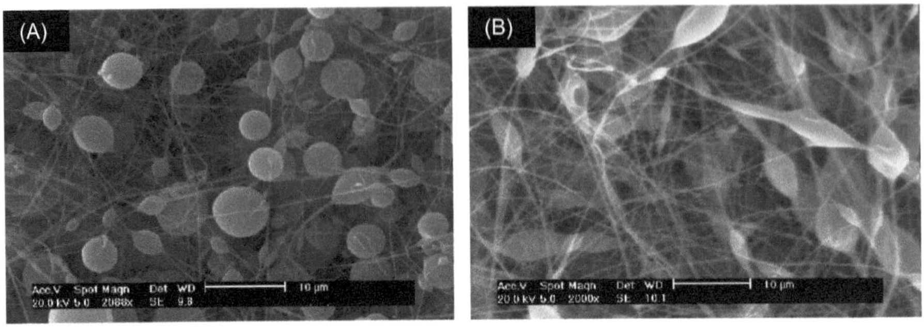

Figure 8.5 Defects in electrospun fibers. (A) Beaded fibers and (B) Spindle-like defects. Controlled morphology and mechanical characterization of electrospun cellulose acetate (CA) fiber webs produced from single and binary solvent systems based on solubility parameters was discussed.

the process driving electrical field and the solution conductivity underpin the charge density carrying capacity of the fiber strand. As charge density increases, there is an observed tendency for bead structures to shrink and become more spindle-like. However, it must be emphasized that bead or spindle morphology is not specific to electrospinning, and is described in most fiber manufacture methods (Walczak, 2002). A reduction in surface tension tends to stop bead formation (Fong et al., 1999). The spinning solution conductivity can be adjusted by adding ionic salts (Stanger et al., 2009). To do this, common salt (NaCl) has been added to a PEO/water solution, increasing the charge density (Angammana and Jayaram, 2008). Zong et al. (2002) examined the effect on fiber morphology of adding salt to poly(D,L-lactic acid) (PDLA). No beads-on-a-string structures were observed using 1% salt solutions to increase the surface charge density of the jet.

Doshi and Reneker (1993) noted that reducing the surface tension of a polymer solution reduces the level of bead formation. The addition of surfactants to the spinnable solution will reduce surface tension and consequently bead formation (Lin et al., 2004; Manee-in et al., 2006). As the charge density (i.e., the charge carried by a unit length) is increased, so does the current flow and the forces causing jet elongation. There is a limit to the degree of charge density on the surface, above which charge dissipates by corona discharge (Cross, 1987). Increasing charge density will affect fiber morphology so that bead size will reduce, more spindle shapes will be observed, and the fiber diameter will decrease (Zong et al., 2002). Similarly, Uyar and Besenbacher (2008) reported that slight variations in the conductivity of the solvent dimethylformamide (DMF) affect the morphology of polystyrene (PS) fibers, and that as the solution conductivity increases, uniform thinner fibers with no beading are observed (Uyar and Besenbacher, 2008). Fibers have been made from carbohydrates with low molecular masses (e.g., maltodextrin, or resistant starch) using the commercial surfactants lecithin, Span 20, and Tween 20 (Pérez-Masiá et al., 2014b). The degree of addition of surfactants alters product structures by affecting the way micelles or surfactant monomers interact with the polymeric components (Kriegel et al., 2009). Cospinning the surfactant as a blend means that the micelles can remain

Table 8.1 Processing parameters in electrospinning

Solution properties	Viscosity Polymer concentration Molecular weight of polymer Electrical conductivity Elasticity Surface tension
Processing conditions	Applied voltage Distance from spinneret to collector Volume feed rate Spinneret diameter
Ambient conditions	Temperature Humidity Atmospheric pressure

intact and survive to become part of the nanofiber product. Micelles can contain lipophilic functional ingredients, this may be a method of functionalizing fibers and increasing the range of application of nanofiber products (Kriegel et al., 2009). Using surfactants in polymer solutions is another method of dictating the properties and morphology and properties at the nanostructure level. In particular, for polyionic polymers such as chitosan ionic surfactants can influence viscosity, conductivity, and surface tension. In turn, this affects Taylor cone formation, jetting onset and nature of the "whipping instability," and thence the nature of the resulting nanostructures products (Kriegel et al., 2009).

Varying the concentration of the polymer and the solvent used to dissolve it alters the solution viscosity and surface tension. As the viscosity increases, so does the size of the beads, and as the beads elongate into spindles, the average distance between them and diameter of the intervening fiber increase. To successfully electrospin from a polymer solution, the processing conditions will need to be optimized (Scampicchio et al., 2015). The main control parameters are general solution properties, machine set points, and the external ambient conditions (Table 8.1).

It is possible to influence fiber morphology by varying these values (Deitzel et al., 2001; Heikkila and Harlin, 2008; Tehrani et al., 2010). Understanding how these variables, both individually and by interaction influence fiber size and morphology underpins the ability to control the process.

8.4.1 Polymer solution properties

The most significant influence over fiber morphology and size is obtained by adjusting the properties of the liquid polymer (Teo and Ramakrishna, 2006). Fiber size and the evolution of beads are influenced by solution viscosity and surface tension (Fong et al., 1999; Spasova et al., 2006). The nature of the solution is defined by the molecular mass of the polymer, the concentration of the polymer, and thence by solution conductivity and surface tension.

8.4.1.1 Polymer concentration, molecular weight, and fiber morphology

Generally polymer solubility increases inversely to molecular mass. The degree of molecular chain entanglement also influences solubility (Shenoy et al., 2005a) tending to increase with molecular weight (Treloar, 1970). The ideal spinnable polymer solution has the maximum polymer content while still being of sufficiently low viscosity to allow jet formation. There is a critical minimum level of polymer concentration for fiber formation (Huang et al., 2003; Shenoy et al., 2005a). Below the threshold level, charging the solution means electrospraying due to the Rayleigh instability—the tipping point between electrostatic repulsion and surface tension, sometimes called capillary wave break up. At lower concentrations of polymer, the degree of network entanglement for polymer chains is too low for fiber formation, and if further increased then a stable fiber will result. There is an upper concentration limit because too high a viscosity prevents continuous electrospinning (Shenoy et al., 2005a). Shenoy et al. (2005b) proposed a semiempirical theory where entanglement number for a particular solution, $(\eta_e)_{soln}$, is critical to whether the formation of electrospray beads or electrospun fiber will happen. The value of $(\eta_e)_{soln}$ is calculated using the following equation:

$$(\eta_e)_{soln} = \frac{\phi M_w}{M_e} \tag{8.2}$$

where, M_w is the molecular weight of the polymer in solution and M_e is the polymer entanglement molecular weight. M_e is related to the topology of the polymer chain (Munir et al., 2009; Shenoy et al., 2005b). An M_e value of two for a polymer solution indicates about one entanglement per chain. For reliable electrospinning, a polymer solution with $M_e > 3.5$ is required. At some point when M_e is between 3 and 2.5, there is a change from beading to continuous fiber. When $M_e < 2$, electrospraying occurs. Electrosprayed product is observed as a deposition of solid beads on the material collector (Munir et al., 2009; Shenoy et al., 2005b). As an example of how M_e influences fiber morphology, Munir et al. (2009) used entanglement number theory to predict the formation of beads, beaded fibers, and continuous fibers for a range of molecular weights and concentrations (Fig. 8.6) (Munir et al., 2009).

Under a concentration of 5 wt.% PVP (in water/ethanol), $(\eta_e)_{soln} \sim 2$, and only the beaded morphology is observed. Raising the solution solids content to increase $(\eta_e)_{soln} \sim 2$ results in beaded fibers. When the concentration of the polymer solution was increased to 10 wt.% and thus $(\eta_e)_{soln} \sim 3.5$, continuous fiber production occurred (Munir et al., 2009). Solutions of polysaccharide in water can be electrospun if they meet two conditions: a low degree of shear thinning at below a 1000 s^{-1} shear rate, and an overlap concentration (i.e., a concentration where it is calculable that adjacent polymer chains will physically overlap) between 10 and 20 g per 100 mL (Stijnman et al., 2011).

Further to the concept of overlap concentrations, Gupta et al. (2005) observe that using a dilute solution only results in polymer droplets, due to an insufficient degree of overlap between adjacent molecules (Gupta et al., 2005). As they increased solids content, a transition between droplets of polymer and fibers with beads on was noted, further concentration increase to a semidilute entangled regime results in beaded fibers.

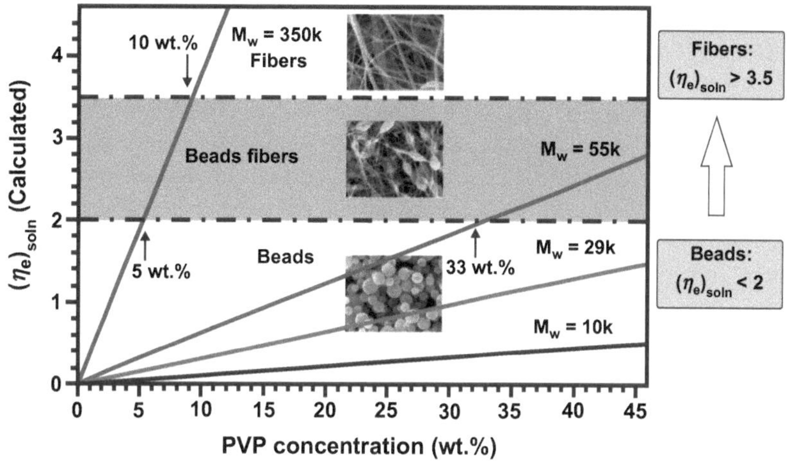

Figure 8.6 Plot of the calculated entanglement number $(\eta_e)_{soln}$ as a function of polymer concentration. In this example, the spinning solution was PVP/(water/ethanol) with various molecular weights.

Further concentration increase results in continuous fiber formation. In addition fiber diameter was also found to increase with solution viscosity (Gupta et al., 2005).

Tan et al. (2005) examined parameters influencing production of ultrafine fibers by electrospinning. The most significant factors were organized into two tranches. The first related to the mass flow rate of polymer through the spinneret (polymer concentration, applied voltage, and flow rate) and the second, those related to electrical force, and the electrical conductivity of the solvent. It was concluded that molecular mass, polymer concentration, and electrical conductivity of the solution are all important in affecting the final fiber morphology.

Demir et al. (2002) consider that viscosity, temperature, and concentration are the most significant properties of the solution, influencing both control of the process and the morphology and properties of the spun product, concluding that the formation of beaded fibers results from dilute solutions, and if the solution contains a higher level of polymer then curly, wavy, and straight fibers are observed. The mean of the distribution fiber diameters was found to be proportional to the cube of the polymer concentration (Demir et al., 2002). Lin et al. (2005) showed that polymer concentration and solution viscosity are directly related (Lin et al., 2005), observing that dilute solutions of polymers electrospin into discrete beads or fibers with beads-on-a-string and very dilute polymer solutions do not produce fibers (Lin et al., 2005). Lim et al. (2008) propose that due dilute solutions are low viscosity, and so polymer chains have a high degree of mobility and are hence will align more easily in the direction of the fiber spinning axis, promoting some degree of increase in orientation at the molecular level (Lim et al., 2008).

Smaller diameter fibers resulting from more dilute solutions do have a higher degree of orientation at the molecular level, and also increased crystallinity, stiffness, and strength at the expense of low ductility (Lim et al., 2008). Zong et al. (2002) also pointed out that the polymer concentration or the viscosity of the polymer solution was one of the most effective fiber morphology control parameters (Zong et al., 2002). Bead size and bead spacing were both found to increase with pinning solution viscosity, and with increased viscosity beads' morphology changed from spherical to spindle shapes. They drew the conclusion that continuous and consistent fibers would result from low electrical field strengths combined with higher polymer concentrations (Zong et al., 2002). Modeling techniques have been demonstrated that relate, say polymer concentration as a process parameter to resulting fiber morphology or diameter (Theron et al., 2004; Thompson et al., 2007; Yördem et al., 2008). Work by Torres-Giner et al. (2008) on biopolymer solutions shows that as pH is decreased, viscosity increases due to the agglomeration of the proteins, and consequently fiber product morphology is altered (Torres-Giner et al., 2008). Dong et al. (2010) examined the effect of variation in concentration, viscosity, and in certain additives, the spinnability of the protein wheat glutenin (WG). When solution concentration was increased PVA/WG fibers had a larger diameter than the equivalent WG product, when dithiothreitol (DTT—also known as Cleland's Reagent and commonly used to reduce protein disulfide bonds, such as those that stabilize the molecular structure of glutenins) was used to make DTT/WG and thiolated poly(vinyl alcohol) (TPVA) was used to make TPVA/WG fibers, thinner beaded morphologies resulted (Dong et al., 2010). A fraction of high molecular weight glutenin was proposed as the responsible factor for the reduction in fiber formation threshold concentration, by comparison with wheat extract which did not possess the component of high molecular mass glutenin. Denaturing the wheat protein to disrupt this previously observed balance of properties prevented fiber formation (Woerdeman et al., 2005).

Verdugo et al. (2014) looked at electrospinning fibers from a biomass protein concentrate made from a microalgae residue. Protein concentrate made from *Botryococcus braunii* microalgae from a biorefinery process was used to develop ultrafine fibers through electrospinning process. *B. braunii* is a source of triterpene oils which are said to have potential as fossil fuel replacements. The results indicated microalgal protein concentrate (MPC) fibers made at pH 1 had smoother surfaces and smaller diameters than those made at pH 12. pH and MPC concentration were identified as the most significant factors influencing the diameter and morphology of the fiber product. Fourier transform infrared analysis (FTIR) indicates a slight frequency shift for the secondary structure of MPC with change in pH. Likely this change in the protein structure increases the physical chain entanglement in the polymer blend (Verdugo et al., 2014).

Vega Lugo and Lim examined how changes in pH can alter the secondary structure of proteins and thence alter viscosity, conductivity, and dynamic surface tension of the protein solution, which in turn were observed to significantly affect the electrospinning behavior of whey protein isolate (WPI). As a globular protein, WPI is difficult to spin on its own, so PEO may be used as an admixture to improve spinnability. WPI/PEO (10% w/w WPI; 0.4% w/w PEO) solutions at low pH resulted in smooth surfaced fibers while neutral solutions produced spheres linked by ultrafine fibers. High pH solutions resulted in fibers that with spindles. The electrospinning-enabling properties of PEO were due to the degree of physical chain entanglement between the two polymers, rather than any specific polymer–polymer interaction. The WPI was found to be uniformly distributed within WPI/PEO electrospun fibers. The electrospinning of beadless fibers was attributed to increase at the molecular level of random coils combined with an α-helix secondary structure in the WPI (Vega Lugo, 2012; Vega-Lugo and Lim, 2012).

8.4.1.2 Electrical conductivity

Raising electrical conductivity and selecting a higher molecular weight polymer while lowering polymer concentration will result in a decrease in spun fiber diameter. The solution electrical conductivity is related to the charge density limit, and so to the amount of elongation of the jet through an increase in electrostatic force. Therefore, for given processing conditions, a more electrically conductive spinning solution may stretch more, resulting in a smaller diameter fiber product (Tan et al., 2005). Plant and animal origin polymers are commonly polyelectrolytic, with each monomer unit bearing an electrolyte group, the ability to form ions in solution increases the ability to carry charge of the solution, and subjects the fiber forming jet to greater electrical origin tensile force: in comparison to synthetic polymers, this does not help the process of fiber formation (Zong et al., 2002).

As mentioned, ionic salts can be added to increase solution conductivity, and by functioning as charge carriers this will increase the production rate (Chaobo et al., 2006). Higher solution conductivities are obtained with an organic acid as the solvent: (Chaobo et al., 2006) used formic acid in conjunction with a small amount of pyridine to dissolve an aramid (Nylon), and then used the solution to electrospin ultrathin (3 nm) nylon fibers.

8.4.1.3 Effect of solvents

The electrical and physical properties of the stock solution from which the fibers are spun, are significantly influenced by the majority component which is the solvent, this in turn affects the morphology of the electrospun fiber product (Eda et al., 2007). A stock polymer solution will consist of typically 80–90 wt.% solvent and 10–20 wt.% polymer, so their influence on fiber morphology is high, and hence the selection of solvents for electrospinning is clearly important (Ghorani et al., 2013; Lu et al., 2006). Wannatong et al. (2004) demonstrated that choice of solvent influences the fiber density of an electrospun mat and the formation of beaded morphologies. Uyar and Besenbacher (2008) showed that the dryness of the spun product is related to the boiling point and density of the polymer solvent. The level of retained solvent in the fiber product increased in inverse relationship to the density and boiling point of the solvent (Wannatong et al., 2004). Similar effects were reported by Jarusuwannapoom et al. (2005).

Solvent choice will also affect fiber product morphology. Yang et al. (2004) examined the effect of using *N*,*N*,-dimethylformamide (DMF), dichloromethane (DCM), ethanol, and mixtures of these solvents when electrospinning uniform polyvinylpyrrolidone (PVP) fibers, observing significant variation in the fiber product morphologies from helical textured to smooth surface textures (Yang et al., 2004). Van der

Schueren et al. (2011) demonstrated that fiber diameter can be substantially decreased by using a binary solvent mixture. Blended solvent systems have been used to improve electrospinning performance, for example, to couple high productivity with reduced fiber diameter. The morphology of electrospun cellulose acetate (CA) fibers produced from single and binary solvent systems was compared using solubility parameters to predict processing conditions for defect-free CA electrospun fibers (Ghorani et al., 2013). Both the Hildebrand solubility parameter (δ) and the radius of the sphere defined as the Hansen space ($D_{(s\text{-}p)}$) of acetone, acetic acid, water, *N*,*N*-dimethylacetamide (DMAc), methanol, and chloroform were examined for suitability for electrospinning CA fibers. The Hildebrand solubility parameter (δ) of acetone and DMAc are within an appropriate range for the solvating CA. The suitability of a binary solvent system comprising acetone:DMAc in a mass ratio of 2: was predicted and confirmed experimentally by continuous electrospinning of high quality CA fibers (Ghorani et al., 2013).

Moomand and Lim (2015) studied the effects of using aqueous ethanol, and isopropanol as electrospinning solvents on the physicochemical properties of zein fibers to encapsulate fish oil (Moomand and Lim, 2015). The electrospun zein fibers made from 70% (w/w) ethanol were ribbon-shaped (still liquid on impact), and those prepared from isopropanol were beaded. The average fiber diameter increased with the level of fish oil loading. This was attributed to the use of a water–ethanol solvent mix and in conjunction with fish oil promoting the formation of an α-helix protein secondary structure, whereas the isopropanol caused the formation of intermolecular and intramolecular β-sheet protein structures (Moomand and Lim, 2015).

Olaru and Olaru (2010) electrospun cellulose acetate phthalate (CAP) from an 85:15 (v:v) acetone–water mixture, 2-methoxyethanol, and a 50:42.5:7.5 (v:v:v) 2-methoxyethanol–acetone–water mixture (Olaru and Olaru, 2010). At low polymer concentrations (12.5%, w:v), the acetone–water mixture 85:15 (v:v) allowed the production of uniform fibers with diameters less than 500 nm, whereas using 2-methoxyethanol, fine hemispherical particles were observed (Olaru and Olaru, 2010). Denis et al. (2015) used a binary solvent system of acetic acid (AA): formic acid (FA) (90:10), which has the advantage of low toxicity, to electrospin polycaprolactone/gelatin (PCL/Gt) fibers (Denis et al., 2015). Electrospinning was possible even though the mixture of PCL and gelatin dissolved in AA/FA was apparently an emulsion. From a practical perspective, PCL/Gt fibers using AA/FA mixture have a morphological structure similar to that of fibers spun from hexafluoroisopropanol (HFIP) solutions, but using a much more benign solvent (Denis et al., 2015).

8.4.2 Processing conditions

Other parameters with a significant effect on the electrospinning process are the applied voltage, the distance from spinneret to target [tip-to-collector distance (TCD)], feed rate, spinneret orifice diameter and also the ambient processing conditions of temperature, absolute humidity and pressure exerted by the atmosphere (Mazoochi and Jabbari, 2011; Pham et al., 2006; Ramakrishna, 2005).

8.4.2.1 Voltage

The most basic part of the electrospinning process is the use of a high applied voltage to charge the polymer solution. Alteration of the applied voltage sets the strength of the electrical field formed between the spinneret and the target, and thence the magnitude of the force drawing the fiber. The "electrospinning number" is defined as the ratio of electrical energy (Vq) to the surface energy (γR^2), $Vq/\gamma R^2$, where V is the voltage, q is the charge, γ is the surface tension, and R is the average radius of the Taylor cone droplet (Shenoy et al., 2005a). When $Vq/\gamma R^2 > 1$, a jet is produced from the Taylor cone. As the degree of application of voltage is increased, the forces of electrostatic repulsion acting on the polymer droplet also increase, which will eventually overcome the counter forces of viscoelasticity and surface tension exerted by the polymer solution. Thus, a higher electrospinning number is expected to increase the degree of elongation experienced by the polymer jet and thus lead to a decrease in fiber diameter (Subramanian et al., 2010).

The amount of electrical charge contained in the solution is critical to the onset of the electrospinning process as electrostatic repulsion overcomes the surface tension of the solution—Rayleigh shattering (Mazoochi and Jabbari, 2011). Depending on the solution feed rate, more current flow driven by a higher applied may be required to replenish this charge keeping the Taylor cone stable (Zong et al., 2002). Negative or positive voltages above about 6 kV will distort a small sessile liquid polymer droplet into the iconic onion shape of the Taylor cone as the jet is initiated (Taylor, 1964). The balance between surface tension and electrical charge determines the shape of the polymer droplet and the Taylor cone at the spinneret (Zong et al., 2002). As the applied voltage is increased, the length of the straight part of the jet (before the "whipping instability") tends to decrease slightly and increasing the critical apex angle of the Taylor cone. There is a high level of acceleration in a straight line close to the tip of cone (Buer et al., 2001). If the voltage applied is increased, the jet will accelerate faster and a greater amount of solution will be drawn from the spinneret. This is all due to the increase in amount of charge present and the consequent increase in the force experienced by the jet. This increase may result in a reduction in size and stability of Taylor cone (Mazoochi and Jabbari, 2011).

If viscosity is increased, usually by raising the polymer solution concentration, a higher level of force is needed to overcome both surface tension and viscoelastic forces to initiate the jet and to stretch it into a fiber (Doshi and Reneker, 1993). The threshold voltage needed to eject a charged stream of polymer solution from the spinneret depends mainly on the polymer concentration in the solution. The voltage supplied and the resultant electrical field both effect the acceleration and thus the degree of stretch of the jet, they will also affect the morphology and diameter of the fiber product (Buer et al., 2001; Megelski et al., 2002; Zhang et al., 2005; Zhuo et al., 2008). It is accepted that the degree of in-flight bifurcation of the fiber jet is related to high levels of applied voltage (Soviet practice is to use in excess of 100 kV) leading to a higher internal electrostatic force and the bifurcation can be seen as a method of decreasing the charge density to a sustainable level by increasing the net surface area of the fibers (Filatov et al., 2007), this results in a decrease in the average

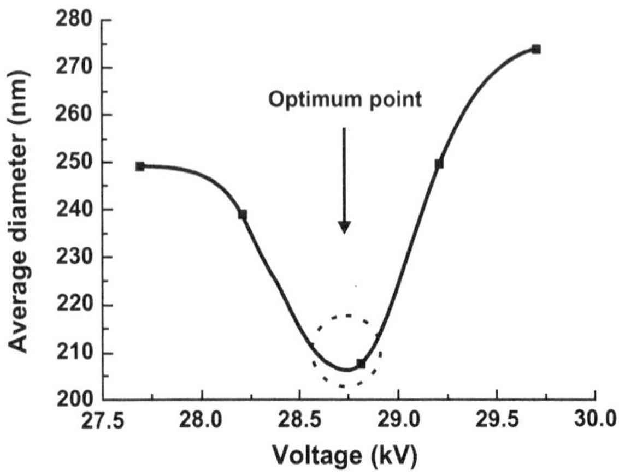

Figure 8.7 Example of the relationship between the applied voltage and the average fiber diameter during electrospinning.

diameters of the electrospun fibers, and an increase in the spread of the distribution of fiber sizes (Heikkila and Harlin, 2008; Lee et al., 2003; Mazoochi and Jabbari, 2011; Tan et al., 2005; Wannatong et al., 2004).

Lin et al. (2008) explain the influence of applied voltage , and spinneret-collector distance (sometimes called the tip to collector distance or TCD) on the fiber product diameter (Lin et al., 2008), suggesting that the mean fiber diameter reaches a minima after the applied voltage is increased before increasing as the applied voltage rises further (Fig. 8.7) (Lin et al., 2008).

This observation is attributed to the shorter time taken for the strand to reach the target and the consequent reduction in the duration of the phase of fiber stretch. This is often referred to as the electrospinning "flight time" (Heikkila and Harlin, 2008; Lin et al., 2008). At higher applied voltages, the production of large diameter fibers with a tendency for bead formation has been reported (Deitzel et al., 2001; Heikkila and Harlin, 2008; Zong et al., 2002). The fiber deposition pattern and degree of crystallinity are also thought to be influenced by the voltage applied (Linnemann et al., 2005; Zhao et al., 2004). At higher applied voltages, the fiber deposition pattern becomes less regular because jet instability onsets earlier in a high strength electrical field (Linnemann et al., 2005; Yuan et al., 2004). Similarly, fiber crystallinity increases with voltage until a critical maximum voltage is reached after which crystallinity decreases (Zhao et al., 2004).

8.4.2.2 Flow rate

If a spinning machine uses a pump, the delivery flow rate governs the quantity of solution at the spinneret. If the machine uses a header tank, then the replenishment rate is effectively matched to the removal rate. Thus for pump-powered machine, at a given applied voltage, there will be corresponding flow rate set point to maintain a stable

Taylor cone (Rutledge et al., 2001). A power law relation is observed between flow rate set point and the necessary level of applied voltage—the flow rate is proportional to the cube of the applied voltage (Demir et al., 2002). Thus, a change in flow rate in a constant electrical field may be expected to significantly affect fiber product morphology (Jalili et al., 2005). However, there is a body of experimental evidence that does not support this conjecture. Zhang et al. (2005) showed that the varying flow rate has only a slight effect on product morphology and that this was the least significant of the control parameters (Zhang et al., 2005).

High volume flow rates tend to be associated with high fiber diameters and conversely, a low flow rate gives uniform low fiber diameters (Ballengee and Pintauro, 2011; Dotti et al., 2007; Megelski et al., 2002; Yuan et al., 2004; Zong et al., 2002). An explanation for this is proposed by Zong et al. (2002) based on variations in the average velocity of the jet and the consequent effect on the drying time of the polymer (Zong et al., 2002). When the droplet formed at the end of the spinneret is large and there is a high feed rate, a larger volume of solvent must be lost prior to fiber impact on the collector plate and the high velocity of the stream means that the time for evaporation is very small. This results in the formation of large beads and fused fiber–fiber intersections observed in the product (Zong et al., 2002). Low flow rates provide more flight time for evaporation, leading to decreased bead formation (Yuan et al., 2004). It has also been noted that the mean pore size in electrospun webs increases with polymer flow rate (Megelski et al., 2002), an observation that is perhaps related to variation of the fiber diameter.

8.4.2.3 Diameter of the spinneret orifice

Some laboratory electrospinning systems use a blunt hypodermic syringe needle as a spinneret. A blunt needle is preferred, as a sharp point encourages loss of power by coronal discharge. This kind of loss can be minimized by using a nonconducting spinneret such as a micropipette tip. In any event, the internal diameter of the spinneret orifice does influence the subsequent spinning process. As the orifice diameter increases, spinneret clogging may happen (Mo et al., 2004) which is said to be an effect of surface tension. A smaller droplet radius increases the effective surface tension, and this effect can readily be achieved by reducing the spinneret diameter. Assuming the electrostatic field strength is constant, the initial acceleration of the jet decreases with increasing polymer surface tension and the average velocity is therefore decreased. Effectively, this means the jet flight time from the spinneret to the collector increases. This increases the time for solvent evaporation, jet attenuation, and splitting. As a result, fibers with smaller diameter and a narrower size distribution can be obtained by using a small diameter spinneret. However, if the orifice diameter is reduced too much, it may be difficult for a satisfactory polymer droplet to form, which also depends on the surface tension of the solution (Zhao et al., 2004). Although it might be expected that thinner electrospun fibers will result from a smaller diameter spinneret, it has been demonstrated that there is no correlation between the spinneret diameter and the fiber diameter (Yamashita et al., 2006).

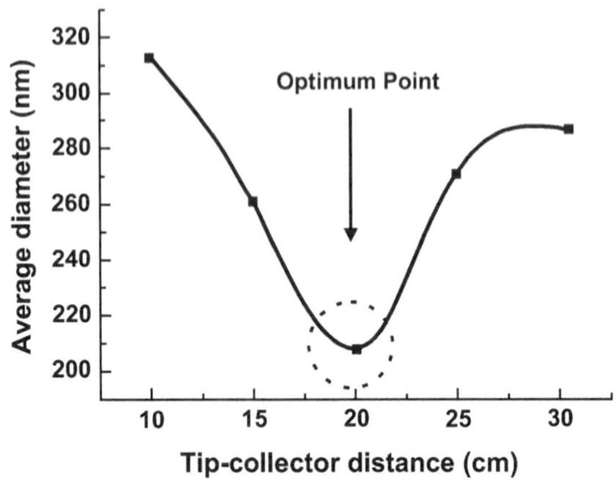

Figure 8.8 Example of the relationship between the tip–collector distance (cm) and the average fiber diameter during electrospinning.

8.4.2.4 Spinneret tip to collector distance

The elongation of the polymer jet into a solid fiber ideally takes place between the spinneret, as the site of formation of the Taylor cone and the earthed collector plate. The TCD is related to the electrostatic field strength and thence to the time of flight of the jet. The deposition of wet fibers can be attributed to an insufficient distance between the spinneret and the collector, so that the drying time is too short to evaporate the solvent before the fibers are deposited on the collector and consequently partially dried fibers with fused fiber–fiber intersections are observed as well as part of a densely packed structure (Barhate et al., 2006). Increasing the flight time by lengthening the TCD results in a longer time for solvent evaporation, which can decrease both the rate of bead formation and the mean fiber diameter (Ahn et al., 2006; Doshi and Reneker, 1993; Heikkila and Harlin, 2008; Jarusuwannapoom et al., 2005; Yuan et al., 2004). However, at a fixed voltage, an increase in the TCD also means a decrease in electrostatic field strength reducing both the chance of jet splitting and the degree of fiber attenuation. Therefore, as the TCD is increased from zero, the average fiber diameter reaches a critical minimum value before increasing again (Fig. 8.8) (Lin et al., 2008).

Decreasing the TCD has a similar effect on fiber morphology and fiber diameter as increasing the voltage because both increase the strength of the electrostatic field—often expressed in kV/cm (Chowdhury and Stylios, 2010; Frenot and Chronakis, 2003; Lee et al., 2003; Zhao et al., 2004). Ying et al. (2006) tell us that if the ambient humidity is low, increasing the TCD, and time in flight may lead to a reduction in fiber diameter as more solvent can evaporate during the increased flight time (Ying et al., 2006). When the ambient humidity is high, the increased distance does not increase the degree of evaporation but does reduce the field strength. Selecting an insulator or a conductor for the earthed collector plate will affect the degree of surface charge built-up during the electrospinning process (Chowdhury and Stylios, 2010). To increase the density of the deposited electrospun fibers, a more highly conductive collector has been suggested, because a conductive plate should prevent charge build-up by removing the incoming charge from the fibers, and the consequent effective reduction in file strength (Chowdhury and Stylios, 2010). However, it should be noted that the general nonconductive nature of polymers means that as the deposited layer of fiber thickens, newly arrived fibers are insulated from the earthed plate by earlier depositions of fiber.

8.4.3 Ambient conditions

Ambient parameters particularly humidity (best measured as absolute humidity rather than relative humidity, as the absolute measure is more directly related to the conductivity of air) and temperature are said to influence fiber morphology and the productivity of the electrospinning process (Chen and Yu, 2010). This is because a direct relationship exists between solvent evaporation rate liquid solvent conductivity and temperature. Both will affect the electrospinnability of a polymer (Chen and Yu, 2010). The viscosity and surface tension of a polymer solution are also affected by temperature (Bhardwaj and Kundu, 2010; De Vrieze et al., 2009). De Vrieze et al. (2009) showed that the average fiber diameter of electrospun fibers is affected by humidity but these trends are difficult to quantify (De Vrieze et al., 2009). In the case of electrospun cellulose acetate fibers, an increase in diameter was observed with increasing humidity, while for water soluble PVP there was a decrease in diameter. It has also been suggested that an increase in humidity can effect pore size distribution and diameter (Casper et al., 2003). Demir et al. (2002) reported differences in the uniformity of fiber diameters with temperature. As a result, electrospinning polyurethane at high temperatures led to greater uniformity in diameter with a noticeable reduction in the effect of the magnitude of the applied electrostatic field (Demir et al., 2002).

8.5 The physical elements of electrospinning and typical apparatus

For a typical lab-based electrospinning setup, a syringe pump is used to dispense the spinning liquid through a spinneret made from a blunt hypodermic syringe needle, and the charge is applied to the electrospinning solution by electrical connection to the needle. To increase the output of the setup, more spinnerets can be used (Fig. 8.9). However, the configuration of the spinneret arrangement needs to be optimized to reduce charge interference between them (Khorshidi et al., 2015; Theron et al., 2004).

With just one more spinneret, the maximum field strength will drop significantly compared to a single spinneret (Angammana and Jayaram, 2008). For spinnerets arranged in arrays, the electrical field gets progressively weaker from the edge to the center. The weakening in electric field becomes more significant when the distance between the spinnerets is reduced. Even at interspinneret distance of 5 cm, there is still discrepancy in the electric field across the surface (Kruncinska et al., 2009). This

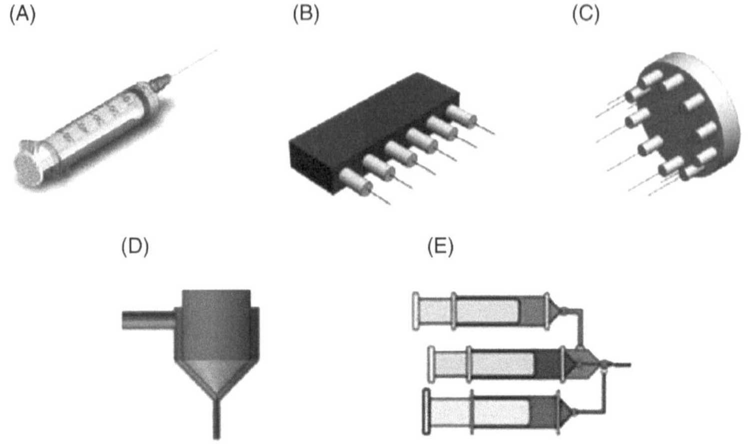

Figure 8.9 Possible geometries for multispinneret injection systems. (A) Single needle, (B) multi needle, (C) multi needle, (D) coaxial, and (E) triaxial.

may result in less stretching of the solution drops closer to the center (Kruncinska et al., 2009). It was shown that the mutual distance of deposition becomes larger with increases in working distance and spinneret spacing, but the influence of voltage is not particularly apparent (Han et al., 2015). When multiple spinnerets are arranged such that there is a central spinneret surrounded by other spinnerets, the coulombic forces exerted by the surrounding jets forces the inner jet to travel in a more direct trajectory toward the collector. This may result in wet deposited fibers from the inner jets necessitating an increase in the TCD to allow sufficient flight time to get dry fibers (Varesano et al., 2010). Fluctuations in the electric field at the spinneret tip due to interference with the adjacent jets may also temporarily arrest electrospinning from the central spinneret before it restarts. When this happens, a larger solution stream may deposit thick wet fibers on the collector (Varesano et al., 2010).

In addition, the phase change of a polymer in a strong electrical field—in this case freezing—results in the formation of an electret. This is the electrostatic equivalent of a permanent magnet. This effect has been known in electrospun fiber formation, and means in addition to the transient surface charge, there is a permanent (in the order of years—Filatov et al., 2007) residual charge intrinsic in the fiber. The potential applications of this phenomenon on the adhesion to the fiber of say, proteins, has not been widely examined, but is likely to be of significant use in constructing nutraceutical delivery systems, for example.

The most popular collector used in electrospinning is a stationary metal plate or piece of foil placed some known distance from the tip. The conical post "whipping instability" expanding helical flight path impinging on the target results in a symmetric circular patch of fiber. As the collector is grounded (Fig. 8.10), the residual surface charges on the as-spun fibers can be rapidly removed, allowing the fibers to form a mat of high areal density. The current flowing to earth can be measured using a microammeter—typically about 6 μA. Using a rotating drum collector allows an increase

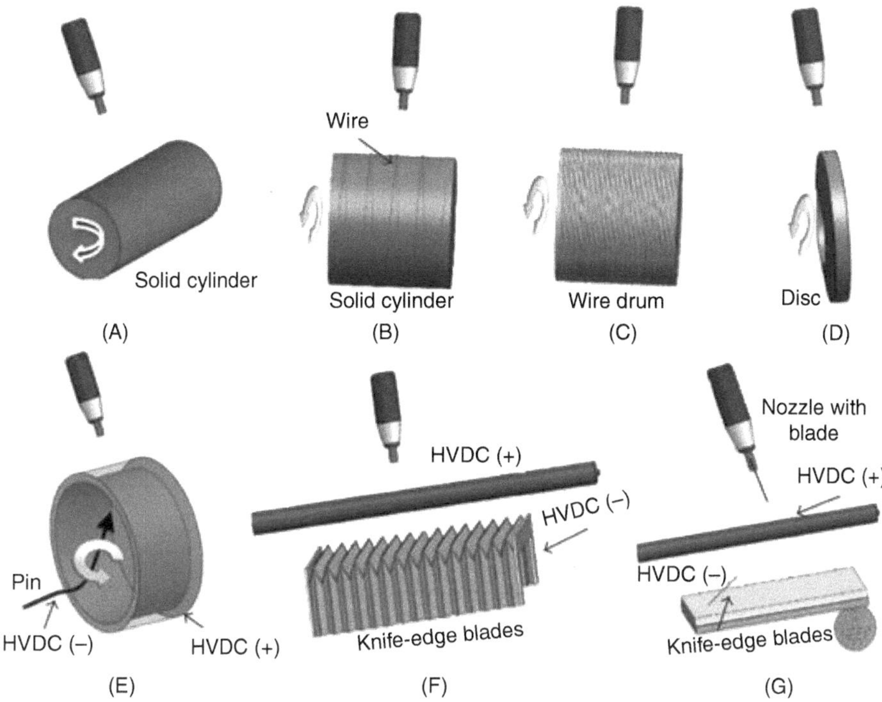

Figure 8.10 Category of electrospinning setups based on rotating device. These configurations include (A) solid or hollow cylinder, (B) wire wounded on an insulated cylinder, (C) wired drum, (D) disc collector, (E) sharp pin inside the rotating collector, (F) knife-edged electrodes, and (G) knife-edged electrode and spinneret system.

in the area of even deposition of fibers, and also allows for some degree of velocity-dependent alignment of nanofibers. Note however that the typical arrival velocity of the fiber strand at the target is in the range of 60–100 m/s, and so high rotation speeds are required to achieve alignment. Different collector geometries reported in the electrospinning literatures are comprehensively reviewed elsewhere (Sahay et al., 2011; Teo and Ramakrishna, 2006).

For a full history of the development of commercial electrospinning plant, the reader is referred to Tucker (2015). For the purposes of this chapter, methods of spinning and spraying that are adaptable for coaxial spraying or spinning to produce encapsulated products are of particular interest. If a fine powder particulate can be dispersed in the spinning dope, then a strand of encapsulated particulates will be the result, but if a liquid is to be encapsulated then the technical problem is more severe. A coaxial process will be required. The concept of coaxial spinning was first proposed by Cooley in 1900, but has not been widely commercialized.

The principle difficulty in the industrial production of electrospun materials can be the rate-limiting step of the use of electrostatic force to draw the fiber or particulate stream from the spinneret. Despite this there are a significant number of companies producing commercial quantities of electrospun fiber. For example, Donaldson (www.donaldson.com) manufacture a 0.2–0.3 μm electrospun fiber turbine system air filter sold under the Spider-Web brand. The company holds in excess of 2000 product patents, but protect their manufacturing technology intellectual property as a rigorously enforced trade secret.

Elmarco (www.elmarco.com) use a needleless spinning system sold under the Nanospider brand. The process uses a rotating cylinder charged to 32–43 kV and half immersed in the spinning dope (Petrik and Maly, 2009). The curved surface of the cylinder puts an element of tension onto the film of liquid on the upper surface of the cylinder allowing the film to rupture into a linear version of the Taylor cone of an individual spinneret. The resulting fiber mass is collected on a belt collector positioned above the tubular emitter. This system neatly steps around the problem of maintaining an even flow to banks of spinnerets, and detecting and clearing with spinneret blockages, but is not currently used in the production of multilayer products.

Walter Simms of Bayer Aktiengesellschaft, Leverkusen, Germany patented (Simm, 1976) a rotating ring emitter where the tension induced by the presence of liquid on a curved surface is assisted by accelerating the surface by means of rotation—centrifugal force to produce a continuous stream of multistrand fibers. Simm proposes the addition of a finely powdered "free flowing" ingredient at this point, and it is conceivable, given a suitable choice of fiber and powder that encapsulated particulates may be produced.

Electrospinning was first proposed as a spinning technology to produce manmade fibers with the aim of producing yarn for the production of convention woven materials. Due in all likelihood to the difficulties of manipulating a fiber stock invisible to the naked eye and with a strong intrinsic electrostatic charge, the process did not take off as a volume production method until work by Fuchs, Petryanov, and Rosenblum (Filatov et al., 2007) lead to the exploitation of these properties of electrospun fiber as a nonwoven filter element. Fiber handling difficulties were eliminated by the direct deposition of the fibers onto a handleable porous substrate. At the height of production this material was being produced at 21 million m^2/year. The Soviet method of production did away with the rate limiting step of using electrostatic attraction to draw the fiber from a spinneret. The spinning dope was charged using exceptionally high voltages (in excess of 100 kV), and then forcibly ejected through the spinneret orifice at rates of up to 1 cm^3/min. During flight the polymer strand repeated bifurcates, reducing in diameter and hence increasing in surface area, until the surface charge reaches a stable equilibrium level. This method produces a high throughput compared to simple electrostatic fiber drawing, although with a wider distribution of fiber sizes. This technique can clearly be extended to use a coaxial spinneret system of the Cooley type (Tucker et al., 2012).

The electrostatic technology used to produce particulates and fibers are broadly similar. The difference between the two being essentially the degree of molecular cohesion in the raw material. It is clear that the technology to produce encapsulated nanoscale encapsulated bioactives is already extant, and available for exploitation.

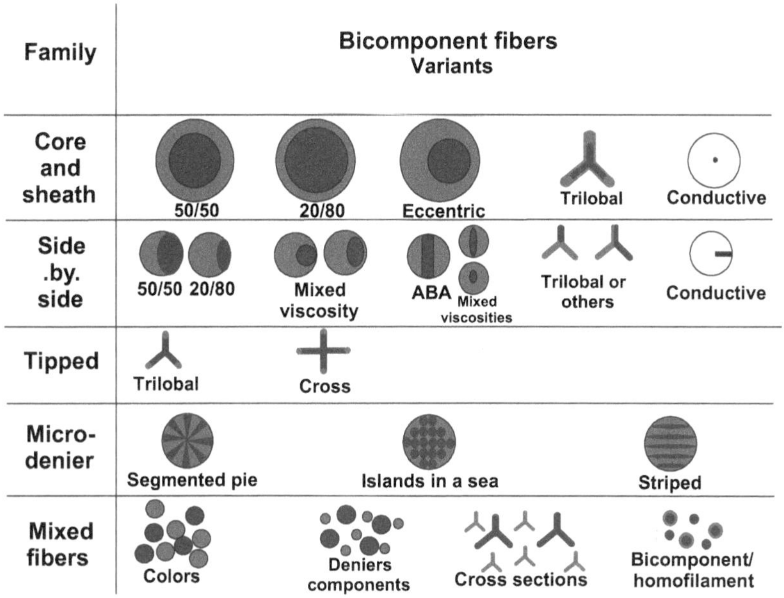

Figure 8.11 Chart on bicomponent fibers.

8.5.1 *Coaxial electrospinning*

Bicomponent fibers can be defined as extruding two polymers from the same spinneret to produce a single fiber strand typically with a skin–core structure. The main objective of this practice is to extend the range of fiber properties beyond those existing in either polymer alone. In Fig. 8.11, skin–core bicomponent fibers where one of the components (core) is fully surrounded by the second component (skin) can be seen (Sun et al., 2004).

The most common way to produce skin–core is coaxial electrospinning where two spinnerets are used with one spinneret positioned inside the other on the same coaxis. This allows the simultaneous feeding of two polymer solutions. Fig. 8.12 shows coaxial electrospinning setup and how the fabrication process works (Loscertales et al., 2002).

The two-layer coaxial jet is formed at the spinneret, and the skin–core structure is retained throughout the drawing process. To make this process work, the feed flow rates of the skin and core material must be closely controlled. This is a process where the use of syringe pumps will have a distinct advantage over header tanks. In addition to matching delivery rates, the ratio of solution viscosities is important, with a slightly higher viscosity being desirable for the inner core material. The injecting speeds and ratio of inner–outer fluid speed will require fine tuning (Li et al., 2010).

Hollow capsules are formed from a coaxial electrified jet of a sol–gel formulation (outside gel) and a regular nonstructured liquid inside (inner liquid), when the jet undergoes varicose break-up, charged liquid capsules result. In-flight evaporation of the solvent from the outer liquid solidifies the capsule outer shell. After collection, the

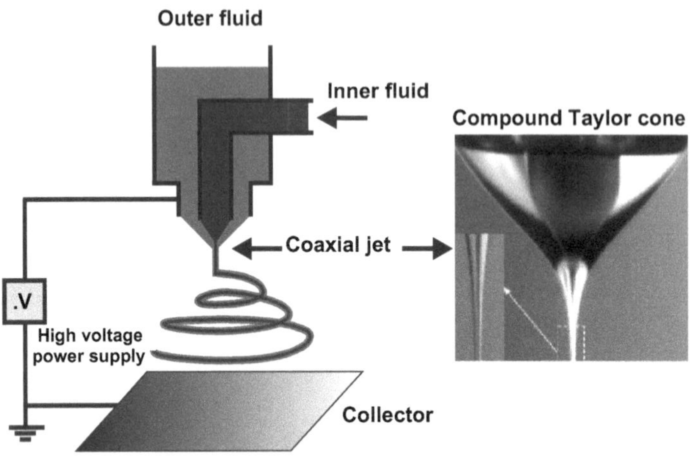

Figure 8.12 The basic setup for coaxial electrospinning and fabrication process of common core–shell nanofibers.

inner liquid solvent is removed by evaporation and diffusion through the outer shell (Li and Xia, 2004; Li et al., 2010).

Coaxial electrospinning is of particular interest to expand the range of application to include core materials that will not form fibers by themselves. The shell fluid must be a spinnable composition: the core liquid need not be an electrospinnable fluid (Yu et al., 2004). For successful spinning, the two fluids need not be miscible; however, the outer skin layer must be capable of forming first a jet, and then a fiber. Clearly, materials selection is key to the success of the whole affair (Greiner and Wendorff, 2007).

Encapsulation of active ingredients provide improved stability both in the use of the final product and during the manufacturing process (Nedovic et al., 2011). For food applications, encapsulation by coaxial electrospinning can reduce damaging due to direct contact with organic solvents or harsh processing conditions that can occur during conventional emulsification. The shell layer can also act as a barrier to prevent the premature release of water-soluble core material. By varying the structure and composition of the nanofibers, it is possible to control the release of the encapsulated agents (Jiang et al., 2014). This may be useful if, for example, a nutraceutical material is to survive contact with salivary amylase in the mouth, and then acid digestion in the stomach, to be finally released from encapsulation and beneficially absorbed in the gut.

A one-step, benign low temperature physical procedure based on coaxial electrospinning was developed for incorporation and controlled release of two model proteins, the serum albumin protein bovine serum albumin (BSA) and the glycoside hydrolase enzyme lysozyme, using biodegradable polycaprolactone (PCL) as the shell and containing the protein load in a polyethylene glycol (PEG) core. The relative thickness of the core and shell were controlled via the feed rate of the core liquid, which in turn influences the release profile of the protein payload. After release the lysozyme maintained both structure and bioactivity (Jiang et al., 2005). Similarly, Jiang et al.

(2006) made biodegradable core–shell fibers with poly(epsilon-caprolactone) as shell and a core of bovine serum albumin (BSA)-containing polysaccharide dextran using coaxial electrospinning. Controlled release of the payload proteins was demonstrated (Jiang et al., 2006).

Coaxial electrospinning has also been used to encapsulate living *Bifidobacterium* bacteria in poly(vinyl alcohol) (PVOH) electrospun fibers with a mean diameter of 150 nm while retaining their viability (López-Rubio et al., 2009). PVOH was used as the encapsulating material because it is generally recognized as safe (GRAS) and so can be used as a food additive, and in addition has good oxygen barrier properties as a dry film, and is water soluble. This final property allows the bacteria to be recovered for viability testing. The specific incorporation of *Bifidobacterium animalis* Bb_{12} caused a decrease in melting point and degree of crystallinity of the PVOH fibers and an increase in the glass transition temperature. The viability tests, carried out at ambient temperature, 4 and $-20°C$ showed that after encapsulation, *B. animalis* Bb_{12} remained viable for 40 days at room temperature and for 130 days when refrigerated, unencapsulated bacteria were used as a control to demonstrate that encapsulation does improve viability (López-Rubio et al., 2009). Yang et al. (2008) investigated lysozyme encapsulated within skin–core fibers prepared by electrospinning from an emulsion precursor, as a model material to examine structural integrity, retention of bioactivity, and release patterns. The skin–core structure cuts down the initial burst release to achieve a more even rate of delivery. Their results show that neither in-process ultrasonification nor electrospinning caused any significant denaturation of the protein: FTIR analysis showed that electrospinning had much less effect on the secondary structure of protein than ultrasonication. The skin–core structured fibers protected the biochemical integrity of the protein during incubation in a liquid medium. Bioactivity assays showed a specific activity loss of $\approx 16\%$ during emulsification to produce the spinning stock solution, and the confirmed the protective effect of the shell materials by examining the activity of protein after release from encapsulation. In vitro degradation studies showed that protein entrapment led to an increased mass loss and reduction in molecular weight (Yang et al., 2008).

8.6 Base encapsulating materials for electrospinning

Electrospinning has the potential to produce products for the emerging noncommodity functional foods market. Functional foods are said to benefit consumer health at a level beyond that achieved by conventional good nutritional practices, and consequently attract a premium price from affluent consumers. Currently available functional foods include omega-3 fatty acid enriched margarines, yogurt and milk to reduce the risk of cardiovascular disease (Ghorbanzade et al., 2017), cholesterol lowering spreads, and cream cheeses, vitamin D enriched fruit juices to militate the onset of osteoporosis (Ghorani and Tucker, 2015). These materials are not always of themselves, spinnable. It is desirable to be able to spin fibers and particulates that contain these materials to protect biochemically complex molecules from damage in storage or during the early stages of ingestion, and also it may be necessary to mask the flavor of a biochemically active material. To produce a spinnable raw material, the payload material can simply be mixed with the spinnable carrier material to produce a fiber or particulate formed of a solid solution of both carrier and payload (Tucker and Ghorani, 2015). Alternatively, as described in the previous section, the core ingredient can be encapsulated within a resistant outer coating (Livney, 2008).

The first step in applying this technology to food manufacture is scoping the range of materials that can be used for electrospinning: fortunately the range of polymers which can be applied to entrap, coat, or encapsulate substances is wide (Table 8.2).

For foodstuffs it is clearly appropriate to develop functional matrices from natural animal or plant origin substances, namely proteins, carbohydrates, and fats (Jafari et al., 2015). Synthetic or nonbiodegradable materials may have perceived environmental and health risks, and will sensibly be rejected by informed consumers (Aceituno-Medina et al., 2013). Materials intended for the encapsulation of food ingredient should themselves be foodstuff origin and GRAS (Freiberg and Zhu, 2004). The use of food-grade polymers and biopolymers as bioactive delivery devices has been proposed and evaluated (Chen et al., 2006; Jones and McClements, 2011). Inventories of suitable food-grade materials (Kriegel et al., 2008), bio-polymers (Schiffman and Schauer, 2008) are available.

Biopolymers are often biodegradable, biocompatible and may have antibacterial properties. These useful characteristics have fueled demand for these products, not least in the area under discussion for encapsulation and delivery systems (Fathi et al., 2014). Examples of the practical application of these materials are active packaging and the preservation of nutrients against challenging environmental conditions (Anu Bhushani and Anandharamakrishnan, 2014). Another prospect for the take-up of electrospun food ingredients is for the control of texture and mouthfeel (Nieuwland et al., 2013). To date, the main application of electrospun foodstuffs is in the protection and delivery of active compounds (Alborzi et al., 2013). Incorporating bioactive compounds into fibers can be achieved by simply mixing the active compounds into the polymer spinning solution. To avoid the use of toxic solvents in food, ethanol, and water soluble polymers are the preferred choice (Kayaci and Uyar, 2012). As an appropriate carrier for complex bioactive compounds, the selected polymer should ideally be plant or animal origin, edible, soluble in edible solvents, as well as being reliably electrospinnable (López-Rubio and Lagaron, 2012). In the biomedical and pharmaceutical sectors polymer-based delivery systems are well established and widely used (Liechty et al., 2010), although there is likely to be less consumer resistance to the presence of a synthetic polymer in a medicine, as the material is only ingested in small quantities, and is not generally perceived as being equivalent to a foodstuff. Proposals for the use of food-grade polymers and biopolymers as bioactive molecule delivery devices in food systems are becoming more widely accepted (López-Rubio and Lagaron, 2012; Nieuwland et al., 2013). Naturally occurring biopolymers in the form of proteins and carbohydrates are obvious candidates for encapsulation, and have been demonstrated to give controlled and sustained release properties to deliver the payload compound in the right place and at the right time (Fathi et al., 2014) (Table 8.2). In this section we have introduced base encapsulating materials suitable for spinning, and surveyed their possible application in the field of food science.

Table 8.2 Examples of applied and potential polymers for encapsulation

Applied electrospinnng polymers for encapsulation			
Based encapsulating materials		***Ingredient***	***References***
Protein	Whey protein concentrate (WPC)	*Bifidobacterium strains*	López-Rubio et al. (2012)
	Whey protein isolate (WPI) and WPC	β-Carotene β-Lactoglobulin Curcumin	López-Rubio and Lagaron (2012) Brahatheeswaran et al. (2012)
	Zein	Gallic acid	Neo et al. (2013)
		β-Carotene	Fernandez et al. (2009)
		(−)-Epi-gallocatechin-gallate (EGCG)	Li et al. (2009)
		α-Tocopherol	Wongsasulak et al. (2014)
		Ferulic acid	Yang et al. (2013)
		Fish oil	Moomand and Lim (2015)
		Tannin	de Oliveira Mori et al. (2014)
	Cellulose	Natural pigment from red cabbage	Devarayan and Kim (2015)
Carbohydrate	Cylodextrins	Vanilla	Kayaci and Uyar (2012)
		Menthol	Kayaci and Uyar (2012)
		Eugenol	Kayaci et al. (2013)
	Sodium alginate/pectin/poly(ethylene oxide) (PEO)	Pure folic acid (Vitamin B_9)	Alborzi et al. (2013)
	Pullulan	*Bifidobacterium strains*	López-Rubio et al. (2012)
	Pullulan/cyclodextrin	Perillaldehyde aroma Limonen	Mascheroni et al. (2013) Anu Bhushani and Anandharamakrishnan (2014)
Synthetic polymer	Cellulose acetate	Vitamin A and E	Taepaiboon et al. (2007)
	PEO/cellulose diacetate (CDA)	T4 phage	Korehei and Kadla (2014)
	PEO/chitosan	α-Tocopherol	Wongsasulak et al. (2014)
	PEO/WPI	Rhodamine B (RhB)	Sullivan et al. (2014)
	PEO/WPI	β-Lactoglobulin	Sullivan et al. (2014)
	Polyvinyl alcohol (PVA)	*B. animalis* subsp. *lactis* Bb_{12}	López-Rubio et al. (2009)
		Curcumin	Sun et al. (2013)
		Escherichia coli, S. albus	Salalha et al. (2006)
		Casein/lipase	Xie and Hsieh (2003)
		Cellulase	Wu et al. (2005)
	PVA/regenerated silk fibroin	Glucose oxidase (GOD)	Liu et al. (1996)

Table 8.2 **Examples of applied and potential polymers for encapsulation *(cont.)***

Applied electrospinnng polymers for encapsulation			
Based encapsulating materials		***Ingredient***	***References***
	Poly(acrylonitrile-*co*-methyl methacrylate)	Beta-galactosidase	El-Aassar (2013)
	Polystyrene	Chymotrypsin	Jia et al. (2002)
	Poly(D,L-lactide) (PDLLA) and PEO	*Bacteriocins Plantaricin 423/ST4SA*	Heunis et al. (2011)
	Polyamide 6.6	pH-Indicator dyes	Van der Schueren et al. (2010)

Potential electospun polymers for encapsulating		
Electrospun polymers		
Protein	Amaranth protein isolate (API)	Aceituno-Medina et al. (2013)
	Casein	Nieuwland et al. (2013)
	Collagen	Bürck et al. (2012)
	Egg albumen	Wongsasulak et al. (2010)
	Gelatin	Songchotikunpan et al. (2008)
	Soy protein isolate	Xu et al. (2012)
	Wheat gluten	Woerdeman et al. (2005)
Carbohydrate	Alginate	Alborzi et al. (2013)
	Carrageenans	Stijnman et al. (2011)
	Chitosan	Desai et al. (2008)
	Dextrans	Sun et al. (2013)
	Galactomannans	Stijnman et al. (2011)
	Glucomannans	Stijnman et al. (2011)
	Guar-gum	Lubambo et al. (2013)
	Inulin	Jain et al. (2014)
	Cellulose	Ghorani and Tucker (2015)
	Starch	Kong and Ziegler (2012)
	Tragacanth	Ranjbar-Mohammadi et al. (2016)
Lipid	Lecithin	McKee et al. (2006)
	Asolectin	Mendes et al. (2015)
Synthetic polymer	PVA/chitosan	Huang et al. (2009)
	PVA/gelatin	Huang et al. (2009)
	PVA/wheat gluten	Dong et al. (2010)
	PEO/WPI	Colín-Orozco et al. (2015)
	Cellulose acetate	Ghorani et al. (2013)
	Cellulose acetate phthalate	Olaru and Olaru (2010)
	Cellulose derivatives	Frenot et al. (2007)
	Poly(vinyl pyrrolidone) (PVP)/poly(L-lactide-*co*-ε-caprolactone) (PLCL)	Li et al. (2008)

8.6.1 Protein-based encapsulating materials used in electrospinning

Proteins are preferred over synthetic polymers (and carbohydrate polymers) for biochemical reasons because proteins form a significant part of the human body, and are often used as dietary supplements and functional foods (Nieuwland et al., 2013). Proteins are notoriously difficult to electrospin as simple solutions, because of their complex secondary and tertiary structures. It is necessary to not only dissolve a protein into solution, but also to generate a random coil molecular conformation to get a spinnable material. Globular proteins have relatively low internal interactions and a low degree of intramolecular attraction, that as we have seen, is required to form a fiber strand during spinning (Nieuwland et al., 2013). As an example of this (Pitkowski et al., 2008) reported that spinning of pure caseinate was not possible from an aqueous solution due to clustering of the molecules.

Common protein based encapsulating materials used are soy protein isolate (Xu et al., 2012), WPI (López-Rubio et al., 2012), whey protein concentrate (WPC) (López-Rubio and Lagaron, 2012), casein (Nieuwland et al., 2013), egg albumen (Wongsasulak et al., 2010), gelatin (Songchotikunpan et al., 2008), collagen (Bürck et al., 2012), zein (Brahatheeswaran et al., 2012), and wheat gluten (Woerdeman et al., 2005). López-Rubio et al. (2012) demonstrated an electrospun functional food by encapsulating *Bifidobacterium* strains in WPC and a carbohydrate (pullulan). As a result, WPC demonstrated greater protective ability than the pullulan, effectively prolonging the survival of the cells in storage even in conditions of high relative humidity (López-Rubio et al., 2012). The efficacy of WPC and WPI was also demonstrated by encapsulating the antioxidants β-carotene and β-lactoglobulin (López-Rubio and Lagaron, 2012; Sullivan et al., 2014). Amaranth (amaranth protein isolate or API) is electrospinnable and the morphology and texture encapsulated functional foods has been studied (Aceituno-Medina et al., 2013).

Zein is a prolamine protein dissolved out of maize (*Zea mays* sp.) with ethanol. This naturally hydrophobic material has long been used in the food sector as food coating where it is sometimes known as "confectioners glaze"(Torres-Giner et al., 2008). Zein, has also been demonstrated to enhance the stability and bioavailability of gallic acid (3,4,5-trihydroxybenzoic acid) (Neo et al., 2013), *β*-carotene (Fernandez et al., 2009), curcumin ((1E,6E)-1,7-bis(4-hydroxy-3-methoxyphenyl) hepta-1,6-diene-3,5-dione) (Brahatheeswaran et al., 2012), (−)-epi-gallocatechin-gallate (EGCG) (Li et al., 2009) α-tocopherol (Wongsasulak et al., 2014), ferulic acid((E)-3-(4-hydroxy-3-methoxy-phenyl)prop-2-enoic acid) (Yang et al., 2013), and the astringent, polyphenol tannin (de Oliveira Mori et al., 2014) for a variety of purposes. Chitosan and gelatin are highly biocompatible so have a popular choice as potential encapsulants. These materials are also biodegradable, exhibit bacteriostatic properties (Chiung et al., 2009). Huang et al. (2004) electrospun gelatin, and quantified the relationship between solution concentration and mechanical properties (Huang et al., 2004). Neo et al. (2012) investigated the impact of solution and processing parameters on the fabrication of zein electrospun fibers. They also examined the interactions between gallic acid and zein, and to assay the effect of electrospinning on the antioxidant properties of gallic acid (Neo, 2014; Neo et al., 2012).

8.6.2 Electrospinning carbohydrate-based encapsulants

All plant and animal origin materials are necessarily biocompatible and biodegradable, and so it is with the polysaccharides. In addition, the biochemistry of these materials means that they are readily modifiable to achieve a particular set of properties. As such polysaccharide materials show promise for nano- and microencapsulation of bioactive food ingredients (Schiffman and Schauer, 2008). As delivery systems, carbohydrates have functional groups that allow them to readily interact with other bioactive compounds, binding a number of hydrophilic and phobic bioactive substances (Assadpour et al., 2016a,b). In addition, these material are stable at high temperatures, in this respect outperforming lipid or proteins which may require an additional stage of melting or denaturation to be usable (Fathi et al., 2014). Carbohydrate monosaccharide, oligosaccharide, and polysaccharides structures are further categorized according to origin: from higher plants (e.g., starch, cellulose, pectin, and guar gum); from animals (e.g., chitosan); algal origin (e.g., alginate, and carrageenan), and from microbes (e.g., xanthan, dextran, and cyclodextrins), although it should be noted that chitosans are the principle structural component of fungi as well (Fathi et al., 2014). Chitosan (Desai et al., 2008), alginate (Alborzi et al., 2010), cellulose and its derivatives (Ghorani et al., 2013), starch (Kong and Ziegler, 2012), pullulan (Fabra et al., 2014; Karim et al., 2009), inulin (Jain et al., 2014), guar-gum (Lubambo et al., 2013), tragacanth (Ranjbar-Mohammadi et al., 2016), and dextrans (Sun et al., 2013) have all been successfully electrospun.

Stijnman et al. (2011) electrospun fibers from aqueous polysaccharide solutions of plant origin (galactomannans and glucomannans), seaweeds (carrageenans and alginate), microbial origin—pullulan, nonbranched, and dextran brush-like structures, in addition to the vegetable-origin carbohydrates starch and methyl cellulose (Stijnman et al., 2011). Electrospun fiber encapsulation using a sodium alginate-pectin-poly(ethylene oxide) blend has been used to protect folic acid (vitamin B_9). This material is susceptible to degradation when exposed to light or stored in acidic conditions. Postencapsulation almost 100% of the folic acid is retained after 41 days of storage at pH 3, this was compared to an unencapsulated control where only 8% of the folic acid was recovered after 1 day of storage at pH 3 (Alborzi et al., 2013).

CDs are cyclic oligosaccharides consisting at the molecular level of α-(1,4)-linked glucopyranose units with a truncated cone-shaped structure, because of this structure CDs can form supramolecular noncovalent host–guest inclusion complexes with a range sizes from small- to macromolecules (Manasco et al., 2012). Electrospun fibers can be made from nonpolymeric systems such as CDs and glucose (Pablo et al., 2015); the method of molecular cohesion is thought to be hydrogen bonding rather than chain entanglement. Like polymers, the concentration of the solution, solvent, and the mechanisms for intermolecular interactions are important for initiation and control of fiber formation. Hydroxypropyl-β-CD, hydroxypropyl-γ-CD, and methyl-β-CD have also been electrospun (Celebioglu and Uyar, 2010, 2012; Manasco et al., 2012). Cyclodextrins can form inclusion complexes with functional food substances such as antioxidants, as well as with antibacterial agents, and flavorings and aromatics. The formation of cyclodextrin complexes has been demonstrated to improve the stability, solubility, retained reactivity and the sustained release encapsulated bioactives (Kayaci and Uyar, 2012). Electrospinning polyvinyl alcohol fibers containing inclusion complexes entraining vanilla, menthol, and eugenol flavors within cyclodextrins (α-CD, β-CD, γ-CD) makes the product more temperature stable and increases shelf life (Kayaci and Uyar, 2012). Perillaldehyde is used in perfumery and is an aromatic monoterpenoid containing an aldehyde functional group extracted from *Perilla frutescens* a plant used in Korean and Japanese cuisine, it has been successfully incorporated into an edible polysaccharide nanofibrous matrix made from a pullulan–cyclodextrin emulsion (Mascheroni et al., 2013). Similarly, emulsion electrospinning has been used to encapsulate another volatile fragrance, limonene (1-methyl-4-(1-methylethenyl)-cyclohexene) into a fiber. Systematic optimization of carbohydrate polymer blends to improve solution properties and resulting in higher tensile strength has recently been reported (Anu Bhushani and Anandharamakrishnan, 2014).

The organic compound cellulose is a polysaccharide consisting of a linear chain with a length varying between hundreds to thousands of β-linked D-glucose units. It is plentiful, renewable, cheap, readily available biopolymer and has a long history in fiber manufacturing, from the use of woven plant fibers to the use of regenerated cellulose as a spun monofilament (Cuculo et al., 2001). Electrospun cellulose nanofibers have physicochemical attributes that make them a desirable choice for food applications such as immobilization of bioactives, nutraceutical delivery, filtration, and applications requiring film formation (Fernandez et al., 2009; Xie and Hsieh, 2003; Yu et al., 2013). Cellulose acetate (CA) has been used as a spinning adjunct to make edible nanofiber films by electrospinning a solution consisting of a mixture of cellulose acetate (CA) in 85% acetic acid and egg albumen (EA) in 50% formic acid (Wongsasulak et al., 2010).

8.6.3 Electrospinning lipid-based encapsulants

Lipids are chemically fatty acids or fatty acid derivatives. The carbon-to-hydrogen bonds in the molecular structure are nonpolar covalent bonds, which means that lipids are soluble in fats and some organic solvents such as ether and acetone, but are not water soluble. The principle lipid groups are triglycerides or neutral fats, phospholipids, steroids, and waxes (Boyle, 2008; McClements and Li, 2010). Phospholipids have been electrospun, to exploit their potential to form films with specific permeability to ions and molecules by self-assembling into highly organized structures that are similar to cell membranes (Mendes et al., 2015). Phospholipids have a polar head group and two hydrophobic hydrocarbon tail, usually a fatty acid, and varying in length between 14 and 24 carbon atoms. One tail is usually unsaturated, possessing more *cis*-double bonds, while the other tail is saturated. These differences in the length and degree of saturation of the fatty acid tails influence the ability of phospholipid molecules to pack tightly, affecting the fluidity of the membrane (Boyle, 2008).

The potential of phospholipids for electrospinning is underpinned by studies which indicate that as we have seen with glucose and cyclodextrins, a high molar mass polymeric structure is not method of producing solutions with sufficient intramolecular interaction to spin into fibers (McKee et al., 2006). For instance, the strong quadruple hydrogen bonding capabilities displayed by certain polymers give them electrospinning performance similar to unfunctionalized polymers with a significantly higher molar mass. Phospholipids can form entangled micelle structures in solution under appropriate conditions. As the concentration of phospholipids is increased, the spherical micellar morphology initially observed changes first to cylindrical, and then as the cylindrical micelles overlap and entangle forming sheets to similar effect to the formation of semicrystalline polymer structures (McKee et al., 2006). A solution of phospholipids (lecithin) above 35% w/w in dimethylformamide:chloroform [3:2 v/v] is capable of being electrospun into a continuous fiber with an average fiber diameter of ≈3.3 μm (from a 45% w/w solution) (McKee et al., 2006). Similarly, Mendes et al. (2015) demonstrate that isooctane is an efficient solvent for electrospinning asolectin (a material containing approximately equal proportions of lecithin, cephalin, and phosphatidylinositol combined with minor amounts of other phospholipids and polar lipids) fibers at a phospholipid concentration of 60% w/w to produce the fibers with average diameter of 6.1±2.7 μm (Mendes et al., 2015). The application of lipids to electrospinning is a current issue which is still under development.

8.6.4 Synthetic encapsulating materials used in electrospinning

Some synthetic polymers, for example, polyvinyl alcohol (PVA), and especially poly(ethylene oxide) (PEO) are combined with biopolymers as spinning adjuncts, as a blended solution to improve the fiber forming ability (Anu Bhushani and Anandharamakrishnan, 2014).

PEO is slightly unusual as a water soluble, and aerobically biodegradable thermoplastic polymer. PEO is biocompatible and of very low toxicity, so the potential for using PEO in biomedical and food applications is obvious (Colín-Orozco et al., 2015). Alborzi et al. (2010) electrospun both low and medium viscosity sodium alginate, pectin, and PEO fibers for use as a carrier to stabilize folic acid. They examined degree of PEO addition and how it affected electrospinning behavior and the consequent morphologies of the alginate–pectin electrospun fibers produced (Alborzi et al., 2010).

Wongsasulak et al. (2010) demonstrated that by itself egg albumen protein is an unsatisfactory polymer for fiber formation, however by blending it with PEO or cellulose acetate the generation of fibers by entrapping the protein in an entangled matrix was achieved (Wongsasulak et al., 2007, 2010).

Colín-Orozco et al. (2015) employed solutions containing WPI and PEO electrospin fibers, varying the PEO/WPI ratio to make fibers with different concentrations of homopolymers (Colín-Orozco et al., 2015). The group looked at surface tension, viscosity, morphology, conductivity, and thermal and vibrational properties as function of the PEO/WPI ratio. Their FTIR results showed that cross-linking assists the formation of fibers. Rosemary (*Rosmarinus officinalis*) extract was used as a model material to test the use of the fibers as a delivery system (Colín-Orozco et al., 2015).

Sullivan et al. (2014) studied electrospinning ability and heat treatment of WPI and β-lactoglobulin to form composite nanofibers in conjunction with PEO. In addition, they showed that the mats swell in water and retain a fibrous quality which makes them desirable for application in regenerative medicine. Finally, they incorporated a small hydrophobic molecule—Rhodamine B (RhB)—as a model flavonoid into WP/PEO nanofiber mats (Sullivan et al., 2014). Furthermore, Wongsasulak et al. (2013, 2014) entrapped α-tocopherol and investigated the effect on mucoadhesivity quantifying the release, degradation, and swelling characteristics of composite fibers electrospun from a of zein–chitosan mixture (Wongsasulak et al., 2013, 2014). α-tocopherol in mucoadhesive fibers (mucoadhesive materials exhibit adhesion between materials, at least one of which is a mucosal surface) were encapsulated using a mixture of zein, PEO, and chitosan, and were found to enhance the property of mucoadhesivity thereby improving bioaccessibility and bioavailability of nutrients (Wongsasulak et al., 2014). In addition, biocompatible fibers were electrospun from egg albumen and PEO (Wongsasulak et al., 2007).

PVA is a highly biocompatible, biodegradable, and nontoxic synthetic material. Chiung et al. (2009) synthesized electrospun PVA/chitosan and PVA/gelatin nanofiber composite membranes (Chiung et al., 2009). Soluble fiber fractions made by alkaline treatment of cereal wastes (corncob and wheat straw) have been electrospun using 6% (w/v) PVA as a spinning adjunct obtaining fibers in the range of 200–800 nm (Kuan et al., 2011). The use of PVA was noted as being crucial in improving the rheological profiles of the agar/PVA blends and thence solution spinnability (Sousa et al., 2015). Additionally, López-Rubio et al. (2009) demonstrated that the viability of *B. animalis* subsp. *lactis* Bb_{12} was boosted by incorporating them into PVA electrospun nanofibers. Dong et al. (2010) investigated the molecular weight distribution (MWD), rheology which they related to various electrospinning traits in a series of wheat gluten mixtures with PVA, dithiothreitol (DTT), and thiolated poly(vinyl alcohol) (TPVA) in a 50:50 ration of water to 1-propanol using size-exclusion chromatography, steady-shear viscosity measurements and examination by scanning electron microscopy (Dong et al., 2010).

Cellulose cannot be applied as practical delivery systems in food without significant modification due to its low solubility in water. Physical, chemical, or biochemical modifications are all available to make cellulose suitable for use in the food industry (Chauhan and Chakrabarti, 2012). Ghorani et al. (2013) both evaluated the morphology and mechanically characterized cellulose acetate (CA) electrospun fiber webs to understand how to vary the morphology of electrospun CA fibers in single and binary solvent systems based on evaluation of solubility parameters and to identify suitable solvent systems and optimum processing conditions consistent with the production of defect-free CA fibrous webs (Ghorani et al., 2013). Edible nanofibrous thin films have also been fabricated by electrospinning from blended solutions of CA in 85% acetic acid and egg albumen in 50% formic acid (Wongsasulak et al., 2010). Vitamins A and E were immobilized onto electrospun cellulose acetate giving product with a smooth surface and a circular cross section (Taepaiboon et al., 2007). This exercise resulted in controlled steady release of vitamins compared to the initial burst release observed in the control of conventionally cast films of cellulose acetate (Taepaiboon et al., 2007).

Olaru and Olaru (2010) demonstrated that cellulose acetate phthalate could be electrospun in a number of solvent systems. Huang et al. (2012) electrospun cellulose acetate phthalate fibers to examine the suitability of the material for drug delivery applications (Huang et al., 2012). Other cellulose derivatives, namely hydroxypropylmethyl cellulose, ethyl cellulose, methyl cellulose, hydroxypropylmethyl cellulose phthalate, and ethyl–cyanoethyl cellulose have been examined for similar applications (Frenot et al., 2007; Lim et al., 2010; Wang et al., 2007; Yu et al., 2013; Zhao et al., 2004).

Khadka et al. (2011) electrospun fiber mats from a synthetic anionic copolypeptide of L-glutamic acid, and L-tyrosine (PLEY), characterizing the in vitro physical, chemical, and biological properties of the fibers (Khadka et al., 2011). Li et al. (2008) coaxially electrospun core/shell poly(vinylpyrrolidone) (PVP)/poly(L-lactide-*co*-ε-caprolactone) (PLCL) fibrous membranes and examined in vitro cytocompatibility. They noted that, in phosphate buffer solution, the PVP/PLCL membranes had the highest water absorption due to both introduction of the PVP component and the core/shell structure (Li et al., 2008).

8.6.5 Applying electrospinning in the food industry

We have shown that both electrospraying and electrospinning are practical methods for the manufacture of food-safe encapsulated delivery systems, and hence that there is much potential to design and develop encapsulated products for high-margin improved functionality products with matching delivery systems (Ghorani and Tucker, 2015).

The developed electrospun structures show suitable results in terms of bioactive protection for application in functional foods (Aceituno-Medina et al., 2015). For example, inclusion complexes such as menthol, vanilla, and eugenol flavorings. Compounding with cyclodextrins (α-CD, β-CD,γ-CD) and electrospinning into polyvinyl alcohol (PVA) nanofibers makes the preparation more temperature stable and increases shelf life. It was reported that γ-CD is more efficient in the stabilization and controlled release of flavors than α-CD and β-CD (Kayaci and Uyar, 2012; Kayaci et al., 2013; Uyar et al., 2009).

Aceituno-Medina et al. (2015) used food-grade electrospun fiber encapsulation to improve the antioxidant capacity of quercetin (2-(3,4-dihydroxyphenyl)-3,5,7-trihydroxy-4H-chromen-4-one) and ferulic acid((E)-3-(4-hydroxy-3-methoxyphenyl)prop-2-enoic acid) after subjection to in vitro digestion (Aceituno-Medina et al., 2015). Protein isolated the amaranth plant (*Amaranthus* sp.), namely amaranth protein isolate (API) was electrospun to examine the importance of morphology and texture when used for bioactive encapsulation matrices for functional food applications (Aceituno-Medina et al., 2013). Curcumin ((1E,6E)-1,7-bis(4-hydroxy-3-methoxyphenyl)hepta-1,6-diene-3,5-dione) or its β-CD inclusion complex were mixed with polyvinyl alcohol (PVA) and subsequently electrospun and assessed for their potential as carriers. PVA-curcumin-cyclodextrin complexed fibers have better thermal stability and sustained release properties than PVA/curcumin fiber alone (Sun et al., 2013). Electrospun hybrid encapsulation structures containing the antioxidant β-carotene loaded as nanoliposomes as a payload have been shown to improve photostability. These are novel materials and show promise for allowing the incorporation of intrinsically hydrophobic β-carotene into water-based food formulations (de Freitas Zômpero et al., 2015). In a recent comprehensive work, Moomand (2014) evaluated the physicochemical aspects of using electrospun zein fibers to encapsulate fish oil. Electrospinning fish oil into electrospun zein improved product stability against oxidative degradation (Moomand, 2014).

In addition to protecting biochemically fragile bioactive compounds, maintaining the viability of probiotic bacteria and bacteriocins during food processing and storage by using proteins and carbohydrates as encapsulants is an additional market sector for electroprocessing (Burgain et al., 2011). Using prebiotic substances encapsulate and thence to protect and enhance the survival rate bioactive compounds and probiotic bacteria within while passing upper gastrointestinal (GI) tract and during food processing and storage is an area of great research and development interest, as a clear route toward a high added value product (López-Rubio et al., 2012). The main problem with common encapsulation techniques are the requirement for processes using high temperatures and/or organic agents, leading an unacceptably low yield of the biochemically fragile nutrients as well difficulties with residual organic materials that will almost certainly be regarded as unacceptable by the target market (López-Rubio and Lagaron, 2012; López-Rubio et al., 2012). For example, using spray drying to encapsulate will significantly lower the viability of bacteria or may damage the molecular structure of the payload (López-Rubio et al., 2012). Electroprocessing is kind to biochemical complexes because it does not involve extremes of temperature or pressure and does not require chemical reactions. A useful range of food-grade, biodegradable, biocompatible, or even conducting polymeric substances can be used as encapsulants (Anu Bhushani and Anandharamakrishnan, 2014). The feasibility and potential for use of soluble dietary fiber (SDF) plant coproducts using okara (a pomace co-product from the production of soybean milk), oil palm trunk (OPT), and oil palm frond (OPF) obtained via alkali treatment as source materials to encapsulate *Lactobacillus acidophilus* has been demonstrated (Fung et al., 2011). Following electrospinning and refrigeration temperatures at storage temperatures 78.6%–90% bacterial survivability was reported (Fung et al., 2011).

Salalha et al. (2006) electrospun a PVA encapsulation in an attempt to improve the viability of the bacteria *E. coli* and *Staphylococcus albus* (Salalha et al., 2006). After 3 months stored at −20 and −55°C the bacteria did not show any significant fall in viability (Salalha et al., 2006). Heunis et al. (2011) electrospun fibers using poly(D,L-lactide) (PDLLA) and PEO as a bacteriocin carrier matrix. Bacteriocins are proteinaceous bacterial toxins produced that inhibit the growth of similar or closely related bacterial strains. *Plantaricin 423*, produced by *Lactobacillus plantarum*, and ST4SA produced by *Enterococcus mundtii*) were used as model systems (Heunis et al., 2011). It is asserted that this technology will find application as an antimicrobial delivery system (Heunis et al., 2011). Core/shell structured coaxially electrospun fibers were also produced from PEO and cellulose diacetate (CDA) for the shell containing T4 phages suspended in a buffer solution (Korehei and Kadla, 2014). In vitro the water-soluble PEO fibers showed a burst release of about 100% of the phage payload over a period of 30 minutes (Korehei and Kadla, 2014). Thus there is significant scope for electrospun fibers designed as delivery systems for functional products

(Anu Bhushani and Anandharamakrishnan, 2014). Solution blending biomolecules as supplementary components within a polymeric matrix prior to electrospinning is an emerging technique to increase the performance of functional foods (Ghorani, 2012). A clear understanding of the mechanisms of electrospinning is essential if production conditions and throughput are to be optimized (Ghorani et al., 2015a).

To overcome difficulties in recovering product from a chemical reaction, sensitivity to temperature, limitation in acceptable range of operating pH, and the risk of inactivation by external agents, enzymes can be protected by immobilization enzymes. Enzyme immobilization is achieved by either affixing on a physical support or encapsulating in solid matrix while still retaining their catalytic powers (Brena et al., 2013; Keyes and Saraswathi, 1985). At the time of publication, using electrospraying for enzyme immobilization is less explored than coelectrospinning and coaxial electrospinning (Wang et al., 2009). In addition to encapsulation, enzymes can be physically adsorbed or covalently attached to nanofibers—known as surface attachment. The fibers act as a support structure for enzyme immobilization. If enzymes are cross-linked to the electrospun substrate the high level of surface area and porosity means greater residual activity: minimizing the size of the substrate improves the efficiency of immobilized enzymes (Anu Bhushani and Anandharamakrishnan, 2014). Liu et al. (1996) immobilized glucose oxidase (GOD) onto a membrane made of poly(vinyl alcohol) and regenerated silk fibroin (Liu et al., 1996). To decrease the resistance to material transport through the substrate, a porous membrane was prepared by adding poly(ethylene glycol) to the spinning solution. A glucose biosensor was made by enzyme immobilization coupled with an oxygen electrode by means of immobilizing GOD in electrospun PVA (Liu et al., 1996). Glucose amperometric biosensors were fabricated by electrospinning GOD immobilized in PVA fibers and successfully characterized for both response time and output current (Anu Bhushani and Anandharamakrishnan, 2014). The deoxidization properties of GOD can also be used in food packaging and preservation (Ge et al., 2012). Wu et al. (2005) immobilized cellulase in electrospun PVA membranes. The enzyme cellulase is commonly used in fruit and vegetable juice processing. If PVA and cellulase are codissolved in a pH 4.6 acetic acid buffer they can be electrospun into fibers with diameter of +200 nm. The fibers were cross-linked using glutaraldehyde vapor and the subsequent catalytic efficiency was determined. The activity of immobilized cellulase in electrospun PVA was ≥65% of that of the free enzyme: electrospun fibers worked better than casting films. The activity of immobilized cellulase was in inverse proportion to cross-linking time and enzyme loading efficiency, which retained 36% of initial activity after six use cycles (Wu et al., 2005). El-Aassar (2013) electrospun poly(acrylonitrile-*co*-methyl methacrylate) polymer activated by glutaraldehyde as support structure immobilizing beta-galactosidase obtained from the filamentous fungus *Aspergillus oryzae* (El-Aassar, 2013). Beta-galactosidase is used in the dairy industry to a make lactose-free "milk" products. Immobilization of these enzymes allows for their use in continuous enzymatic reactions (Anu Bhushani and Anandharamakrishnan, 2014).

Chymotrypsin is a digestive enzyme that breaks down proteins and it is naturally produced by the pancreas, the enzyme is used in both the meat and dairy industries as an industrial product, but is also taken as an enzyme health supplement and as a treatment for certain diseases. The enzyme is selective for peptide bonds with aromatic or large hydrophobic side chains on the carboxyl side of the bond, it also catalyzes the hydrolysis of ester bonds. Jia et al. (2002) electrospun enzyme-carrying biocatalyst fibers (Jia et al., 2002). The immobilization of chymotrypsin in these polystyrene fibers gave structural stability limiting denaturation and increased 18-fold the material half-life in methanol (Jia et al., 2002). Enzymes are expensive and this limits their uptake by the food industry. Immobilizing the enzyme effectively reduces the cost of use. For example, lipase is too expensive to use to hydrolyze the lipids which are the ester bonds in triglycerides, to form fatty acids and glycerol which are widely used in the diary and bakery industries (Sahlström and Bråthen, 1997). Structural support or encapsulation matrices for immobilization of lipase extracted from *Candida rugosa* have been electrospun (Huang et al., 2007; Wang et al., 2006; Ye et al., 2006). Xie and Hsieh (2003) electrospun PVA and a mixture of PEO and casein to encapsulated lipase. The degree of enzyme catalyzed hydrolysis shown by the PVA/lipase membranes was higher than that of the PEO/casein/lipase membrane (Xie and Hsieh, 2003). Lipase activity in the electrospun membrane was 6 times greater than that in the control cast membrane (Xie and Hsieh, 2003). Similarly, Huang et al. (2011) electrospun immobilized *C. rugosa* lipase using cellulose acetate (Huang et al., 2011). When operating conditions were optimized a biocatalyst enzymatic activity of 29.6 U/g was achieved. The immobilized lipase exhibited better thermal stability and durability than at free enzyme control (Huang et al., 2011).

Filtration is an effective method of separating out undesired coproducts, commonly referred to as dregs or sediment. Removal is necessary because the residues may be strongly flavored and thus adversely affect product quality. If the aim is to improve the quality of the foodstuff, then the operator must be careful not to overshoot and remove desirable flavors or texture (Tragardh, 1991). Filtration techniques were classified as an entirely physical process, but at the more advanced end of the process they are now considered as biochemical techniques (Tragardh, 1991). For commercial success membranes with both a high degree of controlled selectivity and long life between services are needed. So, for practical applications permselectivity it is critical to be able to incorporate a required molecule into the membrane. To create sites capable of molecular recognition, molecular imprinting technologies (MITs) are a possible technology (Yoshikawa et al., 2011). Materials with specific binding or catalytic abilities can be synthesized and used to produce polymers with molecular recognition or template polymerization properties (Alexander et al., 2006). Molecular recognition is a well-known natural world process where rapid recognition of antibodies, enzymes, and nucleic acids is essential (Alexander et al., 2006; Kondo et al., 2000; Park et al., 2005). Processes for the manufacture of synthetic polymers with binding sites for specified molecules are the next phase toward the practical use of molecular recognition (Ghorani et al., 2015b). Molecularly imprinted polymer (MIP) synthesis achieves high selectivity from the imprinted material. Molecular imprinting is divided into covalent molecular imprinting and noncovalent molecular imprinting (Locatelli et al.,1998; Park et al., 2005). In covalent molecular imprinting removing of template after polymerization is difficult (Yoshikawa et al., 1999). This is because highly cross-linked polymers are intrinsically resistant to dissolution. Noncovalent imprinting

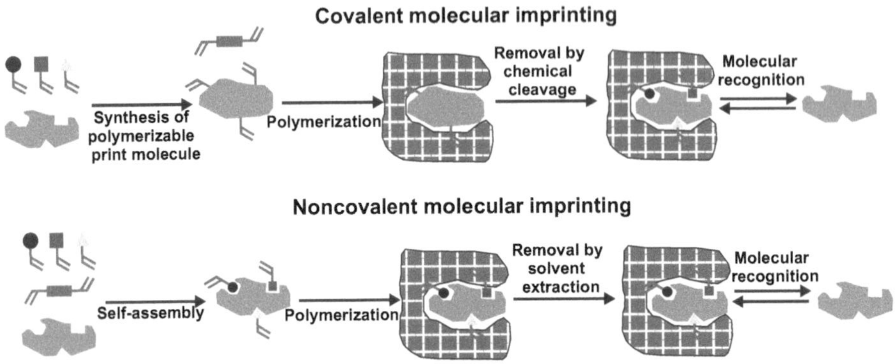

Figure 8.13 General scheme of the principle of the manufacture of molecularly imprinted polymers (MIPs).

(hydrogen bonding, electrostatic interaction, and coordination-bond formation) does not have this problem and is applicable to a wider range of templates than the covalent method. For these reasons noncovalent is generally the method of first choice (Dauwe and Sellergren, 1996; Sellergren, 1989; Sellergren and Shea, 1993).

The molecular imprint process starts with a cross-linking reaction to copolymerize the complex of a functional monomer with the template. The template is then removed from the polymerized matrix forming the binding sites that are compatible with the desired template (Fig. 8.13) (Kim and Chang, 2011).

Compared to the natural biological receptors, the synthetic molecularly imprinted equivalents have better physicochemical stability and perhaps, more importantly are cheaper. There are disadvantages however, the greatest of which are poor site accessibility and heterogeneous distribution of the binding sites which combine to lower system capacity when compared with natural equivalents (Ye, 2013; Yilmaz et al., 1999). Nanoscale membrane fabrication may overcome these disadvantages and make practical application of the method possible (Acton, 2013; Lee et al., 2007; Long et al., 2012). Electrospun materials are an attractive possibility because of their high specific surface area and because it is entirely possible to modify the bulk properties of a fabric by using a multilayer manufacturing process (Ghorani et al., 2013). An example of the application of this technology would be the combined accurate detection and sequential removal by filtration of food pathogens (Ghorani et al., 2015b). Suitable functional polymers with appropriate binding groups are made practically usable by being directly imprinted onto electrospun fibers (Fig. 8.14) (Chakraborty et al., 2009).

In summary, both template molecule and functional polymer are mixed into the spinning solution. During electrospinning, noncovalent interaction between the template and the carrier polymer occurs—effectively a polymer cast is taken of the template. The template is removed from the fiber product by solvent extraction thus exposing the imprinted sites for use (Chronakis et al., 2005; Ghorani, 2012). Alternatively, MIP nanoparticles are made by precipitation polymerization can be easily encapsulated into nanofiber membranes within the size range of 100 nm to 3 μm in

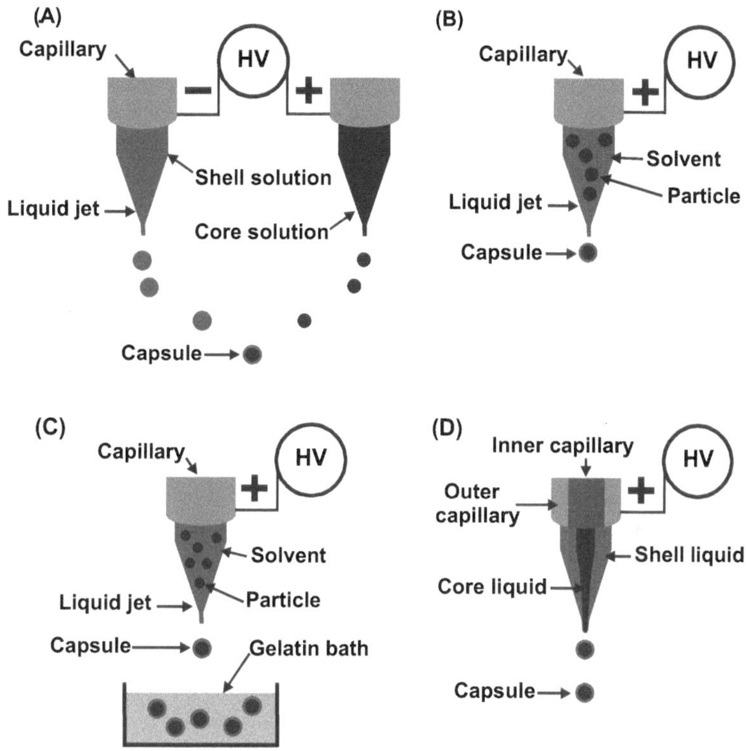

Figure 8.14 Schematics of various electrospinning processes. (A) Collision of droplets of opposite polarity; (B) electrospinning of a template/suspended in polymer with solidification by evaporation; (C) electrospinning of a template dissolved/suspended in polymer with solidification by a chemical or ionic cross-linker; and (D) coaxial electrospinning.

diameter, the MIP nanoparticles are not damaged by the encapsulation process and still provide efficient extraction of trace analytes from real samples. (Chronakis and Ye, 2013; Chronakis et al., 2006; Piperno et al., 2011; Yoshimatsu et al., 2008).

Noncovalent molecularly imprinted fiber membranes improve efficiency of both flux and permselectivity (Sueyoshi et al., 2010; Yoshikawa et al., 2007). Preferred detection methods are either enzymatic, microbiological, chromatographic, plasma analytic, or immunological. Lok and Son (2009), Saini and Kaur (2013), and Song et al. (2014) confirm the desirability of making better use of MIPs when checking food for contaminants, MIPs are especially suitable for this type of work because of their high degree of analyte selectivity. A noncovalently lactose imprinted polymer has been shown to differentiate between various mono- and disaccharides in aqueous solutions (Hadizadeh et al., 2013). In a commercial project, a polymer "trap" with fissures to capture vitamin B_2 (riboflavinor 7,8-Dimethyl-10-[(2S,3S,4R)-2,3,4,5-tetrahydroxy-pentyl]benzo[g]pteridine-2,4-dione) was developed. The project aim was to increase the shelf life of beer. Unfortunately, due to the general biochemical complexity of foodstuffs, a very high degree of selectivity is needed if either the overall composition or the sensory perception of the product is not to be significantly affected. In economic terms, extracting a low level but desirable entity as a fine chemical product is the technical challenge. A generic and economically viable approach to surmount the problems of this nature is required for successful commercial take-up. An example is the extraction of crocin from saffron crocus (*Crocus sativus*) stamens. The use of MIP solid-phase extraction has been demonstrated using a template based on gentiobiose (a glycoside moiety in crocin structure) (Mohajeri et al., 2010). Dubbed Gent-MIP, it has much higher affinity for crocin than other candidate materials and successful extraction of crocin with at a recovery rate of 84% was achieved, effectively by using a part of a big molecule to make an MIP retaining sufficient selectivity for the main structure (Mohajeri et al., 2010). Assembly of molecularly imprinted electrospun nanofiber membranes with example applications in the detection of microbial toxins, antimicrobials, and chemical and dye residues have been demonstrated (Ghorani et al., 2015b).

Molecular imprinting technique are well established, and when combined with the technology of electrospun fiber structures and indeed with other industrially produced fibers, lead the way to the development of new nano- and microstructured smart assemblies, and opportunities for a range of highly novel applications in the food manufacturing (Ghorani et al., 2015b). At the nanoscale, the influence of the interface between the solid and liquid phases extends significantly into the liquid phase. This is why usually only a fraction of a volume percent of a nanoscale solid needs to be added to produce a significant effect on the bulk properties of a composite. There are clearly opportunities for the food industry that can be deduced from this observation. Electrospun nanofibers are more acceptable to consumers as food ingredients than nanoparticulates because they have a significant nonnano dimension, and the method of absorption into the digestive tract is unlikely to be able to bypass the normal mechanism of digestion, as may be the case with insoluble nanoparticulates. The formation of a significant interphase zone has been observed in the liquid phase as well as the solid—a fraction of a volume percent of an electrospun nanofiber can significantly affect viscosity or mouthfeel. MITs—as discussed in this chapter, show promising potential to enhance flavors so that a level of sensory satisfaction could be achieved at lower concentrations. This is important if the ingredients are expensive, or have a health effect, for example, sugar or salt. Additionally, the detection of biochemical indicators for the degree of ripeness or the onset of spoilage may also develop into a significant commercial opportunities (Ghorani et al., 2015b).

Electrospinning also can be used in so-called "smart packaging": active packaging where control of the point of delivery retail packaging storage microenvironment is improved to, say, keep moisture within a tight tolerance to maximize shelf life, and intelligent packaging, where features that indicate internal status or communicate product changes such as degree of ripeness or staling are available. These intelligent packaging aids communicate directly to the consumers or retailer information concerning the quality of the packed food (Butler, 2013). Intelligent packaging is of great significance in the food and beverage industry pH-sensitive nanofiber-based sensors are already available as a part of packaging material. Workers from Ghent university (Van der Schueren et al., 2010) developed halochromic nano nonwovens mixing the pH-indicator dyes Bromocresol Purple and Brilliant Yellow in conjunction with polyamide 6.6 polymer to make an electrospinning solution. The addition of the pH-indicator did not affect the average fiber diameter. If the dyes were not properly dissolved instabilities were observed in the formation of the spinning droplet. The electrospun product showed a clear pH related color change. Dye–fiber interactions did alter the nature of the color change from that which occurred in conventional solution, and so the sensor was only suitable operation over a narrow pH window (Van der Schueren et al., 2010). The pH limitation was overcome by tuning the combination of indicators dyes (Agarwal et al., 2012), the final result was a membrane capable of detecting pH values from 1–10 by giving a unique identifiable color for each pH. The sensors survived extended storage and had good environmental stability over wide range of temperatures, moisture levels, and pH (Agarwal et al., 2012; Anu Bhushani and Anandharamakrishnan, 2014). Electrospun pH-sensitive polycaprolactone (PCL) fibers and PCL/chitosan fiber blends with up to 20% chitosan have been made and characterized (Van der Schueren et al., 2013). Biocompatible polymers are important if certification for medical use is to be obtained. The 3-h time lag associated with PCL electrospun fiber sensors can be improved upon by using blends with chitosan; these have better sensitivity and a response time of only 5 min. This improvement is said to be due to the more hydrophilic nature of chitosan (Van der Schueren et al., 2013). Natural pigments from red cabbage have been used to make electrospun cellulose fibers. The pigment is adsorbed and chemically cross-linked into place (Devarayan and Kim, 2015), this resulted revealed in a biocomposite a universal pH sensor working reversibly over the full 1–14 pH range (Devarayan and Kim, 2015).

8.7 Conclusions and future trends

It is extremely likely that electrospun nanoscale fibers can be used as a delivery system nutrients to protect them during processing and storage or during passage of the components to the target site in body. The advantages of using electrospun nanofibers to encapsulate food bioactives are a combination of sustained and controlled payload release, room temperature processing, reduced protein denaturation, and efficient encapsulation. All these enhancements can be accomplished by the use of food-grade polymers or biopolymers. There is also scope to regulate the size of the final by manipulating processing parameters.

This chapter presented a comprehensive and detailed discussion concerning the basic story of how to electrospin polymer nanofibers including sections on processing, structure, property characterization, and applications. A critical degree of molecular entanglement and increased relaxation times resulting from increased polymer concentration are thought to be the dominant mechanisms for fiber formation, although hydrogen bonding allows certain small molecules to form spinnable solutions. A bead-rich morphology is typically observed when below a minimum critical degree of chain entanglement. When the molecular cohesion (by whatever means) of the solution is too low, and this is usually perceivable as low viscosity, electrospraying (liquid atomization) will occur and discrete droplets are formed rather than fibers.

Processing conditions and the efficacy of control parameters for electrospinning were also examined. Applications of electrospinning and its potential for food manufacture at the laboratory scale have been recognized; however, there is still work to be done on industrial scale manufacture.

The main reason why industrialist have held back on wide adopting of the science and technology under discussion may be attributed to the current primacy of work on synthetic polymers rather than on biopolymers or naturally occurring food components. Certainly, electrospun fibers must be available as practical industrial scale processes for the proposed applications to be taken up. The electroprocessing of novel products for food use particularly in food formulas, ingredients, packaging, sensors, biosensors, and delivery systems has clear commercial potential. The perceived low throughput of electrospinning perhaps restricts commercial take-up and therefore increasing the production efficiency must be a development priority. However, it should be noted that the high surface-to-volume ratio of nanomaterials means that surface energy effects are much more significant than at the macroscale. The formation of an interphase zone, or zone of influence at the phase change boundary means that only small fractions (typically <1 wt.%) are needed to produce highly significant effects. The effect of the interphase on food properties such as mouthfeel of taste has not been significantly investigated and may well open up new areas of say flavor intensification. Finally, many food-grade polymer systems (e.g., natural food hydrocolloids) are difficult to optimize for electrospinning due to their poor viscoelastic behavior, low degree of molecular entanglement, limited solubility, and, more generally, because only a few product parameters can be controlled directly as some of the control parameters are either highly interdependent or derived from the intrinsic physical properties of the polymer solution. Industrially produced electrospun fibers reach a dimensional process capability spread of the order of 10% or worse between nominally identical samples. Therefore, a significant effort is required to bring the mass production of nanofiber fabrics under control and thence to increase of the process throughput and product reproducibility and thence to extend the classes of useable materials.

References

Aceituno-Medina, M., Lopez-Rubio, A., Mendoza, S., Lagaron, J.M., 2013. Development of novel ultrathin structures based in amaranth (*Amaranthus hypochondriacus*) protein isolate through electrospinning. Food Hydrocoll. 31, 289–298.

Aceituno-Medina, M., Mendoza, S., Rodríguez, B.A., Lagaron, J.M., López-Rubio, A., 2015. Improved antioxidant capacity of quercetin and ferulic acid during in-vitro digestion through encapsulation within food-grade electrospun fibers. J. Funct. Foods 12, 332–341.

Acton, Q.A., 2013. Advances in Molecular Nanotechnology Research and Application: 2013 Edition, ScholarlyEditions, USA.

Agarwal, A., Raheja, A., Natarajan, T.S., Chandra, T.S., 2012. Development of universal pH sensing electrospun nanofibers. Sens. Actuators B 161, 1097–1101.

Ahn, Y.C., Park, S.K., Kim, G.T., Hwang, Y.J., Lee, C.G., Shin, H.S., Lee, J.K., 2006. Development of high efficiency nanofilters made of nanofibers. Curr. Appl. Phys. 6, 1030–1035.

Alborzi, S., Lim, L.-T., Kakuda, Y., 2010. Electrospinning of sodium alginate-pectin ultrafine fibers. J. Food Sci. 75, C100–C107.

Alborzi, S., Lim, L.T., Kakuda, Y., 2013. Encapsulation of folic acid and its stability in sodium alginate-pectin-poly(ethylene oxide) electrospun fibres. J. Microencapsul. 30, 64–71.

Alexander, C., Andersson, H.S., Andersson, L.I., Ansell, R.J., Kirsch, N., Nicholls, I.A., O'mahony, J., Whitcombe, M.J., 2006. Molecular imprinting science and technology: a survey of the literature for the years up to and including 2003. J. Mol. Recogn. 19, 106–180.

Almería, B., Deng, W., Fahmy, T.M., Gomez, A., 2010. Controlling the morphology of electrospray-generated PLGA microparticles for drug delivery. J. Colloid Interface Sci. 343, 125–133.

Angammana, C.J., Jayaram, S.H., 2008. Analysis of the effects of solution conductivity on electro-spinning process and fiber morphology. Industry Applications Society Annual Meeting, IEEE, pp. 1–4.

Anu Bhushani, J., Anandharamakrishnan, C., 2014. Electrospinning and electrospraying techniques: potential food based applications. Trends Food Sci. Technol. 38, 21–33.

Assadpour, E., Maghsoudlou, Y., Jafari, S.M., Ghorbani, M., Aalami, M., 2016a. Optimization of folic acid nano-emulsification and encapsulation by maltodextrin-whey protein double emulsions. Int. J. Biol. Macromol. 86, 197–207.

Assadpour, E., Maghsoudlou, Y., Jafari, S.-M., Ghorbani, M., Aalami, M., 2016b. Evaluation of folic acid nano-encapsulation by double emulsions. Food Bioprocess Technol. 9, 2024–2032.

Ballengee, J.B., Pintauro, P.N., 2011. Morphological control of electrospun nafion nanofiber mats. J. Electrochem. Soc. 158, B568–B572.

Barhate, R.S., Loong, C.K., Ramakrishna, S., 2006. Preparation and characterization of nanofibrous filtering media. J. Membr. Sci. 283, 209–218.

Barrero, A., Loscertales, I.G., 2007. Micro- and nanoparticles via capillary flows. Annu. Rev. Fluid Mech. 39, 89–106.

Bhardwaj, N., Kundu, S.C., 2010. Electrospinning: a fascinating fiber fabrication technique. Biotechnol. Adv. 28, 325–347.

Bock, N., Dargaville, T.R., Woodruff, M.A., 2012. Electrospraying of polymers with therapeutic molecules: state of the art. Prog. Polym. Sci. 37, 1510–1551.

Bock, N., Woodruff, M.A., Hutmacher, D.W., Dargaville, T.R., 2011. Electrospraying, a reproducible method for production of polymeric microspheres for biomedical applications. Polymers 3, 131–149.

Bohr, A., Wan, F., Kristensen, J., Dyas, M., Stride, E., Baldursdottír, S., Edirisinghe, M., Yang, M., 2015. Pharmaceutical microparticle engineering with electrospraying: the role of mixed solvent systems in particle formation and characteristics. J. Mater. Sci. 26, 1–13.

Boyle, J., 2008. Molecular biology of the cell, 5th edition by B. Alberts, A. Johnson, J. Lewis, M. Raff, K. Roberts, and P. Walter. Biochem. Mol. Biol. Educ. 36, 317–318.

Brahatheeswaran, D., Mathew, A., Aswathy, R.G., Nagaoka, Y., Venugopal, K., Yoshida, Y., Maekawa, T., Sakthikumar, D., 2012. Hybrid fluorescent curcumin loaded zein electrospun nanofibrous scaffold for biomedical applications. Biomed. Mater. 7, 045001.

Brena, B., González-Pombo, P., Batista-Viera, F., 2013. Immobilization of enzymes: a literature survey. Methods Mol. Biol. 1051, 15–31.

Buer, A., Ugbolue, S.C., Warner, S.B., 2001. Electrospinning and properties of some nanofibers. Text. Res. J. 71, 323–328.

Bürck, J., Heissler, S., Geckle, U., Ardakani, M.F., Schneider, R., Ulrich, A.S., Kazanci, M., 2012. Resemblance of electrospun collagen nanofibers to their native structure. Langmuir 29, 1562e72.

Burgain, J., Gaiani, C., Linder, M., Scher, J., 2011. Encapsulation of probiotic living cells: from laboratory scale to industrial applications. J. Food Eng. 104, 467–483.

Butler, P., 2013. Smart and interactive packaging developments for enhanced communication at the packaging/user interface. In: Farmer, N. (Ed.), Trends in Packaging of Food, Beverages and Other Fast-Moving Consumer Goods (FMCG). Woodhead Publishing Ltd., UK, (Chapter 10).

Casper, C.L., Stephens, J.S., Tassi, N.G., Chase, D.B., Rabolt, J.F., 2003. Controlling surface morphology of electrospun polystyrene fibers: effect of humidity and molecular weight in the electrospinning process. Macromolecules 37, 573–578.

Celebioglu, A., Uyar, T., 2010. Cyclodextrin nanofibers by electrospinning. Chem. Commun. 46, 6903–6905.

Celebioglu, A., Uyar, T., 2012. Electrospinning of nanofibers from non-polymeric systems: polymer-free nanofibers from cyclodextrin derivatives. Nanoscale 4, 621–631.

Chakraborty, S., Liao, I.C., Adler, A., Leong, K.W., 2009. Electrohydrodynamics: a facile technique to fabricate drug delivery systems. Adv. Drug Deliv. Rev. 61, 1043–1054.

Chaobo, H., Shuiliang, C., Chuilin, L., Darrell, H.R., Haiyan, Q., Ying, Y., Haoqing, H., 2006. Electrospun polymer nanofibres with small diameters. Nanotechnology 17, 1558.

Chauhan, V.S., Chakrabarti, S.K., 2012. Use of nanotechnology for high performance cellulosic and papermaking products. Cellul. Chem. Technol. 46, 389.

Chen, L., Remondetto, G.E., Subirade, M., 2006. Food protein-based materials as nutraceutical delivery systems. Trends Food Sci. Technol. 17, 272–283.

Chen, H.-M., Yu, D.-G., 2010. An elevated temperature electrospinning process for preparing acyclovir-loaded PAN ultrafine fibers. J. Mater. Process. Technol. 210, 1551–1555.

Chiung, H.C., Lin, C., Lu, C., Wen, L.C., Chao, C., 2009. Evaluation of the electrospinning manufacturing process based on the preparation of pva composite fibres. Fibres Text. East. Eur. 3 (74), 34–37.

Chowdhury, M., Stylios, G., 2010. Effect of experimental parameters on the morphology of electrospun Nylon 6 fibres. Int. J. Basic Appl. Sci. 10, 116–131.

Chronakis, I.S., Jakob, A., Hagström, B., Ye, L., 2006. Encapsulation and selective recognition of molecularly imprinted theophylline and 17β-estradiol nanoparticles within electrospun polymer nanofibers. Langmuir 22, 8960–8965.

Chronakis, I.S., Milosevic, B., Frenot, A., Ye, L., 2005. Generation of molecular recognition sites in electrospun polymer nanofibers via molecular imprinting. Macromolecules 39, 357–361.

Chronakis, I.S., Ye, L., 2013. Molecularly imprinted nano- and microstructures by electrospinning. Molecular Imprinting: Principles and Applications of Micro- and Nanostructure Polymers, Singapore. DOI: 10.1201/b15678-7.

Colín-Orozco, J., Zapata-Torres, M., Rodríguez-Gattorno, G., Pedroza-Islas, R., 2015. Properties of poly(ethylene oxide)/whey protein isolate nanofibers prepared by electrospinning. Food Biophys. 10, 134–144.

Cross, J., 1987. Electrostatics, Principles, Problems and Applications. Taylor & Francis, UK.

Cuculo, J.A., Aminuddin, N., Frey, M.W., 2001. Solvent spun cellulose fibers. Hanser Gardner Publications, Munich, Germany, pp. 296, 328.

Dabirian, F., Hosseini, Y., Ravandi, S.A.H., 2007. Manipulation of the electric field of electrospinning system to produce polyacrylonitrile nanofiber yarn. J. Text. Inst. 98, 237–241.

Dauwe, C., Sellergren, B., 1996. Influence of template basicity and hydrophobicity on the molecular recognition properties of molecularly imprinted polymers. J. Chromatogr. A 753, 191–200.

de Freitas Zômpero, R.H., López-Rubio, A., de Pinho, S.C., Lagaron, J.M., de La Torre, L.G., 2015. Hybrid encapsulation structures based on β-carotene-loaded nanoliposomes within electrospun fibers. Colloids Surf. B 134, 475–482.

De La Mora, J.F., Loscertales, I.G., 1994. The current emitted by highly conducting Taylor cones. J. Fluid Mech. 260, 155–184.

de Oliveira Mori, C.L.S., Dos Passos, N.A., Oliveira, J.E., Mattoso, L.H.C., Mori, F.A., Carvalho, A.G., de Souza Fonseca, A., Tonoli, G.H.D., 2014. Electrospinning of zein/tannin bio-nanofibers. Ind. Crops Prod. 52, 298–304.

De Vrieze, S., Van Camp, T., Nelvig, A., Hagström, B., Westbroek, P., de Clerck, K., 2009. The effect of temperature and humidity on electrospinning. J. Mater. Sci. 44, 1357–1362.

Deitzel, J.M., Kleinmeyer, J., Harris, D., Beck Tan, N.C., 2001. The effect of processing variables on the morphology of electrospun nanofibers and textiles. Polymer 42, 261–272.

Demir, M.M., Yilgor, I., Yilgor, E., Erman, B., 2002. Electrospinning of polyurethane fibers. Polymer 43, 3303–3309.

Denis, P., Dulnik, J., Sajkiewicz, P., 2015. Electrospinning and structure of bicomponent polycaprolactone/gelatin nanofibers obtained using alternative solvent system. Int. J. Polym. Mater. Polym. Biomater. 64, 354–364.

Desai, K., Kit, K., Li, J., Zivanovic, S., 2008. Morphological and surface properties of electrospun chitosan nanofibers. Biomacromolecules 9, 1000–1006.

Devarayan, K., Kim, B.-S., 2015. Reversible and universal pH sensing cellulose nanofibers for health monitor. Sens. Actuators B 209, 281–286.

Ding, B., Li, C., Miyauchi, Y., Kuwaki, O., Shiratori, S., 2006. Formation of novel 2D polymer nanowebs via electrospinning. Nanotechnology 17, 3685–3691.

Dong, J., Asandei, A.D., Parnas, R.S., 2010. Aqueous electrospinning of wheat gluten fibers with thiolated additives. Polymer 51, 3164–3172.

Doshi, J., Reneker, D.H., 1993. Electrospinning process and applications of electrospun fibers. In: Industry Applications Society Annual Meeting, 1993, Conference Record of the 1993 IEEE, Oct 2–8, vol. 3, pp. 1698–1703.

Doshi, J., Reneker, D.H., 1995. Electrospinning process and applications of electrospun fibers. J. Electrostat. 35, 151–160.

Dotti, F., Varesano, A., Montarsolo, A., Aluigi, A., Tonin, C., Mazzuchetti, G., 2007. Electrospun porous mats for high efficiency filtration. J. Ind. Text. 37, 151–162.

Eda, G., Liu, J., Shivkumar, S., 2007. Solvent effects on jet evolution during electrospinning of semi-dilute polystyrene solutions. Eur. Polym. J. 43, 1154–1167.

El-Aassar, M., 2013. Functionalized electrospun nanofibers from poly(AN-*co*-MMA) for enzyme immobilization. J. Mol. Catalysis B 85, 140–148.

Eltayeb, M., Bakhshi, P.K., Stride, E., Edirisinghe, M., 2013. Preparation of solid lipid nanoparticles containing active compound by electrohydrodynamic spraying. Food Res. Int. 53, 88–95.

Enayati, M., Chang, M.-W., Bragman, F., Edirisinghe, M., Stride, E., 2011. Electrohydrodynamic preparation of particles, capsules and bubbles for biomedical engineering applications. Colloids Surf. A 382, 154–164.

Fabra, M.J., López-Rubio, A., Lagaron, J.M., 2014. On the use of different hydrocolloids as electrospun adhesive interlayers to enhance the barrier properties of polyhydroxyalkanoates of interest in fully renewable food packaging concepts. Food Hydrocoll. 39, 77–84.

Faridi Esfanjani, A.F., Jafari, S.M., 2016. Biopolymer nano-particles and natural nano-carriers for nano-encapsulation of phenolic compounds. Colloids Surf. B 146, 532–543.

Fathi, M., Martin, A., McClements, D.J., 2014. Nanoencapsulation of food ingredients using carbohydrate based delivery systems. Trends Food Sci. Technol. 39, 18–39.

Fernandez, A., Torres-Giner, S., Lagaron, J.M., 2009. Novel route to stabilization of bioactive antioxidants by encapsulation in electrospun fibers of zein prolamine. Food Hydrocoll. 23, 1427–1432.

Filatov, Y., Budyka, A., Kirichenko, V., 2007. Electrospinning of Micro- and Nanofibers: Fundamentals and Applications in Separation and Filtration Processes. Begell House Ltd, New York, NY, USA.
Fong, H., Chun, I., Reneker, D.H., 1999. Beaded nanofibers formed during electrospinning. Polymer 40, 4585–4592.
Freiberg, S., Zhu, X., 2004. Polymer microspheres for controlled drug release. Int. J. Pharm. 282, 1–18.
Frenot, A., Chronakis, I.S., 2003. Polymer nanofibers assembled by electrospinning. Curr. Opin. Colloid Interface Sci. 8, 64–75.
Frenot, A., Henriksson, M.W., Walkenström, P., 2007. Electrospinning of cellulose-based nanofibers. J. Appl. Polym. Sci. 103, 1473–1482.
Fung, W.-Y., Yuen, K.-H., Liong, M.-T., 2011. Agrowaste-based nanofibers as a probiotic encapsulant: fabrication and characterization. J. Agric. Food Chem. 59, 8140–8147.
Gañán-Calvo, A.M., 1999. The surface charge in electrospraying: its nature and its universal scaling laws. J. Aerosol Sci. 30, 863–872.
Ge, L., Zhao, Y.-S., Mo, T., Li, J.-R., Li, P., 2012. Immobilization of glucose oxidase in electrospun nanofibrous membranes for food preservation. Food Control 26, 188–193.
Ghorani, B., 2012. Production and Properties of Electrospun Webs for Therapeutic Applications. University of Leeds (School of Design), Leeds.
Ghorani, B., Hebden, A., Russell, S., Goswami, P., 2015a. Single step assembly of biomolecule-loaded sub-micron polysulphone fibres. Text. Res. J. doi: 10.1177/0040517516629148.
Ghorani, B., Russell, S.J., Goswami, P., 2013. Controlled morphology and mechanical characterisation of electrospun cellulose acetate fibre webs. Int. J. Polym. Sci. 2013, 12.
Ghorani, B., Tucker, N., 2015. Fundamentals of electrospinning as a novel delivery vehicle for bioactive compounds in food nanotechnology. Food Hydrocoll. 51, 227–240.
Ghorani, B., Tucker, N., Yoshikawa, M., 2015b. Approaches for the assembly of molecularly imprinted electrospun nanofibre membranes and consequent use in selected target recognition. Food Res. Int. 78, 448–464.
Ghorbanzade, T., Jafari, S.M., Akhavan, S., Hadavi, R., 2017. Nano-encapsulation of fish oil in nano-liposomes and its application in fortification of yogurt. Food Chemistry 216, 146–152.
Greiner, A., Wendorff, J.H., 2007. Electrospinning: a fascinating method for the preparation of ultrathin fibers. Angew. Chem. Int. Ed. Engl. 46, 5670–5703.
Grigoriev, D., Edirisinghe, M., 2002. Evaporation of liquid during cone-jet mode electrospraying. J. Appl. Phys. 91, 437–439.
Grimm, R.L., Beauchamp, J.L., 2005. Dynamics of field-induced droplet ionization: time-resolved studies of distortion, jetting, and progeny formation from charged and neutral methanol droplets exposed to strong electric fields. J. Phys. Chem. B 109, 8244–8250.
Gupta, P., Elkins, C., Long, T.E., Wilkes, G.L., 2005. Electrospinning of linear homopolymers of poly(methyl methacrylate): exploring relationships between fiber formation, viscosity, molecular weight and concentration in a good solvent. Polymer 46, 4799–4810.
Hadizadeh, F., Zakerian, A., Mohajeri, S., 2013. Non-covalently lactose imprinted polymers and recognition of saccharides in aqueous solutions. J. Iran. Chem. Soc. 10, 207–212.
Han, W., Minhao, L., Xin, C., Junwei, Z., Xindu, C., Ziming, Z., 2015. Study of deposition characteristics of multi-spinneret near-field electrospinning in electric field crossover interference conditions. AIP Adv. 5, 041302.
Hao, S., Wang, Y., Wang, B., Deng, J., Liu, X., Liu, J., 2013. Rapid preparation of pH-sensitive polymeric nanoparticle with high loading capacity using electrospray for oral drug delivery. Mater. Sci. Eng. C 33, 4562–4567.
Hartman, R., Brunner, D., Camelot, D., Marijnissen, J., Scarlett, B., 2000. Jet break-up in electrohydrodynamic atomization in the cone-jet mode. J. Aerosol Sci. 31, 65–95.

Heikkila, P., Harlin, A., 2008. Parameter study of electrospinning of polyamide-6. Eur. Polym. J. 44, 3067–3079.

Heunis, T., Bshena, O., Klumperman, B., Dicks, L., 2011. Release of bacteriocins from nanofibers prepared with combinations of poly(D,L-lactide) (PDLLA) and poly(ethylene oxide) (PEO). Int. J. Mol. Sci. 12, 2158–2173.

Huang, X.-J., Chen, P.-C., Huang, F., Ou, Y., Chen, M.-R., Xu, Z.-K., 2011. Immobilization of *Candida rugosa* lipase on electrospun cellulose nanofiber membrane. J. Mol. Catalysis B 70, 95–100.

Huang, X.-J., Ge, D., Xu, Z.-K., 2007. Preparation and characterization of stable chitosan nanofibrous membrane for lipase immobilization. Eur. Polym. J. 43, 3710–3718.

Huang, C.C., Lin, C.K., Lu, C.T., Lou, C.W., Chao, C.Y., Lin, J.H., 2009. Evaluation of the electrospinning manufacturing process based on the preparation of PVA composite fibres. Fibres. Text East. Eur. 74 (3), 34–37.

Huang, C., Soenen, S.J., Van Gulck, E., Vanham, G., Rejman, J., Van Calenbergh, S., Vervaet, C., Coenye, T., Verstraelen, H., Temmerman, M., 2012. Electrospun cellulose acetate phthalate fibers for semen induced anti-HIV vaginal drug delivery. Biomaterials 33, 962–969.

Huang, Z.-M., Zhang, Y.Z., Kotaki, M., Ramakrishna, S., 2003. A review on polymer nanofibers by electrospinning and their applications in nanocomposites. Compos. Sci. Technol. 63, 2223–2253.

Huang, Z.-M., Zhang, Y., Ramakrishna, S., Lim, C., 2004. Electrospinning and mechanical characterization of gelatin nanofibers. Polymer 45, 5361–5368.

Jafari, S.M., Fathi, M., Mandala, I., 2015. Emerging product formation. Food Waste Recovery: Processing Technologies and Industrial Techniques. Elsevier Inc., Academic Press, San Diego, CA, pp. 293–317.

Jain, E., Scott, K.M., Zustiak, S.P., Sell, S.A., 2015. Fabrication of Polyethylene Glycol-Based Hydrogel Microspheres Through Electrospraying. Macromolecular Materials and Engineering 300, 823–835.

Jain, A.K., Sood, V., Bora, M., Vasita, R., Katti, D.S., 2014. Electrosprayed inulin microparticles for microbiota triggered targeting of colon. Carbohydr. Polym. 112, 225–234.

Jalili, R., Hosseini, S.A., Morshed, M., 2005. The effects of operating parameters on the morphology of electrospun polyacrilonitrile nanofibres. Iran. Polym. J. 14, 1074–1081.

Jarusuwannapoom, T., Hongrojjanawiwat, W., Jitjaicham, S., Wannatong, L., Nithitanakul, M., Pattamaprom, C., Koombhongse, P., Rangkupan, R., Supaphol, P., 2005. Effect of solvents on electro-spinnability of polystyrene solutions and morphological appearance of resulting electrospun polystyrene fibers. Eur. Polym. J. 41, 409–421.

Jaworek, A., 2007. Micro- and nanoparticle production by electrospraying. Powder Technol. 176, 18–35.

Jaworek, A., Krupa, A., 1999. Classification of the modes of EHD spraying. J. Aerosol Sci. 30, 873–893.

Jaworek, A., Krupa, A., Lackowski, M., Sobczyk, A., Czech, T., Ramakrishna, S., Sundarrajan, S., Pliszka, D., 2009. Nanocomposite fabric formation by electrospinning and electrospraying technologies. J. Electrostat. 67, 435–438.

Jaworek, A., Sobczyk, A.T., 2008. Electrospraying route to nanotechnology: an overview. J. Electrostat. 66, 197–219.

Jia, H., Zhu, G., Vugrinovich, B., Kataphinan, W., Reneker, D.H., Wang, P., 2002. Enzyme-carrying polymeric nanofibers prepared via electrospinning for use as unique biocatalysts. Biotechnol. Prog. 18, 1027–1032.

Jiang, H., Hu, Y., Li, Y., Zhao, P., Zhu, K., Chen, W., 2005. A facile technique to prepare biodegradable co-axial electrospun nanofibers for controlled release of bioactive agents. J. Control. Release 108, 237–243.

Jiang, H., Hu, Y., Zhao, P., Li, Y., Zhu, K., 2006. Modulation of protein release from biodegradable core–shell structured fibers prepared by co-axial electrospinning. J. Biomed. Mater. Res. B 79, 50–57.

Jiang, H., Wang, L., Zhu, K., 2014. Co-axial electrospinning for encapsulation and controlled release of fragile water-soluble bioactive agents. J. Control. Release 193, 296–303.

Jones, O.G., McClements, D.J., 2011. Recent progress in bio-polymer nanoparticle and microparticle formation by heat-treating electrostatic protein–polysaccharide complexes. Adv. Colloid Interface Sci. 167, 49–62.

Karim, M.R., Lee, H.W., Kim, R., Ji, B.C., Cho, J.W., Son, T.W., Oh, W., Yeum, J.H., 2009. Preparation and characterization of electrospun pullulan/montmorillonite nanofiber mats in aqueous solution. Carbohydr. Polym. 78, 336–342.

Katouzian, I., Jafari, S.M., 2016. Nano-encapsulation as a promising approach for targeted delivery and controlled release of vitamins. Trends Food Sci. Technol. 53, 34–48.

Kayaci, F., Ertas, Y., Uyar, T., 2013. Enhanced thermal stability of eugenol by cyclodextrin inclusion complex encapsulated in electrospun polymeric nanofibers. J. Agric. Food Chem. 61, 8156–8165.

Kayaci, F., Uyar, T., 2012. Encapsulation of vanillin/cyclodextrin inclusion complex in electrospun polyvinyl alcohol (PVA) nanowebs: prolonged shelf-life and high temperature stability of vanillin. Food Chem. 133, 641–649.

Kessick, R., Tepper, G., 2004. Microscale polymeric helical structures produced by electrospinning. Appl. Phys. Lett. 84, 4807–4809.

Keyes, M.H., Saraswathi, S., 1985. Immobilized Enzymes, Bioactive Polymeric Systems. Springer, US.

Khadka, D.B., Cross, M.C., Haynie, D.T., 2011. Asynthetic polypeptide electrospun biomaterial. ACS Appl. Mater. Interfaces 3, 2994–3001.

Khan, M.K.I., Schutyser, M.A.I., Schroën, K., Boom, R., 2012. The potential of electrospraying for hydrophobic film coating on foods. J. Food Eng. 108, 410–416.

Khorshidi, S., Solouk, A., Mirzadeh, H., Mazinani, S., Lagaron, J.M., Sharifi, S., Ramakrishna, S., 2015. A review of key challenges of electrospun scaffolds for tissue-engineering applications. J. Tiss. Eng. Regen. Med. 10, 715–738.

Kim, W.J., Chang, J.Y., 2011. Molecularly imprinted polyimide nanofibers prepared by electrospinning. Mater. Lett. 65, 1388–1391.

Kondo, Y., Yoshikawa, M., Okushita, H., 2000. Molecularly imprinted polyamide membranes for chiral recognition. Polym. Bull. 44, 517–524.

Kong, L., Ziegler, G.R., 2012. Role of molecular entanglements in starch fiber formation by electrospinning. Biomacromolecules 13, 2247–2253.

Koombhongse, S., Liu, W., Reneker, D.H., 2001. Flat polymer ribbons and other shapes by electrospinning. J. Polym. Sci. B 39, 2598–2606.

Korehei, R., Kadla, J.F., 2014. Encapsulation of T4 bacteriophage in electrospun poly(ethylene oxide)/cellulose diacetate fibers. Carbohydr. Polym. 100, 150–157.

Koski, A., Yim, K., Shivkumar, S., 2004. Effect of molecular weight on fibrous PVA produced by electrospinning. Mater. Lett. 58, 493–497.

Kriegel, C., Arecchi, A., Kit, K., McClements, D.J., Weiss, J., 2008. Fabrication, functionalization, and application of electrospun bio-polymer nanofibers. Crit. Rev. Food Sci. Nutr. 48, 775–797.

Kriegel, C., Kit, K., McClements, D.J., Weiss, J., 2009. Electrospinning of chitosan–poly(ethylene oxide) blend nanofibers in the presence of micellar surfactant solutions. Polymer 50, 189–200.

Kruncinska, I., Komisarczyk, A., Chrzanowski, M., Gliścińska, E., Wrzosek, H., 2009. Electrostatic field in electrospinning with a multicapillary head—modelling and experiment. Fibers Text. East. Eur. 74, 38–44.

Kuan, C.-Y., Yuen, K.-H., Bhat, R., Liong, M.-T., 2011. Physicochemical characterization of alkali treated fractions from corncob and wheat straw and the production of nanofibres. Food Res. Int. 44, 2822–2829.

Lee, K.H., Kim, H.Y., Bang, H.J., Jung, Y.H., Lee, S.G., 2003. The change of bead morphology formed on electrospun polystyrene fibers. Polymer 44, 4029–4034.

Lee, K.H., Kim, D.J., Min, B.G., Lee, S.H., 2007. Polymeric nanofiber web-based artificial renal microfluidic chip. Biomed. Microdevices 9, 435–442.

Li, Y., Lim, L.T., Kakuda, Y., 2009. Electrospun zein fibers as carriers to stabilize (−)-epigallocatechin gallate. J. Food Sci. 74, C233–C240.

Li, F., Song, Y., Zhao, Y., 2010. Core-Shell Nanofibers: Nano Channel and Capsule by Coaxial Electrospinning. INTECH Open Access Publisher.

Li, S., Sun, B., Li, X., Yuan, X., 2008. Characterization of electrospun core/shell poly(vinyl pyrrolidone)/poly (L-lactide-*co*-ε-caprolactone) fibrous membranes and their cytocompatibility in vitro. J. Biomater. Sci. Polym. Ed. 19, 245–258.

Li, D., Xia, Y., 2004. Direct fabrication of composite and ceramic hollow nanofibers by electrospinning. Nano Lett. 4, 933–938.

Liechty, W.B., Kryscio, D.R., Slaughter, B.V., Peppas, N.A., 2010. Polymers for drug delivery systems. Annu. Rev. Chem. Biomol. Eng. 1, 149–173.

Lim, Y.-M., Gwon, H.-J., Jeun, J.P., Nho, Y.-C., 2010. Preparation of Cellulose-Based Nanofibers Using Electrospinning. INTECH Open Access Publisher.

Lim, C.T., Tan, E.P.S., Ng, S.Y., 2008. Effects of crystalline morphology on the tensile properties of electrospun polymer nanofibers. Appl. Phys. Lett. 92, 141908.

Lin, T., Wang, H., Wang, H., Wang, X., 2004. The charge effect of cationic surfactants on the elimination of fibre beads in the electrospinning of polystyrene. Nanotechnology 15, 1375–1381.

Lin, T., Wang, H., Wang, H., Wang, X., 2005. Effects of polymer concentration and cationic surfactant on the morphology of electrospun polyacrylonitrile nanofibres. J. Mater. Sci. Technol. 21, 9–12.

Lin, Y., Yao, Y., Yang, X., Wei, N., Li, X., Gong, P., Li, R., Wu, D., 2008. Preparation of poly(ether sulfone) nanofibers by gas-jet/electrospinning. J. Appl. Polym. Sci. 107, 909–917.

Linnemann, B., Rana, A., Gries, T., 2005. Electrospinning: nanofiber from polycaprolactone (PCL). Chem. Fibres Int. 55, 370–372.

Liu, Y., Zhang, X., Liu, H., Yu, T., Deng, J., 1996. Immobilization of glucose oxidase onto the blend membrane of poly(vinyl alcohol) and regenerated silk fibroin: morphology and application to glucose biosensor. J. Biotechnol. 46, 131–138.

Livney, Y.D., 2008. Complexes and conjugates of bio-polymers for delivery of bioactive ingredients via food. In: Garti, N. (Ed.), Delivery and Controlled Release of Bioactives in Foods and Nutraceuticals. Woodhead Publishing, Cambridge, UK, (Chapter 9).

Locatelli, F., Gamez, P., Lemaire, M., 1998. Molecular imprinting of polymerised catalytic complexes in asymmetric catalysis. J. Mol. Catal. A 135 (1), 89–98.

Lok, C.M., Son, R., 2009. Application of molecularly imprinted polymers in food sample analysis—a perspective. Int. Food Res. J. 16, 127–140.

Long, Y., Chen, H., Wang, H., Peng, Z., Yang, Y., Zhang, G., Li, N., Liu, F., Pei, J., 2012. Highly sensitive detection of nitroaromatic explosives using an electrospun nanofibrous sensor based on a novel fluorescent conjugated polymer. Anal. Chim. Acta 744, 82–91.

López-Rubio, A., Lagaron, J.M., 2012. Whey protein capsules obtained through electrospraying for the encapsulation of bioactives. Innov. Food Sci. Emerg. Technol. 13, 200–206.

López-Rubio, A., Sanchez, E., Sanz, Y., Lagaron, J.M., 2009. Encapsulation of living bifidobacteria in ultrathin PVOH electrospun fibers. Biomacromolecules 10, 2823–2829.

López-Rubio, A., Sanchez, E., Wilkanowicz, S., Sanz, Y., Lagaron, J.M., 2012. Electrospinning as a useful technique for the encapsulation of living bifidobacteria in food hydrocolloids. Food Hydrocoll. 28, 159–167.

Loscertales, I.G., Barrero, A., Guerrero, I., Cortijo, R., Marquez, M., Ganan-Calvo, A.M., 2002. Micro/nano encapsulation via electrified co-axial liquid jets. Science 295, 1695–1698.

Lu, C., Chen, P., Li, J., Zhang, Y., 2006. Computer simulation of electrospinning. Part I. Effect of solvent in electrospinning. Polymer 47, 915–921.

Lubambo, A.F., de Freitas, R.A., Sierakowski, M.-R., Lucyszyn, N., Sassaki, G.L., Serafim, B.M., Saul, C.K., 2013. Electrospinning of commercial guar-gum: effects of purification and filtration. Carbohydr. Polym. 93, 484–491.

Luo, C.J., Loh, S., Stride, E., Edirisinghe, M., 2012. Electrospraying and electrospinning of chocolate suspensions. Food Bioprocess Technol. 5, 2285–2300.

Manasco, J.L., Saquing, C.D., Tang, C., Khan, S.A., 2012. Cyclodextrin fibers via polymer-free electrospinning. RSC Adv. 2, 3778–3784.

Manee-in, J., Nithitanakul, M., Supaphol, P., 2006. Effects of solvent properties, solvent system, electrostatic field strength, and inorganic salt addition on electrospun polystyrene fibres. Iran. Polym. J. 15, 341–354.

Mascheroni, E., Fuenmayor, C.A., Cosio, M.S., di Silvestro, G., Piergiovanni, L., Mannino, S., Schiraldi, A., 2013. Encapsulation of volatiles in nanofibrous polysaccharide membranes for humidity-triggered release. Carbohydr. Polym. 98, 17–25.

Mazoochi, T., Jabbari, V., 2011. Chitosan nanofibrous scaffold fabricated via electrospinning: the effect of processing parameters on the nanofiber morphology. Int. J. Polym. Anal. Charac. 16, 277–289.

McClements, D.J., Li, Y., 2010. Structured emulsion-based delivery systems: controlling the digestion and release of lipophilic food components. Adv. Colloid Interface Sci. 159, 213–228.

McKee, M.G., Layman, J.M., Cashion, M.P., Long, T.E., 2006. Phospholipid nonwoven electrospun membranes. Science 311, 353–355.

Megelski, S., Stephens, J.S., Chase, D.B., Rabolt, J.F., 2002. Micro- and nanostructured surface morphology on electrospun polymer fibers. Macromolecules 35, 8456–8466.

Mendes, A.C., Nikogeorgos, N., Lee, S., Chronakis, I.S., 2015. Nanomechanics of electrospun phospholipid fiber. Appl. Phys. Lett. 106, 223108.

Mo, X.M., Xu, C.Y., Kotaki, M., Ramakrishna, S., 2004. Electrospun P(LLA-CL) nanofiber: a biomimetic extracellular matrix for smooth muscle cell and endothelial cell proliferation. Biomaterials 25, 1883–1890.

Mohajeri, S.A., Hosseinzadeh, H., Keyhanfar, F., Aghamohammadian, J., 2010. Extraction of crocin from saffron (*Crocus sativus*) using molecularly imprinted polymer solid-phase extraction. J. Sep. Sci. 33, 2302–2309.

Mohr, J.G., Rowe, W.P., 1978. Fiber Glass. Van Nostrand Reinhold, New York, NY.

Moomand, K., 2014. Physicochemical Aspects of Ultrathin Electrospun Zein Fibres for Encapsulation of Fish Oil. The University of Guelph, Guelph.

Moomand, K., Lim, L.-T., 2015. Effects of solvent and $n-3$ rich fish oil on physicochemical properties of electrospun zein fibres. Food Hydrocoll. 46, 191–200.

Munir, M.M., Suryamas, A.B., Iskandar, F., Okuyama, K., 2009. Scaling law on particle-to-fiber formation during electrospinning. Polymer 50, 4935–4943.

Nedovic, V., Kalusevic, A., Manojlovic, V., Levic, S., Bugarski, B., 2011. An overview of encapsulation technologies for food applications. Procedia Food Sci. 1, 1806–1815.

Neo, Y.P., 2014. Electrospinning as a Novel Encapsulation Method for Food Applications. ResearchSpace@ Auckland.

Neo, Y.P., Ray, S., Easteal, A.J., Nikolaidis, M.G., Quek, S.Y., 2012. Influence of solution and processing parameters towards the fabrication of electrospun zein fibers with sub-micron diameter. J. Food Eng. 109, 645–651.

Neo, Y.P., Ray, S., Jin, J., Gizdavic-Nikolaidis, M., Nieuwoudt, M.K., Liu, D., Quek, S.Y., 2013. Encapsulation of food grade antioxidant in natural bio-polymer by electrospinning technique: a physicochemical study based on zein–gallic acid system. Food Chem. 136, 1013–1021.

Nieuwland, M., Geerdink, P., Brier, P., Van Den Eijnden, P., Henket, J.T.M.M., Langelaan, M.L.P., Stroeks, N., Van Deventer, H.C., Martin, A.H., 2013. Food-grade electrospinning of proteins. Innov. Food Sci. Emerg. Technol. 20, 269–275.

Ogata, S., Hara, Y., Shinohara, H., 1978. Break-up mechanism of a charged liquid jet. Int. Chem. Eng. 18, 482–488.

Olaru, N., Olaru, L., 2010. Electrospinning of cellulose acetate phthalate from different solvent systems. Ind. Eng. Chem. Res. 49, 1953–1957.

Pablo, L.G.T., Tucker, N., Simmons, L., Watson, A.J.A., Fairbanks, A.J., Staiger, M.P., 2015. Sub-micron sized saccharide fibres via electrospinning. Electrospinning 1, 1–9.

Park, H.R., Yoon, S.D., Bang, E.Y., Rogers, K.R., Chough, S.H., 2005. Molecular imprinting polymers for the separation of toluic acid isomers. J. Appl. Polym. Sci. 96, 650–654.

Pérez-Masiá, R., Lagaron, J., López-Rubio, A., 2014a. Development and optimization of novel encapsulation structures of interest in functional foods through electrospraying. Food Bioprocess Technol. 7, 3236–3245.

Pérez-Masiá, R., Lagaron, J.M., López-Rubio, A., 2014b. Surfactant-aided electrospraying of low molecular weight carbohydrate polymers from aqueous solutions. Carbohydr. Polym. 101, 249–255.

Petrik, S., Maly, M., 2009. Production spinneret-less electrospinning nanofiber technology. MRS Online Proceedings Library Archive 1240, 1240-WW03-07 (12 pages).

Pham, Q.P., Sharma, U., Mikos, A.G., 2006. Electrospinning of polymeric nanofibers for tissue engineering applications: a review. Tiss. Eng. 12, 1197–1211.

Piperno, S., Tse Sum Bui, B., Haupt, K., Gheber, L.A., 2011. Immobilization of molecularly imprinted polymer nanoparticles in electrospun poly(vinyl alcohol) nanofibers. Langmuir 27, 1547–1550.

Pitkowski, A., Durand, D., Nicolai, T., 2008. Structure and dynamical mechanical properties of suspensions of sodium caseinate. J. Colloid Interface Sci. 326, 96–102.

Ramakrishna, S., 2005. An Introduction to Electrospinning and Nanofibers. World Scientific, Hackensack, NJ.

Ranjbar-Mohammadi, M., Zamani, M., Prabhakaran, M., Bahrami, S.H., Ramakrishna, S., 2016. Electrospinning of PLGA/gum tragacanth nanofibers containing tetracycline hydrochloride for periodontal regeneration. Mater. Sci. Eng. C 58, 521–531.

Reneker, D.H., Yarin, A.L., Fong, H., Koombhongse, S., 2000. Bending instability of electrically charged liquid jets of polymer solutions in electrospinning. J. Appl. Phys. 87, 4531–4547.

Rezaei, A., Nasirpour, A., Fathi, M., 2015. Application of cellulosic nanofibers in food science using electrospinning and its potential risk. Compr. Rev. Food Sci. Food Saf. 14, 269–284.

Rutledge, G.C., Li, Y., Fridrikh, S., Warner, S.B., Kalayci, V.E., Patra, P., 2001. Electrostatic spinning and properties of ultrafine fibers. National Textile Center, Annual Report.

Sahay, R., Thavasi, V., Ramakrishna, S., 2011. Design modifications in electrospinning setup for advanced applications. J. Nanomater. 2011, 17.

Sahlström, S., Bråthen, E., 1997. Effects of enzyme preparations for baking, mixing time and resting time on bread quality and bread staling. Food Chem. 58, 75–80.
Salalha, W., Kuhn, J., Dror, Y., Zussman, E., 2006. Encapsulation of bacteria and viruses in electrospun nanofibres. Nanotechnology 17, 4675.
Scampicchio, M., Mannino, S., Cosio, M.S., 2015. Development of food nanostructures by electrospinning. Springer, US.
Schiffman, J.D., Schauer, C.L., 2008. A review: electrospinning of bio-polymer nanofibers and their applications. Polym. Rev. 48, 317–352.
Sellergren, B., 1989. Molecular imprinting by noncovalent interactions. Enantioselectivity and binding capacity of polymers prepared under conditions favoring the formation of template complexes. Die Makromolekulare Chemie 190, 2703–2711.
Sellergren, B., Shea, K.J., 1993. Influence of polymer morphology on the ability of imprinted network polymers to resolve enantiomers. J. Chromatogr. A 635, 31–49.
Shenoy, S.L., Bates, W.D., Frisch, H.L., Wnek, G.E., 2005a. Role of chain entanglements on fiber formation during electrospinning of polymer solutions: good solvent, non-specific polymer-polymer interaction limit. Polymer 46, 3372–3384.
Shenoy, S.L., Bates, W.D., Wnek, G., 2005b. Correlations between electrospinnability and physical gelation. Polymer 46, 8990–9004.
Shin, Y., Hohman, M., Brenner, M., Rutledge, G., 2001. Electrospinning: a whipping fluid jet generates submicron polymer fibers. Appl. Phys. Lett. 78, 1149–1151.
Saini, S.S., Kaur, A., 2013. Molecularly imprinted polymers for the detection of food toxins: a mini review. Adv. Nanopart. 2, 60–65.
Simm, W., 1976. Apparatus for the production of filters by electrostatic fiber spinning. Google Patents.
Smith, K., Alexander, M., Stark, J., 2006. Voltage effects on the volumetric flow rate in cone-jet mode electrospraying. J. Appl. Phys. 99, 064909.
Song, X., Xu, S., Chen, L., Wei, Y., Xiong, H., 2014. Recent advances in molecularly imprinted polymers in food analysis. J. Appl. Polym. Sci. 131, 40766.
Songchotikunpan, P., Tattiyakul, J., Supaphol, P., 2008. Extraction and electrospinning of gelatin from fish skin. Int. J. Biol. Macromol. 42, 247–255.
Sousa, A.M.M., Souza, H.K.S., Uknalis, J., Liu, S.-C., Gonçalves, M.P., Liu, L., 2015. Electrospinning of agar/PVA aqueous solutions and its relation with rheological properties. Carbohydr. Polym. 115, 348–355.
Spasova, M., Mincheva, R., Paneva, D., Manolova, N., Rashkov, I., 2006. Perspectives on: criteria for complex evaluation of the morphology and alignment of electrospun polymer nanofibers. J. Bioact. Compat. Polym. 21, 465–479.
Stanger, J., Tucker, N., Kirwan, K., Staiger, M.P., 2009. Effect of charge density on the Taylor cone in electrospinning. Int. J. Modern Phys. B 23, 1956–1961.
Stijnman, A.C., Bodnar, I., Hans Tromp, R., 2011. Electrospinning of food-grade polysaccharides. Food Hydrocoll. 25, 1393–1398.
Subramanian, C., Weiss, R.A., Shaw, M.T., 2010. Electrospinning and characterization of highly sulfonated polystyrene fibers. Polymer 51, 1983–1989.
Sueyoshi, Y., Fukushima, C., Yoshikawa, M., 2010. Molecularly imprinted nanofiber membranes from cellulose acetate aimed for chiral separation. J. Membr. Sci. 357, 90–97.
Sullivan, S.T., Tang, C., Kennedy, A., Talwar, S., Khan, S.A., 2014. Electrospinning and heat treatment of whey protein nanofibers. Food Hydrocoll. 35, 36–50.
Sun, X.-Z., Williams, G.R., Hou, X.-X., Zhu, L.-M., 2013. Electrospun curcumin-loaded fibers with potential biomedical applications. Carbohydr. Polym. 94, 147–153.
Sun, C., Zhang, D., Liu, Y., Xiao, J., 2004. Bicomponent meltblown nonwovens and fibre splitting. J. Ind. Text. 34, 17–26.

Taepaiboon, P., Rungsardthong, U., Supaphol, P., 2007. Vitamin-loaded electrospun cellulose acetate nanofiber mats as transdermal and dermal therapeutic agents of vitamin A acid and vitamin E. Eur. J. Pharm. Biopharm. 67, 387–397.

Tan, S.H., Inai, R., Kotaki, M., Ramakrishna, S., 2005. Systematic parameter study for ultra-fine fiber fabrication via electrospinning process. Polymer 46, 6128–6134.

Taylor, G., 1964. Disintegration of water drops in an electric field. Proc. R. Soc. Lond. A 280, 383–397.

Tehrani, A.H., Zadhoush, A., Karbasi, S., Khorasani, S.N., 2010. Experimental investigation of the governing parameters in the electrospinning of poly(3-hydroxybutyrate) scaffolds: structural characteristics of the pores. J. Appl. Polym. Sci. 118, 2682–2689.

Teo, W.E., Ramakrishna, S., 2006. A review on electrospinning design and nanofibre assemblies. Nanotechnology 17, 89–106.

Theron, S.A., Zussman, E., Yarin, A.L., 2004. Experimental investigation of the governing parameters in the electrospinning of polymer solutions. Polymer 45, 2017–2030.

Thompson, C.J., Chase, G.G., Yarin, A.L., Reneker, D.H., 2007. Effects of parameters on nano-fiber diameter determined from electrospinning model. Polymer 48, 6913–6922.

Tomita, Y., Ishibashi, Y., Yokoyama, T., 1986. Fundamental studies on an electrostatic ink jet printer: 1st report, electrostatic drop formation. Bull. JSME 29, 3737–3743.

Torres-Giner, S., Gimenez, E., Lagarón, J.M., 2008. Characterization of the morphology and thermal properties of zein prolamine nanostructures obtained by electrospinning. Food Hydrocoll. 22, 601–614.

Torres-Giner, S., Martinez-Abad, A., Ocio, M.J., Lagaron, J.M., 2010. Stabilization of a nutraceutical omega-3 fatty acid by encapsulationin ultrathin electrosprayed zein prolamine. J. Food Sci. 75, N69–N79.

Tragardh, G., 1991. Membrane applications in the food industry. Polym. J. 23, 521–529.

Treloar, L.R.G., 1970. Introduction to Polymer Science. Wykeham Publications, Springer-Verlag, London, New York.

Tucker, N., 2015. The development of electrospinning technologies for commercial application. The Royal Society of Chemistry, UK, (Chapter 3).

Tucker, N., Ghorani, B., 2015. Delivery of nano-tech food ingredients. Food Sci. Technol. 29 (4), 24–26.

Tucker, N., Stanger, J.J., Staiger, M.P., Razzaq, H., Hofman, K., 2012. The history of the science and technology of electrospinning from 1600 to 1995. J. Eng. Fabrics Fibers 7, 63–73.

Uyar, T., Besenbacher, F., 2008. Electrospinning of uniform polystyrene fibers: the effect of solvent conductivity. Polymer 49, 5336–5343.

Uyar, T., Hacaloglu, J., Besenbacher, F., 2009. Electrospun polystyrene fibers containing high temperature stable volatile fragrance/flavor facilitated by cyclodextrin inclusion complexes. React. Funct. Polym. 69, 145–150.

Van der Schueren, L., de Meyer, T., Steyaert, I., Ceylan, Ö., Hemelsoet, K., Van Speybroeck, V., de Clerck, K., 2013. Polycaprolactone and polycaprolactone/chitosan nanofibres functionalised with the pH-sensitive dye Nitrazine Yellow. Carbohydr. Polym. 91, 284–293.

Van der Schueren, L., de Schoenmaker, B., Kalaoglu, Ö.I., de Clerck, K., 2011. An alternative solvent system for the steady state electrospinning of polycaprolactone. Eur. Polym. J. 47, 1256–1263.

Van der Schueren, L., Mollet, T., Ceylan, Ö., de Clerck, K., 2010. The development of polyamide 6.6 nanofibres with a pH-sensitive function by electrospinning. Eur. Polym. J. 46, 2229–2239.

Varesano, A., Rombaldoni, F., Mazzuchetti, G., Tonin, C., Comotto, R., 2010. Multi-jet spinneret electrospinning on textile substrates: observations on process and nanofibre mat deposition. Polym. Int. 59, 1606–1615.

Vega Lugo, A.C., 2012. Understanding the role of poly(ethylene oxide) in the electrospinning of whey protein isolate fibers.

Vega-Lugo, A.C., Lim, L.T., 2012. Effects of poly(ethylene oxide) and pH on the electrospinning of whey protein isolate. J. Polym. Sci. B 50, 1188–1197.

Verdugo, M., Lim, L.-T., Rubilar, M., 2014. Electrospun protein concentrate fibers from microalgae residual biomass. J. Polym. Environ. 22, 373–383.

Walczak, Z.K., 2002. Processes of Fibre Formation. Elsevier, Oxford, United Kingdom.

Wang, Z.-G., Wan, L.-S., Liu, Z.-M., Huang, X.-J., Xu, Z.-K., 2009. Enzyme immobilization on electrospun polymer nanofibers: an overview. J. Mol. Catalysis B 56, 189–195.

Wang, M., Wang, L., Huang, Y., 2007. Electrospun hydroxypropyl methyl cellulose phthalate (HPMCP)/erythromycin fibers for targeted release in intestine. J. Appl. Polym. Sci. 106, 2177–2184.

Wang, Z.-G., Wang, J.-Q., Xu, Z.-K., 2006. Immobilization of lipase from *Candida rugosa* on electrospun polysulfone nanofibrous membranes by adsorption. J. Mol. Catalysis B 42, 45–51.

Wannatong, L., Sirivat, A., Supaphol, P., 2004. Effects of solvents on electrospun polymeric fibers: preliminary study on polystyrene. Polym. Int. 53, 1851–1859.

Woerdeman, D.L., Ye, P., Shenoy, S., Parnas, R.S., Wnek, G.E., Trofimova, O., 2005. Electrospun fibers from wheat protein: investigation of the interplay between molecular structure and the fluid dynamics of the electrospinning process. Biomacromolecules 6, 707–712.

Wongsasulak, S., Kit, K.M., McClements, D.J., Yoovidhya, T., Weiss, J., 2007. The effect of solution properties on the morphology of ultrafine electrospun egg albumen–PEO composite fibers. Polymer 48, 448–457.

Wongsasulak, S., Patapeejumruswong, M., Weiss, J., Supaphol, P., Yoovidhya, T., 2010. Electrospinning of food-grade nanofibers from cellulose acetate and egg albumen blends. J. Food Eng. 98, 370–376.

Wongsasulak, S., Pathumban, S., Yoovidhya, T., 2014. Effect of entrapped α-tocopherol on mucoadhesivity and evaluation of the release, degradation, and swelling characteristics of zein–chitosan composite electrospun fibers. J. Food Eng. 120, 110–117.

Wongsasulak, S., Puttipaiboon, N., Yoovidhya, T., 2013. Fabrication, gastromucoadhesivity, swelling, and degradation of zein–chitosan composite ultrafine fibers. J. Food Sci. 78, N926–N935.

Wu, L., Yuan, X., Sheng, J., 2005. Immobilization of cellulase in nanofibrous PVA membranes by electrospinning. J. Membr. Sci. 250, 167–173.

Xie, J., Hsieh, Y.-L., 2003. Ultra-high surface fibrous membranes from electrospinning of natural proteins: casein and lipase enzyme. J. Mater. Sci. 38, 2125–2133.

Xu, X., Jiang, L., Zhou, Z., Wu, X., Wang, Y., 2012. Preparation and properties of electrospun soy protein isolate/polyethylene oxide nanofiber membranes. ACS Appl. Mater. Interfaces 4, 4331–4337.

Yamashita, Y., Tanaka, A., Ko, F., 2006. Electrospinning for the industrial nanofiber technology. In: Proceedings of International Fiber Conference, Seoul, Korea.

Yang, Q., Li, Z., Hong, Y., Zhao, Y., Qiu, S., Wang, C., Wei, Y., 2004. Influence of solvents on the formation of ultrathin uniform poly(vinyl pyrrolidone) nanofibers with electrospinning. J. Polym. Sci. B 42, 3721–3726.

Yang, Y., Li, X., Qi, M., Zhou, S., Weng, J., 2008. Release pattern and structural integrity of lysozyme encapsulated in core-sheath structured poly(DL-lactide) ultrafine fibers prepared by emulsion electrospinning. Eur. J. Pharm. Biopharm. 69, 106–116.

Yang, J.-M., Zha, L.-S., Yu, D.-G., Liu, J., 2013. Co-axial electrospinning with acetic acid for preparing ferulic acid/zein composite fibers with improved drug release profiles. Colloids Surf. B 102, 737–743.

Yarin, A., Koombhongse, S., Reneker, D., 2001. Bending instability in electrospinning of nanofibers. J. Appl. Phys. 89, 3018–3026.

Ye, L., 2013. Molecular Imprinting: Principles and Applications of Micro- and Nanostructure Polymers. Pan Stanford Publishing, Singapore.

Ye, P., Xu, Z.-K., Wu, J., Innocent, C., Seta, P., 2006. Nanofibrous poly(acrylonitrile-*co*-maleic acid) membranes functionalized with gelatin and chitosan for lipase immobilization. Biomaterials 27, 4169–4176.

Yilmaz, E., Mosbach, K., Haupt, K., 1999. Influence of functional and cross-linking monomers and the amount of template on the performance of molecularly imprinted polymers in binding assays. Anal. Commun. 36, 167–170.

Ying, Y., Zhidong, J., Qiang, L., Zhicheng, G., 2006. Experimental investigation of the governing parameters in the electrospinning of polyethylene oxide solution. IEEE Trans. Dielect. Elect. Insulation 13, 580–585.

Yördem, O.S., Papila, M., Menceloglu, Y.Z., 2008. Effects of electrospinning parameters on polyacrylonitrile nanofiber diameter: an investigation by response surface methodology. Mater. Design 29, 34–44.

Yoshikawa, M., Nakai, K., Matsumoto, H., Tanioka, A., Guiver, M.D., Robertson, G.P., 2007. Molecularly imprinted nanofiber membranes from carboxylated polysulfone by electrospray deposition. Macromol. Rapid Commun. 28, 2100–2105.

Yoshikawa, M., Ooi, T., Izumi, J.-I., 1999. Alternative molecularly imprinted membranes from a derivative of natural polymer, cellulose acetate. J. Appl. Polym. Sci. 72, 493–499.

Yoshikawa, M., Tanioka, A., Matsumoto, H., 2011. Molecularly imprinted nanofiber membranes. Curr. Opin. Chem. Eng. 1, 18–26.

Yoshimatsu, K., Ye, L., Lindberg, J., Chronakis, I.S., 2008. Selective molecular adsorption using electrospun nanofiber affinity membranes. Biosens. Bioelectron. 23, 1208–1215.

Yu, J.H., Fridrikh, S.V., Rutledge, G.C., 2004. Production of submicrometer diameter fibers by two-fluid electrospinning. Adv. Mater. 16, 1562–1566.

Yu, D., Wang, X., Li, X., Chian, W., Li, Y., Liao, Y., 2013. Electrospun biphasic drug release polyvinylpyrrolidone/ethyl cellulose core/sheath nanofibers. Acta Biomater. 9, 5665–5672.

Yuan, X., Zhang, Y., Dong, C., Sheng, J., 2004. Morphology of ultrafine polysulfone fibers prepared by electrospinning. Polym. Int. 53, 1704–1710.

Zhang, W., Chen, M., Zha, B., Diao, G., 2012. Correlation of polymer-like solution behaviors with electrospun fiber formation of hydroxypropyl-β-cyclodextrin and the adsorption study on the fiber. Phys. Chem. Chem. Phys. 14, 9729–9737.

Zhang, S., Kawakami, K., 2010. One-step preparation of chitosan solid nanoparticles by electrospray deposition. Int. J. Pharm. 397, 211–217.

Zhang, C., Yuan, X., Wu, L., Han, Y., Sheng, J., 2005. Study on morphology of electrospun poly(vinyl alcohol) mats. Eur. Polym. J. 41, 423–432.

Zhao, S., Wu, X., Wang, L., Huang, Y., 2004. Electrospinning of ethyl–cyanoethyl cellulose/tetrahydrofuran solutions. J. Appl. Polym. Sci. 91, 242–246.

Zhuo, H., Hu, J., Chen, S., 2008. Electrospun polyurethane nanofibres having shape memory effect. Mater. Lett. 62, 2074–2076.

Zong, X., Kim, K., Fang, D., Ran, S., Hsiao, B.S., Chu, B., 2002. Structure and process relationship of electrospun bioabsorbable nanofiber membranes. Polymer 43, 4403–4412.

9 Nanocapsule formation by electrospraying

José A. Tapia-Hernández, Francisco Rodríguez-Félix*, Iman Katouzian**,†*
*University of Sonora, Sonora, Mexico; **Gorgan University of Agricultural Sciences and Natural Resources, Gorgan, Iran; †Nano-encapsulation in the Food, Nutraceutical, and Pharmaceutical Industries Group (NFNPIG), Universal Scientific Education and Research Network (USERN), Tehran, Iran

9.1 Introduction

The field of food science and technology is of great interest not only for food scholars and nutritionists but also for other scientists to develop the novel ideas in this area. Currently, nanotechnology has entered the food industry presenting its unique superiority and applications. The employment of nanotechnology within the food sector provides many advantages as it comes to support the food industry in fabricating functional and enriched products, improving the quality through assuring the food safety (bacteria-free packaging and coatings), encapsulating biological molecules; for example, carbohydrates and proteins and imposing therapeutic effects with the aim of prolonged and controlled release for the agrofood sector (Jafari et al., 2015).

There are several techniques utilized to prepare nanomaterials within the food industry. The electrospraying technique, also known as electrohydrodynamic atomization has found its way in food nanotechnology as a simple and effective method (Katouzian and Jafari, 2016; Faridi Esfanjani and Jafari, 2016). The spheres gained from the equipment may differ due to parameters, such as flow rate, electrostatic potential, collector distance plus the solution parameters, such as viscosity, density, and concentration. Environmental factors, such as temperature and pressure will also affect the characteristics of the yielding nanocapsules.

Currently, many studies support the benefits of electrospraying technique in improving the quality of food along with the encapsulation of biological compounds. For this reason, this chapter seeks to highlight the electrospraying technique as a useful tool for the nanoencapsulation of nutraceuticals and to incorporate them in food and beverage products.

9.2 Electrospraying: an overview

Initial investigations of voltage application to obtain micro and nanomaterials were called electrical atomization. Later, this technique was modified by applying a constant flow rate to achieve particles of smaller diameters. Thereafter, this technique was termed as electrohydrodynamic spraying. Currently, this approach is named electrospraying, which basically includes applying a high voltage, flow rate, and variation of particle collector distance; in addition to the properties of polymer solution, that is, concentration, viscosity, and density.

Zeleny (1914) pioneered electrospraying as a process of droplet formation. He injected a liquid into a capillary and applied a high electric field. Finally, the electrically charged droplets formed a cone which could be divided into smaller droplets. Vonnegut and Neubauer (1952) utilized a certain voltage to obtain micron-sized spheres via spraying. They concluded that with the help of a high electric potential (5–10 kV), a liquid will yield micron-sized particles through implementing a low electrical conductivity and positive charge.

Drozin (1955) evaluated the effects of dielectric constants in inorganic and organic solutions on the formation of the Taylor cone. Overall, nonpolar organic solutions had lower dielectric constants compared to inorganic solutions, exhibiting that the formation of Taylor cone was unstable.

Jones and Thong (1971) investigated the formation of Taylor cone in a solution of kerosene, reporting that the drops resulted from the cone were uniformly sized, simply as monodisperse droplets confirming the previous studies. On the other hand, Hayati et al. (1986) explained the stability of the Taylor cone by means of the existing mechanisms. In their study, they presented different Taylor cones for various applications, such as paints, electrostatic emulsions, and fumigation of agricultural crops. On the whole, the formation of a stable Taylor cone is necessary to achieve an electrodynamic spray.

Back in 1986, the term electrohydrodynamic atomization of liquids was based on the function of a constant conical meniscus steadily dispersed in a medium of aerosol (Grace and Marijnissen, 1994; Smith, 1986). In 1997, mathematical models were utilized to predict the current flow through the charge of the particle, as well as, the particle size coming out of an aerosol via electrical atomization. These models correlate the surface tension, electrical conductivity, and permittivity of the polymer solution (Ganan-Calvo et al., 1997). Hartman et al. (1999) described three stages in electrohydrodynamic atomization process required to obtain particles with smaller diameter:

1. First, the fluid must be dynamic with an initial acceleration so that the Taylor cone is formed. The Taylor cone is generated through a balance of forces as surface tension, gravity, liquid surface and electric tension, inertia, and viscous stress.
2. In the second phase, the fluid is disintegrated within the cone jet forming tiny droplets.
3. Last, the droplets are sprayed onto a collector surface.

Today, the term "electrohydrodynamic atomization" has been left a bit outdated, although some scientists still use it. At the moment, the formation of particles in the scale of micro or nanometer is nominated as electrospraying. This procedure is based on the fundamentals of applying an electric field to deflect the particle with variants, such as electrospraying by deposition (Ali et al., 2012; Altmann et al., 2014; Fornari et al., 2015; Fukuda et al., 2014; Kilo et al., 2014; Soliwoda et al., 2015; Swarbrick et al., 2006), in solution (Almería and Gomez, 2014; Lee et al., 2010), and coaxial electrospraying (Davoodi et al., 2015; Lee et al., 2011; Moghaddam et al., 2015; Xie et al., 2008), as well as, modifying environmental parameters (pressure, temperature, and humidity), system parameters (flow rate, potential electrical, and distance), and parameters of the polymer solution (viscosity, density, and thickness).

9.3 Types of electrospraying

Today, this technique has been modified in order to obtain better results in the formation of micro and nanoparticles. Moreover, the use of various materials, mainly natural biopolymers has also been investigated. The variations of the technique are electrospraying in solution, coaxial electrospraying, and electrospraying by deposition on a substrate.

Conventional electrospraying techniques and its variants are described in the following text.

9.3.1 Conventional electrospraying

Electrospraying can be exerted as an efficient method to produce micro and nanoparticles in different areas, such as nanomedicine and lately offering various applications within the food sector. The theory of electrospraying technique is based on the ability of an electric field to crush a droplet and transform them to the micron or nanometer scale depending on the control parameters (Jaworek and Sobczyk, 2008). When an electric field is applied to a drop, an electric charge is generated within the droplet, named as Coulomb force; this force competes with the cohesion force of the particle and when it dominates over the cohesive force, a diminution in the surface tension occurs and finally nanoparticles are obtained (Bock et al., 2011). One of the methods to determine the rupture of the drop is to determine a parameter known as Rayleigh Limit (LR). This limit occurs when the surface tension of the drop is elevated by the electrostatic force (Xie et al., 2006). Eq. (9.1) represents the Rayleigh Limit.

$$\begin{aligned} L_R &= q(64\pi^2 \varepsilon \gamma R^3) \\ q &= 8\pi\sqrt{\varepsilon_0 \gamma R^3} \end{aligned} \tag{9.1}$$

Where q is the charge in the droplet surface, γ is the surface tension of fluid, ε is permittivity of the medium, and R is the radius of the droplet.

Fig. 9.1A illustrates the general diagram of a conventional electrospraying device, where the particles are collected on a plate or metallic substratum. Aluminum plate (Wang et al., 2012) is best preferred in this process since the particles are likely to be deposited. Furthermore, according to their electrical conductivity, it is suitable to employ it as a counter electrode. All in all, it is an inexpensive and easily manageable device (Tang and Gomez, 2015).

The collector plate must be vertical to the discharge needle where the Taylor cone and ejected droplets are formed (Bodnár and Rosell-Llompart, 2013). Some authors recommend that the dimensions of this collection plate should be 10 × 10 cm, however, the size depends on system conditions to obtain the desired micro or nanoparticles.

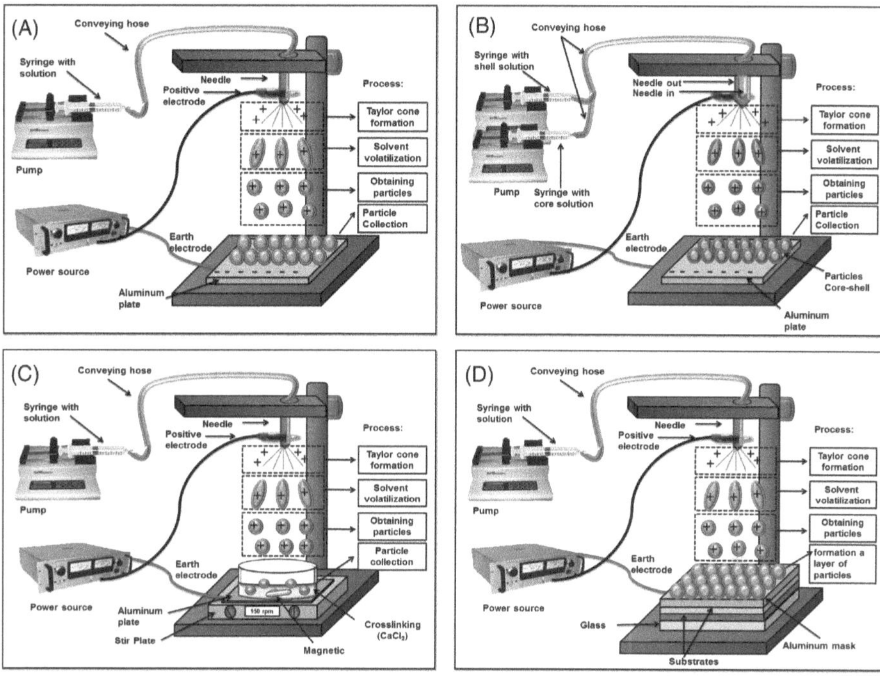

Figure 9.1 (A) General diagram of an equipment of conventional electrospraying. (B) Diagram of the electrospraying technique in solution. (C) Diagram of coaxial electrospraying technique. (D) Diagram of electrospraying by deposition on a substrate.

In a study, Kawakami (2012) used the conventional electrospraying technique to encapsulate carbamazepine and prednisolone as model drugs in a matrix of poly vinyl pyrrolidone. In their research, the sprayed solid particles (400 nm) were collected on an aluminum plate, representing the characteristics of a fine powder.

9.3.2 Electrospraying in solution

The electrospray technique in solution is the first variant of the conventional electrospraying, present in a collection of substances immiscible in a solution. It consists of dropping the electrosprayed polymer solution into an immiscible solution (Fig. 9.1B).

The solution obtained may contain a cross-linking agent, which forms more rigid particles and inhibits the deformation or agglomeration of these particles due to centrifugation or evaporation of the solvent.

The characteristics of the ideal cross-linking agent are:

- Low molecular weight
- Miscible at the solution of collection
- Biocompatible/non-toxic

The main cross-linking agent generally is calcium chloride ($CaCl_2$) as it reinforces the matrices of polysaccharides (Yang et al., 2014) like alginates (Xie and Wang, 2007) and pectins (Aguilar et al., 2015). Commonly, calcium salts are implemented to improve the thermal stability and the protein aggregation (Kuhn et al., 2010). The role of calcium is creating a network through electrostatic interactions with the negatively charged polymers (Uematsu et al., 2004). In the case of proteins, calcium binds to the amino acids present in proteins via the free carboxylic groups, triggering cross-linking (Kinekawa et al., 1998).

Vicini et al. (2015) analyzed the usage of Ca^{+2} as a cross-linking agent in alginate matrices for pharmaceutical applications. In this exploration, the effect of concentration of Ca^{+2} was examined, as well as, the contact time needed for cross-linking. The required criteria for the production particles with smaller diameter were: a concentration of 2% (w/w), electrical potential of 20 kV, flow rate 0.0.1 mL/min and a distance of 5 cm. The results manifested that the higher the concentration and contact time for cross-linking, the greater the particles abundant in Ca^{+2}, due to the higher interaction of anionic groups.

Moghaddam et al. (2015) used the technique of electrospray coaxial for alginate microparticles cross-linked with $CaCl_2$. This study supported the theory that divalent cations bind to the COO^- groups in the alginate to form more stable particles. The alginate concentration used to get smaller diameter particles was 1% (315 ± 33 μm) and the distance of separation was 5 cm between the needle and collector.

9.3.3 Coaxial electrospraying

Coaxial electrospray technique comprises the formation of particles from two diverse polymer solutions, which are added in separate syringes and a needle is inside the other yielding a core-shell structure (Chen et al., 2005; Ghayempour and Mortazavi, 2013). The use of two syringes plus the arrangement of needles is the only difference existing in the system of electrospraying (Bhushani and Anandharamakrishnan, 2014). One of the characteristics of coaxial electrospray is that one needle must be larger in diameter than the other in order that the two solutions are released simultaneously. Concerning the type of solutions contained in both syringes, they can be mixed or simply formed in two phases in the discharge needle (Davoodi et al., 2015).

This technique is feasible for structures with two layers of different polymer or two phases as nanoemulsions (Marín et al., 2007), or when encapsulating an active compound into a polymer is preferred (Hwang et al., 2008; Xie et al., 2008) (Fig. 9.1C). There are two approaches to encapsulate a bioactive compound. In the first case, Two solutions are implemented, where one contains the active ingredient (polymer + active ingredient) and the other only the polymer. Regarding the second case, a polymer is present in the first solution and the bioactive compound in the second solution.

The first case is used when it is not feasible to implement electrospraying on the active compounds, due to their intrinsic properties; for this reason, a polymer is required to serve as a carrier and to be immobilized so as to yield a polymer/polymer + active compound/polymer structure. The second case is used when the bioactive compound can be electrosprayed, obtaining a structure polymer/active compound/polymer (structure core-shell) (Cao et al., 2014). In both cases, an interaction via hydrogen bonds may take place mainly between the bioactive compound and the polymer. Utilizing either of these types depend on the application to be given to micro and nanoparticles by, that is, biomedical, food, or agricultural applications (Zhang et al., 2012). Examples of coaxial electrospray investigations are manifested below:

Wang et al. (2013) encased bovine serum albumin (BSA) as a drug delivery system. In this study, two solutions were utilized to obtain a nanodelivery system; The first solution containing BSA with deionized water was added into a mixture of poly lactic-*co*-glycolic acid (PLGA) in 2,2,2-trifluoroethanol (TFE), as the core material. The second solution was PLGA-TFE, as shell. Both solutions were sprayed with conditions of flow rate 0.8 mL/h, collector distance of 20 cm and 15 kV voltage. The ultimate particle diameter was 2.8 ± 0.60 μm, noting that the coaxial electrospraying technique was possible for encapsulating drugs as controlled release system.

Also, Pérez-Masiá et al. (2015) encapsulated lycopene in edible polymer matrices of dextran, chitosan, and whey protein concentrate by coaxial electrospraying technique. The objective of this work was to incorporate a hydrophobic agent in a hydrophilic matrix. Regarding the three studied matrices, whey protein concentrate showed higher protection to lycopene, increasing its resistance to heat, compared to chitosan and dextran; in addition, the pH was the main factor to degrade the matrix.

Yuan et al. (2015) encapsulated curcumin, a vegetable phenolic agent displaying antiseptic and antiinflammatory features, within a matrix of PLGA. The encapsulation of curcumin is necessary due to its low solubility and poor bioavailability. The method used for encapsulation was coaxial electrospraying, where the internal needle contained curcumin and the external the PLGA. Therefore, with this encapsulation procedure, the bioavailability and optimum drug release of curcumin was improved.

Currently, there are several ongoing studies considering the encapsulation of active compounds in both cases, where promising results has been obtained for the use and protection of active compounds, as well as, their controlled-release.

9.3.4 Electrospraying deposition technique

This variant makes it possible to obtain solid particles not only on a plate but also on a solid substrate, which is covered with the electrospraying material (Fig. 9.1D) (Casula et al., 2015; Zubair et al., 2014) mainly for applications in solar cells (Kim et al., 2012). Rezvanpour and Wang (2011) used PLGA in dichloromethane (DCM) solution to generate particles via electrospraying deposition through a mask of finely-meshed aluminum nets. To sum up, it was realized that a high electrical potential difference between the mask and the solution, as well as, low solution flow rate and short process time, can yield a homogenous particle pattern on the mask.

Kim et al. (2012) fabricated particles acting as organic solar cells (OSC). In this work, several solutions were made to elaborate a multilayered solar cell, the first layer comprised indium tin oxide (ITO), followed by poly(3,4-ethylenediox ythiophene):polystyrene sulfonate (PEDOT:PPS), lastly the third layer contained poly(3-hexylthiophene):[6,6]-phenyl C61-butyric acid methyl ester (P3HT:PCBM) and aluminum (Al). In best conditions, OSC was produced via e-spray method (as-prepared) ending up with the efficiency of 2.98%. Regarding the power conversion efficiency (PCE), this value was increased to 3.08% by exerting a postannealing treatment.

Recently, the coating substrates are being utilized to obtain solar cells, moreover there are several methods by which the coating substrates could be acquired, such as thermal evaporation, chemical vapor deposition, spin on glass, electron beam evaporation, and sputtering. However, electrospraying technique has several advantages over these methods mainly because of its high-speed fabrication, simplicity, producing no waste materials, and being cost-effective.

9.4 Parameters for obtaining micro- and nanoparticles

In order to obtain desired micro and nanoparticles, the parameters which influence the morphology of the particles should be carefully monitored. Three groups of parameters comprise the equipment, solution, and environmental factors (Table 9.1). These factors are described below.

9.4.1 Equipment parameters

The equipment-based factors determining the size and morphology of the produced particles are flow rate regulated by a syringe pump, the electric potential provided by a source of power and distance between the collector and the needle.

9.4.1.1 Electric potential

This is the main parameter of electrospraying process since it is the driving force determining the morphology and size of particles. A higher electrical charge is able to overcome the attraction of the particles on the surface of the polymer solution, thus generating nano or microparticles (Wang and Stark, 2010). Jain et al. (2015) established two conditions for exerting the electrical potential:

1. At low electric potential (less than 10 kV), particles of larger diameter (243 ± 38 μm) and higher size distribution (>0.3%) are formed. This is because the electric potential is still insufficient to dominate over the cohesive forces of the particle and to break the surface tension, in the absence of particle fissions.
2. At high electrical potential (greater than 10 kV), particles of smaller diameter (50–150 μm) with less particle size dispersion (<0.3) are generated. This phenomenon occurs due to the electrical potential, which results in an electric force (Coulomb force) within the droplet. Consequently, this force is able to overcome the surface tension, breaking the drop into smaller droplets.

Moghadam et al. (2010) utilized the electrospraying method to analyze the voltage effect (0–9 kV) on the particle size. It was reported that in the voltage range of 4.5–6.5 kV, the Taylor cone changed the dripping mode at the jet, favoring the production of particles. In brief, at higher voltage (9 kV), micron-sized particles were obtained (400 μm).

Table 9.1 Electrospraying parameters and their influence on the particle size

Parameters		Characteristics	
		Increases	*Decreases*
Equipment	Electric potential	High electrical potential (greater than 10 kV) results in the formation of particles of smaller diameter	Low electric potential (less than 10 kV). There is no formation of Taylor Cone or particles of larger diameter
	Flow rate	At a high flow rate, (≥1 mL/h) it can lead to the formation of large size particles (≥1 μm)	At low flow rate (<1 mL/h) and increasing the electrical conductivity can lead to smaller particles (<1 μm)
	Collector distance	A large distance (15–20 cm or greater) provides more spherical morphology and more volatilization of solvent	Short distance (5–14 cm) results in collapse of the particle
Solution	Concentration	High concentrations tend to form larger particles	Low concentrations tend to form smaller particles
	Viscosity	Increases if the concentration of the polymer is increased and larger particles are obtained	By increasing the voltage and the flow rate, the viscosity decreases and smaller particles are obtained
	Density	Thicker Taylor cone, larger inertial force, therefore droplets are larger	Smaller diameter Taylor cone decreasing fluid resistance, small particles are formed
	Electrical conductivity	If is high, exceeds surface tension and smaller particle diameter will be obtained	If is low, higher surface tension, generates larger particles
Environmental	Relative humidity	If ≤30%, all the solvent is volatilized and completely dry polymer reaches the collector. Smaller particles are obtained	If >30%, no uniform particle diameter and not defined morphology. Larger particles are obtained

Fanzheng et al. (2009), Hartman et al. (2000), and Nguyen et al. (2014) came to a conclusion that higher voltage (15 kV) develops a more stable cone-jet, furthermore particles with smaller diameters will be formed. However, this voltage should not exceed 15 kV since the Taylor cone will become unstable. After all, this voltage value might be different according to the properties of material.

9.4.1.2 Flow rate (Q)

This parameter enables the prediction of material's behavior so as to determine the particle size. Chan et al. (2010) analyzed the preparation of polymethyl silsesquioxane (PMSQ) microspheres and used it as a model wall material entrapping a payload of perfluorohexane (PFH) with a particle size varying between 420 to 1000 nm and the flow rate of 250–650 μL/min. Ghayempour and Mortazavi (2013) fabricated micro and nanoparticles consisting of sodium alginate cross-linked with calcium chloride encapsulating essential oils between 250 and 540 μm in size and the flow rate of 0.1–150 mL/h. These two works determined how the flow rate affects the particle size, obtaining the following results:

1. High flow rate, (≥1 mL/h) can lead to the formation of coarse particles (≥1 μm). This phenomenon occurs because the rate of spraying and polymer concentration are high and therefore the solvent does not volatilize completely.
2. Low flow rate (<1 mL/h) and high electrical conductivity can yield fine particles (<1 μm) since the rate of spraying is low, resulting in greater volatilization of the solvent.

9.4.1.3 Collector distance (L)

In order to fabricate the intended particles and to monitor the uniformity, evaporation of the solvent must be considered and this will depend on the distance between collector and solvent (Jun et al., 2008). Inappropriate distance affects the evaporation of solvent, which may alter the particle formation (morphology and size) (Guarino et al., 2015). Generally, two ranges of distances are considered:

1. A short distance (5–14 cm) causes the collapse of particles as they will be wet at the collector (plate). Thus, particles of different diameters with high polydispersity will be formed.
2. A large distance (15–20 cm or greater) provides more spherical morphology since the polymer chains have enough time to diffuse into the droplets. Therefore, the size and polydispersity are reduced.

Recent studies established the influence of collector distance and the evaporation of the solvent; if a rapid evaporation occurs, particles will display porous or hollow morphologies and even inducing fragmentation of final particles. On the contrary, if a slow evaporation is carried out, particles will be more spherical because there is more time for the solvent to be evaporated (Chul and Jonghwi, 2009).

9.4.2 Solution parameters

These parameters are related to the polymer-solvent interactions, which will affect the constants, such as concentration, viscosity, and density. Each parameter is related to one another, as certain polymer concentrations correspond to certain density and thickness, which influence the morphology and size of the particles being in parallel with the equipment parameters.

9.4.2.1 Concentration

The polymer concentration is an important factor for obtaining micro- and nanoparticles, moreover it is relevant to the viscosity. Khan et al. (2012) established the following guidelines for nano/microparticle formation regarding the concentration parameter:

1. High concentrations tend to form larger particles due to aggregation, therefore it is difficult to overcome the surface tension.
2. Low concentrations tend to form smaller particles since at low concentrations the surface tension of the polymer declines more rapidly due to electrostatic forces generated by the applied voltage, which renders the particles less agglomerated (Khan et al., 2012).

Wang et al. (2015) analyzed the chitosan concentration effect on the properties of obtained microparticles via electrospraying method. They evaluated four concentrations ranging from 1% to 4% and noticed that at a concentration of 4% a high viscosity was obtained (1.68 mPa·s) compared to 2% sample (1.10 mPa·s). Namely, a higher concentration was associated with higher viscosity. The particle size of the prepared solutions varied between 6 to 10 μm, slightly increasing with the enhancement of chitosan concentration, mainly because higher viscosity forms larger droplets in the Taylor cone.

9.4.2.2 *Viscosity*

Viscosity is a parameter that depends on the molecular weight of polymer, concentration and the implemented solvent. At high concentrations, the viscosity of the solution will be larger and vice versa. If the viscosity is too low, the bioactive compound may not be encapsulated (Ghayempour and Mortazavi, 2013). Another parameter affecting the viscosity is the polymer molecular weight. As a result, at higher molecular weights and concentrations coarse particles will be formed (Hazeri et al., 2012; Moghadam et al., 2008).

Moghadam et al. (2010) reported that if the concentration of sodium alginate is elevated from 10 to 30 g/L and also the flow rate is enhanced (20–250 mL/h), viscosity tends to decrease. This phenomenon is more evident at concentration of 30 g/L unlike the concentration of 10 g/L, where there is not a significant change.

9.4.2.3 *Density*

This parameter is directly proportional to the polymer concentration and viscosity. A higher concentration and viscosity tends to increase the density and vice versa (Gianibelli et al., 2001).

The density also has a crucial impact on the development of the Taylor cone. At a higher density, the Taylor cone tends to form drops, while at a lower density Taylor cone shifts to jetting or multijet mode (Jaworek and Sobczyk, 2008). This phenomenon is explained by relating the concept of mass and inertial force (F_ρ) of different solutions. The inertial force is defined as the fluid resistance to changes in conditions either at repose or in motion (Jaworek and Sobczyk, 2008). Greater mass results in the augmentation of density, therefore causing higher resistance or inertial force. In this way, a high polymer concentration (above 6%) is achieved, leading to a cone-type drop, which forms irregular and polydisperse particles. On the other hand, solutions with the lowest concentration (2%) are low in both mass and density, therefore, more uniform particle size and morphology will be obtained considering the low inertial force. Eq. (9.2) relates the density and inertial force in the Taylor cone:

$$F_\rho = \frac{\rho v_i}{dt} \tag{9.2}$$

Where F_ρ = inertial force of the medium, ρ = density of the medium, v_i = Taylor cone rate, and dt = time.

9.4.2.4 Electric conductivity

Electrical conductivity is another parameter, which like viscosity is influenced by the flow rate and electrical potential, furthermore it is defined as the tendency of an agent to pass the electric charge (in this case a polymeric material) (Coll et al., 2015). The results obtained from investigations, relating electrical conductivity with the applied potential, the flow rate, and concentration of polymer are shown as:

1. If the electric potential is high, the flow rate is low and the concentration is high, a greater electrical conductivity will be acquired. Therefore, the higher the electrical conductivity is, the lower surface tension and particle diameter will become.
2. If the electric potential is low, high flow rate, low concentration and lower electrical conductivity are obtained. Lower conductivity and higher surface tension generate larger particles.

Khan et al. (2012) predicted the relation of conductivity and flow rate, suggesting that the increase in conductivity resulted in the decrease of surface tension, yielding smaller particles, nevertheless if flow rate and conductivity are increased, the particle size will be developed (Jaworek and Sobczyk, 2008; Khan et al., 2012).

9.4.3 Environmental parameters

These parameters are indirectly related to the production of micro and nanoparticles since they are not as important as the controllable parameters of equipment or system. Humidity and pressure are incorporated within this category. Moisture is considered as the most important factor, as it is directly related to the ease of solvent evaporation (Salata, 2005).

9.4.3.1 Relative humidity (RH)

Relative humidity is a crucial environmental parameter, along with temperature and pressure. It is basically related to the particle formation (morphology and size). By measuring the relative humidity, it is possible to determine if you can employ the electrospray method or not. Studies have compared the development of micro and nanoparticles at different moisture contents suggesting the following results (Bonadies et al., 2015; Uchida et al., 2015):

1. If the relative humidity is below 30%, it is suitable to work with the equipment, moreover it is assured that all the solvent is volatilized and the completely dried-polymer reaches the collector.

2. If the relative humidity is above 30%, it is not recommended to employ the electrospraying technique, because the particles are not fully dry. As a result, nonuniform particle diameter and various morphologies are obtained.

Another parameter related to the relative humidity is the flow rate. If the humidity is high, it is recommended to provide very low flow rate for that the solvent can be volatilized; otherwise, the solvent will be deposited along with the polymer onto the collecting plate.

9.5 Obtaining materials by electrospraying for the food and nutraceutical industries

The electrospraying technique was use first in medicine for encapsulating drugs with the aim of prolonged and/or controlled release. A research performed by Lee et al. (2010) describes the synthesis of drug particles coated with PLGA via the electrospraying method. They noted that the studied system can reach almost 100% efficiency regarding the encapsulation of drugs utilizing this method. In addition, the changes in the values of PLGA concentration and the electrical conductivity of processed solutions, the gathered particle sizes varied from 165 nm to 1.2 μm.

Recently, the application of electrospraying has expanded beyond food and nutraceutical technology, transforming to a novel technique that is developing rapidly. In this area, the electrospraying can be used for three purposes: primarily, as an application in the area of inedible food, being used to develop intelligent packaging that can resist the entry of pathogens (Guillaume et al., 2010). Second, it is employed to fabricate nanostructures based in food polymeric materials, that is, to produce particles encapsulating bioactive substances or food ingredients (Musyanovych and Landfester, 2014; Pan et al., 2013); and finally, is applicated in the use of polymeric matrices for encapsulation of nutraceuticals (Pérez-Masiá et al., 2014).

9.5.1 Obtaining intelligent packaging and edibles films

The application of nanotechnology to obtain nanospheres has revolutionized the agrifood department. The practices include the development of new materials in food packaging to the systems of nano release and targeted release in the food chain (Granda et al., 2009). In some studies, nanoparticles are used as containers preservatives, which are released when food starts to deteriorate, thus prolonging the shelf life and increasing their bioavailability (Shrivastava and Dash, 2012). Biofilms along with bioactive materials portray a beneficial practice in the food industry. Thus, these technologies can be implemented by mass-producers in order to prolong the product shelf life and fabricate high-quality food (Sánchez-González et al., 2015).

The packaging material directly influences the shelf-life as well as the quality of food products available in the supermarket (Kadam et al., 2015). Currently, these packages contain added molecules, which maintain the quality of food during storage. These molecules may remain in the packaging or migrate into the food and keep the system free of deteriorative microorganisms (Mateescu et al., 2015). Devices used in plastic packaging are termed as biosensors, which offer the potential for the detection of many pathogens and to characterize the food toxicity so as to provide the food security. Polydiacetylene (PDA), is a perfect candidate to be used as a sensor in regard to its peculiar optical features. Changes in different factors, such as temperature, pH and response to biological molecules induce the color changes in films derived from PDA (Santos et al., 2010). Munteanu et al. (2014) used chitosan coating on polyethylene (PE) as packaging material in food by electrospraying, provide some level of protection to fruits and vegetables as decreased gas permeability and antibacterial properties. Antimicrobial test of films depicted the inhibition of 100% for *Salmonella enteritidis* (ATCC 25922), 96.43% for *Escherichia coli* (ATCC 25922) and 90.74% for *Listeria monocytogenes* (ATCC 25922) at 48 h. The mass and thickness of the chitosan layer increased with the augmentation of concentration and spraying time. This electrospraying approach is suitable for obtaining composite packering compared with other techniques, such as immersion or spreading.

Fabricating edible films and coatings are another applications of electrospraying in order to create a bright surface for aesthetic purposes (Milda and Kerry, 2009). A film is different from a coating in a way that films may also be implemented as food contact materials to improve the safety and quality of foods (Acevedo-Fani et al., 2015). Films may form a material independently to control the exchange of carbon dioxide, oxygen, and ethylene involved in the process of respiration, moreover, it can be impemented to prevent the loss of moisture; besides, a coating provides sterility to the surface of food and helps avoiding the loss of important components. Polysaccharides (Zimoch-Korzycka et al., 2015), proteins (Barba et al., 2015), and lipids (Galus and Kadzińska, 2015) are examples of materials used to develop coatings and edible films (Milda and Kerry, 2009).

The electrospraying is favorable for obtaining a uniform film coating in micro or nanoscale since it is done in a single step and being a low cost approach. This technique makes it possible to obtain uniform particles as well as having a film without cracks and voids in it. Besides, electrospraying can be a perfect alternative for the usual spray coating technique (Bhushani and Anandharamakrishnan, 2014).

Pareta and Edirisinghe (2006) elaborated edible films of corn starch by electrospraying. The conditions for the generation of these films were concentration of starch 5% (w/v) with a temperature of gelatinization of 120°C and a time of 30 min. Later, ethanol was added to the small scattered starch particles and subjected to ultrasonic disruption to make a stable modified starch solution. Spraying time is a variable of the thickness of the edible films so as more time passes (2–8 h), the longer and thicker the film becomes. Electrospraying is capable of depositing starch films and coatings. Du et al. (2008) evaluated the storage and antimicrobial activity of edible films emanated from apples containing carvacrol to avoid contamination by *E. coli* O157: H7.94. In another study, Ravishankar et al. (2009) developed edible films from apples with concentrations of 0.5, 1.5, and 3% of cinnamaldehyde or carvacrol that significantly inhibited the growth of pathogens in meat and poultry products. This study showed that the film contributing best biocidal activity was the concentration of 3% carvacrol and cinnamaldehyde with a thickness of 0.130 ± 0.004 and 0.134 ± 0.004, respectively.

Therefore, carvacrol and cinnamaldehyde are good antimicrobial agents added to edible apple films by electrospraying in order to keep food products, such as meat. Morsy et al. (2014) incorporated essential oils and pullulan nanoparticles in films in order to control hazardous bacteria in meat and poultry products. Kurt and Kahyaoglu (2014) characterized a new edible biodegradable film made of glucomannan and analyzed the rheological properties of the solution. Khazaei et al. (2014) prepared and characterized an edible film from basil seed gum (*Ocimum basilicum* L.) and employed glycerol as a plasticizer to obtain favorable results.

9.5.2 Obtaining of nanostructures based on food polymeric materials

Nano- and microparticles produced from food polymeric materials have the characteristic of being biodegradable and can be made by the electrospraying technique. These polymers are commonly used in the pharmaceuticals industry, where they serve as drug delivery systems and to deliver other compounds to the targeted area. It has been shown that particles can also serve as carriers of antimicrobial components; for example, the nanoparticles containing nisin exhibit higher activity against *E. coli* O157: H7 compared to particles without nisin. The most commonly used polymers are starch (Atyabi et al., 2006), polycaprolactone (Wu and Clark, 2007) cellulose, chitosan, alginate, (Bhowmik et al., 2006), gelatin, zein, wheat gluten, and gliadins.

Chitosan is a polymer obtained from chitin, which is widely used to develop nano- and microcapsules (Arya et al., 2009; Zhang and Kawakami, 2010) chitin can be found in marine animals as in the shell of shrimp and some terrestrial insects (Eladlani et al., 2015). Some studies have shown that chitosan is widely used, and can serve as a basis to produce particles and to encapsulate of drugs and bioactive compounds. Xu et al. (2013) used the coaxial electrospraying method to prepare and characterize nanoparticles of folate–chitosan–gemcitabine, obtaining nanoparticles from 200 to 300 nm with activity against pancreatic cancer. They came to a conclusion that it is simple to develop nanocomposites of chitosan via this method.

Starch is a natural polymer obtained from a variety of vegetables, formed from simpler compounds (amylose and amylopectin) (Le Corre et al., 2010). Starch has been widely used to induce sustained release in drug and food delivery systems and other natural compounds. Ghaeb et al. (2015) elaborated starch nanoparticles by the electrospraying technique for its wide use in food products. Studied variables were: flow rate (0.026, 0.039, and 0.052 mL/h), concentration (0.75, 1.00, and 1.25% (*w/w*)), electrical potential (12, 16, and 20 kV) and separation between needle and collector (10, 17.5, and 25 cm). They noted that these conditions resulted in particle sizes smaller than 100 nm, then starch is a feasible material for the production of nanoparticles.

Wu and Clark (2007) elaborated polycaprolactone (PCL) to yield polymeric particles with diverse morphologies via the electrospraying technique. The microstructure of PCL particles is controlled during the electrospraying by modifying the evaporation of solvents.

The microspheres obtained with PCL in chloroform 2% (w/v) showed diameters in the range of 1–5 μm with pore sizes in a range of 0.3–1.3 μm. Midhun et al. (2011)

used electrospraying for the encapsulation of budesonide in a matrix of polycaprolactone in order to overcome the limitations of rapid elimination of this drug and induce a prolonged release. The parameters studied in the investigation were flow rate and electrical potential, as well as the viscosity of the solution. All in all, the encapsulation efficiency was about 75% and it was concluded that this technique is a good candidate for inducing sustained release in the body.

Alginate is a polysaccharide present in the cell walls of marine algae. This polysaccharide is used to develop micro- and nanoparticles as it forms the wall material used for encapsulating active compounds, furthermore it can be used individually or with other materials to form composites. Park et al. (2012) used the electrospraying technique for preparing particles for encapsulation of adenovirus in an alginate matrix cross-linked with $CaCl_2$. The best parameters were shown alginate 0.7% since lower concentrations were nonspherical particles, the flow rate was detected as 2–40 mL/h, still more aspherical 2 mL/h. Consequently, the flow rate, the amount of the used alginate solution plus the applied electrical potential imparted a significant share in altering the size and morphology of alginate particles and that adenovirus can be successfully encased by alginate particles.

Tabeei et al. (2012) explored the non-Newtonian and thixotropic flow behaviors of alginate solution prepared by the electrospraying method. Shi et al. (2016) incorporated lactic probiotic (*Enterococcus faecalis*) in alginate microparticles to overcome the problematic environmental conditions, such as stomach pH in organisms. They concluded it can protect and deliver *Enterococcus faecalis* HZNU P2 under simulated gastrointestinal conditions. Microspheres (800 ± 10 μm) used in foods, such as yogurt exhibit a higher size. However, there is no limitation for their application in nutraceutical and pharmaceutical practices. Zhao et al. (2016) encapsulated *Ganoderma lucidum* spores (GLS) by means of electrospraying. Bioactive compounds are normally susceptible to acidic conditions and oxidative circumstances during processing, storage, and consumption. The selected matrix was alginate and the particle size was dependent on the system parameters. Also, $CaCl_2$ implemented the crosslinker with a level of 0.5 M and at lower concentrations it did not induce intersections. In essence, The encapsulation of GLS via the electrospraying procedure developed an effective delivery system with the aim of prolonged release in the body.

Gelatin is a polymeric substance (polysaccharides) with a complex structure, being colorless and contains amino acids (Chen et al., 2014; Ofokansi et al., 2010). Gelatin is obtained from collagen found in connective tissues of animals, whereas agar-agar is obtained from sea plants (Rhim et al., 2014).

Wool keratin is another naturally-occurring protein, which has plenty of applications in cosmetics, filtration, hygiene, tissue, and particularly in engineering delivery systems (Cardamone, 2010). Rad et al. (2012) used chicken feather for the extraction of quercetin with applications in cosmetics, food, coatings, and biodegradable films. The objective of this investigation was to obtain keratin nanoparticles from electrospraying solution of coated feather keratin. Trifluoroacetic acid was chosen as the appropriate solvent and the average diameter of particles were of 53 nm. The optimal parameters were: electrical potential of 15 kV, a distance of

15 cm between the nozzle and collector, flow rate of 0.047 mL/h and keratin concentration of 0.2% (w/v). In conclusion, it is feasible to achieve keratin nanoparticles by electrospraying method through using higher electrical potential, longer separation distances between the needle and collector, lower flow rates plus lower concentrations. Ebrahimgol et al. (2014) successfully elaborated wool keratin nanoparticles through electrospraying. They suggested that the appropriate parameters for obtaining nanoparticles of keratin were: concentration of 0.5% (w/v) in formic acid, flow rate of 0.02 mL/h and separation distance of 20 cm between the needle and collector. The nanoparticles average diameter in this study was approximately 36–72 nm.

Zein is an abundant source of protein in corn (45%–65% of the whole protein). It was identified in 1987 concerning its solubility in aqueous alcohol solutions. Unfortunately, this protein cannot be used individually due to its low solubility in water (Shukla and Cheryan, 2001). Gomez-Estaca et al. (2012) elaborated curcumin-zein nanoparticles with spherical structure through electrospraying. Zein concentration was about 2.5%–15% (w/w) and the diameter of fabricated particles was of 175–900 nm, furthermore the polymer concentration scaled directly with the diameter of particles. Altogether, the encapsulation efficiency was estimated to be 85%–90% and the electrospraying technique was considered as a practical procedure to zein nanoparticles. Li et al. (2013) prepared microcomposites zein as shell and ibuprofen as core by traditional electrospraying and modified electrospraying. The particles obtained by traditional electrospraying had 1.84 ± 0.64 μm in diameter, while that the particles by modified electrospraying showed 1.78 ± 0.31 μm of diameter. A modified electrospraying process was carried out in this study, where an EP-coated spraying head was implemented for obtaining ibuprofen/zein microparticles. Gaona-Sánchez et al. (2015) investigated the effect of the electrospraying parameters on the formation of zein films. Overall, the electrospraying technique was recognized to be better than the casting method since it leaves the water vapor barrier capacity untouched. Also, monolithic films were obtained via exerting a zein concentration of 11.4%, an electrical potential of 7.8–8.7 kV and a separation between the nozzle and collector of 1.5 ± 0.5 cm. In addition, Fabra et al. (2016) sprayed zein microparticles made with antioxidants, such as α-tocopherol on films of wheat gluten by electrospraying, this material was used as a novel packaging to improve and preserve the characteristics of food.

Gliadins are proteins belonging to wheat gluten and have been used in the preparation of useful materials, such as fibers, particles, and films. Gluten is a polymer network formed mainly by two proteins (the gliadins and glutenins) attached to the starch granules within the endosperm of wheat grain. Liao et al. (2012) developed nanocapsules of wheat gluten deaminated with succinic acid for the encapsulation of fish oil with particles of 527 nm. Gulfam et al. (2012) used electrospraying deposition for the development of anticancer drugs carriers at the nanoscale level. The optimal concentration of zein was 7% aqueous ethanol, with a particle size of 218.66 ± 5.1. The loading efficiency was 72.02 ± 5.6% of cyclophosphamide. The encapsulation of cyclophosphamide in matrix of zein was efficiently done by electrospraying.

9.6 Encapsulation of nutraceuticals

The electrospraying technique is a feasible method for the entrapment of nutraceutical compounds. Some studies conducted for encapsulating nutraceuticals are described below and are summarized in Table 9.2.

Torres-Giner et al. (2010), encapsulated docosahexaenoic acid (DHA) from the omega-3 family by electrospraying. One of the disadvantages of these compounds is that they degrade or get rancid as their polyunsaturated acids react with oxygen. At the same time, rancidity generates unpleasant odors. Pure zein nanocapsules had diameters of approximately 530 ± 215 nm plus the value of 490 ± 200 nm for zein-DHA. Encapsulation of DHA in matrices of zein by electrospraying prevents the reduction of the fatty acid's property as a functional ingredient.

Jung et al. (2011) used the electrospraying technique to lower the size of natural extract particles (*Sophora flavescens*) for utilization in medical applications. This technique is useful to disperse and reduce the particles by charges, increasing contact surface of the extract. The applied equipment conditions were: electrical potential of 3.8 kV and low flow rate of 30 μL/h. The produced particles were within the size of

Table 9.2 Studies of nutraceutical encapsulation by electrospraying technique

Wall material	Nutraceutical	Functión	References
Zein	Docosahexaenoic acid (DHA)	Avoiding degradation or rancidity due to polyunsaturated acids reaction with oxygen	Torres-Giner et al. (2010)
	Extract of *Sophora flavescens*	Antibacterial, antiinflammatory, and anticancer effects	Jung et al. (2011)
WPC	β-Carotene	Antioxidant activity	López-Rubio and Lagaron, 2012
Sodium caseinate	Curcumin	Antiproliferation activity	Pan et al. (2014)
Alginate-pectin	Peppermint oil	Perfumes, cosmetics, medicines, and the flavoring industry	Koo et al. (2014)
Gelatin	Curcumin	Antioxidant activity	Gómez-Estaca et al. (2015)
Gelatin	Epigallocatechin gallate	Antioxidant activity	Gómez-Mascaraque et al. (2015)
Kafirins	Extract of sorghum condensed tannins	Inhibiting α-amylase	Links et al. (2015)
Gelatin, WPC and soybean protein	α-Linolenic acid	Nutraceuticals	Gómez-Mascaraque and López-Rubio (2016)

WPC, Whey protein concentrate.

50–250 nm, representing that the technique is a good candidate for the production of particles from a suspension of the extract.

López-Rubio and Lagaron (2012) encapsulated β-carotene in a matrix of whey protein concentrate (WPC) via electrospraying technique. β-carotene is a bioactive compound displaying antioxidant activity. The conditions in this test were using a voltage of 0–30 kV, flow rate 10^{-3} of 10 mL/h and a separation of 7 cm concerning the needle plus the collector. The best capsules contained 40% WPC. It was suggested that Whey proteins are practical components for the production of micro/nanocapsules by the electrospraying method.

Also, Pan et al. (2014) incorporated curcumin in a lipophilic phyto-cosmetic medium to cover its low water solubility and bioavailability, and then Curcumin was encapsulated in a matrix of sodium caseinate. By studying the solubility under the effects of curcumin pH and self-assembly characteristics of sodium caseinate, it was possible to obtain an encapsulation technology with the advantages of organic solvent-free, low-energy, and being low-cost. The studied encapsulation method was an efficient technique, in the entrapment of lipophilic compounds.

Koo et al. (2014) investigated the microencapsulation of peppermint oil in alginate–pectin network using a coaxial electrospraying system. The experiment conditions were as follows: electrical potential of 5 kV, flow rate of peppermint oil plus the alginate-pectin 50 μL/min and 130 μL/min, respectively together with the separation length of 10 cm between the needle and the collector. The minimum size was 1.54 μm and the maximum was 3.24 μm. Also, the polydispersity scaled directly with the pectin concentration (0.13 ± 0.01–0.89 ± 0.06).

Gómez-Estaca et al. (2015) used electrospraying to encapsulate curcumin in a gelatin network. In spite of curcumin's numerous advantages, the disadvantages of this bioactive compound are attributed to its poor solubility in water, which decrease its dispersion in food matrices, as well as, the bioaccessibility. In this study, two flow rates were used 0.15 and 0.5 mL/h and electrical potential was 14 kV, furthermore the mean diameter of microparticles was 1.2 μm. Overall, the encapsulation of curcumin by employing an aqueous polymer was a practical plan to be exerted in the food industry.

Gómez-Mascaraque et al. (2015) employed the electrospraying technique to obtain food-grade gelatin capsules so as to protect bioactives. The electrosprayed matrices were applied to incorporate epigallocatechin gallate, known as an antioxidant molecule. The conditions considered in the experiment were: gelatin concentration 5%–20% (w/v), voltage 0–30 kV and the flow rate of 0.15–0.2 mL/h. The results showed that particle size distribution decreased with the particle diameter. In addition, as gelatin concentration declined, a greater heterogeneity of capsule sizes was obtained.

Links et al. (2015) used an appropriate therapeutic approach to inhibit enzymes that digest carbohydrates in the intestine, via the polyphenols of plants. They prepared a crude extract of sorghum condensed tannins, afterwards the extract was encapsulated in kafirin microparticles for inhibiting alpha-amylase. The results showed an encapsulation efficiency of 48%, and the in vitro release test represented that these particles were digested partially during the simulated gastrointestinal via pepsin and trypsin-chymotrypsin. To sum up, the system was effective against amylase present in type II diabetes, compared to bare condensed tannins in sorghum.

Gómez-Mascaraque and López-Rubio, 2016 utilized the electrospraying technique in order to encapsulate α-linolenic acid in gelatin matrices (combined with whey protein concentrate) and soybean protein. The results revealed an encapsulation efficiency of 70%, moreover, a delayed degradation of α-linolenic acid was observed at 80°C in both matrices. In conclusion, electrospraying was chosen as an effective technique for encasing nutraceuticals since encapsulation maintains the vital characteristics of the bioactive compound.

9.7 Conclusions

Electrospraying technique and its variants are considered a novel and promising approach for fabricating micro and nanoparticles. This method is performed in a single step without waste production. Through controlling the parameters of this technique (equipment, solution and environment), appropriate micro or nanoparticles are produced according to the desired application. Electrospraying technique can be applied to food sector in various practices, such as packaging, edible films, encapsulation of nutraceutical compounds, and promoting the organoleptic properties. In recent years, this practical approach has been used for encapsulation of nutraceuticals in order to maintain or improve the beneficial effects of bioactive compounds as antioxidant, antiinflammatory, and antimutagenic activities.

References

Acevedo-Fani, A., Salvia-Trujillo, L., Rojas-Graü, M.A., Martín-Belloso, O., 2015. Edible films from essential-oil-loaded nanoemulsions: physicochemical characterization and antimicrobial properties. Food Hydrocolloids 47, 168–177.

Aguilar, K.C., Tello, F., Bierhalz, A.C., Romo, M.G.G., Flores, H.E.M., Grosso, C.R., 2015. Protein adsorption onto alginate-pectin microparticles and films produced by ionic gelation. J. Food Eng. 154, 17–24.

Ali, M., Abbas, M., Shah, S.K., Tuerhong, R., Generosi, A., Paci, B., Gunnella, R., 2012. Realization of solution processed multi-layer bulk heterojunction organic solar cells by electrospray deposition. Org. Electron. 13 (10), 2130–2137.

Almería, B., Gomez, A., 2014. Electrospray synthesis of monodisperse polymer particles in a broad (60 nm–2 μm) diameter range: guiding principles and formulation recipes. J. Colloid Interface Sci. 417, 121–130.

Altmann, K., Schulze, R.D., Friedrich, J., 2014. Polymer deposition morphology by electrospray deposition-modifications through distance variation. Thin Solid Films 564, 269–276.

Arya, N., Chakraborty, S., Dube, N., Katti, D.S., 2009. Electrospraying: a facile technique for synthesis of chitosan-based micro/nanospheres for drug delivery applications. J. Biomed. Mater. Res. B Appl. Biomater. 88, 17–31.

Atyabi, F., Manoochehri, S., Mohamad, S.H., Dinarvand, R., 2006. Cross-linked starch microspheres: effect of cross-linking condition on the microsphere characteristics. Arch. Pharm. Res. 29, 1179–1186.

Barba, C., Eguinoa, A., Maté, J.I., 2015. Preparation and characterization of β-cyclodextrin inclusion complexes as a tool of a controlled antimicrobial release in whey protein edible films. LWT-Food Sci. Technol. 64 (2), 1362–1369.
Bhowmik, B.B., Sa, B., Mukherjee, A., 2006. Preparation and in vitro characterization of slow release testosterone nanocapsules in alginates. Acta Pharm. 56, 417–429.
Bhushani, J.A., Anandharamakrishnan, C., 2014. Electrospinning and electrospraying techniques: potential food based applications. Trends Food Sci. Technol. 38 (1), 21–33.
Bock, N., Woodruff, M.A., Hutmacher, D.W., Dargaville, T.R., 2011. Electrospraying, a reproducible method for production of polymeric microspheres for biomedical applications. Polymers 3 (1), 131–149.
Bodnár, E., Rosell-Llompart, J., 2013. Growth dynamics of granular films produced by electrospray. J. Colloid Interface Sci. 407, 536–545.
Bonadies, I., Cimino, F., Carfagna, C., Pezzella, A., 2015. Eumelanin 3D architectures: electrospun PLA fiber templating for mammalian pigment microtube fabrication. Biomacromolecules 16 (5), 1667–1670.
Cao, L., Luo, J., Tu, K., Wang, L.Q., Jiang, H., 2014. Generation of nano-sized core–shell particles using a coaxial tri-capillary electrospray-template removal method. Colloids Surf. B Biointerfaces 115, 212–218.
Cardamone, J.M., 2010. Investigating the microstructure of keratin extracted from wool: peptide sequence (MALDI-TOF/TOF) and protein conformation (FTIR). J. Mol. Struct. 969 (1), 97–105.
Casula, G., Cosseddu, P., Busby, Y., Pireaux, J.J., Rosowski, M., Szczesna, B.T., Banerjee, R., 2015. Air-stable, non-volatile resistive memory based on hybrid organic/inorganic nanocomposites. Org. Electron. 18, 17–23.
Chan, M.-W., Stride, E., Edirisinghe, M., 2010. Controlling the thickness of hollow polymeric microspheres prepared by electrohydrodynamic atomization. J. R. Soc. Interface 7 (Suppl. 4), S451–S460.
Chen, X., Jia, L., Yin, X., Cheng, J., Lu, J., 2005. Spraying modes in coaxial jet electrospray with outer driving liquid. Phys. Fluids 17 (3), 032101.
Chen, L., Siemiarczuk, A., Hai, H., Chen, Y., Huang, G., Zhang, J., 2014. Development of biocompatible and proton-resistant quantum dots assembled on gelatin nanospheres. Langmuir 30, 1893–1899.
Chul, H.P., Jonghwi, L., 2009. Electrosprayed polymer particles: effect of the solvent properties. J. Appl. Polym. Sci. 114, 430–437.
Coll, A., Bermejo, S., Martin, I., Castaner, L., 2015. Effect of nanofluid conductivity and hu midity on the self-assembly of nanoparticles deposited by electrospray. In: Electron Devices (CDE), Tenth Spanish Conference on IEEE, pp. 1–3.
Davoodi, P., Feng, F., Xu, Q., Yan, W.C., Tong, Y.W., Srinivasan, M.P., Wang, C.H., 2015. Coaxial electrohydrodynamic atomization: microparticles for drug delivery applications. J. Control. Release 205, 70–82.
Drozin, V.G., 1955. The electrical dispersion of liquids as aerosols. J. Colloid Sci. 10 (2), 158–164.
Du, W.X., Olsen, C.W., Avena-Bustillos, R.J., McHugh, T.H., Levin, C.E., Friedman, M., 2008. Storage stability and antibacterial activity against *Escherichia coli* O157:H7 of carvacrol in edible apple films made by two different casting methods. J. Agric. Food Chem. 56, 3082–3088.
Ebrahimgol, F., Tavanai, H., Alihosseini, F., Khayamian, T., 2014. Electrosprayed recovered wool keratin nanoparticles. Polym. Adv. Technol. 25 (9), 1001–1007.
Eladlani, N., Rhazi, M., Taourirte, M., 2015. Nanoparticles and whiskers-based chitosan films for Cr complexation. Environ. Chem. Lett. 13 (1), 105–110.

Fabra, M.J., López-Rubio, A., Lagaron, J.M., 2016. Use of the electrohydrodynamic process to develop active/bioactive bilayer films for food packaging applications. Food Hydrocoll. 55, 11–18.

Fanzheng, M., Yi, J., Zhihui, S., Yizi, Y., Yanyan, L., 2009. Electrohydrodynamic liquid atomization of biodegradable polymer microparticles: effect of electrohydrodynamic liquid atomization variables on microparticles. J. Appl. Polym. Sci. 113, 526–534.

Faridi Esfanjani, A.F., Jafari, S.M., 2016. Biopolymer nano-particles and natural nano-carriers for nano-encapsulation of phenolic compounds. Colloids Surf. B 146, 532–543.

Fornari, E., Roberts, C.J., Temperton, R.H., O'Shea, J.N., 2015. Electrospray deposition in vacuum as method to create functionally active protein immobilization on polymeric substrates. J. Colloid Interface Sci. 453, 252–259.

Fukuda, T., Suzuki, A., Liao, Y., Suzuki, K., 2014. Influence of spray conditions on droplet charge per unit volume for electrospray deposition. J. Aerosol Sci. 77, 38–49.

Galus, S., Kadzińska, J., 2015. Food applications of emulsion-based edible films and coatings. Trends Food Sci. Technol. 45 (2), 273–283.

Ganan-Calvo, A.M., Davila, J., Barrero, A., 1997. Current and droplet size in the electrospraying of liquids. Scaling laws. J. Aerosol Sci. 28 (2), 249–275.

Gianibelli, M., Larroque, O., MacRitchie, F., Wrigley, C., 2001. Biochemical, genetic, and molecular characterization of wheat glutenin and its component subunits. Cereal Chem. 78, 635–646.

Gaona-Sánchez, V.A., Calderon-Dominguez, G., Morales-Sanchez, E., Chanona-Perez, J.J., Velazquez-De La Cruz, G., Mendez-Mendez, J.V., Farrera-Rebollo, R.R., 2015. Preparation and characterisation of zein films obtained by electrospraying. Food Hydrocoll. 49, 1–10.

Ghaeb, M., Tavanai, H., Kadivar, M., 2015. Electrosprayed maize starch and its constituents (amylose and amylopectin) nanoparticles. Polym. Adv. Technol. 26, 917–923.

Ghayempour, S., Mortazavi, S.M., 2013. Fabrication of micro–nanocapsules by a new electrospraying method using coaxial jets and examination of effective parameters on their production. J. Electrostat. 71 (4), 717–727.

Gomez-Estaca, J., Balaguer, M.P., Gavara, R., Hernandez-Munoz, P., 2012. Formation of zein nanoparticles by electrohydrodynamic atomization: Effect of the main processing variables and suitability for encapsulating the food coloring and active ingredient curcumin. Food Hydrocoll. 28 (1), 82–91.

Gómez-Estaca, J., Gavara, R., Hernández-Muñoz, P., 2015. Encapsulation of curcumin in electrosprayed gelatin microspheres enhances its bioaccessibility and widens its uses in food applications. Innov. Food Sci. Emerg. Technol. 29, 302–307.

Gómez-Mascaraque, L.G., López-Rubio, A., 2016. Protein-based emulsion electrosprayed micro-and submicroparticles for the encapsulation and stabilization of thermosensitive hydrophobic bioactives. J. Colloid Interface Sci. 465, 259–270.

Gómez-Mascaraque, L.G., Lagarón, J.M., López-Rubio, A., 2015. Electrosprayed gelatin submicroparticles as edible carriers for the encapsulation of polyphenols of interest in functional foods. Food Hydrocoll. 49, 42–52.

Grace, J.M., Marijnissen, J.C.M., 1994. A review of liquid atomization by electrical means. J. Aerosol Sci. 25 (6), 1005–1019.

Granda, V.M., Valdés, G.A.C., García, C.J.A., Díaz-García, M.E., 2009. Analytical nanotechnology for food analysis. Microchim. Acta 166, 1–19.

Guarino, V., Altobelli, R., Cirillo, V., Cummaro, A., Ambrosio, L., 2015. Additive electrospraying: a route to process electrospun scaffolds for controlled molecular release. Polym. Adv. Technol. 26 (12), 1359–1369.

Guillaume, C., Pinte, J., Gontard, N., Gastaldi, E., 2010. Wheat glutencoated papers for bio-based food packaging: structure, surface and transfer properties. Food Res. Int. 1, 1395.

Gulfam, M., Kim, J.E., Lee, J.M., Ku, B., Chung, B.H., Chung, B.G., 2012. Anticancer drug-loaded gliadin nanoparticles induce apoptosis in breast cancer cells. Langmuir 28 (21), 8216–8223.

Hartman, R.P.A., Brunner, D.J., Camelot, D.M.A., Marijnissen, J.C.M., Scarlett, B., 1999. Electrohydrodynamic atomization in the cone–jet mode physical modeling of the liquid cone and jet. J. Aerosol Sci. 30 (7), 823–849.

Hartman, R.P.A., Brunner, D.J., Camelot, D.M.A., Marijnissen, J.C.M., Scarlett, B., 2000. Jet break-up in electrohydrodynamic atomization in the cone-jet mode. J. Aerosol Sci. 31, 65–69.

Hayati, I., Bailey, A.I., Tadros, T.F., 1986. Mechanism of stable jet formation in electrohydrodynamic atomization. Nature 319 (6048), 41–43.

Hazeri, N., Tavanai, H., Moradi, A.R., 2012. Production and properties of electrosprayed sericin nanopowder. Sci. Technol. Adv. Mater. 13 (3), 35010.

Hwang, Y.K., Jeong, U., Cho, E.C., 2008. Production of uniform-sized polymer core-shell microcapsules by coaxial electrospraying. Langmuir 24 (6), 2446–2451.

Jain, E., Scott, K.M., Zustiak, S.P., Sell, S.A., 2015. Fabrication of polyethylene glycol-based hydrogel microspheres through electrospraying. Macromol. Mater. Eng. 300, 823–835.

Jafari, S.M., Fathi, M., Mandala, I., 2015. Emerging product formation. Food Waste Recovery: Processing Technologies and Industrial TechniquesElsevier Inc., Academic Press, San Diego, CA, pp. 293–317.

Jaworek, A.T.S.A., Sobczyk, A.T., 2008. Electrospraying route to nanotechnology: an overview. J. Electrostatics 66 (3), 197–219.

Jones, A.R., Thong, K.C., 1971. The production of charged monodisperse fuel droplets by electrical dispersion. J. Physics D Appl. Phys. 4 (8), 1159.

Jun, Y., Liang, K., Jingwei, X., Jinsong, H., Chi-Hwa, W., 2008. Characterization of electrospraying process for polymeric particle fabrication. J. Aerosol Sci. 39, 987–1002.

Jung, J.H., Park, S.Y., Lee, J.E., Nho, C.W., Lee, B.U., Bae, G.N., 2011. Electrohydrodynamic nano-spraying of ethanolic natural plant extracts. J. Aerosol Sci. 42 (10), 725–736.

Kadam, A.A., Karbowiak, T., Voilley, A., Debeaufort, F., 2015. Techniques to measure sorption and migration between small molecules and packaging. A critical review. J. Sci. Food Agric. 95 (7), 1395–1407.

Katouzian, I., Jafari, S.M., 2016. Nano-encapsulation as a promising approach for targeted delivery and controlled release of vitamins. Trends Food Sci. Technol. 53, 34–48.

Kawakami, K., 2012. Miscibility analysis of particulate solid dispersions prepared by electrospray deposition. Int. J. Pharm. 433 (1), 71–78.

Khan, M.K.I., Schutyser, M., Schroën, K., Boom, R., 2012. The potential of electrospraying for hydrophobic films coating on food. J. Food Eng. 108, 410–416.

Khazaei, N., Esmaiili, M., Djomeh, Z.E., Ghasemlou, M., Jouki, M., 2014. Characterization of new biodegradable edible film made from basil seed (*Ocimum basilicum* L.) gum. Carbohydr. Polym. 102, 199–206.

Kilo, O., Jabbour, J., Habchi, R., Abboud, N., Brouche, M., Khoury, A., Zaouk, D., 2014. Electrospray deposition and characterization of cobalt oxide thin films. Mater. Sci. Semicond. Process. 24, 57–61.

Kim, Y., Kim, G., Lee, J., Lee, K., 2012. Morphology controlled bulk-heterojunction layers of fully electro-spray coated organic solar cells. Solar Energy Mater. Solar Cells 105, 272–279.

Kinekawa, Y., Fuyuki, T., Kitabatake, N., 1998. Effects of salts on the properties of sols and gels prepared from whey protein isolate and process whey protein. J. Dairy Sci. 81, 1532–1544.

Koo, S.Y., Cha, K.H., Song, D.G., Chung, D., Pan, C.H., 2014. Microencapsulation of peppermint oil in an alginate–pectin matrix using a coaxial electrospray system. Int. J. Food Sci. Technol. 49 (3), 733–739.

Kuhn, K.R., Cavallieri, A.L.F., Cunha, R.L., 2010. Cold-set whey protein gels induced by calcium or sodium salt addition. Int. J. Food Sci. Technol. 45, 348–357.

Kurt, A., Kahyaoglu, T., 2014. Characterization of a new biodegradable edible film made from salep glucomannan. Carbohydr. Polym. 104, 50–58.

Le Corre, D., Bras, J., Dufresne, A., 2010. Starch nanoparticles: a review. Biomacromolecules 11, 1139–1153.

Lee, Y.H., Mei, F., Bai, M.Y., Zhao, S., Chen, D.R., 2010. Release profile characteristics of biodegradable-polymer-coated drug particles fabricated by dual-capillary electrospray. J. Control. Release 145 (1), 58–65.

Lee, Y.H., Bai, M.Y., Chen, D.R., 2011. Multidrug encapsulation by coaxial tri-capillary electrospray. Colloids Surf. B Biointerfaces 82 (1), 104–110.

Li, W., Yu, D.G., Chen, K., Wang, G., Williams, G.R., 2013. Smooth preparation of ibuprofen/zein microcomposites using an epoxy-coated electrospraying head. Mater. Lett. 93, 125–128.

Liao, L., Luo, Y., Zhao, M., Wang, Q., 2012. Preparation and characterization of succinic acid deamidated wheat gluten microspheres for encapsulation of fish oil. Colloid Surf. B 92, 305–314.

Links, M.R., Taylor, J., Kruger, M.C., Taylor, J.R., 2015. Sorghum condensed tannins encapsulated in kafirin microparticles as a nutraceutical for inhibition of amylases during digestion to attenuate hyperglycaemia. J. Funct. Foods 12, 55–63.

López-Rubio, A., Lagaron, J.M., 2012. Whey protein capsules obtained through electrospraying for the encapsulation of bioactives. Innov. Food Sci. Emerg. Technol. 13, 200–206.

Marín, Á.G., Loscertales, I.G., Marquez, M., Barrero, A., 2007. Simple and double emulsions via coaxial jet electrosprays. Phys. Rev. Lett. 98 (1), 014502.

Mateescu, A.L., Dimov, T.V., Grumezescu, A.M., Gestal, M.C., Chifiriuc, M.C., 2015. Nanostructured bioactive polymers used in food-packaging. Curr. Pharm. Biotechnol. 16 (2), 121–127.

Midhun, B.T., Shalumon, K.T., Manzoor, K., Jayakumar, R., Nair, S.V., Deepthy, M., 2011. Preparation of budesonide-loaded polycaprolactone nanobeads by electrospraying for controlled drug release. J. Biomater. Sci. 22 (18), 2431–2444.

Milda, E.E., Kerry, C.H., 2009. Edible Films and Coatings for Food ApplicationsSpringer, Dordrecht Heidelberg London New York, Chapter 1, p 1.

Moghadam, H., Samimi, M., Samimi, A., Khorram, M., 2008. Electrospray of high viscous liquids for producing mono-sized spherical alginate beads. Particuology 6, 271–275.

Moghadam, H., Samimi, M., Samimi, A., Khorram, M., 2010. Electrospray modeling of highly viscous and non-newtonian liquids. J. Appl. Polym. Sci. 118, 1288–1296.

Moghaddam, M.K., Mortazavi, S.M., Khayamian, T., 2015. Preparation of calcium alginate microcapsules containing *n*-nonadecane by a melt coaxial electrospray method. J. Electrostatics 73, 56–64.

Morsy, M.K., Khalaf, H.H., Sharoba, A.M., El-Tanahi, H.H., Cutter, C.N., 2014. Incorporation of essential oils and nanoparticles in pullulan films to control foodborne pathogens on meat and poultry products. J. Food Sci. 79, 675–684.

Munteanu, B.S., Paslaru, E., Fras Zemljic, L., Sdrobis, A., Pricope, G.M., Vasile, C.O.R.N.E.L.I.A., 2014. Chitosan coatings applied to polyethylene surface to obtain food-packaging materials. Cell Chem Technol. 48, 565–575.

Musyanovych, A., Landfester, K., 2014. Polymer micro- and nanocapsules as biological carriers with multifunctional properties. Macromol. Biosci. 14, 458–477.

Nguyen, T.K., Nguyen, V.D., Seong, B., Hoang, N., Park, P., Byun, D., 2014. Control and improvement of jet stability by monitoring liquid meniscus in electrospray and electrohydrodynamic jet. J. Aerosol Sci. 71, 29–39.

Ofokansi, K., Winter, G., Fricker, G., Coester, C., 2010. Matrix-loaded biodegradable gelatin nanoparticles as new approach to improve drug loading and delivery. Eur. J. Pharm. Biopharm. 76, 1–9.

Pan, K., Zhong, Q., Baek, S.J., 2013. Enhanced dispersibility and bioactivity of curcumin by encapsulation in casein nanocapsules. J. Agric. Food. Chem. 61, 6036–6043.

Pan, K., Luo, Y., Gan, Y., Baek, S.J., Zhong, Q., 2014. pH-driven encapsulation of curcumin in self-assembled casein nanoparticles for enhanced dispersibility and bioactivity. Soft matter 10 (35), 6820–6830.

Pareta, R., Edirisinghe, M.J., 2006. A novel method for the preparation of starch films and coatings. Carbohydr. Polym. 63 (3), 425–431.

Park, H., Kim, P.H., Hwang, T., Kwon, O.J., Park, T.J., Choi, S.W., Kim, J.H., 2012. Fabrication of cross-linked alginate beads using electrospraying for adenovirus delivery. Int. J. Pharm. 427 (2), 417–425.

Pérez-Masiá, R., Lagaron, J.M., López-Rubio, A., 2014. Development and optimization of novel encapsulation structures of interest in functional foods through electrospraying. Food Bioprocess Technol. 7 (11), 3236–3245.

Pérez-Masiá, R., Lagaron, J.M., Lopez-Rubio, A., 2015. Morphology and stability of edible lycopene-containing micro-and nanocapsules produced through electrospraying and spray drying. Food Bioprocess Technol. 8 (2), 459–470.

Rad, Z.P., Tavanai, H., Moradi, A.R., 2012. Production of feather keratin nanopowder through electrospraying. J. Aerosol Sci. 51, 49–56.

Ravishankar, S., Zhu, L., Olsen, C.W., McHugh, T.H., Friedman, M., 2009. Edible apple film wraps containing plant antimicrobials inactivate foodborne pathogens on meat and poultry products. J. Food Sci. 74, 440–445.

Rezvanpour, A., Wang, C.H., 2011. Computational and experimental studies of electrospray deposition process in pharmaceutical micro-pattern formation. Chem. Eng. Sci. 66 (17), 3836–3849.

Rhim, J.W., Wang, L.F., Lee, Y., Hong, S.I., 2014. Preparation and characterization of bionanocomposite films of agar and silver nanoparticles: laser ablation method. Carbohydr. Polym. 103, 456–465.

Salata, O.V., 2005. Tools of nanotechnology: electrospray. Curr. Nanosci. 1 (1), 25–33.

Sánchez-González, L., Arab-Tehrany, E., Cháfer, M., González-Martínez, C., Chiralt, A., 2015. Active edible and biodegradable starch films. Polysacch. Bioactivity Biotechnol., 717–734.

Santos, P.A., Ferreira, S.A., Mendes, S.L., Andrade, N., Araujo, S.M., Fernandes, C.A., 2010. Polydiacetylene as a biosensor: fundamentals and applications in the food industry. Food Bioprocess. Technol. 3, 172–181.

Shi, L.E., Zheng, W., Zhang, Y., Tang, Z.X., 2016. Milk-alginate microspheres: protection and delivery of *Enterococcus faecalis* HZNU P2. LWT-Food Sci Technol. 65, 840–844.

Shrivastava, S., Dash, D., 2012. Nanotechnology in food sector and agriculture. Proc. Natl. Acad. Sci. India Sect. B: Biol. Sci. 82, 29–35.

Shukla, R., Cheryan, M., 2001. Zein: the industrial protein from corn. Ind. Crops Prod. 13 (3), 171–192.

Smith, D.P., 1986. The electrohydrodynamic atomization of liquids. IEEE Trans. Ind. Appl. (22), 527–535.

Soliwoda, K., Rosowski, M., Tomaszewska, E., Tkacz-Szczesna, B., Celichowski, G., Grobelny, J., 2015. Electrospray deposition of gold nanoparticles from aqueous colloids on solid substrates. Colloids Surf. A 486, 211–217.

Swarbrick, J.C., Taylor, J.B., O'Shea, J.N., 2006. Electrospray deposition in vacuum. Appl. Surf. Sci. 252 (15), 5622–5626.

Tabeei, A., Samimi, A., Khorram, M., Moghadam, H., 2012. Study pulsating electrospray of non-Newtonian and thixotropic sodium alginate solution. J. Electrostat. 70, 77–82.

Tang, J., Gomez, A., 2015. Control of the mesoporous structure of dye-sensitized solar cells with electrospray deposition. J. Mater. Chem. A 3 (15), 7830–7839.

Torres-Giner, S., Martinez-Abad, A., Ocio, M.J., Lagaron, J.M., 2010. Stabilization of a nutraceutical omega-3 fatty acid by encapsulation in ultrathin electrosprayed zein prolamine. J. Food Sci. 75 (6), N69–N79.

Uchida, K., Higashi, K., Hishida, K., Hotta, A., Miki, N., 2015. Electrospray formation of ring-shaped silica nanoparticles. Jpn. J. Appl. Phys. 54 (2), 020302.

Uematsu, I., Matsumoto, H., Morota, K., Minagawa, M., Tanioka, A., Yamagata, Y., Inoue, K., 2004. Surface morphology and biological activity of protein thin films produced by electrospray deposition. J. Colloid Interface Sci. 269 (2), 336–340.

Vicini, S., Castellano, M., Mauri, M., Marsano, E., 2015. Gelling process for sodium alginate: new technical approach by using calcium rich micro-spheres. Carbohydr. Polym. 134, 767–774.

Vonnegut, B., Neubauer, R.L., 1952. Production of monodisperse liquid particles by electrical atomization. J. Colloid Sci. 7 (6), 616–622.

Wang, K., Stark, J.P., 2010. Voltage effects on the nanoelectrospray characteristics in fully voltage-controlled atomisation of gold nanocolloids. Anal. Chim. Acta 679, 81–84.

Wang, M., Rutledge, G.C., Myerson, A.S., Trout, B.L., 2012. Production and characterization of carbamazepine nanocrystals by electrospraying for continuous pharmaceutical manufacturing. J. Pharm. Sci. 101 (3), 1178–1188.

Wang, Y., Yang, X., Liu, W., Zhang, F., Cai, Q., Deng, X., 2013. Controlled release behaviour of protein-loaded microparticles prepared via coaxial or emulsion electrospray. J. Microencapsul. 30 (5), 490–497.

Wang, X.-X., Ju, X.-J., Sun, S.-X., Xie, R., Wang, W., Liu, Z., Chu, L.-Y., 2015. Monodisperse erythrocyte-sized and acid-soluble chitosan microspheres prepared via electrospraying. RSC Adv. 5, 34243–34250.

Wu, Y., Clark, R.L., 2007. Controllable porous polymer particles generated by electrospraying. J. Colloid Interface Sci. 310 (2), 529–535.

Xie, J., Wang, C.H., 2007. Electrospray in the dripping mode for cell microencapsulation. J. Colloid Interface Sci. 312 (2), 247–255.

Xie, J., Lim, L.K., Phua, Y., Hua, J., Wang, C.H., 2006. Electrohydrodynamic atomization for biodegradable polymeric particle production. J. Colloid Interface Sci. 302 (1), 103–112.

Xie, J., Ng, W.J., Lee, L.Y., Wang, C.H., 2008. Encapsulation of protein drugs in biodegradable microparticles by co-axial electrospray. J. Colloid Interface Sci. 317 (2), 469–476.

Xu, S., Xu, Q., Zhou, J., Wang, J., Zhang, N., Zhang, L., 2013. Preparation and characterization of folate-chitosan-gemcitabine coreshell nanoparticles for potential tumor-targeted drug delivery. J. Nanosci. Nanotechnol. 13, 129–138.

Yang, N., Luan, J., Ashton, J., Gorczyca, E., Kasapis, S., 2014. Effect of calcium chloride on the structure and in vitro hydrolysis of heat induced whey protein and wheat starch composite gels. Food Hydrocoll. 42, 260–268.

Yuan, S., Lei, F., Liu, Z., Tong, Q., Si, T., Xu, R.X., 2015. Coaxial electrospray of curcumin-loaded microparticles for sustained drug release. PloS One 10 (7), e0132609.
Zeleny, J., 1914. The electrical discharge from liquid points, and a hydrostatic method of measuring the electric intensity at their surfaces. Phys. Rev. 3, 69–91.
Zhang, S., Kawakami, K., 2010. One-step preparation of chitosan solid nanoparticles by electrospray deposition. Int. J. Pharm. 397, 211–217.
Zhang, L., Huang, J., Si, T., Xu, R.X., 2012. Coaxial electrospray of microparticles and nanoparticles for biomedical applications. Expert Rev. Med. Devices 9 (6), 595–612.
Zhao, D., Li, J.S., Suen, W., Chang, M.W., Huang, J., 2016. Preparation and characterization of *Ganoderma lucidum* spores-loaded alginate microspheres by electrospraying. Mater. Sci. Eng. C 62, 835–842.
Zimoch-Korzycka, A., Ambrozik-Haba, J., Kulig, D., Jarmoluk, A., 2015. Modification effect of cellulase on the physicochemical characteristic of polysaccharides edible films. J. Polym. Sci. 2015, 7, Article ID 184616.
Zubair, M., Mustafa, M., Lee, K., Yoon, C., Doh, Y.H., Choi, K.H., 2014. Fabrication of CdSe/ZnS quantum dots thin film by electrohydrodynamics atomization technique for solution based flexible hybrid OLED application. Chem. Eng. J. 253, 325–331.

Further Reading

Çelik, S.E., Özyürek, M., Güçlü, K., Apak, R., 2015. Antioxidant capacity of quercetin and its glycosides in the presence of β-cyclodextrins: influence of glycosylation on inclusion complexation. J. Incl. Phenom. Macrocycl. Chem. 83 (3–4), 309–319.
Chen, L., Remondetto, G.E., Subirade, M., 2006. Food protein-based materials as nutraceutical delivery systems. Trends Food Sci. Technol. 17 (5), 272–283.
Deladino, L., Teixeira, A.S., Navarro, A.S., Alvarez, I., Molina-García, A.D., Martino, M., 2015. Corn starch systems as carriers for yerba mate (*Ilex paraguariensis*) antioxidants. Food Bioprod. Process. 94, 463–472.
Kalra, E.K., 2003. Nutraceutical-definition and introduction. AAPS PharmSci. 5 (3), 27–28.
Kanmani, P., Rhim, J.W., 2014. Antimicrobial and physical-mechanical properties of agar-based films incorporated with grapefruit seed extract. Carbohydr. Polym. 102, 708–716.
Sanna, V., Lubinu, G., Madau, P., Pala, N., Nurra, S., Mariani, A., Sechi, M., 2015. Polymeric nanoparticles encapsulating white tea extract for nutraceutical application. J. Agric. Food Chem. 63 (7), 2026–2032.

10 Nanocapsules formation by nano spray drying

Cordin Arpagaus, Philipp John**, Andreas Collenberg**, David Rütti***
*NTB University of Applied Sciences of Technology Buchs, Institute for Energy Systems, Buchs, Switzerland; **BÜCHI Labortechnik AG, Flawil, Switzerland

10.1 Introduction

Spray drying is defined as the transformation of a fluid from a liquid state into a dried particulate form by spraying the fluid into a hot drying medium (Masters, 1991). It is a suitable one-step process for the conversion of various liquid formulations (e.g., aqueous and organic solutions, emulsions, and suspensions) into dry powders.

Spray drying is a simple, fast, and scalable drying technology (Arpagaus and Schwartzbach, 2008) that is well-established in the chemical, food, and pharmaceutical industries (Wang et al., 2005). Spray drying equipment is commercially available and the production cost is lower compared to that of other drying technologies, such as freeze drying (Gharsallaoui et al., 2007). The powders produced are high in quality and have low moisture content, resulting in high shelf stability (Anandharamakrishnan and Padma Ishwarya, 2015). The cooling effect of the evaporating solvent conserves the droplet temperature relatively low; therefore heat-sensitive products can be dried with negligible degradation (Masters, 1991). Spray drying offers high flexibility to control particle size and morphology by optimizing the process parameters and feed formulation (Nandiyanto and Okuyama, 2011).

The dried powder form has higher stability, better protection from the environment (e.g., oxidation, light, and temperature), easier handling and storage, and redispersibility in aqueous solutions (Murugesan and Orsat, 2012; Okuyama et al., 2006). Spray drying is able to result in several types of particle shapes, such as smooth and spherical, collapsed, dimpled, wrinkled, raisin-like, and highly crumpled or folded particles (Nandiyanto and Okuyama, 2011).

Fig. 10.1 shows a traditional laboratory-scale spray dryer. The spray drying process consists of three fundamental steps,

- atomization of the liquid feed,
- drying of the sprayed droplets in the drying gas and formation of dry particles, and
- separation and collection of the dry product from the drying gas.

The nozzle atomizes the feed into droplets. The reduction in particle size leads to a large increase in the surface area. In the drying chamber, the solvent, which can be water, organic solvents, or mixtures thereof, is quickly removed by the continuous flow of the hot drying gas (usually air or inert gas). The dried solid particles are separated from the gas stream by a cyclone and collected in a collection vessel.

Spray drying is the most frequently used encapsulation technique for food products (Jafari et al., 2008b; Mahdavi et al., 2014). It is applied for food materials, such as flavors (Esfanjani et al., 2015; Gharsallaoui et al., 2007; Rajabi et al., 2015), vitamins, minerals, salts, colorants (Assadpour and Jafari, 2017; Assadpour et al., 2017; Mahdavee Khazaei et al., 2014), oils, lipids, spices (Jafari et al., 2007a), polyphenols, proteins, carotenoids, antioxidants, probiotic living cells (Celli et al., 2015), probiotic bacteria, enzymes (Bürki et al., 2011; Dahili and Feczkó, 2015), peptides, and many more.

Encapsulation is accomplished by dissolving, emulsifying, or dispersing the core substance in a solution of the carrier material and then spraying the mixture into the hot drying chamber. The encapsulated material may be present in either liquid or solid form. Encapsulation protects the bioactive compound from the surrounding environment, increases the product storage stability (Anandharamakrishnan, 2014; Celli et al., 2015) and preserves their health-promoting properties (Quintanilla-Carvajal et al., 2010). Moreover, encapsulation is applied to regulate the release of the bioactive compound, to target to specific sites, to mask the taste of an ingredient, and to improve bioavailability (Whelehan, 2014).

Choosing the right material for the encapsulating wall is very important to reach high encapsulation efficiency and capsule stability (Estevinho et al., 2013). Wall materials can be selected from a wide variety of natural (Mahdavi et al., 2014) and synthetic polymers (Gharsallaoui et al., 2007). Natural food-grade compounds are are frequently used to minimize the potential toxic effects associated with ingestion. Some examples of common natural polymers are,

- gums (e.g., arabic gum, guar gum, and carrageenans),
- carbohydrates (e.g., lactose, maltodextrins, cyclodextrins, dextrose, sodium alginate, cellulose derivatives, chitosan, and starches), and
- proteins (e.g., whey, egg, soy, casein, gelatine, collagen, and albumin).

The addition of surfactants is often required to stabilize the polymer particles. Further details about the selection of appropriate wall materials can be found in several other works (Ezhilarasi et al., 2013; Fathi et al., 2014; Ngan et al., 2014; Rajabi et al., 2015; Sabliov and Astete, 2015). Furthermore, excellent studies on the microencapsulation of food ingredients by spray drying exist in the scientific literature (Gharsallaoui et al., 2007; Mahdavi et al., 2014).

The sizes of the particles formed through encapsulation are classified into macro (>5 mm), micro (1 μm to 5 mm), and nano (<1 μm) (Fang and Bhandari, 2012). Capsules smaller than 1 μm in diameter are known as nanocapsules (Couvreur et al., 1995; Jafari et al., 2008b). Jafari et al. (2007a,b, 2008a,b) proposed the term "nanoparticle encapsulation" for cases where the core material is within the nano-size range and is encapsulated into a matrix of micron-sized powder particles. Nanocapsules, due to their reduced size and large specific surface, offer higher absorption rates and bioavailability of the bioactive compounds compared to microcapsules. The smaller the particles, the better is the solubility of the encapsulated nutraceuticals.

Fig. 10.1 illustrates the various morphologies which can be prepared by spray drying encapsulation, such as single core, irregular, multiwall, multicores, composite, and pure bioactives.

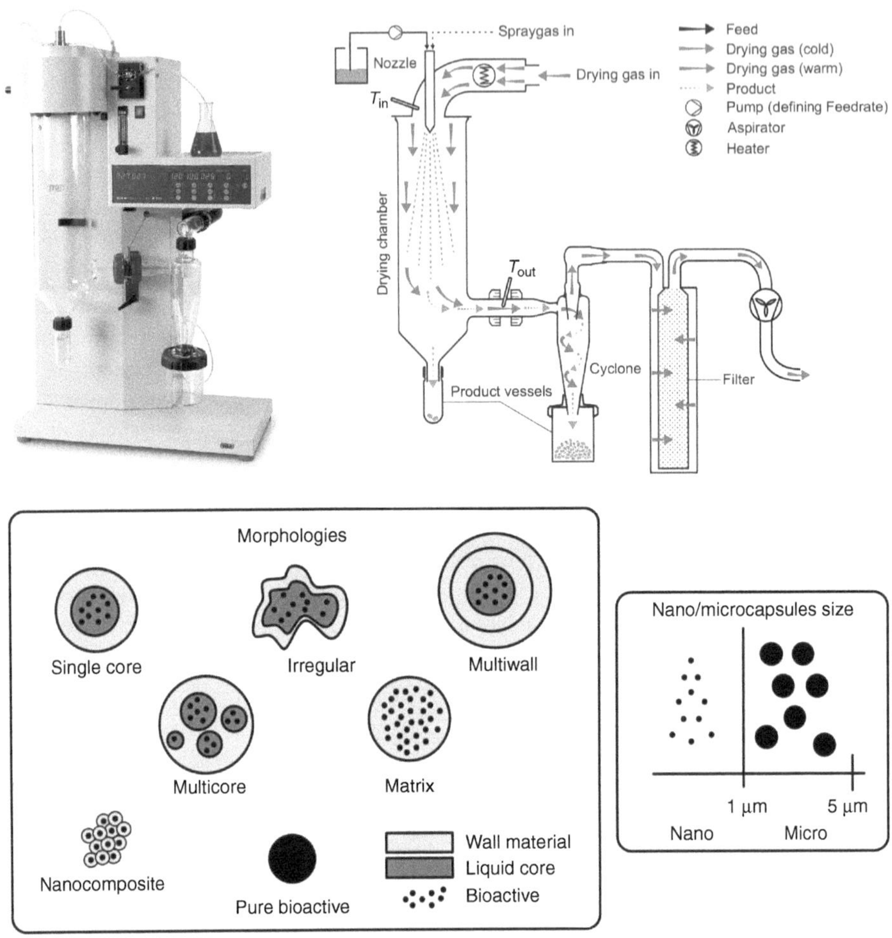

Figure 10.1 Top: Traditional spray dryer—Mini Spray Dryer B-290 with principle flow diagram (Büchi Labortechnik AG, 2015; Wong and John, 2015). Bottom: Morphologies and average size range of different types of particles prepared by encapsulation and nano spray drying.
Source: Adapted from Estevinho, B.N., Rocha, F., Santos, L., Alves, A., 2013. Microencapsulation with chitosan by spray drying for industry applications—a review. Trends Food Sci. Technol. 31, 138–155; Fang, Z., Bhandari, B., 2012. Encapsulation techniques for food ingredient systems. In: Bhandari, B., Roos, Yrjö, H. (Eds.). Food Mater. Sci. Engin., Blackwell Publishing Ltd., Oxford, UK, pp. 320–348 (Chapter 12); Gibbs, B.F., Kermasha, S., Alli, I., Mulligan, C.N., 1999. Encapsulation in the food industry: a review. Int. J. Food Sci. Nutr. 50, 213–224; Haggag, Y.A., Faheem, A.M., 2015. Evaluation of nano spray drying as a method for drying and formulation of therapeutic peptides and proteins. Front. Pharmacol. 6, 1–5.

Single-core capsules consist of a defined core containing the bioactive, surrounded by a uniform shell structure. Multiplecore capsules are typically generated for providing stability. In practical application, the core may be an emulsion, a suspension of solid particles, a solution, or a suspension of smaller capsules. The capsule may even have multiple walls. In matrix particles, the encapsulated material is evenly distributed;

therefore there are no distinctive core and shell (Whelehan and Marison, 2011). The encapsulation of nanoemulsions of oils via emulsifying techniques connected upstream of spray drying has received a great attention within the subject area of food research (Jafari et al., 2007a,b; 2008a,b; Pourashouri et al., 2014). The smaller the emulsion size, the better is the encapsulation efficiency. In principle, the process of nanoencapsulation by nano spray drying is similar to that of microencapsulation (Gharsallaoui et al., 2007). The following list summarizes the main benefits of spray drying in the context of encapsulation applications (Arpagaus et al., 2013; Wong and John, 2015).

- Control of particle size, shape, and morphology (amorphous/crystal form, porosity).
- One-step semicontinuous process to directly convert various feeds into dry powders.
- Process simplicity and ease of operation.
- Low operating costs, energy-efficient technology, and fast process.
- Scale-up capability.
- Open or closed cycle design for aqueous or organic solvents.
- Processing of heat-sensitive substances with low risk of degradation.
- Design of particles with controlled release properties.
- High encapsulation efficiency and extended shelf life.
- Versatile technique for the formation of nanocapsules for numerous encapsulating agents.

The recent boost in nanomedicine and -therapeutics has increased the need for spray dryer systems required to fabricate a high yield of nanoscale particles with narrow distribution of size (Faridi Esfanjani and Jafari, 2016; Katouzian and Jafari, 2016). So far, several expert opinions have been published about the nano spray drying technology (Arpagaus, 2012; Heng et al., 2011; Lee et al., 2011). A search in the online databases of Web of Science and Scopus (accessed on December 31, 2015) with the key word "nano spray drying" gives 472 and 541 hits respectively from the years 1993 to 2015. Fig. 10.2 illustrates this increasing trend of the research activity in this field. Apparently, the potential of nano spray drying has not yet been tapped fully. Research activities in nano spray drying are taking place worldwide, mainly in the fields of pharmaceuticals, materials engineering, and food sciences, although other new applications are expected to be found in the coming years. Several excellent reviews have

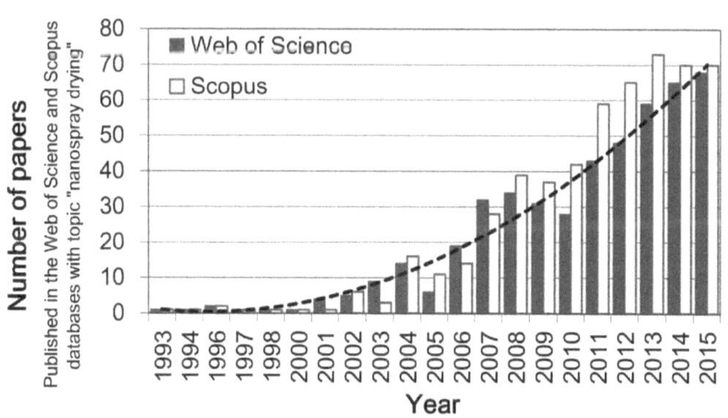

Figure 10.2 Increasing trend of the research activity in this field. Number of articles published since 1993 with the key words "nano spray drying" found in the online databases Web of Science and Scopus (visited on December 31, 2015).

already been published on spray drying technology and the formation of nanoparticles (Arpagaus, 2012; Nandiyanto and Okuyama, 2011; Okuyama et al., 2006; Sosnik and Seremeta, 2015; Wang et al., 2005).

In order to produce nanoscale particles via spray drying technology, some modifications on the experimental set-up of traditional spray dryers are necessary. A major constraint is the limited efficiency of separation and collection of submicron particles by cyclone separators. Typical cyclones are unable to collect particles below 2 μm (Bowen et al., 2013; Chan and Kwok, 2011; Heng et al., 2011; Maa et al., 1998; Maltesen et al., 2008; Schmid et al., 2009). Even with a high-performance glass cyclone (Brandenberger, 2003), the median particle size cannot be reduced below 1.4 μm (Maltesen et al., 2008). In other words, submicron-sized particles cannot be collected using traditional spray dryers. The only feasible option for collection of nanoparticles is to use electrostatic particle collectors (Arpagaus, 2011). Another limitation is the turbulent gas flow in the drying chamber, which results in particle depositions on the chamber wall. Furthermore, traditional atomizers do not allow the fine droplets being generated to reach submicron particle sizes. With the recent advances in nano spray drying technology, particles <1 μm can now be generated.

10.2 Nano spray drying

The Nano Spray Dryer B-90 was introduced in 2009 to extend spray drying to the submicron scale. This laboratory spray dryer is based on a fundamentally new concept of spray drying technology, involving the fabrication of submicron particles from a solution, nanoemulsion, or nanosuspension. Its technological novelty lies in the gentle flow of laminar drying gas, the vibrating mesh spray technology to form fine droplets, and the highly efficient electrostatic precipitator to collect nanoparticles.

The modular glass assembly allows modification of the length of the drying chamber and is easy to clean. An integrated outlet gas filter works as a safety measure to protect the operator and the environment from the finest particles. Fig. 10.3 shows the Nano Spray Dryer B-90and illustrates its functional principle, which is not fundamentally different from traditional spray drying. The process consists of the following basic steps - heating of the drying gas, droplet generation, drying of the droplets, and particle collection.

Table 10.1 compares the major characteristics of the Nano Spray Dryer B-90 to those of the traditional Mini Spray Dryer B-290 developed by Büchi Labortechnik AG. Overall yields are limited in the traditional spray dryer, at best reaching 50–70% (Arpagaus and Meuri, 2010; Li et al., 2010; Schmid et al., 2011). Moreover, at least 30 mL of the liquid sample is required for a feasibility test (Schmid et al., 2011). In the nano spray dryer, a very small amount (e.g., a few milligrams of solid) are sufficient. This is especially relevant when expensive materials are used, such as peptides under clinical investigations in the early stages of drug development (Büchi Labortechnik, 2010; Lee et al., 2011; Schmid et al., 2011). On the other hand, the traditional spray dryer allows a more straightforward scale-up, from laboratory scale to pilot and industrial production, as described by Arpagaus and Schwartzbach (2008).

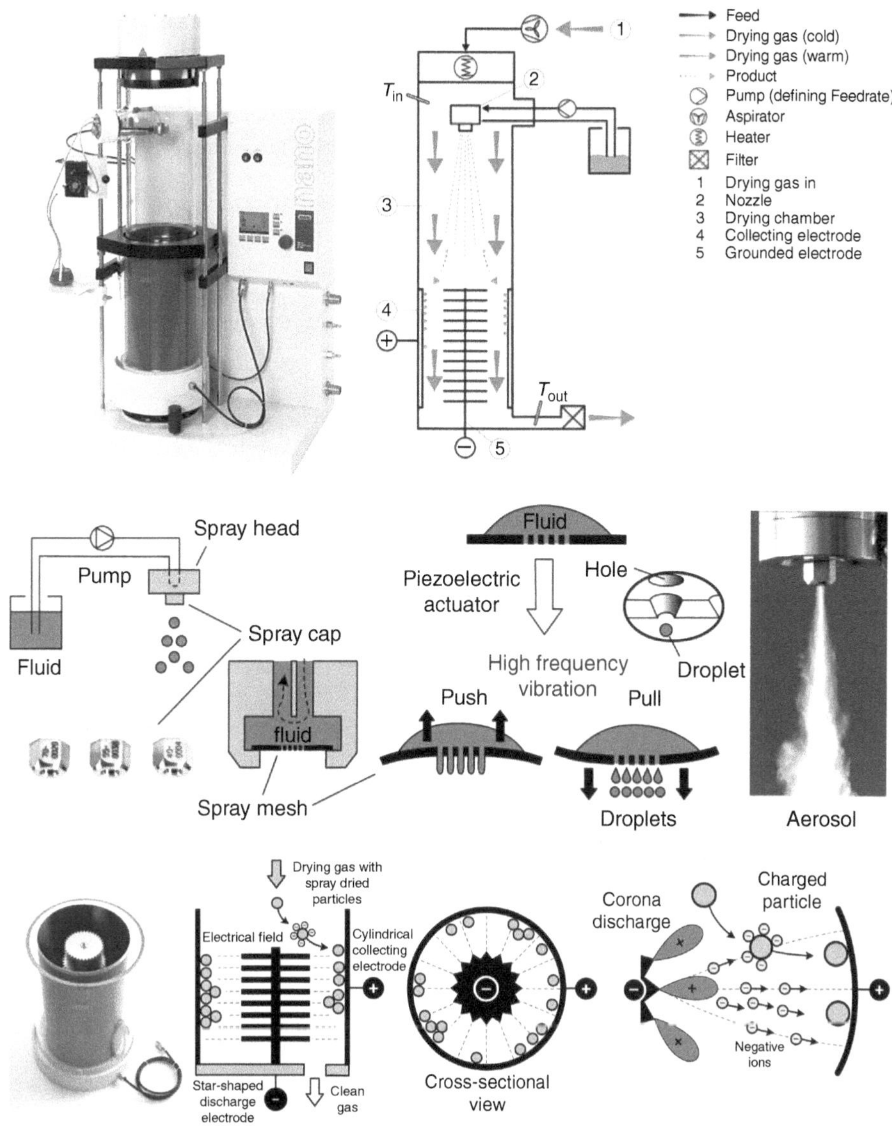

Figure 10.3 Top: Nano Spray Dryer B-90 and a schematic of its functional principle (Arpagaus et al., 2010a; Büchi Labortechnik AG, 2015, 2010). Middle: Droplet generation by vibration mesh technology in the Nano Spray Dryer B-90 (Büchi Labortechnik AG, 2015). Bottom: Functional principle of the electrostatic particle collector.
Source: Bottom, Picture from Büchi Labortechnik, AG, 2013. Nano Spray Dryer B-90 Technical data sheet, en 1311 A. Available from: http://static2.buchi.com/sites/default/files/technical-data-pdf/B-90_Data_Sheet_en_A_0.pdf

Table 10.1 **Major characteristics of the Nano Spray Dryer B-90 in comparison to the traditional Mini Spray Dryer B-290**

Characteristics	Nano Spray Dryer B-90	Mini Spray Dryer B-290
Major benefits	Small sample quantities, fine particles, high yields	Traditional spray drying, established process
Minimum sample amount	2 mL	30 mL
Maximum sample viscosity	<10 cps	<300 cps
Spray generation device	Vibrating mesh	Two-fluid nozzle (optional: ultrasonic nozzle)
Spray generation energy	Piezoelectric actuator (ultrasonic frequency)	Pressurized gas (two-fluid nozzle), ultrasonic vibration (ultrasonic nozzle)
Nozzle diameter	4.0, 5.5, and 7.0 μm (spray mesh hole size)	0.7, 1.4, and 2.0 mm (nozzle tip diameter)
Water droplet size range	3–8 μm (4.0 μm spray mesh), 4–12 μm (5.5 μm), 5–15 μm (7.0 μm)	5–30 μm (0.7 mm tip)
Water droplet size distribution	Narrow (span 1.2–1.6)	Broader (span 1.8) (span = $(d_{90}-d_{10})/d_{50}$)
Dried particle size range	0.3–5 μm	2–25 μm
Maximum drying temperature	120°C	220°C
Maximum drying gas flow	10 m^3/h (laminar)	35 m^3/h (turbulent)
Drying chamber diameter	0.18 m	0.165 m
Water evaporation capacity	200 mL/h	1 L/h
Water feed rate (depends on application, parameter settings, and spray head)	10–20 mL/h (4.0 μm spray mesh) 25–50 mL/h (5.5 μm) 80–150 mL/h (7.0 μm)	Depends on peristaltic pump settings, 50% corresponds to about 900 mL/h
Particle separation device	Electrostatic particle collector	Cyclone and glass vessel
Yield	Up to 90%	Up to 70%
Scale-up capability	Limited by vibrating mesh technology and electrical particle collector	Possible to upscale to kg and ton scale

Source: Data summarized from Arpagaus (2011); Arpagaus et al. (2010a,b); Büchi Labortechnik AG (2010, 2013).

10.2.1 Droplet generation

The droplet generation in the nano spray dryer is based on vibrating mesh technology. Fig. 10.3 illustrates the mode of operation of this technology. The underlying physical principle of droplet formation is utilized in ink-jet printers and vibrating mesh nebulizers for inhalation therapy (Knoch and Keller, 2005; Lass et al., 2006; Smart et al., 2002; Vecellio, 2006). An overview of electronic nebuilizers that use vibrating mesh technology to generate fine low-velocity aerosols has been given by Dhand (2002).

The vibrating mesh technology is incorporated in a spray head for nano spray drying. It includes a piezoelectric actuator driven by an electronic circuit. A small exchangeable spray cap is screwed onto the lower part of the spray head at a defined torque. The spray cap is made of a thin stainless steel mesh which is perforated with an array of precisely micron-sized holes. Spray meshes are available with orifice diameters of 4.0, 5.5, and 7.0 μm.

The large number of holes are made in the thin membrane by laser drilling. These holes have a conical shape with the larger diameter toward the liquid supply side and the narrow diameter facing the droplet release side (Fig. 10.3). By appropriate adjustment of the parameters during the drilling process, the mesh can be optimized for the production of droplets of a certain size (Lass et al., 2006). The drilling procedure determines the number and position of the nozzles, their tapered shape, and orifice size (Lass et al., 2006). The shape of the holes has been optimized for the creation of individual droplets.

Plastic tubes connect the spray head to the feed reservoir. A peristaltic pump with variable speed circulates the fluid uniformly from the reservoir through the spray head to the spray mesh and back to the reservoir. This circulation mode enables efficient and continuous atomization.

Electricity drives the piezoelectric actuator at an ultrasonic frequency of about 60 kHz, which is close to the resonance frequency of the spray cap, thus minimizing electrical energy consumption. The piezoelectric crystal contracts and expands resulting in an upward and downward oscillation of the spray mesh. The mesh deforms by a few microns into the fluid side, thus pumping and loading the holes with fluid. The deformation on the other side of the fluid ejects the droplets through the holes into the drying chamber. The rapid vibration causes each hole to act as a micropump that draws the fluid through the holes to form identically sized droplets. The output is a fine low-velocity aerosol optimized for nano spray drying. Millions of precisely sized droplets with narrow size distribution are created every second. The droplet size depends on the mesh hole size and the physicochemical properties of the fluid, such as viscosity and surface tension. The output droplet size of a 4-μm spray mesh is about 3 to 8 μm (span 1.2 to 1.6) (Schmid et al., 2011). Liquids that are too viscous may result in intermittent droplet generation or even cessation of droplet generation. A major benefit of the vibrating mesh technology is the ability to aerosolize very small amounts of fluids in the millilitre range. The power unit controls the amplitude and frequency of the vibration. The spray rate intensity can be adjusted by the user by setting the percentage value of the ON time.

10.2.2 Drying of droplets

The Nano Spray Dryer B-90 uses a patented system to heat up the laminar gas flow (EP 2056037 A1; Schön and Baumgartner, 2009). The drying gas flows through a compact porous metal disc that enables efficient heat transfer from the hot metal surface to the drying gas. The porous structure is embedded with an electrical heating coil. This coil ensures optimal energy transfer to the porous metal and uniform heat distribution in the entire heater volume. The heating system takes only a few minutes to raise the inlet temperature from ambient to 120°C. It also ensures a laminar flow of gas in the spray chamber to dry the fine aerosol. This is crucial because turbulence would lead to uncontrolled spray formation and particle loss on the drying chamber walls. The laminar flow enables gentle drying, which makes the system suitable for drying heat-sensitive products.

The aerosol droplets are injected from the top of the nano spray dryer in the same downward direction as the flow of drying gas. The aspirator speed maintains the drying gas flow rate within a range of 80 to 160 L/min. The vertical configuration guides the submicron particles toward the electrostatic particle collector while minimizing particle adhesion to the sidewalls of the drying chamber, thus resulting in high collection efficiency. Thanks to these features, i.e. mild drying conditions, laminar gas flow, and co-current operation, the product is treated with care and drying takes place gently. Nano spray drying is thus a very suitable technology for heat-sensitive products as it has low risk of degradation or activity loss.

The typical system configuration for aqueous feeds is the open mode with air used as drying gas. Closed mode operation with an Inert Loop B-295 equipment can be used for oxygen-sensitive substances and flammable organic solvents. In such cases, inert gases, usually nitrogen and carbon dioxide, need to be used as the drying medium to prevent the formation of an explosive gas mixture. The oxygen concentration is kept below 4% and the inert gas is circulated throughout the entire circuit. An admixture of CO_2 is necessary to prevent a dielectric breakdown at the collecting electrode of the electrostatic particle collector.

10.2.3 Particle collection

The particle collection mechanism in the nano spray dryer is based on electrostatic charging, which, unlike cyclones, is independent of particle mass (Arpagaus, 2012; Lee et al., 2011). Fig. 10.3 shows the electrostatic particle collector and illustrates its principal function. The working principle consists of the following basic steps:

1. generation of an electrical field of about 15 kV voltage between the discharge and collecting electrodes,
2. generation of negatively charged ions in the gas,
3. charging of the spray dried particles, and
4. deflection of the particles to the collection electrode.

The electrostatic particle collector in the Nano spray dryer has a tubular shape. It is comprised of a stack of star-shaped discharge electrodes and a smooth cylindrical colleting electrode. A high voltage of about 15 kV is applied to the metallic cylinder to generate an electrical field between the discharge and the collecting electrodes. As the electrical potential difference increases, the gas near the sharply curved tips of the star-shaped center electrodes begins to ionize. A corona discharge appears at the surface and extends into the gas. This creates a luminous glow discharge around the electrode that is visible in the dark.

The glow discharge ionizes the drying gas, forming negative gas ions that move toward the collecting electrode. As they drift toward the collecting electrode, the gas ions stick to the suspended spray-dried particles in the drying gas, giving them a negative polarity. The Coulomb force moves the charged particles along the path of invisible electric field lines toward the positively charged collecting electrode, where they attach and build up a layer of particles. On coming in contact with the collecting electrode, the particle charge is dissipated. The drying gas exits the collector in a purified form. Further details about the principle function of electrostatic particle collectors (also called electrical precipitators) can be found in the works of Riehle (1997), Mizuno (2000), and Masters (1991).

The electrostatic particle collector can capture submicron particles (< 1 μm) at a separation efficiency greater than 99% for small solid batches of 30 to 500 mg (Bürki et al., 2011; Lee et al., 2011; Li et al., 2010; Schmid et al., 2011). The electrostatic precipitator can even collect thin-walled particles without breaking (Feng et al., 2011; Sun et al., 2011). The particles are gently removed from the internal surface of the collecting electrode cylinder by utilizing the particle scraper and particle collecting paper that are included in the delivery of the laboratory instrument. The particles are filled in airtight glass vials and normally stored in a desiccator until further usage and analysis. Commonly applied analytical methods to characterize the produced spray-dried powders are scanning electron microscopy (SEM) to determine particle size and morphology, laser diffraction to measure particle size, and X-ray diffraction or differential scanning calorimetry to identify the amorphous/crystalline state. The analysis by laser diffraction typically results in higher mean particle diameters compared to SEM, which is attributed to the different measurement principles (Beber et al., 2014). Thanks to the modular assembly of the collecting electrode, the set-up is simple to clean, thus saving time.

10.3 Optimizing the Nano Spray Drying Process Parameters

10.3.1 Influences of process parameters

Before the spray drying application work begins, it is advantageous to conduct feasibility tests at the laboratory bench using a small sample of the material to determine the feasibility of the application. Nano spray drying is a technology in which the properties of the fabricated powder depend on the correlation and interdependency of the process parameters and formulation variables. Fig. 10.4 illustrates the main adjustable process parameters and formulation variables for a nano spray dryer (marked black), as well as the outputs (marked grey).

Table 10.2 provides an overview of the main nano spray drying process parameters and presents their influence on the final powder characteristics. The thickness of each arrow illustrates the strength of the related influence. Some guidelines for the

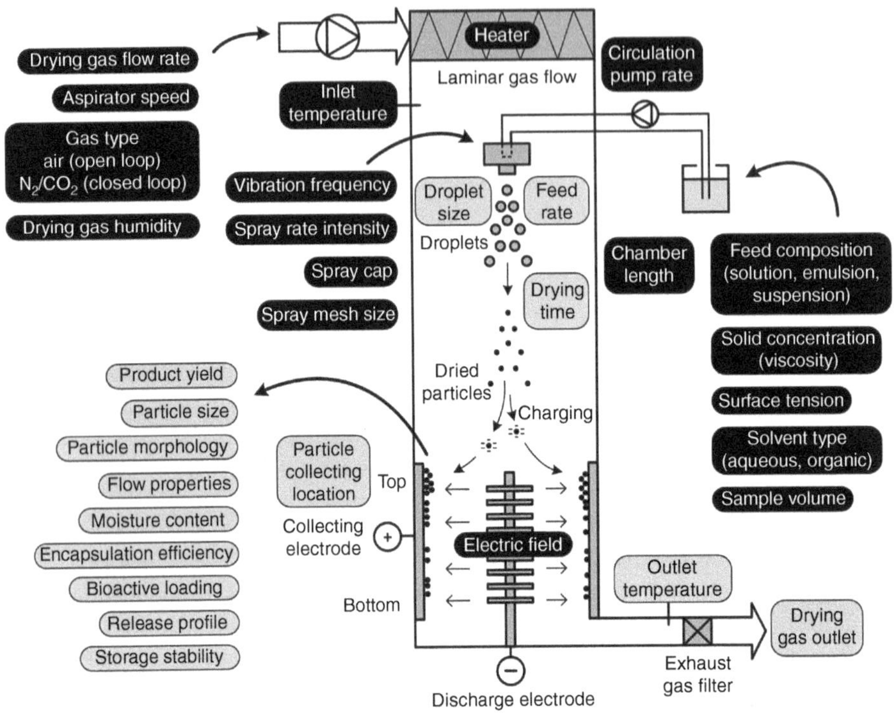

Figure 10.4 The main adjustable process parameters and formulation variables for a nano spray dryer *(black)*, as well as the outputs *(gray)*.

optimization of the process of formation of nanocapsules have been identified and are described in later sections. Some initial process conditions are listed, which can be used for nano spray drying identical or similar products.

The optimization of the process parameters is usually done by trial and error. However, design of experiment (DoE) studies are a useful approach to optimize the nano spray drying process. Several authors have investigated the effects of the main input process parameters and formulation variables by using factorial experimental designs. Table 10.3 gives a summary of a few such cases. The designs allow us to determine the optimal working conditions with fewer experiments, thereby reducing costs and material usage.

The most important input parameters identified are the spray mesh size, the spray rate intensity, the solid concentration, the polymer and surfactant concentrations, the drying gas inlet temperature, the solvent type, and the drying gas flow rate. The effects on the particle size, product yield, and the amount of output produced per second are of primary importance. Other output parameters of interest are particle morphology, bioactive loading, controlled release profile, encapsulation efficiency, and stability. Overall, DoE studies provide valuable information that helps in predicting the outputs of future experiments.

Table 10.2 Nano spray drying process parameter table

Parameter	Drying gas flow rate ↑	Drying gas humidity ↑	Inlet temperature ↑	Spray mesh size ↑	Spray rate intensity ↑	Circulation pump rate ↑	Solid concentration (viscosity) ↑	Surfactant/stabilizer in feed ↑	Solvent instead of water
Outlet temperature	⬆	↑	⬆	↑	⬇	—	↑	—	⬆
Droplet size	—	—	—	⬆	↑	↑	—	↓	↓
Particle size	—	—	↑	⬆	↑	↑	⬆	↓	↓
Feed rate	—	—	—	⬆	⬆	↑	⬇	↑	↑
Moisture content	⬇	⬆	↓	—	↓	—	↓	—	⬇
Yield	—	↓	↑	—	—	—	↑	↑	↑
Stability	—	—	⬇	↑	⬇	↑	—	⬆	—

⬆/⬇, Strong increasing/decreasing influence; ↑/↓, weak increasing/decreasing influence; —, minimal or no influence.

Table 10.3 **Design of experiment (DoE) studies with regard to nano spray drying**

Process & Formulation Parameters								Output Parameters											References and Applications
Drying gas inlet temperature	Spray mesh size	Spray rate intensity/pump rate	Solid concentration	Polymer concentration/type	Surfactant concentration/type	Solvent concentration/type	Drying gas flow rate	Particle size	Particle size distribution	Particle shape/morphology	Crystallinity	Yield	Product produced per time	Bioactive loading	Encapsulation efficiency	Moisture content	Enzyme activity/shelf life	Number of experiments	
■		■		■	■	■		□	□			□						56	Draheim et al. (2015) *PLGA*
■	■					■		□	□	□	□	□					□	27	Bürki et al. (2011) *β-Galactosidase in trehalose*
■	■		■					□		□		□	□					27	Littringer et al. (2013) *Salbutamol sulfate*
■	■		■					□	□			□						25	Schoubben et al. (2014) *Inhalable capreomycin*
			■	■				□	□			□						21	Li et al. (2010) *Different polymers*
■			■	■			■					□	□					20	Gu et al. (2015) *Model polymer*
■	■			■		■	■	□				□			□			18	Lee et al. (2011) *Bovine serum albumin*
■	■	■	■	■				□			□	□	□	□	□			9	Schafroth et al. (2012) *Cyclosporin in PLGA*
					■	■		□		□	□	□		□		□		9	Durli et al. (2014) *Dexamethasone in PLGA*

■; □, Process parameters investigated.

10.3.2 Drying gas flow rate, humidity and temperature

Fig. 10.5 illustrates the typical temperature profile of the drying gas and the resulting droplets in the nano spray dryer. The gas temperature falls rapidly on coming in contact with the aerosol droplets. The decrease in temperature becomes slower with an increase in drying time and distance as the outlet temperature for the dryer is gradually reached. The droplet temperature rises initially from the feed temperature to the saturated surface temperature and afterward it remains constant during the first period of drying (constant evaporation at wet-bulb temperature). Adequate migration of moisture from within the droplet to the surface maintains the droplet surface saturation. During drying, the droplets can shrink, agglomerate, and lose sphericity as the moisture gets evaporated. As calculated by Feng et al. (2011), the drying time periods for the droplets are in the order of 10 ms, assuming an initial water droplet diameter of 7 μm, drying temperature of 75°C, and drying air flow rate of 100 L/min.

At the end of the drying period, the particle surface temperature (see droplet curve in Fig. 10.5) approaches the temperature of the surrounding gas in the dryer outlet, which determines the final outlet temperature. The outlet gas temperature in the nano spray dryer is mainly controlled by the drying gas flow rate, the inlet drying gas temperature, and the fluid feed rate (which in turn is primary controlled by the spray rate intensity and spray mesh size). Moreover, the outlet temperature is a derivative of the evaporation enthalpy of the solvent, the concentration of the solid in the liquid, and heat loss through the drying chamber wall. The outlet temperature is the highest temperature to which the product may be heated.

Fig. 10.6 shows the range of achievable outlet temperatures for the nano spray dryer at different inlet temperatures, when applied to aqueous solutions. From the setting of the inlet temperature, it is possible to predict the outlet temperature by using a

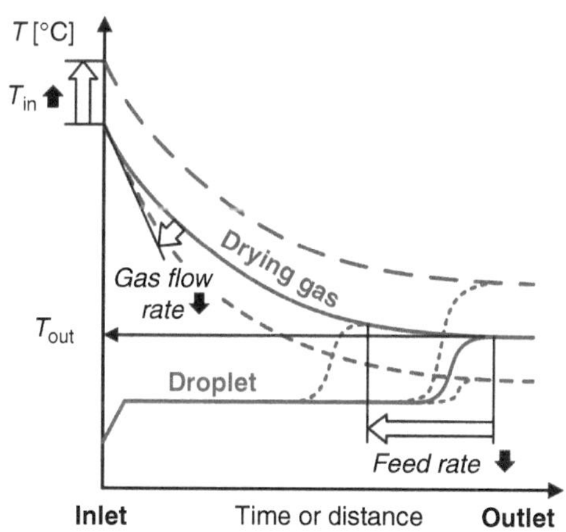

Figure 10.5 Typical temperature profile of the drying gas and the resulting droplets in the nano spray dryer The gas temperature falls rapidly on coming in contact with the aerosol droplets.

Source: Adapted from Arpagaus, C., Rütti, D., Meuri, M., 2013. Enhanced solubility of poorly soluble drugs via spray drying. In: Douroumis, D., Fahr, A. (Eds.), Drug Delivery Strategies for Poorly Water-Soluble Drugs, John Wiley & Sons, Ltd., pp. 551–585.

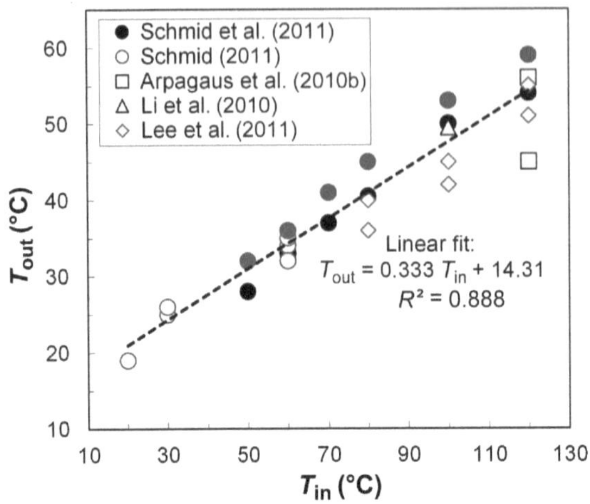

Figure 10.6 The range of achievable outlet temperatures for the nano spray dryer at different inlet temperatures, when applied to aqueous solutions. From the setting of the inlet temperature, it is possible to predict the outlet temperature by using a linear fit approximation.

linear fit approximation. The outlet temperature lies between 28°C and 59°C and thus is lower compared to traditional spray drying technology. This makes nano spray drying a very suitable process for heat-sensitive biological materials (Aquino et al., 2014; Heng et al., 2011). A higher inlet temperature causes a proportional increase in the outlet temperature. A higher inlet temperature also reduces the relative humidity of the drying gas, resulting in a higher yield and drier, less sticky powder as output. Schmid (2011) observed a residual moisture content lower than 0.5% in mannitol powders and 2–5% in trehalose, indicating that the product would be amorphous. To achieve a final product with low residual moisture, the inlet temperature has to be set as high as possible and the difference in temperatures of the inlet and the outlet has to be as small as possible (Arpagaus et al., 2013).

A smaller feed rate increases the outlet temperature because less fluid is evaporated over the same period of time. The spray rate intensity correlates to the time during which the vibration of the spray mesh is active, and it is adjustable between 10% and 100% by the operator. A low spray intensity results in a reduced throughput and the overall process takes longer. An increase in the spray rate intensity causes more droplets to form per unit of time; however, the size of the droplets is only marginally changed, if at all (Littringer et al., 2013). Decreasing the spray rate intensity from 100% to 25% reduces the particle size from about 7.7 μm to 5.3 μm and increases the span from about 1.8 to 2.1 (Beck-Broichsitter et al., 2015b).

A lower drying gas flow rate reduces the slope of the gas temperature profile (i.e. it decreases the drying energy) and leads to a lower outlet temperature (see the drying gas curve in Fig. 10.5). The drying gas flow rate in the nano spray dryer can be adjusted to a value between 80 L/min and 160 L/min by setting the aspirator speed. This range corresponds to gas velocities of about 0.05–0.1 m/s inside the drying chamber and a laminar gas flow with Reynolds numbers between 345 and 690. The gas residence time in the drying chamber is about 12 to 20 s.

A higher degree of humidity in the drying gas means a lower capacity for vapor uptake, resulting a slightly higher outlet temperature. The humidity of the final product increases; this could lead to moist particles that stick to the glasswall and consequently to a lower yield.

The drying gas flow rate and the inlet temperature have very little impact on the particle size, as shown by Lee et al. (2011) and Baba and Nishida (2012). Interestingly, when the temperature is as high as 120°C, hollow sphere particles can be seen on SEM micrographs, as shown in Fig. 10.14 (Suryaprakash et al., 2014). Vehring (2008) describes this phenomenon with the help of the dimensionless Peclet number (Pe), which is the ratio of the droplet evaporation rate to the diffusion coefficient of the solutes inside the droplet. If Pe is less than one, drying is slow enough for the solute to distribute evenly inside the evaporating droplet by diffusion. The solute precipitates instantly throughout the droplet, leading to dense dried particles. If Pe is greater than one, the evaporation of the solvent takes place faster than the diffusion. The particle surface becomes enriched with the solutes, which can form a shell or skin at high drying temperatures, impeding the further shrinking of the droplet. Since this shell formation takes place when the droplet volume is still large, it results in larger particles.

10.3.3 Droplet size

To control the size of the droplets, the properties of the atomized fluid, such as viscosity and surface tension, need to be considered. Hagen-Poiseuille's law governs the rate of flow through small holes. A higher fluid viscosity leads to a lower flow rate. Hence, a lower portion of fluid volume is delivered during the mesh vibration, leading to the formation of smaller droplets, as shown experimentally in a study of vibrating mesh nebulization by Ghazanfari et al. (2007). Surface tension and hydrodynamic effects cause the liquid jets to break into a cloud of precisely controlled droplets Smart et al. (2002). The detachment of smaller droplets is favored by a higher viscosity, a lower surface tension, and a smaller spray mesh diameter (Beck-Broichsitter et al., 2012). Organic solvents typically generate smaller droplets because they have lower surface tension, viscosity, and density. Suryaprakash et al. (2014) found that increasing the methanol concentration in an aqueous feed results in smaller initial droplets.

Schmid et al. (2011) found a strong correlation between the size of the spray mesh used and the resulting droplet size by spraying water directly into the beam of a laser diffraction instrument. Fig. 10.7 illustrates these findings. Mean droplet sizes of 4.8, 6.0, and 7.2 μm were obtained with ± 0.5 μm reproducibility using the 4.0, 5.5, and 7.0 μm spray meshes respectively. A smaller spray mesh size leads to smaller droplets and consequently to a lower dried particle size.

In addition, (Schmid et al., 2011) measured the droplet sizes of trehalose solutions (at 0.1%, 1%, and 10% concentration) with and without surfactant. Droplets with a mean size of approximately 5 μm were obtained with a 4.0 μm spray mesh, irrespective of the trehalose concentration or the addition of surfactant. This implies that the total solid concentration and the viscosity or the surface tension of the spray solution

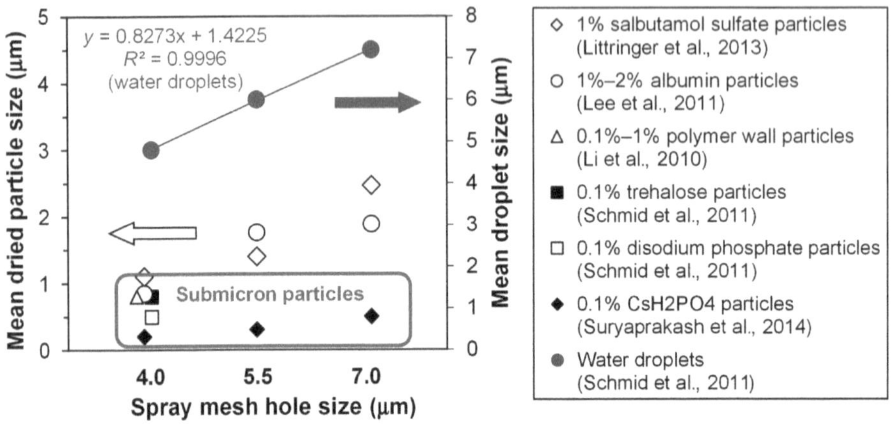

Figure 10.7 Correlations between droplet and particle sizes. The submicron particle size is reached when using a 4.0 μm spray cap and diluted solid concentrations of about 0.1%–1% (w/v). Mean droplet sizes of 4.8, 6.0, and 7.2 μm were obtained with ± 0.5 μm reproducibility using the 4.0-, 5.5-, and 7.0-μm spray meshes respectively. A smaller spray mesh size leads to smaller droplets and consequently to a lower dried particle size.

do not greatly influence the size of droplets formed with the vibrating mesh technology (Schmid et al., 2011). The spray mesh size primarily determines the droplet size. However, the output feed rate decreases with an increase in the concentration of the solution.

Compared to a traditional Mini Spray Dryer B-290, the Nano Spray Dryer B-90 produces droplets of a smaller size. The traditional spray dryer generates water droplets in the range of about 9 to 29 μm (Maltesen et al., 2008) with a mean size of approximately 14 μm (Schmid et al., 2011), depending on the gas and feed flow rates in the two-fluid nozzle. In addition, the droplets from the nano spray dryer have a narrower size distribution (span values of about 1.2 to 1.6) than those from the Mini Spray Dryer B-290 (span values of approximately 1.8).

10.3.4 Particle size

The spray mesh size determines directly the size of the droplets and consequently that of the dried solid particles. Fig. 10.7 and Table 10.4 illustrate various correlations for different applications. The submicron particle size region is typically reached when using a 4.0 μm spray mesh and diluted solutions at 0.1–1% (w/v) solid concentrations. Further reduction of particle size is possible by further dilution.

Fig. 10.8 shows some SEM images of the bovine serum albumin particles Lee et al. (2011) obtained with a 4.0, 5.5, and 7.0 μm spray mesh respectively. Particles of approximately 0.7, 1.7, and 2.6 μm in size respectively were achieved at 1% (w/v) solid concentration.

Table 10.4 summarizes the literature data demonstrating the influence of the spray mesh size and the solid concentration on the particle size. As can be seen, particles as small as 100 nm can be reached when the solution is highly diluted. Assuming a one to one transformation of atomized droplets to dried particles, the final particle

Table 10.4 **Influence of spray mesh size on particle size (in nanometers)**

Substance	Solvent	Concentration (% w/v)	Mesh size 4.0 μm	5.5 μm	7.0 μm	References
Chitosan (low-density)	0.5% acetic acid	0.025	95	215	265	Ngan et al. (2014)
Ethambutol	Water	1	220	—	—	Ahmad et al. (2014)
Gentamicin sulfate in alginate/pectin	Water	0.1	310	520	850	De Cicco et al. (2014)
		0.25	345	550	980	
		0.5	405	610	1000	
Arabic gum	Water	0.1	355	—	—	Li et al. (2010)
		1	580	—	—	
Whey protein	Water	0.1	420	—	—	Li et al. (2010)
		1	595	—	—	
Sodium chloride	Water	0.1	515	—	—	Li et al. (2010)
		1	995	—	—	
Disodium phosphate	Water	0.1	500	—	—	Schmid et al. (2011)
Trehalose	Water	0.1	800	—	—	Schmid et al. (2011)
Bovine serum albumin (with surfactant)	Water	0.1	460	—	—	Lee et al. (2011)
		1	700	1700	2600	
Sodium alginate	Water	0.13	760	—	—	Blasi et al. (2010)
Vancomycin in chitosan	0.5% acetic acid	0.2	450	—	—	Cerchiara et al. (2015)
Curcumin in albumin	Water	0.5	715	—	—	Jain (2014)
Salbutamol sulfate	Water	1	1000	1600	3100	Littringer et al. (2013)
Galactosidase in trehalose	Water	5	1800	4500	4900	Bürki et al. (2011)

(*Continued*)

Table 10.4 **Influence of spray mesh size on particle size in nanometer (*cont.*)**

Substance	Solvent	Concentration (% w/v)	Mesh size			References
			4.0 μm	5.5 μm	7.0 μm	
CsH_2PO_4	Methanol:water (56:44)	0.1	200	300	500	Suryaprakash et al. (2014)
		1	540	—	—	
Calpain inhibitor steroid nano-crystals	Ethanol	0.05	420	605	845	Baba and Nishida (2012)
		0.5	995	—	—	
Fluorometholone	Ethanol	0.1	620	795	855	Baba and Nishida (2013)
Dexamethasone	Ethanol	1	835	1100	1300	Baba and Nishida (2013)
PLGA	DCM	1	—	3200	4200	Dahili and Feczkó (2015)
Ethyl cellulose	DCM	1	1500	2700	3400	Dahili et al. (2015)
PLGA	DCM	0.5	—	570	—	Beck-Broichsitter et al. (2012)
		1	—	760	—	
		2	685	830	930	
		5	—	1100	—	

—, Data not available.

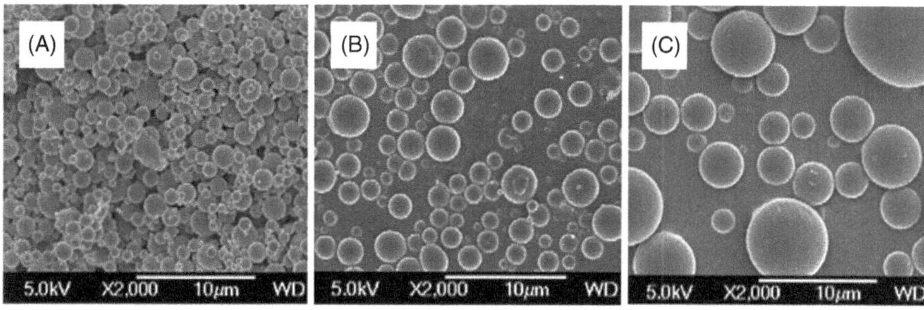

Figure 10.8 Scanning electron microscopy (SEM) images of the bovine serum albumin particles that Lee et al. (2011) obtained with a (A) 4.0-, (B) 5.5-, and (C) 7.0-μm spray mesh, respectively. Particles of approximately 0.7, 1.7, and 2.6 μm in size respectively were achieved at 1% (w/v) solid concentration.

size is directly correlated to the droplet size and to the solid concentration in the feed. The relationship can be approximated the equation $D_P \sim C^{1/3} D_D$ (Nandiyanto and Okuyama, 2011) where D_D and D_P are droplet and particle sizes and C is the solid concentration in the feed, assuming that droplets are spherical and have a homogeneous composition.

The results from a study by Beck-Broichsitter et al. (2015a) agree with this relationship. Similar trends have been observed by Beck-Broichsitter et al. (2012) in an experiment where aqueous nanosuspensions were spray dried. Higher nanoparticle concentrations and the use of a wider spray mesh were found to generate larger composite particles. As expected, particles size as well as distribution increased with the spray mesh size. The particles size distribution also became wider at a higher feed concentration (Beck-Broichsitter et al., 2012). Suryaprakash et al. (2014) found a cube-root fit function to the data, which showed the particle diameter increasing monotonically with increasing solid concentration. This indicates that the initial droplet size does not change significantly with the solid concentration. As a consequence, Beck-Broichsitter et al. (2015a) coated a 4.0 μm spray mesh with a polymer by means of chemical vapor deposition. The uniform polymer layer reduced the mesh hole cross-section by 50% from 4.0 μm to about 2.0 μm. As a result, finer droplets with narrower size distribution could be generated. The new spray mesh resulted in significantly smaller particles (1.3 μm vs. 3.6 μm) with narrower distribution (span of 1.4 vs. 1.8). This coating technique could enable easy prototyping and optimization of vibrating mesh designs in the future.

10.3.5 Solid concentration

The solid concentration influences the feed rate, particle size, and outlet temperature. A higher solid concentration implies that there is less liquid for vaporization, which in turn increases the outlet temperature. A higher solid concentration in a droplet increases the particle size according to the previous relationship. Li et al. (2010) spray dried some examples of polymeric wall materials, such as arabic gum and whey protein solutions at concentrations of 0.1%, 1%, and 10% (w/w). At 0.1% concentration, very small particles were achieved with peaks of 353 and 421 nm for arabic gum and whey protein respectively. An increase in sample concentration increases the particle size and the distribution (span). These findings are in agreement with the results of other studies reported in literature (Baba and Nishida, 2013). Beck-Broichsitter et al. (2012) investigated the influence of the molecular weight of poly(lactic-co-glycolic) acid (PLGA) on particles size. A higher molecular weight of PLGA was found to raise the mean particle size. An increase in the solid concentration of PLGA results in an increase in the kinematic viscosity and density of the feed solution (Beck-Broichsitter et al., 2015b). Ngan et al. (2014) observed a similar effect with chitosan at a highly diluted concentration of 0.025% (w/v). The average particle size increased with molecular weight. A direct correlation between viscosity and the size of dexamethasone-loaded poly(caprolactone) particles was also discovered by Fontana et al. (2014).

10.3.6 Feed rate

The feed rate (amount of fluid sprayed per unit of time) in the nano spray dryer depends on the spray mesh size, the set value of the relative spray rate intensity, the recirculation pump rate, and the feed formulation (i.e. the substance, the solid concentration, the solvent type, and the addition of surfactant). Table 10.5 shows the influence of the spray mesh size on the feed rate for various substances.

The feed rate increases with mesh size, but also depends on the application, the parameter settings, and the spray head used. According to the manufacturing information from Büchi Labortechnik, 2010, the water feed rate is in the ranges of 10–20, 25–50, and 80–150 mL/h for a 4.0, 5.5, and 7.0 μm spray mesh respectively. When using an aqueous substance and a 4.0 μm spray mesh, the feed rates may vary in a range of 10 mL/h (Arpagaus et al., 2010a) to 16 mL/h (Sun et al., 2011) or 3–25 mL/h (Li et al., 2010). With a 5.5 μm spray mesh and chitosan in 1% acetic acid, feed rates up to about 50 mL/h were measured by Gautier et al. (2010). The addition of organic solvents to the water tends to increase the feed rates. This can be attributed to the lower surface tension. Pretreating the mesh with surfactant solutions increases the feed rate, as shown by Aquino et al. (2014). A 4.0 μm mesh was dressed with a thin film of *n*-hexane as surfactant prior to nano spray drying. This led to an increase in throughput from about 8 to 24 mL/h. The further admixture of a small amount of surfactant directly into the feed also improved the throughput of trehalose spray solution, as reported by Schmid (2011). Draheim et al. (2015) obtained smaller particles by adding surfactant.

The spray rate intensity is another adjustable parameter. The operator can set the spray rate intensity to any value in a range between 10 and 100%. A lower spray rate intensity reduces the throughput of the spray solution almost linearly, as shown by (Beck-Broichsitter et al., 2015b) for sildenafil loaded PLGA in 5% (w/v) DCM using a 5.5 μm spray mesh.

The circulation rate of the peristaltic pump also influences the droplet size and the feed rate. If the feed is pumped faster during circulation, the pressure on the spray mesh (inside the spray cap) rises, thus increasing the initial droplet size and consequently

Table 10.5 Influence of spray mesh size on feed rate for various substances at 100% spray rate intensity

Substance	Concentration (% w/v)	Solvent	Mesh size (μm)	Feed rate (mL/h)	Reference
—	—	Water	4.0	10–20	Büchi Labortechnik (2010)
			5.5	25–50	
			7.0	80–150	
Arabic gum, maltodextrin, polyvynyl alcohol, modified starch	1	Water	4.0	3–25	Li et al. (2010)
Salbutamol sulphate in lactose	10	Water	4.0	10	Arpagaus et al. (2010a)
Lithium carbonate (Li_2CO_3)	10	Water	4.0	16	Sun et al. (2011)
Levofloxacin in chitosan and leucine	4	Water	5.5	20	Merchant et al. (2014)
Chitosan (Mw 30,000 Da)	0.1	1% acetic acid	5.5	~ 50	Gautier et al. (2010)
Ketoprofen lysinate in leucine	1–5	30% isopropyl Alcohol	4.0	2–8	Aquino et al. (2014)
			5.5	20	
			7.0	67	
Sildenafil in PLGA	5	Acetone	4.0	18 ± 5	Beck-Broichsitter et al. (2015a)
Sildenafil in PLGA	5	DCM	4.0	23 ± 6	Beck-Broichsitter et al. (2015b)
			5.5	64 ± 15	
			7.0	163 ± 26	
Dexamethasone in PLGA	0.5	Acetone/water	4.0	34	Fontana et al. (2014)
Dexamethasone in PLGA and different surfactants	0.5	Acetone, ethanol, or propanol	4.0	3–25	Durli et al. (2014)

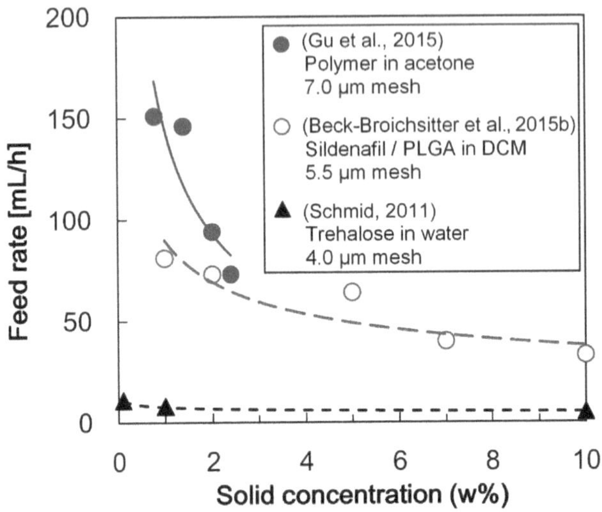

Figure 10.9 Influence of the solid concentration on the feed rate. Generally, a higher solid concentration slows down the discharge of droplets from the vibrating mesh holes (Hagen–Poiseuille's law), resulting in a lower feed rate.

the feed rate. The number of the droplets generated per unit of time remains constant. As shown by Suryaprakash et al. (2014), the mean size of CsH_2PO_4 particles can be increased from about 150 to 650 nm by increasing the circulation pump rate from 2 to 25 mL/min.

The solid concentration is another important factor that influences the feed rate. Generally, a higher solid concentration slows down the discharge of droplets from the vibrating mesh holes (Hagen–Poiseuille's law), resulting in a lower feed rate (Beck-Broichsitter et al., 2015b; Gu et al., 2015; Schmid, 2011) as shown in Fig. 10.9 and Table 10.6. Table 10.6 also presents the feed rate values of concentrated capreomycin suspensions from the experiments of Schoubben et al. (2013). Flow rates of 27 and 28 mL/h in ethanol–water mixtures, and 35 mL/h in a mixture of acetone and water were achieved by using a 7.0 µm mesh.

10.3.7 Product yield

The product yield is calculated based on the initial solid mass in the feed and the resulting powder weighed after nano spray drying. Under optimized conditions, very high yields of 76%–96% can be achieved for 100 mg sample amounts of various substances, as shown in Table 10.7.

The ability to process small sample amounts makes the nano spray dryer suitable for experimenting with valuable biological materials, such as monoclonal antibodies and recombinant proteins (Heng et al., 2011; Schmid, 2011). The efficient electrostatic particle collector is able to capture most particles. Gu et al. (2015) showed that the yield increases with increase in the polymer concentration and the inlet temperature.

Table 10.6 Influence of solid concentration on feed rate

Substance	Solvent	Spray mesh (μm)	Concentration (% w/v)	Feed rate (mL/h)	References
Trehalose	Water	4.0	0.1	11	Schmid (2011)
			1	8	
			10	5	
Sildenafil-loaded PLGA	DCM	5.5	1	81 ± 21	Beck-Broichsitter et al. (2015b)
			2	73 ± 12	
			5	64 ± 15	
			7	40 ± 9	
			10	33 ± 12	
Hypromellose acetate succinate-LF (model polymer)	Acetone	7.0	0.8	151	Gu et al. (2015)
			1.4	146	
			2.0	94	
			2.4	73	
Capreomycin oleate and linoleate suspension	Ethanol/water	7.0	21	27	Schoubben et al. (2013)
			12	28	
Capreomycin linolenate suspension	Acetone/water	7.0	11	35	Schoubben et al. (2013)

The polymer solution is more easily dried by the heated air, resulting in more convenience during particle collection. However, results from Littringer et al. (2013) implied that the yield is independent of the process parameters and there is no correlation between the initial mass of salbutamol sulfate used for the preparation of the feed solution and the product yield.

Variations in the yield may be due to particle depositions around the spray cap and the chamber wall (Brinkmann-Trettenes et al., 2014; Bürki et al., 2011; Lee et al., 2011; Schmid et al., 2011), or the manual collection of the powder with a scraper. It is recommended to recover the powder from the collecting electrode cylinder in a controlled and dry atmosphere, and to store the samples afterward in glass vials inside a desiccator (Schmid, 2011) to prevent crystallization and moisture uptake.

Studies from Li et al. (2010); Brinkmann-Trettenes et al. (2014); and Suryaprakash et al. (2014) indicate that the electrostatic particle collector captures bigger particles more rapidly than smaller ones, revealing a particle segregation effect over the cylinder length. Powder samples captured on the lower part of the collecting electrode were found to be slightly smaller than the particles from the upper part. These findings correspond to the basic laws of electrostatic physics. The electrostatic force (Coulomb force) acting on the particles increases with surface charge. Larger particles have a bigger surface charge and migrate to the collecting electrode faster than smaller particles. Brinkmann-Trettenes et al. (2014) demonstrated that particles with the same size are deposited at the same horizontal position of the collection cylinder, implying

Table 10.7 Optimized product yields achieved by nano spray drying

Product yield (%)	Solid sample amount (mg)	Substance	References
Food and nutraceutical applications			
Up to 95	30–300	Arabic gum, maltodextrin, polyvinyl alcohol, and modified starch	Li et al. (2010)
60–94	500	β-galactosidase with trehalose	Bürki et al. (2011)
>90	n.a.	Sodium alginate (polysaccharide)	Blasi et al. (2010)
65–80	210	Dexamethasone (steroid) in PLGA	Fontana et al. (2014)
75–91	800	Hypromellose (enteric film coating, emulsifier)	Gu et al. (2015)
68–76	n.a.	Bovine serum albumin (protein)	Lee et al. (2011)
Drug delivery applications			
96	100	Carbopol loaded vildagliptin (antidiabetic drug)	Harsha et al. (2013a)
91–93	~1000	Nicergoline (blood pressure regulating drug)	Martena et al. (2012)
64–92	500	Amoxicillin (antibiotic) in gelatine	Harsha et al. (2013b)
50–78	10–50	Trehalose, mannitol, or disodium phosphate	Schmid et al. (2009, 2011)
72–90	100	Dexamethasone (steroid) in poly(ε-caprolactone)	Beber et al. (2014)
41–90	0.25–2500	Salbutamol sulfate (asthma drug)	Littringer et al. (2013)
71–88	300	Resveratrol in poly(ε-caprolactone)	Dimer et al. (2015b)
74–88	400	Vancomycin (antibiotic)in chitosan	Cerchiara et al. (2015)
62–87	n.a.	Ketoprofen (antiinflammatory drug) in leucine	Aquino et al. (2014)
81–85	2000	Vildagliptin (antidiabetic drug) in gelatine	Harsha et al. (2015a)
70–82	20–200	Bacitracin (polypeptide antibiotic)	Schoubben et al. (2014)
51–79	525	Levofloxacin (antibiotic) in chitosan	Merchant et al. (2014)
77	2000	Fluticasone propionate (hormone) in lactose	Arpagaus et al. (2010a)

n.a., Data not available.

that the nano spray dryer is axially symmetrical. Suryaprakash et al. (2014) measured the deposition rate of particles on carbon paper substrates positioned at various heights along the length of the particle-collecting cylinder. The deposition rate decreased monotonically downstream, in agreement with the earlier considerations.

10.3.8 Organic solvent instead of water

Typical organic solvents generally used in nano spray drying applications are ethanol (Baba and Nishida, 2013; Zellnitz et al., 2015a), methanol (Öztürk et al., 2015; Schmid, 2011; Suryaprakash et al., 2014), acetone (Beber et al., 2014; Beck-Broichsitter et al., 2015b; Durli et al., 2014; Fontana et al., 2014; Li et al., 2010; Perecin et al., 2015), dichloromethane (DCM) (Bege et al., 2013; Panda et al., 2014; Schafroth et al., 2012), and mixtures thereof with water (Aquino et al., 2014; Dimer et al., 2015a; Kanogwan and Srichana, 2016; Martena et al., 2012; Schoubben et al., 2015, 2013; Son et al., 2013b).

The selection of the solvent is based on the solubilization of the bioactive compound and the encapsulating wall materials (e.g., leucine, chitosan, ethyl cellulose, PLGA, and polycaprolactone polymer). Acetone–water mixtures dissolve for example dexamethasone well and the low viscosity of the acetone leads to higher feed rates through the vibrating mesh, thus reducing the processing time (Fontana et al., 2014). The mixing ratio is adjusted to allow the complete dissolution of the compounds (Schoubben et al., 2013). Organic solvents require less energy to vaporize compared to water, which increases the temperature of the drying gas at the outlet. Therefore, the inlet temperature setting needs to be optimized according to the solvent type to achieve complete evaporation. DCM is widely used as a solvent in drug encapsulation applications, as it offers a low boiling point and a high PLGA solubility. PLGA requires low glass transition temperatures and needs to be processed at low inlet temperatures to avoid problems related to sticky output. Optimum inlet temperatures for DCM were found to be in the range of 29–32°C (Schafroth et al., 2012). Therefore, the DCM evaporation and particle formation can develop under near-ambient conditions, which is very suitable for temperature-sensitive compounds.

10.3.9 Particle morphologies

SEM is normally used to examine nano spray dried powders, as it provides important information regarding particle morphology, such as size, shape, and surface structure of particles. Optical microscopy by fluorescence is used to confirm proper encapsulation, as shown by Bege et al. (2013) with fluorescent dye-loaded PLGA particles and (Pérez-Masiá et al., 2015) with encapsulated folic acid in whey protein and resistant starch.

The morphology of particles prepared by the nano spray drying method include dense, hollow, porous, and encapsulated structures with spherical, wrinkled, shriveled, or doughnut-like shapes. The particle shape depends on the feed properties (e.g., material type, solid concentration, solvent, and surfactant) and drying conditions (e.g., temperatures). In general, slow drying yields more compact particles, while fast drying leads to the formation of hollow particles (Nandiyanto and Okuyama, 2011). Figs. 10.10–10.15 show the different particle shapes prepared by nano spray drying.

A spherical shape is the most stable form for a droplet. Therefore, the formation of compact spherical particles can be expected in nano spray drying. Several researchers have optimized the smoothness and sphericity of particles by adding surfactants

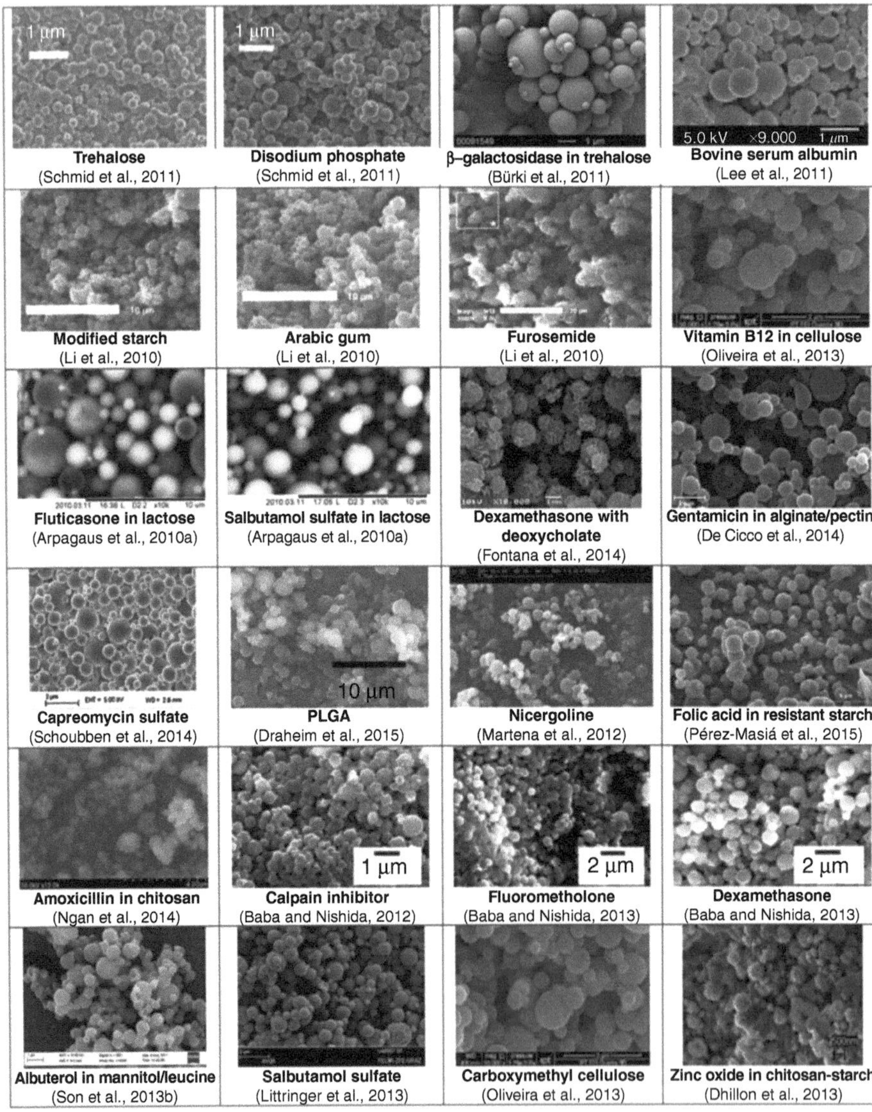

Figure 10.10 Spherical particles prepared by nano spray drying.

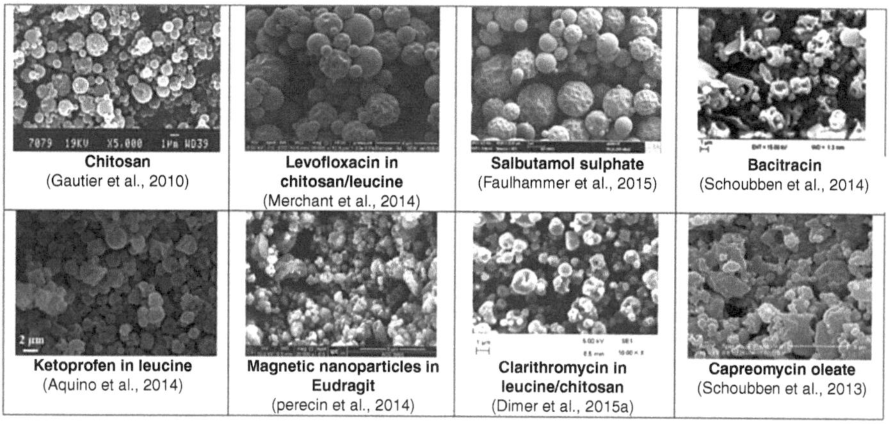

Figure 10.11 Shriveled/wrinkled particles prepared by nano spray drying.

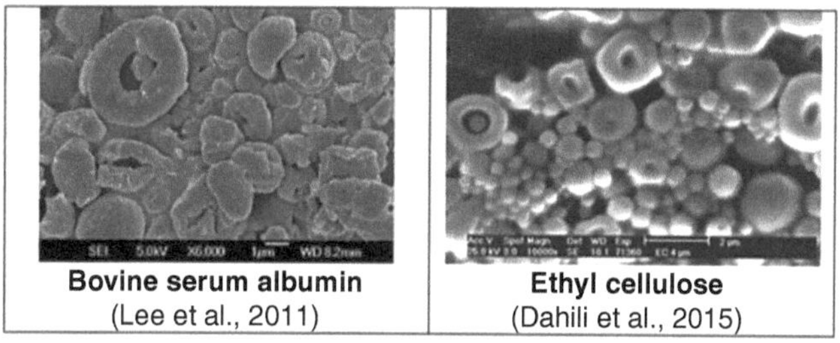

Figure 10.12 Particles with doughnut shape prepared by nano spray drying.

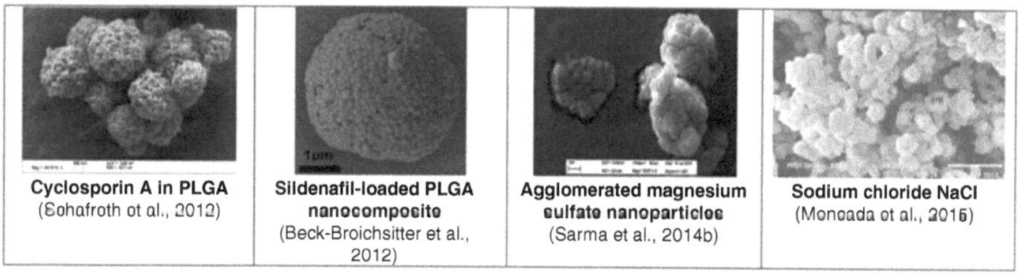

Figure 10.13 Composite particles prepared by nano spray drying.

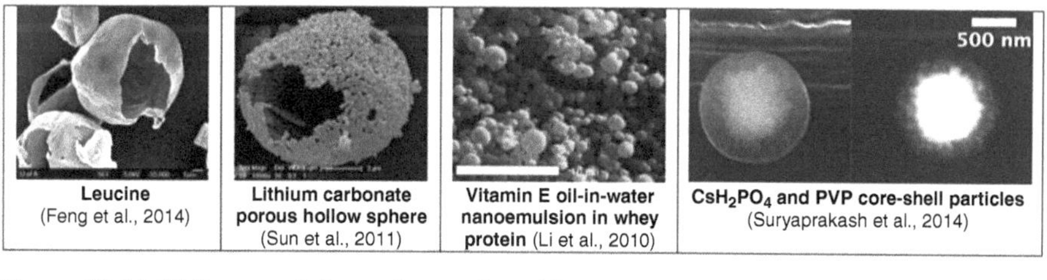

Figure 10.14 Hollow particles and capsules with encapsulated liquid prepared by nano spray drying.

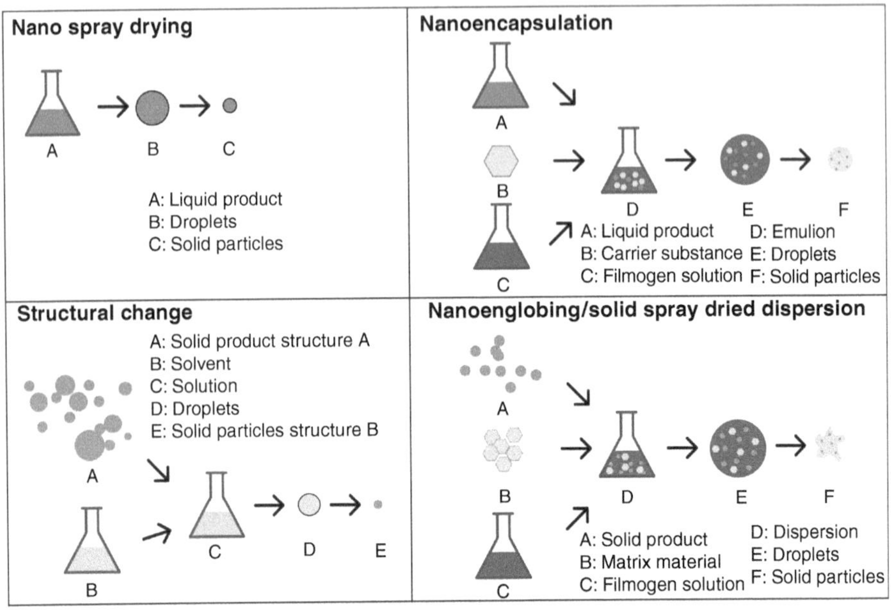

Figure 10.15 Application areas for a nano spray dryer—nano spray drying, structural change, nanoencapsulation, and nanoenglobing of solid dispersions.

(Bürki et al., 2011; Li et al., 2010; Schmid et al., 2009, 2011). The surfactant balances the surface-to-viscous forces inside the drying droplet and enables the formation of a smooth spherical surface on the dry particle. More corrugated surfaces are observed for dexamethasone without the encapsulating polymer (Fontana et al., 2014).

Shriveled/wrinkled particles are obtained from surfactant-free formulations (Lee et al., 2011). Larger particles are more likely to have shriveled surfaces (Bürki et al., 2011).

Doughnut-like particles are the result of hydrodynamic effects and the loss of structural stability of the sprayed droplets (Dahili et al., 2015; Lee et al., 2011). It may occur due to high viscosity, which affects the morphology of the particles. However, a doughnut shape is not preferred in encapsulation, since it results in a higher specific surface area compared to a spherical shape.

Composite particles are due to nanoparticle suspensions forming prior to the nano spray drying. The composites consist of agglomerated nanoparticles, which give rise to a high specific surface area (Beck-Broichsitter et al., 2012). The nanoparticles preserve their integrity, specific properties, redispersibility, enhanced solubility, and targeting capacity (Sarma et al., 2014b).

Hollow particles with a thin shell are formed due to high temperatures and fast evaporation (Sun et al., 2011). Use of surface-active compounds, such as leucine (Feng et al., 2011) and the addition of organic solvents in the formulations promote the formation of hollow particles. Rapid drying forms the particle structure early and does not allow the particles to shrink during drying (Kanogwan and Srichana, 2016).

10.3.10 Encapsulation efficiency and active compounds loading

The selection of a suitable wall material for nanoencapsulation by nano spray drying is based on factors similar to that for microencapsulation (Gharsallaoui et al., 2007). The relevant criteria are mechanical strength, compatibility with the encapsulated active compound, appropriate release properties, high encapsulation efficiency, and stability under storage conditions. Carbohydrates, gums, and proteins are commonly used. PLGA, poly(ε-caprolactone), chitosan, carbopol, and L-leucine are other wall materials used frequently in delivery applications. Encapsulation efficiency (ratio of actual to theoretical active compound content) and active compound loading (ratio between active compound amount in particles and total amount of particles) are other relevant parameters for nanocapsules. Table 10.8 summarizes some representative literature data. Nanocapsules with encapsulation efficiencies of over 95% and adjustable active compound loading are obtained.

Table 10.8 Encapsulation efficiencies and bioactive compound loadings achieved by nano spray drying

Bioactive compound	Wall material	Bioactive compound loading (%)	Encapsulation efficiency (%)	References
Albuterol	Mannitol, leucine	28	93	Son et al. (2013a)
Amoxicillin	Gelatine	89	—	Harsha (2013)
Clozapine	PLGA	13	95	Panda et al. (2014)
Dexamethasone	PLGA	27	81	Schafroth et al. (2012)
Dexamethasone	Poloxamer, surfactants	83	91	Durli et al. (2014)
Dexamethasone	Poly(ε-caprolactone)	47	94	Beber et al. (2014)
Dexamethasone	Poly(ε-caprolactone)	43	94	Fontana et al. (2014)
Folic acid (vitamin B9)	Resistant starch	—	53	Pérez-Masiá et al. (2015)
Folic acid (vitamin B9)	Whey protein	—	84	Pérez-Masiá et al. (2015)
Gentamicin sulphate	Alginate/pectin	28	83	De Cicco et al. (2014)
Insulin-like growth factor	Trehalose	74	93	Schultz et al. (2015)
Metformin	Carbopol	89	—	Harsha (2015)
Nimodipine	PLGA	38	96	Bege et al. (2013)
Resveratrol	Poly(ε-caprolactone)	31	99	Dimer et al. (2015b)
Vancomycin	Chitosan	55	76	Cerchiara et al. (2015)
Vildagliptin	Gelatine	76	—	Harsha et al. (2015a)
Vildagliptin	Carbopol	89	—	Harsha et al. (2013a)

—, Not known.

10.3.11 Controlled release of active compounds

The type of encapsulating polymer, its molecular weight, and the particle size primarily influence the release mechanism. Smaller particles exhibit a higher speed of active compound release compared to larger ones, due to the difference in exposed surface area, as evidenced by (Beck-Broichsitter et al., 2015a) with sildenafil-loaded polymeric particles. The release kinetics of active compounds encapsulated in PLGA polymers with a low molecular weight are faster compared to those with a high molecular weight, due to the higher hydrolysis and degradation rate (Panda et al., 2014). A PLGA polymer with low molecular weight can provide sustained release for four to 6 weeks (Schafroth et al., 2012). Release profiles typically exhibit two phases. In the first burst phase, about 10%–30% of the active compound is released due to free binding at the surface of the particles (Harsha et al., 2013a, 2015a). The rest of the active compound is released in a second phase for a longer duration (sustained release) (Harsha, 2015, 2013; Harsha et al., 2013a).

10.3.12 Stability of active compounds during nano spray drying

During the nano spray drying process, the active compound is exposed to several potential stress factors, such as fluid pumping, atomization into droplets, heat, and particle collection and recovery. Several studies show that proteins can be nano spray dried with only marginal activity loss, if any. Schultz et al. (2015) investigated the activity of the insulin-like growth factor encapsulated in trehalose. The activity remain unaffected and the spray drying process has no impact on product degradation or aggregation.

(Schmid, 2011) studied protein stability using a monoclonal antibody and the sensitive enzyme L-lactic dehydrogenase in trehalose. A slight reduction in enzyme activity was detected during pump circulation. In addition, a marginal aggregate formation of antibody was found, which can be attributed to the vibrating mesh temperature of about 80°C (at 100% spray rate intensity) and 59°C (at 25% spray rate intensity). This temperature impact is deemed to be critical because the feed is in contact with the spray mesh multiple times and has to pass through its holes for atomization. To minimize the risk of activity loss, it is recommended to keep the feed temperature low during the whole spraying process by cooling the feed vessel in ice (Schoubben et al., 2013). As shown by Bürki et al. (2011), the activity loss of β-galactosidase enzyme can also be minimized by using a larger spray mesh (i.e., 7.0 μm) to reduce the mechanical shear of atomization, pure aqueous solutions (low ethanol content), and a low inlet temperature of around 80°C.

10.3.13 Storage stability

Several studies show that the activity of the encapsulated product is preserved after nano spray drying if the powder is stored under controlled conditions and a stabilizer is added into the feed formulation. Most nano spray dried powders tend to be amorphous because the drying time is too short to form crystalline structures. Examples include, salbutamol (Chan and Kwok, 2011; Littringer et al., 2013), albuterol in mannitol

(Son et al., 2013a), trehalose (Schmid, 2011), and clozapine and risperidone in PLGA (Panda et al., 2014).

Controlling the storage conditions ensures that the particles remain in their amorphous state (Zellnitz et al., 2015a). Nano spray dried powders should be stored over silica gel in glass vials inside a desiccator to prevent crystallization. For example, Littringer et al. (2013) prepared amorphous salbutamol, which was kept in stable conditions at a relative humidity lower than 60% and temperature of 25°C. Nano spray dried β-galactosidase enzyme in trehalose lost only 3% of its activity during 3 weeks of storage at 70°C (Bürki et al., 2011). Dahili and Feczkó (2015) enhanced the activity of peroxidase enzyme by immobilizing it in ethylcellulose and PLGA. Pérez-Masiá et al. (2015) observed bioactive stability of folic acid (vitamin B_9) in whey protein after 60 days under dry storage conditions and in darkness. Merchant et al. (2014) found no evidence of change in the crystallinity of chitosan powder after storage for 60 days at room temperature and ambient humidity. De Cicco et al. (2014) conducted accelerated storage stability tests of gentamicin in alginate/pectin powders at 40°C and 75% relative humidity. The drug was preserved for 6 months. Finally, Harsha (2013) kept nanoparticle formulations of amoxicillin stable for 24 months at 25°C and 60% relative humidity.

10.3.14 Challenges in nano spray drying

Some fluids may be too viscous to pass through the mesh system and suspensions may restrict the mesh holes, which could result in intermittent droplet generation or even cessation of droplets generation. The processing time may gradually increase. Sometimes the product may even build-up deposits on the vibrating mesh and reduce the product yield (Schmid, 2011; Schmid et al., 2011). However, as stated by Schmid et al. (2011), virtually any substance can be atomized successfully with the nano spray dryer, as long as the right process parameters and formulation variables are used. There are several options to remedy nozzle blockage:

- cleaning of the spray mesh in an ultrasonic bath (Brinkmann-Trettenes et al., 2014)
- sonication of dispersions prior to nano spray drying (Brinkmann-Trettenes et al., 2014)
- constantly stirring of the feed while pumping to avoid agglomerations
- replacing the spray mesh with a new one (when the feed rate is greatly reduced)
- changing to a wider spray mesh to avoid obstruction (Dhillon et al., 2014; Schoubben et al., 2013)
- feed filtration or centrifugation to remove impurities (Gu et al., 2015; Harsha et al., 2015a)
- feed dilution to avoid mesh blockage (Pérez-Masiá et al., 2015)
- addition of surfactant or dispersion enhancers to optimize the performance of the vibrating mesh performance (Draheim et al., 2015; Son et al., 2013a)
- modification of the solvent mixture (Dimer et al., 2015b)

As stated by Heng et al. (2011), the nano spray drying technology is not expected to be widely adopted for large-scale industrial applications at present. The Nano Spray Dryer B-90 has been designed primarily for the early stages of product development and the processing of small batches of powder in the milligram or gram scale

(Arpagaus, 2010). Processing these batches may take a few hours (Durli et al., 2014; Haggag and Faheem, 2015).

Equipment designed for large-scale nano spray drying is yet to be developed and commercialized. Further studies are needed to investigate the scalability of the process parameters (Gu et al., 2015). A possible scale-up solution for droplet generation could be using multiple vibrating mesh atomizers in parallel arrangements or a larger nozzle unit (Heng et al., 2011). Industrial-scale electrostatic particle collectors like a mechanical rapping device could be installed with an autocleaning system to remove the deposited particles from the collecting electrode continuously before it reaches its saturation.

10.4 Nano spray drying applications

Various feasibility tests have been performed with the Nano Spray Dryer B-90 between 2010 and 2015. The collected experience of the published works resulting from such tests is reviewed in the following sections. Our focus is on food applications, drug delivery, and material science. Applications can be divided into four distinct areas—nano spray drying, nanoization/structural change, nanoencapsulation, and nanoenglobing, as shown in Fig. 10.15.

Nano spray drying is suitable for most solutions, as long as it yields a dried product that behaves like a solid. Practical applications include polymeric wall materials, like arabic gum, whey protein, polyvinyl alcohol, modified starch, and maltodextrin (Li et al., 2010, 2015). A solution of the product (A) is dispersed into fine droplets (B). The solvent evaporates immediately and a solid powder is formed. The final product (C) is a fine powder that can be amorphous or crystalline. Feed dilution results in a final product with smaller particle size.

Nanoization or structural change is used to change the particle morphology, for example, if a very fine powder is required. This improves the solubility of the final product. Typical examples are drugs with poor solubility in water (Li et al., 2010; Martena et al., 2012; Schmid et al., 2011) and salts (Moncada et al., 2015). The main advantage of nanoization is that particles of very regular size are achieved. A crystalline product (A) is dissolved in a solvent (B) and this solution (C) is dispersed into tiny droplets (D). The result is a final product (E) similar to the final product described in nano spray drying.

In nanoencapsulation, a (liquid) product is embedded inside a solid matrix. A typcial example is encapsulated oil-in-water nanoemulsion (Beck-Broichsitter et al., 2012; Li et al., 2010, 2015; Panda et al., 2014). An emulsion (D) is created from the liquid product to be treated (A), a carrier substance (B), such as maltodextrin, and a filmogen solution (C), such as gum arabic in water. Organic binders like polyvinylalcohol, maltodextrin, and gelatine are also used as polymeric wall materials to provide efficient encapsulation. The emulsion is then sprayed into small droplets (E). The solvent evaporates, leaving a solid matrix around the dispersed second phase (F). The result is that small droplets of the product (A) are stored in the carrier substance (B), and embedded in the filmogen (C).

Nanoenglobing is similar to the nanoencapsulation process. It involves a solid material being embedded in another solid or liquid product. Examples include, carotenoids in gelatine or magnetic iron oxide nanoparticles in Eurdragit and maltodextrin (Perecin et al., 2014). A nanodispersion (D) is created from the product to be treated (A), a matrix (B), water, and added filmogen (C). This suspension is then sprayed into small droplets (E). The matrix and the filmogen lead to an agglomeration or coating of the suspended particles (F), resulting in nanocomposite particles.

10.4.1 Food and nutraceutical applications

Since the market introduction in 2009, the Nano Spray Dryer B-90 has been employed predominantly in drug delivery applications and material science (Arpagaus, 2011, 2012). Nanoencapsulation of active food ingredients is still at a nascent stage (Anandharamakrishnan and Padma Ishwarya, 2015) but is constantly developing. This section reviews the published works on food applications carried out with the nano spray dryer. Table 10.9 shows the results under optimized experimental conditions, which can be used as first-guess values for applications of the Nano Spray Dryer B-90 on similar substances.

Nano spray drying has been successfully used with a variety of representative polymeric wall materials, including, for example, sodium alginate (Blasi et al., 2010), trehalose (Schmid et al., 2009), arabic gum, whey protein, polyvinyl alcohol, modified starch, and maltodextrin dissolved in water at various concentrations (Li et al., 2010, 2015).

Submicron particles with a spherical shape were achieved, as shown by SEM pictures in Fig. 10.10. Low-viscosity sodium alginate is used as an emulsifier and immobilization agent. Spherical particles with smooth surfaces were obtained and the product yield was higher than 90% (Blasi et al., 2010).

Moncada et al. (2015) produced salt particles with sizes ranging from of 0.5 to 1.9 μm from a 3% (w/v) sodium chloride solution. Fig. 10.10 illustrates the spherical morphology of the fine salt particles. The effect of the sensory and microbiological characteristics of surface-salted cheese crackers was further investigated. Nano spray dried salts resulted in a significantly higher level of preferred saltiness than the control. The particles, being smaller in size than regular salt, would reduce sodium intake by 25%–50% if used in surface-salted cheese crackers. NaCl solutions were also spray dried by Li et al. (2010) to demonstrate the capacity of nanoization and the structural change of water-soluble salt substances. Particle sizes from 0.5 to 1.0 μm were observed by increasing the salt concentration from 0.1 to 1.0% (w/v).

The same authors Li et al. (2010) encapsulated Vitamin E acetate nanoemulsions (smaller than 100 nm) in arabic gum, modified starch, maltodextrin, and whey protein. Vitamin E acetate was chosen as a model for the oily phase. The nanoemulsions were obtained through low-energy emulsification and were mixed with 1% (w/v) aqueous wall material solutions at a weight ratio of 1:4. Optimal drying conditions (100°C inlet temperature and 40–60°C outlet temperature) ensured that the food ingredients were not damaged during the drying process. Nonaggregated solid particles of submicron size were produced with almost 100% encapsulation efficiency. Redissolving of the

Table 10.9 Food and nutraceutical applications established by nano spray drying

Substance	Application	Wall material	Solvent	*T* in (°C)	*T* out (°C)	Drying gas (L/min)	Particle size (μm)	References
Nano spray drying								
Sodium alginate	Polymeric wall material	—	Water	110	—	100	0.4–1.2	Blasi et al. (2010)
Arabic gum, whey protein, polyvinyl alcohol, modified starch, maltodextrin	Polymeric wall materials	—	Water	100	38–60	100	0.2–1.1	Li et al. (2010)
Trehalose	Protein stabilizer	Surfactant, Polysorbate	Water	60–100	30–45	115	0.3–3.0	Schmid et al. (2009)
Chitosan	Bioresorbable biopolymer, fat blocker	—	Water acetic acid	120	55	130	0.6–1.6	Gautier et al. (2010)
Hydroxypropyl methylcellulose	Excipient for oral medicaments	—	Acetone	56	—	120	2–10	Gu et al. (2015)
Poly(lactic-*co*-glycolic acid)	Biodegradable polymer	Surfactants, poloxamer, sorbitan monostearate	Acetone, ethyl acetate DCM	50–90	29–36	115 (N_2/CO_2)	2.4–8	Draheim et al. (2015)
PLGA suspensions	Biodegradable polymer	—	DCM	30–50	20–30	100	2.8–4.4	Beck-Broichsitter et al. (2012)
Hydroxypropyl methylcellulose (hypromellose)	Enteric polymer film coating	—	Acetone	56	—	120	2–10	Gu et al. (2015)

Nanoization/Structural Change								
Sodium chloride	Surface-salted cheese crackers	—	Water	95	—	125	0.5–1.9	Moncada et al. (2015)
Sodium chloride	Model of water soluble salt	—	Water	100	—	100	0.5–1.0	Li et al. (2010)
Nanoencapsulation								
Vitamin B_{12}	Water-soluble vitamin	Arabic gum, cashew nut gum, sodium alginate, carboxymethyl cellulose, eudragit RS100	Water	120	50–60	130	0.2–5.5	Oliveira et al. (2013)
Folic acid (synthetic Vitamin B_9)	Water-soluble vitamin	Guar gum, whey protein, resistant starch	Water	90	45	140	0.2–4.5	Pérez-Masiá et al. (2015)
Vitamin E acetate (nanoemulsion)	Vitamin supplement	Arabic gum, modified starch, maltodextrin, whey protein	Water	100	38–60	100	0.4–1.1	Li et al. (2010)
Curcumin	Antioxidant Food additive	Human serum albumin	Water	120	—	150	0.2–0.7	Jain (2014)
Leucine	Aminoacid, Nutritional supplement	Trehalose	Water	75	45	100	2.1–5.4	Feng et al. (2011)
Bovine serum albumin	Model protein derived from cows	Surfactant, polyoxyethylene, sorbitan monoleate	Water	80–120	36–55	150	0.5–2.6	Lee et al. (2011)
β-galactosidase	Enzyme catalyzing hydrolysis of saccharides	Trehalose	Water	80	36–53	100–110	1–5	Bürki et al. (2011)

—, Data not known.

spray-dried nanoparticles in water led to nanoemulsions without any significant degradation or size increase (Li et al., 2010, 2015). It was concluded that encapsulation of nanoemulsions by nano spray drying preserved the integrity of the nanoemulsion system. This is considered as a new and promising solution for the delivery of nanoemulsions (Li et al., 2015).

Furthermore, a recent application presented by Pérez-Masiá et al. (2015) involves the encapsulation of folic acid (synthetic vitamin B_9) in whey protein and resistant starch. Folic acid is an essential water-soluble vitamin of nutraceutical importance. Spherical submicron and micron capsules were obtained. Whey protein displayed a higher encapsulation efficiency than resistant starch, which could be attributed to the interactions between the bioactive and the protein matrix. It was further observed that whey protein conserved the folic acid better against degradation during storage in an aqueous solution, as well as dry storage. The capsules made of whey protein were able to preserve to bioactive stability at almost 100% in dark conditions, while 40% of the nonencapsulated folic acid degraded.

Oliveira et al. (2013) encapsulated Vitamin B_{12} in arabic gum, cashew nut gum, sodium alginate, and carboxymethyl cellulose. The influence of the different feed properties were evaluated. Aqueous solutions with viscosity of 6.3–9.2 mPa s were successfully atomized and spray dried. SEM micrographs displayed spherical submicron particles clustered together, as shown in Fig. 10.10. To prevent the irreversible aggregation of the nanoparticles after spray drying, "spacer" excipients, such as sucrose, trehalose, and leucine are proposed.

Jain (2014) encapsulated curcumin in cross-linked human serum albumin particles. Curcumin is a highly active antioxidant; it is also a food additive and preservative. The particles were smooth, spherical, and submicron-sized in a range of 0.2–0.7 μm. Cross-linking of albumin produced spray dried particles with slower curcumin release, which followed a first-order release profile.

In conclusion, it is expected that other food-related encapsulation applications will use the nano spray drying technique. Potential applications in the future are foreseen in the dairy and food industry (Murugesan and Orsat, 2012; Patel and Upadhyay, 2014), as well as for the nanoencapsulation of vitamins, flavors, lipids, dyes, milk and egg products, coffee, tea, cereal products, spices, and so on.

In addition, experimental results from drug delivery studies may provide useful clues as to how the nano spray drying process can be adapted for macromolecules. The nano spray drying method is suitable for various solutions, including emulsions, suspensions, heat-sensitive materials, biological products, proteins, and enzymes. Further research studies also need to concentrate on the development of new polymeric wall materials with good encapsulation properties.

10.4.2 Drug delivery applications

Nano spray drying is widely spread for various drug delivery applications. Thanks to its unique ability to process small sample amounts in a range of 10 mg to 2.5 g at optimized product yields of up to 76 to 96%, as shown in Table 10.7. Typical nano spray dried pharmaceutical excipients applied include trehalose as a stabilizer to increase shelf life (Schmid et al., 2009), L-leucine amino acid as a dispersing agent (Feng et al., 2011), hydroxypropyl methylcellulose as a viscoelastic polymer (Gu et al., 2015), and chitosan (Gautier et al., 2010) or PLGA (Beck-Broichsitter et al., 2012; Draheim et al., 2015) as a biodegradable polymers.

Several studies demonstrate that nano spray drying is also suitable for directly converting pure drugs into dry nanopowders, because of the associated benefits of nanoization and structural change (Ahmad et al., 2014; Baba and Nishida, 2012, 2013; Lee et al., 2013; Li et al., 2010; Littringer et al., 2013; Martena et al., 2012; Schmid, 2011; Zellnitz et al., 2015a). Table 10.10 summarizes some published works related to nanoized and nanoencapsulated drugs, including the experimental process parameters for nano spray drying and the achieved particle size.

This list includes drugs for the treatment of tuberculosis through inhalation (Ahmad et al., 2014; Kanogwan and Srichana, 2016; Schoubben et al., 2013, 2014, 2015), bronchial asthma (Arpagaus et al., 2010a; Baba and Nishida, 2013; Littringer et al., 2013; Son et al., 2013a; Zellnitz et al., 2015a), inflammation (Aquino et al., 2014; Baba and Nishida, 2013; Beber et al., 2014; Durli et al., 2014; Fontana et al., 2014; Hindle et al., 2015; Lee et al., 2013; Öztürk et al., 2015), high blood pressure (Bege et al., 2013; Martena et al., 2012), diabetes (Harsha et al., 2013a, 2015a), Alzheimer's and Parkinson's diseases (Baba and Nishida, 2012), fungal affection (Schmid, 2011), ophthalmic disorders, congestive heart failure and edema (Li et al., 2010), and others.

The nanosize dimensions and the amorphous state provide pronounced improvement in drug release and bioavailabilty, and encourage future developments in this research field. Typical organic solvents used to dissolve the drugs that are often poorly soluble in water are ethanol, methanol, and acetone.

Nano spray drying enables the encapsulation of drugs in polymers with high efficiency and drug loading capacity, as shown in Table 10.8. Nanocapsules, with their reduced size and large specific surface area, offer better absorption rates and higher bioavailability. This enables the generation of target drug delivery systems (Arpagaus, 2011; Harsha et al., 2015b; Li et al., 2010).

Various polymers have been applied as wall materials for encapsulation. This includes water-soluble polymers, such as chitosan, alginate, gelatine, arabic gum, whey protein, cellulose derivatives, lactose, modified starch, maltodextrin, polyvinylpyrrolidone, and polyvinylalcohol, as well as hydrophobic polymers like poly(ε-caprolactone), poly(lactic-*co*-glycolic acid), and Kollidon.

Organic solvents are commonly used to adjust the solubility of the polymer and the drug, as well as the boiling point, resulting in particles that have the desired properties. The low boiling points of DCM (40°C) or acetone (56°C) lead to fast drying and prevent particles from sticking to the walls or agglomerating. The evaporation temperature is lower than the melting temperature or the glass transition temperature of certain polymers. PLGA is one of the most commonly used materials for biomedical applications due to its well-established biocompatibility and biodegradability (Arpagaus and Schafroth, 2007, 2009).

Furthermore, the drug-loaded particles obtained by nano spray drying are administered through different routes, including oral, pulmonary (Beck-Broichsitter

Table 10.10 **Drug delivery applications established by nano spray drying**

Substance	Application	Wall material	Solvent	*T* in (°C)	*T* out (°C)	Drying gas (L/min)	Particle size (μm)	References
Nano spray drying								
Calpain inhibitor	Drug to cure Alzheimer's and Parkinson's diseases	—	Ethanol	50	35	100 (N_2/CO_2)	0.1–1.3	Baba and Nishida (2012)
Ciprofloxacin and gatifloxacin	Antiinflammatory, antibacterial		Water	120	40–45	120 (N_2)	0.8–3.6	Lee et al. (2013)
Dexamethasone	Antiinflammatory steroid, asthma drug	—	Ethanol	50	35	100 (N_2/CO_2)	0.4–2.2	Baba and Nishida (2013)
Ethambutol	Antituberculosis drug for inhalation	—	Water	110	—	—	0.2	Ahmad et al. (2014)
Fluorometholone	Antiinflammatory steroid, asthma drug	—	Ethanol	50	35	100 (N_2/CO_2)	0.4–1.2	Baba and Nishida (2013)
Furosemide	Drug to treat congestive heart failure and edema	—	Acetone	100	—	100	0.7–1.7	Li et al. (2010)
Griseofulvin	Antifungal drug	—	Methanol, acetone	70	40–45	120 (N_2)	3.4–6.5	Schmid (2011)
Nicergoline	Blood pressure regulating drug	—	Ethanol, water	50	—	90	0.1–1.6	Martena et al. (2012)
Salbutamol	Asthma drug	—	Water	80–120	40–50	110	1.0–6.4	Littringer et al. (2013)
Salbutamol	Asthma drug	—	Water ethanol	100–120	—	110	1–5	Zellnitz et al. (2015a)

Nanoencapsulation								
Amoxicillin	Antibacterial drug to treat stomach infections	Gelatine	Water	80	—	—	0.6	Harsha (2013)
Amoxicillin	Antibacterial drug to treat stomach infections	Chitosan	Water acetic acid	120	80	130	0.1–0.4	Ngan et al. (2014)
Azithromycin	Antiinflammatory drug to treat cystic fibrosis	Sodium chloride, mannitol, leucine	Water	80	45	120	1–5	Hindle et al. (2015)
Capreomycin	Antituberculosis drug for inhalation	Leucine	Ethanol, acetone, water	80–120	—	100	1–5	Schoubben et al. (2015, 2014, 2013)
Clarithromycin	Antibiotic to treat lung infections	Leucine, chitosan	Water ethanol	80	—	120	0.3–2.5	Dimer et al. (2015a)
Clozapine and risperidone	Antipsychotic drugs to treat Schizophrenia	PLGA, polyvinylalcohol	DCM water	60	31–38	110–118 (N_2)	0.2–0.4	Panda et al. (2014)
Coumarin	Fluorescent dye suspended in fibrin glue for surgery	PLGA	DCM	45	—	100	0.8–3.0	Bege et al. (2013)
Cyclosporin	Immune system suppressant drug	PLGA	DCM ethanol	29–32	28–32	102–132 (N_2/CO_2)	0.9–2.2	Schafroth et al. (2012)
Dexamethasone	Steroid, antiinflammatory drug	Polycapro-lactone, deoxy-cholate	Acetone water	55	34	110	0.6–1.6	Fontana et al. (2014)
Dexamethasone	Steroid, antiinflammatory drug	Surfactants	Ethanol acetone propanol	55	—	110	0.4–1.2	Durli et al. (2014)

(Continued)

Table 10.10 **Drug delivery applications established by nano spray drying (*cont.*)**

Substance	Application	Wall material	Solvent	*T* in (°C)	*T* out (°C)	Drying gas (L/min)	Particle size (μm)	References
Dexamethasone	Anti-inflammatory drug to treat skin diseases	Polycaprolactone, deoxycholate	Acetone water	55	—	110	0.4–2.3	Beber et al. (2014)
Dexketoprofen	Antiinflammatory drug to control analgesic delivery	Kollidon	Methanol	120	54	—	0.1–0.4	Öztürk et al. (2015)
Fluticasone propionate	Asthma drug	Lactose	Water	120	45–56	130	0.5–3.0	Arpagaus et al. (2010a)
Gentamicin	Antibacterial drug for wound dressing	Alginate, pectin	Water	90	—	100	0.3–1	De Cicco et al. (2014)
Insulin-like growth factor I	Anabolic protein similar to insulin	Trehalose, silk-fibroin	Water	70	—	115–130	0.3–3.2	Schultz et al. (2015)
Ketoprofen	Antiinflammatory drug to treat cystic fibrosis	Leucine	Water isopropyl-alcohol	60–110	—	100	2.4–6.6	Aquino et al. (2014)
Levofloxacin	Antibiotic to treat lung infections	Chitosan, L-leucine	Water	120	—	133	2.5–4.6	Merchant et al. (2014)
Methotrexate	Drug to treat lung cancer	Carbopol	Water	85–115	—	—	6.8	Harsha et al. (2015b)

Nimodipine	Dilatation regulator in brain arterioles	PLGA	DCM	45	—	100	0.7–3.7	Bege et al. (2013)
Phospholipid dispersions	Pulmonary drug delivery	Trehalose, egg, lecitin	Water	60–120	44–47	90–140	0.5–4	Brinkmann-Trettenes et al. (2014)
Pyrazinamide	Antituberculosis drug	L-leucine	Water ethanol	90–115	—	100	2.8–3.8	Kanogwan and Srichana (2016)
Resveratrol	Drug against pulmonary arterial hypertension	Polycaprolactone, deoxy-cholate, trehalose	Acetone water	55	—	110 (N_2, CO_2)	1–5	Dimer et al. (2015b)
Salbutamol	Asthma drug	Lactose	Water	120	52–56	130	0.5–3.0	Arpagaus et al. (2010a)
Salbutamol (Albuterol)	Asthma drug	Mannitol, L-leucine, poloxamer	Water ethanol	70	40	120	<1	Son et al. (2013a)
Sildenafil	Drug against pulmonary arterial hypertension	PLGA	Acetone	45	25–30	100	1.2–3.9	Beck-Broichsitter et al. (2015a)
Vancomycin (nanoparticles)	Antibiotic for colon drug delivery	Chitosan	Water	80	40	—	0.4–0.5	Cerchiara et al. (2015)
Vildagliptin	Antidiabetic oral drug (mucoadhesive)	Gelatine, carbopol	Water	120	27–35	100–110	0.1–0.6	Harsha et al. (2015a, 2013a)

—, Data not known.

et al., 2015a; Dimer et al., 2015b; Brinkmann-Trettenes et al., 2014; Harsha et al., 2015b; Aquino et al., 2014), ophthalmic, nasal, intravenous, or even cerebral Bege et al. (2013), thus underlining the versatility of the nano spray drying method.

Several studies have been conducted on nano spray dried oral dosage forms for immediate and sustained drug release. Harsha et al. (2013a, 2015a) encapsulated antidiabetic vildagliptin in gelatin and carbopol polymer for mucoadhesive oral administration with sustained drug release. Spherical submicron particles with smooth surfaces were obtained with up to 90% drug loading and high product yields. In another study, Harsha (2013) prepared the oral dosage forms of amoxicillin-loaded gelatin nanoparticles to treat helicobacter pylori infection in stomach. Complete drug release was delayed by up to 12 h. Öztürk et al. (2015) prepared dexketoprofen-loaded Kollidon-SR nanoparticles for controlled analgesic delivery. The powder formulations seem to be promising for the oral use of dexketoprofen as a nonsteroidal antiinflammatory drug. Cerchiara et al. (2015) prepared chitosan particles for the colon-specific delivery of vancomycin through oral administration. The antibiotic drug vancomycin is administered to treat serious colon infections. The nanoparticles were produced at high yield and high encapsulation efficiency, and demonstrated a prolonged drug release at a low pH resembling the stomach environment. In addition, the particles showed an efficient and uniform bactericidal activity.

Nano spray drying has emerged as an attractive technology to produce dry powders within the respirable size range of 1–5 μm, as demonstrated by many studies listed in Table 10.10. In order to reach the deep lung regions, the particles have to be in the optimized aerodynamic size range. Particles that are too large will deposit in the throat and are not able to access the deep bronchiolar region, whereas particles that are too small will be exhaled. The high bioavailability in the lung permits the drug doses to be lowered compared to oral administration, thus resulting in reduced systemic side effects and benefits to compliance of the patient (Kanogwan and Srichana, 2016).

A main benefit of nano spray drying is the optimization of the size and morphology of the particles so that they are suitable for inhalation therapy (Arpagaus and Meuri, 2010; Arpagaus et al., 2010b; Bürki et al., 2011; Lee et al., 2013; Schoubben et al., 2014). There has been significant research activity in the formulation of dry powder aerosols for the treatment of lung diseases, such as asthma, inflammation, hypertension, cystic fibrosis, cancer, tuberculosis, diabetes, and bacterial infection. The results open up a fresh perspective for the development of new nano spray dried powders for pulmonary administration.

Research activity to improve the flowability and dosing accuracy of dry powder inhalers is ongoing. One strategy is to use glass beads or coarse lactose particles as drug carriers. Zellnitz et al. (2014, 2015b) prepared surface-modified glass beads (with sizes between 400 and 600 μm) and covered them with nano spray dried salbutamol. The particles adhered to the glass beads with up to 30% surface coverage. This way, fine particle fractions (<5 μm) of 25%–35% could be achieved. (Faulhammer et al., 2015) investigated the effect of coarse lactose (particle size of 63–227 μm) carriers and measured fine particle fractions of 8%–12%.

A newly proposed strategy combines the delivery of spray driedsubmicron particles with minimal losses due to mouth–throat deposition. This method utilizes the hygroscopic growth of the inhaled particles within the airways to achieve nearly complete pulmonary deposition. The applied nano spray dried powders are made of albuterol sulfate, mannitol, L-leucine, and poloxamer. Son et al. (2013b) concluded that it would be possible to aerosolize this optimized dry powder formulation with a mass median aerodynamic diameter of about 1.5 μm. Fine particle fractions (<5 μm) of 95% were reached (Son et al., 2013a). Other studies with terbutaline sulfate (Behara et al., 2014b), antibiotics, such as ciprofloxacin (Farkas et al., 2015; Longest et al., 2015), and azithromycin (Hindle et al., 2015) demonstrate the feasibility of this enhanced excipient-growth approach.

Further research to improve the efficiency of dry powder inhalers should be directed toward the optimization of the wall coating in the devices (Behara et al., 2014a), and the utilization of fluidized bed approaches to increase the aerosolized powder masses (Farkas et al., 2015).

Besides the oral and pulmonary drug administration routes, nano spray dried applications have created nanoencapsulated drugs that can be transferred through the skin or during surgery. De Cicco et al. (2014) loaded alginate-pectin nanopowders with 25% antibacterial gentamicin. This formulation turns into a gel when administered locally on a wound. Sustained penetration was obtained, achieving a total drug permeation period of 3–6 days. Ngan et al. (2014) prepared amoxicillin-loaded chitosan nanoparticles with improved antibacterial activity. This study confirms that these nanoparticles strongly inhibit bacterial growth.

Bege et al. (2013) encapsulated nimodipine in PLGA particles. Nimodipine is a substance that regulates the dilatation of blood vessels, particularly in the brain. The particles were suspended in fibrin sealant as a depot system for treatments related to brain surgery. The fibrin glue is biocompatible, biodegradable, and stable in the cerebrospinal fluid, that is, the fluid in the brain and the spine. To visualize the homogeneity of the suspended particles in the adhesive, PLGA particles were also colored with encapsulated green fluorescent coumarin dye. The drug loading and encapsulation efficiency could be clearly examined by fluorescence spectroscopy.

Beber et al. (2014) obtained submicron polymeric particles containing dexamethasone and prepared creams (oil-in-water emulsions) for local administration to the skin. The drug release in the skin was observed to have the ability to control the drug release kinetics by optimizing the powder properties.

Panda et al. (2014) entrapped the antipsychotic drugs clozapine and risperidone within PLGA nanoparticles with high-drug loading and encapsulation efficiency for sustained delivery to treat Schizophrenia. Nanoparticles, as small as around 250 nm were prepared by using PLGA polymers with low molecular weight. The drugs were evenly dispersed within the PLGA particles without any crystalline content. Sustained drug-release profiles were observed over several days.

10.4.3 Material science applications

Although much of the literature has focused on biological samples, interesting applications of the nano spray dryer are possible in the materials sector, as summarized in Table 10.11.

Table 10.11 Material science applications established by nano spray drying

Substance	Application	Wall material	Solvent	*T* in (°C)	*T* out (°C)	Drying gas (L/min)	Particle size (μm)	References
Nano spray drying								
Lithium carbonate ($LiHCO_3$) (nanosuspension)	Material for rechargeable lithium batteries	—	Water	90	—	130	0.2 (primary) 4–9 (composites)	Sun et al. (2011)
Magnesium sulfate ($MgSO_4$) nanoparticles	Supplement for algae to produce lipid from waste glycerol	—	Water	120	—	120	0.1 (individual) 4–5 (aggregates)	Sarma et al. (2014b)
Ferric citrate ($C_6H_5FeO_7$) nanoparticles	Supplement for microorganism to produce hydrogen	—	Water	120	—	120	0.5–1.7	Sarma et al. (2014a)
Nanoencapsulation and nanoenglobing								
Horseradish peroxidase enzyme	Enzymatic removal of phenols from wastewaters	Ethyl cellulose	DCM	40	—	100 (N_2)	1.5–3.4	Dahili et al. (2015)
CsH_2PO_4 nanoparticles	Solid acid fuel cells	Polyvinyl-pyrrolidone	Methanol water mixture	30–120	—	60–150 (N_2, CO_2)	0.1–0.8	Suryaprakash et al. (2014)
Zinc oxide powder	Antimicrobial and antibiofilm agent	Chitosan citric acid whey starch glycerol	Water acetic acid	120	—	150–160	0.1–0.4	Dhillon et al. (2014)
Magnetite nanosuspension	Cancer therapy by magnetic hyperthermia	Polycaprolactone	Acetone	65	—	—	0.6	Perecin et al. (2015)
Magnetic iron oxide nanosuspension	Cancer therapy by magnetic hyperthermia	Eudragit maltodextrin	Water	120	54	130	0.3–1	Perecin et al. (2014)

Sun et al. (2011) prepared battery-grade lithium carbonate (Li_2CO_3) nanocrystals from purified lithium bicarbonate ($LiHCO_3$) precursor. SEM pictures reveal self-assembled hollow spheres of lithium carbonate in the size range of 4–9 μm scale, that are composed of 200 nm crystalline particles (Fig. 10.13). The achieved surface area of the nanosized product is promising for application in the lithium-battery industry.

Sarma et al. (2014a) synthesized nano spray driedferric citrate particles as supplements for Enterobacter aerogenes microorganism to produce hydrogen by bioconversion of crude glycerol, which is a side-product of the biodiesel production process. The large specific surface area of the nanoparticles lead to a significantly higher hydrogen yield, comparable to that from dark fermentative hydrogen production.

Sarma et al. (2014b) prepared $MgSO_4$ nanoparticles of around 100 nm in size as supplements for enhanced lipid production by crude glycerol bioconversion with the algae *Chlorella vulgaris*. The addition of 1 g/L of nanoparticles was found to improve the lipid production by 118%. Unless dispersed in water, the nanoparticles tended to aggregate and form larger particles of around 4–5 μm (Fig. 10.13).

Dahili et al. (2015) cross-linked horseradish peroxidase enzyme with ethylcellulose particles formed by nano spray drying. The enzyme catalyzes the removal of phenolic compounds from wastewater. As expected, the residual activity of the enzyme increased upon decreasing the spray-dried particle size, thanks to the higher specific surface area. The immobilization also improved the pH tolerance up to a range of 4–10, as well as the storage stability, compared to free enzyme. The immobilized enzyme preserved over 50% of its activity after 4 weeks of storage at room temperature.

Suryaprakash et al. (2014) explored nano spray drying as a new synthesis method for CsH_2PO_4 nanoparticles for application in solid acid-fuel-cell electrodes. They produced nanoparticles of size 0.1–0.8 μm, which were deposited directly inside the nano spray dryer onto a dense CsH_2PO_4 electrolyte fuel-cell pellet with a deposition rate of 1 mg/h per cm^2. Polyvinylpyrrolidone was used as stabilizing component for the deposited CsH_2PO_4 nanostructure. The particles formed a core–shell structure, as can be seen in Fig. 10.14.

Another strategy to produce nanoencapsulated formulations is to use nanocomposite particles composed of encapsulated (or englobed) nanosuspensions (see principle in Fig. 10.15). These particles are more physically stable than single primary nanoparticles and offer efficient delivery.

Dhillon et al. (2014) encapsulated zinc oxide nanoparticles in chitosan by nano spray drying using various encapsulating stabilizers, such as citric acid, glycerol, starch, and whey. The particles showed significant antimicrobial activity and biofilm growth inhibition against pathogenic microorganisms. This study indicates that there is a promising potential for englobing metal oxide nanoparticles in polymeric wall materials.

Perecin et al. (2014) provided the example of encapsulation of magnetic iron oxide nanoparticles (about 30 nm size, prepared by coprecipitaiton) in Eudragit and maltodextrin. The nano spray drying of the nanosuspension created polymer capsules of size between 320 and 955 nm with encapsulated magnetic nanoparticles. Potential applications are foreseen in cancer therapy by magnetic hyperthermia. The particles have the necessary biocompatibility and magnetization to promote heating and effect cytotoxicity when applied in an alternating magnetic field.

In a subsequent study, (Perecin et al., 2015) encapsulated magnetite iron nanoparticles into polycaprolactone, forming particles with a size of approximately 600 nm. This type of powder formulation with increased magnetite loading is promising for the treatment of tumors by magnetic hyperthermia.

10.5 Conclusions

Nano spray drying is a valuable technique to convert various liquid feeds into dry powders, particularly in the fields of food ingredients, drug delivery, and material science. The number of publications associated with nano spray drying and its applications in nanoencapsulation has increased rapidly over the last 5 years.

The introduction of the Nano Spray Dryer B-90 by BÜCHI Labortechnik AG in 2009 has made nano spray drying a reality at the laboratory scale. This dryer makes it possible to create particles of an order of magnitude smaller than can be created by traditional spray dryers. This represents a major benefit of this novel technology. The ability to effectively produce spray-dried nanocapsules in the submicron scale opens new possibilities in particle engineering in the lab for the coming decade and beyond.

The technological novelty of the Nano Spray Dryer B-90 includes the gentle laminar drying gas flow, the vibrating mesh spray technology that creates the finest droplets, and the highly efficient electrostatic particle collector to separate and collect the dried nanoparticles. The combination of the innovative droplet generation principle and resourceful particle separation technology opens the way to apply nano spray drying for a variety of new applications.

The technique demonstrates the capability to encapsule nanoemulsions containing vitamins in different food-grade polymeric wall materials, including sodium alginate, arabic gum, whey protein, polyvinyl alcohol, modified starch, and maltodextrin. Nano spray drying is also successfully employed for the nanoization and structural change of pure drugs and water-soluble salts. Tiny spray-dried particles, as small as 100 nm in size are created, depending on the spray mesh size, solid concentration, surface tension of the feed, and spray rate intensity. Various particle morphologies can be achieved, including dense, hollow, and porous, with spherical, wrinkled, shriveled, or doughnut shapes. Future food applications are foreseen in terms of nanoencapsulation of flavors, lipids, dyes, and milk and egg products.

Nano spray drying is widely spread for different drug delivery applications thanks to its unique ability to process small sample amounts in a range of 10 mg to 2.5 g. The electrostatic particle collector highly extends the size spectrum of separable particles toward the nanoscale. Uniquely high yields of 76%–96% are obtained, thus enabling the economical use of expensive pharmaceutical ingredients and pure drugs. Heat-sensitive peptides, proteins, hormones, and amino acids can be dried and stabilized in nanopowder forms at optimized product yields. Biocompatible polymers are utilized as wall materials encapsulating different bioactives providing controlled release and higher bioavailability. High encapsulation efficiencies of over 95% are reached with adjustable drug loadings.

The gentle nano spray drying process, with outlet temperatures between 28 and 59°C, prevents significant degradation of the encapsulated compounds and enables the preservation of the activity during storage. The drug-loaded particles are administered through different routes, including pulmonary, oral, ophthalmic, nasal, intravenous, skin, or even cerebral, thus underlining its great versatility of applications. The majority of current literature has focused on inhalation therapy with optimized respirable particles size of 1–5 μm.

Nanoencapsulated particles also provide interesting applications in the material science sector, as antimicrobial agents, supplements for microorganism to produce fuels and lipids, or materials for rechargeable lithium batteries and fuel cells.

The Nano Spray Dryer B-90 offers the possibility to adjust the process parameters and formulation variables to optimize the properties of the final product. The spray mesh size and the solid concentration are the key parameters controlling the final particle size. Other important inputs parameters for nanoencapsulation are the drying temperature, the drying gas flow rate, the spray rate intensity, and the solvent type. Of primary interest are the effects on the particle size, morphology, yield, number of product produced per unit of time, bioactive loading, release profile, encapsulation efficiency, and stability. Depending on the application, an optimized set of process parameters is found. The experimental conditions presented in this chapter can be used as first-guess values for applications of the nano spray dryer on similar substances.

At present, the application of nano spray drying for the formation of nanoencapsulates is confined to laboratory scale operations. Experimental studies are on the level of small batches in the milligram to gram scale and require a few hours of processing times, due to the use of diluted feeds with 0.1%–1% (w/v) solid concentrations to produce nanoparticles. Thus, future research work needs to focus on the further commercialization of the nano spray drying technology to the industrial scale. Increased customer demands for the laboratory devices, coupled with promising new applications, are expected to eventually encourage and stimulate the development of more industrially relevant nano spray dryer equipment.

Another research trend is directed to further decreasing the particle size toward the nanoscale. The smallest spray mesh currently available on the nano spray dryer has 4 μm holes and generates droplets of about 5 μm. It would be interesting to see the effect on the particle size on using even smaller spray meshes, while changing the frequency of vibration. Smaller orifices would enable more efficient feed atomization, but would also be accompanied by a lower production rate.

Moreover, new developments in computational fluid dynamics and heat transfer modeling of droplets drying and particles formation, as well as temperature and humidity profiling inside the spray dryer could be beneficial to further controlling the nano spray drying technique.

Application trends for nano spray drying lie in the fields of pulmonary drug delivery, the nanotherapeutics, the encapsulation of nanoemulsions containing food ingredients, and the formulation of nanocrystals of poorly water-soluble drugs. The reduction in spray-dried particle size offers various new prospects for the application in functional foods, such as carotenoids, polyphenols, minerals, flavors, antioxidants, and vitamins. There are areas where the full potential of the nano spray drying process is yet to be fully exploited.

References

Ahmad, M.I., Ungphaiboon, S., Srichana, T., 2014. The development of dimple-shaped chitosan carrier for ethambutol dihydrochloride dry powder inhaler. Drug Dev. Ind. Pharm. 41, 791–800.

Anandharamakrishnan, C., 2014. Drying techniques for nanoencapsulation. Techniques for Nanoencapsulation of Food Ingredients. Springer, New York, pp. 51–60.

Anandharamakrishnan, C., Padma Ishwarya, S., 2015. Spray drying for nanoencapsulation of food components. Spray Drying Techniques for Food Ingredient Encapsulation. John Wiley & Sons, Ltd., Chichester, UK, pp. 180–197, (Chapter 8).

Aquino, R.P., Stigliani, M., Del Gaudio, P., Mencherini, T., Sansone, F., Russo, P., 2014. Nanospray drying as a novel technique for the manufacturing of inhalable NSAID powders. Hindawi Publ. Corp. Sci. World J., 1–7.

Arpagaus, C., 2010. Spray Drying R&D Solutions—BÜCHI's Nano Spray Dryer: A world novelty in laboratory scale. ONdrugDelivery, pp. 30–31. Available from: www.ondrugdelivery.com

Arpagaus, C., 2011. Nano Spray Dryer B-90: Literature review and applications, best@buchi Information Bulletin, Number 63/2011.

Arpagaus, C., 2012. A novel laboratory-scale spray dryer to produce nanoparticles. Dry. Technol. 30, 1113–1121.

Arpagaus, C., Meuri, M., 2010. Laboratory scale spray drying of inhalable particles: a review. Respir. Drug Deliv. 2, 469–476.

Arpagaus, C., Schafroth, N., 2007. Spray Dried Biodegradable Polymers as target material for Controlled Drug Delivery, best@buchi Information Bulletin, Number 46/2007.

Arpagaus, C., Schwartzbach, H., 2008. Scale-up from the Büchi Mini Spray Dryer B-290 to the Niro MOBILE MINOR, best@buchi Information Bulletin, Number 52/2008.

Arpagaus, C., Schafroth, N., 2009. Laboratory scale spray drying of biodegradable polymers. Respir. Drug Deliv. Eur. 2, 269–274.

Arpagaus, C., Schafroth, N., Meuri, M., 2010a. Laboratory scale spray drying of lactose: a review, best@buchi Information Bulletin, Number 57/2010.

Arpagaus, C., Schafroth, N., Meuri, M., 2010b. Laboratory scale spray drying of inhalable drugs: a review, best@buchi Information Bulletin, Number 59/2010.

Arpagaus, C., Rütti, D., Meuri, M., 2013. Enhanced solubility of poorly soluble drugs via spray drying. In: Douroumis, D., Fahr, A. (Eds.), Drug Delivery Strategies for Poorly Water-Soluble Drugs. John Wiley & Sons, Ltd., pp. 551–585.

Assadpour, E., Jafari, S.M., Maghsoudlou, Y., 2017. Evaluation of folic acid release from spray dried powder particles of pectin-whey protein nano-capsules. International J. Biol. Macromol. 95, 238–247.

Assadpour, E., Jafari, S.M., 2017. Spray drying of folic acid within nano-emulsions; optimization by taguchi approach. Drying Technol., doi:10.1080/07373937.2016.1242016.

Baba, K., Nishida, K., 2012. Calpain inhibitor nanocrystals prepared using Nano Spray Dryer B-90. Nanoscale Res. Lett. 7, 1–9.

Baba, K., Nishida, K., 2013. Steroid nanocrystals prepared using the nano spray dryer B-90. Pharmaceutics 5, 107–114.

Beber, T.C., Andrade, D.F., Kann, B., Fontana, M.C., Coradini, K., Windbergs, M., Beck, R.C.R., 2014. Submicron polymeric particles prepared by vibrational spray-drying: semisolid formulation and skin penetration/permeation studies. Eur. J. Pharm. Biopharm. 88, 602–613.

Beck-Broichsitter, M., Schweiger, C., Schmehl, T., Gessler, T., Seeger, W., Kissel, T., 2012. Characterization of novel spray-dried polymeric particles for controlled pulmonary drug delivery. J. Control. Release 158, 329–335.
Beck-Broichsitter, M., Paulus, I.E., Greiner, A., Kissel, T., 2015a. Modified vibrating-mesh nozzles for advanced spray-drying applications. Eur. J. Pharm. Biopharm. 92, 96–101.
Beck-Broichsitter, M., Strehlow, B., Kissel, T., 2015b. Direct fractionation of spray-dried polymeric microparticles by inertial impaction. Powder Technol. 286, 311–317.
Bege, N., Renette, T., Endres, T., Beck-Broichsitter, M., Hänggi, D., Kissel, T., 2013. In situ forming nimodipine depot system based on microparticles for the treatment of posthemorrhagic cerebral vasospasm. Eur. J. Pharm. Biopharm. 84, 99–105.
Behara, S.R.B., Farkas, D.R., Hindle, M., Longest, P.W., 2014a. Development of a high efficiency dry powder inhaler: effects of capsule chamber design and inhaler surface modifications. Pharm. Res. 31, 360–372.
Behara, S.R.B., Longest, P.W., Farkas, D.R., Hindle, M., 2014b. Development and comparison of new high-efficiency dry powder inhalers for carrier-free formulations. J. Pharm. Sci. 103, 465–477.
Blasi, P., Schoubben, A., Giovagnoli, S., Rossi, C., Ricci, M., 2010. Alginate micro- and nanoparticle production by spray drying. Proc. Meet. Lact. as a Carr. Inhal. Prod. Parma, Italy, Sept. 26–28, 2010, pp. 137–138.
Bowen, M., Turok, R., Maa, Y.-F., 2013. Spray drying of monoclonal antibodies: investigating powder-based biologic drug substance bulk storage. Dry. Technol. 31, 1441–1450.
Brandenberger, H., 2003. Development of a novel high-performacne cyclone to increase the yield in a Mini Spray Dryer, best@buchi Information Bulletin, Number 27/2003.
Brinkmann-Trettenes, U., Barnert, S., Bauer-Brandl, A., 2014. Single step bottom-up process to generate solid phospholipid nano-particles. Pharm. Dev. Technol. 19, 326–332.
Büchi Labortechnik, AG, 2010. Nano Spray Dryer B-90 Brochure, Flawil, Switzerland, 11592236 en 1012. Available from: www.buchi.com
Büchi Labortechnik, AG, 2013. Nano Spray Dryer B-90 Technical data sheet, en 1311 A. Available from: http://static2.buchi.com/sites/default/files/technical-data-pdf/B-90_Data_Sheet_en_A_0.pdf.
Büchi Labortechnik, AG, 2015. Spray Drying & Encapsulation Solutions–Particle formation for lab scale. Brochure, No. 11592837 en 15/09 E 24 pp.
Bürki, K., Jeon, I., Arpagaus, C., Betz, G., 2011. New insights into respirable protein powder preparation using a nano spray dryer. Int. J. Pharm. 408, 248–256.
Celli, G.B., Ghanem, A., Brooks, M.S. L., 2015. Bioactive encapsulated powders for functional foods—a review of methods and current limitations. Food Bioproc. Technol. 8, 1825–1837.
Cerchiara, T., Abruzzo, A., di Cagno, M., Bigucci, F., Bauer-Brandl, A., Parolin, C., Vitali, B., Gallucci, M.C., Luppi, B., 2015. Chitosan based micro- and nanoparticles for colon-targeted delivery of vancomycin prepared by alternative processing methods. Eur. J. Pharm. Biopharm. 92, 112–119.
Chan, H.-K., Kwok, P.C.L., 2011. Production methods for nanodrug particles using the bottom-up approach. Adv. Drug Deliv. Rev. 63, 406–416.
Couvreur, P., Dubernet, C., Puisieux, F., 1995. Controlled drug delivery with nanoparticles—current possibilities and future trends. Eur. J. Pharm. Biopharm. 41, 2–13.
Dahili, L.A., Feczkó, T., 2015. Cross-linking of horseradish peroxidase enzyme to fine particles generated by Nano Spray Dryer B-90. Period. Polytech. Chem. Eng. 59, 209–214.
Dahili, L.A., Kelemen-Horváth, I., Feczkó, T., 2015. 2,4-Dichlorophenol removal by purified horseradish peroxidase enzyme and crude extract from horseradish immobilized to nano spray dried ethyl cellulose particles. Process Biochem. 50, 1835–1842.

De Cicco, F., Porta, A., Sansone, F., Aquino, R.P., Del Gaudio, P., 2014. Nanospray technology for an in situ gelling nanoparticulate powder as a wound dressing. Int. J. Pharm. 473, 30–37.

Dhand, R., 2002. Nebulizer that use a vibrating mesh or plate with multiple apertures to generate aerosol. Respir. Care 47 (12), 1406–1418.

Dhillon, G.S., Kaur, S., Brar, S.K., 2014. Facile fabrication and characterization of chitosan-based zinc oxide nanoparticles and evaluation of their antimicrobial and antibiofilm activity. Int. Nano Lett. 4, 107.

Dimer, F., de Souza Carvalho-Wodarz, C., Haupenthal, J., Hartmann, R., Lehr, C.-M., 2015a. Inhalable clarithromycin microparticles for treatment of respiratory infections. Pharm. Res. 32, 3850–3861.

Dimer, F., Ortiz, M., Pohlmann, A.R., Guterres, S.S., 2015b. Inhalable resveratrol microparticles produced by vibrational atomization spray drying for treating pulmonary arterial hypertension. J. Drug Deliv. Sci. Technol. 29, 152–158.

Draheim, C., de Crécy, F., Hansen, S., Collnot, E.-M., Lehr, C.-M., 2015. A design of experiment study of nanoprecipitation and nano spray drying as processes to prepare plga nano- and microparticles with defined sizes and size distributions. Pharm. Res. 32, 2609–2624.

Durli, T.L., Dimer, F.A., Fontana, M.C., Pohlmann, A.R., Beck, R.C.R., Guterres, S.S., 2014. Innovative approach to produce submicron drug particles by vibrational atomization spray drying: influence of the type of solvent and surfactant. Drug Dev. Ind. Pharm. 40, 1011–1020.

Esfanjani, A.F., Jafari, S.M., Assadpoor, E., Mohammadi, A., 2015. Nano-encapsulation of saffron extract through double-layered multiple emulsions of pectin and whey protein concentrate. J. Food Eng. 165, 149–155.

Estevinho, B.N., Rocha, F., Santos, L., Alves, A., 2013. Microencapsulation with chitosan by spray drying for industry applications—a review. Trends Food Sci. Technol. 31, 138–155.

Ezhilarasi, P.N., Karthik, P., Chhanwal, N., Anandharamakrishnan, C., 2013. Nanoencapsulation techniques for food bioactive components: a review. Food Bioproc. Technol. 6, 628–647.

Fang, Z., Bhandari, B., 2012. Encapsulation techniques for food ingredient systems. In: Bhandari, Bhesh, Roos, Yrjö H. (Eds.), Food Materials Science and Engineering. Blackwell Publishing Ltd., Oxford, UK, pp. 320–348, (Chapter 12).

Farkas, D.R., Hindle, M., Longest, P.W., 2015. Characterization of a new high-dose dry powder inhaler (DPI) based on a fluidized bed design. Ann. Biomed. Eng. 43, 2804–2815.

Faridi Esfanjani, A.F., Jafari, S.M., 2016. Biopolymer nano-particles and natural nano-carriers for nano-encapsulation of phenolic compounds. Colloids Surf. B. 146, 532–543.

Fathi, M., Martin, A., McClements, D.J., 2014. Nanoencapsulation of food ingredients using carbohydrate based delivery systems. Trends Food Sci. Technol. 39, 18–39.

Faulhammer, E., Wahl, V., Zellnitz, S., Khinast, J.G., Paudel, A., 2015. Carrier-based dry powder inhalation: Impact of carrier modification on capsule filling processability and in vitro aerodynamic performance. Int. J. Pharm. 491, 231–242.

Feng, A.L., Boraey, M.A., Gwin, M.A., Finlay, P.R., Kuehl, P.J., Vehring, R., 2011. Mechanistic models facilitate efficient development of leucine containing microparticles for pulmonary drug delivery. Int. J. Pharm. 409, 156–163.

Fontana, M.C., Durli, T.L., Pohlmann, A.R., Guterres, S.S., Beck, R.C.R., 2014. Polymeric controlled release inhalable powder produced by vibrational spray-drying: one-step preparation and in vitro lung deposition. Powder Technol. 258, 49–59.

Gautier, S., Arpagaus, C., Schafroth, N., Meuri, M., Deschamps, A., Maquet, V., 2010. Very fine chitosan microparticles with narrow & controlled size distribution using spray-drying technologies. Drug Deliv. Technol. 10, 30–37.

Gharsallaoui, A., Roudaut, G., Chambin, O., Voilley, A., Saurel, R., 2007. Applications of spray-drying in microencapsulation of food ingredients: an overview. Food Res. Int. 40, 1107–1121.

Ghazanfari, T., Elhissi, A.M.A., Ding, Z., Taylor, K.M.G., 2007. The influence of fluid physico-chemical properties on vibrating-mesh nebulization. Int. J. Pharm. 339, 103–111.

Gu, B., Linehan, B., Tseng, Y.-C., 2015. Optimization of the Büchi B-90 spray drying process using central composite design for preparation of solid dispersions. Int. J. Pharm. 491, 208–217.

Haggag, Y.A., Faheem, A.M., 2015. Evaluation of nano spray drying as a method for drying and formulation of therapeutic peptides and proteins. Front. Pharmacol. 6, 1–5.

Harsha, S., 2013. Pharmaceutical suspension containing both immediate/sustained-release amoxicillin-loaded gelatin nanoparticles: Preparation and in vitro characterization. Drug Des. Devel. Ther. 7, 1027–1033.

Harsha, S., 2015. In vitro and in vivo evaluation of nanoparticles prepared by nano spray drying for stomach mucoadhesive drug delivery. Dry. Technol. 33, 1199–1209.

Harsha, S., Aldhubiab, B.E., Alhaider, I.A., Attimarad, M., Nair, A., 2013a. Carbopol 934-P loaded with vildagliptin for diabetic delivery: in vitro and in vivo evaluation of nanoparticles. Curr. Nanosci. 9, 642–647.

Harsha, S., Attimard, M., Khan, T.A., Nair, A.B., Aldhubiab, B.E., Sangi, S., Shariff, A., 2013b. Design and formulation of mucoadhesive microspheres of sitagliptin. J. Microencapsul. 30, 257–264.

Harsha, S., Aldhubiab, B.E., Nair, A.B., Alhaider, I.A., Attimarad, M., Venugopala, K.N., Srinivasan, S., Gangadhar, N., Asif, A.H., 2015a. Nanoparticle formulation by Büchi B-90 Nano Spray Dryer for oral mucoadhesion. Drug Des. Devel. Ther. 9, 273–282.

Harsha, S., Al-Dhubiab, B.E., Nair, A.B., Al-Khars, M., Al-Hassan, M., Rajan, R., Attimarad, M., Venugopala, K.N., Asif, A.H., 2015b. Novel drying technology of microsphere and its evaluation for targeted drug delivery for lungs. Dry. Technol. 33, 502–512.

Heng, D., Lee, S.H., Ng, W.K., Tan, R.B.H., 2011. The Nano Spray Dryer B-90. Expert Opin. Drug Deliv. 8, 965–972.

Hindle, M., Kaviratna, A., Yamarthi, D., Longest, P.W., 2015. Characterization of spray dried azithromycin dry powders for inhalation. J. Aerosol Med. Pulmonary Drug Deliv. 4, A9.

Jafari, S.M., He, Y., Bhandari, B., 2007a. Encapsulation of Nanoparticles of d-limonene by spray drying: role of emulsifiers and emulsifying techniques. Dry. Technol. 25, 1069–1079.

Jafari, S.M., He, Y., Bhandari, B., 2007b. Role of powder particle size on the encapsulation efficiency of oils during spray drying. Dry. Technol. 26, 1081–1089.

Jafari, S.M., Assadpoor, E., Bhandari, B., He, Y., 2008a. Nano-particle encapsulation of fish oil by spray drying. Food Res. Int. 41, 172–183.

Jafari, S.M., Assadpoor, E., He, Y., Bhandari, B., 2008b. Encapsulation efficiency of food flavours and oils during spray drying. Dry. Technol. 26, 816–835.

Jain, I., 2014. Crosslinking albumin for drug release from spray dried particles, Master Thesis, Electronic Thesis and Dissertations, Paper 674. University of Louisville, USA.

Kanogwan, K., Srichana, T., 2016. Nano spray-dried pyrazinamide-L-leucine dry powders, physical properties and feasibility used as dry powder aerosols. Pharm. Dev. Technol. 21 (1), 68–75.

Katouzian, I., Jafari, S.M., 2016. Nano-encapsulation as a promising approach for targeted delivery and controlled release of vitamins. Trends Food Sci. Technol. 53, 34–48.

Knoch, M., Keller, M., 2005. The customised electronic nebuliser: a new category of liquid aerosol drug delivery systems. Expert Opin. Drug Deliv. 2, 377–390.

Lass, J.S., Sant, A., Knoch, M., 2006. New advances in aerosolised drug delivery: vibrating membrane nebuliser technology. Expert Opin. Drug Deliv. 3, 693–702.

Lee, S.H., Heng, D., Ng, W.K., Chan, H.-K., Tan, R.B.H., 2011. Nano spray drying: a novel method for preparing protein nanoparticles for protein therapy. Int. J. Pharm. 403, 192–200.

Lee, S.H., Teo, J., Heng, D., Ng, W.K., Chan, H.K., Tan, R.B.H., 2013. Synergistic combination dry powders for inhaled antimicrobial therapy: formulation, characterization and in vitro evaluation. Eur. J. Pharm. Biopharm. 83, 275–284.

Li, X., Anton, N., Arpagaus, C., Belleteix, F., Vandamme, T.F., 2010. Nanoparticles by spray drying using innovative new technology: the Büchi Nano Spray Dryer B-90. J. Control. Release 147, 304–310.

Li, X., Anton, N., Vandamme, T.F., 2015. Spray-drying of nano- and microcapsules of nutraceuticals. In: Shi, J. (Ed.), Functional Food Ingredients and Neutraceuticals—Processing Technologies. CRC Press, Taylor & Francis Group, Boca Raton, pp. 455–480.

Littringer, E.M., Zellnitz, S., Hammernik, K., Adamer, V., Friedl, H., Urbanetz, N.A., 2013. Spray drying of aqueous salbutamol sulfate solutions using the nano spray dryer b-90 - the impact of process parameters on particle size. Dry. Technol. 31, 1346–1353.

Longest, P.W., Golshahi, L., Behara, S.R.B., Tian, G., Farkas, D.R., Hindle, M., 2015. Efficient nose-to-lung (N2L) aerosol delivery with a dry powder inhaler. J. Aerosol Med. Pulm. Drug Deliv. 28, 189–201.

Maa, Y.F., Nguyen, P.A., Sit, K., Hsu, C.C., 1998. Spray-drying performance of a bench-top spray dryer for protein aerosol powder preparation. Biotechnol. Bioeng. 60, 301–309.

Mahdavee Khazaei, K., Jafari, S.M., Ghorbani, M., Hemmati Kakhki, A., 2014. Application of maltodextrin and gum Arabic in microencapsulation of saffron petal's anthocyanins and evaluating their storage stability and color. Carbohydr. Polym. 105 (1), 57–62.

Mahdavi, S.A., Jafari, S.M., Ghorbani, M., Assadpoor, E., 2014. Spray-drying microencapsulation of anthocyanins by natural biopolymers: a review. Dry. Technol. 32, 509–518.

Maltesen, M.J., Bjerregaard, S., Hovgaard, L., Havelund, S., van de Weert, M., 2008. Quality by design—spray drying of insulin intended for inhalation. Eur. J. Pharm. Biopharm. 70, 828–838.

Martena, V., Censi, R., Hoti, E., Malaj, L., Di Martino, P., 2012. A new nanospray drying method for the preparation of nicergoline pure nanoparticles. J. Nanoparticle Res. 14, 1–10.

Masters, K., 1991. Spray drying handbook, 5th edition Longman Scientific & Technical, Harlow, USA.

Merchant, Z., Taylor, K.M.G., Stapleton, P., Razak, S.A., Kunda, N., Alfagih, I., Sheikh, K., Saleem, I.Y., Somavarapu, S., 2014. Engineering hydrophobically modified chitosan for enhancing the dispersion of respirable microparticles of levofloxacin. Eur. J. Pharm. Biopharm. 88, 816–829.

Mizuno, A., 2000. Electrostatic precipitation. IEEE Trans. Dielectr. Electr. Insul. 7, 615–624.

Moncada, M., Astete, C., Sabliov, C., Olson, D., Boeneke, C., Aryana, K.J., 2015. Nano spray-dried sodium chloride and its effects on the microbiological and sensory characteristics of surface-salted cheese crackers. J. Dairy Sci. 98, 5946–5954.

Murugesan, R., Orsat, V., 2012. Spray drying for the production of nutraceutical ingredients—a review. Food Bioproc. Technol. 5, 3–14.

Nandiyanto, A.B.D., Okuyama, K., 2011. Progress in developing spray-drying methods for the production of controlled morphology particles: from the nanometer to submicrometer size ranges. Adv. Powder Technol. 22, 1–19.

Ngan, L.T.K., Wang, S.-L., Hiep, Đ.M., Luong, P.M., Vui, N.T., Đinh, T.M., Dzung, N.A., 2014. Preparation of chitosan nanoparticles by spray drying, and their antibacterial activity. Res. Chem. Intermed. 40, 2165–2175.

Okuyama, K., Abdullah, M., Lenggoro, I.W., Iskandar, F., 2006. Preparation of functional nanostructured particles by spray drying. Adv. Powder Technol. 17, 587–611.

Oliveira, A.M., Guimarães, K.L., Cerize, N.N., Tunussi, A.S., Poço, J.G., 2013. Nano spray drying as an innovative technology for encapsulating hydrophilic active pharmaceutical ingredients (API). J. Nanomed. Nanotechnol. 4, 1–6.

Öztürk, A.A., Yenilmez, E., Yazan, Y., 2015. Preparation and Characterization of Dexketoprofen Tromethamol Loaded Kollidon-SR Nanoparticles. In: ISOPS 11th International Symposium on Pharmaceutical Sciences, June 9–12, 2015, Ankara, Turkey.

Panda, A., Meena, J., Katara, R., Majumdar, D.K., 2014. Formulation and characterization of clozapine and risperidone co-entrapped spray-dried PLGA nanoparticles. Pharm. Dev. Technol. 7450, 1–11.

Patel, R., Upadhyay, J.B., 2014. Nano Spray Dryer—A Technique to Handle Specific Products Range. In: National Seminar on "Indian Dairy Industry—Opportunities and Challenges", Technical Articles. pp. 181–183.

Perecin, C.J., Yoshioka, S.A., De Oliveira, A.M., Chitta, V.A., Cerize, N.N.P., 2014. Nanopartículas Superparamagnéticas Encapsuladas com Polímeros Para Aplicação No Tratamento De Câncer Por Hipertermia. In: XXIV, Congresso Brasiliero de Engenharia Biomédica, CBEB, pp., 2703–2706.

Perecin, C.J., Yoshioka, S., Chitta, V., Gratens, X., Léo, P., De Oliveira, A., Cerize, N., 2015. Magnetite Nanoparticles Encapsulated with PCL and Poloxamer by Nano Spray Drying technique. In: XIV, Brazil, MRS (Materials Research Society) Meeting, Rio, Awarded Abstract: ASEE, Symposium B.

Pérez-Masiá, R., López-Nicolás, R., Periago, M.J., Ros, G., Lagaron, J.M., López-Rubio, A., 2015. Encapsulation of folic acid in food hydrocolloids through nanospray drying and electrospraying for nutraceutical applications. Food Chem. 168, 124–133.

Pourashouri, P., Shabanpour, B., Razavi, S.H., Jafari, S.M., Shabani, A., Aubourg, S.P., 2014. Impact of wall materials on physicochemical properties of microencapsulated fish oil by spray drying. Food Bioproc. Technol. 7, 2354–2365.

Quintanilla-Carvajal, M.X., Camacho-Díaz, B.H., Meraz-Torres, L.S., Chanona-Pérez, J.J., Alamilla-Beltrán, L., Jimenéz-Aparicio, A., Gutiérrez-López, G.F., 2010. Nanoencapsulation: a new trend in food engineering processing. Food Eng. Rev. 2, 39–50.

Rajabi, H., Ghorbani, M., Jafari, S.M., Mahoonak, S.A., Rajabzadeh, G., 2015. Retention of saffron bioactive components by spray drying encapsulation using maltodextrin, gum Arabic and gelatin as wall materials. Food Hydrocoll. 51, 327–337.

Riehle, C., 1997. Basic and theoretical operation of electrostatic precipitators. In: Parker, K.R. (Ed.), Applied Electrostatic Precipitation. Blackie Academic & Professional, An Imprint of Chapman & Hall, London, UK, (Chapter 3).

Sabliov, C.M., Astete, C.E., 2015. Polymeric nanoparticles for food applications. In: Sabliov, C.M., Chen, H., Y., Y.R. (Eds.), Nanotechnology and Functional Foods: Effective Delivery of Bioactive Ingredients. John Wiley & Sons, Ltd., Chichester, UK, pp. 272–296, (Chapter 17).

Sarma, S.J., Brar, S.K., Reigner, J., Le Bihan, Y., Buelna, G., 2014a. Enriched hydrogen production by bioconversion of biodiesel waste supplemented with ferric citrate and its nano-spray dried particles. RSC Adv. 4, 49588–49594.

Sarma, S.J., Das, R.K., Brar, S.K., Le Bihan, Y., Buelna, G., Verma, M., Soccol, C.R., 2014b. Application of magnesium sulfate and its nanoparticles for enhanced lipid production by mixotrophic cultivation of algae using biodiesel waste. Energy 78, 16–22.

Schafroth, N., Arpagaus, C., Jadhav, U.Y., Makne, S., Douroumis, D., 2012. Nano and microparticle engineering of water insoluble drugs using a novel spray-drying process. Colloids Surf. B Biointerfaces 90, 8–15.

Schmid, K., 2011. Spray drying of protein precipitates and Evaluation of the Nano Spray Dryer B-90. PhD Thesis, Ludwig-Maximilians-University, Munich.
Schmid, K., Arpagaus, C., Friess, W., 2009. Evaluation of a vibrating mesh spray dryer for preparation of submicron particles. Respir. Drug Deliv. Eur. 2, 323–326.
Schmid, K., Arpagaus, C., Friess, W., 2011. Evaluation of the nano spray dryer B-90 for pharmaceutical applications. Pharm. Dev. Technol. 16, 287–294.
Schön, M., Baumgartner, R., 2009. Method for Heating and Laminating, Electrostatic Separator, Spray Drier, Separating Device and Method for Separating Particles. Patent EP 2056037 (A1), priority date: 2007-10-30.
Schoubben, A., Blasi, P., Marenzoni, M.L., Barberini, L., Giovagnoli, S., Cirotto, C., Ricci, M., 2013. Capreomycin supergenerics for pulmonary tuberculosis treatment: preparation, in vitro, and in vivo characterization. Eur. J. Pharm. Biopharm. 83, 388–395.
Schoubben, A., Giovagnoli, S., Tiralti, M.C., Blasi, P., Ricci, M., 2014. Capreomycin inhalable powders prepared with an innovative spray-drying technique. Int. J. Pharm. 469, 132–139.
Schoubben, A., Blasi, P., Giontella, A., Giovagnoli, S., Ricci, M., 2015. Powder, capsule and device: an imperative ménage à trois for respirable dry powders. Int. J. Pharm. 494, 40–48.
Schultz, I., Vollmers, F., Lühmann, T., Rybak, J.-C., Wittmann, R., Stank, K., Steckel, H., Kardziev, B., Schmidt, M., Högger, P., Meinel, L., 2015. Pulmonary insulin-like growth factor I delivery from trehalose and silk-fibroin microparticles. ACS Biomater. Sci. Eng. 1, 119–129.
Smart, J., Berg, E., Nerbrink, O., Zuban, R., Blakey, D., New, M., Mastersizer, M., 2002. TouchSpray technology: comparison of the droplet size with cascade impaction and laser diffraction. Respir. Drug Deliv. VIII 2, 525–528.
Son, Y.-J., Longest, P.W., Hindle, M., 2013a. Aerosolization characteristics of dry powder inhaler formulations for the excipient enhanced growth (EEG) application: effect of spray drying process conditions on aerosol performance. Int. J. Pharm. 443, 137–145.
Son, Y.-J., Longest, P.W., Tian, G., Hindle, M., 2013b. Evaluation and modification of commercial dry powder inhalers for the aerosolization of a submicrometer excipient enhanced growth (EEG) formulation. Eur. J. Pharm. Sci. 49, 390–399.
Sosnik, A., Seremeta, K.P., 2015. Advantages and challenges of the spray-drying technology for the production of pure drug particles and drug-loaded polymeric carriers. Adv. Colloid Interface Sci. 223, 40–54.
Sun, Y., Song, X., Wang, J., Yu, J., 2011. Preparation of lithium carbonate hollow spheres by spray pyrolysis. Cryst. Res. Technol. 46, 173–177.
Suryaprakash, R.C., Lohmann, F.P., Wagner, M., Abel, B., Varga, A., 2014. Spray drying as a novel and scalable fabrication method for nanostructured CsH_2PO_4, Pt-thin-film composite electrodes for solid acid fuel cells. RSC Adv. 4, 60429–60436.
Vecellio, L., 2006. The mesh nebulizer: a recent technical innovation for aerosol delivery. Breathe 2, 252–260.
Vehring, R., 2008. Pharmaceutical particle engineering via spray drying. Pharm. Res. 25, 999–1022.
Wang, B., Zhang, W., Zhang, W., Mujumdar, A.S., Huang, L., 2005. Progress in drying technology for nanomaterials. Dry. Technol. 23, 7–32.
Whelehan, M., 2014. Encapsulator B-390/ B-395 Pro—Laboratory Guide, Büchi Labortechnik AG, No. 11595052 en 1408 A.
Whelehan, M., Marison, I.W., 2011. Microencapsulation using vibrating technology. J. Microencapsul. 28, 669–688.
Wong, T.W., John, P., 2015. Advances in spray drying technology for nanoparticle formation. In: Aliofkhazraei, M. (Ed.), Handbook of Nanoparticles. Springer International Publishing, Cham, Switzerland, pp. 1–16.

Zellnitz, S., Schroettner, H., Urbanetz, N.A., 2014. Surface modified glass beads as model carriers in dry powder inhalers—influence of drug load on the fine particle fraction. Powder Technol. 268, 377–386.
Zellnitz, S., Narygina, O., Resch, C., Schroettner, H., Urbanetz, N.A., 2015a. Crystallization speed of salbutamol as a function of relative humidity and temperature. Int. J. Pharm. 489, 170–176.
Zellnitz, S., Schroettner, H., Urbanetz, N.A., 2015b. Influence of surface characteristics of modified glass beads as model carriers in dry powder inhalers (DPIs) on the aerosolization performance. Drug Dev. Ind. Pharm. 41, 1710–1717.

Further Reading

Feng, S.-S., Mu, L., Win, K.Y., Huang, G., 2004. Nanoparticles of biodegradable polymers for clinical administration of paclitaxel. Curr. Med. Chem. 11, 413–424.

Part Four

Nanoencapsulation Technologies Based on Biopolymer Nanoparticles

11 Nanocapsule formation by individual biopolymer nanoparticles

*Rohollah Sadeghi**,†, *Laleh Mehryar***, *Mahsan Karimi**, *Jozef Kokini*†
*Islamic Azad University, Kermanshah, Iran; **University of Urmia, Urmia, Iran; †Purdue University, West Lafayette, IN, United States

11.1 Introduction

Recently, nanoparticles have risen to fame for encapsulating and target delivery of drug and food compounds. Several techniques have been successfully used to fabricate nano- and microparticles such as desolvation using a desolvating agent, emulsification, precipitation, gelation using thermal treatment, nanospray drying, nab-technology and self-assembly (Elzoghby et al., 2012a; Faridi Esfanjani and Jafari, 2016; Jafari et al., 2015; Katouzian and Jafari, 2016). Among these, desolvation and/or nanoprecipitation has gained attraction, mostly due to the advantages it offers in terms of the procedure and the applied materials, including simplicity of the procedure, wide range of the biopolymers that could be used, the ability to employ different solvents and desolvating agents, feasibility of producing relatively uniform, stable, and spherical particles, and so on. Therefore, it could be beneficial to use biocompatible and biodegradable materials to prepare nanoparticles (Elzoghby et al., 2012b). Proteins are an example of such biodegradable polymers.

In this chapter, formation of the nanocapsule by either single protein or carbohydrate biopolymers via desolvation and precipitation methods will be discussed in two main sections. In the next chapter, nanocapsules biosynthesized through biopolymers complexation along with their applications in the food and nutraceutical sciences would be considered.

11.2 Protein nanoparticles (desolvation method)

According to Elzoghby et al. (2012b), protein-based nanocarriers have gained lots of interest mainly because they are generally recognized as safe (GRAS), and biodegradable, as well as they have nonantigen nature, great profile of nutritional requirement, plentiful sustainable sources, significant binding capacity, and great potential to produce bioactive peptides by enzymatic reactions. Other advantages, which highlight the usage of proteins as delivery vehicles, are related to their self-protection ability, provided by the reticuloendothelial system (RES) toward opsonization which is done via aqueous steric barrier. They also possess great ability to form emulsions, gels, and foams along

Nanoencapsulation Technologies for the Food and Nutraceutical Industries with the improved water binding capacity (Elzoghby et al., 2011, 2012b). Due to the various functional groups and sites present in the proteins' sequences and structures as well as in their nanoparticles, they are capable of offering binding sites to a variety of active compounds, which lead to target delivery of the active molecules. In recent years, albumin and gelatin nanoparticles are being vastly employed as drug delivery carriers (Kaul and Amiji, 2005; Kommareddy and Amiji, 2005; Langer et al., 2003).

11.2.1 Principles of the desolvation process

According to Sahoo et al. (2015), desolvation is defined as a self-sorting process with the minimum of Gibbs free energy to prepare nanoparticles from biopolymers, such as proteins, that is, based on a theory, that replacing some or all the water molecules (solvent) around protein molecules with a desolvating agent (another solvent) causes conformational changes in protein structure and significant decreases in the solubilitywhich result in the phase separation and precipitation of proteins. This makes them ready to interact with each other and form nanoparticles (Farrugia and Groves, 1999; Lohcharoenkal et al., 2014). As Kinsella and Whitehead (1989) have mentioned, unfolding of a protein structure exposes its interactive groups such as thiols and amines, which facilitates interaction of protein molecules. When a globular protein unfolds, amine and thiol groups and hydrophobic pockets relocate to the surface of the molecule, which increases intramolecular bindings and decreases hydrophobic interactions (Clark et al., 1981; Ko and Gunasekaran, 2006). It seems this theory completely works for nanoparticulation of the serum albumins during desolvation.

In this method, which is also known as drawing-out precipitation, reaching to the desired nanoparticulation basically depends on balancing the attractive and repulsive forces and decreasing intramolecular hydrophobic interactions (Lohcharoenkal et al., 2014). In general, a desolvating agent is a solvent of different polarity and hydrogen-bond-forming capacity, which is added dropwise to the protein solution (Fig. 11.1). The commonly used agents are ethanol, methanol, acetone, propanol, etc. and also a mixture of them which have been investigated by many researchers as aforementioned, the mechanism upon which the desolvating agent acts is to reduce protein solubility by substituting the solute (Farrugia and Groves, 1999). Gradual adding of ethanol to protein solution unfolds the protein structure at first, followed by

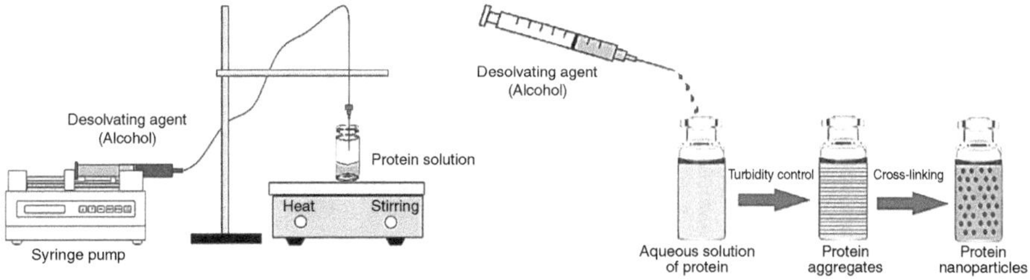

Figure 11.1 Schematic representation of the desolvation process for the fabrication of polymeric nanoparticles.

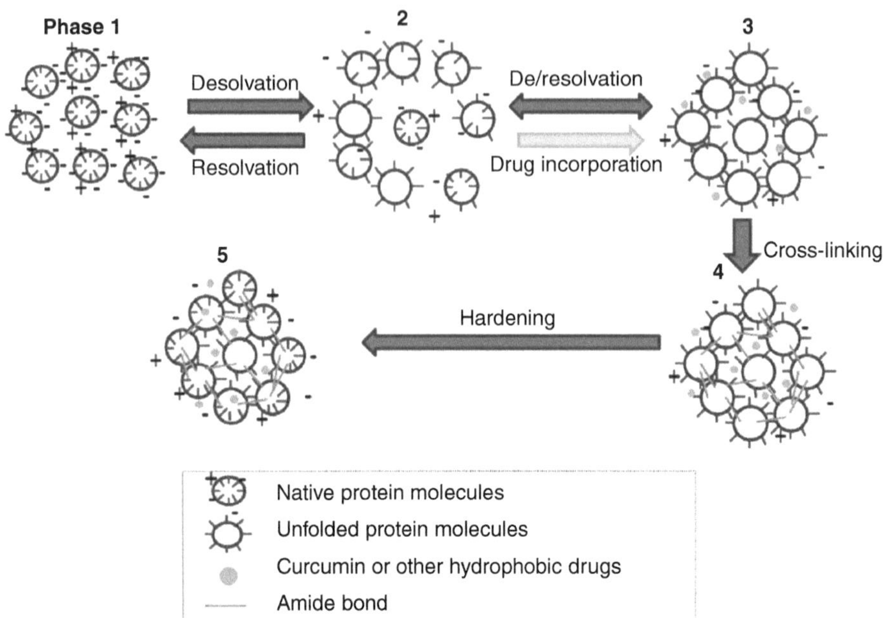

Figure 11.2 Schematic of protein nanoparticulation and role of cross-linking agents. *Source:* Reprinted with permission from Teng, Z., Luo, Y., Wang, Q., 2012. Nanoparticles synthesized from soy protein: preparation, characterization, and application for nutraceutical encapsulation. J. Agric. Food Chem. 60 (10), 2712–2720. Copyright (2012), with permission from American Chemical Society.

proteins folding and the possible increase of peptide–peptide interaction and finally, specific nanostructures will be assembled to form nanoparticles. The new structure has more hydrophobic surface, which leads to the lower solubility of the produced aggregates or nanoparticles in comparison with the native proteins. Subsequently, produced nanoparticles should be cross-linked to avoid aggregation. Cross-linking has proved to be an important step in nanoparticulation process mainly due to its role in biodegradability and release of the encapsulated material. It also prevents aggregation and agglomeration of the resulted nanoparticles (Niknejad and Mahmoudzadeh, 2015). The most commonly used cross-linking agent is GTA[a] which has been proved to have toxic effects. Figs. 11.2 and 11.3 present the schematic of protein nanoparticulation and its cross-linking procedure.

Oppenheim et al. (1978) first developed the desolvation method to fabricate gelatin nanoparticles. They added desolvating agents to the gelatin solution till the solution reached the desolvating region (turbid) in which the intensity of the scattered light noticeably increased. Then, an aldehyde (GTA) solution was added as the hardening agent to stabilize the produced nanoparticles. Oppenheim et al. (1978) used polyvalent anions SO_4^{2-} or cations (Ca^{2+} and Mg^{2+}) and an alcohol (ethanol or isopropanol) as

[a] Glutaraldehyde

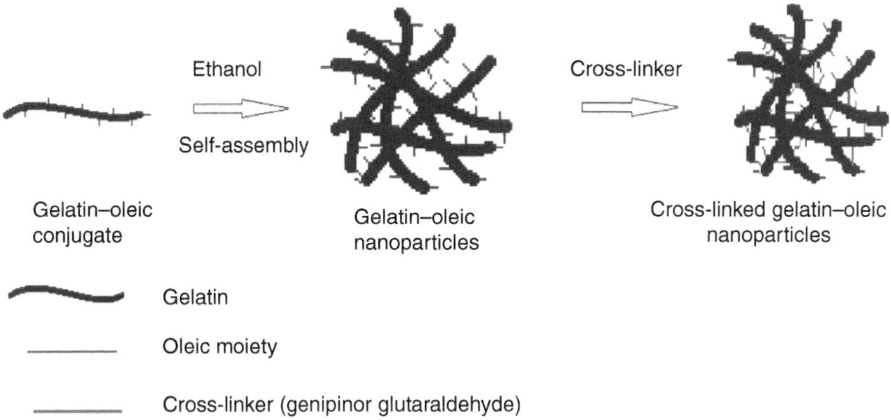

Figure 11.3 Schematic of gelatin–oleic nanoparticle (GON) structure and its preparation process.
Source: Reprinted from Park, C., Le-Ngoc Vo, C., Kang, T., Oh, E., Lee, B.-J., 2015. New method and characterization of self-assembled gelatin–oleic nanoparticles using a desolvation method via carbodiimide/*N*-hydroxysuccinimide (EDC/NHS) reaction. Eur. J. Pharm. Biopharm. 89, 365–373. Copyright (2015), with permission from Elsevier.

desolvating agents. Later Oppenheim and Stewart (1979) reported the fabrication of gelatin and HSA[b] nanoparticles using salting out process by adding sodium sulfate and then isopropanol and GTA. They also conjugated gelatin and albumin nanoparticles with fluorescein isothiocyanate (FITC) and reported that the gelatin nanoparticles were uptaken by mouse mammary tumor EMT 6.

Lin et al. (1993) successfully prepared HSA nanoparticles by desolvation method with acetone. Farrugia and Groves (1999) optimized the gelatin nanoparticles' preparation conditions with ethanol and found that pH and ethanol to protein solution ratio had significant effects on the particulation of gelatin. Ultimately, they reported the particle size of 200–250 nm for the gelatin nanoparticles with the PDI[c] of less than 0.2. Furthermore, Yoshikawa et al. (2012) showed that 17 disulfide bonds in BSA[d] structure could affect the BSA behavior in the presence of ethanol, while SH-modified BSA did not show the same trend. In a study conducted by Arroyo-Maya et al. (2014), the conformational changes of α-lactalbumin (α-LA) during nanoparticulation using desolvation method were studied. They used acetone as a desolvating agent, and the CD[e] results showed significant configurational changes during particulation; β-strand increased ~35%, while α-helix decreased 26%. They finally proposed that acetone induces denaturation of α-LA, and they may interact with each other to form nanoparticles.

[b] Human serum albumin
[c] Polydispersity index
[d] Bovine serum albumin
[e] Circular dichroism

The schematic of desolvation process for producing nanoparticles is shown in Fig. 11.1. Also, in Figs. 11.2 and 11.3 the schematic of protein nanoparticulation and its cross-linking is presented.

11.2.2 Effect of the operational parameters on the nanoparticles characteristics prepared by desolvation method

The effect of the operational parameters on the characterization of the nanoparticles fabricated by the desolvation method is very critical to fabricate small and uniform nanoparticles. Langer et al. (2003) investigated the effect of some parameters such as, protein type and concentration, cross-linking process, environmental conditions, and pH of the protein solution on the characterization of HSA nanoparticles using ethanol as the desolvating agent. Rahimnejad et al. (2012) assessed the effect of pH values, temperature, BSA concentration, ethanol adding rate, agitation speed, and GTA concentration on the defined ranges of particle size and PDI.

11.2.2.1 Upscaling

Wacker et al. (2011) developed the upscaling of desolvation process using ethanol by combining the paddle stirring systems along with a peristaltic pump to optimize the nanoparticulation of HSA up to 2000 mg. They proved that the nanoparticles characteristics were not negatively affected by increasing the desolvating volume up to 100 mL. They reported the particle size of 251.2 ± 27.0 and 234.1 ± 1.5 nm for different volumes (10–100 mL), and the polydispersity indices of lower than 0.2 for all the studied conditions. Later, Yedomon et al. (2013) studied the upscaling process of producing BSA nanoparticles (up to 10 g) by membrane contactor. First, the particles (139 nm, PI: 0.046) were produced at small scale via syringe pump at optimal conditions (pH 8.2, BSA concentration of 100 mg/mL, and desolvating agent flow rate of 1 mL/min), thereafter they were upscaled. Upscaling was performed by infusing ethanol through the pores of the membrane. It was found that increasing the ethanol pressure (1–2.7 bars) would decrease the size of the particles and lead to a high nanoparticulation production yield (94%–96%). It was also concluded that at the optimal conditions (pH 8.2, BSA concentration of 100 mg/mL, ethanol pressure of 2.7 bars, flow rate of 30.7 mL/s for the circulating phase), there was not any difference between the BSA nanoparticles prepared by the two different method. The results showed the shelf stability of 2 months for nanoparticles at 4°C.

Preparation of uniform and small protein nanoparticles with acceptable functionality; such as high amounts of available amino groups on the particle surface or specific carrier capability is still a challenging task, which is gaining more interest. There are different parameters affecting the nanoparticle size; such as type and amount of desolvating agent along with the rate of addition, stirring rate, protein type and concentration, cross-linking agent type, cross-linking conditions, temperature, pH values, and so on.

11.2.2.2 Amount and addition rate of the desolvating agent

Weber et al. (2000) showed that the amount of the added desolvating agent could mainly determine the particle size of HSA nanoparticles. Based on the photon correlation spectroscopy (PCS) measurements, adding ethanol up to 1.5 times of HSA solution increased particle size, nevertheless at volumes more than 1.5-folds, no changes were observed. In light of these findings, Langer et al. (2003) noted that the rate of ethanol addition did not have any effect on the average diameter of HSA nanoparticles (280 nm); however based on PCS data, the rate of ethanol addition greatly affected the particle size distribution, for example, at lower rates (0.5 mL/min), PDI were higher (0.10) but at rates between 1.0 and 2.0 mL/min, more uniform nanoparticles were obtained (PDI: 0.01–0.02). Rahimnejad et al. (2012) studied different ethanol addition rates (0.5, 1, 1.5, and 2 mL/min) on BSA particle size and stated that there was no significant difference in the particle size; indeed different rates of ethanol addition, accelerated its transition phase.

Recently, Sadeghi et al. (2014) reported that increasing the ratio of desolvating agentis resulted in larger BSA nanoparticles with lower PDI, subsequently after reaching an exact ratio both particle size and PDI would be constant. They also stated that by elevating the ethanol ratio from 0X to 3X, the particle size increased while they could not find any significant changes in particle size from 3X to 5X. For polydispersity, addition of 1X ethanol increased PDI, while additions of 2X, 3X, and 4X decreased the PDI. Addition of 4X and 5X ethanol did not show any significant effect on PDI. It seems that adding 1X desolvating agent, initiates the nucleation and adding more desolvating agent leads to nuclei growth. Won and Yong-Hee (2008) studied the effect of ethanol concentration on the particle size of the recombinant human gelatin and noted that increasing the ethanol from 59% to 62% (v/v) significantly increased the nanoparticle size (49–85 nm, with a PDI less than 0.10). They also suggested that nanoparticles were not stable at ethanol concentrations less than 59% or greater than 62%. In the most of nanoparticulation studies, the appropriate nanoparticles were manufactured in 70%–80% ethanol (Sadeghi et al., 2014).

Teng et al. (2012) stated that increasing the ethanol concentration from 0 to 30%–40%, gives bigger soy protein isolate (SPI) nanoparticles, which could be attributed to the interaction of hydrophobic groups and pockets in the protein structure, rendering to the aggregation. Weber et al. (2000) reported similar results. The peak particle size at SPI concentrations of 4 and 12 mg/mL was reported 606 and 862 nm, respectively, which may be accredited to the formation of aggregates at higher protein concentrations. Subsequent addition of ethanol up to 80%, considerably decreased the particle size. According to the authors' claim, the decreasing trend in particle size by increasing the desolvating agent concentration hadn't been stated before and they were pioneers in stating this fact. They related this process to either the collapse of larger aggregates or the fabrication of smaller particles. It was also shown that increasing the ethanol concentration (0%–80%) produces higher percentage of smaller particles.

11.2.2.3 Type of the desolvating agent

Sadeghi et al. (2014) analyzed the usage of acetone as desolvating agent; the trend of particulation was the same as ethanol except that by adding more than 3X acetone, BSA precipitated, representing that the particle size and PDI were higher for acetone. They also compared ethanol, acetone, and a mixture of ethanol and acetone as desolvating agents for preparation of BSA nanoparticles via desolvation method.

They came into conclusion that acetone was a more efficient desolvating agent to prepare BSA nanoparticles compared to others; however more uniform and spherical particles were produced using ethanol. Mixture of ethanol and acetone resulted in more spherical nanoparticles compared to acetone, yet polydispersity of the particles was lower than those produced by ethanol. SEM[f] images confirmed that ethanol was the best desolvating agent to prepare uniform and spherical BSA nanoparticles. On the other hand, Arroyo-Maya et al. (2012) used acetone, ethanol, and isopropanol to prepare nanoparticles from α-lactalbumin using desolvation method. They concluded that acetone was the best desolvating agent to manufacture smaller, spherical, and more uniform nanoparticles. They also applied different pretreatments such as, heat pretreatment (60°C, 30 min), and high pressure pretreatment (600 MPa, 55°C, 30 min) and compared with the samples with no treatment, and concluded that in all of the cases, the smallest nanoparticles were obtained with acetone from non-treated protein solutions and the order of the desolvating agents were reported to be as acetone (152.3 nm, PDI = 0.166), ethanol (246 nm, PDI = 0.214), and isopropanol (324.9 nm, PDI = 0.179). Different conditions and parameters might lead to these antithetic results, such as different protein configurations, nanoparticles' preparation methods, measurement methods, and so on. Sadeghi et al. (2014) measured the particle size, distribution and PDI after particulation and before GTA treatment, while Arroyo-Maya et al. (2012) measured the mentioned parameters after GTA treatment.

Kaur and Singh (2015) developed BSA nanoparticles with acetone, chloroform, and ethanol as desolvating agents. According to their experiments, acetone was selected as the best agent in producing nanoparticles compared to the other agents, mostly owing to their effect in producing aggregates and macrosized particles. In another study, Mehryar et al. (2016) investigated the effect of ethanol and methanol on the particle size of the sunflower protein isolate nanoparticles and found that ethanol (192.6 nm at 4°C and 169.1 nm at 50°C) produces smaller particles compared to methanol (218.6 nm at 4°C and 194.44 nm at 50°C) at 4 and 50°C. However, at 25°C, no difference was observed between the size of the particles gained by the different desolvating agent. Recently, Etorki et al. (2016) investigated the effect of three desolvating agents of ethanol, methanol, and acetone on nanoparticulation of α-LA and OVA and found significant differences in their application. Methanol proved to produce smaller particles of both α-LA and OVA than ethanol and acetone. They suggested that the polarity of the desolvating agent has a significant effect on desolvation process and nanoparticle size.

11.2.2.4 Concentration of the desolvating agent

Yoshikawa et al. (2012) performed a study to understand how ethanol concentration causes folding and unfolding of BSA using CD method. The far-UV CD spectra of BSA in water showed two minima peaks at 209 and 222 nm, suggesting the presence of approximately 65% α-helix structure in the proteins. Adding 0%–20% ethanol did not cause significant changes in the CD spectra. However, by adding 30%–50% ethanol,

[f] Scanning electron microscopy

significant changes were observed in the secondary structure of proteins; in this range increasing the ethanol concentration led to a gradual decrease in α-helix structure of protein. Yoshikawa et al. (2012) also found that by increasing the concentration of ethanol above 50%, BSA molecules refold and new α-helix structure forms. In the range of 60%–80% ethanol, a significant and gradual folding in the protein structure was observable, but by increasing the ethanol concentration to 90 and 99%, a remarkable folding was observed. At low concentrations of ethanol (less than 50%), solvation of hydrophobic chains weakens the structure of BSA and unfolds the protein structure, while higher concentrations (higher than 60%) yield more hydrophobic environment and drive hydrophilic groups closer to each other, thus stabilizing BSA molecules by hydrogen bonds; in the next step protein will be folded again (Yoshikawa et al., 2012).

11.2.2.5 Stirring rate

According to Rahimnejad et al. (2012), stirring rate (200, 400, 500, and 700 rpm) of the protein solution could vastly affect the size of BSA nanoparticles. They noted that the particle size had a decreasing trend (from 110 to 91 nm at 4°C) by increasing stirring rates; however after a definite rate (500 rpm), the particle size did not change. It seems that more uniformity in the preparation would be obtained via vial, which leads to better particulation; furthermore it also depends on the factors such as, vial size and shape, stirring method, stirrer geometry, and so on. Unfortunately, there is not enough scientific literature on the effect of stirring on the particulation of proteins using the desolvation method. Von Storp et al. (2012) employed various agitating speeds, pre-agitating and the constant adding of desolvating agent to the HSA solution to produce HSA nanoparticles. The results implied that all the aforementioned factors played important roles in determining the shape and size of the nanoparticles; spherical nanoparticles were achieved with predictable and reproducible sizes between 50 and 300 nm while a mixtures of methanol and ethanol produced spherical HSA nanoparticles as small as 50–80 nm.

11.2.2.6 Cross-linking conditions (type, concentration, and time)

Cross-linking agents are chemical substances (e.g., GTA, DTSSP,[g] BS[h]) which are necessary for obtaining uniform and stable particles and therefore lead to more and better release conditions (Rahimnejad et al., 2012). The amount and the density of the cross-linking agent is therefore an important factor. Cross-linking may improve the protein solubility at higher temperatures. Weber et al. (2000) found that increasing the amounts of GTA (40%–100%) reduces the amount of surface accessible amino groups of HSA nanoparticles and the zeta potential of the carrier system as well. They suggested that heating to the denaturation temperature would be useful in stabilizing the nanoparticles and therefore would increase the amount of surface accessible amino groups of the nanoparticles. Rahimnejad et al. (2012) studied the effect of GTA concentrations (100, 150, 200, and 300 mg/L) on BSA nanoparticles

[g] 3,3′ Dithiobis (sulfosuccinimidyl propionate)
[h] Bis (sulfosuccinimidyl) suberate

and found no significant difference in the particle sizes. Similar findings have been stated by Langer et al. (2003), which indicated that 8% GTA in water (in the range of 0.235–1.175 μL/mg HSA) did not influence HSA particle size but reduced the isoelectric point of HSA from 5.3 (no cross-linking agent) to 4.91 (for GTA concentration of 1.175 μL/mg HSA).

Estrada et al. (2014) examined the effect of 3 diverse cross-linking agents (8% GTA; DTSSP, and BS at a ratio of 1:2.2 cross-linking to lysine) on the particle size of the protein carriers. After 2 h of cross-linking, they concluded that GTA, produces larger particles with wider size distribution, in comparison with DTSSP or BS; moreover BS led to smaller particles compared to DTSSP. They also showed that the lower the cross-linking times (20 min compared to 2 h), the smaller the particle sizes would be. Arroyo-Maya et al. (2014) considered the effect of protein–GTA ratio on nanoparticle properties and reported that applying half of the concentration of the cross-linking agent increases the amount of the assembled α-LA particles by c. 40% considering that the size of the nanoparticles remained unchanged yet; at lower concentrations of GTA (one-tenth or less) aggregates were formed.

Won and Yong-Hee (2008) modified the desolvation method and produced recombinant human gelatin (rHG) nanoparticles via genipin as a natural cross-linking agent. Their results revealed that the obtained nanoparticles had the potential to be used in drug delivery systems because they are safe and offer controlled release and erupt less at the early stages. They also studied the effect of genipin concentration (0.01%–0.1%, w/v) on the particle size and PDI of gelatin nanoparticles and showed that 0.05% genipin was the least concentration that stable gelatin nanoparticles could be formed. At 3 h of cross-linking, increasing the genipin concentration enhanced the particle size (229–267 nm) and decreased PDI (0.154–0.022). Dreis et al. (2007) found that GTA concentrations (0.294–1.175 μL of 8% GTA solution per mg of HSA) could not significantly change the HSA nanoparticles size which were loaded by doxorubicin (particle size and zeta potential of 404 ± 14 nm and −56.1 ± 1.4 mV, respectively). They also noticed an increased doxorubicin loading efficiency (70%–96.3%) by increasing GTA concentration (0.294–1.175 μL of 8% GTA solution per milligram of HSA).

According to Park et al. (2015) findings, decreasing the concentration of cross-linking agent (from 1.5 to 0.15 μL and 10 to 2 mg for 25% w/v GTA and genipin, respectively) led to an increase in particle size (173.5 nm to unstable and 171 nm to unstable for 25% w/v GTA and genipin, respectively) and PDI (0.026–0.097 and 0.013–0.087 for 25% w/v GTA and genipin, respectively) and a decrease in zeta potential (−21.78 to −24.13 mV and −23.91 to −20.03 mV for 25% w/v GTA and genipin, respectively) of the gelatin–oleic nanoparticles. It was also concluded that the type of cross-linking agent did not significantly affect on the particle size, but at very low concentrations particles became unstable.

Chemical cross-linking agents have shown harmful, carcinogenic, and toxic effects therefore should be removed from the sample before consumption. Furthermore, they can cause environmental pollution. Other substitute for harmful cross-linking agents include physical cross-linking methods, that is, drying, heating, ultraviolet (UV) irradiation, and chemical techniques like glucose (Niknejad and Mahmoudzadeh, 2015). Safe cross-linking agents (carbodiimides and genipin) have rather low cross-linking efficiency and are costly. For example, genipin as a natural cross-linking agent causes very few toxic effects (~10,000 times less than GTA) (Elzoghby et al., 2012b; Liang et al., 2003). Liang et al. (2003), studied the effect of genipin on the cross-linking of gelatin microspheres and reported that increasing the swelling diameter of genipin-cross-linked microspheres was considerably less than GTA-cross-linked microspheres. Moreover, it was found that the breakdown of genipin cross-linked microspheres occurs at smaller rates than GTA-cross-linked ones. Therefore, obtaining a safe and of course efficient cross-linking agent is still a challenging and important task.

Among cross-linking agents, those with the natural origin are becoming very popular. For instance, citric acid has gained more interest in recent years because it is GRAS and guarantees human health (Reddy et al., 2012). Xu et al. (2015b) were the first to report the mechanism and kinetics of citric acid cross-linking for proteins. They studied the mechanism in neutral solutions (pH 6.8) under low temperatures (50 and 75°C) and stated that the number of responsible carboxyl groups in citric acid molecule for reacting with proteins seems to be more than one. They also noted that citric acid is a proper choice in enzyme immobilization and protein-based materials, which starts its cross-linking effect in relatively alkaline pH values. It was concluded that after 2-h treatment at 75°C and 3 h at 50°C, the cross-linking process was completed. Xu et al. (2015a) applied citric acid (9 wt.% at citric acid: protein ratio of 1:1) as a cross-linking agent in target delivery of hollow corn protein nanoparticles. The mechanism was based on the nucleophilic substitution between zein protein and citric acid (the amine group of zein reacts with the carboxylic groups of citric acid). They suggested that cross-linked nanoparticles possessed higher surface electrical negativity and their molecular weight increased markedly. The cross-linked nanoparticles were found stable at pH 7.4 for 6 days.

Qazvini and Zinatloo (2011) applied two nontoxic cross-linking agents (carbodiimide (CDI) and *N*-hydroxysuccinimide (NHS) to produce gelatin nanoparticles through two step desolvation technique via acetone. They showed that CDI/NHS cross-linked nanoparticles were more homogeneous (PDI of 0.141 and 0.45 for CDI/NHS and GTA, respectively) and smaller (184 and 280 nm for CDI/NHS and GTA, respectively) than those prepared with GTA, which could be due to the cross-linking mechanisms. Another nontoxic alternative to the chemical and harmful cross-linking agents, that is, GTA, is the enzyme transglutaminase. Fuchs et al. (2010) evaluated the effect of transglutaminase (60 U) on the cross-linking efficiency of gelatin nanoparticles prepared by two-step desolvation method with acetone. The enzyme was added at pH 7.4 and at 16, 25, or 37°C. The produced nanoparticles were below 250 nm. The results of process optimization indicated that water was the preferred cross-linking medium, besides the best incubation time and temperature were 48 h and 25–37°C, respectively. They reported that transglutaminase may be a better cross-linking agent in comparison with GTA.

In a recent study conducted by Varca et al. (2016), BSA nanoparticles were developed via desolvation method followed by irradiation as cross-linking treatment. Irradiation was performed in phosphate buffer (pH = 7.2) by means of a gamma cell with or without ethanol (30%, v/v) or methanol (40%, v/v). The studied irradiated doses were 2.5, 5, 7.5, and 10 kGy and the optimized properties were obtained at 10 kGy.

The mechanism of cross-linking was attributed to the formation of bityrosine and it was concluded that controllable irradiation dose would produce particles of controlled sizes resulting in cross-linking. By increasing the irradiation dose (0–10 kGy), bigger particles were produced in all of the experiments (6.7–16.6 nm, 14.3–25.1 nm, and 10.6–15.7 nm for native BSA, ethanol, and methanol, respectively).

Niknejad and Mahmoudzadeh (2015) compared different cross-linking methods (UV, glucose, and their combination) for preparation of docetaxel-loaded albumin nanoparticles via desolvation method. Their results showed that applying UV along with glucose, was less cytotoxic than GTA alone.

11.2.2.7 Temperature

Weber et al. (2000) stated that the temperature and heat denaturation time affected the stability of nanoparticles but were effectless on the particle size and the amount of amino groups. They could not find a definite trend in temperature dependency of particle size; however, particulation at 4°C led to smaller particles compared to higher temperatures (20, 30, and 40°C). At temperatures of 20 and 30°C, there were no significant change in particle size and at 40°C, smaller particles were formed. Due to the lower HSA solubility at higher temperatures (30 and 40°C), more particles were formed. Ko and Gunasekaran (2006) produced uniform β-lactoglobulin (BLG) nanoparticles of 59 ± 5 nm by desolvation method. Their study revealed that heating the solution to 60°C (at pH = 9.0) unfolded the protein molecules and therefore suppressed hydrophobic interactions, reduced self-aggregation and formed smaller particles. In conclusion, BLG was suggested as a promising material for preparing nanoparticles based on its smaller molecular weight, extremely unfolding trend and more hydrophilic structure (Ko and Gunasekaran, 2006).

Rahimnejad et al. (2012) studied the effect of temperature (4, 14, and 24°C) on the size of BSA nanoparticles and found that it increases (from 89.3 to 163.6 nm at 20 mg/mL of BSA solution) with temperature elevation. Arroyo-Maya et al. (2014) studied thermal stability of α-LA nanoparticles by DSC[i] and reported that nanoparticles at pH 7.0 showed a broad endothermic transition followed by a large increase in heat capacity. The absence of endothermic regions in previously denatured nanoparticle samples was an indication of the irreversibility of the denaturation process. Arroyo-Maya et al. (2014) concluded that endothermic curves are a reflection of the structural changes of protein. The obtained curves were comparable at pH 9.0 and 7.0, but at pH 3.0, the endothermic region was hardly noticeable and had moved to the lower temperatures. In acidic samples, spectra of low and high temperatures were somewhat the same; however; the ellipticity under 220 nm at pH 7.0 and 9.0, showed to be rather negative during heating.

A thermosensitive aggregation method ("gentle heating," "quick heating," and "heat surprise") was applied to produce stable nanoparticles from the elastin model polypeptide $(GVGVP)_{251}$, by Fujimoto et al. (2010), followed by gamma-ray cross-linking. It was concluded that the only successful process in achieving stable nanoparticles (less than c. 150 nm with a narrow size distribution) was "heat surprise" process; however other processes led to the larger particles. Bagheri et al. (2013) indicated

[i] Differential scanning calorimetry

that preheating (60°C for 30 min) would result in smaller whey protein nanoparticles (280–183 nm when dispersed in aqueous ethanol and 41–212 nm and 103 nm when dispersed in water). This phenomenon is attributed to the unfolding of whey proteins under heat treatment, which causes hidden groups to be exposed and produce smaller particles as a consequence. Similarly, Mehryar et al. (2016) produced sunflower protein isolate nanoparticles via desolvation method at three temperature levels and two desolvating agents including ethanol and methanol. They found that increasing desolvation temperature would lead to a decrease in particle size of ethanol treated samples (192.6, 174.6, and 169.1 nm at 4, 25, and 50°C, respectively). In the case of methanol, the same trend was obvious; however the optimum temperature was 25°C (218.6, 173.4, and 194.4 nm at 4, 25, and 50°C, respectively). Recently, Etorki et al. (2016) investigated the effect of heat pretreatment (at 25, 50, and 80°C) on nanoparticulation of α-LA and OVA. They reported that protein denaturation especially at 80°C was led to the fabrication of smaller nanoparticles for both proteins.

11.2.2.8 The effect of pH value on the desolvation process

Lin et al. (1993) produced HSA nanospheres via desolvation method using acetone. They developed smaller nanospheres (from 199 to 105.8 nm) by increasing the pH value (7–9) of the HSA solution. Langer et al. (2003) stated that the initial pH of the HSA solution could influence the particle size. At different pH values (7–9), the particle mean diameters varied between 280 and 150 nm, respectively. It was concluded that increasing pH values results in the smaller particle sizes, which is in agreement with Lin et al. (1993). In contrast, particle yield presented a decreasing trend (95%–66%) by increasing pH values (Langer et al., 2003). Findings of Lin et al. (1993) was comparable to the results obtained by Langer et al. (2003), and it indicated that the higher the pH values, the smaller the produced nanoparticles, which could be as a result of the effect of higher pH values on the protein charge and solubility. At higher pH values, protein molecules are charged negatively on their surface and precipitation would be repressed. The secondary structure would also change these pH values and beta sheet formations increase while small hydrophobic interactions (thiol–disulfide interchange) result in the smaller particles (Gunasekaran et al., 2007). Rahimnejad et al. (2012) considered the effect of pH ranges from 4.0 to 9.0 at BSA concentrations of 5 mg/mL and found that different pH values did not change particle sizes in a regular manner; moreover particles tend to be larger at isoelectric pH values due to the instability of protein particles.

Arroyo-Maya et al. (2014) measured spectroscopic properties of α-LA nanoparticles in pH values from 3.0 to 9.0 and revealed minor changes in the far-UV CD spectrum of the nanoparticles. A light minimum shoulder at 210 nm is vivid, which relates to the α-helix spectrum. This point somewhat intensifies at pH 9.0, and disappears at pH 3.0. According to the spectral results, faint increase of 2% was observed in α-helix structure at pH values of 3.0–9.0. Farrugia and Groves (1999) prepared nanoparticles by desolvation method utilizing ethanol and found that increasing pH values (from 5.5 to 8.5) decreases the gelatin particle size from 228.5 to 177.3 nm at ethanol concentration of 67.5% (w/w). Podaralla and Perumal (2012) reported that the particle size and PDI of 6,7-dihydroxycoumarin (DHC) loaded zein nanoparticles were higher at pH values of 2.0 (~385 nm and 0.39) and 12.0

(~378 nm and 0.44, respectively). However, smaller nanoparticles were produced at pH 7.4 with the particle size and PDI of 143 nm and 0.26, respectively. According to the literature, the higher pH values around 9 lead to the formation of smaller and more uniform nanoparticles via desolvation method.

11.2.2.9 Protein concentration

Langer et al. (2003) measured the particle size of HSA nanoparticles as a function of protein concentration (25–100 mg/mL) and suggested that concentration may influence the particle diameter. By increasing HSA concentration, PDI was reduced. According to Rahimnejad et al. (2012), BSA nanoparticle sizes decreased (from 110.3 to 91 nm at 4°C) along with the augmentation of concentration (5–30 mg/mL). In contrast, Won and Yong-Hee (2008) reported that by increasing the concentration of the recombinant human gelatin (rHG) from 0.05% to 0.1%, the size of nanoparticles increased (49–220 nm). These antithetical results may be related to the DLS measurements, yet concentration of protein and protein nanoparticles show a significant effect on DLS results. According to the literature and previous experiences, it seems that enhancing the concentration could increase the particle size; however the measurement should be done in the same concentration of nanoparticles. According to Galisteo-González and Molina-Bolívar (2014), increasing BSA concentration (from 12 to 100 mg/mL), slowly increased particle size (from ~74.5 to 139.5 nm) and led to a decrease in PDI (from ~0.157 to 0.037). It was also stated that at BSA concentrations above 25 mg/mL, particles are considered to be monodispersed.

The ascending trend of particle size by increasing the concentration may be described by the classical theory of nucleation in which, increasing the BSA concentration, rises the viscosity of the solution and reduces the rate of protein transport or nucleation rate in water and ethanol and finally larger particles are formed. Based on another theory, higher protein concentrations may cause supersaturation. Subsequently, there will be more nuclei, which could be agglomerated by particle collision forming bigger particles (Galisteo-González and Molina-Bolívar, 2014). It was also concluded that the number of nanoparticles would nearly remain constant by increasing the protein concentration. Similarly, Podaralla and Perumal (2012) found that increasing zein concentration (0.09–0.5% w/v), would produce larger nanoparticles (~223–384 nm). Also, Kaur and Singh (2015) showed that by increasing BSA concentrations (15, 20, and 30 mg/mL) bigger particles were formed while the smallest particles were produced at concentrations of 15 mg/mL.

11.2.2.10 Buffer type

Podaralla and Perumal (2012) used different buffers to see if they could affect the size of DHC loaded zein nanoparticles before and after freeze drying. According to their results, obtained particles were of equal size in phosphate and citrate buffers (pH = 7.4) before drying, yet, freeze drying led to particle size increasing in phosphate buffer (pH = 7.4) by twofold, nevertheless the particle size didn't change in citrate buffer (pH = 7.4). PDI was reported to increase from 0.5 to 0.72 and 0.5 to 0.6 in phosphate and citrate buffers, respectively.

11.2.2.11 Ionic strength

According to Podaralla and Perumal (2012), increasing the ionic strength of aqueous phase of DHC-loaded zein nanoparticles (0.1–1 M) decreased particle size (309–269 nm) and PDI (0.54–0.45); however the alterations were not meaningful ($P < 0.05$). Zeta potential was also suggested to show an increasing trend by intensifying ionic strength (from about +24 to −32 mV).

11.2.3 Production of nanoparticles from different protein sources

11.2.3.1 Nanoparticles from animal proteins

Animal proteins are known to be nontoxic, abundant, and healthy (Elzoghby et al., 2012b).

11.2.3.1.1 Gelatin

Gelatin is formed in the hydrolysis process of collagen in extremely low or high pH values or through the thermal/enzymatic degradation of collagen. Its functional groups are available to be chemically bounded with bioactive compounds and therefore may be used in target delivery purposes. Gelatin rose to fame for delivery of active compounds, which dates back to 1978, regarding investigation of Oppenheim et al. Gelatin has been recognized as GRAS by FDA and it has proven to be successful in the delivery of anticancer, antimicrobial, antiinflammatory agents, and so on.

Stability of gelatin nanoparticles increases in biological fluids in comparison to nonbiological environments, which can offer controlled and continuous release of the encapsulated compounds (Sahoo et al., 2015). Gelatin constitutes about one-third of all the proteins of vertebrates which is the major part of the extracellular protein in the tendon and bone and more than half of the skin protein. Considering the problems of large particle size and instability of gelatin nanoparticles related to the current desolvation methods, Coester et al. (2000) developed a new bistage desolvation strategy to form gelatin nanoparticles to overcome the mentioned deficiencies. Gelatin nanoparticles could be produced by different methods including desolvation (Coester et al., 2000), coacervation phase separation (Yeh et al., 2005), emulsification-solvent evaporation, reverse phase evaporation, and nanoprecipitation. Yeh et al. (2005) developed paclitaxel (one of the most important anticancer drugs) loaded gelatin nanoparticles by coacervation-phase separation method; an aqueous solution of 20% sodium sulfate was added dropwise to gelatin solution encompassing Tween 20. Then, a paclitaxel–isopropanol solution was added. Afterward, aqueous solution of GTA (25%) was used for cross-linking. These particles could bioaccumulate more successfully in different parts of the body (average of 3.2-fold) compared to the conventional paclitaxel formulation which applies cremophor/ethanol as solubilizes. Lee et al. (2012b) investigated gelatin nanoparticles containing drug molecules prepared by nanoprecipitation while water was used as solvent and ethanol as nonsolvent. Surface activated heparin-functionalized gelatin nanoparticles (GNPs) (developed by bistage desolvation strategy) containing cisplatin were also developed by Jain et al. (2012). GNPs had the particle size of

173 nm and after coupling with heparin (H-GNPs), reached to 189 nm. Results offer hope for the use of H-GNPs as drug delivery vehicles for solid tumor.

In a recent study, Park et al. (2015) produced gelatin–oleic conjugate (GOC), with the ability to self-assemble into nanoparticles by using a new method. Oleic acid (OA) binding to gelatin was completed through carbodiimide/*N*-hydroxysuccinimide (EDC/NHS) reaction in a medium containing water–ethanol as solvents. According to these findings, particle size and zeta potentials of GONs loaded with IRT were reported to be less than 250 nm and 20 mV, respectively. The obtained particles proved to be stable against aggregation. Incubation method had the advantage of producing particles with a small mean diameter; thus, were applied to evaluate loading and encapsulation efficiencies. It was found that DL[j] and EE[k] of the GONs were 11.28 and 38.20%, respectively.

11.2.3.1.2 Collagen

Collagen constitutes the major part of the vertebrates body which accounts for the 20%–30% of the total body proteins of mammalians (Lee et al., 2001). Due to some biological characteristics including biodegradability, weak antigenicity, and great biocompatibility, collagen has been regarded as one of the most fundamental materials in medical research applications. Because of the formation of very robust and durable fibers via self-aggregation and cross-linking, the application of collagen in biomedical research has been restricted (Lee et al., 2001). Collagen systems have also showed some disadvantages; including resource deficiencies, mechanical weakness and the presence of some infected parts that needs to be managed properly. Nicklas et al. (2009) produced marine (*Chondrosia reniformis*) sponge collagen nanoparticles (developed by organized hydrolysis in pH > 7) and applied it for the transdermal transfer of 17β-estradiol-hemihydrate through the skin. They could tract higher amounts of estradiol in saliva samples of postmenopausal patients treated with the gel containing nanoparticles. It was concluded that, products containing estradiol-loaded collagen nanoparticles could release estradiol for a more extended period of times and higher rates of adsorption compared to common gels in the market.

11.2.3.1.3 Albumin

Serum albumin protein has been widely used for nanoparticulation; it offers several known benefits; such as high solubility, binding ability, compact structure, and so on.

Albumin comprises most of the plasma proteins (35–50 g/L human serum) with a molecular weight of 66.5 kDa (Elzoghby et al., 2012a). It has proved to be harmless, nonstimulator of immune system, environmental friendly, and could be degraded into safe materials (Elzoghby et al., 2012b). It is also believed that since endogenous albumin only requires 19 days to reduce to half of its initial amount in blood system, it can be suitable in developing the pharmacokinetic profile and targeting characteristics of many novel drugs (Elzoghby et al., 2012b). Egg white albumin (ovalbumin), BSA, and HSA are the common albumins (Elzoghby et al., 2012b).

[j] Drug loading content
[k] Encapsulation efficiency

There are many methods to produce nanoparticles from albumin proteins including desolvation, self-assembly, emulsification, thermal gelation, spray drying, and nanoparticle albumin bound technology (NAB-technology) (Elzoghby et al., 2012b). Market approved products of albumin nanoparticles; for treating metastatic breast cancer, the taxol albumin nanoparticle (Abraxane) has been successfully used (Elsadek and Kratz, 2012). Moreover, a great number of drug formulations based on albumin proteins have been applied in clinical trials. For example, antibody fusion proteins (MM-111) for treating HER2/neu positive breast cancer (phase I), a camelid albumin-binding nanobody anti-HSA-anti-TNF-α (ATN-103) in phase II studies for treating rheumatoid arthritis, and a few more examples have been explained by Elsadek and Kratz (2012). In the past decades, albumin based nanocarriers were vastly established in many fields of research, that is, drug targeting, enhancement of the pharmacokinetics, and medicine and food technology (Kratz, 2008). These are mostly due to the high affinity of albumin proteins to efficiently bind to a wide range of bioactives and drugs, mainly as a result of the multiple functional binding sites. The presence of these multifunctional sites on the surface of the albumin nanoparticles gives them the potential to be used for target delivery application as well as applying surface modification (Elzoghby et al., 2012b). Albumin has become known as a great carrier for drug delivery purposes due to the benefits it offers; such as having acidic nature, high solubility in aqueous media, versatility, its stability in a wide range of pH, solubility in ethanolic media, heating resistance, preferential absorption in target tissues, lack of toxicity, and immunogenicity (Kratz, 2008; Lohcharoenkal et al., 2014). Fig. 11.4 presents the four main techniques to fabricate drug delivery systems based on albumin.

There are many studies on HSA nanoparticle production by desolvation method. For example, Dreis et al. (2007) successfully produced HSA nanoparticles having size range of 150–500 nm. Wartlick et al. (2004) developed HSA nanoparticles (290–330 nm with PDI of 0.0489 and 0.0818) loaded with several antisense oligonucleotides (ASO) by desolvation method. The GTA (4.7–10.6 μL of 8% GTA solution, equal to 40%–90% of the theoretic value) cross-linking method plays a major role in determining the biodegradability and drug release profile of the nanoparticles.

11.2.3.1.4 Milk proteins

11.2.3.1.4.1 Casein

Casein is known as one of the mostly found proteins of milk. It is inexpensive, nontoxic, and highly stable. Due to the structural and physicochemical properties of caseins, they offer beneficial properties as drug delivery vehicles. As explained in Chapter 5, caseins are rich in hydrophobic or hydrophilic amino acid residues (amphiphilic proteins). According to this fact, caseins strongly tend to form globular casein micelles (50–500 nm) (Shapira et al., 2010a,b). They have also unique surface activities and properties, outstanding emulsification properties and self-organization characteristics plus great gelation and water holding and binding abilities (Elzoghby et al., 2012b). Another unique characteristic of caseins is that they are heat-stable, oppose to the susceptible whey proteins that denature at temperatures above 70°C (Livney, 2010).

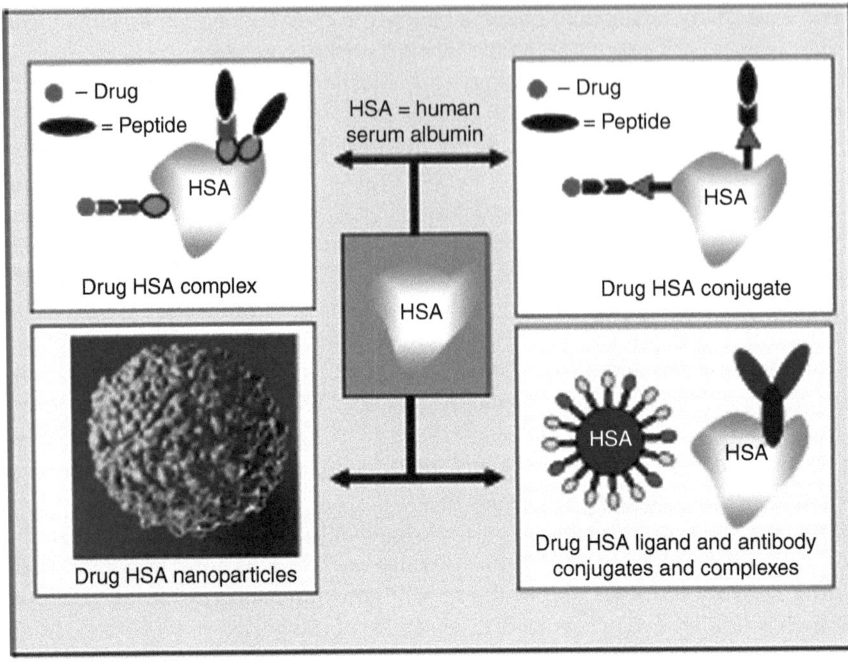

Figure 11.4 The main technologies that use albumin as a drug carrier.
Source: Reprinted from Elsadek, B., Kratz, F., 2012. Impact of albumin on drug delivery—new applications on the horizon. J. Control. Release 157 (1), 4–28. Copyright (2012), with permission from Elsevier.

11.2.3.1.4.2 Whey proteins

Whey proteins are composed of two main dominant proteins including beta-lactoglobulin (BLG), which constitutes the major part, and alpha-lactalbumin (αLA), the second chief one. Due to its globular structure, BLG shows resistance to low pH values and proteolytic degradation (Ko and Gunasekaran, 2006). Gunasekaran et al. (2007) produced BLG nanoparticles (131 nm) using acetone as the desolvating agent and GTA cross-linking to stabilize the nanoparticles. The experiments showed that preheating (60°C) of BLG solution then setting pH to 9.0, resulted in particles of 59 nm and acceptable uniformity. Compared with BSA nanoparticles, BLG nanoparticles have proved to be more stable (at pH ≤7) and smaller in size with less polydispersity, which is related to the lower hydrophobic interactions of BLG. The enhanced stability was ascribed to the denser matrix configuration and lower amounts of basic amino acid in BLG (Elzoghby et al., 2012b). Figs. 11.4 and 11.5 present different technologies and functionalities of proteins in delivery applications.

Encapsulation and delivery tasks are accomplished by several mechanisms including ion binding, binding hydrophobic and other molecules, surface activity, self-assembly and coassembly, gelation, targeting and reactive gel swelling for pH-triggered release, interaction ability with other polymers, covalent conjugation, noncovalent interactions, etc. (Livney, 2010).

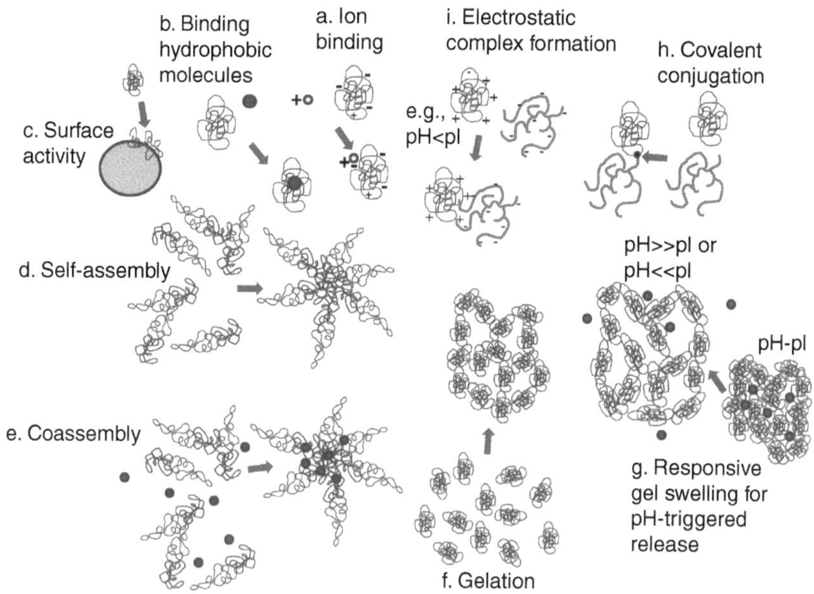

Figure 11.5 Several effective functionalities of milk proteins in delivery applications. *Source:* Reprinted from Livney, Y.D., 2010. Milk proteins as vehicles for bioactives. Curr. Opin. Colloid Interface Sci. 15 (1), 73–83. Copyright (2010), with permission from Elsevier.

11.2.3.1.4.3 Silk proteins

According to Numata and Kaplan (2010), the proteins which are produced and rotated into fibers by lepidoptera larvae, such as silkworms, spiders, scorpions, mites, and flies are expressed as silk. Silk proteins are regarded as promising carriers in delivery purposes owing to their biocompatibility, low rate of breakdown, self-aggregation, great mechanical strength, processing flexibility, their potential to be processed into films, hydrogels, nanofibers, nanoparticles, three-dimensional porous scaffold, controllable structure, and morphology.

Altman et al. (2003) stated that compared to the other polymers, silk is less inflammatory and is made of fibroin (a fibrous protein) and sericin (a glue-like protein). Fibroin incorporates the core of silk while sericin wraps the fibroin fibers together. In a similar study, Kundu et al. (2010) developed silk fibroin nanoparticles (150–170 nm) by desolvation method with dimethyl sulfoxide as a desolvating agent. The results revealed stable, spherical, negatively charged, and nontoxic particles, which were able to release VEGF (vascular endothelial growth factor) during 3 weeks.

11.2.3.2 Nanoparticles from plant proteins

Plant proteins are other probable carriers in terms of their low cost, reduced risk of transmitting ailments (bovine spongiform encephalitis, mad cow disease) and

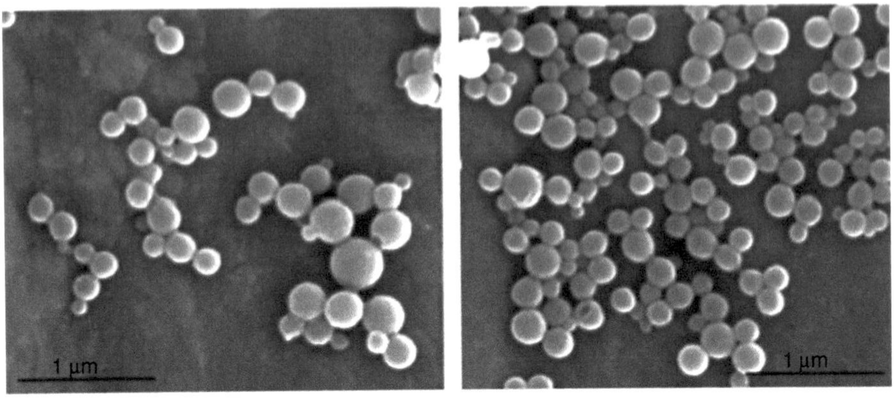

Figure 11.6 Scanning electron microscopy (SEM) of folic acid–loaded zein nanoparticles. *Source:* Reprinted from Peñalva, R., Esparza, I., González-Navarro, C.J., Quincoces, G., Peñuelas, I., Irache, J.M., 2015a. Zein nanoparticles for oral folic acid delivery. J. Drug Deliv. Sci. Technol. 30, 450–457. Copyright (2015), with permission from Elsevier.

hydrophobic nature (e.g., zein and gliadin), which induce sustained release. Owing to this fact, the obtained nanoparticles may be produced without any need of additional chemical or physical treatments to strengthen them, thus the use of toxic chemical cross-linking agents is avoided (Elzoghby et al., 2012b). Plant proteins hold functional groups that enable them to adsorb or covalently combine with the molecules having the ability to alter the targeting properties of nanoparticles.

11.2.3.2.1 Zein

Peñalva et al. (2015b) investigated the encapsulation of resveratrol via zein nanoparticles. Based on the obtained results, the resveratrol-loaded nanoparticles showed to be of ~310 nm in size and had a negative zeta potential (−51 mV). Their PDI which shows the homogeneity of the obtained particles was also reported to be less than 0.2. Peñalva et al. (2015a) also developed zein nanoparticles for the oral delivery of folic acid. Zein was ultrafiltered and spray dried and then desolvation method was used to prepare nanoparticles, and the folic acid-–loaded zein nanoparticles size was approximately 200 nm, with a negative zeta potential of −24 mV and a low PDI (0.2). Fig. 11.6, presents SEM images of folic acid–loaded zein nanoparticles.

11.2.3.2.2 Gliadin

Duclairoir et al. (2003) produced gliadin nanoparticles (450–475 nm) via desolvation method. Affinity of gliadin to interact with compounds with the similar polarity facilitates its drug coupling process. In a study, Ezpeleta et al. (1996) produced gliadin nanoparticles (around 500 nm in size) by desolvation method using ethanol as the desolvating agent. It was found that the gliadin nanoparticles were stable in phosphate-buffered saline (PBS) up to 4 days and cross-linking with GTA significantly improved their stability.

11.2.4 Encapsulation of different food components within protein nanoparticles

Owing to the poor diet and therefore low intake of micronutrients in daily food servings, enrichment seems quite indispensable. As mentioned by Onwulata (2012), protected delivery or encapsulation has been considered as one of the most important ways of delivering nutrients, mainly because it enhances nutrient bioavailability, flavor retention, and stabilizes the food ingredients.

Biodegradable polymers have been increasingly used over the past decade to encapsulate pharmaceuticals and biomedical agents. They are mainly used as the controlled drug delivery systems (Elzoghby et al., 2011; Langer and Chasin, 1990). The aforementioned biodegradable polymers may be used either in natural or synthetic forms. It is obvious that synthetic polymers are more advantageous than natural polymers due to their extended applications. Yet, concerning the toxicity of the synthetic polymers, applying safe carriers such as natural polymers are quite necessary (Elzoghby et al., 2011). Food proteins are promising nanocarriers for the delivery purposes mainly due to their unique characteristics as described before (Elzoghby et al., 2011). On the other hand, therapeutically relevant peptides and proteins increasingly took part in curing the viral, malignant, and autoimmune ailments. The aforementioned nanocarriers suffer from their labile nature, high manufacturing expenses, provoking immunogenic and allergic responses, along with the low bioavailability and sensitivity toward peptidases (Kratz, 2008).

The potential applications of protein nanoparticles produced by the desolvation method for the various food ingredients and nutraceuticals has been summarized in Table 11.1, and a further discussion is presented in this section.

11.2.4.1 Curcumin

Sadeghi et al. (2014) encapsulated curcumin in BSA NPs using desolvation method. The encapsulation efficiency (EE) was about 20%. They merely reported the entrapped curcumin as encapsulated curcumin; however after freeze drying of the formulation, free curcumin molecules could physically attach to the surfaces of the NPs. Curcumin-loaded BSA NPs had a great redispersibility and showed that the encapsulation technique could change the releasing profile. Controlled release study showed that about 70% of the curcumin released from NPs after 3 days, then reached to a constant rate. This part of release profile is similar to the physical mixture of curcumin and BSA. They proposed that around 70% of the added curcumin molecules may interact and attach to the NPs surfaces. On the other hand, around 30% of the added curcumin was entrapped within the structure of BSA NPs. The stability study of the curcumin-loaded BSA NPs was done to prove that the nanoparticles were stable in the targeted conditions and curcumin molecules could not be released. It was also concluded that sonication for 10 s decreased the particle size and PDI of the redispersed nanoparticles. The same release profiles were suggested by Kolluru et al. (2013) for doxorubicin and indocyanine green. They also used freeze drying after particulation in the presence of the dye or drug. Sneharani et al. (2010) stated that

Table 11.1 **A summary of nanoencapsulation and delivery of bioactive components using protein nanoparticles prepared by desolvation method**

Protein type	Particle size or range (nm)		Encapsulated material	EE[a] (%)	LE[b]	Release trend	Achievements	References
	Unloaded	Loaded						
BSA	175.4 ± 0.5	—	Resveratrol	88.7	39.4	72 h	Resveratrol-BSA NPs are promising drugs	Li et al. (2012)
Zein coated with carboxymethyl chitosan	86–200	—	Vitamin D3	52.2–87.9	1.7%–3.9%	1.5 h (burst release) to 7 h (sustained release)	Calcium addition to cross-link with CMCS was found to affect the release results	Luo et al. (2012)
Whey protein	41–212	—	Date palm pit extract	70.26–78.33	—	—	Higher entrapment efficiency of date palm pit extract	Bagheri et al. (2013)
Soy protein conjugated with folic acid	109–168.7	160.7–338.8	Curcumin	32.5–89.0	2.89%–5.37%	8 h	Folic acid improved the final properties	Teng et al. (2013)
BSA	—	88 ± 4	Curcumin	19.8	—	(~70%) 72 h	Controlled release—sonication for 10 s decreases particle size and PDI of the redispersed nanoparticles	Sadeghi et al. (2014)
Zein	—	164, 193	Folic acid	57	—	SGF: no release SIF: biphasic pattern, 70% (burst release, 2 h), followed by sustained release (24 h)	No release was observed in simulated gastric fluid	Peñalva et al. (2015a)

Gliadin	—	450–475	Hydrophobic (VE), amphiphilic (linalool and LLA) drugs, cationic amphiphilic BZC	Higher in VE (79.2%), linalool and LLA (82.4%)	972.0 VE μg gliadins mg – 1980.0 LLA μg gliadins mg – 1	Burst release in 1 h followed by sustained release	Suitable controlled release systems	Duclairoir et al. (2003)
Gliadin	~500	~500	*trans*-Retinoic acid (RA)	85%–97%	76.4 μg drug/mg protein	20% (burst release in 15 min)	Stability over 4 days in PBS-controlled release (3 h-release rate of 0.065 mg RA/h)	Ezpeleta et al. (1996)
Gelatin	300–900		Paclitaxel	70%	2%	More than 55% (15 min), 87% (2 h), and 92% (3 h)	Higher accumulation	Yeh et al. (2005)
HSA	150–500	—	Doxorubicin	96.3%	70%–95%	Decreases at higher concentrations of GTA	Drug-loaded nanoparticles showed higher anticancer activity	Dreis et al. (2007)
Gelatin	184–280		Paracetamol	27% (CDI/NHS) and 10% (GA)	7.3% (CDI/NHS) and 4% (GA)	A little bit faster in GA-GNPs	CDI/NHS cross-linking agent was introduced as a choice for drug delivery applications	Qazvini and Zinatloo (2011)
HSA	290–330	—	Antisense oligonucleotides	20.8%	23.2%–94.7%	Sustained release during 24 h	A rapid uptake of nanoparticles by different breast cancer cells	Wartlick et al. (2004)

(Continued)

Table 11.1 A summary of nanoencapsulation and delivery of bioactive components using protein nanoparticles prepared by desolvation method (*cont.*)

Protein type	Particle size or range (nm)		Encapsulated material	EE[a] (%)	LE[b]	Release trend	Achievements	References
	Unloaded	Loaded						
Surface-functionalized gelatin	173, 189 (coupled)		Cisplatin	71 (GNPs) and 70.8 (H-GNPs)	0.5 mg/mL	8.41 (from GNPs) and 7.8% (H-GNPs) (burst release in 2 h)	Promising carriers for tumor-specific delivery	Jain et al. (2012)
BSA	101–325	—	A drug (doxorubicin) and a dye (indocyanine green)	88.80 (doxorubicin), 96.47 (indocyanine green)	2.22% (doxorubicin), 0.19% (indocyanine green)	9 days 3 days	Targeted drug delivery to tumor	Kolluru et al. (2013)
HSA	150	—	Docetaxel	80.5	17.2	26.8% (8 h), 57.5% (28 h)	Specific cancer cell targeting	Jiang et al. (2015)
BSA	156	—	Amphotericin B (Amb)	Over 82	—	48% over 5 days	Prolonged AmB release	Casa et al. (2015)
BSA	189.9	—	Serratiopeptidase	82.64	—	~6 h	Promising formulation to treat rheumatoid arthritis	Kaur and Singh (2015)
BSA	418.2	—	Abacavir sulfate	—	1.2–5.9	24 h	Enhances the bioavailability of the drug	Wilson et al. (2015)
BSA	164.1 ± 5.40	—	Hirudin	85.14	66.38	48 h	Good encapsulation and sustained-release capacity	Jing et al. (2015)
BSA	215.8 ± 3.2	—	Etoricoxib	72	—	72 h	Targeting prospective, spatial delivery, and amplified bioavailability	Bilthariya et al. (2015)

BZC, Benzalkonium chloride; LLA, linalyl acetate; VE, vitamin E.
[a]Encapsulation efficiency.
[b]Loading efficiency.

the stability of the curcumin–βLG was improved 6.7 times compared to the curcumin solution. The efficiency of βLG nanoparticles for encapsulating curcumin was reported to be >96%. In another study, folic acid–conjugated soy protein nanoparticles were produced by Teng et al. (2013) to investigate drug-targeted delivery; meanwhile curcumin was applied as a hydrophobic bioactive. They reported the particle size, EE, and loading efficiency as 109–168.7 nm, 32.5%–89.0%, and 2.89%–5.37%, respectively; furthermore according to the release data, there was a fast and complete release in PBS/Tween 20 buffer. First, a rapid release (1 h) happened, in which 24 and 29% of the SPI and folic acid conjugated soy protein isolate (FA-SPI) were released initially, followed by a sustained release during 7 h. After 8 h, SPI and FA-SPI nanoparticles released 39 and 58% of the encapsulated curcumin. Curcumin-loaded OVA nanoparticles were prepared by desolvation method using acetone as the desolvating agent (Delfiya et al., 2016a). Based on the optimization analysis, curcumin loading, entrapment efficiency, yield, and particle size of 4.125%, 55.23%, 72.85%, and 232.6 nm were reported. A similar work was done by ethanol as the desolvating agent and the obtained results for the aforementioned parameters were reported to be 5.78%, 56.78%, 71.65%, and 263.5 nm, respectively (Delfiya et al., 2016b).

11.2.4.2 Folic acid

Hydrophobic proteins have a great potential to be used as encapsulating systems. Peñalva et al. (2015a) reported the ability of zein nanoparticles developed by desolvation method as an oral delivery system for folic acid. In this system, zein was soluble in 70% ethanol while water was a desolvating agent. Zein was able to prepare globular and smooth nanoparticles which could be loaded by folic acid; moreover the results depicted that folic acid loading capacity and EE were 54 ± 7 μg/mg and 57 ± 6%, respectively. Both amphiphilic nature of zein and perfect nanoparticulation by desolvation method led to a new formulation to encapsulate folic acid in zein NPs. It was also found that the release profile changed by adjusting the pH values. At low pH values (simulated gastric condition), no significant release was observed; however at higher pH values (simulated intestinal condition), 70% of the loaded folic acid was released from NPs. In vivo study on the oral administration of folic acid loaded NPs showed an increase in folic acid concentration in the serum versus time. The mechanism suggested for improving the bioavailability of the folic acid was based on preventing precipitation of folic acid during crossing the low pH part of the digestive system.

11.2.4.3 Carvacrol

Carvacrol is one of the known phenolic compounds, which shows a high protective effect against oxidation and tumor formation. Keshavarzi et al. (2015) encapsulated carvacrol in HSA nanoparticles using two techniques. Their results revealed that the entrapment capacity of the nanoparticles prepared by desolvation was higher than those prepared by emulsion/desolvation technique. The sizes of carvacrol-loaded nanoparticles were also reported to be 132 ± 42 nm and 230 ± 38 nm for the aforementioned techniques, respectively. Based on the results gained by release studies (PBS at pH 7.4), the burst release during the first 3 h was higher in nanoparticles prepared by the emulsion/desolvation method (26.8%) than the one prepared by the desolvation method

Figure 11.7 SEM of resveratrol-loaded BSA nanoparticles.
Source: Republished from Li, J.L., Zu, Y.G., Zhao, X.H., Zhao, D.M., Chen, X.Q., An, Z.G., Wang, K.L., 2012. Preparation, characterization, and in vitro evaluation of resveratrol-loaded bovine serum albumin nanoparticles. Adv. Mater. Res. 345, 349–354, with permission from Scitec Publications.

(21.4%). Cytotoxicity of the loaded carvacrol on gastric cancer cells line showed that the IC50 indices of free carvacrol, unloaded, and carvacrol-loaded nanoparticles were 30, 1070, and 120 μg/mL, respectively. It was finally concluded that carvacrol loaded HSA nanoparticles are promising nanocarriers against gastric cancer.

11.2.4.4 Resveratrol

Several advantageous effects of resveratrol in human body are related to its anticancer, antioxidant, antiinflammatory activity along with combating cardiovascular diseases. Li et al. (2012) produced resveratrol-loaded BSA nanoparticles via desolvation method followed by cross-linking using GTA with the particle size, entrapment, and loading efficiencies of 175.4 ± 0.5 nm, 88.7, and 39.4%, respectively. According to the release data, in the first 36 h an initial burst release happened while sustained release was continued up to 72 h. Fig. 11.7 shows SEM images of the produced nanoparticles, which represent the spherical and narrow size distribution of the obtained nanocapsules (Li et al., 2012).

According to Peñalva et al. (2015b), loading capacity of resveratrol loaded zein nanoparticles was 80 μg/mg (with the EE of 82%) and in vitro studies showed that the release profile was not pH dependent, which was well fitted to the Peppas–Salin kinetic model. It was also reported that oral bioavailability of resveratrol increased to 50% in zein nanoparticles compared to the control sample.

11.2.4.5 Retinoic acid

Ezpeleta et al. (1996) suggested that the entrapment efficiency of retinoic acid (RA) loaded gliadin nanoparticles was 85%–97%. The release profiles followed two-stage pattern, which shows an early rapid release and a zero-order diffusion. The loading efficiency was 76.4 μg/mg; and it was reported that the nanoparticles were stable over 4 days in PBS along with the controlled release.

11.2.4.6 Vitamin D_3

Due to its vital roles, vitamin D_3 has gained great interest in recent studies. Luo et al. (2012) investigated carboxymethyl chitosan (CMCS) coated zein nanoparticles for encapsulating vitamin D_3 (VD_3). The obtained nanoparticles (86–200 nm) showed the encapsulation and loading efficiency values of 52.2%–87.9% and 1.7%–3.9%, respectively. According to the kinetic release studies, a burst release occurred in the first 1.5 h followed by a sustained release in the following 7 h.

11.2.4.7 Date extract

Bagheri et al. (2013) investigated the ability of desolvated whey protein nanoparticles to encapsulate date palm pit extract. According to their results, EE was about ~78%. Loaded particles were 163 nm in alcoholic dispersion and 92 nm in water dispersion. It was also proved that the date extract did not bind covalently.

11.2.4.8 Drugs and others bioactive compounds

There have been a great number of research papers regarding the tumor targeting properties of the labeled albumins since 1990. Wunder et al. (2003) reported that albumin tends to gradually increase in the arthritic paws of mice affected with collagen-initiated arthritis. This was affirmed by applying scintigraphy and laser-optical techniques with albumin labeled by either fluorescein or radioactive materials. It was concluded that albumin may actively deliver the drugs to the inflamed tissues. With a diameter of nearly 130 nm, nab-paclitaxel was studied comprehensively, and held promise as an efficient antitumor agent and quite suitable to cure the metastatic breast cancer (Desai et al., 2006; Kratz, 2008). Tumor combating capabilities of the paclitaxel and nab-paclitaxel were evaluated in the mamma carcinoma MX-1 model and it was found that nab-paclitaxel could be taken up 33% more than paclitaxel in the subcutaneously growing MX-1 tumors. Thus nab-paclitaxel was reported to be a much more stronger weapon against the tumors in mice than paclitaxel (Kratz, 2008).

Other recent developments in the preclinical scale consist of attaching high-affinity peptides for albumins to antibody fragments, the utilization of albumin-binding gadolinium contrast agents and attaching antiviral, antibacterial, and anticancer drugs to albumin as albumin-binding prodrugs (Elsadek and Kratz, 2012). Kolluru et al. (2013) prepared BSA nanoparticles using desolvation method and loaded them by a drug (doxorubicin) and a dye (indocyanine green). They reported entrapment efficiency of 88.80 and 96.47% for doxorubicin and indocyanine green, respectively. According to the DSC results, doxorubicin showed an amorphous nature, which is the sign of the enhanced stability of the drug in the nanoparticle. That is why it required longer time for release (burst release of ~ 9% in the first 2 h followed by 50% during 9 days) than indocyanine green (burst release of 30% in the first 2 h and constant release in 3 days). It was also mentioned that quasi-Fickian diffusion mechanism could be a good model to explain the release profile of indocyanine green and doxorubicin from BSA nanoparticles. To sum up, the studied nanoparticles showed promising results in targeting breast tumors in the light of the previous in vivo studies.

Casa et al. (2015) applied desolvation method to produce amphotericin B (AmB) loaded BSA nanoparticles of 156 nm in size. The EE of AmB was estimated to be over 82%. According to their experiments, it was found that 18% of AmB was released in the first 24 h while 48% was released constantly in the next 72 h. Moreover, no burst release was observed throughout the experiment, which indicates that the drug was not attached on the nanoparticle surface. Results also showed that encapsulation did not affect on the antifungal activity of AmB against *Candida albicans* and *Candida tropicalis*. Lastly, the kinetic analysis study of AmB release profile from BSA nanoparticles showed that Higuchi model could best fit the release data ($R^2 = 0.9960$).

In another study, Kaur and Singh (2015) developed BSA nanoparticles via desolvation method using acetone as desolvating agent. Based on the optimization results, treatments with 15 mg BSA, 5 mL acetone, 50 μL and 15 mg BSA, 5 mL acetone, and 66.81 μL GTA, produced the smallest nanoparticles (194.4 and 189.9 nm, respectively). It was also determined that the mentioned treatments had good entrapment efficiencies (85 and 82.64%, respectively). In vitro release studies showed that percentage cumulative values of albumin nanoparticles and pure drug (serratiopeptidase) were 79.3 and 92.3, respectively.

BSA nanoparticles loaded with Abacavir sulfate (an antiviral drug) were prepared by Wilson et al. (2015) via desolvation method using ethanol to maintain the antiviral effects of encapsulated drug for HIV treatment. Drug loading to the particles was at the range of 1.2%–5.9% w/w, while the particle size and zeta potential were reported to be 418.2 nm and −40.8 mV, respectively. In vitro drug release studies by dialysis bag, suggested a biphasic pattern in which an initial burst followed by slow sustained release were observed. The initial release lasted for 15 min for 2.4%–7.6% of the drug and the drug release continued for 38.73 and 51.36% w/w for 24 h. Evaluation of models proved that the Higuchi model obtained the highest coefficient of determination thus could best fit the experimental data, which supports diffusion controlled release of the drug. According to the n (release exponent) value of Korsmeyer–Peppas model (0.425), Fickian release mechanism was attributed to the data. It was concluded that the studied nanoparticles could be considered as promising carriers mainly due to the high loading capacity, relatively high zeta potential, sustained release and enhanced bioavailability of the target drug.

Since hirudin is a potent harmless antithrombotic that does not provoke allergic responses; Jing et al. (2015) decided to develop hirudin-BSA nanoparticles to overcome the drawbacks of heparin and improve its antithrombotic effect. Some of the present drawbacks include: rapid distribution outside the cell after venous injection, short half-life (only about 1 h), need to repeat injections and high price. The aforementioned nanoparticles were produced by desolvation method with the average size of 164.1 nm and average zeta potential of −20.41 mV. Moreover, EE and loading capacity were 85.14 and 66.38%, respectively. Release studies in phosphate buffer solution via dialysis tubes showed that the acquired data fitted well to bioexponential model, besides, in vitro release studies indicated that the burst release occurred in the first 0.5 h, continued by gentle constant release during 48 h. On the other hand, in vivo release studies supported the nanoencapsulation theory which was able to release the drug over time. This could overcome its quick elimination caused by the drug's short half-life. It was concluded that the produced nanoparticles showed good encapsulation and sustained-release capacity.

Folic acid was used as an overexpressed folate receptor ligand for activated macrophages in rheumatoid arthritis-targeted delivery to fabricate etoricoxib-loaded albumin nanoparticles (Bilthariya et al., 2015). The desolvation procedure was applied to produce etoricoxib-loaded BSA globular nanoparticles (215.8 ± 3.2 nm). Based on the results, the entrapment efficiency of etoricoxib was 72 ± 1.3%. Controlled release data specified that 93.1 ± 2.2% of etoricoxib was released in 72 h and etoricoxib concentration in inflamed joints after 24 h was 9.67 ± 0.34 mg/g which indicates the targeting potential of the loaded nanoparticles. It was also concluded that data followed zero-order kinetics.

Docetaxel rose to fame as a useful antitumor drug. It has also shown many therapeutic applications especially in the tumor treatment of breast, ovarian, gastric, prostate, and lung. Jiang et al. (2015) investigated antitumor activities of folate-decorated docetaxel-loaded HSA nanoparticles. Nanoparticles were prepared via desolvation method, while the carboxylic groups of *N*-hydroxysuccinimide ester of folate (NHS-folate) linked to the amine groups of the HSA nanoparticles. Nanoparticles were spherical and 150 nm with a zeta potential of −30 mV. Based on the drug loading studies, drug loading and entrapment efficiencies were estimated to be 17.2 and 80.5%, respectively; however, according to the in vivo investigations, human hepatoma cell line was targeted successfully by folate-dressed nanoparticles. Antitumor activities of the docetaxel-loaded nanoparticles and docetaxel control sample were found somewhat similar in in vitro studies, while 66.2% and 59.5% of inhibition ratios were reported in in vivo investigations, respectively. Release studies showed that the drug release from the HSA nanoparticles was complex, which has been ascribed by the drug diffusion through the nanoparticles' bodies. In the first 8 h, 26.8% was released then, 57.5% of burst release occurred in 28 h. This sustained release could lead to a successful cancer cell death.

Dreis et al. (2007) produced HSA nanoparticles via desolvation method to bind doxorubicin to it. Binding of these drugs can decrease their toxic effects, optimize their distribution pattern and/or may solve multidrug resistance issue. Doxorubicin could attach to the surface of the HSA nanoparticles or locate into the interior parts of the nanoparticle with a loading efficiency of 70%–95%. To produce small nanoparticles with a narrow size distribution, doxorubicin should be at the concentrations of 0.5 and 1.0 mg/mL. Loading efficiencies of more than 95% was also obtained at low doxorubicin concentrations. Efficiency of 96.3% was achieved at 200% cross-linking agent concentration which may be caused by the development of covalent bonds between drug and matrix protein. It was also concluded that drug release decreased at higher concentrations of GTA. They finally reported that the drug-loaded nanoparticles have higher anticancer activity compared to the doxorubicin solution.

Qazvini and Zinatloo (2011) investigated gelatin nanoparticles produced from two step desolvation method via acetone for encapsulating paracetamol (a model drug). CDI/NHS cross-linked nanoparticles acted more efficiently in entrapping (27% and 10% for CDI/NHS and GA, respectively) and loading of drug (7.3% and 4% for CDI/NHS and GA, respectively); although GA-GNPs had a little bit faster release kinetics.

Duclairoir et al. (2003) studied the entrapment of vitamin E (VE), mixture of linalool, linalyl acetate (LLA), and benzalkonium chloride (BZC) in gliadin nanoparticles because of the beneficial effects including "antioxidant and antiaging," "antibacterial and antifungal," and "antiseptic, bactericide, spermicide, or virucide" properties of the aforementioned compounds. These particles could be well applied as the controlled delivery device for aforementioned hydrophobic and amphiphilic drugs. According to the obtained results, loading efficiency of gliadin nanoparticles to entrap VE and the combination of linalool and LLA were higher compared to BZC. Similar polarity of the gliadin and the drug explains their high binding capacity which facilitates better drug loading.

Yeh et al. (2005) reported the particle size, entrapment, and drug loading efficiencies of paclitaxel loaded gelatin nanoparticles at the range of 300–900 nm, 70 and 2%, respectively. Considering the release studies, a rapid drug release at more than 55 (15 min), 87 (2 h), and 92% (3 h) occurred. According to Wartlick et al. (2004) study, loading and entrapment efficiencies of HSA nanoparticles (290–330 nm, and PDI of 0.0489 and 0.0818) loaded with different ASO were reported to be in the range of 23.2%–94.7% and 20.8%, respectively. Drug loading studies showed a sustained release during 24 h and therefore a rapid uptake of nanoparticles by different breast cancer cells happened. In another study, cisplatin loaded surface-functionalized gelatin nanoparticles (GNPs) showed entrapment efficiencies of 71 and 70.8% for GNPs and H-GNPs, respectively. In cell uptake studies, calcein was applied in 0.5 mg/mL instead of the drug and the release studies showed initial burst release in which 8.41% (from GNPs) and 7.8% (from H-GNPs) of the drug was released in the first 2 h.

11.3 Polysaccharide nanoparticles (nanoprecipitation method)

A similar technique of desolvation, named nanoprecipitation method has been developed for the fabrication of nanoparticles from other biopolymers including polysaccharides. Particulation of biopolymers using nanoprecipitation was reported by Fessi et al. (1989) for the first time, which was built upon interfacial deposition of biopolymer molecules by the displacement of solvents such as water and miscible organic solvents like ethanol, acetone, and so on.

11.3.1 Principles of the nanoprecipitation method

Fessi et al. (1989) prepared poly-(D,L-lactide) polymer (PLA) particles by adding a solution of PLA in acetone to water and used amphiphilic constituents in both acetone and water phases, then reported the loading of indomethacin in PLA nanoparticles. Polymeric nanoparticles were prepared by combining the hydrophobic cellulose acetate (CA) and hydrophilic polysaccharides such as hydroxyethyl cellulose (HEC), carboxymethyl cellulose (CMC), low-molecular-weight chitosan (L-CHI), and amino cellulose (AC) using nanoprecipitation method (Kulterer et al., 2012). Spherical, narrow-size-distribution with various size and zeta potential, functionalized, and highly stable polymeric composite nanoparticles could be formed via this technique. Meanwhile, Heinze and Hornig (2009) stated the advantages of the nanoprecipitation technique as: (1) no need to use large amounts of toxic solvents; (2) obtaining monodispersed nanoparticle and (3) no need to use the external energy sources.

There are many parameters determining the particle size such as concentration of the starch solution; in which the higher the concentration, the larger the size of formed particles. Starch molecules are in better dispersed state in lower concentrations which enables the preparation of smaller nanoparticles during nanoprecipitation (Tan et al., 2009). Wu et al. (2016) also fabricated larger particles by increasing the concentration of starch solution. Higher molecular weight accompanied by the lower degree of substitution would favor the preparation of smaller size nanoparticles. Morphology and particle size of starch nanoparticles in different concentrations are shown in scanning electron microscopy (SEM) images in Fig. 11.8. Another effective parameter in fabrication of composite nanoparticles using nanoprecipitation technique is pH. In the study conducted by Kulterer et al. (2012), relatively smaller composite nanoparticles were prepared at high pH media (≥ 7) compared to low pH conditions (<7). It was also suggested that at the border of CA-rich areas and its adjacent solution, the concentration of the hydrophilic polysaccharides increases, and this phenomenon can separate these domains based on the electrosteric repulsion (combination of electrostatic forces which are related to charge density and steric repulsion which is related to polymer structure, swelling, and hydrophilicity). Pyrene was used as a hydrophobic model component to show the potential of the CA/CMC composite nanoparticles to encapsulate different compounds (Kulterer et al., 2012). Kulterer et al. (2012) also reported that 75% of pyrene was loaded by the composite nanoparticles which were located in the CA matrix of the nanoparticles.

The "ouzo" region is the range of the polymer concentration that the dilution of the solution is appropriate and metastable (Schubert et al., 2011). Fig. 11.9 represents a phase diagram of the poly(ε-caprolactone) (PCL) nanoparticle formation in solvent/nonsolvent (acetone/water), where the "ouzo" effect can be observed and nanoprecipitation could occur (Schubert et al., 2011). The "ouzo" region is indicated by BI zone in Fig. 11.9. Literature review showed that, in the preparation of the nanoparticles from different molecular weight of waxy maize starch acetate with various degrees of substitution (by dissolving the starch acetate in acetone, and adding distilled water to the starch solution drop by drop), the ratio of water to organic solvent (acetone) would play an important role in the particulation profile of starch, and the mean diameter of the nanoparticles decreases from 596 to 460 nm when the ratio of water to acetone increases from 1 to 3.5 (Tan et al., 2009). Strengthening the interfacial interaction between acetone and water compresses the starch nanoparticles and results in the smaller nanoparticles (Tan et al., 2009). On the other hand, Wu et al. (2016) applied acid hydrolysis method to decrease the starch molecular weight and found that the higher the extent of the hydrolysis, the smaller the starch nanoparticles would be. In another study, Ma et al. (2008) fabricated nanoparticles from corn starch by adding ethanol to hot gelatinized starch solution and fabricated citric acid modified starch nanoparticles (CASNs) in the size range of 50–100 nm.

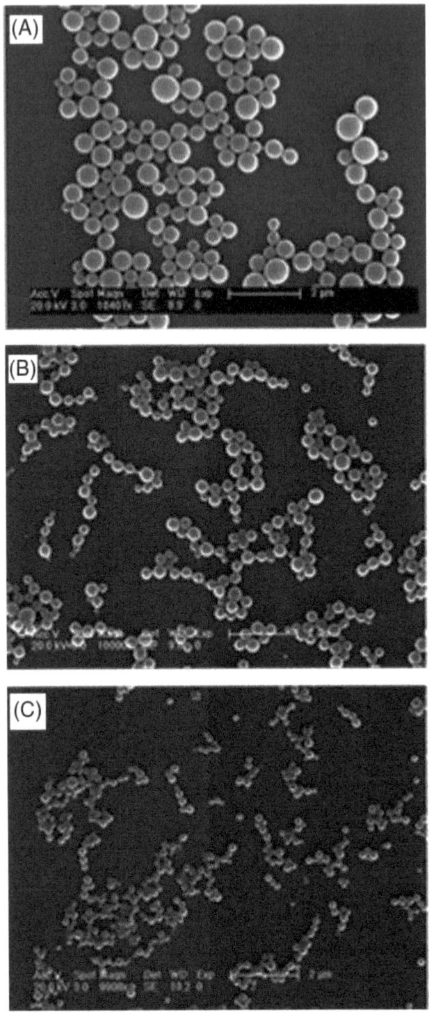

Figure 11.8 SEM images of SA nanospheres (DS 2.97) prepared by the dropwise addition of water into a polymer acetone solution with concentrations of (A) 10 mg/mL, (B) 5 mg/mL, and (C) 1 mg/mL.
Source: Reprinted from Tan, Y., Xu, K., Li, L., Liu, C., Song, C., Wang, P., 2009. Fabrication of size-controlled starch based nanospheres by nanoprecipitation. ACS Appl. Mater. Interfaces 1 (4), 956–959. Copyright (2009), with permission from American Chemical Society.

Chin et al. (2011) stated the preparation process of native sago starch nanoparticles with the controlled sizes by nanoprecipitation method in which, they used the combination of NaOH and urea (0.8 and 1 wt.%) to solubilize starch in water. At that moment, sago starch solutions with and without surfactant were added dropwise into the absolute ethanol under stirring. Spherical nanoparticles were formed when the starch solution to ethanol ratio reached 1:20. It was concluded that the presence

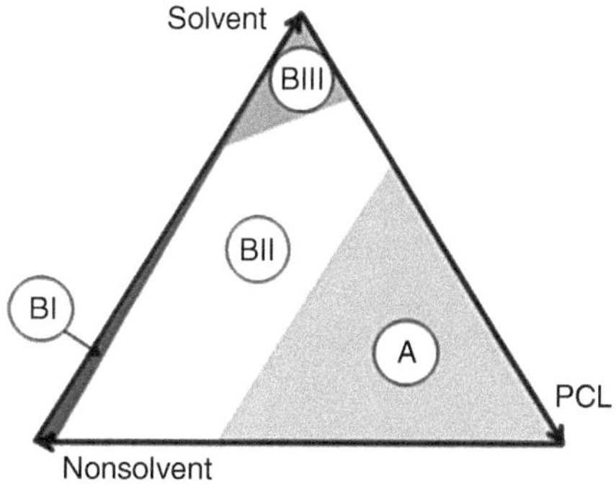

Figure 11.9 Ternary phase diagram of PCL in acetone/water (solvent/nonsolvent) shows the domain of nanoprecipitation (BI), of flocculation (BII), of dissolution of PCL (BIII), and of insolubility of (A).
Source: Reproduced from Schubert, S., Delaney, Jr., J.T., Schubert, U.S., 2011. Nanoprecipitation and nanoformulation of polymers: from history to powerful possibilities beyond poly(lactic acid). Soft Matter 7 (5), 1581–1588, with permission from The Royal Society of Chemistry.

Figure 11.10 SEM micrographs of precipitated starch nanoparticles prepared by addition of 1 mL of 1% starch into 20 mL of ethanol (A) without surfactant, (B) with 4% of CTAB, and (C) with 4% of Tween 80.
Source: Reprinted from Chin, S.F., Pang, S.C., Tay, S.H., 2011. Size controlled synthesis of starch nanoparticles by a simple nanoprecipitation method. Carbohydr. Polym. 86 (4), 1817–1819. Copyright (2011), with permission from Elsevier.

of surfactant in the starch solution formed smaller nanoparticles; however, in the presence of 4% hexadecyl (cetyl) trimethylammonium bromide (CTAB) and Tween 80, particle size range decreased to 250–300 and 150–200 nm, respectively (Chin et al., 2011). Fig. 11.10 illustrates SEM images of sago starch nanoparticles with and without the surfactants (Chin et al., 2011). Since Tween-80 is more hydrophilic compared to CTAB, its interactions with starch molecules are stronger, which forms smaller starch nanoparticles (Chin et al., 2011).

In a recent study, Wu et al. (2016) used dimethylsulfoxide (DMSO) to disperse starch molecules in solvent instead of alkaline and/or urea. They proved that since some of the active components are sensitive to alkaline condition, using DMSO as an alternative, may open a new horizon for the encapsulation of active components within starch nanoparticles. For this purpose, dispersed starch solution in DMSO was added dropwise into different alcohols (methanol, ethanol, *n*-propanol, isopropanol, *n*-butanol, and *n*-pentanol) as nonsolvent moieties for nanoparticulation, and the mean diameter of the obtained nanoparticles were reported to be 207.9, 209.2, 210.3, 212.8, 216.4, and 475.6 nm, respectively. Increasing the ratio of nonsolvent to solvent from 5 to 30 formed smaller particles, and SEM images showed the morphology of the particles was identical when the ratio of nonsolvent to starch solution was 10 or higher.

11.3.2 Encapsulation of different bioactive components within polysaccharide nanoparticles

A summary on the nanoparticulation of polysaccharides using nanoprecipitation method and their application as nanocarrier delivery systems is presented in Table 11.2. Acetylated corn starch nanoparticles (ACSNPs) were fabricated to encapsulate ciprofloxacin (Najafi et al., 2016) and ibuprofen (IBU) (Han et al., 2013), for which acetone and water were used as solvent and nonsolvent, respectively. EE of ciprofloxacin in ACSNPs varied from 20.9% to 89.7% depending on the degree of substitution and water-to-starch solution ratio. The highest EE was obtained with the higher degree of substitution (2.66%) and higher water-to-starch solution ratio (3:1) (Najafi et al., 2016). Han et al. (2013) showed that the weight ratio of ACS to IBU significantly affected the particle properties and encapsulation ability, in which by an increase in weight ratio from 1:1 to 4:1, the particle size and the EE increased from 189 nm and 32.7% to 975 nm and 69.3%, respectively (Han et al., 2013). They also reported that the ratio of water to acetone plays an important role in determining the particle size and the EE; the higher the ratio of the water to acetone, the lower the particle size and the EE. They stated that the interfacial turbulence and thermal inequalities coming from the broaden difference between surface tensions decrease the mean particle diameters. On the other hand, one of the explanations for the lower EE was the quicker precipitation of nanoparticles and shorter time necessary to entrap the drug molecules (Han et al., 2013). Higher degree of substitution (DS) indicates that higher hydrophobic acetyl groups are present in the starch molecules thus the solubility of the ACS changes in both acetone and water medium. Han et al. (2013) reported that larger particle size, lower polydispersity and higher EE could all be gained by increasing the degree of substitution from 1.82% to 2.85%. The releasing profile studies showed that increasing the degree of substitution, accompanied by elevated starch to drug and water to acetone ratios improve the control on the drug release (Han et al., 2013).

Thereafter, El-Naggar et al. (2015) used nanoprecipitation technique to fabricate diclofenac sodium loaded nanoparticles from cross-linked starch. They homogenized native starch dispersion in sodium hydroxide solution, Tween 80 and diclofenac sodium then, sodium tripolyphosphate was added as a cross-linking agent to the aqueous solution. After 2 h of cross-linking, ethanol was added to precipitate diclofenac sodium loaded cross-linked starch nanoparticles (El-Naggar et al., 2015). They reported that this technique would be able to fabricate relatively smaller nanoparticles compared to other methods (less than 60 nm), while EE was calculated to be between 73.32%–91.94%.

Table 11.2 A summary of polysaccharides nanoparticulation using nanoprecipitation method and some of encapsulation and delivery applications

Polysaccharide type	Particle size (nm)		Encapsulated material	EE[a]	LE[b]	Release trend	Achievements	References
	Unloaded	Loaded						
Composite of CA and CMC[c]	89–159	—	Pyrene	75	—	—	The pyrene is localized in the CA matrix of the nanoparticles	Kulterer et al. (2012)
Sago starch	150–400	—	—	—	—	—	Particle size can be modulated using surfactant	Chin et al. (2011)
Amphiphilic starch mixed ester	c. 270	—	—	—	—	—	Pickering emulsions as potential delivery systems	Tan et al. (2012)
Cross-linked maize starch NPs	—	21–68	Diclofenac sodium	73.32–91.94	—		Enhancement of the controlled release for the transdermal delivery	El-Naggar et al. (2015)
Acetylated corn starch	—	189–975	Ibuprofen	32.7–71.1	—	Larger particle size and higher degree of substitution enhance the release profile	An feasible technique for encapsulation of hydrophobic drug	Han et al. (2013)
Cellulose stearoyl ester	50–550	—	—	—	—	—	Fabrication of temperature-responsive superhydrophobic surfaces using the NPs	Geissler et al. (2014)

(*Continued*)

Table 11.2 A summary of polysaccharides nanoparticulation using nanoprecipitation method and some of encapsulation and delivery applications (*cont.*)

Polysaccharide type	Particle size (nm)		Encapsulated material	EE[a]	LE[b]	Release trend	Achievements	References
	Unloaded	Loaded						
Pullulan acetate	~80	128–253	Paclitaxel		6.7–13.1	Sustained in vitro release properties	A higher reduction in tumor volume (in vivo study)	Lee et al. (2012a)
Acetylated starch	—	221–324	Ciprofloxacin	20.9–89.7	—	—	Encapsulation property is related to degree of substitution	Najafi et al. (2016)
Chitosan-coated polycaprolactone	—	114–125	Curcumin	>99	—	—	Strong ability to interact with mucin	Mazzarino et al. (2012)
Chitosan-poly(ε-caprolactone)	98–215	—	—	—	—	—	The graft copolymer NPs have a great potential in delivery systems	Guan et al. (2007)
Cress seed gum	24–520	—	—	—	—	—	Optimization of fabrication parameters by RSM	Taheri and Razavi (2015)
Acetylated cashew gum	108–314	70–170	Indomethacin			An initial burst in the first 2 h and a controlled release up to 72 h	Perfectly stable nanoparticles for 1 year	Pitombeira et al. (2015)
α-Tocopherol succinate conjugated dextran[d]	—	295–325	Doxorubicin hydrochloride	58.9–87.9	4.21–8.12	An initial burst release for about 12 h and subsequent sustained release	The micelles have potential as drug delivery vehicles	Yu et al. (2015)

Fatty acid chloride-starch esters[d]	316–420	—	—	—	—	—	Reduction in particle size by increasing the dialysis time	Namazi et al. (2011)
Poly-(D,L-lactide)	—	229	Indomethacin	100	—	—	Very high drug loading capacity	Fessi et al. (1989)
Dextran, pullulan, or starch esters	50–200	—	Fluorescent dye	—	—	—	The dye could be encapsulated in the nanoparticles	Aschenbrenner et al. (2013)
Sago starch	—	50–80	Curcumin	48–78	—	No initial burst release. Sustained release from the swollen starch NPs	Cost-effective delivery nanocarriers	Pang et al. (2015)
Starch-Maleate	—	30–120	Curcumin	—	15[e]	Enhancement of curcumin solubility in water	Cost-effective delivery nanocarriers	Pang et al. (2015)
PLA-grafted dextran copolymer	141–224	—	—	—	—	—	PLA core and dextran-enriched surface nanoparticles showed good colloidal stability in water even at high ionic strengths	Gavory et al. (2011)

[a]Encapsulation efficiency.
[b]Loading capacity (%w/w).
[c]Cellulose acetate (CA), carboxymethyl cellulose (CMC)
[d]Dialysis method.
[e]Unit mg/g.

The results of FTIR, X-ray diffraction and DSC experiments, showed that the drug structure would convert from crystalline to amorphous form during encapsulation, and according to the optimization data, the best conditions for the fabrication of smaller and more uniform particles with the highest EE and sustained release profile were obtained. Their findings presented evidences that the cross-linked starch nanoparticles have a great potential to encapsulate targeted components such as diclofenac sodium (El-Naggar et al., 2015). Paclitaxel-incorporated nanoparticles were successfully prepared from pullulan acetate (PA), a hydrophobic polysaccharide, using nanoprecipitation method (Lee et al., 2012a). This technique was successfully applied for the fabrication of spherical and small pullulan acetate nanoparticles (~80 nm). It was found that, loading the nanoparticles with paclitaxel increased the particle size (>100 nm) (Lee et al., 2012a). Paclitaxel-incorporated nanoparticles showed lower antitumor activity in in vitro studies compared to free paclitaxel; however in in vivo study tumor volume was reported to be smaller using HCT116 human colon carcinoma cells (Lee et al., 2012a). Cellulose alkyl esters nanoparticles were fabricated via this method as well (Geissler et al., 2014; Zhang et al., 2015), in which the size range of cellulose stearoyl esters (CSEs) nanoparticles was 50–550 nm according to the CSEs solution concentration ranges (0.1–10 mg/mL) (Geissler et al., 2014). Crystalline structure of the nanoparticles and their surface hydrophobicity were highly influenced by the characteristics of alkyl groups and the degree of substitution as well; therefore, the feasibility of the crystallization decreases with short alkane chains (Zhang et al., 2015). Cellulose stearoyl esters nanoparticles can be fabricated in crystalline, amorphous, or the intermediate status which is temperature driven due to temperature-responsive property of CSEs (Zhang et al., 2015).

Emulsions are through the best delivery systems developed to carry lots of bioactive components, especially using Pickering emulsions enables the elimination of synthetic surfactants (Marto et al., 2015). Starch nanoparticles prepared by nanoprecipitation have been reported as an effective stabilization system for Pickering emulsions (Tan et al., 2012).

11.4 Future trends

Among the various methods employed for the preparation of nanoparticles, desolvation/nanoprecipitation has been well regarded as the best techniques due to the simplicity of the method and the beneficial properties that the fabricated nanoparticles offer. More systematic studies on the optimization of nanoparticulation parameters are necessary to increase our knowledge and ability to predict the properties of the nanoparticle. Novel techniques for improving the stability of the nanoparticles which are acceptable for food product and also being capable of stabilizing the nanoparticles can dramatically increase the use of the method for encapsulation of nutraceuticals and pharmaceuticals. Functionalization and bioconjugation of the original proteins or polysaccharides before particulation can be used to bind a specific bioactive or targeted component to the biopolymers. Profound studies on the encapsulation, releasing profile, and bioavailability of the bioactive components along with the reviewed research works presented in this chapter might eventuate in commercialization of products for consumers.

References

Altman, G.H., Diaz, F., Jakuba, C., Calabro, T., Horan, R.L., Jingsong, C., Lu, H., Richmond, J., Kaplan, D.L., 2003. Silk-based biomaterials. Biomaterials 24 (3), 401–416.

Arroyo-Maya, I.J., Hernández-Sánchez, H., Jiménez-Cruz, E., Camarillo-Cadena, M., Hernández-Arana, A., 2014. α-Lactalbumin nanoparticles prepared by desolvation and cross-linking: structure and stability of the assembled protein. Biophys. Chem. 193, 27–34.

Arroyo-Maya, I.J., Rodiles-López, J.O., Cornejo-Mazon, M., Gutierrez-Lopez, G.F., Hernández-Arana, A., Toledo-Núñez, C., Barbosa-Cánovas, G.V., Flores-Flores, J.O., Hernandez-Sanchez, H., 2012. Effect of different treatments on the ability of α-lactalbumin to form nanoparticles. J. Dairy Sci. 95 (11), 6204–6214.

Aschenbrenner, E., Bley, K., Koynov, K., Makowski, M., Kappl, M., Landfester, K., Weiss, C.K., 2013. Using the polymeric ouzo effect for the preparation of polysaccharide-based nanoparticles. Langmuir 29 (28), 8845–8855.

Bagheri, L., Madadlou, A., Yarmand, M., Mousavi, M.E., 2013. Nanoencapsulation of date palm pit extract in whey protein particles generated via desolvation method. Food Res. Int. 51 (2), 866–871.

Bilthariya, U., Jain, N., Rajoriya, V., Jain, A.K., 2015. Folate-conjugated albumin nanoparticles for rheumatoid arthritis-targeted delivery of etoricoxib. Drug Dev. Ind. Pharm. 41 (1), 95–104.

Casa, D.M., Karam, T.K., Alves, S., Cristo, A.d., Zgoda, A.A., Khalil, N.M., Mainardes, R.M., 2015. Bovine serum albumin nanoparticles containing amphotericin B: characterization, cytotoxicity and in vitro antifungal evaluation. J. Nanosci. Nanotechnol. 15 (12), 10183–10188.

Chin, S.F., Pang, S.C., Tay, S.H., 2011. Size controlled synthesis of starch nanoparticles by a simple nanoprecipitation method. Carbohydr. Polym. 86 (4), 1817–1819.

Clark, A.H., Judge, F.J., Richards, J.B., Stubbs, J.M., Suggett, A., 1981. Electron microscopy of network structures in thermally-induced globular protein gels. Int. J. Peptide Protein Res. 17 (3), 380–392.

Coester, C.J., Langer, K., Von Briesen, H., Kreuter, J., 2000. Gelatin nanoparticles by two step desolvation a new preparation method, surface modifications and cell uptake. J. Microencapsul. 17 (2), 187–193.

Delfiya, D.S.A., Thangavel, K., Amirtham, D., 2016a. Preparation of curcumin loaded egg albumin nanoparticles using acetone and optimization of desolvation process. Protein J. 35 (2), 124–135.

Delfiya, D.S.A., Thangavel, K., Natarajan, N., Amirtham, D., 2016b. Preparation of curcumin-loaded egg albumin nanoparticles using ethanol as desolvation agent. Asian J. Chem. 28 (7), 1536.

Desai, N., Trieu, V., Yao, Z., Louie, L., Ci, S., Yang, A., Tao, C., et al., 2006. Increased antitumor activity, intratumor paclitaxel concentrations, and endothelial cell transport of cremophor-free, albumin-bound paclitaxel, ABI-007, compared with cremophor-based paclitaxel. Clin. Cancer Res. 12 (4), 1317–1324.

Dreis, S., Rothweiler, F., Michaelis, M., Cinatl, J., Kreuter, J., Langer, K., 2007. Preparation, characterisation and maintenance of drug efficacy of doxorubicin-loaded human serum albumin (HSA) nanoparticles. Int. J. Pharm. 341 (1), 207–214.

Duclairoir, C., Orecchioni, A.-M., Depraetere, P., Osterstock, F., Nakache, E., 2003. Evaluation of gliadins nanoparticles as drug delivery systems: a study of three different drugs. Int. J. Pharm. 253 (1), 133–144.

El-Naggar, M.E., El-Rafie, M.H., El-Sheikh, M.A., El-Feky, G.S., Hebeish, A., 2015. Synthesis, characterization, release kinetics and toxicity profile of drug-loaded starch nanoparticles. Int. J. Biol. Macromol. 81, 718–729.

Elsadek, B., Kratz, F., 2012. Impact of albumin on drug delivery—new applications on the horizon. J. Control. Release 157 (1), 4–28.
Elzoghby, A.O., El-Fotoh, W.S.A., Elgindy, N.A., 2011. Casein-based formulations as promising controlled release drug delivery systems. J. Control. Release 153 (3), 206–216.
Elzoghby, A.O., Samy, W.M., Elgindy, N.A., 2012a. Albumin-based nanoparticles as potential controlled release drug delivery systems. J. Control. Release 157 (2), 168–182.
Elzoghby, A.O., Samy, W.M., Elgindy, N.A., 2012b. Protein-based nanocarriers as promising drug and gene delivery systems. J. Control. Release 161 (1), 38–49.
Estrada, L.H., Chu, S., Champion, J.A., 2014. Protein nanoparticles for intracellular delivery of therapeutic enzymes. J. Pharm. Sci. 103 (6), 1863–1871.
Etorki, A.M., Gao, M., Sadeghi, R., Maldonado-Mejia, L.F., Kokini, J.L., 2016. Effects of desolvating agent types, ratios, and temperature on size and nanostructure of nanoparticles from α-lactalbumin and ovalbumin. J. Food Sci. 81, E2511–E2520.
Ezpeleta, I., Irache, J.M., Stainmesse, S., Chabenat, C., Gueguen, J., Popineau, Y., Orecchioni, A.-M., 1996. Gliadin nanoparticles for the controlled release of all-trans-retinoic acid. Int. J. Pharm. 131 (2), 191–200.
Faridi Esfanjani, A.F., Jafari, S.M., 2016. Biopolymer nano-particles and natural nano-carriers for nano-encapsulation of phenolic compounds. Colloids Surf. B 146, 532–543.
Farrugia, C.A., Groves, M.J., 1999. Gelatin behaviour in dilute aqueous solution: designing a nanoparticulate formulation. J. Pharm. Pharm. 51 (6), 643–649.
Fessi, H., Puisieux, F., Devissaguet, J.P., Ammoury, N., Benita, S., 1989. Nanocapsule formation by interfacial polymer deposition following solvent displacement. Int. J. Pharm. 55 (1), R1–R4.
Fuchs, S., Kutscher, M., Hertel, T., Winter, G., Pietzsch, M., Coester, C., 2010. Transglutaminase: new insights into gelatin nanoparticle cross-linking. J. Microencapsul. 27 (8), 747–754.
Fujimoto, M., Hara, M., Hayashi, T., Furuta, M., 2010. Effect of heating process on the formation of nanoparticles of elastin model polypeptide, (GVGVP) 251, by gamma-ray cross-linking. Polym. Bull. 64 (7), 707–716.
Galisteo-González, F., Molina-Bolívar, J.A., 2014. Systematic study on the preparation of BSA nanoparticles. Colloids Surf. B 123, 286–292.
Gavory, C., Durand, A., Six, J.-L., Nouvel, C., Marie, E., Leonard, M., 2011. Polysaccharide-covered nanoparticles prepared by nanoprecipitation. Carbohydr. Polym. 84 (1), 133–140.
Geissler, A., Biesalski, M., Heinze, T., Zhang, K., 2014. Formation of nanostructured cellulose stearoyl esters via nanoprecipitation. J. Mater. Chem. A 2 (4), 1107–1116.
Guan, X., Quan, D., Shuai, X., Liao, K., Mai, K., 2007. Chitosan-graft-poly(∈-caprolactone) s: an optimized chemical approach leading to a controllable structure and enhanced properties. J. Polym. Sci. A 45 (12), 2556–2568.
Gunasekaran, S., Ko, S., Xiao, L., 2007. Use of whey proteins for encapsulation and controlled delivery applications. J. Food Eng. 83 (1), 31–40.
Han, F., Gao, C., Liu, M., 2013. Fabrication and characterization of size-controlled starch-based nanoparticles as hydrophobic drug carriers. J. Nanosci. Nanotechnol. 13 (10), 6996–7007.
Heinze, T., Hornig, S., 2009. Versatile concept for the structure design of polysaccharide-based nanoparticles. ACS Symposium Series, 1017Oxford University Press, pp. 169–183.
Jafari, S.M., Fathi, M., Mandala, I., 2015. Emerging product formation. Food Waste Recovery: Processing Technologies and Industrial Techniques. Elsevier Inc., Academic Press, San Diego, CA, pp. 293–317.
Jain, A., Gulbake, A., Jain, A., Shilpi, S., Hurkat, P., Jain, A., Jain, S.K., 2012. Development of surface-functionalised nanoparticles for FGF2 receptor-based solid tumour targeting. J. Microencapsul. 29 (1), 95–102.

Jiang, S., Gong, X., Zhao, X., Zu, Y., 2015. Preparation, characterization, and antitumor activities of folate-decorated docetaxel-loaded human serum albumin nanoparticles. Drug Deliv. 22 (2), 206–213.

Jing, F., Xu, W., Liu, D., Wang, C., Sui, Z., 2015. Enhanced antithrombotic effect of hirudin by bovine serum albumin nanoparticles. J. Exp. Nanosci. 11 (8), 619–629.

Katouzian, I., Jafari, S.M., 2016. Nano-encapsulation as a promising approach for targeted delivery and controlled release of vitamins. Trends Food Sci. Technol. 53, 34–48.

Kaul, G., Amiji, M., 2005. Tumor-targeted gene delivery using poly(ethylene glycol)-modified gelatin nanoparticles: in vitro and in vivo studies. Pharm. Res. 22 (6), 951–961.

Kaur, H., Singh, A., 2015. Design, development and characterization of serratiopeptidase loaded albumin nanoparticles. J. Appl. Pharm. Sci. 5 (02), 103–109.

Keshavarzi, M., Shakeri, S., Kiani, K., 2015. Preparation and in vitro investigation of antigastric cancer activities of carvacrol-loaded human serum albumin nanoparticles. IET Nanobiotechnol. 9 (5), 294–299.

Kinsella, J.E., Whitehead, D.M., 1989. Proteins in whey: chemical, physical, and functional properties. Adv. Food Nutr. Res. 33 (C), 343–438.

Ko, S., Gunasekaran, S., 2006. Preparation of sub-100-nm β-lactoglobulin (BLG) nanoparticles. J. Microencapsul. 23 (8), 887–898.

Kolluru, L.P., Rizvi, S.A.A., D'Souza, M., D'Souza, M.J., 2013. Formulation development of albumin based theragnostic nanoparticles as a potential delivery system for tumor targeting. J. Drug Target. 21 (1), 77–86.

Kommareddy, S., Amiji, M., 2005. Preparation and evaluation of thiol-modified gelatin nanoparticles for intracellular DNA delivery in response to glutathione. Bioconjugate Chem. 16 (6), 1423–1432.

Kratz, F., 2008. Albumin as a drug carrier: design of prodrugs, drug conjugates and nanoparticles. J. Control. Release 132 (3), 171–183.

Kulterer, M.R., Reichel, V.E., Kargl, R., Köstler, S., Sarbova, V., Heinze, T., Stana-Kleinschek, K., Ribitsch, V., 2012. Functional polysaccharide composite nanoparticles from cellulose acetate and potential applications. Adv. Funct. Mater. 22 (8), 1749–1758.

Kundu, J., Chung, Y.-I., Kim, Y.H., Tae, G., Kundu, S.C., 2010. Silk fibroin nanoparticles for cellular uptake and control release. Int. J. Pharm. 388 (1), 242–250.

Langer, K., Balthasar, S., Vogel, V., Dinauer, N., Von Briesen, H., Schubert, D., 2003. Optimization of the preparation process for human serum albumin (HSA) nanoparticles. Int. J. Pharm. 257 (1), 169–180.

Langer, R., Chasin, M., 1990. Biodegradable polymers as drug delivery systems. Marcel Dekker, New York, p. 43–70.

Lee, S.J., Hong, G.-Y., Jeong, Y.-I., Kang, M.-S., Oh, J.-S., Song, C.-E., Lee, H.C., 2012a. Paclitaxel-incorporated nanoparticles of hydrophobized polysaccharide and their antitumor activity. Int. J. Pharm. 433 (1), 121–128.

Lee, E.J., Khan, S.A., Park, J.K., Lim, K.-H., 2012b. Studies on the characteristics of drug-loaded gelatin nanoparticles prepared by nanoprecipitation. Bioprocess Biosyst. Eng. 35 (1–2), 297–307.

Lee, C.H., Singla, A., Yugyung, L., 2001. Biomedical applications of collagen. Int. J. Pharm. 221 (1), 1–22.

Li, J.L., Zu, Y.G., Zhao, X.H., Zhao, D.M., Chen, X.Q., An, Z.G., Wang, K.L., 2012. Preparation, characterization, and *in vitro* evaluation of resveratrol-loaded bovine serum albumin nanoparticles. Adv. Mater. Res. 345, 349–354.

Liang, H.-C., Chang, W.-H., Lin, K.-J., Sung, H.-W., 2003. Genipin-crosslinked gelatin microspheres as a drug carrier for intramuscular administration: in vitro and in vivo studies. J. Biomed. Mater. Res. A 65 (2), 271–282.

Lin, W., Coombes, A., Davies, M., Davis, S., Illum, L., 1993. Preparation of sub-100 nm human serum albumin nanospheres using a pH-coacervation method. J. Drug Target. 1 (3), 237–243.

Livney, Y.D., 2010. Milk proteins as vehicles for bioactives. Curr. Opin. Colloid Interface Sci. 15 (1), 73–83.

Lohcharoenkal, W., Wang, L., Chen, Y.C., Rojanasakul, Y., 2014. Protein nanoparticles as drug delivery carriers for cancer therapy. BioMed Res. Int. 2014, 1–12.

Luo, Y., Teng, Z., Wang, Q., 2012. Development of zein nanoparticles coated with carboxymethyl chitosan for encapsulation and controlled release of vitamin D3. J. Agric. Food Chem. 60 (3), 836–843.

Ma, X., Jian, R., Chang, P.R., Yu, J., 2008. Fabrication and characterization of citric acid-modified starch nanoparticles/plasticized-starch composites. Biomacromolecules 9, 3314–3320.

Marto, J., Gouveia, L., Jorge, I.M., Duarte, A., Gonçalves, L.M., Silva, S.M.C., Antunes, F., et al., 2015. Starch-based Pickering emulsions for topical drug delivery: a QbD approach. Colloids Surf. B 135, 183–192.

Mazzarino, L., Travelet, C., Ortega-Murillo, S., Otsuka, I., Pignot-Paintrand, I., Lemos-Senna, E., Borsali, R., 2012. Elaboration of chitosan-coated nanoparticles loaded with curcumin for mucoadhesive applications. J. Colloid Interface Sci. 370 (1), 58–66.

Mehryar, L., Esmaiili, M., Zeynali, F., Sadeghi, R., Imani, M., 2016. Production of nanoparticles from sunflower seed protein isolate and usage feasibility studies of the nanoparticles for encapsulation of functional components. PhD thesis, Urmia University.

Najafi, S.H.M., Baghaie, M., Ashori, A., 2016. Preparation and characterization of acetylated starch nanoparticles as drug carrier: ciprofloxacin as a model. Int. J. Biol. Macromol. 87, 48–54.

Namazi, H., Fathi, F., Dadkhah, A., 2011. Hydrophobically modified starch using long-chain fatty acids for preparation of nanosized starch particles. Sci. Iranica 18 (3), 439–445.

Nicklas, M., Schatton, W., Heinemann, S., Hanke, T., Kreuter, J., 2009. Preparation and characterization of marine sponge collagen nanoparticles and employment for the transdermal delivery of 17β-estradiol-hemihydrate. Drug Dev. Ind. Pharm. 35 (9), 1035–1042.

Niknejad, H., Mahmoudzadeh, R., 2015. Comparison of different crosslinking methods for preparation of docetaxel-loaded albumin nanoparticles. Iran. J. Pharm. Res. 14 (2), 385.

Numata, K., Kaplan, D.L., 2010. Silk-based delivery systems of bioactive molecules. Adv. Drug Deliv. Rev. 62 (15), 1497–1508.

Onwulata, C.I., 2012. Encapsulation of New Active Ingredients*. Annu. Rev. Food Sci. Technol. 3, 183–202.

Oppenheim, R.C., Marty, J.J., Speiser, P., 1978. Injectable compositions, nanoparticles useful therein, and process of manufacturing same. U.S. Patent 4,107,288, issued August 15, 1978.

Oppenheim, R.C., Stewart, N.F., 1979. The manufacture and tumour cell uptake of nanoparticles labelled with fluorescein isothiocyanate. Drug Dev. Ind. Pharm. 5 (6), 563–571.

Pang, S.C., Chin, S.F., Nadirah, A., Tay, S.H., Yazid, S.N.A.M., 2015. Fabrication of polysaccharide-based nanoparticles as drug delivery nanocarriers. ECS Trans. 66 (37), 15–32.

Park, C., Le-Ngoc Vo, C., Kang, T., Oh, E., Lee, B.-J., 2015. New method and characterization of self-assembled gelatin–oleic nanoparticles using a desolvation method via carbodiimide/*N*-hydroxysuccinimide (EDC/NHS) reaction. Eur. J. Pharm. Biopharm. 89, 365–373.

Peñalva, R., Esparza, I., González-Navarro, C.J., Quincoces, G., Peñuelas, I., Irache, J.M., 2015a. Zein nanoparticles for oral folic acid delivery. J. Drug Deliv. Sci. Technol. 30, 450–457.

Peñalva, R., Esparza, I., Larraneta, E., González-Navarro, C.J., Carlos, G., Irache, J.M., 2015b. Zein-based nanoparticles improve the oral bioavailability of resveratrol and its anti-inflammatory effects in a mouse model of endotoxic shock. J. Agric. Food Chem. 63 (23), 5603–5611.

Pitombeira, N.A.O., Veras Neto, J.G., Silva, D.A., Feitosa, J.P.A., Paula, H.C.B., de Paula, R.C.M., 2015. Self-assembled nanoparticles of acetylated cashew gum: characterization and evaluation as potential drug carrier. Carbohydr. Polym. 117, 610–615.

Podaralla, S., Perumal, O., 2012. Influence of formulation factors on the preparation of zein nanoparticles. AAPS PharmSciTech 13 (3), 919–927.

Qazvini, N.T., Zinatloo, S., 2011. Synthesis and characterization of gelatin nanoparticles using CDI/NHS as a non-toxic cross-linking system. J. Mater. Sci. 22 (1), 63–69.

Rahimnejad, M., Najafpour, G., Bakeri, G., 2012. Investigation and modeling effective parameters influencing the size of BSA protein nanoparticles as colloidal carrier. Colloids Surf. A 412, 96–100.

Reddy, N., Jiang, Q., Yang, Y., 2012. Preparation and properties of peanut protein films cross-linked with citric acid. Ind. Crops Prod. 39, 26–30.

Sadeghi, R., Moosavi-Movahedi, A.A., Emam-jomeh, Z., Kalbasi, A., Razavi, S.H., Karimi, M., Kokini, J., 2014. The effect of different desolvating agents on BSA nanoparticle properties and encapsulation of curcumin. J. Nanopart. Res. 16 (9), 1–14.

Sahoo, N., Sahoo, R.K., Biswas, N., Guha, A., Kuotsu, K., 2015. Recent advancement of gelatin nanoparticles in drug and vaccine delivery. Int. J. Biol. Macromol. 81, 317–331.

Schubert, S., Delaney, Jr., J.T., Schubert, U.S., 2011. Nanoprecipitation and nanoformulation of polymers: from history to powerful possibilities beyond poly(lactic acid). Soft Matter 7 (5), 1581–1588.

Shapira, A., Assaraf, Y.G., Epstein, D., Livney, Y.D., 2010a. Beta-casein nanoparticles as an oral delivery system for chemotherapeutic drugs: impact of drug structure and properties on co-assembly. Pharm. Res. 27 (10), 2175–2186.

Shapira, A., Markman, G., Assaraf, Y.G., Livney, Y.D., 2010b. β-Casein–based nanovehicles for oral delivery of chemotherapeutic drugs: drug-protein interactions and mitoxantrone loading capacity. Nanomedicine 6 (4), 547–555.

Sneharani, A.H., Karakkat, J.V., Singh, S.A., Rao, A.G.A., 2010. Interaction of curcumin with β-lactoglobulin—stability, spectroscopic analysis, and molecular modeling of the complex. J. Agric. Food Chem. 58 (20), 11130–11139.

Stainmesse, S., Orecchioni, A., Nakache, E., Puisieux, F., Fessi, H., 1995. Formation and stabilization of a biodegradable polymeric colloidal suspension of nanoparticles. Colloid Polym. Sci. 273 (5), 505–511.

Taheri, A., Razavi, S.M.A., 2015. Fabrication of cress seed gum nanoparticles, an anionic polysaccharide, using desolvation technique: an optimization study. BioNanoSci. 5 (2), 104–116.

Tan, Y., Xu, K., Li, L., Liu, C., Song, C., Wang, P., 2009. Fabrication of size-controlled starch-based nanospheres by nanoprecipitation. ACS Appl. Mater. Interfaces 1 (4), 956–959.

Tan, Y., Xu, K., Liu, C., Li, Y., Lu, C., Wang, P., 2012. Fabrication of starch-based nanospheres to stabilize Pickering emulsion. Carbohydr. Polym. 88 (4), 1358–1363.

Teng, Z., Luo, Y., Wang, Q., 2012. Nanoparticles synthesized from soy protein: preparation, characterization, and application for nutraceutical encapsulation. J. Agric. Food Chem. 60 (10), 2712–2720.

Teng, Z., Luo, Y., Wang, T., Zhang, B., Wang, Q., 2013. Development and application of nanoparticles synthesized with folic acid conjugated soy protein. J. Agric. Food Chem. 61 (10), 2556–2564.

Varca, G.H.C., Queiroz, R.G., Lugão, A.B., 2016. Irradiation as an alternative route for protein cross-linking: cosolvent free BSA nanoparticles. Rad. Phys. Chem.

Von Storp, B., Engel, A., Boeker, A., Ploeger, M., Langer, K., 2012. Albumin nanoparticles with predictable size by desolvation procedure. J. Microencapsul. 29 (2), 138–146.

Wacker, M., Zensi, A., Kufleitner, J., Ruff, A., Schütz, J., Stockburger, T., Marstaller, T., Vogel, V., 2011. A toolbox for the upscaling of ethanolic human serum albumin (HSA) desolvation. Int. J. Pharm. 414 (1), 225–232.

Wartlick, H., Spänkuch-Schmitt, B., Strebhardt, K., Kreuter, J., Langer, K., 2004. Tumour cell delivery of antisense oligonuclceotides by human serum albumin nanoparticles. J. Control. Release 96 (3), 483–495.

Weber, C., Coester, C., Kreuter, J., Langer, K., 2000. Desolvation process and surface characterisation of protein nanoparticles. Int. J. Pharm. 194 (1), 91–102.

Wilson, B., Paladugu, L., Priyadarshini, S.R.B., Jenita, J.J.L., 2015. Development of albumin-based nanoparticles for the delivery of abacavir. Int. J. Biol. Macromol. 81, 763–767.

Won, Y.-W., Yong-Hee, K., 2008. Recombinant human gelatin nanoparticles as a protein drug carrier. J. Control. Release 127 (2), 154–161.

Wu, X., Chang, Y., Fu, Y., Ren, L., Tong, J., Zhou, J., 2016. Effects of non-solvent and starch solution on formation of starch nanoparticles by nanoprecipitation. Starch/Staerke 68, n/a–n/a.

Wunder, A., Muller-Ladner, U., Stelzer, E., Neumann, E., Sinn, H., Gay, S., Fiehn, C., 2003. Albumin-based drug delivery as novel therapeutic approach for rheumatoid arthritis. Arthritis Res. Ther. 5 (Suppl. 3), 1.

Xu, H., Shen, L., Xu, L., Yang, Y., 2015a. Controlled delivery of hollow corn protein nanoparticles via non-toxic cross-linking: in vivo and drug loading study. Biomed. Microdevices 17 (1), 1–8.

Xu, H., Shen, L., Xu, L., Yang, Y., 2015b. Low-temperature cross-linking of proteins using non-toxic citric acid in neutral aqueous medium: mechanism and kinetic study. Ind. Crops Prod. 74, 234–240.

Yedomon, B., Fessi, H., Charcosset, C., 2013. Preparation of Bovine Serum Albumin (BSA) nanoparticles by desolvation using a membrane contactor: a new tool for large scale production. Eur. J. Pharm. Biopharm. 85 (3), 398–405.

Yeh, T.K., Lu, Z., Wientjes, M.G., Au, J.L.-S., 2005. Formulating paclitaxel in nanoparticles alters its disposition. Pharm. Res. 22 (6), 867–874.

Yoshikawa, H., Hirano, A., Arakawa, T., Shiraki, K., 2012. Effects of alcohol on the solubility and structure of native and disulfide-modified bovine serum albumin. Int. J. Biol. Macromol. 50 (5), 1286–1291.

Yu, J., Zhou, Y., Chen, W., Ren, J., Zhang, L., Lu, L., Luo, G., Huang, H., 2015. Preparation, characterization and evaluation of α-tocopherol succinate-modified dextran micelles as potential drug carriers. Materials 8 (10), 6685–6696.

Zhang, K., Geissler, A., Heinze, T., 2015. Reversibly crystalline nanoparticles from cellulose alkyl esters via nanoprecipitation. Part. Part. Syst. Charac. 32 (2), 258–266.

Nanocapsule formation by complexation of biopolymers

12

Seyed M.H. Hosseini, Fatemeh Ghiasi, Mastaneh Jahromi
Shiraz University, Shiraz, Iran

12.1 Introduction

Main biopolymeric components of foods include polysaccharides and proteins. These two ingredients are required to provide an equilibrated caloric diet for the survival of organisms (Schmitt et al., 2009). As they are large molecular structures synthesized by an organism or a plant, they are broadly known as biopolymers (Goh et al., 2009). The chemical structure of biopolymers has more complexity than synthetic polymers. Generally, the natural polymers are nontoxic, biocompatible, and biodegradable (Esfanjani and Jafari, 2016). Enderle et al. (2005) extended the term biopolymers to bioplastics, such as polylactic acid, polyglycolic acid, and their block copolymer. Biopolymers are utilized as essential functional ingredients in a broad range of industrial applications (i.e., in foods, pharmaceuticals, and personal care utilities) (Corredig et al., 2011; McClements, 2006; Rodríguez Patino and Pilosof, 2011). In food products, they have critical functions in characterizing the shelf life, stability, viscoelastic and sensory properties, and in developing the building blocks of structure and texture through rheological control of the aqueous medium caused by aggregation and gelation (Corredig et al., 2011; Dickinson, 1998, 2008; Tolstoguzov, 2003).

Polysaccharides, such as long-chain carbohydrate molecules, are extensively applied as thickeners, emulsifiers, and bulk gel formation agents to modify the texture and to manufacture edible packaging materials (Corredig et al., 2011; Dickinson, 1998; Goh et al., 2009; Hosseini et al., 2009). Owing to the interfacial activity, proteins synthesized from amino acids are involved in the formation of highly stable foams and emulsions. In addition, they can impart excellent colloidal stability to the emulsion-dispersed phase by combining electrostatic repulsion and steric hindrance (Dickinson, 2008; McClements, 2004). In addition, an overprotection from the phase separation may be obtained by polysaccharides, which modifies the rheological properties of a continuous phase network structure and also engineers the interface of dispersed droplets (Dickinson, 2008). Based on what was explained in Chapter 11, naturally occurring food proteins are utilized as carriers for other molecules [e.g., retinol entrapment inside the calyx of β-lactoglobulin (β-lg) present in milk] to provide effective protection against degradation both in milk and stomach after ingestion (Fox, 2009).

The physical and chemical features of individual biopolymers as a function of intrinsic and extrinsic molecular key parameters (e.g., molecular weight, steric conformation, concentration, charge amount and distribution, size homogeneity, ionic strength, ion type, pH, solvent quality, temperature, and various molecular interactions within or between different biopolymers) have been extensively explored throughout the past few decades (Goh et al., 2009). Compared to single biopolymers, formation of biopolymer complexes offers different functional and promising properties due to the synergistic combination between the functional features of various biopolymers (McClements, 2006; Schmitt et al., 1998, 2009). Engineered functionalities, such as encapsulation, emulsion, and foam stabilization; formation of lightening agents by fat mimicry; creation of novel mixed electrostatic gels at concentrations well below those required for common gel forming agents; formation of new structures at various (nano-, meso-, or macro-) scales, and waste management may be developed using covalent linkages, Coulombic interactions, hydrogen bonding, and hydrophobic interactions in biopolymer blends. The new functionalities resulting from self-assembling structures are generally the focus of researchers studying soft condensation in material science (de Vries and Stuart, 2006; Schmitt et al., 2009; van der Sman, 2012). The ability to control and modify involved interactions can assist food technologists to engineer supra-macromolecular structures to develop food products having more desirable structural properties (Schmitt et al., 2009; Turgeon et al., 2003).

The application of biopolymers and their mixtures for designing advanced engineered delivery systems to entrap bioactive components and to enable subsequent fortification of food products is a highly desirable and promising technique (Assadpour and Jafari, 2017; Assadpour et al., 2017; Esfanjani et al., 2017; Mehrnia et al., 2017). In addition to the vast food applications, biopolymers can be used as promising replacements for synthetic polymers in nonfood applications (e.g., wound dressings, drug delivery, and scaffold engineering). Biopolymers are capable of encapsulating substances of different hydrophilicity ranging from completely the water soluble to water insoluble. In addition, mild preparation process without involving harsh chemicals and poisonous solvents applied in the formation of structured vehicles is of paramount importance in green (organic solvent–free) industries (Huang et al., 2007; Mukhopadhyay and Kundu, 2015; Pal et al., 2009). Therefore, natural individual biopolymers and their mixtures introduce new possibilities for developing multifunctional nanometric encapsulation systems for protected and targeted release of sensitive health-promoting substances (Hu and Huang, 2013; Esfanjani and Jafari, 2016; Katouzian and Jafari, 2016).

12.2 Molecular forces between biopolymers and factors affecting them

For careful designing and fine-tuning of structured food products with specific functional properties, it is necessary to be aware of various attractive and repulsive forces involved in ingredient interactions (McClements et al., 2009). Interaction strength and specificity differ broadly among different attractive interactions. The net (i.e., repulsive or attractive) interaction between bio-macromolecules is a consequence of various forces acting between different segments of two adjacent macromolecular species (Dickinson, 1998; Goh et al., 2009; Schmitt et al., 1998). A brief introduction of the main molecular forces responsible for the creation of the supramolecular assemblies is given in Table 12.1.

Table 12.1 **Different types of molecular forces between biopolymers**

Type of interaction	Characteristics	Methods of generation	References
Covalent bonding	Specific and very strong (25–250 kcal/mol) Insensitive to changes in pH and salt concentration	MRs	Caillard et al. (2010); Chen et al. (2012)
		Cross-linking between different phenolic residues using oxidoreductase enzymes	Boeriu et al. (2004)
		Cross-linking using TG	Gan et al. (2008); Heidebach et al. (2009)
		Using genipin as a natural cross-linker	Devi and Maji (2010); Song et al. (2009)
Coulombic interactions	Very strong (25–250 kcal/mol) Repulsive or attractive Unstable toward changes in pH and salt concentration	Altering the pH and/or salt concentration	Aberkane et al. (2010); Girard et al. (2003); Hosseini et al. (2013a,b)
Hydrogen bond	Strong (2–10 kcal/mol) attractive collective interaction with a specific orientational ionic character	A ubiquitous bonding between polar groups particularly at low temperatures	Braudo and Antonov (1993)
Hydrophobic interactions	Relatively strong (1–2 kcal/mol) attractive collective interaction between hydrophobes	Conformational and structural modifications of biopolymers resulting in chain unfolding and exposing of hydrophobic regions	Pan et al. (2007)
van der Waals forces	Universal short-range, weak attractive interactions of electrostatic character between neutral molecules	Acting between all groups to some extent	Dickinson (1998)

MR, Maillard-type reaction; TG, transglutaminase.

In a specific system, several factors, including those related to intrinsic characteristics (e.g., molecular mass and conformation, charge density and distribution, flexibility, and hydrophilicity); solvent (e.g., ion type, ionic strength, pH, dielectric constant, and solvent quality); and processing (e.g., pressure, temperature, mixing conditions, and method of acidification) and external characteristics dictate the predominant form of interactions between biopolymers. The aforementioned parameters can be modulated for designing and controlling interactions among novel structures (McClements et al., 2009).

In aqueous dispersions, biopolymers experience two different types of demixing phenomena: segregative and associative phase separations (Fig. 12.1). These two phenomena are mainly dependent on the charges of polymers, which are influenced by external and internal factors (Fig. 12.2) (Nakagawa and Nagao, 2012; Schmitt and Turgeon, 2011; Schmitt et al., 2009; Turgeon and Laneuville, 2009). The resultant phase separation phenomenon is the consequence of long- or short-range (low- or high-amplitude) intermolecular interactions between contributors, as well as different affinities of biopolymers toward the solvent (Schmitt et al., 2009).

Cosolubility takes place in the presence of similarly charged species. This phenomenon can be observed in a biopolymer mixture containing very low concentrations or in mixed systems containing biopolymer subunits (i.e., monosaccharides and amino acids). When biopolymers carry electrical charges of a similar sign (or when one partner is neutral), macromolecules are phase separated into two separate immiscible biopolymer-rich phases (Fig. 12.1A). High concentrations and ionic strengths are essential to observe this phenomenon (also known as thermodynamic incompatibility) (Dickinson, 1998; Doublier et al., 2000; Goh et al., 2009; Turgeon and Laneuville, 2009; Turgeon et al., 2003; Ye, 2008). Under these conditions, the net interactions between biopolymers are repulsive. These interactions, which mainly result from excluded volume effects, are nonspecific and generally of temporary duration (Dickinson, 1998). Individual mixtures of structurally different polysaccharides or proteins may also undergo segregative phase separation (e.g., mixture of alcohol-soluble prolamins and acid-soluble glutelins, branched and linear polysaccharides, and intact and heat-denatured forms of the same proteins) (Tolstoguzov, 2002).

When both partners carry electrical charges of the opposite sign, thermodynamic compatibility (also known as complex coacervation) occurs at a relatively low concentration (in most cases <4 wt.%) and low ionic strengths (<400 mM). The associative phase separation results in the spontaneous formation of an upper macromolecule-depleted (water-rich) phase (also called the equilibrium solution) together with a lower phase rich in complexed contributors (Doublier et al., 2000; Eleya and Turgeon, 2000). For attractive columbic interactions, the pH of solution should be between the isoelectric point (p*I*) of protein and the pK_a of polysaccharide. Electrostatic interactions are the major molecular forces present in the biopolymer-rich phase. In the bottom phase, the resultant structure derived from the complexed biopolymers can appear as coacervates, interpolymeric (insoluble) complexes (or precipitates), and electrostatic gels (Turgeon and Laneuville, 2009) depending on the biopolymer type (Fig. 12.1B). When one partner (particularly polysaccharide) is in an excess amount, soluble nanocomplexes can be obtained instead of precipitates (Fig. 12.1C).

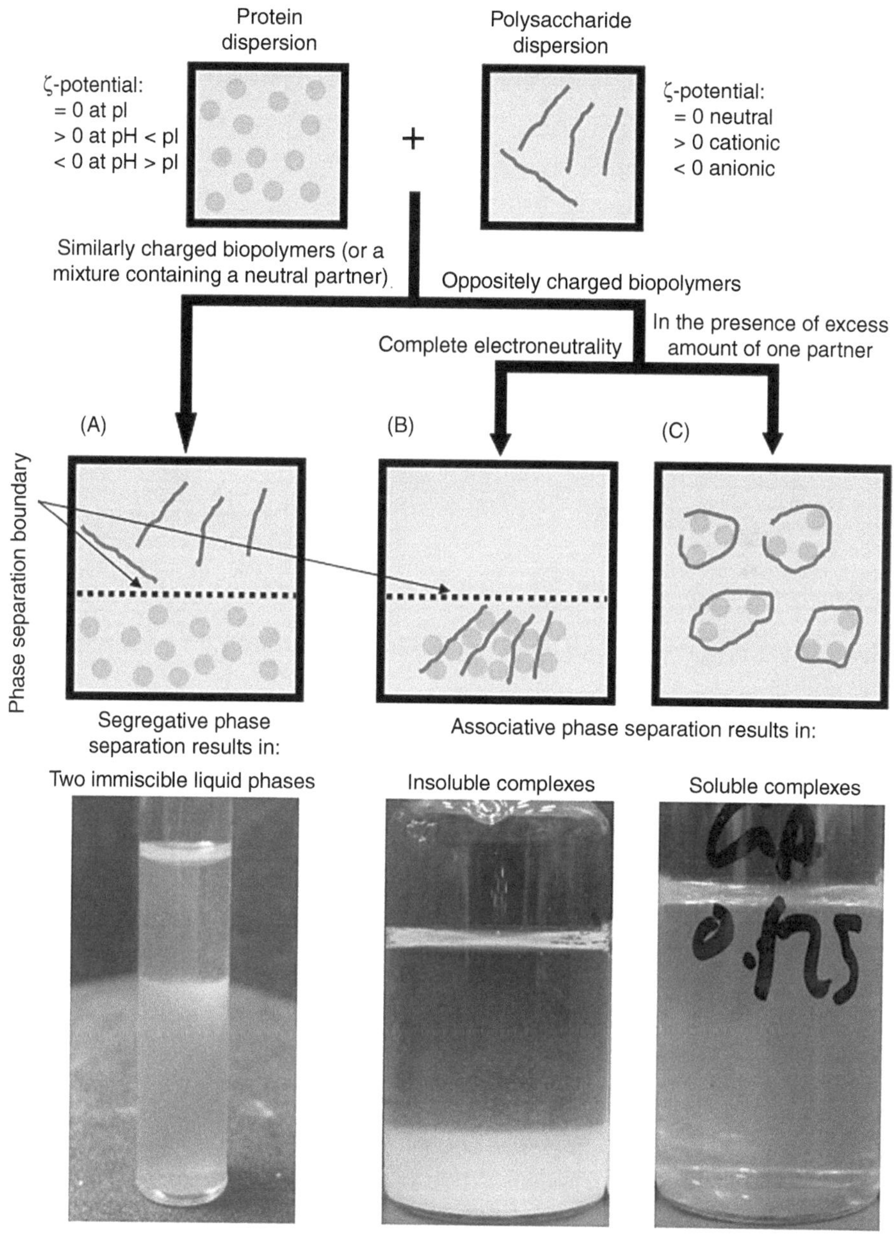

Figure 12.1 Two different types of phase separation phenomena occurring in a mixture of biopolymers (protein + polysaccharide).

Complex coacervation of various biopolymer pairs (especially those of protein and polysaccharide) under the influence of different factors has been extensively studied previously. The interested readers are referred to the excellent reviews available in this research topic for additional details (de Kruif et al., 2004; Kizilay et al., 2011;

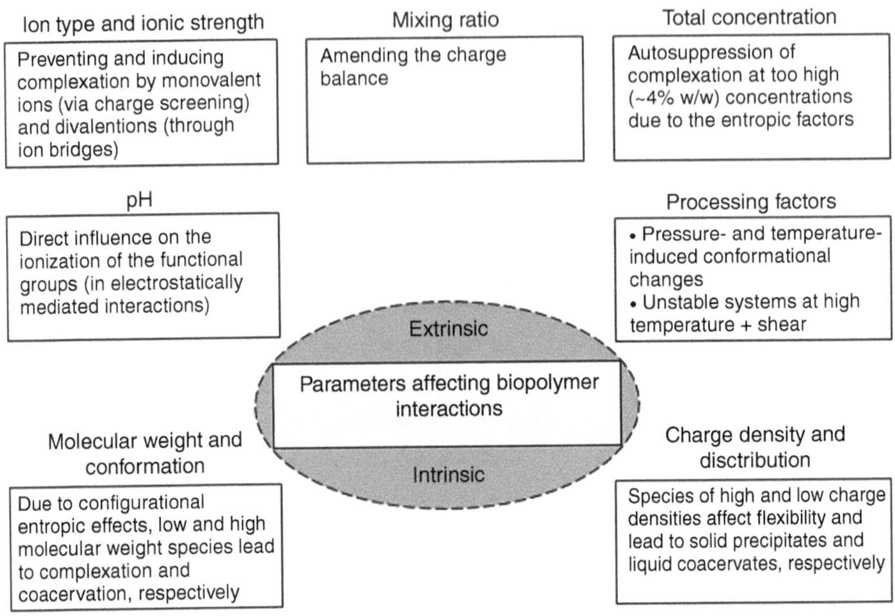

Figure 12.2 A brief review of the important parameters influencing the interaction of biopolymers.

Lapitsky, 2014; Rodríguez Patino and Pilosof, 2011; Schmitt and Turgeon, 2011; Turgeon and Laneuville, 2009; Turgeon et al., 2007; van der Sman, 2012).

12.3 Application of biopolymer complexes in nanoencapsulation technology

Generally, physical entrapment and chemical complexation of bioactive molecules (particularly hydrophobic ones) are the main strategies applied in the development of supramolecular biopolymer–based carrier systems (Matalanis et al., 2011). Actually, different types of biopolymer mixtures are capable of binding nutraceuticals of interest and creating molecular complexes. The active molecules can be encapsulated and protected using protein–polysaccharide, polysaccharide–polysaccharide, and protein–protein (nano)complexes.

12.3.1 Nanoencapsulation using structures derived from protein and polysaccharide interactions

Complexes and coacervates derived from protein and polysaccharide interactions are ideal tools in nanoencapsulation technology. A large spectrum of different combinations of proteins + polysaccharides applied for (nano)encapsulation of different components can be observed in Table 12.2 (showing milk proteins) and Table 12.3

Table 12.2 Summary of the milk protein + polysaccharide complexes/coacervates used for encapsulating different components and their remarkable results

Encapsulants	Encapsulated compounds	Remarkable results	References
WPI–CMC WPI–GA WPI–CS	Procyanidin-rich extract	• Average size of particles (μm): 10 • Polydispersity index: 0.8–1.2 • Encapsulation efficiency (%): 72–80 • Controlled release	Berendsen et al. (2015)
WPI–grafted dextran/ chondroitin sulfate (Electrostatic complexes)	Lutein	• Average size of spherical particles with a smooth surface (nm): 50–100 • Zeta potential (mV): −36.8 to +15.8 • Encapsulation efficiency (%): 94.07 • Loading content (% w/w): 8.03 • Stable nanoparticles at a wide pH range (1–8) and in the presence of high salt concentration (200 mM NaCl)	Dai et al. (2015)
WPI/GA	β-Carotene	• Average size of spherical particles with a smooth surface (μm): 149.7–160.9 • Encapsulation efficiency (%): 45.9–77.3 • Zeta potential (mV): −0.138 • Initial burst release followed by sustained release • Effective protection during storage	Jain et al. (2015)

WPI/beet pectin	Anthocyanin	• Average size of particles (nm): <200
		• Loading efficiency (%):
		  • 55 (Anthocyanin addition before heating the complex solution)
		  • 35 (Anthocyanin addition after heating the complex solution)
		• Negatively charged complexes at pH 4–8 became less anionic at lower pH values
		• Improved heat stability
		• Decreased antioxidant activity
			Arroyo-Maya and McClements (2015)

(Continued)

Table 12.2 Summary of the milk protein + polysaccharide complexes/coacervates used for encapsulating different components and their remarkable results (*cont.*)

Encapsulants	Encapsulated compounds	Remarkable results	References

WPI/GA	Probiotic bacteria	• Average size of particles (μm):
		  • 10–15 (at pH 4)
		  • 1–10 (at pH 2)
		• Encapsulation efficiency (%): >94.3
		• Enhanced viability of the microorganisms after being exposed to an environment of low pH (pH 2.0)
		• Survival in simulated gastric juice (%): >73
		• Viability after 60-day refrigerated storage at pH 4 (%): >86
			Bosnea et al. (2014)

WPI/sugar beet pectin	Caffeine	• Average size of spherical particles (nm): 200 • Polydispersity index: 0.40 • Physical entrapment within particles • Lower transition temperatures of conjugate as compared to the parent biopolymers as determined by DSC	Gazme and Madadlou (2014)
WPI/CMC	β-Pinene	• Average size of particles with a compact or a "spongy-like" structure ($D_{3,2}$, μm): 0.165–0.560 • Encapsulation efficiency (%): 13.05–36.90 • Loading capacity (%): 1.07–30.10	Koupantsis et al. (2014)

WPI/GA	Omega-3 fatty acids	• Optimum condition:
		  • pH: 3.75
		  • WPI/GA ratio: 3:1
		• Better stability against oxidation
		• Higher oil microencapsulation efficiency and lower surface oil content, making spray-dried solid microcapsules suitable for production
		• Micropore-free surface of spray-dried microcapsules
		• Microporous surface of freeze-dried microcapsules
			Eratte et al. (2014)

WPI/LMP	Thiamine	• Average size of particles (μm):
		  • Increased with decreasing protein/pectin ratio using postblending acidification method
		  • Increased with decreasing pH using preblending acidification method
		  • Remained constant with decreasing pH using postblending acidification method
		• Loading efficiency (%):
		  • Decreased with decreasing pH using postblending acidification method
		  • Had no specific pattern in preblending acidification
		• Encapsulation efficiency (%):
		  • 6.99 Using preblending acidification (pH 3.5)
		  • 4.50 Using postblending acidification (pH 3.0–3.5)
			Bédié et al. (2008)

WPC/MD	Folic acid	• Optimum formulation:
		  • Dispersed phase volume (%): 12
		  • Water to surfactant ratio: 0.9
		  • Folic acid content (mg/mL): 3
		• Final nanoemulsion droplet size (nm): <100
		• Encapsulation efficiency (%): > 86
			Assadpour et al. (2016)

WPC/pectin	Olive leaf phenolic compounds	• Average droplet size after one day of production (nm):
		  • W/O/W emulsions stabilized by WPC: 675.7
		  • W/O/W emulsions stabilized by WPC–pectin: 1443
		• Average droplet size after a 20-day storage (nm):
		  • W/O/W emulsions stabilized by WPC: 347.7
		  • W/O/W emulsions stabilized by WPC–pectin: 1992.4
		• Encapsulation efficiency after 1 day of production (%):
		  • W/O/W emulsions stabilized by WPC: 93.34
		  • W/O/W emulsions stabilized by WPC–pectin: 96.64
		• Encapsulation efficiency after 20-day storage (%):
		  • W/O/W emulsions stabilized by WPC: 72.73
		  • W/O/W emulsions stabilized by WPC–pectin: 88.81
			Mohammadi et al. (2016a,b)

(*Continued*)

Table 12.2 Summary of the milk protein + polysaccharide complexes/coacervates used for encapsulating different components and their remarkable results (*cont.*)

Encapsulants	Encapsulated compounds	Remarkable results	References
		• Controlled release in WPC–pectin emulsions and fastest release rate in WPC emulsion • Achieving higher antioxidant activity of encapsulated olive leave extract in soybean oil with increasing solubility and controlled release of olive leaf phenolic compounds through their nanoencapsulation	

WPC/pectin
WPC/MD
	Saffron extract (crocin, picrocrocin, and safranal)	• Emulsion droplet size (Z-average, nm):
		  • 536.3 (WPC/pectin; sequential adsorption)
		  • 482.3 (WPC/pectin; simultaneous adsorption)
		• Encapsulation efficiency (%):
		  • For picrocrocin:
		    – 56.51 (MD/WPC matrix)
		    – 96.82 (WPC/pectin; sequential adsorption)
		    – 95.96 (WPC/pectin; simultaneous adsorption)
		  • For saffranal:
		    – 51.57 (MD/WPC matrix)
		    – 95.94(WPC/pectin; sequential adsorption)
		    – 93.02 (WPC/pectin; simultaneous adsorption)
		  • For crocin:
		    – 62.55 (MD/WPC matrix)
		    – 96.66 (WPC/pectin; sequential adsorption)
		    – 95.96 (WPC/pectin; simultaneous adsorption)
		• Encapsulated powder morphology:
		  • Dent and wrinkle surface (MD/WPC matrix)
		  • Spherical particles with a smooth surface (WPC/pectin)
			Esfanjani et al. (2015, 2017)

WPC/mesquite gum
WPC/GA
	Chia essential oil	• Average size of dimpled spherical crack-free particles ($D_{4,3}$, μm):
		  • 2.32–3.35 (before spray-drying)
		  • 13.17–28.20 (after spray-drying)
		• Encapsulation efficiency (%): >70 (after drying)
		• Droplet coalescence rate ($s^{-1} \times 10^{-8}$): 1.54–7.66
			Rodea-González et al. (2012)

Cyclodextrin/WPC/xanthan gum	Fish oil	• Average size of particles (nm): 120–700 • Surface oil (%): 8.04–40.53 • Total oil (%): 43.79–53.37 • Encapsulation efficiency (%): 7.45–84.03 • Loading efficiency (%): 3.26–42.30 • 30% Decrease in odor intensity of the encapsulated fish oil compared to the original value • Improved oxidative stability	Na et al. (2011)
Caseinate/GE	Polyunsaturated lipids	• Average size of irregularly shaped particles (μm): 7 • Temperature-triggered release of bioactive agents under oral processing • Stable filled hydrogel particles at pH 5.8	Zhang et al. (2015)
Sodium caseinate/GA (Multilayered interfacial membranes)	EPA/DHA (Fish oil)	• Average size of aggregated irregularly shaped particles (nm): 232.3 • Encapsulation efficiency (%): 78.88 • Bioaccessibility within enriched fruit juice (%): EPA (56.16), DHA (36.25), and EPA + DHA (47.37)	Ilyasoglu and El (2014)

TG–cross-linked LMP/caseinate	Polyunsaturated lipids	• Average size of particles ($D_{4,3}$, μm):
		  • 4.6
		  • 4.3 (cross-linked)
		• Zeta potential (mV):
		  • −37.2
		  • −36.7(cross-linked)
		• Stability toward lipid oxidation as compared to conventional emulsions
		• Release after ingestion
			Zhang et al. (2014a)

(*Continued*)

Table 12.2 Summary of the milk protein + polysaccharide complexes/coacervates used for encapsulating different components and their remarkable results (*cont.*)

Encapsulants	Encapsulated compounds	Remarkable results	References
Sodium caseinate/ CMC	β-Pinene	• Average size of particles with a compact or a "spongy-like" structure ($D_{3,2}$, μm): 0.213–0.408 • Encapsulation efficiency (%): 17.80–29.02 • Loading capacity (%): 2.15–25.09	Koupantsis et al. (2014)
Casein/pectin (Complex coacervation)	Acetaminophen	• Average size of spherical particles (μm): 4.68–5.18 • Encapsulation efficiency (%): 32.9–57.0 • Controlled release • Larger release in simulated gastric fluid than in simulated intestinal fluid • Resistant to the spray-drying process	Baracat et al. (2012)

Casein-*grafted*-dextran (MR)	β-Carotene	• Average size of spherical particles (nm): 200
		• Encapsulation efficiency (%):
		  • 54 (Weight ratio of β-carotene in feed to casein in feed: 1:10)
		  • 10.5 (Weight ratio of β-carotene to casein in feed: 1:1)
		  • 4.7 (Weight ratio of β-carotene to casein in feed: 5:1)
		• Loading efficiency (%):
		  • 5.4 (Weight ratio of β-carotene in feed to casein in feed: 1:10)
		  • 10.5 (Weight ratio of β-carotene to casein in feed: 1:1)
		  • 18.8 (Weight ratio of β-carotene to casein in feed: 5:1)
		• Stability in aqueous solution against dilution, pH and ionic strength changes, and $FeCl_3$-induced oxidation and longtime storage
			Pan et al. (2007)

β-Lg/alginate	β-Carotene, folic acid, curcumin, and ergocalciferol	• Intensity-weighted average size of particles (nm):
		  • 278 nm for curcumin-soluble complexes
		  • >1000 for folic acid–soluble complexes
		• Electrophoretic mobility (10^{-8} m^2/Vs):
		  • −4.30 for curcumin-soluble complexes
		  • −3.95 for folic acid–soluble complexes
		• β-Carotene revealed the highest binding affinity and the lowest binding stoichiometry at both pH values (4 and 7)
		• Colloidal stability of low water–soluble nutraceuticals even after 30 days of production
			Hosseini et al. (2015)

β-Lg–AG	Sunflower oil	• Increasing emulsion stability with ultrasound treatment
		• Optimum conditions to improve the emulsion's stability:
		  • AG/β-lg ratio of 12.5:1
		  • 10-min ultrasound treatment with an amplitude of 25%
		  • Thermal treatment: 50°C
		  • pH 3
		• The higher the AG/ β-lg ratio, the bigger the particle size and lower the zeta potential
		• The more the ultrasound amplitude and time, the less the emulsion's particle size and zeta potential
			Shamsara et al. (2015)

β-Lg/GA	Ubiquinone (coenzyme Q_{10})	• Average size of spherical particles (μm):
		  • 3.6 (Core/wall ratio of 1)
		  • 14.4 (Core/wall ratio of 1.875)
		• Postdrying encapsulation efficiency (%): 40.01–92.45
		• Predrying encapsulation efficiency (%): 89.17–99.01
		• Oil loading (%): 62.5–1250
		• Controlled release
			Ahmadi et al. (2015)

β-Lg/HMP	Linoleic acid	• Average size of particles (nm):
		  • 316 (Protein/polysaccharide ratio of 1)
		  • 385 (Protein/polysaccharide ratio of 2)
		• Enhanced protein–ligand complexation after a moderate heat treatment (60°C, 10 min)
			Perez et al. (2015)

(*Continued*)

Table 12.2 Summary of the milk protein + polysaccharide complexes/coacervates used for encapsulating different components and their remarkable results (*cont.*)

Encapsulants	Encapsulated compounds	Remarkable results	References

CS/denatured β-lg microcapsules	Brilliant blue	• Average size of spherical particles (μm):
		  • 20–30 (2% CS)
		  • 25–40 (4% CS)
		• Zeta potential (mV):
		  • −23.18 (2% CS)
		  • −30.16 (4% CS)
		• Encapsulation efficiency (%):
		  • 39.51 (2% CS)
		  • 38.61 (4% CS)
		• Protein coating resistance to simulated stomach conditions
		• Hydrolysis of protein coating under simulated small intestine conditions resulted in the release of dye-loaded CS into the intestinal fluid at a constant rate
			Lee et al. (2012)

β-Lg/LMP	Vitamin D_2	• Average size of particles (nm): 50–70 • Appropriate for enrichment of clear liquid foods • High affinity of vitamin to the nanoparticles • Increased protection against oxidation as compared to the protection provided by protein alone	Ron et al. (2010)
β-Lg/LMP	DHA	• Average size of spherical particles (nm): 100 • An increase in the absolute value of zeta potentials (mV) with increasing pectin concentration • Transparent dispersions • High protection against chemical degradation during an accelerated shelf-life stress test	Zimet and Livney (2009)

AG, Apricot gum; CMC, carboxymethyl cellulose; CS, chitosan; DHA, docosahexaenoic acid; DSC, differential scanning calorimetry; EPA, eicosapentaenoic acid; GA, gum Arabic; GE, gelatin; HMP, high-methoxyl pectin; β-lg, β-lactoglobulin; LMP, low-methoxyl pectin; MD, maltodextrin; MP, whey microparticles; W/O/W, water in oil in water; WPC, whey protein concentrate; WPI, whey protein isolate;

Table 12.3 Summary of the nonmilk protein + polysaccharide complexes/coacervates used for encapsulating different components and their remarkable results

Encapsulants	Encapsulated compounds	Remarkable results	References

GE/GA
GE/GA cross-linked with TG
MR-modified GE/GA
	Stearidonic acid soybean oil	• Volume-weighted average size of spheroid particles (μm, dependent on the homogenization level):
		  • 56.4–296.5 for GE/GA
		  • 55.4–287.4 for TG–cross-linked GE/GA
		  • 37.9–286.6 for MR-modified GE/GA
		• Encapsulation efficiency (%): >99
		• Enhanced antioxidant activity using MR
		• Reduced antioxidant activity after cross-linking by TG
		• Minimum lipid oxidation and maximum thermal stability in MR-modified microcapsules
			Ifeduba and Akoh (2015)

GE/GA	Poppy seed oil	• Average size of spherical particles with a smooth surface (μm): <5 • Encapsulation efficiency (%): 76.8 • Loading capacity (%): 18.2 • Enhanced storage stability	Yang et al. (2015)
Imprinted-like biopolymer micelles developed through the coassembly of GE–dextran conjugates using hydrophilic tea polyphenol followed by cross-linking with genipin	Curcumin	• Average size of spherical particles (nm): 160–220 • Encapsulation efficiency (%): >99 • 5×10^4-fold enhanced aqueous solubility • Controlled release • Increased cellular uptake • Improved cellular toxicity in HeLa cells • Improved oral bioavailability	Zhang et al. (2014c)
Genipin-cross-linked GE/GA	Mustard seed essential oil	• Average size of spherical particles (μm): 5–10 • Strong chemical stability under different conditions of relative humidity and temperature • Controlled release	Peng et al. (2014)

(*Continued*)

Table 12.3 **Summary of the nonmilk protein + polysaccharide complexes/coacervates used for encapsulating different components and their remarkable results (*cont.*)**

Encapsulants	Encapsulated compounds	Remarkable results	References

GE/GA	Xylitol and menthol	• Average size of spherical particles (μm):
		  • 98.57[a]
		  • 98.99[b]
		• Encapsulation efficiency (%):
		  • Menthol:
		    – 93.75 [a]
		    – 101.62[b]
		  • Xylitol:
		    – 41.85[a]
		    – 45.54[b]
		• Controlled release
			Santos et al. (2014)

GE/GA	Aspartame	• Average size of spherical particles (μm): 84.22–102.38 • Encapsulation efficiency (%): 45.15–71.70 • Hygroscopicity (g/100 g): 10.86–38.13 • Solubility (%): 9.46–95.70 • Controlled release characteristic	Rocha-Selmi et al. (2013)
GE/GA	Ascorbic acid	• Average size of spherical particles (μm): 51.67–83.82 • Encapsulation efficiency (%): >90 • Controlled release characteristic • Decreased hygroscopicity and solubility • Increased stability	Comunian et al. (2013)

GE/GA (Complex coacervation)	Turmeric oleoresin	• Average size of spherical multinucleated particles (μm):
		  • 46 (2.5% total biopolymer concentration and core:wall material of 1)
		  • 53 (5% total biopolymer concentration and core:wall material of 0.5)
		• Encapsulation efficiency (%): 49–73
		• Coacervation yield (%): 80.1–95.2
		• Solubility (%): 86.5–96.8
		• Increased stability to light
			Zuanon et al. (2013)

GE/dextran (Complex coacervate core micelles synthesized using MR)	Tea polyphenol	• Average size of spherical particles (nm): 86 • Zeta potential (mV): ~0 at pH 5 and more negative at pH > 7 • Encapsulation efficiency (%): 70 • Loading capacity (%): 360 • In vitro sustained release • Cytotoxicity in MCF-7 cells	Zhou et al. (2012)
Glutaraldehyde-cross-linked GE (type A)/sodium alginate	Olive oil	• An increase in the average size of free flowing spherical particles by increasing biopolymer concentration • Oil load (%): 26.84–187.97[c] • Oil content (%): 13.66–58.33[c] • Encapsulation efficiency (%): 62.54–89.37[c] • Improved thermal stability after cross-linking	Devi et al. (2012)

GE/GA	β-Carotene	• Average size of particles (nm): 50–4,000
		• Core–shell nano-microparticles, surrounded by polymeric membranes
		• The membrane properties (in frozen emulsion with a pH of 4.7):
		  • Maximum encapsulation yield in a cooling rate of −1.0°C/min
		  • Limited release rate in cooling rate of −2.0°C/min
			Nakagawa and Nagao (2012)

(Continued)

Table 12.3 Summary of the nonmilk protein + polysaccharide complexes/coacervates used for encapsulating different components and their remarkable results (*cont.*)

Encapsulants	Encapsulated compounds	Remarkable results	References
Glutaraldehyde–cross-linked GE (type A)/sodium CMC	Neem seed oil	• An increase in the average size of spherical microparticles by an increase in the biopolymer concentration • Oil load (%): 36.52–247.11[d] • Oil content (%): 20–59[d] • Encapsulation efficiency (%): 71.63–96.90[d] • Improved thermal stability after cross-linking	Devi and Maji (2011)

Glutaraldehyde–cross-linked GA/GE	Capsaicin	• Average size of rough surface spherical particles (nm): 100
		• Encapsulation efficiency (%): 80.8
		• Optimum conditions
		  • Viscosity (cP): 15–20
		  • Cross-linking time (min): 40–80
		  • Agitation rate (rpm): 15,000
		• Increased melting point
		• Improved thermal stability of capsaicin
			Jincheng et al. (2010)

GE/GA	β-Cypermethrin	• Average size of spherical particles (nm): <100 • Zeta potential (mV): −18 to −28 • Encapsulation efficiency (%): >60 • Polydispersity index: <0.14	Zhu et al. (2009)

Glutaraldehyde and TG–cross-linked GE/GA	Vetiver essential oil (containing KD as the fluorescent compound)	• Average size of particles (μm):
		  • 39 and 33 for wet and freeze-dried microparticles, respectively (cross-linked with GA)
		  • 52 and 45 for wet and freeze-dried microparticles, respectively (cross-linked with TG)
		• Faster release of KD from wet microparticles than from freeze-dried ones
		• Synergistic effect of the drying process with chemical cross-linking
		• More efficient cross-linking with GA
			Prata et al. (2008)

CMCS/SPI (Using simple ionic gelation method)	Vitamin D_3	• Average size of particles (nm): 162–243
		• Zeta potential (mV): −10 to −20
		• CMCS/SPI complex (as compared to CMCS):
		  • Decreased Ca^{2+} requirement
		  • Better particle formation over a broad range of pH
		  • Increased loading efficiency (6.06%)
		  • Increased encapsulation efficiency (up to 96.8%)
		• CMCS/SPI complex (as compared to SPI):
		  • Lower (42.3% vs. 86.1%) release of vitamin D_3 in simulated gastric fluid
		  • Higher (36.0% vs. 8.2%) release under simulated intestinal condition
			Teng et al. (2013)

TG–cross-linked SPI/GA	Omega-3	• Average size of particles with a wrinkled surface (μm): 44.27–54.85[e] • Encapsulation efficiency (%): 19.81–70.77[e]	de Conto et al. (2013)
SPI/GA	Sweet orange oil	• Average size of surface hole–free spherical particles (μm): 7.57 • Optimum conditions (pH 4, SPI/GA ratio: 1, core material loading for the highest encapsulation efficiency, and yield 10%) • Flavor-retention capability (%): up to 90.97 for D-limonene • A decrease in coacervation by an increase in the ionic strength	Jun-xia et al. (2011)
SPI/pectin	Propolis extract	• Average size of spherical particles (μm): 10–90 • Encapsulation efficiency (%): 72.01–66.12 (for formulations with 2.5 and 5.0 g/100 mL of colloids) • Hygroscopicity ($g_{H_2O}/100\ g_{sample}$): 33.46–34.10 • Antimicrobial activity against *Staphylococcus aureus* • Insignificant effect of coacervation and freeze-drying processes on antioxidant activity	Nori et al. (2011)

(*Continued*)

Table 12.3 Summary of the nonmilk protein + polysaccharide complexes/coacervates used for encapsulating different components and their remarkable results (*cont.*)

Encapsulants	Encapsulated compounds	Remarkable results	References
SPI/pectin	Casein hydrolysate	• Average size of spherical particles (μm): 16.24–24.12 • Encapsulation efficiency (%): 78.8–91.62 • Less hygroscopicity ($g_{H_2O}/100\ g_{sample}$): 20.08–26.0 versus 53 for free hydrolysate • Decreased bitter taste	Mendanha et al. (2009)
Alginate–protein (whey proteins, BSA, calcium caseinate, soy proteins, and hemp proteins); microcapsules coated with CS and pectin	Green tea flavan-3-ols and caffeine	• Average particle size ($D_{0.5}$ or median, μm): 573.09–1,124.36 • Encapsulation efficiency of total phenols (%): 60.78–80.17 • Encapsulation efficiency of flavan-3-ols (%): 42.30–82.75 • Caffeine content (mg/g beads): 1.20–11.04 • Burst release of polyphenols (over 50% in the first 5–10 min) followed by sustained release up to 120 min	Belščak-Cvitanović et al. (2015)

Alginate/pectin mixtures coated with whey and egg white proteins	Sunflower oil	• Average size of particles (μm): 259–287
		• Zeta potential (mV): −63.2 to 28.5
		• Polydispersity index: 1.5–2
		• Encapsulation efficiency (%):
		  • 87.6 at pH 3.5
		  • 90.8 at pH 3.75
			Aguilar et al. (2015)

Lipoproteins/CMC	Doxorubicin	• Average size of spherical particles with a smooth surface (μm): 90 • Encapsulation efficiency (%): 98 • Polydispersity index: <0.3 • Zeta potential (mV): −35 • Effective in vitro toxicity in cancer cells	He et al. (2015)

Ovalbumin fibrils/HMP	Limonene	• Average size of spherical particles (μm): 5–6
		• Zeta potential (mV):
		  • +30 (Odd number of layers with protein fibrils as an outer layer)
		  • −20 (Even number of layers with polysaccharide as an outer layer)
		• 3-Fold decreased release rate by increased number of shell layers from four to eight
		• The more layers the shell consists of, the more resistance against the physical drying process
			Humblet-Hua et al. (2011)

CMCS/zein	Indole-3-carbinol	• Average size of spherical particles of uniform surface (nm): 113.5 • Zeta potential (mV): −19.53 • Encapsulation efficiency (%): 77.79 • Polydispersity index: 0.19 • Improved protection against thermal- and ultraviolet-mediated degradation	Luo et al. (2013)
CMCS/zein	3,30-Diindolylmethane	• Average size of spherical particles of uniform surface (nm): 89.1 • Zeta potential (mV): −19.80 • Encapsulation efficiency (%): 78.08 • Polydispersity index: 0.19 • Improved protection against thermal- and ultraviolet-mediated degradation	Luo et al. (2013)
Glutaraldehyde–cross-linked collagen hydrolysate/CS	Lavender oil	• Encapsulation efficiency (%): 36.84–73.73 • Loading efficiency (%): 77.45–615.17 • Microcapsules of spherical shape • No significant interaction between the complex and the core • Initial burst release of about 25%–30% of core (up to 1 h), followed by a gradual and sustained release phase for the following 72 h	Ocak (2012)

(Continued)

Table 12.3 Summary of the nonmilk protein + polysaccharide complexes/coacervates used for encapsulating different components and their remarkable results (*cont.*)

Encapsulants	Encapsulated compounds	Remarkable results	References

Amaranth protein isolate/ pullulan	Quercetin and ferulic acid	• Encapsulation efficiency (%):
		  • 93.6 for quercetin
		  • 83.7 for ferulic acid
		• Loading efficiency (%): 20
		• Smooth and defect-free fibers
		• Controlled release during an in vitro digestion
		• Decreased thermal stability
		• Improved antioxidant capacity
			Aceituno-Medina et al. (2015)

γ-Polyglutamic acid/CS	Doxorubicin (An anticancer drug)	• Average size of particles (nm): 150–630 • Zeta potential (mV): 40.3–51.3 • Encapsulation efficiency (%): 45.5–69.8 • Considerable antiproliferative activity against the human cancer cell line	Hellmers et al. (2013)
Pea protein isolates/MD	Ascorbic acid	• Average size of spherical and shriveled particles (μm): 7 • Ligand retention after the drying process (%): 69	Pereira et al. (2009)
Cowpea protein isolates/MD	Ascorbic acid	• Average size of spherical and shriveled particles (μm): 8 • Ligand retention after drying process (%): 66	Pereira et al. (2009)

BSA, Bovine serum albumin; CMCS, carboxymethyl chitosan; GE, gelatin; KD, khusimyl dansylate; SPI, soy protein isolate.
[a]Xylitol and 2 g of menthol/100 g of a simple emulsion.
[b]Xylitol and 6 g of menthol/100 g of a simple emulsion.
[c]Different combinations of gelatin (1.4–2.8 g), sodium alginate (0.4–1.2 g), water (180 mL), olive oil (1.0–7.0 g), glutaraldehyde (1.25–5.0 mmol), and temperature (60 ± 1°C).
[d]Different combinations of gelatin (0.5–4.0 g), sodium alginate (0.214–1.714 g), water (200 mL), olive oil (1.5–7.0 g), glutaraldehyde (5.0–20.0 mmol), and temperature (45 ± 1°C).
[e]Dependent on the SPI:GA ratio and total concentration, core material, and TG concentrations.

(for other proteins). Among different structures resulting from protein–polysaccharide complexation, the coacervates are superior to others for encapsulation technology. Coacervates have great interfacial properties resulting in utilization for taste masking and controlled release purposes due to their ability to form a protective film (Ducel et al., 2008; Turgeon and Laneuville, 2009). It should be noted that coacervates and complexes are structurally different from each other. However, in literature, the words "complexes" and "coacervates" are sometimes used interchangeably. Structurally, coacervates are in a liquid form, which may then separate as dense and structured liquid spherical droplets (vesicles); interpolymeric complexes (insoluble complexes or amorphous coprecipitates) exhibit fractal aggregates; and gels are 3D interconnected complexes (Dickinson, 1998; Ducel et al., 2008; Turgeon and Laneuville, 2009). Indeed, the structure determination of biopolymer complex coacervation at various hierarchical scales, including molecular (nano-), meso-, and large scales, is one of the most challenging and exciting aspects of these demixing phenomena. Due to the dispersing of coacervate droplets in a continuous phase, systems capable of forming coacervates have sometimes been considered as water in water (W/W) emulsions (Ducel et al., 2008; Turgeon and Laneuville, 2009).

For each binary mixed system composed of various proteins and polysaccharides, there are some unique factors dictating the formation either coprecipitates or coacervates. An increase in the biopolymer charge density and a decrease in the backbone flexibility (indirectly influenced by the charge density) generally result in the formation of complexes instead of coacervates (Turgeon et al., 2007). This phenomenon can be observed in binary mixtures of globular proteins and some highly charged and stiff anionic polysaccharides, such as sodium alginate, carrageenans, and carboxymethyl cellulose (CMC). On the other hand, low and intermediate charge densities, as well as less stiffness lead to the formation of spherical coacervates, as observed in biopolymer mixtures containing gelatin (GE), acacia gum, hyaluronic acid, and some types of pectins (Turgeon and Laneuville, 2009; Turgeon et al., 2007). Spontaneous complex coacervation (having negative Gibbs free energy) may be enthalpically or entropically driven. Therefore, the resultant final structure is influenced by the strength of the interaction. Interactions of a higher strength result in a tight binding of spherical protein molecules to the linear backbones of polysaccharides, inducing extensive release of counterions and hydrating water molecules, both present in the close proximity of the biopolymer molecules, into the bulk phase (entropic effects) (Turgeon and Laneuville, 2009). The extensive release phenomenon results in the dewatering of bio-macromolecules to the point that complexation (but not coacervation) becomes predominant (Kayitmazer et al., 2007). Moreover, Kumar et al. (2007) reported that larger molecular weight is required for the coacervate formation.

Yang et al. (2015) studied the stability of spray-dried microcapsules prepared from complex coacervation between gum Arabic (GA) and GE (as wall materials) in poppy seed oil. The optimum ratio of oil to GA/GE mixture was 1:3. Encapsulation and loading efficiencies of around 77 and 18% were obtained using spherical microcapsules, respectively. Moreover, microcapsules could enhance the storage stability of oil. As mentioned previously, the dominant interaction present within this type of biopolymer complexes is electrostatic in nature; however, other interactions may also be involved in complex formation. Ifeduba and Akoh (2015) compared the effectiveness of two different encapsulation systems, namely classical GE–GA microcapsules (prepared by complex coacervation technique) and a system prepared by Maillard-type reactions (MR) (through dry heating under controlled conditions), for the encapsulation of stearidonic (moroctic) acid soybean oil. After loading, a portion of microcapsules was cross-linked using transglutaminase (TG). Different types of microcapsules were utilized to fortify yogurt as a model food. Microcapsules prepared from MRs showed the highest thermal and oxidative stabilities during processing and storage, demonstrating the importance of binding strength, as well as antioxidant components developed during the MR (Ifeduba and Akoh, 2015). However, it should be noted that MR during conjugation of biopolymers may result in the formation of poisonous compounds.

Developing multilayers is an efficient strategy to improve the stability and to prevent the burst release of bioactive components. Multilayers are synthesized through a process called layer-by-layer self-assembly. Humblet-Hua et al. (2011) utilized alternating layers of ovalbumin and high-methoxyl pectin (HMP) for microcapsules preparation. A 3-time decrease in the release rate was observed due to an increase in the shell thickness through an increase in the number of layers.

The weight ratio of protein to polysaccharide (or vice versa) is a significant factor in determining the morphological characteristics of the resultant complex. Koupantsis et al. (2014) reported that the microstructure of milk protein–CMC complexes were noticeably influenced by the biopolymer mixing ratio; being compact when the ratio was low and spongy like when it was high. Irrespective of the systems under study, the stability provided by the higher protein-to-polysaccharide weight ratios is not generally long lasting, resulting in coacervate growth and deformation. In other words, the coalescence of coacervates results in the formation of specific structures of high sedimentation affinity (Sanchez et al., 2002). Precipitation ruins the visual appearance of the final product (particularly of liquid-type foods) during its shelf life.

In the recent years, many efforts have been made to benefit from the potential of soluble (or insoluble) complexes for the encapsulation of bioactives with different hydrophobicities (Bédié et al., 2008; Fioramonti et al., 2014; Hosseini et al., 2015; Nori et al., 2011; Perez et al., 2015; Ron et al., 2010; Zimet and Livney, 2009). As mentioned previously (Section 12.2), proteins and ionic polysaccharides may form soluble nanocomplexes in the presence of excess amounts of one contributor and at pH levels corresponding to an electrostatic attraction between them (polysaccharide $pK_a < pH <$ protein pI). Zimet and Livney (2009) reported that the ω−3 fatty acid, docosahexaenoic acid (DHA), was entrapped within the supramolecular structures arising from the electrostatic interaction between bovine β-lg and low-methoxyl pectin (LMP). The oxidative stability of DHA increased after encapsulation in protein nanocarriers. Moreover, an overprotection against oxidative degradation was obtained after the electrostatic deposition of LMP onto protein nanocarriers (Zimet and Livney, 2009). This research team also studied the bioprotection conferred to ergocalciferol after entrapment within either β-lg nanovehicles or the same binary complexes. Similarly, the provided protection by soluble mixed biopolymer nanocomplexes was better than that provided by single complexes of vitamin D_2 residing in the protein β-barrel (Ron et al., 2010).

Hosseini et al. (2015) utilized the intrinsic transporting characteristics of β-lg toward some hydrophobic and hydrophilic bioactives to make green-type carrier systems, which could be used in fortification of liquid and clear acidic food products. A secondary outer layer of sodium alginate was placed onto the surface of protein–ligand complexes via Coulombic interactions. The complexation between β-lg and four (β-carotene, vitamin D_2, curcumin, and folic acid) nutraceutical models, as a function of nutraceutical type and pH, was confirmed using fluorimetry studies. Proton nuclear magnetic resonance spectroscopy provided additional information on the binding of hydrophilic ligands to the protein. The findings obtained in this study resulted in the fabrication of nanometric encapsulation systems for the entrapment of bioactives with different hydrophilicities and the subsequent fortification of clear, acidic beverages (Hosseini et al., 2015).

Linoleic acid as a model polyunsaturated fatty acid was successfully entrapped within biopolymer nanoparticles arising from electrostatic interactions of HMP and bovine β-lg (Perez et al., 2015). An improved protein–ligand complexation has been reported after applying a mild thermal processing at 60°C for 10 min (Perez et al., 2015). Recently, TEM micrographs have surprisingly confirmed the deposition of anionic sodium alginate around the cationic β-lg (i.e., the formation of core–shell nanostructures) (Fig. 12.3) (Mirpoor, 2016).

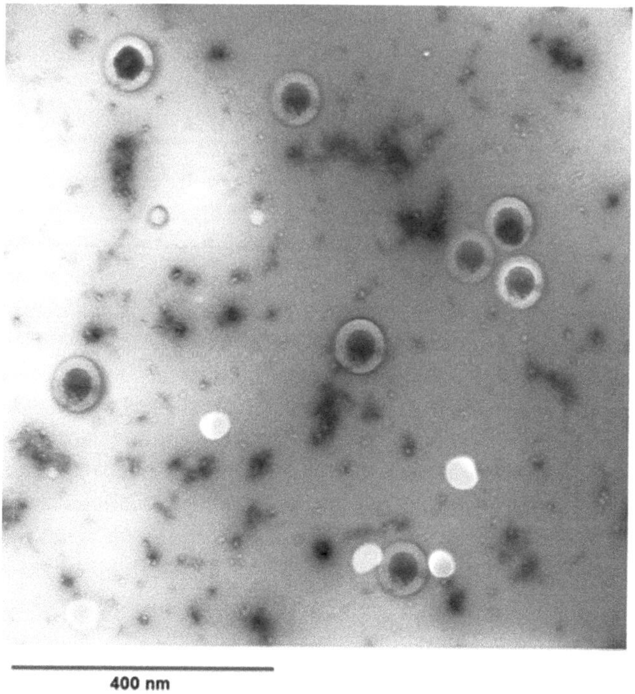

Figure 12.3 Core–shell nanostructures arising from the β-lg (core)–sodium alginate (shell) electrostatic interaction.

Different properties of curcumin were studied after entrapment within the electrostatically complexed bovine serum albumin (BSA) and ι-carrageenan (Yang et al., 2013). The authors concluded that the complex coacervation stabilized the tertiary structure of BSA and, therefore, provided domains of more hydrophobicity for curcumin binding. The binding stoichiometry of curcumin was not affected by the carrier type, suggesting that the anionic polysaccharide was located at the surface of complexes. The presence of the polysaccharide at the outer surface did not disturb the hydrophobic binding sites of the protein (Yang et al., 2013).

12.3.2 Nanoencapsulation using polysaccharide + polysaccharide complexes

Oppositely charged polysaccharides can be used for developing mixed-type biopolymer complexes to encapsulate different functional components (Table 12.4). Non-neutral polysaccharides are almost anionic with some exceptions. Therefore, cationic chitosan (CS) is mainly the ubiquitous biopolymer within complexes, resulting from polysaccharide + polysaccharide electrostatic interactions. CS is a natural, nontoxic, biodegradable, biocompatible, and mucoadhesive cationic amino polysaccharide ($pK_a \approx 6.5$) produced from chitin after partial deacetylation in an alkaline condition (Martins et al., 2012). Among different anionic polysaccharides, sodium alginate (because of having a standard structure, as well as gelation characteristics in the presence of calcium ions) has been widely used for complexation with CS. Alginate is a block copolymer composed of β-D-mannuronate and α-L-guluronate residues (Draget et al., 2006). Similarly to the protein + polysaccharide systems, various internal and external parameters (as mentioned in Section 12.2) are involved in polysaccharide + polysaccharide complexation.

Rahaiee et al. (2015) studied the stability of crocin, a water-soluble carotenoid present in saffron, encapsulated within CS–sodium alginate nanoparticles. Hydrogen bonding was the major driving force for crocin binding to mixed polysaccharides, as confirmed by FTIR spectroscopy. The encapsulation efficiency and loading capacity were reported as 38.16 and 30.96%, respectively. Nanoencapsulation enhanced crocin stability under unfavorable environmental conditions (Rahaiee et al., 2015).

Due to some drawbacks associated with sodium alginate–CS nanoparticles, Nagarwal et al. (2012) first developed sodium alginate–CS nanoparticles and then coated them with CS to nanoencapsulate fluorouracil applied in ocular bioavailability. Ionic gelation method was used to produce alginate–CS nanoparticles followed by suspension in a CS solution. Particle size and morphological characteristics of nanoparticles were significantly influenced by the second step. Moreover, the coating process enhanced mucoadhesiveness and, hence, bioavailability (Nagarwal et al., 2012).

To have sufficient bioavailability, the digestion rate is of importance. Li and McClements (2011) studied the in vitro digestibility (monitored by a pH-stat method) of lipids transformed into filled hydrogels. A protein-stabilized oil-in-water, (O/W) double-layer emulsion was first prepared and then entrapped within coacervates of cationic CS/anionic calcium alginate. Emulsion-filled hydrogels were physically stable at

Table 12.4 Summary of the polysaccharide + polysaccharide complexes/coacervates used for encapsulating different components and their remarkable results

Encapsulants	Encapsulated compounds	Remarkable results	References
CS/alginate	Curcumin diethyl disuccinate	• 0.15:1 as optimum CS/alginate mass ratio • No chemical reaction between ligand and polymers • Higher cellular uptake of encapsulated compound into Caco-2 cells	Bhunchu et al. (2015)
CS-*grafted*-PAMAM/alginate	Insulin	• Average size of particles (nm): 98–150 • Encapsulation efficiency (%): ~97 • Loading efficiency (%): 27 • Relative bioavailability (%): ~11.78 • Improved water solubility, pH responsiveness, and oral bioavailability of insulin • Self-sustained release • Protection against enzymatic deactivation • No sign of acute systemic toxicity	Mukhopadhyay and Kundu (2015)
CS/alginate	Crocin	• Average size of spherical particles with a nonsmooth surface (nm): 165–230 • Zeta potential (mV): −33.52 • Encapsulation efficiency (%): 38.16 • Loading efficiency (%): 30.96 • Enhanced crocin stability under unfavorable conditions	Rahaiee et al. (2015)
CS/alginate	Acetamiprid	• Average size of spherical particles (nm): 201.5 • Polydispersity index: 0.390 • Zeta potential (mV): −32.1 • Encapsulation efficiency (%): 62 • In vitro controlled release at different pHs (4, 7, and 10)	Kumar et al. (2015)

(*Continued*)

Table 12.4 **Summary of the polysaccharide + polysaccharide complexes/coacervates used for encapsulating different components and their remarkable results (*cont.*)**

Encapsulants	Encapsulated compounds	Remarkable results	References

CS/alginate	Quercetin	• Increasing CS concentration led to:
		  • An increase in encapsulation efficiency (%) from 70 to 81.59
		  • A decrease in drug release (%) from 87.2 to 18.57 (at different pH)
		  • A decrease in swelling index from 850.5 to 150.5 (at different pH)
		  • Altering the morphology from pellet-type to spherical-shaped microparticles
			Hazra et al. (2015)

Sodium alginate–coated CS/CS	5-Fluorouracil	• Average size of rough particles (nm): 412–505 • Zeta potential (mV): 18.5–28.9 • Encapsulation efficiency (%): 2.68-18.93 • Loading efficiency (%): 6.19–26.66 • Controlled release • Higher bioavailability • Enhanced mucoadhesiveness	Nagarwal et al. (2012)

CS (containing ascorbic acid for CS dissolution)/alginate	Different polyphenolic antioxidants (raspberry leaf, hawthorn, ground ivy, yarrow, nettle, and olive leaf extracts)	• Average size of particles (median, $D_{0.5}$) (µm):
		  • Raspberry leaf: 1014.52
		  • Hawthorn: 956.94
		  • Ground ivy: 1025.83
		  • Yarrow: 1035.40
		  • -Nettle: 1785.67
		  • Olive leaf: 781.27
		• Encapsulation efficiency (%):
		  • Raspberry leaf: 87.48
		  • Hawthorn: 84.50
		  • Ground ivy: 89.37
		  • Yarrow: 82.39
		  • Nettle: 80.88
		  • Olive leaf: 89.39
		• Diffusion controlled release
		• The highest amount of microelements was present in nettle microparticles
			Belščak-Cvitanović et al. (2011)

CS/alginate (Direct and indirect methods)	Protein-stabilized lipid droplets	• Average size of particles ($D_{3,2}$, μm):
		  • 31.7 for the direct method
		  • 25.3 for the indirect method
		• Zeta potential (mV):
		  • +38.7 for direct method
		  • +40.3 for indirect method
		• Relative stability toward aggregation/dissociation at pH 1–6
		• Particle size dependency of lipid digestion rate
			Li and McClements (2011)

CS/alginate using an extrusion technique	Probiotic bacteria and quercetin	• Average size of particles (μm): 345.4–542.7 • Encapsulation efficiency (%): 19.5–40.2 • Enhanced stability in adverse conditions • Enhanced survival of the probiotics • Reduced porosity of alginate beads and hence decreased leakage of encapsulated probiotics • Enhanced survival of the probiotics • Resistance to the simulated gastric conditions (pH 2.0, 2 h) and bile salt solution (3%, 2 h)	Chávarri et al. (2010)
Alginate/CS/pluronic	Curcumin	• Average size of spherical particles (nm): 100 • Encapsulation efficiency (%): >10 • No sign of toxicity in HeLa (human cervical cancer) cells (at a concentration of 500 (μg/mL) • Controlled release	Das et al. (2010)

Lectin-conjugated CS/alginate	5-Fluorouracil	• Average size of spherical particles (μm):
		  • 6.6 (Before conjugation)
		  • 14.7 (After conjugation)
			Dodov et al. (2009)

(*Continued*)

Table 12.4 **Summary of the polysaccharide + polysaccharide complexes/coacervates used for encapsulating different components and their remarkable results (*cont.*)**

Encapsulants	Encapsulated compounds	Remarkable results	References

		• Zeta potential (mV):
		  • +9.6 (Before conjugation)
		  • +15.3 (After conjugation)
		• Encapsulation efficiency (%):
		  • 75.6 (Before conjugation)
		  • 72.8 (After conjugation)
		• Controlled release
		• Enhanced in vitro mucoadhesiveness	

Layer-by-layer self-assembly of alginate/CS onto liposomes	BSA	• Average size of spherical particles (nm): 383 • Polydispersity index: <0.3 • Zeta potential (mV): +44.6 • Encapsulation efficiency of liposomes coated with 10 alternating layers of polyelectrolytes: >80% • Increased loading capacity and encapsulation efficiency by increased number of layers • Albumin entrapment within both aqueous liposomal core and polyelectrolyte shell • Sustained linear release	Haidar et al. (2008)
Genipin–cross-linked alginate/CS	*Lactobacillus plantarum* Mammalian HepG2	• Average size of spherical particles (μm): 450 • Suppressed capsular swelling (swelling ratio < 10%) • Appropriate permeability in water at a wide range of pH • Capsular integrity and bacterial survival for 6 months	Chen et al. (2007)

Glutaraldehyde–cross-linked CS/κ-carrageenan
CS microspheres covered by κ-carrageenan
	Essential oil of pimento	• Average size of particles (μm):
		  • 1224 for irregular aggregates of CS microspheres
		  • 1172–1211 for irregular-shaped (cross-linked) CS/κ-carrageenan showing a rigid and harsh surface full of holes and cracks
		• Encapsulation efficiency (%):
		  • 98.31 for CS microspheres
		  • 92.16–96.23 for cross-linked CS/κ-carrageenan
		• Higher swelling degree of CS microspheres
		• Antimicrobial activity against *Candida utilis*, *Bacillus cereus*, and *Bacillus subtilis*
			Dima et al. (2014)

CS/carrageenan	Glucose oxidase	• Average size of spherical particles (nm):
		  • 618.4–1288.5[a]
		  • 577.5–1386.7[b]
		• Polydispersity index:
		  • 0.22–0.80[a]
		  • 0.36–0.50[b]
		• Zeta potential (mV):
		  • +34.25 to +35.65[a]
		  • +36.10 to +37.85[b]
		• Controlled release characteristic
			Briones and Sato (2010)

CS/fucoidan (on polystyrene nanoparticles as template)	Poly-L-lysine	• Average size of spherical particles (nm): ~50 (after the core removal)
		• Zeta potential (mV): +26.0
		• Encapsulation efficiency (%):
		  • 45.1 by adsorption
		  • 26.0 by diffusion
		• Loading capacity: >15 by adsorption
		• pH-dependent release
			Pinheiro et al. (2015)

(Continued)

Table 12.4 Summary of the polysaccharide + polysaccharide complexes/coacervates used for encapsulating different components and their remarkable results (*cont.*)

Encapsulants	Encapsulated compounds	Remarkable results	References

HC/CMC
CS/CMC
	MF, MB, TH, and MT	• Highly water-soluble across a wide pH range
		• Encapsulation efficiency (%):
		  • 60–70 (CS/CMC)
		  • 90–95 (HC/CMC)
		• Longer duration of sustained drug release from HC/CMC
		• Lower equilibrium water content of HC/CMC
		• Higher erosion rate and lower swelling capacity of HC/CMC
			Lai and Shum (2015)

CS/sulfated pullulan	BSA	• Average size of particles (nm): 271 • Zeta potential (mV): +50 • Encapsulation efficiency (%): 44 • Loading efficiency (%): 30 • 30% Burst release continued in a controlled manner up to 24 h • No overt sign of toxicity in a respiratory cell model	Dionísio et al. (2013)
CS/chondroitin sulfate	BSA	• Average size of spherical particles (nm): 156.7–195.4 • Polydispersity index: 0.02–0.038 • Zeta potential (mV): +32.1 to + 35.7 • Production yield (%): 27.5–53.3 • Encapsulation efficiency (%): 80 (for 5% w/w BSA) • Controlled release	Santo et al. (2012)

CS/cashew gum	*Lippia sidoides* essential oil	• For cashew gum:CS ratio 10:1 (at different cashew gum concentration and matrix:oil ratio)
		  • Average size of spherical particles (nm): 335–558
		  • Polydispersity index: 0.365–0.613
		  • Encapsulation efficiency (%): 40–70
		  • Loading efficiency (%): 5.5–11
		• For 1% cashew gum, matrix:oil 10:1, and different relative gum:CS ratios
		  • Average size of spherical particles (nm): 335–899
		  • Polydispersity index: 0.373–0.633
		  • Zeta potential (mV): +4 to +49.6
		• Controlled release
			Abreu et al. (2012)

Self-assembled CS/modified lecithin	Dextran–fluorescein isothiocyanate, BSA, and Coomassie blue	• Average size of spherical particles (nm): 268.3–305.2 • Zeta potential (mV): +62.1 to + 65.5 • Encapsulation efficiency (%): 8.7–62.7 • Excellent stability over extended pH (3–6) and ionic strength (≤500 mM NaCl concentration) ranges	Chuah et al. (2009)
Zinc sulfate–cross-linked CS/dextran sulfate	Amphotericin B	• Average size of spherical shaped smooth particles (nm): 600–800 • Zeta potential (mV): −32 • Loading efficiency (%): 65 • Fast-release behavior • Decreased nephrotoxicity	Tiyaboonchai and Limpeanchob (2007)
Aminated pullulan/ carrageenan	BSA	• Average size of particles (nm): 244 • Zeta potential (mV): +58 • Encapsulation efficiency (%): 34 • Loading efficiency (%): 31 • 30% Burst release continued in a controlled manner up to 24 h • No overt sign of toxicity in a respiratory cell model	Dionísio et al. (2013)

HC, Hypromellose-*graft*-chitosan; MB, methylene blue, MF, mometasone furoate; MT, metronidazole; PAMAM, polyamidoamine; TH, tetracycline hydrochloride.
[a]CS:carrageenan ratio of 3.
[b]CS:carrageenan ratio of 5.

pH values from 1 to 6, while instability occurred at higher pH levels. Droplets of larger than 100 μm decreased the digestion rate appreciably (Li and McClements, 2011).

Controlled digestion in the small intestine is important to obtain desired bioavailability; however, in some cases the delivery system should be designed for colon targeting. CS-coated alginate microspheres have been utilized to encapsulate probiotic bacteria, along with the prebiotic quercetin, to enhance their survival during contact with harsh gastric conditions (Chávarri et al., 2010). Low encapsulation yield and survival were observed during the simultaneous encapsulation of probiotics and prebiotics. However, a significant stability was observed when probiotics were separately microencapsulated (Chávarri et al., 2010).

Microspheres incorporated with *Pimenta dioica* essential oil (as an oil in water emulsion) have been prepared via an extrusion technique using CS and/or CS/κ-carrageenan as wall materials (Dima et al., 2014). The swelling degree of CS microspheres was significantly higher than that of CS/κ-carrageenan mixed microspheres (Dima et al., 2014). The decreased swelling was attributed to the electrostatic interaction–mediated disappearance of free amino and carboxyl functional groups in a mixed biopolymer system. The release mechanism was dependent on the CS/κ-carrageenan mixing ratio. A zero-order release kinetic for a weight ratio of 1 and a non-Fickian model for weight ratios of 1:0, 3:1, and 2:1 were reported (Dima et al., 2014).

Protecting the bioactivity of enzymes and proteins is of great importance, particularly when incorporated into delivery systems designed for oral administration. Glucose oxidase, as a model enzyme, was successfully encapsulated within various CS–carrageenans complexes (Briones and Sato, 2010). An increase in the charge density ($\kappa < \iota < \lambda$) of carrageenan decreased the enzyme entrapment efficiency. The mixed-type complexes retained an encapsulated enzyme activity of about 80% in HCl solution (pH = 1.2), 73% in chitosanase dispersion, and 66% in pepsin dispersion previously treated by chitosanase. CS–κ-carrageenan also revealed the lowest release rate (Briones and Sato, 2010).

Modified (aminated and sulfated) pullulans were electrostatically complexed with carrageenan and CS, respectively, to develop positively charged nanocarriers (180–270 nm) for transmucosal delivery of the model, BSA (Dionísio et al., 2013). A fast release of about 30% was observed at pH 7.4, which was maintained up to 24 h. No sign of acute toxicity was observed against respiratory cells (Dionísio et al., 2013).

12.3.3 Nanoencapsulation using protein + protein complexes

Coassembly of proteins through noncovalent bonding is a powerful method for generating ordered supramolecular structures. However, in contrast to the polysaccharide + protein and/or polysaccharide + polysaccharide systems, some limited studies describe the spontaneous complexation between two oppositely charged proteins and their coassembly into nanometer- or micrometer-sized structures composed of at least two different proteins. Under specific physicochemical conditions (the correct mixing ratios and defined pH conditions), proteins of acidic character (i.e., having low p*I* values), such as β-lg, BSA, and caseins, are able to spontaneously self-assemble with proteins containing a strong basic character, such as lysozyme, lactoferrin, and avidin.

There are substantial differences between classical complex coacervation and heteroprotein coacervation. Entropic factors, such as configurational entropy, as well as counterion release, into the bulk phase have minor roles in heteroprotein coacervation (Kizilay et al., 2014). Some proteins, such as caseins and osteopontin, have relatively extended structures and their interactions with other proteins are similar to those of the polysaccharide–protein interactions. Anema and de Kruif (2012, 2013) studied the electrostatic complexation between casein, globular lysozyme, and lactoferrin. The interaction between anionic osteopontin and cationic lactoferrin has been studied from a thermodynamic point of view by Yamniuk et al. (2009). The coacervation between GEs of different types (acidic and basic) and the drug (salbutamol sulfate) release from the GE coacervates has been studied by Tiwari et al. (2009). The interaction between two globular proteins is more complicated. Anema and de Kruif (2014) studied the electrostatic association in a system composed of lactoferrin and β-lg. Their results showed that stoichiometric complexes could be obtained independent of the pH.

To our knowledge, protein–protein complex coacervation is still a live research area. There are a few studies on the potential applications of protein–protein complex coacervates in encapsulation technology. Chapeau et al. (2016) designed an encapsulation system for bioprotection and targeted delivery of vitamin B_9 through heteroprotein coacervation between β-lg and lactoferrin. Vitamin-loaded heteroprotein coacervates obtained at intermediate protein concentrations revealed higher efficacy as a transporter. The entrapment efficiency that was around 6–16 mg of vitamin/g protein could meet the recommended daily intake of vitamin B_9. The presence of vitamin B_9 had no influence on the coassembly of two globular proteins.

β-lg and lysozyme self-assembled coprecipitates of spherical structure were developed and utilized to encapsulate vitamin D_3 (Diarrassouba et al., 2015b). The encapsulation efficiency was around of 91%. In vivo assessment of vitamin D_3 bioavailability was studied in force-feeding rats (Diarrassouba et al., 2015a). An increased stability of vitamin D_3 was observed after encapsulation. The observed release in the intestines was enhanced in the presence of proteolytic enzymes. Analysis of the serum level of 25-OH vitamin D_3 confirmed that the encapsulation resulted in increased bioavailability (Diarrassouba et al., 2015a).

Hydrophobicity and low chemical stability limit the bioavailability of resveratrol after oral administration. To encapsulate resveratrol, prolamine-type protein particles were prepared from gliadin or zein using the antisolvent precipitation method (Davidov-Pardo et al., 2015). A fraction of protein particles was covered with pectin or sodium caseinate. The average sizes of uncovered gliadin and zein nanocarriers were 260 and 120 nm, respectively. The best coatings for gliadin and zein nanovehicles were pectin and sodium caseinate, respectively. The encapsulation efficiencies of resveratrol in coated gliadin and zein nanoparticles were 53 and 86%, respectively. The encapsulation efficiencies of coated particles were higher than uncovered ones (Davidov-Pardo et al., 2015). The physical stability of (covered) protein carriers against heat treatment (90°C; 30 min), as well as an increase in the ionic strength was studied in another work (Joye et al., 2015). Physical stability was observed in the dispersion of covered carriers after heat treatment. Among different samples, only the dispersion containing zein nanocarriers coated with caseinate was stable at high salt concentrations. This sample was completely redispersed after lyophilization and could confer UV stability to the nutraceutical compound (Joye et al., 2015).

Complexed biopolymers can also be utilized for increasing the effectiveness of antimicrobial compounds. Thymol is one of the main antimicrobial components of thyme essential oil, and was incorporated into zein nanocarriers and then coated with sodium caseinate and CS HCl (Zhang et al., 2014b). Both zein nanocarriers and those coated with caseinate were spherical with smooth surface; whereas, rough surfaces with some clumps were observed in CS HCl–stabilized zein nanocarriers. The antimicrobial effect of encapsulated thymol against Gram-positive bacteria was more long lasting than that of unentrapped (free) thymol.

Another important implication of protein + protein complexes is to stabilize emulsions as delivery systems. Whey protein– and (beta) casein-stabilized emulsions undergo destabilization upon Ca^{2+} addition. However, lactoferrin incorporation into milk protein–stabilized emulsions could enhance calcium stability (Ye et al., 2012). Lactoferrin protects calcium-induced flocculation by: (1) binding calcium ions via sialic acid residues on lactoferrin molecules and reducing the number of free ions available to form calcium bridges between protein molecules; (2) electrostatic interaction with negatively charged phosphate (and carboxyl) groups on milk protein molecules resulting in lactoferrin–milk protein interfacial complexes, and hence blocking calcium phosphate interactions; and (3) providing additional stability through increased steric repulsion between interfaces of milk protein/lactoferrin coassemblies, due to the large molecular size of lactoferrin (McCarthy et al., 2014).

In addition to the electrostatic interactions, protein + protein complexes can be obtained through heat treatment. As an example, heat treatment of milk causes interactions mainly between κ-casein and β-lg in a pH-dependent manner. Therefore, the pH of milk before heat treatment has a significant influence on the various types of gelation, including acid- or rennet-induced gelation, which is of considerable importance to the dairy industry (Vasbinder and de Kruif, 2003).

12.4 Conclusions and future trends

The number of "generally regarded as safe" biomaterials (of food-grade quality), which can be used to produce nanoencapsulating systems, is quite limited; therefore, finding suitable delivery systems is of importance. Biopolymers are considered as the key structural components in foodstuffs and their complexes and coacervates are promising tools to develop tailor-made structures. Significant strides have been made, yet ongoing research is directed toward identifying the most important extrinsic and intrinsic parameters controlling complex formation.

The application of biopolymer complexes and coacervates, including protein + polysaccharide, polysaccharide + polysaccharide, and protein + protein as encapsulating systems, for foods and nutraceuticals is very interesting. The large diversity of biopolymer couples facilitates fine-tuning and careful engineering of carrier systems to achieve desired requirements (e.g., the desired structure, including soluble or insoluble complexes, coacervates and gels, storage stability, bioprotection, and controlled release). Biopolymers may also contain bioactive agents (such as bioactive peptides sequences present in a protein).

The main drawback of electrostatic biopolymer complexes and coacervates is that the molecular driving forces are highly dependent on the pH, ionic strength, concentration, and temperature. Therefore, certain conditions are required to avoid agglomeration of the prepared (nano)capsules. The numerous possible combinations of biopolymers may lead to the application of a particular mixture in a specific circumstance (Turgeon and Laneuville, 2009).

Covalently cross-linked biopolymer complexes are the only systems that can properly withstand the changes in physicochemical conditions resulting from nonreversible linkages. Cross-linkers are required to obtain these complexes; however, most of them suffer from high toxicity (such as glutaraldehyde) or an unknown in vivo fate. For the time being, the choice of safe and biocompatible cross-linkers is limited to genipin, TG, oxidoreductase enzymes, and MRs. Low amounts of swelling and having no stimuli-responsive characteristics are the other weaknesses of covalently cross-linked complexes.

At present, our understanding of the structure–function relationships is limited. Further studies should be undertaken to provide more details regarding the morphology of the stable soluble nanocomplexes using cryotransmission electron microscopy and atomic force microscopy. Another important research field is the optimization of multiingredient systems. Designing intelligent or stimuli-responsive nanometric carrier systems using multilayered nanoparticles is another active area of research.

More investigations are also needed to clarify whether these complexes can bioprotect nutraceuticals during heat treatment (industrial pasteurization and cold storage) and from (photo)oxidation. Additional research on the interactions of nanocapsules with food matrices and their influence on the sensorial aspects of the fortified food is also required.

Last but not least, delivery systems must release the nutraceutical in an active form at the specific site of absorption. Therefore, evaluating the bioavailability of nutraceuticals loaded in such complexes, together with toxicological studies, is surely needed to convince the food industry of the usefulness of this encapsulation technology. To apply the nanocapsules on an industrial scale in foods, economically scalable methods are also needed.

References

Aberkane, L., Jasniewski, J., Gaiani, C., Scher, J., Sanchez, C., 2010. Thermodynamic characterization of acacia gum−β-lactoglobulin complex coacervation. Langmuir 26 (15), 12523–12533.

Abreu, F.O., Oliveira, E.F., Paula, H.C., de Paula, R.C., 2012. Chitosan/cashew gum nanogels for essential oil encapsulation. Carbohydr. Polym. 89 (4), 1277–1282.

Aceituno-Medina, M., Mendoza, S., Rodríguez, B.A., Lagaron, J.M., López-Rubio, A., 2015. Improved antioxidant capacity of quercetin and ferulic acid during in-vitro digestion through encapsulation within food-grade electrospun fibers. J. Funct. Foods 12, 332–341.

Aguilar, K.C., Tello, F., Bierhalz, A.C., Romo, M.G.G., Flores, H.E.M., Grosso, C.R., 2015. Protein adsorption onto alginate-pectin microparticles and films produced by ionic gelation. J. Food Eng. 154, 17–24.

Ahmadi, N., Nasirpour, A., Sheikhzeinodin, M., Keramat, J., 2015. Microencapsulation of ubiquinone using complex coacervation for functional yoghurt. Food Sci. Biotech. 24 (3), 895–904.

Anema, S.G., de Kruif, C.G., 2012. Co-acervates of lactoferrin and caseins. Soft Matter 8, 4471–4478.

Anema, S.G., de Kruif, C.G., 2013. Coacervates of lysozyme and β-casein. J. Colloid Interface Sci. 398, 255–261.

Anema, S.G., de Kruif, C.G., 2014. Complex coacervates of lactotransferrin and β-lactoglobulin. J. Colloid Interface Sci. 430, 214–220.

Arroyo-Maya, I.J., McClements, D.J., 2015. Biopolymer nanoparticles as potential delivery systems for anthocyanins: fabrication and properties. Food Res. Int. 69, 1–8.

Assadpour, E., Jafari, S.M., 2017. Spray drying of folic acid within nano-emulsions; optimization by Taguchi approach. Drying Technol. doi: 10.1080/07373937.2016.1242016.

Assadpour, E., Jafari, S.M., Maghsoudlou, Y., 2017. Evaluation of folic acid release from spray dried powder particles of pectin-whey protein nano-capsules. Int. J. Biol. Macromol. 95, 238–247.

Assadpour, E., Maghsoudlou, Y., Jafari, S.M., Ghorbani, M., Aalami, M., 2016. Optimization of folic acid nano-emulsification and encapsulation by maltodextrin-whey protein double emulsions. Int. J. Biol. Macromol. 86, 197–207.

Baracat, M.M., Nakagawa, A.M., Casagrande, R., Georgetti, S.R., Verri, Jr., W.A., de Freitas, O., 2012. Preparation and characterization of microcapsules based on biodegradable polymers: pectin/casein complex for controlled drug release systems. AAPS J. 13 (2), 364–372.

Bédié, G.K., Turgeon, S.L., Makhlouf, J., 2008. Formation of native whey protein isolate–low methoxyl pectin complexes as a matrix for hydro-soluble food ingredient entrapment in acidic foods. Food Hydrocoll. 22 (5), 836–844.

Belščak-Cvitanović, A., ðorđević, V., Sven Karlović, S., Vladimir Pavlović, V., Komes, D., Damir Ježek, D., Bugarski, B., Nedović, V., 2015. Protein-reinforced and chitosan-pectin coated alginate microparticles for delivery of flavan-3-ol antioxidants and caffeine from green tea extract. Food Hydrocoll. 51, 361–374.

Belščak-Cvitanović, A., Stojanović, R., Manojlović, V., Komes, D., Cindrić, I.J., Nedović, V., Bugarski, B., 2011. Encapsulation of polyphenolic antioxidants from medicinal plant extracts in alginate–chitosan system enhanced with ascorbic acid by electrostatic extrusion. Food Res. Int. 44 (4), 1094–1101.

Berendsen, R., Güell, C., Ferrando, M., 2015. A procyanidin-rich extract encapsulated in water-in-oil-in-water emulsions produced by premix membrane emulsification. Food Hydrocoll. 43, 636–648.

Bhunchu, S., Rojsitthisak, P., Rojsitthisak, P., 2015. Effects of preparation parameters on the characteristics of chitosan–alginate nanoparticles containing curcumin diethyl disuccinate. J. Drug Deliv. Sci. Technol. 28, 64–72.

Boeriu, C.G., Oudgenoeg, G., Spekking, W.T.J., Berendsen, L., Vancon, L., Boumans, H., Gruppen, H., Van Berkel, W.J.H., Laane, C., Voragen, A.G.J., 2004. Horseradish peroxidase-catalyzed cross-linking of feruloylated arabinoxylans with beta-casein. J. Agric. Food Chem. 52 (21), 6633–6639.

Bosnea, L.A., Moschakis, T., Biliaderis, C.G., 2014. Complex coacervation as a novel microencapsulation technique to improve viability of probiotics under different stresses. Food Bioprocess Technol. 7 (10), 2767–2781.

Braudo, E.E., Antonov, Y.A., 1993. Non-coulombic complex formation of protein as a structure forming factor in food systems. In: Schwenke, K.D., Mothes, R. (Eds.), Food Proteins: Structure and Functionality. VCH, Weinheim, pp. 210–215.

Briones, A.V., Sato, T., 2010. Encapsulation of glucose oxidase (GOD) in polyelectrolyte complexes of chitosan–carrageenan. React. Funct. Polym. 70 (1), 19–27.

Caillard, R., Mateescu, M.A., Subirade, M., 2010. Maillard-type cross-linked soy protein hydrogels as devices for the release of ionic compounds: an in vitro study. Food Res. Int. 43 (10), 2349–2355.

Chapeau, A.-L., Tavares, G.M., Hamon, P., Croguennec, T., Poncelet, D., Bouhallab, S., 2016. Spontaneous co-assembly of lactoferrin and β-lactoglobulin as a promising biocarrier for vitamin B9. Food Hydrocoll. 57, 280–290.

Chávarri, M., Marañón, I., Ares, R., Ibáñez, F.C., Marzo, F., del Carmen Villarán, M., 2010. Microencapsulation of a probiotic and prebiotic in alginate-chitosan capsules improves survival in simulated gastro-intestinal conditions. Int. J. Food Microbiol. 142 (1), 185–189.

Chen, B., Li, H., Ding, Y., Suo, H., 2012. Formation and microstructural characterization of whey protein isolate/beet pectin coacervations by laccase catalyzed cross-linking. LWT Food Sci. Technol. 47 (1), 31–38.

Chen, H., Ouyang, W., Jones, M., Metz, T., Martoni, C., Haque, T., Cohen, R., Lawuyi, B., Prakash, S., 2007. Preparation and characterization of novel polymeric microcapsules for live cell encapsulation and therapy. Cell Biochem. Biophys. 47 (1), 159–167.

Chuah, A.M., Kuroiwa, T., Ichikawa, S., Kobayashi, I., Nakajima, M., 2009. Formation of biocompatible nanoparticles via the self-assembly of chitosan and modified lecithin. J. Food Sci. 74 (1), N1–N8.

Comunian, T.A., Thomazini, M., Alves, A.J.G., de Matos Junior, F.E., de Carvalho Balieiro, J.C., Favaro Trindade, C.S., 2013. Microencapsulation of ascorbic acid by complex coacervation: protection and controlled release. Food Res. Int. 52 (1), 373–379.

Corredig, M., Sharafbafi, N., Kristo, E., 2011. Polysaccharide–protein interactions in dairy matrices, control and design of structures. Food Hydrocoll. 25 (8), 1833–1841.

Dai, Q., Zhu, X., Abbas, S., Karangwa, E., Zhang, X., Xia, S., Feng, B., Jia, C., 2015. Stable nanoparticles prepared by heating electrostatic complexes of whey protein isolate-dextran conjugate and chondroitin sulfate. J. Agric. Food Chem. 63 (16), 4179–4189.

Das, R.K., Kasoju, N., Bora, U., 2010. Encapsulation of curcumin in alginate-chitosan-pluronic composite nanoparticles for delivery to cancer cells. Nanomedicine 6 (1), 153–160.

Davidov-Pardo, G., Joye, I.J., McClements, D.J., 2015. Encapsulation of resveratrol in biopolymer particles produced using liquid antisolvent precipitation. Part 1: preparation and characterization. Food Hydrocoll. 45, 309–316.

de Conto, L.C., Grosso, C.R.F., Gonçalves, L.A.G., 2013. Chemometry as applied to the production of omega-3 microcapsules by complex coacervation with soy protein isolate and gum Arabic. LWT Food Sci. Technol 53 (1), 218–224.

de Kruif, C.G., Weinbreck, F., de Vries, R., 2004. Complex coacervation of proteins and anionic polysaccharides. Curr. Opin. Colloid Interface Sci. 9 (5), 340–349.

de Vries, R., Stuart, M.C., 2006. Theory and simulations of macroion complexation. Curr. Opin. Colloid Interface Sci. 11 (5), 295–301.

Devi, N., Maji, T.K., 2010. Genipin crosslinked microcapsules of gelatin A and κ-carrageenan polyelectrolyte complex for encapsulation of neem (*Azadirachta indica* A. Juss.) seed oil. Polym. Bull. 65 (4), 347–362.

Devi, N., Maji, T.K., 2011. Study of complex coacervation of gelatin a with sodium carboxymethyl cellulose: microencapsulation of neem (*Azadirachta indica* A. Juss.) seed oil (NSO). Int. J. Polym. Mater. 60 (13), 1091–1105.

Devi, N., Hazarika, D., Deka, C., Kakati, D.K., 2012. Study of complex coacervation of gelatin A and sodium alginate for microencapsulation of olive oil. J. Macromol. Sci. A 49 (11), 936–945.

Diarrassouba, F., Garrait, G., Alvarez, P., Beyssac, E., Subirade, M., 2015a. Food protein-based microspheres for increased uptake of vitamin D3. Food Chem. 173, 1066–1072.

Diarrassouba, F., Remondetto, G., Garrait, G., Alvarez, P., Beyssac, E., Subirade, M., 2015b. Self-assembly of β-lactoglobulin and egg white lysozyme as a potential carrier for nutraceuticals. Food Chem. 173, 203–209.

Dickinson, E., 1998. Stability and rheological implications of electrostatic milk protein polysaccharide interactions. Trends Food Sci. Technol. 9 (10), 347–354.

Dickinson, E., 2008. Interfacial structure and stability of food emulsions as affected by protein–polysaccharide interactions. Soft Matter 4 (5), 932–942.

Dima, C., Cotârlet, M., Alexe, P., Dima, S., 2014. Microencapsulation of essential oil of pimento [*Pimenta dioica* (L) Merr.] by chitosan/k-carrageenan complex coacervation method. Innov. Food Sci. Emerg. Technol. 22, 203–211.

Dionísio, M., Cordeiro, C., Remuñán-López, C., Seijo, B., da Costa, A.M.R., Grenha, A., 2013. Pullulan-based nanoparticles as carriers for transmucosal protein delivery. Eur. J. Pharm. Sci 50 (1), 102–113.

Dodov, M.G., Calis, S., Crcarevska, M.S., Geskovski, N., Petrovska, V., Goracinova, K., 2009. Wheat germ agglutinin-conjugated chitosan–Ca–alginate microparticles for local colon delivery of 5-FU: development and in vitro characterization. Int. J. Pharm. 381 (2), 166–175.

Doublier, J.L., Garnier, C., Renard, D., Sanchez, C., 2000. Protein–polysaccharide interactions. Curr. Opin. Colloid Interface Sci. 5 (3), 202–214.

Draget, K.I., Moe, S.T., Skjåk-Bræk, G., Smidsrød, O., 2006. Alginates. In: Stephen, A.M., Phillips, G.O., Williams, P.A. (Eds.), Food Polysaccharides and Their Applications. second ed. CRC Press, FL, pp. 289–334.

Ducel, V., Pouliquen, D., Richard, J., Boury, F., 2008. 1 H NMR relaxation studies of protein–polysaccharide mixtures. Int. J. Biol. Macromol. 43 (4), 359–366.

Eleya, M.O., Turgeon, S.L., 2000. The effects of pH on the rheology of β-lactoglobulin/κ-carrageenan mixed gels. Food Hydrocoll. 14 (3), 245–251.

Enderle, J.D., Blanchard, S.N., Bronzino, J.D., 2005. Introduction to Biomedical Engineering, second ed Elsevier Academic Press, Burlington, MA.

Eratte, D., Wang, B., Dowling, K., Barrow, C.J., Adhikari, B.P., 2014. Complex coacervation with whey protein isolate and gum Arabic for the microencapsulation of omega-3 rich tuna oil. Food Funct. 5 (11), 2743–2750.

Esfanjani, A.F., Jafari, S.M., Assadpoor, E., Mohammadi, A., 2015. Nano-encapsulation of saffron extract through double-layered multiple emulsions of pectin and whey protein concentrate. J. Food Eng. 165, 149–155.

Esfanjani, A.F., Jafari, S.M., 2016. Biopolymer nano-particles and natural nano-carriers for nano-encapsulation of phenolic compounds. Colloids Surf. B. 146, 532–543.

Esfanjani, A.F., Jafari, S.M., Assadpour, E., 2017. Preparation of a multiple emulsion based on pectin-whey protein complex for encapsulation of saffron extract nanodroplets. Food Chem. 221, 1962–1969.

Fioramonti, S.A., Perez, A.A., Aríngoli, E.E., Rubiolo, A.C., Santiago, L.G., 2014. Design and characterization of soluble biopolymer complexes produced by electrostatic self-assembly of a whey protein isolate and sodium alginate. Food Hydrocoll. 35, 129–136.

Fox, P.F., 2009. Milk: an overview. In: Thompson, A., Boland, M., Singh, H. (Eds.), Milk Proteins: From Expression to Food. Elsevier, New York, NY, pp. 1–54.

Gan, C.Y., Cheng, L.H., Easa, A.M., 2008. Evaluation of microbial transglutaminase and ribose cross-linked soy protein isolate-based microcapsules containing fish oil. Innov. Food Sci. Emerg. Technol. 9 (4), 563–569.

Gazme, B., Madadlou, A., 2014. Fabrication of whey protein–pectin conjugate particles through laccase-induced gelation of microemulsified nanodroplets. Food Hydrocoll. 40, 189–195.

Girard, M., Turgeon, S.L., Gauthier, S.F., 2003. Thermodynamic parameters of β-lactoglobulin-pectin complexes assessed by isothermal titration calorimetry. J. Agric. Food Chem. 51 (15), 4450–4455.

Goh, K.K.T., Sarkar, A., Singh, H., 2009. Milk protein–polysaccharide interactions. In: Thompson, A., Boland, M., Singh, H. (Eds.), Milk Proteins: From Expression to Food. Elsevier, New York, NY, pp. 347–376.

Haidar, Z.S., Hamdy, R.C., Tabrizian, M., 2008. Protein release kinetics for core–shell hybrid nanoparticles based on the layer-by-layer assembly of alginate and chitosan on liposomes. Biomaterials 29 (9), 1207–1215.

Hazra, M., Mandal, D.D., Mandal, T., Bhuniya, S., Ghosh, M., 2015. Designing polymeric microparticulate drug delivery system for hydrophobic drug quercetin. Saudi Pharm. J. 23 (4), 429–436.

He, L., Liang, H., Lin, L., Shah, B.R., Li, Y., Chen, Y., Li, B., 2015. Green-step assembly of low density lipoprotein/sodium carboxymethyl cellulose nanogels for facile loading and pH-dependent release of doxorubicin. Colloids Surf. B 126, 288–296.

Heidebach, T., Först, P., Kulozik, U., 2009. Transglutaminase-induced caseinate gelation for the microencapsulation of probiotic cells. Int. Dairy J. 19 (2), 77–84.

Hellmers, F., Ferguson, P., Koropatnick, J., Krull, R., Margaritis, A., 2013. Characterization and in vitro cytotoxicity of doxorubicin-loaded γ-polyglutamic acid-chitosan composite nanoparticles. Biochem. Eng. J. 75, 72–78.

Hosseini, M.H., Razavi, S.H., Mousavi, M.A., 2009. Antimicrobial, physical and mechanical properties of chitosan-based films incorporated with thyme, clove and cinnamon essential oils. J. Food Process. Preserv. 33 (6), 727–743.

Hosseini, S.M.H., Emam-Djomeh, Z., Razavi, S.H., Moosavi-Movahedi, A.A., Saboury, A.A., Atri, M.S., Van der Meeren, P., 2013b. β-Lactoglobulin–sodium alginate interaction as affected by polysaccharide depolymerization using high intensity ultrasound. Food Hydrocoll. 32 (2), 235–244.

Hosseini, S.M.H., Emam-Djomeh, Z., Razavi, S.H., Moosavi-Movahedi, A.A., Saboury, A.A., Mohammadifar, M.A., et al., 2013a. Complex coacervation of β-lactoglobulin–κ-carrageenan aqueous mixtures as affected by polysaccharide sonication. Food Chem. 141 (1), 215–222.

Hosseini, S.M.H., Emam-Djomeh, Z., Sabatino, P., Van der Meeren, P., 2015. Nanocomplexes arising from protein-polysaccharide electrostatic interaction as a promising carrier for nutraceutical compounds. Food Hydrocoll. 50, 16–26.

Hu, B., Huang, Q.R., 2013. Biopolymer based nano-delivery systems for enhancing bioavailability of nutraceuticals. Chinese J. Polym. Sci. 31 (9), 1190–1203.

Huang, Y.I., Cheng, Y.H., Yu, C.C., Tsai, T.R., Cham, T.M., 2007. Microencapsulation of extract containing shikonin using gelatin–acacia coacervation method: a formaldehyde-free approach. Colloids Surf. B 58 (2), 290–297.

Humblet-Hua, K.N.P., Scheltens, G., van der Linden, E., Sagis, L.M.C., 2011. Encapsulation systems based on ovalbumin fibrils and high methoxyl pectin. Food Hydrocoll. 25 (4), 569–576.

Ifeduba, E.A., Akoh, C.C., 2015. Microencapsulation of stearidonic acid soybean oil in complex coacervates modified for enhanced stability. Food Hydrocoll. 51, 136–145.

Ilyasoglu, H., El, S.N., 2014. Nanoencapsulation of EPA/DHA with sodium caseinate–gum Arabic complex and its usage in the enrichment of fruit juice. LWT Food Sci. Technol. 56 (2), 461–468.

Jain, A., Thakur, D., Ghoshal, G., Katare, O.P., Shivhare, U.S., 2015. Microencapsulation by complex coacervation using whey protein isolates and gum acacia: an approach to preserve the functionality and controlled release of β-carotene. Food Bioprocess Technol. 8, 1635–1644.

Jincheng, W., Xiaoyu, Z., Sihao, C., 2010. Preparation and properties of nanocapsulated capsaicin by complex coacervation method. Chem. Eng. Commun. 197 (7), 919–933.

Joye, I.J., Davidov-Pardo, G., McClements, D.J., 2015. Encapsulation of resveratrol in biopolymer particles produced using liquid antisolvent precipitation. Part 2: stability and functionality. Food Hydrocoll. 49, 127–134.

Jun-xia, X., Hai-yan, Y., Jian, Y., 2011. Microencapsulation of sweet orange oil by complex coacervation with soybean protein isolate/gum Arabic. Food Chem. 125 (4), 1267–1272.

Katouzian, I., Jafari, S.M., 2016. Nano-encapsulation as a promising approach for targeted delivery and controlled release of vitamins. Trends Food Sci. Technol. 53, 34–48.

Kayitmazer, A.B., Strand, S.P., Tribet, C., Jaeger, W., Dubin, P.L., 2007. Effect of polyelectrolyte structure on protein-polyelectrolyte coacervates: coacervates of bovine serum albumin with poly (diallyldimethylammonium chloride) versus chitosan. Biomacromolecules 8 (11), 3568–3577.

Kizilay, E., Kayitmazer, A.B., Dubin, P.L., 2011. Complexation and coacervation of polyelectrolytes with oppositely charged colloids. Adv. Colloid Interface Sci. 167 (1), 24–37.

Kizilay, E., Seeman, D., Yan, Y., Du, X., Dubin, P.L., Donato-Capel, L., Bovetto, L., Schmitt, C., 2014. Structure of bovine β-lactoglobulin–lactoferrin coacervates. Soft Matter 10, 7262–7268.

Koupantsis, T., Pavlidou, E., Paraskevopoulou, A., 2014. Flavour encapsulation in milk proteins–CMC coacervate-type complexes. Food Hydrocoll. 37, 134–142.

Kumar, S., Chauhan, N., Gopal, M., Kumar, R., Dilbaghi, N., 2015. Development and evaluation of alginate–chitosan nanocapsules for controlled release of acetamiprid. Int. J. Biol. Macromol. 81, 631–637.

Kumar, A., Dubin, P.L., Hernon, M.J., Li, Y., Jaeger, W., 2007. Temperature-dependent phase behavior of polyelectrolyte-mixed micelle systems. J. Phys. Chem. B 111 (29), 8468–8476.

Lai, W.F., Shum, H.C., 2015. Hypromellose-graft-chitosan and its polyelectrolyte complex as novel systems for sustained drug delivery. ACS Appl. Mater. Interfaces 7 (19), 10501–10510.

Lapitsky, Y., 2014. Ionically crosslinked polyelectrolyte nanocarriers: recent advances and open problems. Curr. Opin. Colloid Interface Sci. 19 (2), 122–130.

Lee, P.S., Yim, S.G., Choi, Y., Ha, T.V.A., Ko, S., 2012. Physiochemical properties and prolonged release behaviours of chitosan-denatured β-lactoglobulin microcapsules for potential food applications. Food Chem. 134 (2), 992–998.

Li, Y., McClements, D.J., 2011. Controlling lipid digestion by encapsulation of protein-stabilized lipid droplets within alginate–chitosan complex coacervates. Food Hydrocoll. 25 (5), 1025–1033.

Luo, Y., Wang, T.T., Teng, Z., Chen, P., Sun, J., Wang, Q., 2013. Encapsulation of indole-3-carbinol and 3, 3′-diindolylmethane in zein/carboxymethyl chitosan nanoparticles with controlled release property and improved stability. Food Chem. 139 (1), 224–230.

Martins, A.F., Oliveria, D.M., Pereira, A.G.B., Rubira, A.F., Muniz, E.C., 2012. Chitosan/TPP microparticles obtained by microemulsion method applied in controlled release of heparin. Int. J. Biol. Macromol. 51, 1127–1133.

Matalanis, A., Jones, O.G., McClements, D.J., 2011. Structured biopolymer-based delivery systems for encapsulation, protection, and release of lipophilic compounds. Food Hydrocoll. 25 (8), 1865–1880.

McCarthy, A.A., Kelly, A.L., O'Mahony, J.A., Fenelon, M.A., 2014. Sensitivity of emulsions stabilised by bovine β-casein and lactoferrin to heat and $CaCl_2$. Food Hydrocoll. 35, 420–428.

McClements, D.J., 2004. Protein-stabilized emulsions. Curr. Opin. Colloid Interface Sci. 9 (5), 305–313.

McClements, D.J., 2006. Non-covalent interactions between proteins and polysaccharides. Biotechnol. Adv. 24, 621–625.

McClements, D.J., Decker, E.A., Park, Y., Weiss, J., 2009. Structural design principles for delivery of bioactive components in nutraceuticals and functional foods. Crit. Rev. Food Sci. Nutr. 49 (6), 577–606.

Mehrnia, M.A., Jafari, S.M., Makhmal-Zadeh, B.S., Maghsoudlou, Y., 2017. Rheological and release properties of double nano-emulsions containing crocin prepared with Angum gum. Arabic gum and whey protein. Food Hydrocoll. doi: 10.1016/j.foodhyd.2016.11.033.

Mendanha, D.V., Ortiz, S.E.M., Favaro-Trindade, C.S., Mauri, A., Monterrey-Quintero, E.S., Thomazini, M., 2009. Microencapsulation of casein hydrolysate by complex coacervation with SPI/pectin. Food Res. Int. 42 (8), 1099–1104.

Mirpoor, F., 2016. Fortifying Golab Sherbet as a Traditional Herbal Beverage Using Nanoencapsulated Quercetin. M.Sc. thesis. Department of Food Science and Technology, Shiraz University, Shiraz, Iran.

Mohammadi, A., Jafari, S.M., Assadpour, E., Faridi Esfanjani, A., 2016a. Nano-encapsulation of olive leaf phenolic compounds through WPC-pectin complexes and evaluating their release rate. Int. J. Biol. Macromol. 82, 816–822.

Mohammadi, A., Jafari, S.M., Esfanjani, A.F., Akhavan, S., 2016b. Application of nano-encapsulated olive leaf extract in controlling the oxidative stability of soybean oil. Food Chem. 190, 513–519.

Mukhopadhyay, P., Kundu, P.P., 2015. Chitosan-graft-PAMAM–alginate core–shell nanoparticles: a safe and promising oral insulin carrier in an animal model. RSC Adv. 5 (114), 93995–94007.

Na, H.S., Kim, J.N., Kim, J.M., Lee, K.Y., 2011. Encapsulation of fish oil using cyclodextrin and whey protein concentrate. Biotechnol. Bioprocess Eng. 16 (6), 1077–1082.

Nagarwal, R.C., Kumar, R., Pandit, J.K., 2012. Chitosan coated sodium alginate–chitosan nanoparticles loaded with 5-FU for ocular delivery: in vitro characterization and in vivo study in rabbit eye. Eur. J. Pharm. Sci. 47 (4), 678–685.

Nakagawa, K., Nagao, H., 2012. Microencapsulation of oil droplets using freezing induced gelatin–acacia complex coacervation. Colloids Surf. A 411 (5), 129–139.

Nori, M.P., Favaro-Trindade, C.S., de Alencar, S.M., Thomazini, M., de Camargo Balieiro, J.C., Castillo, C.J.C., 2011. Microencapsulation of propolis extract by complex coacervation. LWT Food Sci. Technol. 44 (2), 429–435.

Ocak, B., 2012. Complex coacervation of collagen hydrolysate extracted from leather solid wastes and chitosan for controlled release of lavender oil. J. Environ. Manage. 100, 22–28.

Pal, K., Paulson, A.T., Rousseau, D., 2009. Biopolymers in controlled-release delivery systems. In: Kasapis, S., Norton, I.T., Ubbink, J.B. (Eds.), Modern Biopolymer Science: Bridging the Divide between Fundamental Treatise and Industrial Application. Elsevier, New York, NY, pp. 519–557.

Pan, X., Yao, P., Jiang, M., 2007. Simultaneous nanoparticle formation and encapsulation driven by hydrophobic interaction of casein-graft-dextran and β-carotene. J. Colloid Interface Sci. 315 (2), 456–463.

Peng, C., Zhao, S.Q., Zhang, J., Huang, G.Y., Chen, L.Y., Zhao, F.Y., 2014. Chemical composition, antimicrobial property and microencapsulation of mustard (*Sinapis alba*) seed essential oil by complex coacervation. Food Chem. 165, 560–568.

Pereira, H.V.R., Saraiva, K.P., Carvalho, L.M.J., Andrade, L.R., Pedrosa, C., Pierucci, A.P.T.R., 2009. Legumes seeds protein isolates in the production of ascorbic acid microparticles. Food Res. Int. 42 (1), 115–121.

Perez, A.A., Sponton, O.E., Andermatten, R.B., Amelia, C., Rubiolo, A.C., Santiago, L.G., 2015. Biopolymer nanoparticles designed for polyunsaturated fatty acid vehiculization: protein–polysaccharide ratio study. Food Chem. 188, 543–550.

Pinheiro, A.C., Bourbon, A.I., Cerqueira, M.A., Maricato, É., Nunes, C., Coimbra, M.A., Vicente, A.A., 2015. Chitosan/fucoidan multilayer nanocapsules as a vehicle for controlled release of bioactive compounds. Carbohydr. Polym. 115, 1–9.

Prata, A.S., Zanin, M.H., Ré, M.I., Grosso, C.R., 2008. Release properties of chemical and enzymatic crosslinked gelatin-gum Arabic microparticles containing a fluorescent probe plus vetiver essential oil. Colloids Surf. B 67 (2), 171–178.

Rahaiee, S., Shojaosadati, S.A., Hashemi, M., Moini, S., Razavi, S.H., 2015. Improvement of crocin stability by biodegradeble nanoparticles of chitosan-alginate. Int. J. Biol. Macromol. 79, 423–432.

Rocha-Selmi, G.A., Bozza, F.T., Thomazini, M., Bolini, H.M., Fávaro-Trindade, C.S., 2013. Microencapsulation of aspartame by double emulsion followed by complex coacervation to provide protection and prolong sweetness. Food Chem. 139 (1), 72–78.

Rodea-González, D.A., Cruz-Olivares, J., Román-Guerrero, A., Rodríguez-Huezo, M.E., Vernon-Carter, E.J., Pérez-Alonso, C., 2012. Spray-dried encapsulation of chia essential oil (*Salvia hispanica* L.) in whey protein concentrate-polysaccharide matrices. J. Food Eng. 111 (1), 102–109.

Rodríguez Patino, J.M., Pilosof, A.M.R., 2011. Protein–polysaccharide interactions at fluid interfaces. Food Hydrocoll. 25 (8), 1925–1937.

Ron, N., Zimet, P., Bargarum, J., Livney, Y.D., 2010. Beta-lactoglobulin–polysaccharide complexes as nanovehicles for hydrophobic nutraceuticals in non-fat foods and clear beverages. Int. Dairy J. 20 (10), 686–693.

Sanchez, C., Mekhloufi, G., Schmitt, C., Renard, D., Robert, P., Lehr, C.-M., Lamprecht, A., Hardy, J., 2002. Self-assembly of β-lactoglobulin and acacia gum in aqueous solvent: structure and phase-ordering kinetics. Langmuir 18 (26), 10323–10333.

Santo, V.E., Gomes, M.E., Mano, J.F., Reis, R.L., 2012. Chitosan–chondroitin sulphate nanoparticles for controlled delivery of platelet lysates in bone regenerative medicine. J. Tissue Eng. Regen. Med. 6 (S3), s47–s59.

Santos, M.G., Carpinteiro, D.A., Thomazini, M., Rocha-Selmi, G.A., da Cruz, A.G., Rodrigues, C.E.C., Favaro-Trindade, C.S., 2014. Coencapsulation of xylitol and menthol by double emulsion followed by complex coacervation and microcapsule application in chewing gum. Food Res. Int. 66, 454–462.

Schmitt, C., Turgeon, S.L., 2011. Protein/polysaccharide complexes and coacervates in food systems. Adv. Colloid Interface Sci. 167 (1), 63–70.

Schmitt, C., Aberkane, L., Sanchez, C., 2009. Protein-polysaccharide complexes and coacervates. In: Phillips, G.O., Williams, P.A. (Eds.), Handbook of Hydrocolloids. second ed. CRC Press, FL, pp. 420–476.

Schmitt, C., Sanchez, C., Desobry-Banon, S., Hardy, J., 1998. Structure and technofunctional properties of protein-polysaccharide complexes: a review. Crit. Rev. Food Sci. Nutr. 38 (8), 689–753.

Shamsara, O., Muhidinov, Z.K., Jafari, S.M., Bobokalonov, J., Jonmurodov, A., Taghvaei, M., Kumpugdee-Vollrath, M., 2015. Effect of ultrasonication, pH and heating on stability of apricot gum-β-lactoglobulin two layer nanoemulsions. Int. J. Biol. Macromol. 81, 1019–1025.

Song, F., Zhang, L.M., Yang, C., Yan, L., 2009. Genipin-crosslinked casein hydrogels for controlled drug delivery. Int. J. Pharm. 373 (1), 41–47.

Teng, Z., Luo, Y., Wang, Q., 2013. Carboxymethyl chitosan–soy protein complex nanoparticles for the encapsulation and controlled release of vitamin D3. Food Chem. 141 (1), 524–532.

Tiwari, A., Bindal, S., Bohidar, H.B., 2009. Kinetics of protein-protein complex coacervation and biphasic release of salbutamol sulfate from coacervate matrix. Biomacromolecules 10, 184–189.

Tiyaboonchai, W., Limpeanchob, N., 2007. Formulation and characterization of amphotericin B–chitosan–dextran sulfate nanoparticles. Int. J. Pharm. 329 (1), 142–149.

Tolstoguzov, V., 2002. Thermodynamic aspects of biopolymer functionality in biological systems, foods, and beverages. Crit. Rev. Biotech. 22 (2), 89–174.

Tolstoguzov, V., 2003. Some thermodynamic considerations in food formulation. Food Hydrocoll. 17 (1), 1–23.

Turgeon, S.L., Laneuville, S.I., 2009. Protein + polysaccharide coacervates and complexes: from scientific background to their application as functional ingredients in food products. In: Kasapis, S., Norton, I.T., Ubbink, J.B. (Eds.), Modern Biopolymer Science. Elsevier, New York, NY, pp. 327–363.

Turgeon, S.L., Beaulieu, M., Schmitt, C., Sanchez, C., 2003. Protein–polysaccharide interactions: phase-ordering kinetics, thermodynamic and structural aspects. Curr. Opin. Colloid Interface Sci. 8 (4), 401–414.

Turgeon, S.L., Schmitt, C., Sanchez, C., 2007. Protein–polysaccharide complexes and coacervates. Curr. Opin. Colloid Interface Sci. 12 (4), 166–178.

van der Sman, R.G.M., 2012. Soft matter approaches to food structuring. Adv. Colloid Interface Sci. 176, 18–30.

Vasbinder, A.J., de Kruif, C.G., 2003. Casein–whey protein interactions in heated milk: the influence of pH. Int. Dairy J. 13, 69–677.

Yamniuk, A.P., Burling, H., Vogel, H.J., 2009. Thermodynamic characterization of the interactions between the immunoregulatory proteins osteopontin and lactoferrin. Mol. Immunol. 46, 2395–2402.

Yang, X., Gao, N., Hu, L., Li, J., Sun, Y., 2015. Development and evaluation of novel microcapsules containing poppy-seed oil using complex coacervation. J. Food Eng. 161, 87–93.

Yang, M., Wu, Y., Li, J., Zhou, H., Wang, X., 2013. Binding of curcumin with bovine serum albumin in the presence of ι-carrageenan and implications on the stability and antioxidant activity of curcumin. J. Agric. Food Chem. 61 (29), 7150–7155.

Ye, A., 2008. Complexation between milk proteins and polysaccharides via electrostatic interaction: principles and applications–a review. Int. J. Food Sci. Technol. 43 (3), 406–415.

Ye, A., Lo, J., Singh, H., 2012. Formation of interfacial milk protein complexation to stabilize oil-in-water emulsions against calcium. J. Colloid Interface Sci. 378, 184–190.

Zhang, Z., Decker, E.A., McClements, D.J., 2014a. Encapsulation, protection, and release of polyunsaturated lipids using biopolymer-based hydrogel particles. Food Res. Int. 64, 520–526.

Zhang, Y., Niu, Y., Luo, Y., Ge, M., Yang, T., Yu, L., Wang, Q., 2014b. Fabrication, characterization and antimicrobial activities of thymol-loaded zein nanoparticles stabilized by sodium caseinate–chitosan hydrochloride double layers. Food Chem. 142, 269–275.

Zhang, L., Qi, Z., Huang, Q., Zeng, K., Sun, X., Li, J., Liu, Y.N., 2014c. Imprinted-like biopolymeric micelles as efficient nanovehicles for curcumin delivery. Colloids Surf. B 123, 15–22.

Zhang, Z., Zhang, R., Tong, Q., Decker, E.A., McClements, D.J., 2015. Food-grade filled hydrogels for oral delivery of lipophilic active ingredients: temperature-triggered release microgels. Food Res. Int. 69, 274–280.

Zhou, H., Sun, X., Zhang, L., Zhang, P., Li, J., Liu, Y.N., 2012. Fabrication of biopolymeric complex coacervation core micelles for efficient tea polyphenol delivery via a green process. Langmuir 28 (41), 14553–14561.

Zhu, Y., An, X., Li, S., Yu, S., 2009. Nanoencapsulation of β-cypermethrin by complex coacervation in a microemulsion. J. Surfactants Deterg. 12 (4), 305–311.

Zimet, P., Livney, Y.D., 2009. Beta-lactoglobulin and its nanocomplexes with pectin as vehicles for ω-3 polyunsaturated fatty acids. Food Hydrocoll. 23 (4), 1120–1126.

Zuanon, L.A.C., Malacrida, C.R., Telis, V.R.N., 2013. Production of turmeric oleoresin microcapsules by complex coacervation with gelatin–gum Arabic. J. Food Process Eng. 36 (3), 364–373.

Further Reading

Dev, A., Mohan, J.C., Sreeja, V., Tamura, H., Patzke, G.R., Hussain, F., Weyeneth, S., Nair, S.V., Jayakumar, R., 2010. Novel carboxymethyl chitin nanoparticles for cancer drug delivery applications. Carbohydr. Polym. 79, 1073–1079.

Laneuville, S.I., Sanchez, C., Turgeon, S.L., Hardy, J., Paquin, P., 2005. Small-angle static light-scattering study of associative phase separation kinetics of blactoglobulin-xanthan gum mixtures under shear. In: Dickinson, E. (Ed.), Food Colloids: Interaction, Microstructure and Processing. The Royal Society of Chemistry, Cambridge, pp. 443–465.

Yang, Z., Peng, Z., Li, J., Li, S., Kong, L., Li, P., Wang, Q., 2014. Development and evaluation of novel flavour microcapsules containing vanilla oil using complex coacervation approach. Food Chem. 145, 272–277.

Part Five

Bioavailability, Characterization, and Safety of Nano-Encapsulated Ingredients

13 Bioavailability and release of bioactive components from nanocapsules

Seid Mahdi Jafari, Iman Katouzian*,**, Hamid Rajabi*, Mohammad Ganje**

*Gorgan University of Agricultural Sciences and Natural Resources, Gorgan, Iran; **Nano-encapsulation in the Food, Nutraceutical, and Pharmaceutical Industries Group (NFNPIG), Universal Scientific Education and Research Network (USERN), Tehran, Iran

13.1 Overview of release

Controlled release is defined as the delivery of bioactvies in response to stimulant or time, represented by the European Directive (3AQ19a). In order to make this process possible, carriers should be engineered for the gradual release of a nutrient and to provide exclusive benefits. A targeted delivery system can be natural or engineered leading to the optimized release extent and absorption in the targeted place of the body. Moreover, delivery networks have the potential to protect the bioactive compounds against harsh processing conditions and digestion (Baker, 1987).

The efficiency of a controlled-release formulation is subjected to the balance between the kinetics of its adhesion and the rate of erosion of the mucus. The concentration of the adherent mucus to the epithelial layer is determined by equilibrium between the exudation rate regulated by the epithelium plus the mechanical shear induced by the process of digestion. Therefore, it seems necessary to understand the adhesion mechanism of nanocarriers in order to raise the quality of sustained release. Functional agents like chitosan or lectin can incite enhanced adhesion in the delivery systems. Nutrient and/or drug delivery is done once the bioactive compound gets to the target cells.

13.2 Release mechanisms

In the area of nutraceutical delivery systems and food formulations, it is of utmost importance to employ mass transport equations along with release kinetics of nutrient molecules so as to understand the release mechanism of bioactives (Grassi et al., 2007). Some of the well-known mechanisms for controlled-release are partitioning, dissolution, osmosis, diffusion, swelling, and erosion (Fig. 13.1). The aforementioned mechanisms impose physical transformation of components present in the food matrix when they are put into a biological environment (Raval et al., 2010).

Nanoencapsulation Technologies for the Food and Nutraceutical Industries

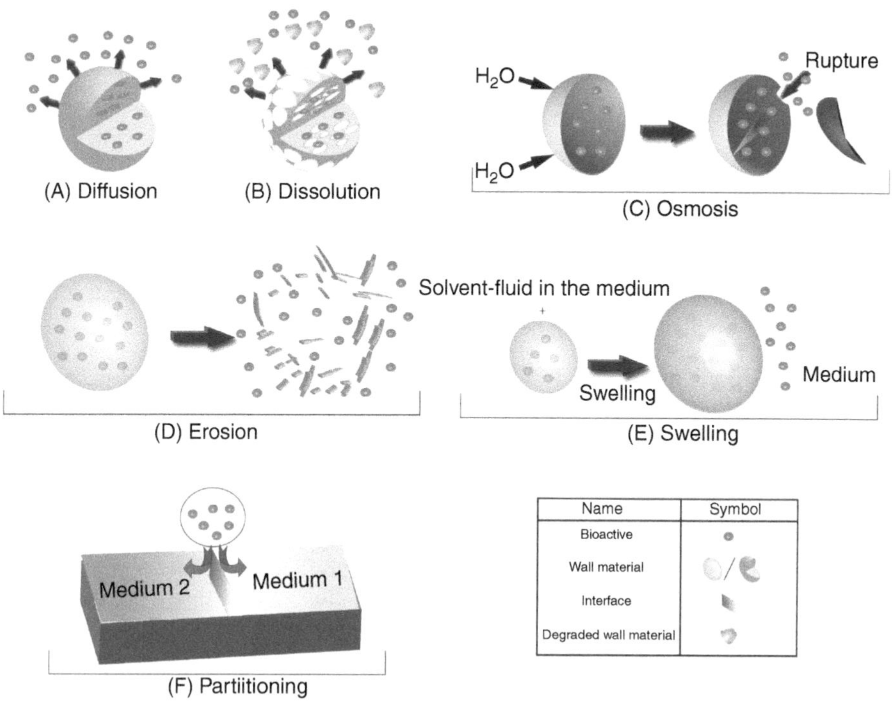

Figure 13.1 Different mechanisms for the release of encapsulated bioactive compounds. (A) Diffusion, (B) dissolution, (C) osmosis, (D) erosion, (E) swelling, (F) partitioning.

Dissolution is defined as the dispersion of the nutrient from its guest molecule to the surrounding medium like tissue or water phase. The extent of a nutrient in a system at saturation is characterized by the solubility of that bioactive compound. During the process of delivery, nutrients often encounter an interface between two materials or phases. The partition coefficient determines the direction of where a nutrient will be heading considering two different phases and is generally estimated by the rate of nutrient solubility in the in the present media. Diffusion is an imperative component of numerous controlled release frameworks. Osmosis is another release phenomenon in which particular solutes (osmolytes) are separated from water, in other words the membrane is permeable to water but will not let other solutions pass through it. In this system, water passes through the selectively permeable membrane until the level of both solutions become even on both sides of the membrane (Santus and Baker, 1995). Water absorption by a polymer network will enhance its volume, which is known as swelling. Swelling is normally a commencement for polymer dissolution. Nevertheless, if the polymer chain is too long or is not compatible with water, swelling may take place without dissolution. Swelling of a polymer may be influenced by nutrients or other compounds acting as osmolytes, which draw water into the polymer. Swelling can be induced via selection of suitable polymers plus the modification of external parameters, such as temperature and pH. In the pharmaceutical field, biodegradable delivery systems are of great interest, especially the therapies that include injection since retrieval is not necessary following the complete release of the drug. Controlled release leads to targeted delivery and will heal cells associated with toxicity.

13.3 Bioavailability of nutraceuticals and their uptake in gut

13.3.1 Bioavailability of nutraceuticals

Several definitions of bioavailability exist, but broadly it refers to the uptake of an ingested bioactive substance that launches the normal body functions (Aggett, 2010; Hurrell and Egli, 2010). Bioavailability captures two essential features: (i) *the absorption rate*—how fast the bioactive agent enters the systemic circulation; and (ii) *the absorption extent*—how much of the bioactive reaches the systematic circulation (Borel et al., 2013; Johnson et al., 2005). According to the FDA,[a] the bioavailability is the absorbance of a bioactive ingredient and further distribution via the circulatory system so that it reaches the active site to impart its functional roles. (Pathak and Raghuvanshi, 2015). Poor bioavailability reflects failure of the ingredient to reach the minimum effective level in the blood to impart its expected biological performance.

Typically, a number of different steps determine the biological fate of bioactive agents after ingestion (Aggett, 2010; Katouzian and Jafari, 2016):

- Release of the bioactive agent from the protective shell,
- Digestion by enzymes within the intestine,
- Adherence and uptake by the mucosal layer of the intestine,
- Moving across the gut layer (passing through and/or between the epithelium cells) to the lymphatic system or portal vein,
- Systemic distribution,
- Systemic deposition (storage),
- Metabolic and chemical reactions in living cells,
- Excretion (via urine or feces).

Various external and internal factors determine the overall bioavailability rate of an ingested bioactive agent. The external factors include the nature of the bioactive agent and the chemical structure of the food matrix, while internal factors are gender, age, wellbeing, nutrient status and life phase (e.g. pregnancy). Factors such as nutrient status determine how much of a nutrient is actually used, stored, or excreted, while some definitions of bioavailability are only centered on the assimilated extent of the nutrient (Heaney, 2001). The bioavailability of naturally-occurring food macromolecules like carbohydrates, proteins and lipids are usually more than 90% of the consumed food product. On the other hand, the bioavailability of micronutrients, such as vitamins and

[a] The US Food and Drug Administration

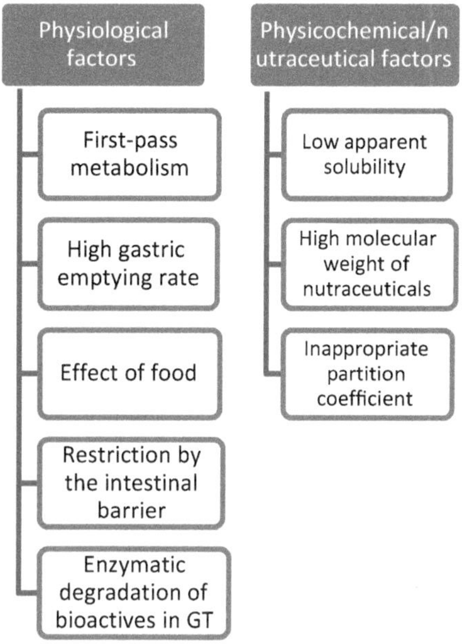

Figure 13.2 Factors determining the bioavailability of nutraceuticals and other bioactive components.
Source: Reprint with permission from Jafari, S.M., McClements, D.J., 2017. Nanotechnology approaches for increasing nutrient bioavailability. In: Advances in Food and Nutrition Research, Boston, MA, vol. 81, Elsevier.

minerals and nutraceuticals (such as flavonoids and carotenoids) can vary greatly depending on their molecular and physicochemical characteristics. For instance, the bioavailability of hydrophobic nutrients is limited by their poor solubility, high melting point, chemical instability, poor bioavailability, and ingredient interactions (McClements, 2013). Overall, several factors determine the bioavailability of nutrients and nutraceuticals as mentioned in Fig. 13.2.

Insights into how to formulate food-grade delivery systems with improved bioavailability characteristics can often be obtained from the pharmaceutical industry. Research has shown that the bioavailability of many poorly water-soluble medications can be dramatically improved via nanoencapsulation (Aboalnaja et al., 2016; Thanki et al., 2013). The relatively high surface area of these systems can lead to the rapid dissolution of drugs within the gastrointestinal tract (Erfanian et al., 2014). Consequently, there is great interest in developing nanoscale delivery systems for nutraceuticals and other bioactive components intended for food use.

To produce biologically effective functional foods it is essential that the encapsulated nutraceuticals or nutrients have a sufficiently high bioavailability, *that is,* the bioactives are absorbed into the blood stream at a level that is high enough to have a beneficial health effect (Assadpour et al., 2016a,b). There are many examples where the potential health benefits of bioactive agents are not fully realized because of their low oral bioavailability. Vitamin D may be consumed in the form of pills or capsules or in food formulations, such as fortified milk and fruit juices (Katouzian and Jafari, 2016). Human studies have shown that the fraction of vitamin D absorbed into the blood is typically relatively low for most oral formulations (Barakat et al., 2013; Borel et al., 2015). Concerning the fact that the dose of nutraceuticals in functional foods is rather low (compared to pharmaceutical products), a biological effect is therefore questionable. The same applies to the various flavonoids, which are becoming increasingly popular as antioxidants, such as rutin, apigenin, hesperidin, hesperetin, and quercetin (Esfanjani and Jafari, 2016). In general, bioactive molecules are best absorbed when they are fully dissolved within the GIT fluids. Many highly hydrophobic bioactives (such as vitamin D, vitamin E, carotenoids, and curcumin) have poor water-solubility but may still dissolve in aqueous GIT fluids because of the presence of colloidal particles that have hydrophobic regions that can solubilize them, that is, mixed micelles and vesicles. In these cases, the bioavailability of the bioactive agent scales directly the type and amount of coingested lipids. Another strategy utilized in the pharmaceutical industry is to produce nanocrystals of the bioactive substance that may be directly absorbed by the epithelium cells (Jafari et al., 2015).

In summary, technologies developed in the food industry have been inspired from the pharmaceutical sector to improve the performance of nutraceuticals. In this chapter, we focus on the utilization of nanotechnology-based approaches, which have been widely implemented in the drug delivery systems and are now finding increasing use in the food industry.

Some of the mechanisms associated with improving the bioavailability of nutraceuticals are based on understanding the uptake of nanoparticles within the GIT, for example, targeting of gut-associated lymphoid tissue (GALT), absorption by Peyer's patches, and improvement in the gastrointestinal solubility (McClements and Xiao, 2012; Pathak and Raghuvanshi, 2015). These mechanisms are briefly described in Fig. 13.3 and schematically in Fig. 13.4.

13.3.2 Enhancers and inhibitors of nutrient bioavailability

Bioactive compounds may interact with each other or their surrounding medium, resulting in the alteration of bioavailability. Enhancers can make a nutrient soluble or to prevent their reaction with inhibitors. For instance, as carotenoids are lipophilic, adding minute amounts of fat or oil to the meal (35 g per meal) will result in the improvement of their bioavailability (van het Hof et al., 2000). It has been suggested that vitamin C is able to enhance the absorption of iron by two- or threefolds (Teucher et al., 2004). This implies, for instance, by consuming a glass of orange juice in addition to a bowl of breakfast grain, our body can be supplied with the most of the iron existing in the food.

Inhibitors obstruct nutrient absorption via; (1) disguise the nutrient by binding it to other elements, or (2) lowering the absorption of a nutrient by rendering it insoluble. Phytic acid is a typical example of inhibitors which are found in certain plant foods

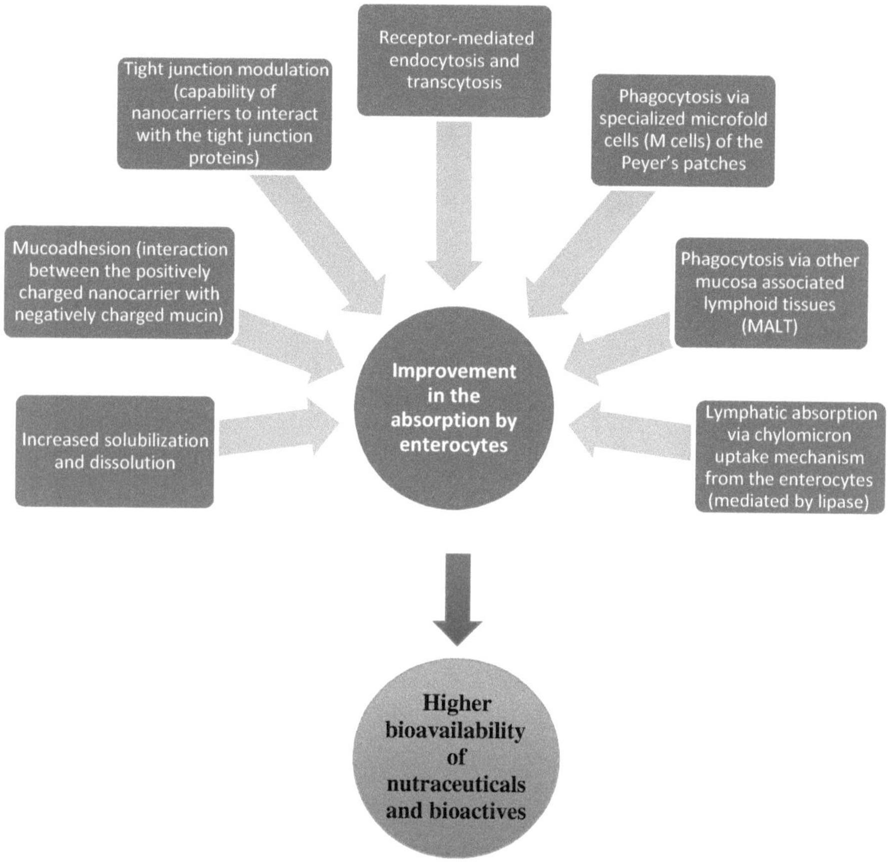

Figure 13.3 Role of nanocarriers in higher bioavailability of nutraceuticals through improvement in the absorption by enterocytes.
Source: Reprint with permission from Jafari, S.M., McClements, D.J., 2017. Nanotechnology approaches for increasing nutrient bioavailability. In: Advances in Food and Nutrition Research, Boston, MA, vol. 81, Elsevier.

(e.g., nuts, seeds, wholegrain cereals). This compound has a robust binding affinity for minerals including calcium, iron, and zinc that make the micronutrients unavailable for absorption (Zhou and Erdman, 1995). Fermentation, soaking, and germination are considered as useful methods to reduce the amount of phytic acid (Hotz and Gibson, 2007). Also, the interaction between calcium and nonhaem iron is a case of competition for the same absorption procedure. Both minerals link to a transporter on the surface of enterocytes, nonhaem iron is able to enter the cells by this trend, while calcium hinders the access of iron as it maintains its position in the doorway. This reaction is obvious, where calcium and/or iron supplements are consumed in lieu of the meal setting (Gibson, 2007).

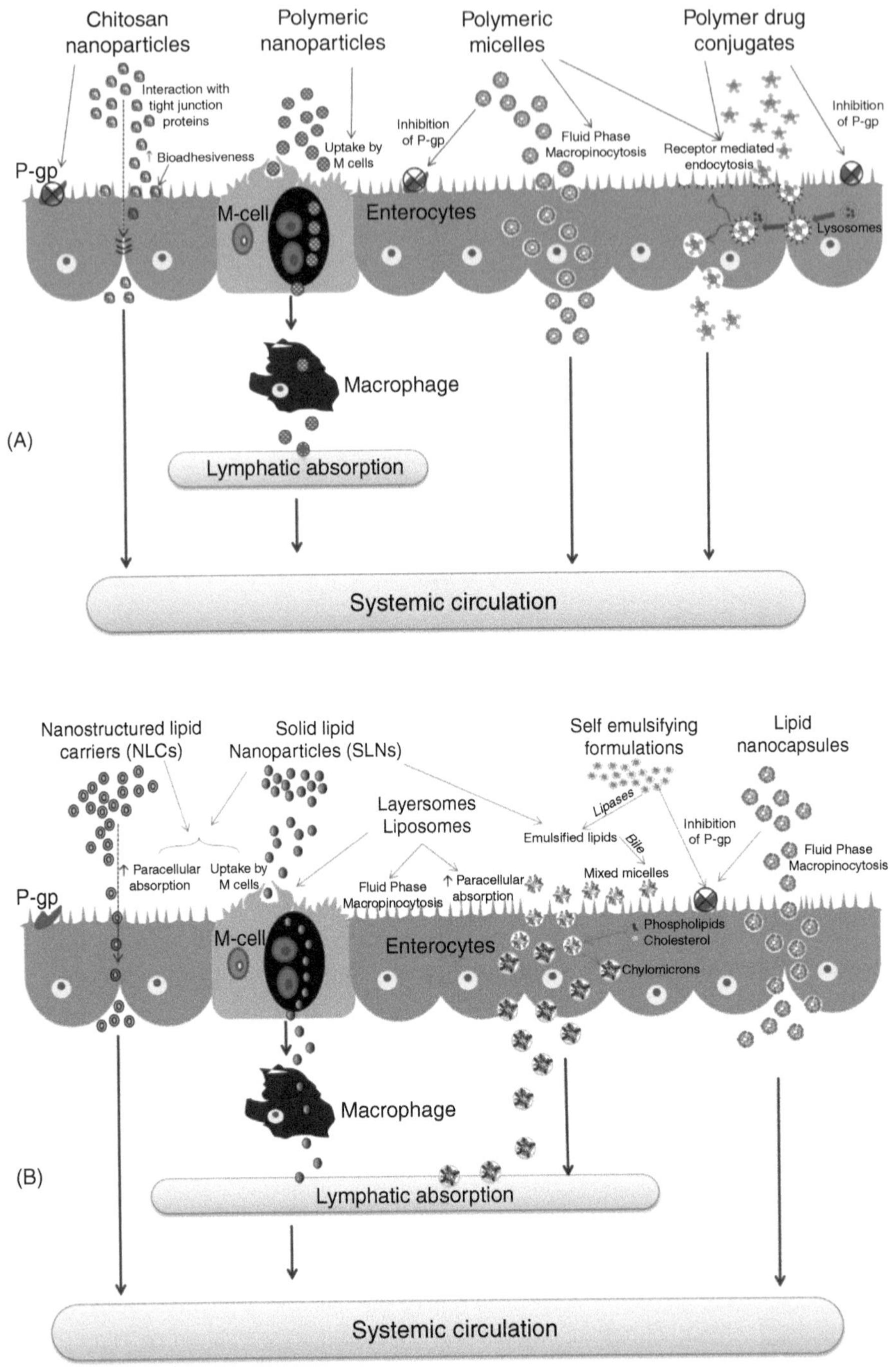

Figure 13.4 Schematic representation of the role of nanodelivery systems in enhancing the absorption mechanisms leading to improved bioavailability. (A) Polymeric nanoparticles, (B) lipid nanoparticles.

Source: Reprint with permission from Thanki, K., Gangwal, R.P., Sangamwar, A.T., Jain, S., 2013. Oral delivery of anticancer drugs: challenges and opportunities. J. Control. Release 170(1), 15–40.

13.3.3 Impact of host factors

Internal factors can be ordered into gastrointestinal and systemic items. Gastrointestinal parameters are represented by the absorption trend of vitamin B12. Furthermore, systemic factors comprise the absence of a particular nutrient or alterations in the physiologic state, for example, pregnancy. Concerning both of the conditions, the uptake and utilization of nutrient will be controlled by the body to maintain the equilibrium of that certain nutrient in body fluids (Gibson, 2007). Calcium and zinc are normally regulated by this procedure. Nevertheless, digestive disorders and inflammation may decrease the absorption rate of the gut. As an example, the absorption of iron is downregulated in infections like the common cold (Lynch, 2007).

13.3.4 Scaling absorption, metabolism, and tissues affected by nutrients

The estimation of nutrients applied in food products may not be a pragmatic practice due to their further interactions in various cells for a couple of reasons. In order to explore the health impacts of food components in human health, first it is necessary to acquire information about the release of nutrients from the matrix they been entrapped in. Moreover, the rate and concentration of the nutrients plus the site of absorption and their health impact on the cell genome should be considered.

If it is intended to measure the amount of the drug or nutrient that has reached the target zone in the body, samples should be provided from the treated tissue or cells to detect the concentration of the administered bioactive compound. Also, a more practical means would be the employment of various doses of the intended groups or volunteers and obtain the efficient dose to consider it as the optimized measurement. For example, concentration after some time in the entire blood, or a part of that, such as plasma, erythrocytes, or a predetermined number of white blood cells. This surrogate approach is defined Schümann et al. (1997) attributed this approach to the proportion of an oral dose of bioactive compound that reaches the circulatory system. However, understanding the ADME (absorption, distribution, metabolism, and excretion) profile of a nutrient requires the modeling of components (Jacquez, 1996) in order to understand the biological effects of the components, for example, gene expression, protein synthesis plus the functionality that describes the health outcome.

13.3.5 Mucoadhesion

Physiological activity in the gastrointestinal (GIT) structure is intense, specifically due to its contractions and propulsive activity, which affect the controlled- release of functional food and limit its efficiency (Fig. 13.5). Mucoadhesion is an intricate mechanism responsible for developing the oral delivery in the gastrointestinal tract and enhancing the adhesion of the controlled-release formulation to the mucus layer. Bioadhesion involves adhesives derived from natural polymeric substances; also mucoadhesion is the process of the adherence of bioadhesives to the mucus (Gu et al., 1987).

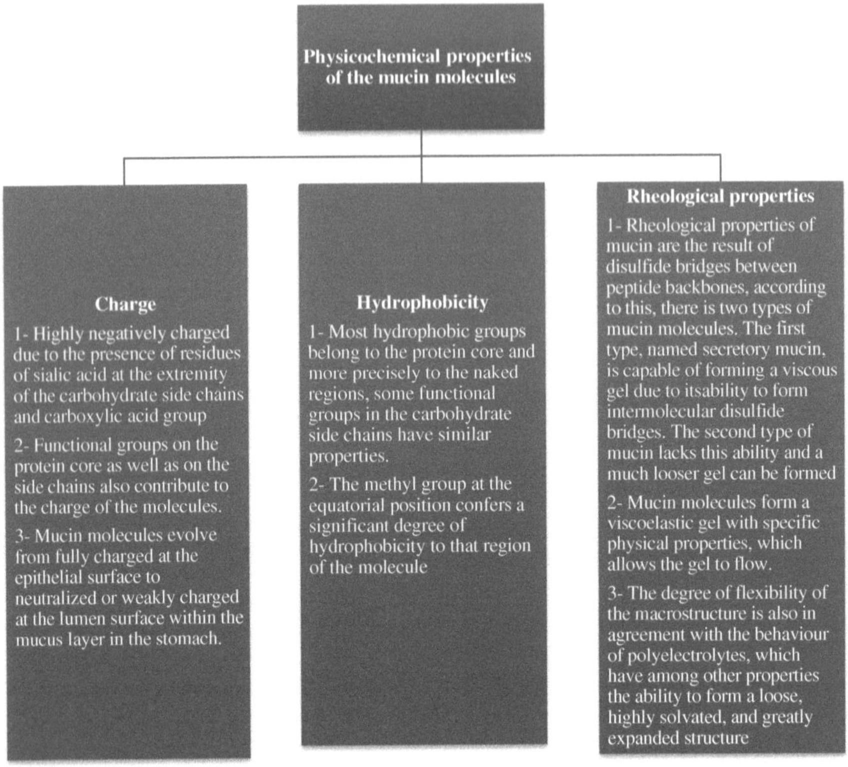

Figure 13.5 Physicochemical properties of mucin molecules.

One single hypothesis cannot thoroughly support such a complicated process, therefore five different theories are proposed concerning this issue. The theories include adsorption, electronic, wetting, fracture, and diffusion. Nonetheless, these theories cannot describe the adhesion mechanisms properly, until they clarify the important parameters for the bioadhesion procedure (Fig. 13.6).

The formation of chemical and mechanical bonds is subjected to many parameters, which can influence their strength. These elements may comprise the polymer structure or the media circumstances. The principal factors are underlined in subsequent sections (Fig. 13.7).

13.3.6 *Molecules influencing the mucoadhesion process*

Several types of molecules have been tested regarding their mucoadhesion. Yet, polymers are classified into two categories known as generations. The first generation is relatively water soluble and is able to yield high viscosity solutions even at low levels. This group is often pH sensitive since their structure is derived from polyacrylic acid. Guggi et al. (2004) suggested that it is possible to enhance the mucoadhesion up to two- to fourfolds by lowering the pH below 5. Indeed, mucoadhesion occurs in

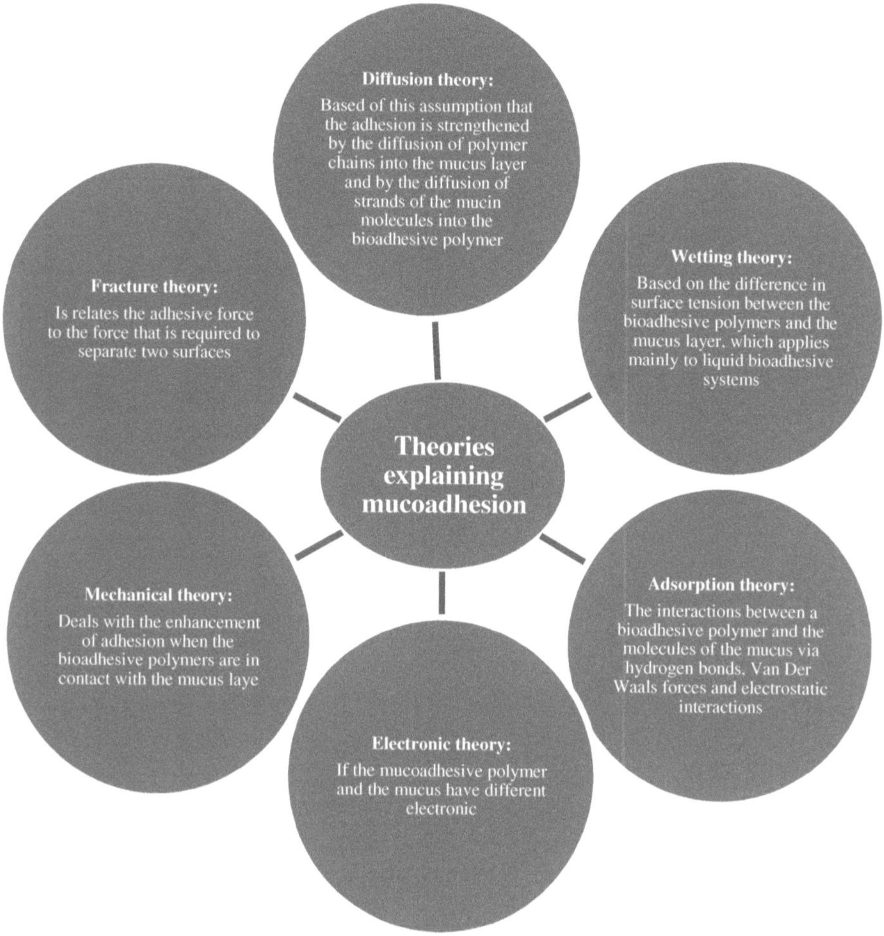

Figure 13.6 Theories explaining mucoadhesion.

the acid form of carboxyls. On the other hand, at pH higher than 6, repulsion forces dominate and restrict the adhesion process. When nearly 80% of the carboxyl groups acquire protons, a favorable adhesion process is accomplished. Chitosan is a typical example of this group. Another generation of mucoadhesive structures is by their specific adherence to mucosal surfaces like lectin.

13.3.7 Nutrient transfer across the mucus layer

The uptake of nutrients is affected by the mucoadhesion in combination with the diffusion of nutrients through the mucosal layer earlier from reaching their target zones. It has been shown that the mucus prevents the higher extent of nutrient compared to an equivalent aqueous layer (Khanvilkar et al., 2001). For instance,

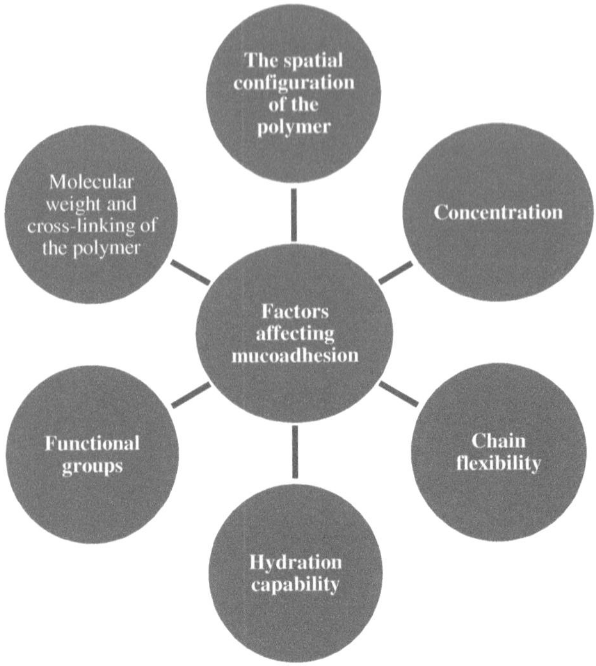

Figure 13.7 Factors affecting mucoadhesion.

structures, such aminoglycosides, fatty acids and some antibodies have a dramatically slow diffusion across the mucus medium. The retardation process can be attributed to the mucus structure. Two components in the mucus structure are of extraordinary significance: First, the mucin molecules, which constitute the main part of the formed gel. Second, the lipids, which scales directly with the diffusion mechanism across the mucus coating.

The mucin molecules generate a viscoelastic gel that slows down the procedure of diffusion since molecules must somehow pass the porous structure. In addition, the amount of water (free water) being trapped in the mucus matrix imparts a significant task in the diffusion of nutrients. Therefore, the reduction in the amount free water equals reduction in the diffusion rate (Bhat et al., 1995).

The mucin molecules comprise hydrophobic parts on the protein structure together with negative charges. Accordingly, positively charged, light molecules have a strong affinity for the negatively charged mucin. Due to the electrostatic interactions, even simpler complexes, such as potassium ions transfer rate are limited (Desai et al., 1992). On the contrary, the ζ-potential does not impart a dramatic effect in the transport of nanospheres across the mucus. Moreover, the electrostatic interactions are able to interrupt the diffusion of molecules. Electrostatic interactions form relatively weak bonds and this deficiency can be covered by implementing other molecules. Eventually, different sorts of interactions can happen when nutrients migrate through the mucus stratum (Larhed et al., 1998).

13.4 Different approaches for studying the release profile

Usually, the description of bioavailability in food products is not a simple task due to the interactions of food matrix during consumption plus the effects of food preparation techniques. Meanwhile, different techniques are used for monitoring and determining the release rate and value of different food ingredients in tests (Assadpour and Jafari, 2017; Assadpour et al., 2017). In vivo and in vitro studies need to be carried out for determining the safety of consuming nanocapsules since currently, no common standard regulations are available for the fabrication and consumption of nanofoods (Rashidi and Khosravi-Darani, 2011). In Chapter 15, safety aspects of food nanocapsules and nanomaterials are explored in detail.

13.4.1 In vivo tests

In vivo studies are basically applied to test animals, such as mice, fish, and rabbits (Goodman et al., 2011; Rubió et al., 2013). These assays provide useful information about the biological and physiological factors including the metabolism of the employed formulations plus their bioavailability (Holst and Williamson, 2008). The in vivo animal models provide valuable information to determine oral bioavailability in a conclusive manner. Unlike the man-made machinery that can be examined step-by-step, in vivo models provide only unilateral end result determining the fractions of ingested dosages that are available to the systemic circulation without providing enough explanation of the events that lead to it.

There are various methodologies to evaluate the bioavailability of different food bioactive ingredients. For example, considering the minerals, in vivo bioavailability can be evaluated by techniques, such as metabolic balance, radioisotopes, stable isotopes, etc. The vital micronutrient minerals in the food industry are iron, copper, zinc, selenium, and manganese. Metabolic balance is a term used for checking the extent of consumed minerals plus the excreted quantity during a specified period of time. This term scales directly with the minerals that are highly absorbed in the GI tract. Moreover, concerns of the measurement are normally lower than the computed difference between the ingested and excreted minerals.

Understanding the level of repletion is another method implemented in lieu of metabolic balance for the minerals with poor absorption profiles. The repletion strategy is related to the deficiency of some minerals in the body so that the need for these minerals is approximately high (Sacks and Houchin, 1978). The plasma concentration method is appropriate for minerals in high doses, where the pace of absorption is higher than excretion. Also, this method is best suited to the nutritional items containing highly-absorbed minerals (Navarro and Wood, 2003). The plasma concentration of minerals is basically assessed via the atomic absorption.

Employing radioisotopes is an efficient approach to pursue the mass balance (rate of metabolization and elimination) of minerals within the ingested food products plus the body retention and imaging studies (Siegel et al., 1999).

The bioavailability of organic agents (water-soluble/insoluble) are like those utilized for minerals. Generally, the absorption rate of lipophilic organic compounds is more than the highly soluble minerals (namely, calcium and iron). Therefore, in this case, it is common to use only the metabolic mass balance (ingested minus excreted) (Williamson and Manach, 2005). Nevertheless, stable isotopes and mass spectrometry are usually preferred to pursue the byproducts of these ingredients (Emken, 2001; Zhang and Zuo, 2004).

In vivo studies have been performed to determine the bioavailability of nanoencapsulated food bioactive components. For instance, Gou et al. (2011) used the nanoprecipitation method to nanoencapsulate curcumin by PEG and PLGA as the wall materials. The nanocapsules particle size was about 27 nm and the encapsulation efficiency was 99%. Furthermore, the entrapped curcumin depicted a stronger antineoplastic effect compared to the free curcumin. As a result, the in vivo assays presented higher encapsulation efficiency, gradual release, and higher rate of bioavailability. In another study, Park et al. (2007) analyzed the impact of interfacial structure on the digestive potential of emulsified and encapsulated lipids by performing in vivo assays. Soybean O/W emulsions were prepared via the layer-by-layer electrostatic deposition technique, which contained coated lipid droplets with the composition of lecithin or lecithin in combination with chitosan. There were no differences in body weights, food intake, major organ weights, or fecal fat contents between all treatments, where total fat absorption was >90%. Unlike previous studies, entrapping lipid molecules by chitosan does not limit the digestion process. Thus, chitosan can be utilized to encase lipids and lipid-soluble materials without any modification in the bioavailability.

Zhou et al. (2010) applied high-pressure homogenization method to prepare a lecithin nanoemulsion with the droplet size of 58–92 nm. In an in vivo study, the hydration potential of a lotion was enhanced as it was formulated with lecithin-loaded nanoemulsions (2.5 times higher with the concentration of 10%). Other researches have suggested that insulin-loaded nanoliposomes comprising chitosan, PEG, or mucin elevate its uptake and remedial impact. These results are obtained via the interaction of engineered nanoliposomes with the mucus stratum along with the retention ability under the harsh circumstances of the GI tract (Iwanaga et al., 1999; Torchilin, 2005; Wu et al., 2004).

Wang et al. (2010) utilized wheat feruloyl oligosaccharides (FOs) to suppress the activity of free radicals in the plasma of treated rats. The oxidative markers and the activity of antioxidant enzymes were assessed in the plasma volume, which had a normal diet fortified with 1% FOs. The antiradical impact of 0.5 mg FOs was evaluated via 2,2 azobis (2-amidinopropane) dihydrochloride as the free radical inducer. As a result, plasma antioxidant enzyme activity was enhanced via the administration of FOs, besides malondialdehyde and oxidized glutathione levels were decreased considering the groups fortified with FOs in comparison to the control. Patrignani et al. (2016) tested the antioxidant performance, antihypertensive impact, and Maillard reaction output (MRO) emanated from biscuits in albino rats. The rats were nourished with dissimilar diets for 6 weeks and then the antioxidant performance was assessed in the serum as well as lipid peroxidation by the end of the test. In conclusion, the dietary MRO declined the growth efficiency, increased the antioxidant activity of serum, enhanced the proportion between lactic and the whole aerobic bacteria, increased the water entrapment of feces along with lowering the blood pressure but the mineral absorption rate did not decrease.

Qingming et al. (2010) tested the antioxidant potential of barley malt extract. Scavenging properties of the biomacromolecule was evaluated on hydroxyl and superoxide radicals plus the protection against reactive oxygen species. In this work, the Kunming mice were administered with D-galactose to determine the antioxidant performance of malt extract. Regarding both in vitro and in vivo assays, the extract displayed great antioxidant potential by scavenging hydroxyl- and superoxide radicals, robust reduction along with preventing the damage of biomacromolecules, such as DNA and proteins. Malt extract maintained the antioxidant activities of enzymes, decreased the malondialdehyde levels in hepatocytes and brain cells plus carbonyl concentrations and elevated the overall antioxidant potential in mice treated with D-galactose.

13.4.2 *In vitro assays*

These tests are normally carried out to assure the effectiveness of a formulation and to get confirmation certificates from the regulatory organizations. At the beginning of production stages, usually in vitro tests are in priority since they are simple and factors that control the dissolution or uptake of a nutrient can be modified. Also, in vitro assays are beneficial quality control tools. Ceratin protocols prepared by regulatory centers should be followed when in vitro methods are chosen as a means for quality control tool or as an alternative for in vivo approaches.

One of the prerequisites for controlled release systems is the retention of the nanocarrier in the intestinal tract as long as the nutrient is released in the small intestine and before it crosses the small intestine (in some cases the colone is the delivery zone and then the system should be designed in a way to be released in that place).

In vitro tests have represented that some carotenoids like lycopene and β-cryptoxanthin trigger the process of ossification (bone generation) and mineralization preventing osteoporosis (Yamaguchi et al., 2005). Hence, cell line models are used to determine the uptake of carotenoids, thus the factors (food network degradation rate and the degree of solubility of carotenoids) with the limitations to be controlled is ignored (Briviba et al., 2001).

In a study, Lee and Heo (2000) simulated the survival of bifidobacteria in the calcium alginate network and presented a model to examine the impact of gel concentration, particle size and number of initial cells on the viability of bifidobacteria. Bulky alginate beads (>1 mm) created a rough texture in the microbial solution, whereas small-sized beads (<100 mm) permit a rapid diffusion of molecules in the matrix (Hansen et al., 2002).

Moreover, Wu et al. (2008) fabricated a nanoencapsulation system with quercetin and aminoalkyl methacrylate copolymers as core and wall materials consecutively. Results indicated that the radical scavenging property of entrapped quercetin against superoxide anions was enhanced up to 883- and 1377 times as compared to the free type.

Neves et al. (2013) examined the behavior of solid lipid nanoparticles (cetyl palmitate and polysorbate 60) loaded with resveratrol in the simulated gastric fluid along with the conditions of initial 3-h incubation at 37°C with a pH of 1.2, proceeded by a

33-h incubation in the pH of 7.4. In effect, nanocarriers were stable within the GI tract environment and depicted a sustained release in the GI mimicked fluids.

In another in vitro study, Niu et al. (2011) suggested that sodium glycocholate nanoliposomes exhibit enhanced stability against trypsin, pepsin and α-chymotrypsin disintegration and increased percolation across the intestine epithelium.

Casein-carbohydrate complexes are promising protective mediums for encapsulating probiotics within an O/W emulsion (Crittenden et al., 2006). In vitro experiments showed that the probiotic cells (*Bifidobacterium infantis*) may resist the ingestion conditions in simulated gastric solution and were liberated in the simulated intestinal liquid.

Iwanaga et al. (2003) explored release profile of insulin from gelatin microspheres prepared with various cross-linking densities via in vitro tests and concluded that the release rate could be regulated by the density of cross-linking.

Furthermore, Tiyaboonchai et al. (2007) formulated SLN loaded with curcumin by the microemulsification technique at the temperature of 75°C. At optimized process conditions, lyophilized curcumin being loaded in SLN illustrated a spherical shape having a mean particle size of 450 nm and incorporation efficacy of 70%. In vitro release studies showed a prolonged release of the curcuminoids from the SLN for up to 12 h. Moreover, lyophilized curcuminoids represented satisfactory physicochemical stability over the storage period of 6 months.

A study reported that Eudragit microcapsules incorporating vitamin C were effective candidates for the treatment of colorectal cancer consumed by the oral administration. The microcapsules were prepared by the spray-drying method. Also, their size and morphology were determined by using SEM and then the release profile was obtained by the dialysis method. To sum up, the implemented microcapsules revealed a favorable size distribution and morphology, which makes them proper candidates for the delivery of vitamin C to treat the colorectal cancer via the oral route.

Altogether, in vitro practices are very useful tools for testing the bioavailability efficacy of nanovehicles and their following interaction with the employed cell lines. Besides, the interaction mechanisms of the nanoliposomal structures and cells should be determined via experiments in order to elevate their uptake, enable a targeted release in the predetermined organelles plus the duration of the process for beneficial in vivo effects (Acosta, 2009).

13.5 Release modeling

The immediate (burst) release of encapsulated nutrients may cause uncontrolled release, decreasing the absorption rate in the gut and defect the technological and pharmaceutical purpose of the whole process. The proplonged release can be defined as release at a longer time immediately after administration. The goal of extended release formulations is the availability of encapsulated nutrients over extended period of time (Gaonkar et al., 2014). Extended release is the programmed release of finite amounts of encapsulated nutrients at different periods of time.

Whenever a new formulation of encapsulated materials is prepared, it should be determined whether the new formulation indicates dissolution in the right manner. Recently pharmaceutical/nutraceutical industries and the registration authorities are centering on the guest molecule dissolution/release studies. Regarding the quantitative studies of core material release properties of any formulation, it is best the correct values to be obtained via suitable mathematical formulas as a function of formulation features (Lakkis, 2007; Assadpour et al., 2017a; Mehrnia et al., 2017). It is likely that theoretical analysis of these mathematical models is attained throughout the process. Meanwhile, some empirical equations are considered suitable, where the theoretical concept does not exist.

Different conditions of media, such as temperature, water activity, and dissolution medium can influence the release profile of encapsulated flavors (Fabra et al., 2012). Besides, storage condition and food structure have undeniable effects on release properties. Normally a programmed release is necessary for an optimized process. Therefore, modeling methods are applied to have a preprogrammed release for the core materials in a desired manner over prolonged periods. With this release procedure, the goal of nutraceutical delivery to the targeted place at a desired value will be reached.

13.5.1 Mathematical modeling

Mathematical modeling of release gives information about mass transfer and chemical reactions occurring in food and nutraceuticals delivery systems as well as the effect of parameters used for designing the nanocarriers, such as the device geometry and bioactive loading, on nutrient release mechanism. Hence, the optimized design needed for a particular bioactive release profile can be predicted via a systematic procedure with a minimum number of experimental investigations. Food systems with complex interfaces are marked with nonlinear constructive equations in accordance with stress-deformation manner dependency. The dynamic behavior of the complex encapsulation system is actually based on the Interfacial Transport Phenomena (Sagis, 2008). Consequently, with The developments of mathematical models may lead to accurate predictions in regard to the delivery performance (Narasimhan and Peppas, 1997) and ameliorate the design of delivery systems.

Mathematical modeling can be very useful in different aspects of encapsulation and release phenomena (Versypt et al., 2013). It can be used in the prediction of nutrients release rate plus their diffusion behavior through polymers, thus laborious experiments are less likely to be employed. Moreover, the optimization of release profile, determining the Physical mechanism of nanomaterials transport by relating release data with mathematical models, Prediction of the effect of design parameters (morphology and size) is obtained via mathematical modeling in a fast and simple trend (Gaonkar et al., 2014).

Indeed, these models describe the concentration of dissolved nutrients (C) as a function of time (t) indicated by $C = f(t)$ equation. Herein, the most common mathematical models applied for this purpose are shown in Table 13.1 and explored subsequently (Shaikh et al., 2015).

1. *Zero-order model*
   This model scales directly with the dissolution of a nutrient and is only a function of time and is independent of the bioactive concentration. This model is used in the systems with low release rate.

Table 13.1 Common used mathematical models for analysis of release

Type of model	Model equation	Model parameters
Zero-Order Model	$C = C_0 - K_0\, t$	• C = Amount of nutrient release or dissolved • C_0 = Initial amount of nutrient in solution • K_0 = Zero order rate constant • t = time
First Order Kinetic Model	$\text{Log } C = \text{Log } C_0 - Kt/2.303$	• C_0 = Initial concentration of nutrient • K = First order constant • t = Time
Higuchi Model	$C = [D\,(2qt - C_s)\, C_s t]^{1/2}$	• C = Total amount of nutrient release per unit area of the matrix • D = Diffusion coefficient for the nutrient in the matrix • qt = Total amount of nutrient in a unit volume of matrix • C_s = Dimensional solubility of nutrient in the polymer matrix • t = Time
Korsemeyer-Peppas Model	$\frac{C_t}{C_\infty} = kt^n$	• $\frac{C_t}{C_\infty}$ = Fraction of nutrient release at time t • k = Rate constant • n = Release exponent
Hixson-Crowell equation	$C_t^{\frac{1}{3}} - C_t^{\frac{1}{3}} = K_{HC}t$	• C_t = Amount of nutrient released in time t • C_0 = Initial amount of nutrient in the tablet • K_{HC} = Rate constant for Hixson-Crowell equation
Weibull Model	$C = C_0[1 - exp\left[\frac{-(t-T)^b}{a}\right]$	• C = Amount of dissolved nutrient as a function of time t • C_0 = Total amount of nutrient being released. • T = Lag time measured as a result of dissolution process parameters • a = Scale parameter that describe the time dependence • b = Shape of dissolution curve.

Table 13.1 Common used mathematical models for analysis of release (*cont.*)

Type of model	Model equation	Model parameters
Hopfenberg Model	$\frac{M_t}{M_\infty} = 1 - \left(1 - \frac{k_0 \cdot t}{C_0 \cdot a}\right)^n$	• M_t is the amount of nutrient dissolved in time t • M_∞ is the total amount of nutrient dissolved when the pharmaceutical dosage form is exhausted • k_0 is the erosion rate constant • C_0 is the initial concentration of nutrient in the matrix • a is the initial radius for a sphere or cylinder or the half thickness for a slab • n is 1, 2 and 3 for a slab, cylinder, and sphere, respectively
Baker–Lonsdale Model	$f = \frac{3}{2}\left[1 - (1 - \frac{M_t}{M_\infty})^{\frac{2}{3}}\right] \frac{M_t}{M_\infty} = k_t$	• M_t is the nutrient release amount at time t M_∞ is the amount of nutrient released at an infinite time • k is the release constant which corresponds to the slope of the graph when plotted as $[d(M_t/M_\infty)]/dt$ with respect to the root of time inverse

2. *First order kinetic model*
   The absorption and excretion of some molecules can be predicted by this model. Also, the first order diffusion depends on the concentration of the nutrient.
3. *Higuchi model*
   In the matrix systems, Higuchi model is the common mathematical equation for the description of nutrient release and usually fits the different geometrics and porous systems. It is suggested that the initial active ingredient concentration is dramatically higher than the solubility in the food network, which leads to the one-dimensional diffusion within the system. Moreover, no swelling or dissolution of the matrix occurs as the diffusivity of the bioactive compound is constant and proper volume of dissolution medium are always attained in the release environment and finally, the active agent particles are much smaller than the microcapsule systems (Higuchi, 1963).
4. *Korsemeyer-Peppas model (the power law)*
   This model describes the release of core molecules from a polymeric network (Korsmeyer et al., 1983) with a simple relationship. In order to illustrate the procedure of nutrient release, in the first place, 60% of the acquired data is adjusted with the Korsemeyer-Peppas model. More importantly, Following theories have to be considered in this model:
   a. The typical equation is useful for short times and where the curve's proportion is $Mt/M\infty < 0.6$, it can be employed to obtain the index n.

Table 13.2 *n* **Values and related release mechanisms**

Release exponent (n)	Nutrient transport mechanism	Rate as a function of time
$n = 0.5$	Fickian diffusion	$t^{-0.5}$
$0.5 < n < 1$	Non Fickian diffusion	t^{n-1}
$n = 1$	Case II transport	Zero order release
n Higher than 1	Super Case II transport	t^{n-1}

b. The core material is released in one dimension.
c. The proportion of length to thickness in the system should be at least 10.

The *n* value describes different liberation profiles of matrices with cylindrical shapes; moreover the value of *n* specifies the release type of a nutrient as shown in Table 13.2.

5. *Hixson–Crowell model*
This model is used in systems with variable surface areas and particle diameters. In the Hixson-Crowell equation, the area of particles scales directly with the cube root of its volume. Furthermore, the release pace is limited by the dissolution rate in this model and diffusion has an inconsiderable effect on the release profile (Ramteke et al., 2014).
6. *Weibull model*
A major application of Weibull model is in the area of bioactive dissolution or dosage release in pharmaceuticals. Since this model is generally empirical-based it has some limitations (Costa and Lobo, 2001). For example, the rate of intrinsic dissolution rate for bioactives is not defined. Besides, there is neither kinetic fundament used to describe the dissolution status, nor the characterization of the dissolution of bioactive kinetic properties. Nonetheless, this model is best suited for comparing the release profiles of matrix type bioactive delivery systems (Ramteke et al., 2014).
7. *Hopfenberg model*
It is exerted to correlate the core material's release from surface eroding polymers so long as the surface area remains constant during the degradation process. This model is used to identify the mechanisms of release from the optimized oil sphere using derived data from the composite profile, which is essentially displayed the site-specific biphasic release kinetics (Mahat, 2010).
8. *Baker–Lonsdale model*
Baker–Lonsdale model is applied to the linear functions for different release formulations. In fact, this is the modified version of the Higuchi model (Lokhandwala et al., 2013).

13.5.2 Intelligent modeling

The implementation of empirical relations for food matrix complex systems is not a simple mathematical operation and is not precise in many cases (Husseini et al., 2007). Thus, intelligent modeling seems to be a better option in such situations. The benefits of intelligent modeling outnumber the advantages of empirical correlations, which are listed below:

- straightforward
- error and noise tolerance
- plasticity (keeping the potential of prediction even when some of the neurons are eliminated in the ongoing process)

- black box modeling procedure
- robustness (being able to cope with process changes)

The complex nature of food systems needs a practical modeling technique that can simulate the release of nutrients in different conditions.

One of these promising techniques is artificial neural networks (ANN) used for inline and in situ evaluations. Moreover, this method is nondestructive and therefore is useful in the food industry (Malekjani et al., 2013; Bahmani et al., 2016). ANNs are actually inspired by the learning process of the human brain. They are can be utilized in different areas of system identification, pattern recognition, classification, etc. (Esfanjani and Jafari, 2016). With the help of this modeling technique release assays can be predicted with high accuracy in near future.

Monge et al. (2004) used principal component analysis (PCA) and ANNs to study the collected data from electronic nose device in accordance with the flavor liberation of a complex essence entrapped in the network of pectin gels. Rheological experiments were carried out to determine the Gelation kinetics of pectin with high degree of methoxylation in presence and absence of the flavor. Flavor release was examined in different days after encapsulation by interpreting the signals emitted from the gas sensor of the e-nose along with the fingerprints of the whole sensors array. The connection with rheological measurements represents that the e-nose is able to identify the gel production onset and to specify the release alterations throughout the aging process by employing PCA. By the means of ANN analysis, samples were classified into three groups according to their flavor formulations and concentration in the gas state. In other words, this intelligent model is able to make different samples distinguishable. Compared to PCA, ANN is a nonparametric method and both linear and nonlinear functions are covered in the working range. The best topology for ANN was the back propagation neural network, including an anonymous layer plus 5 chosen neurons. Regarding the optimization process, 80% of the exemplars were used for training; 10% were used as a cross-validation information, and 10% were used as the testing information.

Husseini et al. (2007) employed ANNs to specify the release procedure of Doxorubicin (Dox) from Pluronic P105 micelles. First, the collected release data were utilized for training, validating, and testing the ANN model. Afterward, ultrasonic frequency, power intensity, pluronic P105 temperature, and concentration were determined by the sensitivity analysis. In fact, ANN models are a means to predict the release profile of Dox in regard to the aforementioned parameters. The accuracy of this model depends on the high number experimental data to match the system predictions with real release measurements in order to approve the efficacy of the ANN model. The presence of one hidden layer is necessary to have an uncomplicated network structure. Also, the number of neurons in the input and output layers matched the number of input and output variables, therefore the input and output layer had 4 and 1 neurons, respectively. For determining the optimum number of neurons in the hidden layer, the performance of different networks was tested. It was observed that few neurons were in the hidden layer, which leads to a great solution forming eight neurons. Based on the expanded model, the release process does not have a considerable connection with the temperature element, thus the release process is not dramatically modified by thermal processing. Finally, sensitivity plots of P105 concentration displayed greater release in low concentrations of the copolymer. These researchers introduce neural network models as a powerful method to predict the release process of Dox from Pluronic P105 micelles. To sum up, this modeling procedure is capable of the experiment conditions and can also be applied to in vivo assays.

In order to unite the self-adaptability and learning ability of the neural network with the fuzzy system and overcome the imprecision of real systems a practical technique was established known as adaptive network-based fuzzy inference system (ANFIS) (Jang, 1993).

Hezaveh et al. (2012) used ANFIS for studying the release process of nutrients. In this study, beta-carotene and -carrageenan/NaCMC hydrogel were used as guest and host molecules consecutively. The MATLAB software was used for programming the ANFIS model. The initial data were obtained from running experiments and used for model validation. Subsequently, hydrogel swelling (Since the aims of this work is to investigate the swelling ratio of hydrogel under different experimental conditions) was chosen as the output variable and pH, crosslinker level, and mixture proportion were set as input data. Besides, the optimal combination of the number and type of membership functions along with the number of epochs were achieved via shooting method. As a result, the model was compatible with the experimental data supporting the ANFIS modeling as an accurate, rapid and straightforward technique for predicting the nutrient liberation processes.

13.6 Targeted release

In general, targeted release can be defined as the release of a bioactive compound in a special area of the body; such as mouth or the digestive tract. The structure of the nanovehicle should be chosen according to the nature of the nutrient as the core material (Katouzian and Jafari, 2016; McClements, 2014; Mohammadi et al., 2016). The target site in which nutrients are delivered plus their release features are directly influenced via the material's size. Therefore, nanoscale carriers possess more surface area compared to their microscale counterparts and are more active so that an improved bioavailability is obtained (Jafari et al., 2007).

Today, a number of companies have started developing their nanofood products with the aim of targeted delivery. Below some of these companies plus their manufactured products are represented.[b]

1. *Tip-Top*: This company is active in the field of bakery products in west Australia. They have fabricated nanocapsules containing fish oil to enrich their products, mask the unpleasant odor of fish oil and to cause targeted delivery in the intestine.
2. *Nutralease*: Self-assembled liquid structures (NSSL) are utilized with the aim of targeted delivery. Up to now, they have incorporated nutrients, such as β-carotene, lutein, lycopene, phytosterols, DHA/EPA, and CoQ10 in their products. Furthermore, Shemen industries have also marketed this method as a means to deliver Canola Activa oil to the cells since it lowers the cholesterol intake.

[b] http://www.azonano.com/

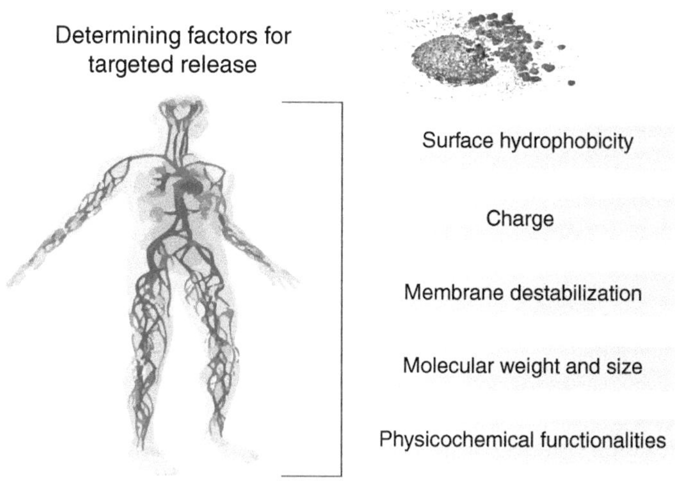

Figure 13.8 Crucial parameters affecting targeted release.

3. *Biodelivery sciences International*: This firm has produced nanochelates with the size of 50 nm which can be employed for targeted delivery of bioactives, such as lycopene, vitamins, and omega fatty acids.
4. *Aquanova*: The members were able to develop a nanocarrier and cause satiety plus fat reduction in food products as an intelligent approach to weight management, which is a challenge in these days. An emulsifier called NovaSOL Sustain causes satiety and fat reduction in food production by using alpha-lipoic acid and CoQ10 respectively.
5. *livethesource*: This company fabricates nanocapsules on the basis of natural plant lipids, thus they are of great interest as nutrient delivery systems. Moreover, it delivers the nutrient in a rapid and efficient way. Some of their products include daily multivitamin, symbiotic diet, and vitamin D booster.
6. *Other companies*: Leading companies in the food sector; such as Nestle, Kraft, and Unilever improve the texture of their food products via nanoemulsions of oils and fats. With this method, the usage of fat for obtaining the desired sensory properties is minimized which results in much healthier processed food.

13.6.1 Factors affecting the targeted release

A physically targeted delivery system is based on the physicochemical characteristics of a nanovehicle so as to deliver the nutrient to the target site. The nature of the delivery systems is of utmost importance for predicting its in vivo behavior but is often neglected. As shown in Fig. 13.8, these features comprise but are not limited to molecular weight, size, surface hydrophobicity, surface charge, and destabilization activity toward biomembranes (Kerr et al., 2000; Takakura et al., 1996).

13.6.2 Ligand-receptor-based interaction and delivery systems

Ligands have the potential to be employed in targeted delivery systems since the ligand-receptor complex acquire a structural complementarity for the persistence of the intended interaction. Herein, The studies related to the typical ligands along with their complementary receptors used for pharmaceutical purposes are summarized in Table 13.3. For more information, please refer to Akinc et al. (2010).

13.7 Conclusions

Here, distinct mechanisms of controlled release were studied. Several mechanisms may be present in a special controlled release process, particularly when there is more than one phase present in this process. This chapter explored the necessity for controlled release so that nutrients are released in the right place. Lately, pharmaceutical/nutraceutical industries are focusing on the quantitative investigations for predicting the release process as optimized values can be obtained by proper mathematical solutions as a function of material formulations. Furthermore, empirical equations also uitlized in case the theoretical concept is not available. In both micro and nanostructured systems, nutrients are released with chemical or physical alterations in the nanovehicle structure, which were explained before. Nevertheless, controlled release is not regulated by only one method. In fact, the release mechanism may initiate by one mechanism and then it is controlled by the complex reaction of other release procedures. Therefore, the intricacy of release procedure yields various applications in the food industry with the aim of improving the public health.

Table 13.3 Some of the most commonly ligands and their complementary receptors (Akinc et al., 2010)

Receptors/ ligands	Description	Examples
Antigens	Anticancer targeting *n* delivery/ tumor-associated antigens for the targeting of antibodies	Carcinoembryonic antigen, prevalent in gastrointestinal (GI), lung and breast tumors
Cadherins	Cadherins are a group of glycoproteins that facilitate Ca^{2+}-dependent cell–cell adhesive interactions/when cadherin function is disrupted, the release of a tumor cell can result	Aggressive metastasis of undifferentiated epithelial carcinoma cells that had lost cell-cell adhesion could be stopped by transfection with E-cadherin cDNA
Selectins	One type of cell adhesion molecule that is responsible for carbohydrate binding/ selectins mediate cell adhesion by recognizing specific carbohydrate ligands arranged on the surfaces of cells	Binding affinity and selectivity of the tetrasaccharide glycolipids sialyl-Lewis X (sLex) and sialyl-Lewis A (sLea)

(Continued)

Table 13.3 Some of the most commonly ligands and their complementary receptors (Akinc et al., 2010) (*cont.*)

Receptors/ ligands	Description	Examples
Integrins	Heterodimeric glycoproteins that consist of α and β subunits/ identified as excellent candidates for the development of target specific cancer therapies.	Adhesion between cell-cell and cell–extracellular matrix occur when integrin recognizes and specifically interacts with minimal peptide core sequences within the extracellular matrix
Vitamins	Essential for normal cellular function and growth	Folic acid, riboflavin, biotin, and vitamin B6 all have been evaluated as potential ligands for targeted delivery of therapeutic agents to specific cells
Transferrin	A glycoprotein responsible for transporting iron into cells	Iron binds to transferrin and enters the cell through a highly specific receptor-mediated endocytosis via the transferrin receptor
Hormones	The presence of hormone receptors in hormone-sensitive cancers presents potential applications of hormone-targeted drug delivery of traditional drugs	When doxorubicin is conjugated to an analogue of the luteinizing hormone-releasing hormone (LH-RH), cancer cell death was observed at subnanomolar concentrations
Low-density lipoprotein (LDL)	Naturally occurring spherical macromolecular particles that transport lipids such as cholesterols and triacylglycerols through the blood–cell membrane barrier of various cells.	Very good candidates for the targeted drug delivery of traditional drugs to various tissues

References

Aboalnaja, K.O., Yaghmoor, S., Kumosani, T.A., McClements, D.J., 2016. Utilization of nanoemulsions to enhance bioactivity of pharmaceuticals, supplements, and nutraceuticals: nanoemulsion delivery systems and nanoemulsion excipient systems. Expert Opin. Drug Deliv., 1–10.

Acosta, E., 2009. Bioavailability of nanoparticles in nutrient and nutraceutical delivery. Curr. Opin. Colloid Interface Sci. 14 (1), 3–15.

Aggett, P.J., 2010. Population reference intakes and micronutrient bioavailability: a European perspective. Am. J. Clin. Nutr. 91 (5), 1433S–1437S.

Akinc, A., Querbes, W., De, S., Qin, J., Frank-Kamenetsky, M., Jayaprakash, K.N., Jayaraman, M., Rajeev, K.G., Cantley, W.L., Dorkin, J.R., 2010. Targeted delivery of RNAi therapeutics with endogenous and exogenous ligand-based mechanisms. Mol. Ther. 18 (7), 1357–1364.

Assadpour, E., Maghsoudlou, Y., Jafari, S.M., Ghorbani, M., Aalami, M., 2016a. Optimization of folic acid nano-emulsification and encapsulation by maltodextrin-whey protein double emulsions. Int. J. Biol. Macromol. 86, 197–207.

Assadpour, E., Maghsoudlou, Y., Jafari, S.M., Ghorbani, M., Aalami, M., 2016b. Evaluation of folic acid nano-encapsulation by double emulsions. Food Bioproc. Technol. 9, 2024–2032.

Assadpour, E., Jafari, S.M., Maghsoudlou, Y., 2017. Evaluation of folic acid release from spray dried powder particles of pectin-whey protein nano-capsules. Int. J. Biol Macromol. 95, 238–247.

Assadpour, E., Jafari, S.M., 2017. Spray Drying of Folic Acid within Nano-Emulsions; Optimization by Taguchi Approach. Drying Technol. doi: 10.1080/07373937.2016.1242016.

Bahmani, A., Jafari, S.M., Shahidi, S.-A., Dehnad, D., 2016. Mass Transfer Kinetics of Eggplant during osmotic dehydration by neural networks. J. Food Process. Preserv. 40 (5), 815–827.

Baker, R., 1987. ontrolled Release of Biologically Active Agents. John Wiley & Sons, New York, 84-131.

Barakat, A., Shegokar, R., Dittgen, M., Müller, R.H., 2013. Coenzyme Q10 oral bioavailability: effect of formulation type. J. Pharm. Invest. 43 (6), 431–451.

Bhat, P.G., Flanagan, D.R., Donovan, M.D., 1995. The limiting role of mucus in drug absorption: drug permeation through mucus solution. Int. J. Pharm. 126 (1), 179–187.

Borel, P., Preveraud, D., Desmarchelier, C., 2013. Bioavailability of vitamin E in humans: an update. Nutr. Rev. 71 (6), 319–331.

Borel, P., Caillaud, D., Cano, N., 2015. Vitamin D bioavailability: state of the art. Crit. Rev. Food Sci. Nutr. 55 (9), 1193–1205.

Briviba, K., Schnäbele, K., Schwertle, E., Blockhaus, M., Rechkemmer, G., 2001. β-Carotene inhibits growth of human colon carcinoma cells in vitro by induction of apoptosis. Biol. Chem. 382 (12), 1663–1668.

Costa, P., Lobo, J.M.S., 2001. Modeling and comparison of dissolution profiles. Eur. J. Pharm. Sci. 13 (2), 123–133.

Crittenden, R., Weerakkody, R., Sanguansri, L., Augustin, M., 2006. Synbiotic microcapsules that enhance microbial viability during nonrefrigerated storage and gastrointestinal transit. Appl. Environ. Microbiol. 72 (3), 2280–2282.

Desai, M., Mutlu, M., Vadgama, P., 1992. A study of macromolecular diffusion through native porcine mucus. Experientia 48 (1), 22–26.

Emken, E., 2001. Stable isotope approaches, applications and issues related to polyunsaturated fatty acid metabolism studies. Lipids 36 (9), 965–973.

Erfanian, A., Mirhosseini, H., Manap, M.Y.A., Rasti, B., Bejo, M.H., 2014. Influence of nanosize reduction on absorption and bioavailability of calcium from fortified milk powder in rats. Food Res. Int. 66, 1–11.

Esfanjani, A.F., Jafari, S.M., 2016. Biopolymer nano-particles and natural nano-carriers for nano-encapsulation of phenolic compounds. Colloids Surf B. Biointerfaces. 146, 532–543.

Fabra, M.J., Chambin, O., Voilley, A., Gay, J.-P., Debeaufort, F., 2012. Influence of temperature and NaCl on the release in aqueous liquid media of aroma compounds encapsulated in edible films. J. Food Eng. 108 (1), 30–36.

Gaonkar, A.G., Vasisht, N., Khare, A.R., Sobel, R., 2014. Microencapsulation in the Food Industry: A Practical Implementation Guide. Academic Press, San Diego, CA.

Gibson, R.S., 2007. The role of diet-and host-related factors in nutrient bioavailability and thus in nutrient-based dietary requirement estimates. Food Nutr. Bull. 28 (Suppl. 1), S77–S100.

Goodman, M., Bostick, R.M., Kucuk, O., Jones, D.P., 2011. Clinical trials of antioxidants as cancer prevention agents: past, present and future. Free Radic. Biol. Med. 51 (5), 1068–1084.

Gou, M., Men, K., Shi, H., Xiang, M., Zhang, J., Song, J., Long, J., Wan, Y., Luo, F., Zhao, X., 2011. Curcumin-loaded biodegradable polymeric micelles for colon cancer therapy in vitro and in vivo. Nanoscale 3 (4), 1558–1567.

Grassi, M., Grassi, G., Lapasin, R., Colombo, I., 2007. Part 1: Gastrointestinal Tract. In Understanding Drug Release and Absorption Mechanisms, a Physical, and Mathematical Approach, first ed. CRC Press, Boca Raton, pp. 29–68.

Gu, J., Robinson, J., Leung, S., 1987. Binding of acrylic polymers to mucin/epithelial surfaces: structure-property relationships. Crit. Rev. Ther. Drug Carrier Sys. 5 (1), 21–67.

Guggi, D., Marschütz, M.K., Bernkop-Schnürch, A., 2004. Matrix tablets based on thiolated poly (acrylic acid): pH-dependent variation in disintegration and mucoadhesion. Int. J. Pharm. 274 (1), 97–105.

Heaney, R.P., 2001. Factors influencing the measurement of bioavailability, taking calcium as a model. J. Nutr. 131 (4), 1344S–1348S.

Hezaveh, H., Muhamad, I.I., Noshadi, I., Shu Fen, L., Ngadi, N., 2012. Swelling behaviour and controlled drug release from cross-linked κ-carrageenan/NaCMC hydrogel by diffusion mechanism. J. Microencapsul. 29 (4), 368–379.

Higuchi, T., 1963. Mechanism of sustained-action medication. Theoretical analysis of rate of release of solid drugs dispersed in solid matrices. J. Pharm. Sci. 52 (12), 1145–1149.

Holst, B., Williamson, G., 2008. Nutrients and phytochemicals: from bioavailability to bioefficacy beyond antioxidants. Curr. Opin. Biotechnol. 19 (2), 73–82.

Hotz, C., Gibson, R.S., 2007. Traditional food-processing and preparation practices to enhance the bioavailability of micronutrients in plant-based diets. J. Nutr. 137 (4), 1097–1100.

Hurrell, R., Egli, I., 2010. Iron bioavailability and dietary reference values. Am. J. Clin. Nutr. 91 (5), 1461S–1467S.

Husseini, G.A., Abdel-Jabbar, N.M., Mjalli, F.S., Pitt, W.G., 2007. Modeling and sensitivity analysis of acoustic release of Doxorubicin from unstabilized pluronic P105 using an artificial neural network model. Technol. Cancer Res. Treat. 6 (1), 49–56.

Iwanaga, K., Ono, S., Narioka, K., Kakemi, M., Morimoto, K., Yamashita, S., Namba, Y., Oku, N., 1999. Application of surface-coated liposomes for oral delivery of peptide: effects of coating the liposome's surface on the GI transit of insulin. J. Pharm. Sci. 88 (2), 248–252.

Iwanaga, K., Yabuta, T., Kakemi, M., Morimoto, K., Tabata, Y., Ikada, Y., 2003. Usefulness of microspheres composed of gelatin with various cross-linking density. J. Microencapsul. 20 (6), 767–776.

Jacquez, J.A., 1996. Compartmental Analysis in Biology and Medicine, third ed Biomedware, Ann Arbor, MI, p. 512.

Jafari, S.M., He, Y., Bhandari, B., 2007. Role of powder particle size on the encapsulation efficiency of oils during spray drying. Drying Technol. 25 (6), 1081–1089.

Jafari, S.M., Fathi, M., Mandala, I., 2015. Emerging product formation. Food Waste Recovery: Processing Technologies and Industrial Techniques. Elsevier Inc., Academic Press, San Diego, CA, 293-317.

Jang, J.-S.R., 1993. ANFIS: adaptive-network-based fuzzy inference system. IEEE Trans. Syst. Man Cybernet. 23 (3), 665–685.

Johnson, J., Mistry, V., Vukovich, M., Hogie-Lorenzen, T., Hollis, B., Specker, B., 2005. Bioavailability of vitamin D from fortified process cheese and effects on vitamin D status in the elderly. J. Dairy Sci. 88 (7), 2295–2301.

Katouzian, I., Jafari, S.M., 2016. Nano-encapsulation as a promising approach for targeted delivery and controlled release of vitamins. Trends Food Sci. Technol. 53, 34–48.

Kerr, J.S., Slee, A.M., Mousa, S.A., 2000. Small molecule αv integrin antagonists: novel anticancer agents. Expert Opin. Invest. Drugs 9 (6), 1271–1279.

Khanvilkar, K., Donovan, M.D., Flanagan, D.R., 2001. Drug transfer through mucus. Adv. Drug Deliv. Rev. 48 (2), 173–193.

Korsmeyer, R.W., Gurny, R., Doelker, E., Buri, P., Peppas, N.A., 1983. Mechanisms of solute release from porous hydrophilic polymers. Int. J. Pharm. 15 (1), 25–35.

Lakkis, J.M., 2007. Encapsulation and controlled release technologies in food systems. Blackwell, Iowa.

Larhed, A.W., Artursson, P., Björk, E., 1998. The influence of intestinal mucus components on the diffusion of drugs. Pharm. Res. 15 (1), 66–71.

Lee, K.-Y., Heo, T.-R., 2000. Survival of bifidobacterium longumImmobilized in calcium alginate beads in simulated gastric juices and bile salt solution. Appl. Environ. Microbiol. 66 (2), 869–873.

Lokhandwala, H., Deshpande, A., Deshpande, S., 2013. Kinetic modeling and dissolution profiles comparison: an overview. Int. J. Pharm. Bio. Sci 4 (1), 729.

Lynch, 2007. Influence of infection/inflammation, thalassemia and nutritional status on iron absorption. Int. J. Vitamin. Nutr. Res. 77 (3), 217–223.

Mahat, B., 2010. Mathematical Models used in Drug Release Studies. Department of Pharmacy, Kathmandu University, Dhulikhel, Nepal.

Malekjani, N., Jafari, S.M., Rahmati, M.H., Zadeh, E.E., Mirzaee, H., 2013. Evaluation of thin-layer drying models and artificial neural networks for describing drying kinetics of canola seed in a heat pump assisted fluidized bed dryer. Int. J. Food Eng. 9 (4), 375–384.

McClements, D.J., 2013. Utilizing food effects to overcome challenges in delivery of lipophilic bioactives: structural design of medical and functional foods. Expert Opin Drug Deliv. 10 (12), 1621–1632.

McClements, D.J., 2014. Nanoparticle- and Microparticle-based Delivery Systems: Encapsulation, Protection, and Release of Active Components. CRC Press, Boca Raton, FL.

McClements, D.J., Xiao, H., 2012. Potential biological fate of ingested nanoemulsions: influence of particle characteristics. Food Func. 3 (3), 202–220.

Mohammadi, A., Jafari, S.M., Assadpour, E., Esfanjani, A.F., 2016. Nano-encapsulation of olive leaf phenolic compounds through WPC–pectin complexes and evaluating their release rate. Int. J. Biol. Macromol. 82, 816–822.

Mehrnia, M.A., Jafarim, S.M., Makhmal-Zadeh, B.S., Maghsoudlou, Y., 2017. Rheological and release properties of double nano-emulsions containing crocin prepared with Angum gum, Arabic gum and whey protein. Food Hydrocolloids. doi: 10.1016/j.foodhyd.2016.11.033.

Monge, M.a.E., Bulone, D., Giacomazza, D., Bernik, D.L., Negri, R.M.n., 2004. Detection of flavour release from pectin gels using electronic noses. Sensors Actuators B Chem. 101 (1), 28–38.

Narasimhan, B., Peppas, N., 1997. The role of modeling studies in the development of future controlled-release devices. Controlled Drug Delivery: Challenges and Strategies, American Chemical Society, Washington, DC, 529–557.

Navarro, M., Wood, R.J., 2003. Plasma changes in micronutrients following a multivitamin and mineral supplement in healthy adults. J. Am. Coll. Nutr. 22 (2), 124–132.

Neves, A.R., Lucio, M., Martins, S., Lima, J., Reis, S., 2013. Novel resveratrol nanodelivery systems based on lipid nanoparticles to enhance its oral bioavailability. Int. J. Nanomedicine. 8 (1), 177–187.

Niu, M., Lu, Y., Hovgaard, L., Wu, W., 2011. Liposomes containing glycocholate as potential oral insulin delivery systems: preparation, in vitro characterization, and improved protection against enzymatic degradation. Int. J. Nanomedicine 6, 1155–1166.

Park, G.Y., Mun, S., Park, Y., Rhee, S., Decker, E.A., Weiss, J., McClements, D.J., Park, Y., 2007. Influence of encapsulation of emulsified lipids with chitosan on their in vivo digestibility. Food Chem. 104 (2), 761–767.

Pathak, K., Raghuvanshi, S., 2015. Oral bioavailability: issues and solutions via nanoformulations. Clin. Pharmacokinet. 54 (4), 325–357.

Patrignani, C., Group, P.D., et al., 2016. Review of particle physics. Chinese Phys. C 40 (10), 100001.

Qingming, Y., Xianhui, P., Weibao, K., Hong, Y., Yidan, S., Li, Z., Yanan, Z., Yuling, Y., Lan, D., Guoan, L., 2010. Antioxidant activities of malt extract from barley (*Hordeum vulgare* L.) toward various oxidative stress in vitro and in vivo. Food Chem. 118 (1), 84–89.

Ramteke, K., Dighe, P., Kharat, A., Patil, S., 2014. Mathematical models of drug dissolution: a review. Sch. Acad. J. Pharm 3, 388–396.

Rashidi, L., Khosravi-Darani, K., 2011. The applications of nanotechnology in food industry. Crit. Rev. Food Sci. Nutr. 51 (8), 723–730.

Raval, A., Parikh, J., Engineer, C., 2010. Mechanism of controlled release kinetics from medical devices. Braz. J. Chem. Eng. 27 (2), 211–225.

Rubió, L., Motilva, M.-J., Romero, M.-P., 2013. Recent advances in biologically active compounds in herbs and spices: a review of the most effective antioxidant and anti-inflammatory active principles. Crit. Rev. Food Sci. Nutr. 53 (9), 943–953.

Sacks, P.V., Houchin, D.N., 1978. Comparative bioavailability of elemental iron powders for repair of iron deficiency anemia in rats. Studies of efficacy and toxicity of carbonyl iron. Am. J. Clin. Nutr. 31 (4), 566–571.

Sagis, L.M., 2008. Dynamics of controlled release systems based on water-in-water emulsions: a general theory. J. Control. Release 131 (1), 5–13.

Santus, G., Baker, R.W., 1995. Osmotic drug delivery: a review of the patent literature. J. Control. Release 35 (1), 1–21.

Schümann, K., Classen, H., Hages, M., Prinz-Langenohl, R., Pietrzik, K., Biesalski, H., 1997. Bioavailability of oral vitamins, minerals and trace elements in perspective. Arzneimittel-Forschung 47 (4), 369–380.

Shaikh, H.K., Kshirsagar, R., Patil, S., 2015. Mathematical models for drug release characterization: a review. WJPPS 4, 324–338.

Siegel, J.A., Thomas, S.R., Stubbs, J.B., Stabin, M.G., 1999. MIRD pamphlet no. 16: techniques for quantitative radiopharmaceutical biodistribution data acquisition and analysis for use in human radiation dose estimates. J. Nucl. Med. 40 (2), 37S.

Takakura, Y., Mahato, R.I., Nishikawa, M., Hashida, M., 1996. Control of pharmacokinetic profiles of drug—macromolecule conjugates. Adv. Drug Deliv. Rev. 19 (3), 377–399.

Teucher, Olivares, Cori, 2004. Enhancers of iron absorption: ascorbic acid and other organic acids. Int. J. Vitamin Nutr. Res. 74 (6), 403–419.

Thanki, K., Gangwal, R.P., Sangamwar, A.T., Jain, S., 2013. Oral delivery of anticancer drugs: challenges and opportunities. J. Control. Release 170 (1), 15–40.

Tiyaboonchai, W., Tungpradit, W., Plianbangchang, P., 2007. Formulation and characterization of curcuminoids loaded solid lipid nanoparticles. Int. J. Pharm. 337 (1), 299–306.

Torchilin, V.P., 2005. Recent advances with liposomes as pharmaceutical carriers. Nat. Rev. Drug Discov. 4 (2), 145–160.

van het Hof, K.H., West, C.E., Weststrate, J.A., Hautvast, J.G., 2000. Dietary factors that affect the bioavailability of carotenoids. J. Nutr. 130 (3), 503–506.

Versypt, A.N.F., Pack, D.W., Braatz, R.D., 2013. Mathematical modeling of drug delivery from autocatalytically degradable PLGA microspheres—a review. J. Control. Release 165 (1), 29–37.

Wang, J., Sun, B., Cao, Y., Wang, C., 2010. Wheat bran feruloyl oligosaccharides enhance the antioxidant activity of rat plasma. Food Chem. 123 (2), 472–476.

Williamson, G., Manach, C., 2005. Bioavailability and bioefficacy of polyphenols in humans. II. Review of 93 intervention studies. Am. J. Clin. Nutr. 81 (1), 243S–255S.

Wu, T.H., Yen, F.L., Lin, L.T., Tsai, T.R., Lin, C.C., Cham, T.M., 2008. Preparation, physicochemical characterization, and antioxidant effects of quercetin nanoparticles. Int. J. Pharm. 346 (1), 160–168.

Wu, Z., Ping, Q., Wei, Y., Lai, J., 2004. Hypoglycemic efficacy of chitosan-coated insulin liposomes after oral administration in mice. Acta Pharmacologica Sinica 25, 966–972.

Yamaguchi, M., Igarashi, A., Morita, S., Sumida, T., Sugawara, K., 2005. Relationship between serum β-crytoxanthin and circulating bone metabolic markers in healthy individuals with the intake of juice (*Citrus unshiu*) containing β-crytoxanthin. J. Health Sci. 51 (6), 738–743.

Zhang, K., Zuo, Y., 2004. GC-MS determination of flavonoids and phenolic and benzoic acids in human plasma after consumption of cranberry juice. J. Agric. Food Chem. 52 (2), 222–227.

Zhou, J.R., Erdman, Jr., J.W., 1995. Phytic acid in health and disease. Crit. Rev. Food Sci. Nutr. 35 (6), 495–508.

Zhou, H., Yue, Y., Liu, G., Li, Y., Zhang, J., Gong, Q., Yan, Z., Duan, M., 2010. Preparation and characterization of a lecithin nanoemulsion as a topical delivery system. Nanoscale Res. Lett. 5 (1), 224–230.

Further Reading

Allen, A., Flemström, G., 2005. Gastroduodenal mucus bicarbonate barrier: protection against acid and pepsin. Am. J. Physiol. Cell Physiol. 288 (1), C1–C19.

Ashwell, M., Lambert, J.P., Alles, M.S., Branca, F., Bucchini, L., Brzozowska, A., de Groot, L.C., Dhonukshe-Rutten, R.A., Dwyer, J.T., Fairweather-Tait, S., 2008. How we will produce the evidence-based EURRECA toolkit to support nutrition and food policy. Eur. J. Nutr. 47 (1), 2–16.

Augustin, M.A., Sanguansri, L., Bode, O., 2006. Maillard reaction products as encapsulants for fish oil powders. J. Food Sci. 71 (2), E25–E32.

Dekker, J., Rossen, J.W., Büller, H.A., Einerhand, A.W., 2002. The MUC family: an obituary. Trends Biochem. Sci. 27 (3), 126–131.

Demonty, I., Ras, R.T., van der Knaap, H.C., Duchateau, G.S., Meijer, L., Zock, P.L., Geleijnse, J.M., Trautwein, E.A., 2008. Continuous dose-response relationship of the LDL-cholesterol–lowering effect of phytosterol intake. J. Nutr. 108, 095125.

Hannon-Fletcher, M.P., Armstrong, N.C., Scott, J.M., Pentieva, K., Bradbury, I., Ward, M., Strain, J., Dunn, A.A., Molloy, A.M., Kerr, M.A., 2004. Determining bioavailability of food folates in a controlled intervention study. Am. J. Clin. Nutr. 80 (4), 911–918.

Hansen, L.T., Allan-Wojtas, P., Jin, Y.-L., Paulson, A., 2002. Survival of Ca-alginate microencapsulated *Bifidobacterium* spp. in milk and simulated gastrointestinal conditions. Food Microbiol. 19 (1), 35–45.

Hong, Z., Chasan, B., Bansil, R., Turner, B.S., Bhaskar, K.R., Afdhal, N.H., 2005. Atomic force microscopy reveals aggregation of gastric mucin at low pH. Biomacromolecules 6 (6), 3458–3466.

Kosaraju, S.L., D'ath, L., Lawrence, A., 2006. Preparation and characterisation of chitosan microspheres for antioxidant delivery. Carbohydr. Polym. 64 (2), 163–167.

Lee, S.J., Rosenberg, M., 2000. Whey protein-based microcapsules prepared by double emulsification and heat gelation. LWT-Food Sci. Technol. 33 (2), 80–88.

Ma, Q.H., Xia, Q., Lu, Y.Y., Hao, X.Z., Gu, N., Lin, X.F., Luo, D., 2007. Preparation of tea polyphenols-loaded solid lipid nanoparticles based on the phase behaviors of hot microemulsions. Solid State Phenomena 121, 705–708.

Madene, A., Jacquot, M., Scher, J., Desobry, S., 2006. Flavour encapsulation and controlled release–a review. Int. J. Food Sci. Technol. 41 (1), 1–21.

Pillai, O., Panchagnula, R., 2001. Polymers in drug delivery. Curr. Opin. Chem. Biol. 5 (4), 447–451.

Rock, C.L., Lovalvo, J.L., Emenhiser, C., Ruffin, M.T., Flatt, S.W., Schwartz, S.J., 1998. Bioavailability of β-carotene is lower in raw than in processed carrots and spinach in women1, 2, 3. J. Nutr. 128 (5), 913–916.

Rosenberg, M., Kopelman, I., Talmon, Y., 1990. Factors affecting retention in spray-drying microencapsulation of volatile materials. J. Agric. Food Chem. 38 (5), 1288–1294.

Round, A., Berry, M., McMaster, T., Stoll, S., Gowers, D., Corfield, A., Miles, M., 2002. Heterogeneity and persistence length in human ocular mucins. Biophys. J. 83 (3), 1661–1670.

Sanders, N.N., De Smedt, S.C., Demeester, J., 2000. The physical properties of biogels and their permeability for macromolecular drugs and colloidal drug carriers. J. Pharma. Sci. 89 (7), 835–849.

Schrooyen, P.M., van der Meer, R., De Kruif, C., 2001. Microencapsulation: its application in nutrition. Proc. Nutr. Soc. 60 (04), 475–479.

Shayeghi, M., Latunde-Dada, G.O., Oakhill, J.S., Laftah, A.H., Takeuchi, K., Halliday, N., Khan, Y., Warley, A., McCann, F.E., Hider, R.C., 2005. Identification of an intestinal heme transporter. Cell 122 (5), 789–801.

Taylor, T.M., Weiss, J., Davidson, P.M., Bruce, B.D., 2005. Liposomal nanocapsules in food science and agriculture. Crit. Rev. Food Sci. Nutr. 45 (7–8), 587–605.

Truswell, A.S., 2007. Vitamin B12. Nutr. Diet. 64 (s4), S120–S125.

Wamsley, A., Jasti, B., Phiasivongsa, P., Li, X., 2004. Synthesis of random terpolymers and determination of reactivity ratios of *N*-carboxyanhydrides of leucine, β-benzyl aspartateand valine. J. Polym. Sci. A Polym. Chem. 42 (2), 317–325.

Winkels, R.M., Brouwer, I.A., Siebelink, E., Katan, M.B., Verhoef, P., 2007. Bioavailability of food folates is 80% of that of folic acid. Am. J. Clin. Nutr. 85 (2), 465–473.

Wright, A., Finglas, P., Dainty, J., Hart, D., Wolfe, C., Southon, S., Gregory, J., 2003. Single oral doses of 13 C forms of pteroylmonoglutamic acid and 5-formyltetrahydrofolic acid elicit differences in short-term kinetics of labelled and unlabelled folates in plasma: potential problems in interpretation of folate bioavailability studies. Br. J. Nutr. 90 (02), 363–371.

Wright, A.J., Finglas, P.M., Dainty, J.R., Wolfe, C.A., Hart, D.J., Wright, D.M., Gregory, J.F., 2005. Differential kinetic behavior and distribution for pteroylglutamic acid and reduced folates: a revised hypothesis of the primary site of PteGlu metabolism in humans. J. Nutr. 135 (3), 619–623.

Wu, Y., Clark, R.L., 2008. Electrohydrodynamic atomization: a versatile process for preparing materials for biomedical applications. J. Biomater. Sci. Polym. Ed. 19 (5), 573–601.

14 Instrumental analysis and characterization of nanocapsules

Seid Mahdi Jafari, Afshin F. Esfanjani***

*Gorgan University of Agricultural Sciences and Natural Resources, Gorgan, Iran; **Ferdowsi University of Mashhad, Mashhad, Iran

14.1 Introduction

Nanocapsules features are directly related to their functional and physicochemical properties, which are crucial for the best fortification of a broad range of formulations, such as foods and other biological systems. Therefore, applying simple and accurate methods based on analytical instruments seems inevitable for the characterization of the nanocapsules.

This chapter contains some of the main analytical methods listed in Table 14.1. Modern and powerful microscopies can be used for studying the morphology of nanocapsules. For instance, electron microscopy is based on electron beams and includes scanning electron microscopy (SEM) and transmission electron microscopy (TEM), which can provide discernible images of nanocapsules (Egerton, 2006). Besides, scanning probe microscopies, such as confocal laser scanning microscopy (CLSM) and atomic force microscopy (AFM) can also provide major information about the structure and morphology of nanocapsules. AFM can provide main information about atomic and molecular structure of materials, furthermore high resolution and depth selective image fluorescence wavelength from nanocapsules can be achieved by CLSM (Aguilera and Stanley, 1999; Yoon et al., 2005).

It is true that microscopic study will provide precise information about the size distribution of nanocapsules, but these methods are time-consuming. Thus, accurate instruments have been designed, which are capable of measuring the size distribution of nanocapsules in a short time. Instruments based on light scattering method include static light scattering (SLS) and dynamic light scattering (DLS) which are the most widely used instruments to characterize nanocapsules particle size (Alexander and Dalgleish, 2006). In addition, laser-induced breakdown detection (LIBD) based on plasma generation is one of the size measurement instruments, which is more sensitive for detecting small particles (<50 nm) compared to conventional light scattering methods (Bundschuh et al., 2001a). There are also some instruments based on gravitational settling and centrifugation that can be used for size measurement of nanocapsules (McClements, 2014).

Surface characteristics of nanocapsules; such as surface charge and surface component can influence their functional and physicochemical properties, as well as their application. X-ray photoelectron spectroscopy (XPS) and zeta-potential measurement can be used for obtaining information about charge and elements of the surface of nanocapsules, respectively, which will be explored in this chapter.

Nanoencapsulation Technologies for the Food and Nutraceutical Industries

Table 14.1 Selected parameters for the characterization of nanocapsules

Parameter	Description	Analytical techniques
Morphology	Information on the morphology and structural characteristics	Electron Microscopy (TEM and SEM) and Scanning Probe Microscopies (AFM and CLSM)
Particle size	Information on primary particle size, size of agglomerates/ aggregates, size range, and number size distribution	Microscopic study (TEM and SEM), AFM, DLS, SLS, gravitational settling, and LIBD
Surface charge	Information on zeta potential	Zeta-potential parameter by Zetasizer
Surface component	Information on surface elements	XPS
Physicochemical	Information about phase transitions, thermal properties, crystallization behavior, melting, molecular structure, interaction bonding, and complex formation	XRD, dilatometry, DSC, FTIR, and NMR
Stability	Information about stability of nanocapsule systems during storage and against environmental stress	Sedimentation, droplet size change, turbidity, viscosity change, centrifugation stability, and environment stress
Image analysis	Information about stability, droplet size, and optical properties	Image processing method
Microenvironment	Information about the Stability and biological fate of nanoencapsulated ingredients	Fluorescence spectroscopy

Measuring the physicochemical properties of nanocapsules, useful information about functional properties, behavior, and application of nanomaterials can be obtained (Mcclements, 2007). Some of the main physicochemical properties of nanocapsules like the phase transitions, thermal properties, crystallization behavior, melting, molecular structure, interaction bonding, and complex formation can be provided fourier transform infrared spectroscopy (FTIR) by, nuclear magnetic resonance (NMR), differential scanning calorimetry (DSC), dilatometry, and X-ray diffraction (XRD).

Awareness of the stability of nanocapsules is necessary for designing the best delivery system. Some of the fundamental stability tests include storage stability, accelerated stability, and stability against environment stress (temperature, mechanical, and pH stress) are common analytical methods to be used for measuring of nanoca=psules stability (McClements, 2004a). Measurement of stability can be performed by image analyzing of droplet changes (Size and distribution) of nanocapsules. In addition, image analyzing is a simple method for measuring optical properties of nanocapsules

(Hosseini et al., 2015; McClements, 2014). The stability of encapsulated compounds in nanocapsules, bioavailability, and microenvironment of nanocapsules can be determined by fluorescence spectroscopy (Sessa et al., 2014).

14.2 Morphology of nanocapsules

The studying of nanocapsules morphology can provide information about the functional and physicochemical properties of them. The microscopic study is a common and easy method for this purpose which can provide information about the structure of nanocapsules by preparing images from them, next the obtained images can be analyzed via related software. In order to achieve a high-quality image from nanocapsules resolution, magnification, and contrast should be present within the images. There are different types microscopes which can be used for analyzing the morphology of nanocapsules. Optical microscopes cannot show particles smaller than 200 nm, therefore modern and powerful microscopes; such as electron microscopy, CLSM, and AFM should be employed for the best results (Aguilera and Stanley, 1999; McClements, 2014; Gupta and Wells, 2004).

14.2.1 Electron microscopy

Electron microscopy as a common method is used to study the structure of nanocapsules (<1000 nm). In electron microscopy, a beam of electrons is used for creating images of particles instead of light, so that it provides as much as a thousand-fold increases in resolving power. As shown in Fig. 14.1A, the electron beams interact with the specimen by passing through an ultrathin specimen. High energy electron beams are emitted from the cathode and accelerated into anode by electron field, then, these electron beams are concentrated via the electromagnetic and electrostatic lenses and are transmitted through a specimen that is to some degree straightforward to electrons and to some extent disseminates them out of the pillar. When it rises up out of the specimen, the electron beam conveys data about the structure of the specimen which is amplified by the target focal point arrangement of the magnifying instrument. The amplified electron picture is given by anticipating onto a fluorescent survey screen covered with a phosphor or scintillator material (Fig. 14.1A) (Egerton, 2006; Gan et al., 2005; Hall, 1954; Hayat, 1974; Hren, 2013).

There are two basic electron microscopy techniques, which are widely used to study nanocapsule systems: TEM and SEM. The fundamental distinction amongst SEM and TEM is that the previous outputs the surface of sample, while the last enters through the specimen and creates a picture of the molecule by cooperating with the electrons and is most appropriate to analyze nanosized particles in biological and live systems. TEM gives preferred determination over SEM in light of the fact that the vitality of the electron beam in TEM is higher; moreover, it is capable of creating images with a magnification of $\times 10^3$ to $\times 10^6$ and a resolution smaller than 1 nm. On the other hand, SEM is able to provide images from the surface of materials with a magnification of $\times 10$ to $\times 500{,}000$ at a resolution of less than 1–20 nm, depending on

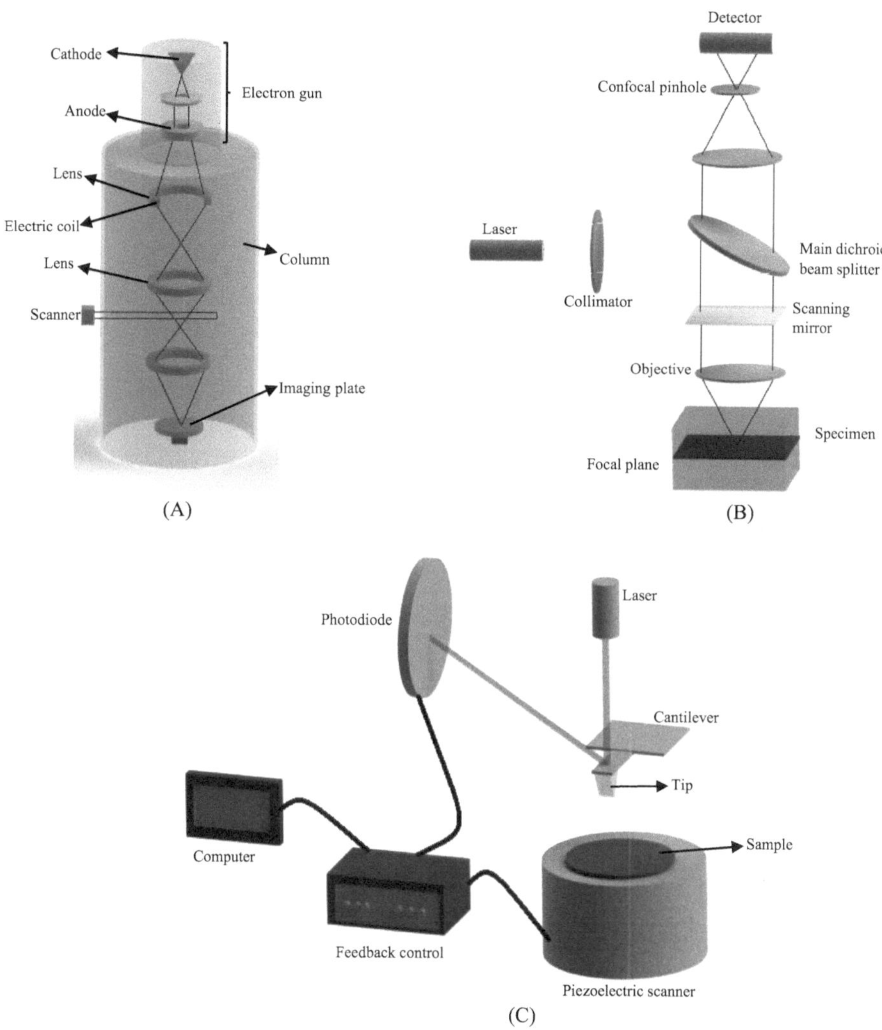

Fig. 14.1 Schematic diagram of analytical instruments for morphology study of nanocapsules. Transmission electron microscope (A), confocal laser scanning microscope (B), atomic force microscope (C).

the type of sample (Aguilera and Stanley, 1999; Egerton, 2006; McClements, 2014). For SEM, the sample should be in a dried form.

14.2.2 Confocal laser scanning microscopy

CLSM is a scanning probe microscopy, which can yield images of nanocapsules with high resolution and selectivity depth. The standard of CLSM depends on identification of concentrated light created by the connection of the enlightening shaft with the little specimen brightening (point enlightenment) in which the pinhole spatially channels permit just engaged light to achieve the identifier (Fig. 14.1B). This system is particular to the fluorescence wavelength of the dye utilized. For studying the morphology of nanocapsules without utilizing fluorescence by CLSM, it is necessary to incorporate a fluorescent dye into them (Alvarez-Román et al., 2004; Yoon et al., 2005).

14.2.3 Atomic force microscopy

The atomic force microscopy (AFM) gives data about the structure of materials at the nuclear and sub-atomic level, furthermore, it can be connected for measuring of molecule size qualities (Jafari et al., 2015). AFM as scanning probe microscopies is based on the scan of the surface by a sharp tip where there is no current drawn between the tip and the specimen. As explained schematically in Fig. 14.1C mounted tip on a cantilevered arm is closely held against the surface of the sample. The ventricular movement of a tip from surface driven by the repulsive forces are detected by the optical system that senses the spacing between the probe and the sample, yielding a correction signal (Fig. 14.1C). By producing an amplified signal of the minute deflections of the cantilever, the image of its structure can be achieved. The AFM offers perception in three measurements. The determination in the vertical pivot, or Z, is restricted by the vibration environment of the instrument, though determination in the level hub, or *X–Y*, is constrained by the measurement of tip used for examining. Normally, AFM instruments have vertical resolutions under 0.1 nm and *X–Y* resolutions around 1 nm (Aguilera and Stanley, 1999; Latterini and Tarpani, 2012; Scalf and West, 2006).

14.3 Size of nanocapsules

It is possible to verify the real size of produced nanocapsules by applying accurate particle size analysis instruments. Furthermore, identification of physicochemical and functional properties of nanocapsules can be predicted via their size. On the other hand, for the fortification of intended formulations (like the food products), evaluation of nanocapsules behavior in fortified systems (such as their release in vivo and in vitro), and also measuring their stability, bioavailability, rheology, and optical properties are necessarily dependent on their size information. Usually, polydispersity index (PDI), particle size distribution (PSD), or the central tendency and width of the distribution are used for the representation of particle dimensions in nanocapsules.

The particle diameter is usually used for describing particle size of a perfectly spherical particle, besides there are common nonspherical nanocapsules with very large aspect ratios (such as nanotubes or nanofibers). Accordingly, some different diameter measures can be used for an equivalent size of a specific type (Mohammadi et al., 2016a; Hassellöv et al., 2008; Madden and Hochella, 2005).or example,

1. Number average: size average of quantities of particles inside a specific size class, $Dn = \frac{\Sigma ni \times di}{\Sigma ni}$
2. Volume average: size average of volume of particles inside a specific size class, $Dv = \frac{\Sigma vi \times di}{\Sigma vi}$
3. *Z*-average size: a power-weighted average ascribed to specific techniques, $Dn = \frac{\Sigma ni \times di^6}{\Sigma ni \times di^5}$

Many different analytical techniques with different principles can be utilized for providing information about PSD of nanocapsules. For example, microscopic methods discussed in the previous section can be applied for measuring the nanocapsule size and their size distribution but these methods are laborious and time-consuming. On the other hand, there are special novel and powerful particle sizers, which can provide rapid and precise information about PSD (Mcclements, 2007; McClements, 2014). Selection of methods for measuring PSD relies on upon numerous parameters; like the type and size range of particles plus the availability and price of technology. This section presents brief information about common and efficient instrumental methods for measuring nanocapsules size.

14.3.1 Dynamic light scattering

DLS (also called photon correlation spectroscopy, laser diffractometry, Fraunhofer diffraction, or Mie scattering) is widely used for size determination of colloid particles especially in nanorange (<1000 nm), which is absolute, noninvasive and nondestructive. When the particles in Brownian motion are exposed to a monochromatic mild beam, including a laser, a doppler shift takes place because the light hits the shifting particles and hence converting the wavelength (typically red light at 633 nm or near-infrared at 830 nm for biomolecular applications) of the approaching light (Fig. 14.2A). This change is related to the size of the particles: smaller particles circulate extra swiftly than large ones and consequently supply more rapid depth fluctuations. It is possible to compute the PSD and describe the particles movement in the medium, measuring the dispersion coefficient of the particles and utilizing the autocorrelation capacity, moreover DLS gives information about particle concentration range and microrheology experienced by small particles (Alexander and Dalgleish, 2006; Arzenšek, 2010).

14.3.2 Static light scattering

Another technique which is most commonly used to provide information about the shape, size, size distribution, and aggregate structure size distribution of nanocapsules is SLS (also called laser diffraction), which is an accurate, rapid, nonintrusive, and in situ characterization method (Alexander and Dalgleish, 2006; Arzenšek, 2010; Kozan, 2007). SLS displays particle size data by measuring the light scattering pattern (angular variation) as a laser beam goes through a weaken scattered particulate specimen. Large particles dissipate light at little points with respect to the laser beam and small particles disseminate light everywhere edges. The particle size is then calculated by analyzing the angular scattering intensity data through a software that contains Mie theory of light scattering (Fig. 14.2B). Indeed, SLS uses Mie theory to ascertain the PSD, accepting a volume proportionate spheres model. Mie hypothesis requires gaining knowledge of the refractive report and fanciful part of optical properties of the instance being measured, alongside the refractive list of the dispersant. Many modern instruments are based on the optical properties of the dispersant found from published data. On the other hand, for samples without a refractive index, optical properties, and

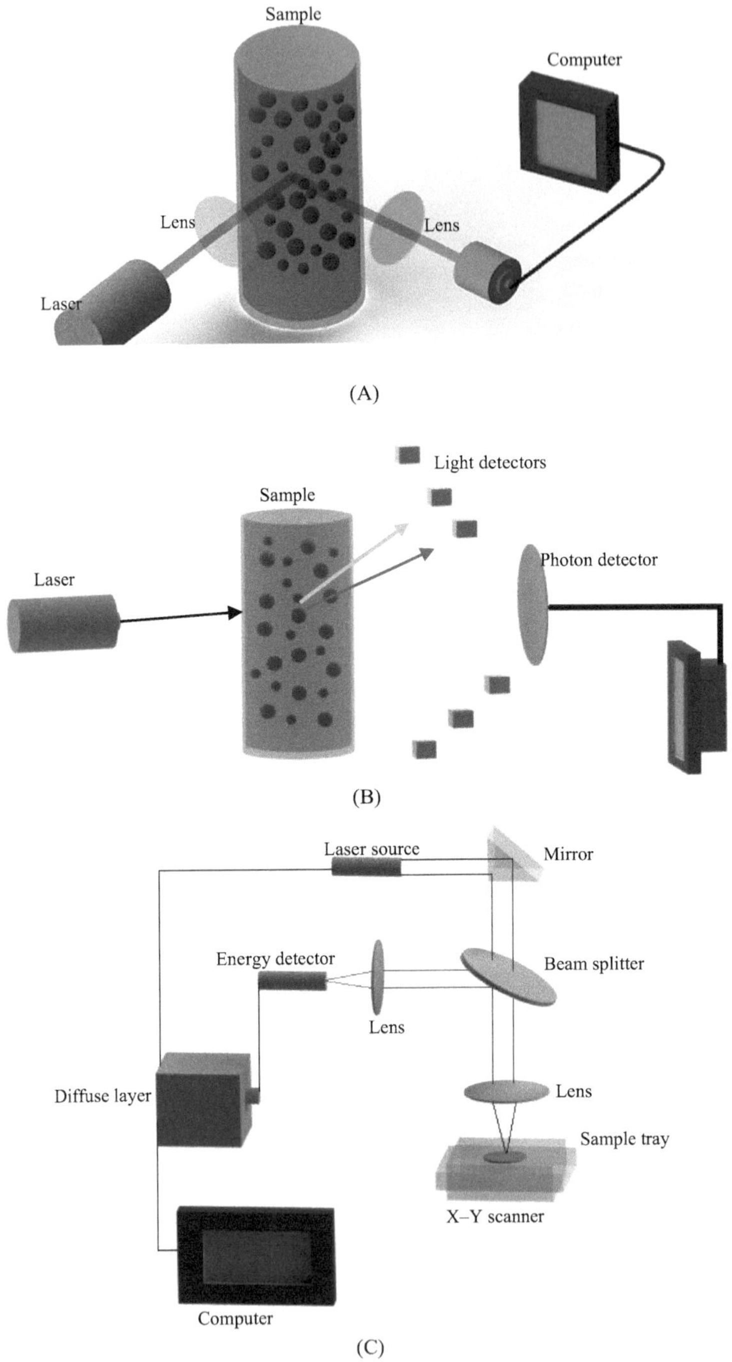

Fig. 14.2 Schematic diagram of analytical instruments for measuring size of nanocapsules. Dynamic light scattering (DLS) (A), static light scattering (SLS) (B), laser-induced breakdown detection (LIBD) (C).

imaginary component can be measured or estimated by utilizing an iterative methodology based upon the integrity of fit between the displayed information and the real information gathered for the sample (Driscoll et al., 2001; Kübart and Keck, 2013).

14.3.3 Gravitational settling and centrifugation

There are a variety of techniques; such as ultrasonic velocity, light scattering, electrical conductivity, X-rays, or NMR that depend on centrifugation and gravitational settling. These methods yield data about the particle length of colloidal dispersions with the aid of measuring the rate at which debris flow in a gravitational or centrifugal field. Large particles circulate greater unexpectedly than smaller ones and consequently measurements of particle speed can be used to determine the particle size the usage of a suitable mathematical model. These methods ordinarily measure the time reliance of the molecule fixation versus test tallness and can measure the size of particles with widths between roughly 1 nm and 1 mm (Mcclements, 2007; McClements, 2014).

14.3.4 Laser-induced breakdown detection

LIBD is another technique that can be several requests of size greater sensitive for figuring out little particles (>50 nm) than routine light scrambling strategies. LIBD depends on the plasma era (or breakdown) on single particles by a centered, pulse laser beam that can be used for detection of colloids as a sensitive method. Plasma formation is occurred by passing nanocapsules through the central volume of the optical cell. A microscope associated with an activated charge-coupled-gadget (CCD) camera or a piezoelectric crystal attached to the cuvette can be used for detection of plasma light emission (Fig. 14.2C). The detection data by CCD or piezoelectric crystal are presented as breakdown probability (BP). Due to the fact that BP for a given laser strength relies upon both on particle attention and size, it's miles necessary to clarify both (Bundschuh et al., 2001a,b).

14.4 Electric charge of nanocapsules

There are many different charged materials, such as ionic surfactants, proteins, polysaccharides, fatty acids, and so on that can be used for stabilization of nanocapsule systems, thus measuring the surface charge of nanocapsules is necessary to determine their stability, physicochemical, and functional properties, encapsulation efficiency, and pathway of their cellular uptake. Regarding the preparation of multilayer nanocapsules (especially polymeric nanocarriers) it is necessary to have information about their surface charge. For example, in layer by layer nanocapsules, the first layer is usually positive (like whey proteins below their isoelectric point) and the second layer is negative (like the polysaccharides) (Dubas et al., 2006; McClements, 2004a, 2014).

Zeta-potential measurement is the common technique for describing the surface charge of nanocapsules due to better representation and simple measurement of the

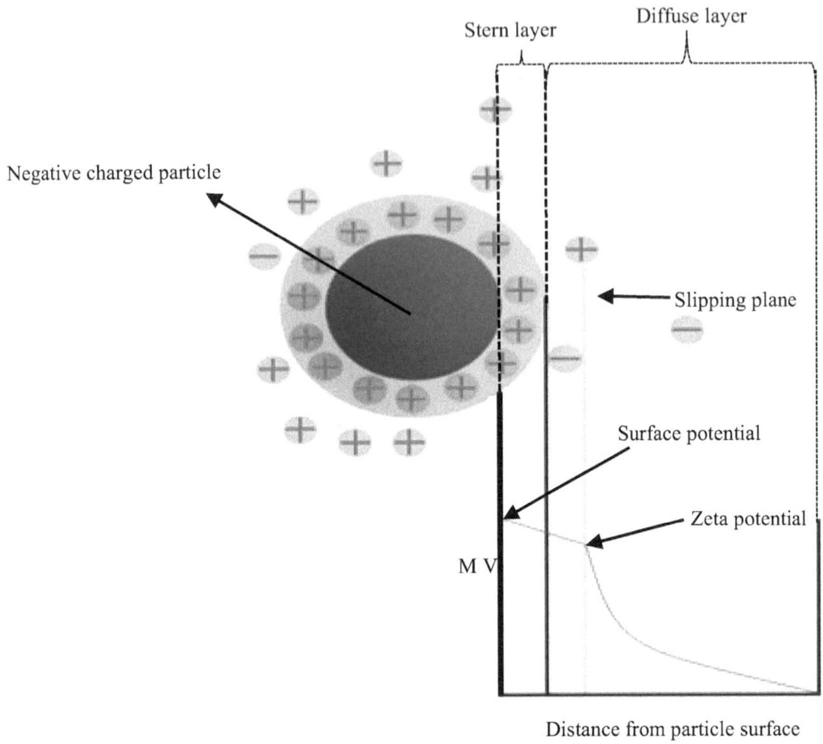

Fig. 14.3 Illustration of a particle, having a negative surface potential, a Stern layer composed of negative ions, a diffuse layer surrounding both the particles and the measured zeta potential is located at the slipping plane.

electrical characteristics. It is characterized as the separation far from the particle surface underneath which any counter particles remain unequivocally joined to the particle when it moves in an electrical field (Clogston and Patri, 2011; Hunter, 2001; McClements, 2004a). The stern layer and diffuse layer surround the particles in the solution which are called "electrical double layers". A layer of ions is bounded to the particles surface, which is a stern layer and the outer layer is the diffuse layer with loosely associated ions. The electrostatic potential in limit isolating the stern layer and diffuse layer is characterized as zeta-potential (Fig. 14.3) (Clogston and Patri, 2011; Mitri et al., 2011; Nguyen et al., 2014; Sze et al., 2003).

Zeta-potential measurement reveals information about the dispersion, aggregation, or flocculation needed for improving the dispersions, emulsions, and suspensions. Regularly, measuring the z-potential versus pH profile describes the electrical properties of a particle. Zeta-potential has values that for the most part range from +100 mV to −100 mV, besides colloidal particles with zeta possibilities of more noteworthy than +30 mV or under −30 mV are considered unequivocally cationic and anionic, respectively (Clogston and Patri, 2011).

Zeta potential measurement is based on analyzing particle movement due to the applied electrical field across the sample and it is measured via the light scattering of the particles; finally, the zeta-potential is calculated by Henry equation. The electrophoresis and laser doppler velocimetry, also called laser doppler electrophoresis are two main techniques that can be used for measuring zeta-potential. These methods measure how speedy a particle actions in a liquid whilst an electrical discipline is applied. The zeta-sizer instrument (Malvern, United Kingdom) is commonly used for measuring zeta-potential of nanocapsules as it conveys data about particle size and molecular weight (Clogston and Patri, 2011; Hunter, 2001; McClements, 2004a).

14.5 Surface component of nanocapsules

Analyzing nanocapsules surface component reveals major information about their functional and chemical properties, the stability of encapsulated ingredients in nanocapsules and applicability of nanocapsules in formulations. Accordingly, an easy and accurate method is needed to study the surface component of nanocapsules. XPS is one of the most sensitive and accurate surface analysis techniques, indeed it is based on creating electron image from samples by scanning of a focused X-ray beam. X-rays are irradiated on the surface of the sample and is absorbed by surface atoms rendering electrons to be reemitted. Identification of sample's composition and the electronic state can be made directly by studying the kinetic energy and number of the emitted electrons (Fig. 14.4). In order To prevent the interactions of X-rays and photoelectrons with the air, XPS has to be performed under high vacuum (Andrade, 1985).

The surface compounds of nanocapsules can be detected by XPS. Leakage of core materials is likely to occur during solvent extraction; accordingly, XPS can analyze this ingredient on the surface of nanocapsules. For example, as an illustration, XPS can be applied for analyzing the unencapsulated component in the surface of produced powders by a spray dryer (Jafari et al., 2008).

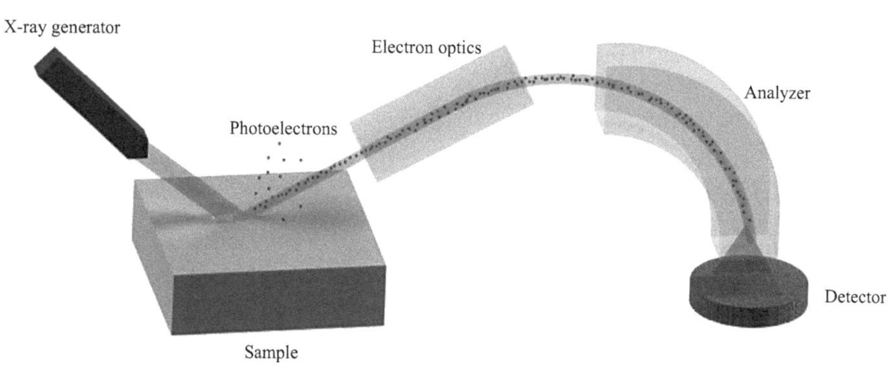

Fig. 14.4 Schematic diagram of X-ray photoelectron spectroscopy (XPS).

14.6 Physicochemical properties of nanocapsules

Nanocapsules can be formulated by different materials; like the lipids, carbohydrates, proteins, and surfactants; these materials undergo many different changes during, after, and before productions or in fortified formulations; for instance, melting of lipids, conformational changes of proteins and polysaccharides, and phase transitions of surfactant and phospholipids. In essence, identification of the physicochemical state of nanocapsules can provide fundamental information about their application, stability, functional properties, and resistance (Faridi Esfanjani and Jafari, 2016; Faridi Esfanjani et al., 2017). For these reasons, it is important to be familiar with instrumental methods, which investigate physicochemical properties of nanocapsules. Here some of these techniques are discussed.

14.6.1 X-ray diffraction

XRD basically provides information about the physical state of the atomic and molecular structure of a crystal which can also be used for determination of changes in the spatial arrangement of molecules within a material like nanocrystals (Qazi et al., 2009; Dehnad et al., 2014).

14.6.2 Dilatometry

In this method, dimension changes of a sample (such as nanocapsules) are measured under fixed load, while the substance is exposed to a controlled temperature program in a specified atmosphere. It can provide information about phase transitions, thermal properties, crystallization behavior, and melting of nanocapsules in the delivery system (Mcclements, 2007; McClements, 2014; Menard, 2008).

14.6.3 Differential scanning calorimetry

DSC as a thermal systematic procedure, ready to set up an association amongst temperature and particular physical properties of samples which are an important technique for measuring changes in enthalpy of samples that are exposed to temperature variations like the heating or cooling process(Dehnad et al., 2014; Jafari et al., 2015). Therefore, this method can provide fundamental information about the glass transition temperature (T_g), heat capacity (C_p), melting point (T_m), and recrystallization times of nanocapsules in delivery systems (Fang et al., 2009; Jenning et al., 2000; Sarmento et al., 2006).

14.6.4 Fourier transform infrared spectroscopy

FT-IR is a favored strategy for infrared spectroscopy in which IR radiation is gone through a sample. A portion of the infrared radiation is consumed by the specimen and some are gone through (transmitted). The ensuing spectrum represents the molecular absorption and transmission, making a molecular fingerprint of the sample(Assadpour et al., 2017). Indeed, FTIR provides an infrared spectrum of absorption or emission of a solid, liquid, or gas and can be used for determination of interaction bonding and complex formation in colloid delivery systems (Esfanjani et al., 2015; Sarmento et al., 2006; Shukla et al., 2003). For instance, when the charged complex of pectin-whey protein is used for formulating double-layered capsules, the FTIR study can provide information about their interaction; exhibiting a sharp band at 1140 cm^{-1} which are related to the C = O stretching and thus the formation of WPC/pectin complex. Moreover, charged complex whey protein-pectin showed a peak at 1553 cm^{-1} which it related to bonds between the amino groups of whey protein and carboxyl groups of pectin (Esfanjani et al., 2015).

14.6.5 Nuclear magnetic resonance

NMR is a physical phenomenon available for the characterization of molecular structure and conformational dynamics with atomic level details. It is based on magnetic properties of certain atomic nuclei. When the nanocapsule samples are exposed to a pulse of radio frequency electromagnetic radiation, the hydrogen nuclei inside a sample changes the resonance frequency and provides information about the electronic structure of the sample. These changes of resonance frequency can determine the physical state of colloid delivery systems, for example, because solids change the resonance solid content of particles in colloidal delivery systems can be determined by these frequency changes frequency more than liquids (Hoch and Stern, 1996; Jores et al., 2003; Mayer, 2005).

14.7 Stability of nanocapsules

Stability of nanocapsules is a fundamental parameter, which is necessary to evaluate the stability of produced nanocapsules with the aim of storage and applying them in targeted formulations, such as food products or other biological systems. For example, high stability of nanocapsules is important for the fortification of milk or beverages during storage and consumption so as to avoid undesirable changes and degradation of encapsulated compounds. There are several reasons for physical instability of nanoemulsions as popular colloid delivery systems illustrated in Fig. 14.5. In the same way, changes of structure, shape, location, or size of particles happens during instability of colloid delivery systems, such as emulsions (Faridi Esfanjani and Jafari, 2016; Esfanjani et al., 2015; Hosseini et al., 2015; Jafari et al., 2012; McClements, 2004a; Solans et al., 2005; Tadros et al., 2004). In this section, different analytical methods for evaluating the stability of nanocapsules have been discussed.

14.7.1 Storage stability testing

Storage stability analysis is the most common technique for determining the stability of nanocapsules in many formulations, such as foods, pharmaceuticals, cosmeceuticals along with other biological and industrial systems. In this method, samples (like nanoemulsions) are transferred into fixed test tubes and then are stored for a required time under specified conditions, such as simulated storage or consumption

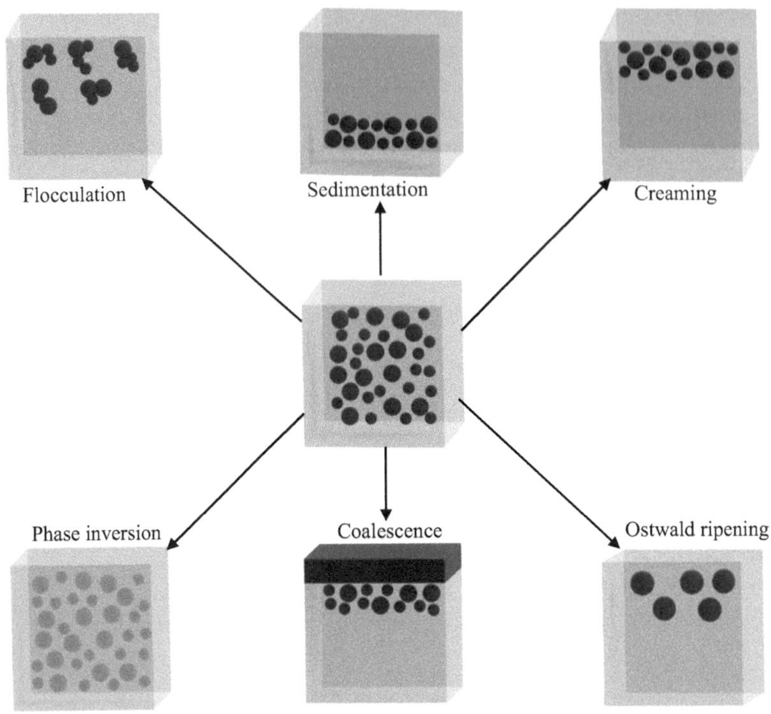

Fig. 14.5 The variety of physical instability mechanisms of nano-emulsions, including creaming, sedimentation, flocculation, coalescence, Ostwald ripening, and phase inversion.

environment of fortified formulations with nanocapsules. Furthermore, it is possible to investigate the influence of ambient conditions on the stability of nanosystems by changing some factors like the temperature and air condition during storage. The stability indices including creaming index, zeta-potential, PSD, encapsulation efficiency, turbidity, centrifugation test, and viscosity are measured during prespecified storage time (e.g., in a month) (McClements, 2004a; Qian et al., 2012).

Creaming or sedimentation stability is a simple and available method applied to provide information about the long period stability of nanocapsule (especially for liquid delivery systems, such as emulsions). The creaming test is performed by transferring freshly prepared colloid delivery systems into test tubes and their storage for a required time. The volume of apparently stable systems without any creaming is monitored daily during storage (ht) and compared to the initial systems height (h_0) to determine emulsion creaming index (CI) (McClements, 2004a; McClements, 2014).

$$\mathrm{CI} = \frac{ht}{h_0} \times 100 \tag{14.1}$$

Droplet size change is measured according to Section 14.2 during storage of nanocapsules systems in test tubes. Stability of nanocapsules systems based on their droplet size change can be measured by stability index (ESI) of the colloid delivery system, which is defined as a relative ratio of the stored colloid delivery system data (e.g., D43, D32, or Z average) to the freshly-made sample data. A sharp change (decreasing or increasing) of ESI shows the instability of colloid delivery systems since particle size variations cause the ESI changes (Jafari et al., 2007).

Turbidity loss amount can measure the cloud stability of colloid delivery systems. This method is commonly based on measuring absorption of diluted colloid delivery systems (with sugar solution) by a UV spectrophotometer: samples are stored in bottles and small amounts are consistently taken from these bottles (usually at a distance from bottom) and daily reading of their absorbance is continued until their absorbance decrease to 10% of the preliminary absorbance value. A first-order model is normally used for fitting the absorbance data with the aim of analyzing turbidity based on the following equation (Buffo et al., 2001; Gharibzahedi et al., 2012; Rao and McClements, 2012):

$$\ln A = \ln A_0 - k_1 t \tag{14.2}$$

Where t is time, A the absorbance at time t, A_0 the absorbance at time 0 and k_1 the first-order rate constant. In this equation, $A = 0.1A_0$ (end-point of the colloid delivery system stability) is followed and, k_1 is measured as the turbidity loss amount; smaller k_1 value refers to the scale back loss amount of turbidity.

Viscosity is a type of rheological property of colloid delivery systems, which can be used for measuring the stability of nanocapsules systems. Two main instruments that are applied for the determination of shear forces and compression dynamic shear rheometers and forces are shear viscometers, and universal testing machines, respectively. Usually, the viscosity of samples can be measured immediately after preparation, after 1 day, and during storage time. Many different parameters can influence the viscosity behavior of colloid delivery systems; for example, physical state and droplet size (liquids, solids, or viscoelastic materials), formulation composition (e.g., gums provide higher viscosity), concentration of different compounds in formulation (when the aqueous phase is large, the viscosity is relatively low), and creaming. Thereupon, measuring the viscosity of samples during storage can provide information about their quality changes and stability. For example, the high and stable viscosity in emulsion systems causes the droplets to be keep away from each other and as a result, their stability is increased. Moreover, creaming depends on viscosity; the phase separation is reduced by the high and stable viscosity of samples during storage (Jafari et al., 2012; Lutz et al., 2009; Mohammadi et al., 2016b; Sakulku et al., 2009).

14.7.2 Accelerated stability tests

The accelerated technique is a method for measuring the stability of the colloid delivery system in a short time, as well as prediction of their long-term stability to gravitational separation. Centrifugation stability as an accelerated technique is a common and easy analytical method for estimating the stability of nanocapsules systems. This index is determined by centrifugation of required amount of samples (such as an emulsion):

$$ES = \frac{F_{ev}}{i_{ev}} \tag{14.3}$$

Where F_{ev} is the final sample volume, and i_{ev} is initial volume (Mou et al., 2008; Sciarini et al., 2009; Shafiq et al., 2007). The high stability of a sample against phase separation is shown by a low ES index.

Also, predicting the stability of colloid systems can be performed by increasing the stresses on the colloidal system as below:

Temperature stress can be applied by low or high temperatures to the colloidal systems. Storage at low temperatures will measure if the colloidal formulation can recuperate its unique structure after defrosting, while at raised temperatures, the maturing of the colloidal formulation is quickened to anticipate its long haul stability. Besides, measuring the stability of colloid formulations against fluctuations in temperature can be performed by cycles of freeze-thaw between high and low temperatures (Aoki et al., 2005; Gu et al., 2005; Güzey and McClements, 2006; Tadros, 2004).

Mechanical stress can be applied by pouring colloid formulations into containers and then subjecting them to fixed mechanical forces like rotation. Mechanical stress stability is very important for colloid formulations during their production process which show their behavior under mechanical devices (like the homogenizer) or transfer in pipes (Heinzelmann and Franke, 1999; McClements, 2004b).

pH stress on colloid formulations can be determined by applying different pH values in preparing the samples. Usually, colloid formulations are subjected to a variety of pH ranges in foods or other biological systems and are produced in different pH ranges. In conclusion, it is important monitoring their stability under a variety of pH values (Gu et al., 2005).

14.8 Image analysis of nanocapsules

The analyzing of color, size, shape, texture, and other characteristics by image processing has been rapidly developed in many different fields, such as biological or food systems (Davies, 2000; Ginesu et al., 2004; Zheng et al., 2006). This is an easy, available, and cheap method that basically comprises of an advanced or camcorder for image procurement, standard settings illuminants and a PC programming for picture examination. This technique has also been established in nanocapsule systems area which can provide information about particle shape, particle size, PSD, stability, and optical properties (Burgess et al., 2004; Du and Sun, 2004). Particle data of nanocapsules can be determined by the microscopic analysis (providing an image by a microscope) furnished with a PC controlled image examination framework. Meanwhile, stability analysis of colloid formulations is possible by microscopic technique and through studying the particle changes (like the size or shape) (Merkus, 2009; Nguyen et al., 2014).

On the other hand, storage and application of nanocapsules in various formulations need measuring the optical properties of colloid formulations. For example, the color of nanoencapsulated ingredients must be compatible with fortified systems (like milk). The opacity, color, and surface sheen are most important optical properties of nanocapsules. The opacity is affected via particle size and distribution, and the absorption properties of any chromophores present.

The image processing setup is composed of a box containing two fluorescent lamps and black walls, a camera which is placed 25 cm in parallel or at the top of the sample, a computer, and an image analysis software (like Image J software or Clemex Vision PE). The color of provided images by the camera are measured by computer vision system (converted from RGB to Lab system by "color space converter" plugin) (changed over from RGB to Lab framework by color space converter module) (Hosseini et al., 2015). Image processing can give data about the optical properties of delivery frameworks with $L^*a^*b^*$ framework. In this framework, the * and b^* qualities are color facilitates, where $+a^*$ is the red bearing, $-a^*$ is the green course, $+b^*$ is the yellow heading, and $-b^*$ is the blue bearing. The mistiness of a colloidal scattering can thusly be portrayed by the delicacy (L^*), while the general shading force can be described by the chroma: $C = (a^{*2} + b^{*2})^{1/2}$ (Hosseini et al., 2015; Assadpour, et al., 2016).

14.9 Fluorescence spectroscopy of nanocapsules

Fluorescence spectroscopy is based on measuring emitted photons (fluorescence) from samples. When the samples absorb beam light (usually ultraviolet light), their electrons excite to higher energy state. Hence, the photons of lower energy (fluorescence) are emitted by the return of these electrons to lower energy state (Fig. 14.6). Indicator probe of fluorescence provides fluorescence spectroscopy and imaging as the main tool for analyzing the samples. These tools provide some fundamental information about nanocapsule properties,such as the critical aggregation concentration, location and distribution of encapsulated compounds within nanocapsules, understanding the microenvironment of the encapsulated compounds, measuring transformation of produced materials across the nanocapsules, studying fate of coated materials in nanocapsules and nanocapsule in vivo systems (Jäger et al., 2007; Kim et al., 2005; Sessa et al., 2014; Sharma and Schulman, 1999; Sabliov et al., 2015; Yu and Huang, 2010).

Stability of encapsulated compounds within nanocapsules is very important for designing the best colloid delivery systems. There may be possible challenges in the destructive reactions of nanocapsule, such as oxidation or degradation by free radicals. These reactions especially occur in sensitive nanocapsule systems (such as O/W nanoemulsions) at inappropriate conditions like high temperature and presence of minor impurities. The produced free radicals then transport the hydrophobic core of nanocapsules and cause numerous reactions. Thus, the resulted instability in nanocapsules should be measured by an easy analytical method. The transportation of free radical

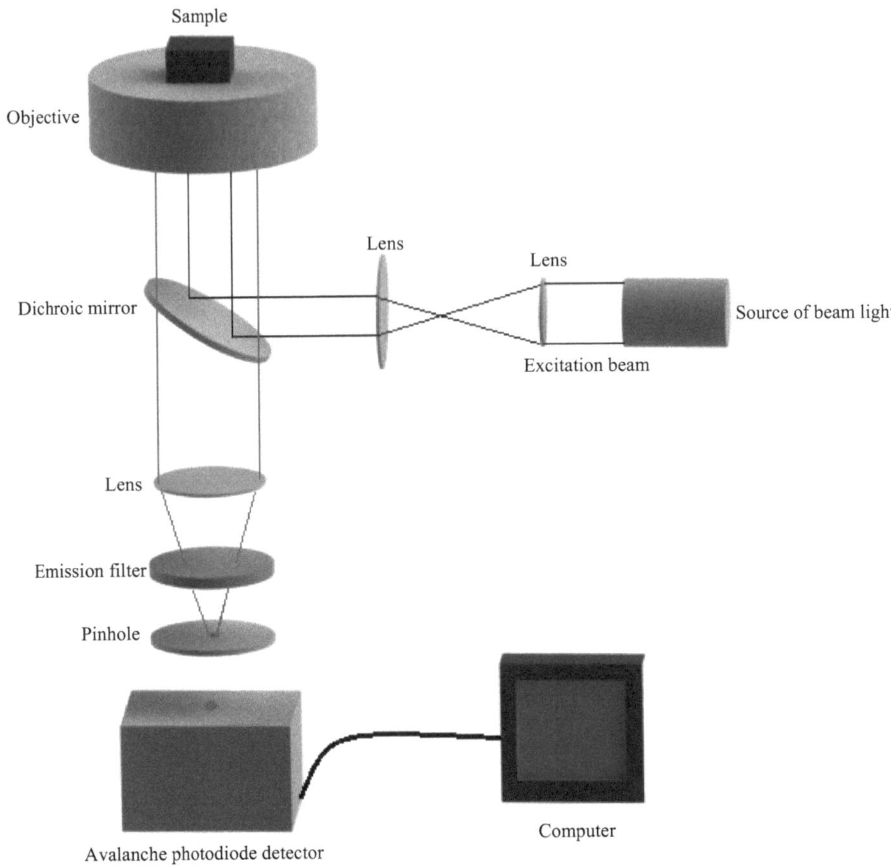

Fig. 14.6 Schematic figure of a setup of fluorescence correlation spectroscopy (FCS).

and oxygen inside nanocapsules can be identified by fluorescence spectroscopy. The principle of this technique is based on measuring fluorescence intensity change of encapsulated fluorescent compound in nanocapsules upon association via oxygen and free radicals. For example, in O/W nanoemulsions containing a fluorescent compound within oil phase, the fluorescence intensity is increased by transport of hydroxyl radicals across the emulsion interface and likewise the loss in fluorescence intensity of the encapsulated fluorescent compound is shown upon interaction with peroxide radicals (Sabliov et al., 2015).

The fluorescence image determines the behavior of nanocapsules in the gastrointestinal tract (GIT) plus uptake and could create essential comprehension of the dissemination of nanocontainers inside in vivo frameworks. Detection of fluorescence discharge from fluorophores in the entire collection of living little creatures is performed by using the sensitive camera for providing fluorescence images. Long emissions at the close infrared (NIR) district are for the most part wanted to beat the photon weakening in living tissues (Goutayer et al., 2010; Rao et al., 2007).

References

Aguilera, J.M., Stanley, D.W., 1999. Microstructural principles of food processing and engineering. Springer Science & Business Media.

Alexander, M., Dalgleish, D.G., 2006. Dynamic light scattering techniques and their applications in food science. Food Biophy. 1, 2–13.

Alvarez-Román, R., Naik, A., Kalia, Y., Fessi, H., Guy, R.H., 2004. Visualization of skin penetration using confocal laser scanning microscopy. Eur. J. Pharm. Biopharm. 58, 301–316.

Andrade, J.D., 1985. X-ray photoelectron spectroscopy (XPS). Surface and interfacial aspects of biomedical polymers. Springer, US, pp. 105–195.

Aoki, T., Decker, E.A., Mcclements, D.J., 2005. Influence of environmental stresses on stability of O/W emulsions containing droplets stabilized by multilayered membranes produced by a layer-by-layer electrostatic deposition technique. Food Hydrocol. 19, 209–220.

Arzenšek, D., 2010. Dynamic light scattering and application to proteins in solutions. Seminar, 4th year. Department of Physics, University of Ljubljana, pp. 1–18.

Assadpour, E., Maghsoudlou, Y., Jafari, S.M., Ghorbani, M., Aalami, M., 2016. Optimization of folic acid nano-emulsification and encapsulation by maltodextrin-whey protein double emulsions. Int. J. Biol. Macromol. 86, 197–207.

Assadpour, E., Jafari, S.M., Maghsoudlou, Y., 2017. Evaluation of folic acid release from spray dried powder particles of pectin-whey protein nano-capsules. Int. J. Biol. Macromol. 95, 238–247.

Buffo, R., Reineccius, G., Oehlert, G., 2001. Factors affecting the emulsifying and rheological properties of gum acacia in beverage emulsions. Food Hydrocol. 15, 53–66.

Bundschuh, T., Knopp, R., Kim, J., 2001a. Laser-induced breakdown detection (LIBD) of aquatic colloids with different laser systems. Colloids Surf. A 177, 47–55.

Bundschuh, T., Yun, J.-I., Knopp, R., 2001b. Determination of size, concentration and elemental composition of colloids with laser-induced breakdown detection/spectroscopy (LIBD/S). Fresenius' J. Anal. Chem. 371, 1063–1069.

Burgess, D.J., Duffy, E., Etzler, F., Hickey, A.J., 2004. Particle size analysis: AAPS workshop report, cosponsored by the Food and Drug Administration and the United States Pharmacopeia. AAPS J. 6, 23–34.

Clogston, J.D., Patri, A.K., 2011. Zeta potential measurement. Methods Mol. Biol. 697, 63–70.

Davies, E.R., 2000. Image processing for the food industry. World Scientific, New York, p. 289.

Dehnad, D., Mirzaei, H., Emam-Djomeh, Z., Jafari, S.M., Dadashi, S., 2014. Thermal and antimicrobial properties of chitosan-nanocellulose films for extending shelf life of ground meat. Carbohydrate polymers 109, 148–154.

Driscoll, D.F., Etzler, F., Barber, T.A., Nehne, J., Niemann, W., Bistrian, B.R., 2001. Physicochemical assessments of parenteral lipid emulsions: light obscuration versus laser diffraction. Int. J. Pharm. 219, 21–37.

Du, C.-J., Sun, D.-W., 2004. Recent developments in the applications of image processing techniques for food quality evaluation. Trends Food Sci. Technol. 15, 230–249.

Dubas, S.T., Kumlangdudsana, P., Potiyaraj, P., 2006. Layer-by-layer deposition of antimicrobial silver nanoparticles on textile fibers. Colloids Surf. A: Physicochem. Eng. Aspects 289, 105–109.

Egerton, R., 2006. Physical principles of electron microscopy: an introduction to TEM, SEM, and AEM. Springer Science & Business Media, New York.

Esfanjani, A.F., Jafari, S.M., Assadpoor, E., Mohammadi, A., 2015. Nano-encapsulation of saffron extract through double-layered multiple emulsions of pectin and whey protein concentrate. J. Food Eng. 165, 149–155.

Faridi Esfanjani, A.F., Jafari, S.M., 2016. Biopolymer nano-particles and natural nano-carriers for nano-encapsulation of phenolic compounds. Colloids Surf. B. Biointerfaces. 146, 532–543.

Faridi Esfanjani, A.F., Jafari, S.M., Assadpoor, E., 2017. Preparation of a multiple emulsion based on pectin-whey protein complex for encapsulation of saffron extract nanodroplets. Food Chem. 221, 1962–1969.

Fang, G., Li, H., Yang, F., Liu, X., Wu, S., 2009. Preparation and characterization of nano-encapsulated n-tetradecane as phase change material for thermal energy storage. Chem. Eng. J. 153, 217–221.

Gan, Q., Wang, T., Cochrane, C., Mccarron, P., 2005. Modulation of surface charge, particle size and morphological properties of chitosan–TPP nanoparticles intended for gene delivery. Colloids Surf. B Biointerfaces 44, 65–73.

Gharibzahedi, S.M.T., Mousavi, S.M., Hamedi, M., Ghasemlou, M., 2012. Response surface modeling for optimization of formulation variables and physical stability assessment of walnut oil-in-water beverage emulsions. Food Hydrocol. 26, 293–301.

Ginesu, G., Giusto, D.D., Märgner, V., Meinlschmidt, P., 2004. Detection of foreign bodies in food by thermal image processing. IEEE Trans. Ind. Electron. 51, 480–490.

Goutayer, M., Dufort, S., Josserand, V., Royère, A., Heinrich, E., Vinet, F., Bibette, J., Coll, J.-L., Texier, I., 2010. Tumor targeting of functionalized lipid nanoparticles: assessment by in vivo fluorescence imaging. Eur. J. Pharm. Biopharm. 75, 137–147.

Gu, Y.S., Regnier, L., Mcclements, D.J., 2005. Influence of environmental stresses on stability of oil-in-water emulsions containing droplets stabilized by β-lactoglobulin–ι-carrageenan membranes. J. Colloid Interface Sci. 286, 551–558.

Gupta, A.K., Wells, S., 2004. Surface-modified superparamagnetic nanoparticles for drug delivery: preparation, characterization, and cytotoxicity studies. IEEE Transact. Nano-Bioscience 3, 66–73.

Güzey, D., McClements, D.J., 2006. Influence of environmental stresses on O/W emulsions stabilized by β-lactoglobulin–pectin and β-lactoglobulin–pectin–chitosan membranes produced by the electrostatic layer-by-layer deposition technique. Food Biophy. 1, 30–40.

Hall, C.E., 1954. Introduction to electron microscopy. Soil Sci. 77, 254.

Hassellöv, M., Readman, J.W., Ranville, J.F., Tiede, K., 2008. Nanoparticle analysis and characterization methodologies in environmental risk assessment of engineered nanoparticles. Ecotoxicology 17, 344–361.

Hayat, M.A., 1974. Principles and techniques of scanning electron microscopy. Biological applications. Volume 1. Van Nostrand Reinhold Company, New York.

Heinzelmann, K., Franke, K., 1999. Using freezing and drying techniques of emulsions for the microencapsulation of fish oil to improve oxidation stability. Colloids Surf. B Biointerfaces 12, 223–229.

Hoch, J.C., Stern, A.S., 1996. NMR data processing. Wiley-Liss, New York.

Hosseini, A., Jafari, S.M., Mirzaei, H., Asghari, A., Akhavan, S., 2015. Application of image processing to assess emulsion stability and emulsification properties of Arabic gum. Carbohydr. Polym. 126, 1–8.

Hren, J., 2013. Introduction to analytical electron microscopy. Springer Science & Business Media, New York.

Hunter, R.J., 2001. Foundations of colloid science. Oxford University Press, New York.

Jafari, S.M., Assadpoor, E., Bhandari, B., He, Y., 2008. Nano-particle encapsulation of fish oil by spray drying. Food Res. Int. 41, 172–183.

Jafari, S.M., Beheshti, P., Assadpoor, E., 2012. Rheological behavior and stability of D-limonene emulsions made by a novel hydrocolloid (Angum gum) compared with Arabic gum. Journal of Food Engineering 109, 1–8.

Jafari, S.M., He, Y., Bhandari, B., 2007. Effectiveness of encapsulating biopolymers to produce sub-micron emulsions by high energy emulsification techniques. Food Res. Int. 40, 862–873.

Jafari, S.M., Khanzadi, M., Mirzaei, H., Dehnad, D., Chegini, F.K., Maghsoudlou, Y., 2015. Hydrophobicity, thermal and micro-structural properties of whey protein concentrate-pullulan-beeswax films. Int. J. Biol. Macromol. 80, 506–511.

Jäger, A., Stefani, V., Guterres, S.S., Pohlmann, A.R., 2007. Physico-chemical characterization of nanocapsule polymeric wall using fluorescent benzazole probes. Int. J. Pharm. 338, 297–305.

Jenning, V., Thünemann, A.F., Gohla, S.H., 2000. Characterisation of a novel solid lipid nanoparticle carrier system based on binary mixtures of liquid and solid lipids. Int. J. Pharm. 199, 167–177.

Jores, K., Mehnert, W., Mäder, K., 2003. Physicochemical investigations on solid lipid nanoparticles and on oil-loaded solid lipid nanoparticles: a nuclear magnetic resonance and electron spin resonance study. Pharm. Res. 20, 1274–1283.

Kim, B.-D., Na, K., Choi, H.-K., 2005. Preparation and characterization of solid lipid nanoparticles (SLN) made of cacao butter and curdlan. Eur. J. Pharm. Sci. 24, 199–205.

Kozan, M., 2007. Characterization of Coloidal Nanoparticle Aggregates Using Light Scattering Techniques. Ph.D. thesis, University of Kentucky. Available from: http://uknowledge.uky.edu/gradschool_diss/567, paper 567.

Kübart, S.A., Keck, C.M., 2013. Laser diffractometry of nanoparticles: frequent pitfalls & overlooked opportunities. J. Pharm. Technol. Drug Res. 2, 17.

Latterini, L., Tarpani, L., 2012. AFM Measurements to Investigate Particulates and Their Interactions with Biological Macromolecules. INTECH Open Access Publisher, Rijeka, pp. 87–98.

Lutz, R., Aserin, A., Wicker, L., Garti, N., 2009. Release of electrolytes from W/O/W double emulsions stabilized by a soluble complex of modified pectin and whey protein isolate. Colloids Surf. B Biointerfaces 74, 178–185.

Madden, A.S., Hochella, M.F., 2005. A test of geochemical reactivity as a function of mineral size: manganese oxidation promoted by hematite nanoparticles. Geochim. Cosmochim. Acta 69, 389–398.

Mayer, C., 2005. NMR studies of nanoparticles. Annual Reports on NMR Spectroscopy 55, 205–258.

McClements, D.J., 2004a. Food emulsions: principles, practices, and techniques. CRC press, Boca Raton.

McClements, D.J., 2004b. Protein-stabilized emulsions. Curr. Opin. Colloid Interface Sci. 9, 305–313.

Mcclements, D.J., 2007. Critical review of techniques and methodologies for characterization of emulsion stability. Crit. Rev. Food Sci. Nutr. 47, 611–649.

McClements, D.J., 2014. Nanoparticle-and microparticle-based delivery systems: Encapsulation, protection and release of active compounds. CRC Press, New York.

Menard, K.P., 2008. Dynamic mechanical analysis: a practical introduction. CRC press, Boca Raton.

Merkus, H.G., 2009. Particle size measurements: fundamentals, practice, quality. Springer Science & Business Media, The Netherlands.

Mitri, K., Shegokar, R., Gohla, S., Anselmi, C., Müller, R.H., 2011. Lipid nanocarriers for dermal delivery of lutein: preparation, characterization, stability and performance. Int. J. Pharm. 414, 267–275.

Mohammadi, A., Jafari, S.M., Assadpour, E., Faridi Esfanjani, A., 2016a. Nano-encapsulation of olive leaf phenolic compounds through WPC-pectin complexes and evaluating their release rate. Int. J. Biol. Macromol. 82, 816–822.

Mohammadi, A., Jafari, S.M., Esfanjani, A.F., Akhavan, S., 2016b. Application of nano-encapsulated olive leaf extract in controlling the oxidative stability of soybean oil. Food Chem. 190, 513–519.

Mou, D., Chen, H., Du, D., Mao, C., Wan, J., Xu, H., Yang, X., 2008. Hydrogel-thickened nanoemulsion system for topical delivery of lipophilic drugs. Int. J. Pharm. 353, 270–276.

Nguyen, M.H., Aripin, N.F.K., Chen, X.G., Park, H.J., 2014. Methodologies Used for the Characterization of Nano-and Microcapsules—Nano-and Microencapsulation for Foods, pp. 65–94.

Qazi, S.J.S., Rennie, A.R., Cockcroft, J.K., Vickers, M., 2009. Use of wide-angle X-ray diffraction to measure shape and size of dispersed colloidal particles. J. Colloid Interface Sci. 338, 105–110.

Qian, C., Decker, E.A., Xiao, H., Mcclements, D.J., 2012. Physical and chemical stability of β-carotene-enriched nanoemulsions: influence of pH, ionic strength, temperature, and emulsifier type. Food Chem. 132, 1221–1229.

Rao, J., Dragulescu-Andrasi, A., Yao, H., 2007. Fluorescence imaging in vivo: recent advances. Curr. Opin. Biotechnol. 18, 17–25.

Rao, J., McClements, D.J., 2012. Food-grade microemulsions and nanoemulsions: role of oil phase composition on formation and stability. Food Hydrocoll. 29, 326–334.

Sabliov, C., Chen, H., Yada, R. (Eds.), 2015. Nanotechnology and Functional Foods: Effective Delivery of Bioactive Ingredients. John Wiley and Sons Ltd, London, UK, p. 285.

Sakulku, U., Nuchuchua, O., Uawongyart, N., Puttipipatkhachorn, S., Soottitantawat, A., Ruktanonchai, U., 2009. Characterization and mosquito repellent activity of citronella oil nanoemulsion. Int. J. Pharm. 372, 105–111.

Sarmento, B., Ferreira, D., Veiga, F., Ribeiro, A., 2006. Characterization of insulin-loaded alginate nanoparticles produced by ionotropic pre-gelation through DSC and FTIR studies. Carbohydr. Polym. 66, 1–7.

Scalf, J., West, P., 2006. Part I: introduction to nanoparticle characterization with AFM. Pacific Nanotechnology, Santa Clara. Available from: www.nanoparticles.org/pdf/Scalf-West.pdf

Sciarini, L., Maldonado, F., Ribotta, P., Pérez, G., León, A., 2009. Chemical composition and functional properties of *Gleditsia triacanthos* gum. Food Hydrocoll. 23, 306–313.

Sessa, M., Balestrieri, M.L., Ferrari, G., Servillo, L., Castaldo, D., D'onofrio, N., Donsì, F., Tsao, R., 2014. Bioavailability of encapsulated resveratrol into nanoemulsion-based delivery systems. Food Chem. 147, 42–50.

Shafiq, S., Shakeel, F., Talegaonkar, S., Ahmad, F.J., Khar, R.K., Ali, M., 2007. Development and bioavailability assessment of ramipril nanoemulsion formulation. Eur. J. Pharm. Biopharm. 66, 227–243.

Sharma, A., Schulman, S.G., 1999. Introduction to fluorescence spectroscopy. Wiley-interscience, New York.

Shukla, N., Liu, C., Jones, P.M., Weller, D., 2003. FTIR study of surfactant bonding to FePt nanoparticles. J. Magn. Magn. Mater. 266, 178–184.

Solans, C., Izquierdo, P., Nolla, J., Azemar, N., Garcia-Celma, M., 2005. Nano-emulsions. Curr. Opin. Colloid Interface Sci. 10, 102–110.

Sze, A., Erickson, D., Ren, L., Li, D., 2003. Zeta-potential measurement using the Smoluchowski equation and the slope of the current–time relationship in electroosmotic flow. J. Colloid Interface Sci. 261, 402–410.

Tadros, T., 2004. Application of rheology for assessment and prediction of the long-term physical stability of emulsions. Adv. Colloid Interface Sci. 108, 227–258.

Tadros, T., Izquierdo, P., Esquena, J., Solans, C., 2004. Formation and stability of nano-emulsions. Adv. Colloid Interface Sci. 108, 303–318.

Yoon, T.J., Kim, J.S., Kim, B.G., Yu, K.N., Cho, M.H., Lee, J.K., 2005. Multifunctional nanoparticles possessing a "magnetic motor effect" for drug or gene delivery. Angew. Chemie 117, 1092–1095.

Yu, H., Huang, Q., 2010. Enhanced in vitro anti-cancer activity of curcumin encapsulated in hydrophobically modified starch. Food Chem. 119, 669–674.

Zheng, C., Sun, D.-W., Zheng, L., 2006. Recent applications of image texture for evaluation of food qualities—a review. Trends Food Sci. Technol. 17, 113–128.

Safety and regulatory issues of nanocapsules

15

Seid Mahdi Jafari, Iman Katouzian*,**, Sahar Akhavan**
*Gorgan University of Agricultural Sciences and Natural Resources, Gorgan, Iran; **Nano-encapsulation in the Food, Nutraceutical, and Pharmaceutical Industries Group (NFNPIG), Universal Scientific Education and Research Network (USERN), Tehran, Iran

15.1 Introduction

Nanotechnology may have beneficial impacts on the food and beverage sector as all the well-known companies are searching more useful methods to improve the quality of their products as well as optimizing their production processes. By the time, over 300 companies are manufacturing their nanobased food and beverage products (Momin and Joshi, 2015). According to a workshop held in the area of food nanotechnology (Yaktine and Pray, 2009), some of the advantages of employing nanotechnology within the food sector are yielding promising nanoprocesses, fabricating biodegradable and green nanopackaging, modifying the texture and taste of cuisines, and formulating low-calorie foods and beverages in order to change lifestyles into healthy ones. Yet, it is not clear if nanomaterials implemented in food products cause harmful biological responses in the body and therefore governmental bodies, as well as international organizations, are disseminating principles in regard to the safety aspects of handling nanoscale materials plus their definitions and deleterious effects on the body and nature. Nevertheless, not many countries have legislated and applied the standards and principles in working with nanomaterials. The US Food and Drug Administration (FDA) is known to be the first public sector to define the concept (Anandharamakrishnan, 2014) of nanotechnology and its application to fabricate nanoproducts. Chau et al. (2007) have gathered a list of different organizations, which are active in the research field of safety prospects of nanofoods. Although the possible toxicity of using nanoparticles is a concern, not much budget and studies are dedicated to this area to date. In 2006, the International Risk Governance Council (IRGC) examined the situation of how nanotechnology is being governed in different areas. According to this report, eleven governments, eleven industrial firms, five research organizations, and nine nongovernment organizations (NGOs) participated in this analysis from different parts of the world. As a result, they suggested that the activities of R&D centers is inevitable as it will lead to potential advantages emanating from the application of nanotechnology (Wei et al., 2009).

However, like other novel technologies, this innovation may also bear some possible disadvantages. Nanomaterials exerted in the food formulations normally enter the body and circulatory system via the oral route; in addition the respiratory system and skin permeation are also other possible pathways of entry. Intestine cells (enterocytes)

Nanoencapsulation Technologies for the Food and Nutraceutical Industries

have their predesigned mechanisms in allowing or rejecting the nutrients or foreign particles tending to pass through them. In this case, the nanosized agents are able to cross epithelium of small intestine, thus there is a possibility in the enhancement of absorption and bioavailability, which may trigger health consequences according to the higher absorption. Furthermore, the absorption rate is different considering the gastrointestinal (GI) tract, like a diseased gut compared to an ordinary one (des Rieux et al., 2006). Yet more assays need to be carried out in order to distinguish the effects of these miniature materials inside the living cells. It was suggested that rate of production and consumers demand may expand in future (Chaudhry et al., 2008).

15.2 Safety and toxicity aspects of food nanoparticles

Although the field of nanotechnology is relatively new, concerns are already arising that nanoparticles may prove toxic to humans or the environment; therefore there is a pressing need to understand how nanoparticles interact with the human body following exposure to consumers. The side effects of nanofood materials are still unclear but their potential hazard cannot be ignored due to the high surface area to volume ratio, which causes them to be highly reactive. Besides, these minute particles simply cross the cell membranes and may accumulate in different locations of biological systems and damage cells. Hence, precise case studies and clinical trials should be conducted to clarify the effects of these miniature materials on body (Chen et al., 2014; Katouzian and Jafari, 2016; Kim et al., 2014; Verma and Stellacci, 2010; Wei et al., 2009).

Safety concerns of nanoscale materials and nanoencapsulated food materials may be evaluated under three categories (1) Nanoparticles as indirect food contaminants; (2) nanoparticles which modify the digestion process as well as the absorption, and (3) nanoparticles with unknown toxicity.

An individual might be exposed directly or indirectly to nanosized materials in the food industry or through other industries, which enters the food chain via agricultural, fisheries, or animal husbandry pathways. Food industry pathways comprise foods being coated with organic or inorganic nanomaterials and nanofilms during fabrication or processing. Also, migration of nanobased food contact materials into the food is another way of becoming exposed to nanomaterials. Fig. 15.1 represents the exposure routes of consumers to nanomaterials in the food products.

The processing of food ingredients into nanoscale creates a completely different physicochemical behavior compared to their original size. (Chaudhry et al., 2008). These novel nanoscale compounds can cause possible toxicity within body as they may find their way into the food chain undetected and enter the cells of living organisms (Baltes, 2008; Dingman and REHS, 2008; Risk Assessment Section, 2010). These nanoparticles are possible factors for causing the disruption of cell's performance leading to their death. Redox balance alteration and tissue inflammation are some examples related to toxic effects of nanoparticulates (Arora et al., 2012).

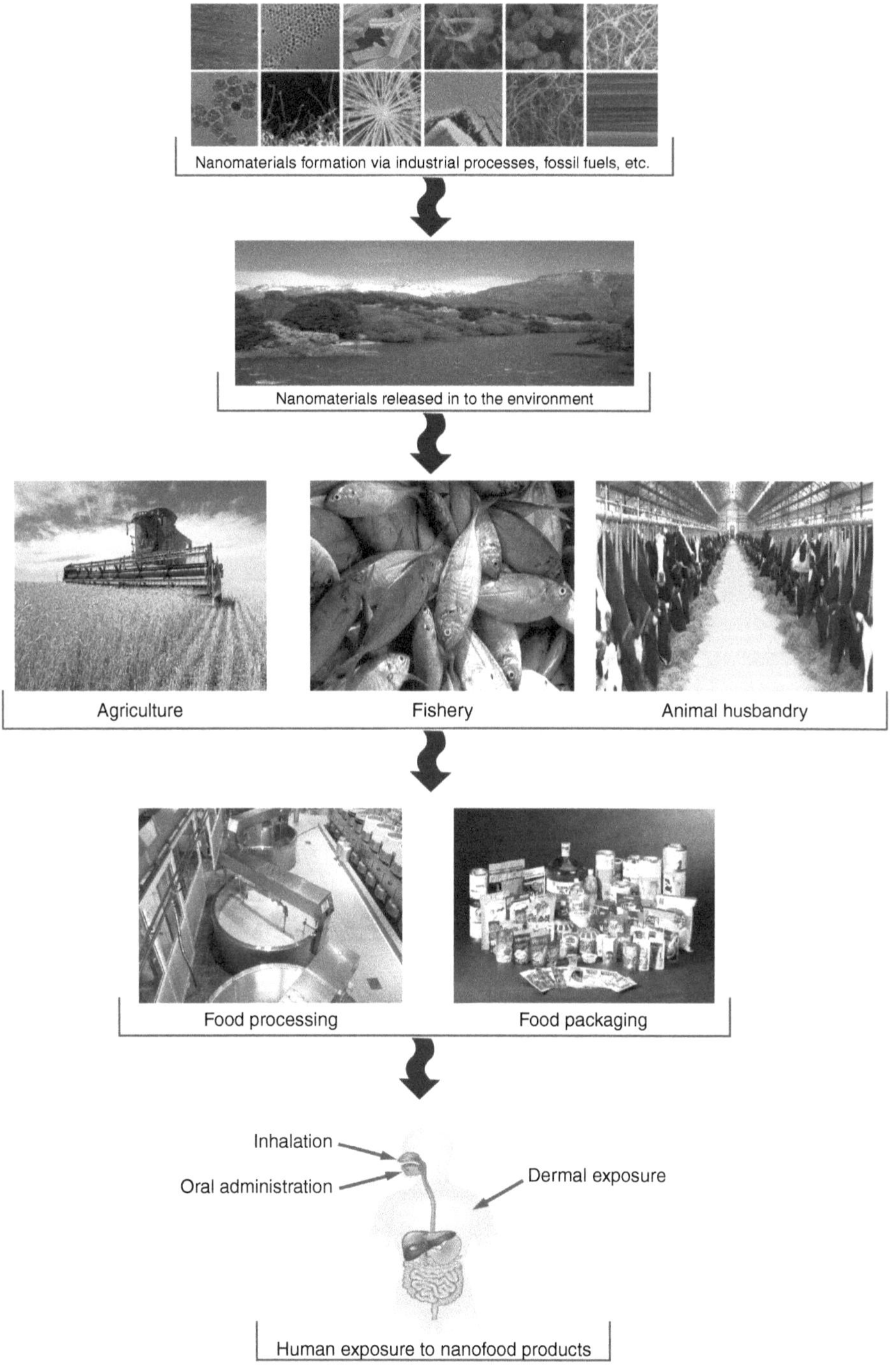

Figure 15.1 Consumers exposure to nanomaterials by using nanofood products and effects of the industry processing and environment on the food chain.
Source: Adapted from Rossi, M., Cubadda, F., Dini, L., Terranova, M., Aureli, F., Sorbo, A., Passeri, D., 2014. Scientific basis of nanotechnology, implications for the food sector and future trends. Trends Food Sci Technol. 40 (2), 127–148..

Generally, after the consumption of nanoparticulates in food matrix, these substances may have three fates in the gastrointestinal (GI) tract which are displayed in Fig. 15.2.

1. The nanoparticulates are fully digested and absorbed in the GIT. Also, the surfactants utilized in the production procedure should be analyzed via safety tests
2. Nanoparticulates are degraded partially and thus the core material is released slightly. On the other hand, conjugates will be generated between core-shell materials, hence various behaviors plus biokinetic action is observed in regard to these conjugates. Moreover, these unknown conjugates may be transferred to different parts of the body and tissues and can pose immunological reactions as they are undefined to the immune system (Dehalu et al., 2012).
3. The nanoparticulate is resistant to digestion, and thus core materials are not released in the GI tract. In this case, two options are considered:

First, the core/shell material will be excreted from the digestive system and second, the core-shell material passes through the epithelium of intestine and access the blood stream. Here, immunological and toxicological assays seem necessary to be done (Katouzian and Jafari, 2016).

15.2.1 Separation and identification of nanocapsules after the process of digestion/absorption

Although several works have been done in recent years in regard to nanotechnology being employed in the food sector, it is not yet clear the exact behavior of nanocapsules after digestion and how their physicochemical characteristics will be modified following absorption. In the last decade, several techniques, such as dynamic light scattering (DLS), scanning electron microscopy (SEM), transmission electron microscopy (TEM), and HPLC have been applied to detect nearly any kind of nanostructures (Linsinger et al., 2013; Peters et al., 2011). Although characterization approaches are available for scholars, isolation, and purification of nanocapsules are not consensual after their transport in the GI tract and also in body fluids. Lately, some studies have centered on the characterization, isolation and purification of nanocapsules within food matrices (Blasco and Picó, 2011; Kowalczyk et al., 2011; Peters et al., 2011; Zattoni et al., 2014). Also, nanostructures in addition to food networks react with body fluids and thus their main properties will be altered. This modification can be associated with the reaction among nanostructures and proteins (for instance, alterations in charge value and agglomeration state) (Lee et al., 2011) plus pH values in GI medium (Pinheiro et al., 2013). Several investigations have focused on the identification of synthetic agents as well as inorganic nanomaterials, such as gold and silver following digestion (Akbulut et al., 2012; Mwilu et al., 2013). Up to now, few studies have explored the methods for the isolation and characterization of nanostructures in the body fluids and tissues. Due to their exclusive properties, like structure, shape, degradation in an acidic or alkaline medium, different methods should be selected for their segregation and purification. Here some investigations are summarized as they seek the absorption and characterization of nanostructures in the GI tract.

In a study flow field flow fractionation (FIFFF) together with inductively coupled plasma mass spectrometry were employed to characterize the size of selenium

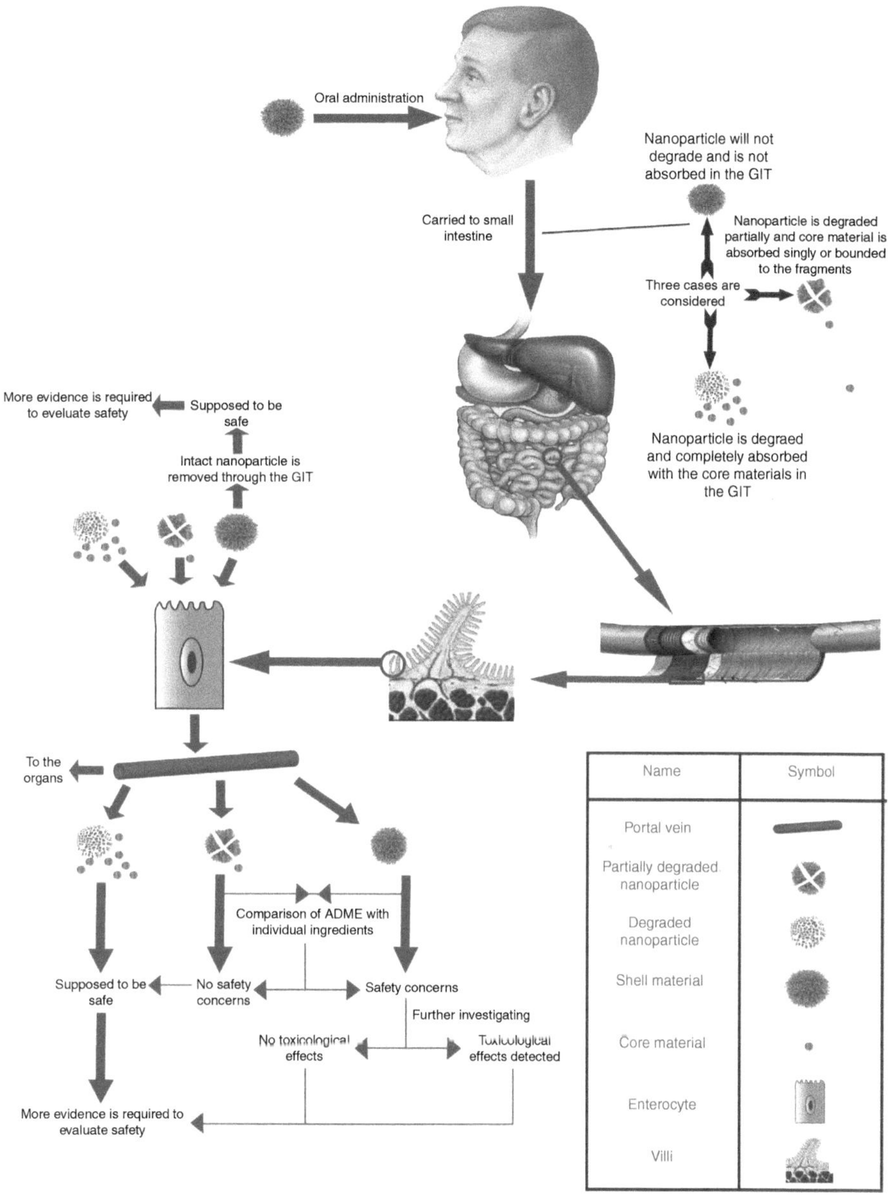

Figure 15.2 Nanoparticles fate and their ADME (adsorption, distribution, metabolism, and excretion) profile within the GIT.
Source: Adapted from Katouzian, I., Jafari, S.M., 2016. Nano-encapsulation as a promising approach for targeted delivery and controlled release of vitamins. Trends Food Sci. Technol. 53, 34–38.

nanomaterials following ingestion within the GI tract, moreover alginate/pectin, ovalbumin and β-lactoglubulin were used as stabilizers in this research(Pornwilard et al., 2014). The results obtained via TEM microscopy and FIFFF supported each other as their results were approximately alike. In another study, polymeric micelle's stability comprising polyethylene glycol (PEG), polylactic acid, and Tween 80 was tested within the human serum via fluorescent method. Finally, the micelles were segregated through flow field flow fractionation (AF4) and were identified via deflection of light (Miller et al., 2012).

Lee et al. (2011) analyzed the integration of chitosan nanoparticles following the GI transit, especially when the reach the Caco-2 cells. Free chitosan was measured (labeled by fluorescein isothiocyanate (FITC)) via ultracentrifugation so as to determine the degradation rate of chitosan nanoparticulates. Furthermore, CLSM was applied to observe and recognize the properties of these nanoparticles present in the Caco-2 cells.

Dandekar et al. (2010) explored the in vitro digestion of hydrogel-based nanoparticles with curcumin as the core material consisted of LMP and HMP, which are low-methoxy pectin and high-methoxy pectin, respectively plus sodium alginate (ALG) and iota carrageenan (CAR). The integrity of nanostructures was monitored during gastric and duodenal ingestion. As a result, the usage of HMP was not successful in protecting nanostructures from gastric proteolysis and changes were observed even during short times of digestion. Also, the digestion of nanostructures resulted in the formation of peptides, representing special SDS-PAGE bands.

Typical problems of consuming nanofood products may include administration of extra doses, which may later alter the ordinary physiological functions, accumulation of nanoparticles in different zones of the body, plus the interactions that would occur between the nanomaterials and food chemical compounds (Sahoo et al., 2007).

15.2.2 Typical mechanisms of nanoparticle toxicity in the GI tract

As stated earlier, when nanofood systems are consumed both the nanoscale network and associated products could deposit in the body and thus their concentration may exceed the natural content present in the body. Moreover, due to partial elimination nanomaterials may be retained in cells and thus normal activities of organs/tissues may be altered and even cases of genotoxicity and immunotoxicity might occur (Katouzian and Jafari, 2016). Therefore, the potential health risks of nanostructures must be assessed before mass production of nanobased foods.

The reaction among the biological environment and nanoparticles and their possible toxicity is related to the nature of materials plus the physicochemical characteristics (size, shape, surface properties, and charge) of nanoscale materials. As an illustration, Loh et al. (2012) suggested that positive surface charges of chitosan nanoparticles had less effect on the in vitro cytotoxicity than of the particle size.

Nanostructures have a high surface area, which renders them extremely reactive compared to their macroscale counterparts. According to these features, nanostructures may impose cytotoxicity and physiological changes within the digestive system, including inflammation, immunological responses (Phagocytosis, supplementary activation of inflammatory cells) (Zolnik et al., 2010), corona generation, mitochondrial disorders, protein denaturation (Simón-Vázquez et al., 2014), and finally DNA alteration and cell death (Lai, 2012; Lozano et al., 2011). Some of these interactions might be advantageous for body, such as employing the nanostructures as possible antigens carriers for vaccines which trigger unique immunological reactions; on the other hand others might be hazardous.

Here some of the physiological mechanisms that are conducted by nanomaterials are described in subsequent sections.

15.2.2.1 Protein corona

Upon the entry of nanomaterials in the circulatory system, they could possibly affect the biomolecules (e.g., DNA, lipids, and proteins of serum) within body and impose toxicity effects. As the mixture of cellular proteins bind to the nanomaterials, protein corona will be generated in which the stability and action of nanostructures are altered. Thereupon, the protein corona modifies the physiological responses (Wolfram et al., 2014). For instance, it can affect endocytosis and exocytosis of the cell membranes as well as the phagocytosis process (Peng et al., 2014; Ruh et al., 2012). Nanostructures are recognized by the cells according to their charge, type of coating, and size (Moros et al., 2012). Moreover, the uptake of these miniature materials is extensively modified by their agglomeration on body fluids (Diaz et al., 2008; Mahmoudi et al., 2014). Besides, nanoparticles can alter the physiological performance of proteins via their conformational modifications. In conclusion, the in vivo fate of nanoparticles will be modified as it is compared with in vitro tests.

It has been suggested that the synthesis of protein corona will affect the biodistribution and the pharmacokinetic profile. Yet, there is a debate about the pros and cons of this protein (Dell'Orco et al., 2012). For example, these proteins can be applied to deliver a nanoparticle to a certain part of body (Dufort et al., 2012). In another work conducted by Peng et al. (2013), it was revealed that the combination of nanoparticles with albumin was an efficient method to increase the circulation time for nanostructures as well as declining their toxicity. On the contrary, as nanoparticles attach to the proteins, they will be rapidly absorbed to liver and spleen, next they will be eliminated via the reticuloendothelial system (Wolfram et al., 2014). When the retention and long life of these nanoparticles are required this characteristic is a negative one. Also, more tests should be done to perceive the effect of nanostructures on different cellular compartments (like proteins and polysaccharides).

15.2.2.2 Oxidative stress

The toxicity of nanomaterials can be recognized by means of measuring the reactive oxygen species generation in cells (Manke et al., 2013). ROS might have hazardous effects on cells in high doses as they cause lipid peroxidation, inflammation and modifications in genes arrangement, which lead to synthesis of irregular proteins (Sharifi et al., 2012). The inflammation process triggered by nanostructured occurs due to the control of transcription factors, such as activating protein 1 (AP-1), external signal linked kinases (ERK), c-Jun protein N-terminal kinases (JNK) plus the protein kinase routes activated via p38 mitogen (Aillon et al., 2009).

ROS emanates from different resources, like low content of antioxidants, phagocytic cell reaction against external agents, Availability of transition metal elements, environmental parameters along with size and shape of nanoparticles (Arora et al., 2012; Karakoti et al., 2006).

It was reported that (Yan et al., 2013) ROS could be formed as nanoparticles alter the balance of biochemical/physiological reactions in cells and consequently organelles, such as mitochondria will be damaged. On the contrary, direct contact of nanoparticles with cellular ingredients is needed for the generation of ROS. Nanostructures with extremely active dangling bonds are able to take electrons from electron donors and thus oxidizing the cellular structures.

15.2.2.3 Stimulating the immune system and developing inflammation

Innate and adaptive defense systems are two common immune types, which activate different cell species. Regarding the innate immune system, neutrophils, monocytes, dendritic cells and macrophages are some of the existing cells in this category. They attack foreign substances via phagocytosis, generation of ROS and antimicrobial peptides (Giese and Marx, 2014). Inflammation and cytotoxicity are unwanted immunological responses triggered by interactions between nanoparticles and innate immune system. It should be noted that the physicochemical structure of the nanoparticle is significant in how the immune system identifies and reacts to them. When nanoparticles become present in blood, the complement system (a component of the innate humoral immune system) will be activated, which will then mark the surface of nanoparticles with C3b and iC3b opsonic complement segments. These combinations will subsequently be eliminated via phagocytic cells (Chang, 2010; Zolnik et al., 2010). Type B and T lymphocytes are responsible for adaptive immune response, moreover innate and adaptive immune systems can both take action either with their released products or directly by their cells. Some of the secreted agents are cytokines responsible for inflammation, such as interleukin (IL)-1β, IL-6 plus antibodies (Elsabahy and Wooley, 2013). The uptake of chitosan nanostructures and PLGA covered with PEG was studied by (Semete et al., 2010) in Balb/C mice during 24 h following oral delivery. According to the released pro-inflammatory cytokines, IL-6, IL-2, IL-12p70, and TNF-α present in plasma were observed to be low in mice fed by nanomaterials of PLGA. Furthermore, the amount of cytokines and chemokines, such as INF-γ, IL4, and IL5 exhibited a steady trend in mice treated with PLGA. Finally, it was concluded that immunological reactions do not occur after the intake of PLGA nanostructures from the in vivo tests.

15.2.3 In vivo and in vitro assays for predicting the safety of food-based nanomaterials

In vitro and in vivo tests are needed to obtain the toxicological effect of food nanomaterials, besides in vitro assays have to be done prior to any in vivo study. Before the toxicity assays, the purity and physicochemical properties of nanomaterials should also be determined in both their dry and wet conditions (Oberdörster et al., 2005; Sayes and Warheit, 2009). Here, some of the recent in vivo plus in vitro studies related to food nanoengineered materials are summarized (Table 15.1).

Table 15.1 **Examples of in vivo and in vitro tests conducted to analyze the safety and toxicity of nanostructures in biological environment**

In vitro assays					
Bioactive agent	**Size (nm)**	**Nanostructure form**	**Nanoencapsulation method**	**Purpose**	**References**
Glucocorticoid budesonide	190–220	Nanoliposome	Emulsion solvent evaporation	Glucocorticoid budesonide as possible cure for inflamed bowel cells	Leonard et al. (2012)
Puerarine	400–500	Nanosuspension	High pressure homogenization (HPH)	Studying the anticancer action of puerarin in HT-29 cell (human colon)	Wang et al. (2013)
Safranal	158.16 ± 6.31	Nanoliposome	Solvent evaporation	Employing saffron as a natural substance for the treatment of cancer	Malaekeh-Nikouei et al. (2013)
TiO_2	460–600			Estimating the cytotoxicity and genotoxicity of commercial TiO_2 nanoparticles	Hamzeh and Sunahara (2013)
SiO_2	15			Evaluating antimicrobial and physical properties of SiO nanoparticles@polymer	Dhapte et al. (2014)
Soy protein isolate	28–179	Nanoparticle	Modified cold gelation	Determining the toxicity of SPI nanoparticles in Caco-2 cells	Zhang et al. (2012)
Doxorubicin hydrochloride	279.3 ± 8.3	Chitosan-based nanogels	Cross-linking	Analyzing the cytotoxicity effect of doxorubicin hydrochloride on Caco-2 cell line	Feng et al. (2014)

(Continued)

Table 15.1 Examples of in vivo and in vitro tests conducted to analyze the safety and toxicity of nanostructures in biological environment (*cont.*)

In vivo assays					
Bioactive agent	**Size (nm)**	**Nanostructure form**	**Nanoencapsulation method**	**Purpose**	**References**
Sio_2	30	FITC-labeled carboxymethyl chitosan (FITC-CMCS)	Cross-linking	Impact of SiO_2 biological system of mice	So et al. (2008)
Quercetin	70–400	Cationic nano-structured lipid carriers	Emulsification and solidification	Biodistribution of entrapped quercetin in healthy male mice	Liu et al. (2014)
Lidocaine	221 ± 99	Nanofiber	Electrospinning	Efficacy of lidocaine nanofibers to be released on a fractured femoral zone	Chou et al. (2016)
Curcumin	110	Nanoemulsion	Nanoemulsification	Analyzing the cytotoxic effects of curcumin on fibroblasts of rats	Vecchione et al. (2016)
TiO_2	20–80	Anatase and rutile nanocomposite	Melt blending	Analyzing the antimicrobial effects of Titanium dioxide as direct contact material for fresh pears	Bodaghi et al. (2013)

15.2.3.1 In vitro studies

In vitro tests can provide useful information for implementing the in vivo assays. In addition, sufficient data can be acquired via the in vitro studies for bulk materials but the in vivo studies are complicated due to the reactions made within the biological systems, furthermore in vitro studies are normally chosen for large-scale production and it is an economical method (Fischer and Chan, 2007). Some of the in vitro studies, which highlight the toxicological properties of nanomaterials, are described further.

In a research, the anticolorectal cancer effects of lecithin and hydroxypropyl methylcellulose (HPMC) in puerarin nanosuspensions were analyzed in cell line of HT-29. Cytotoxicity tests revealed that puerarin nanosuspensions were able to cease the proliferation of HT-29 cells (Wang et al., 2013).

Leonard et al. (2012) induced an inflammation of intestinal mucosa according to the Caco-2 cells, macrophages isolated from blood, and dendritic antigen-presenting cells which are the agents present in the intestinal innate immune system. Inflammation was triggered by the IL-1β cytokine. Next, polymeric PLGA nanoparticles and liposomes were applied as possible vehicles to load glucocorticoid budesonide and further treat the inflamed bowel cells. Glucocorticoid budesonide exhibited high efficacy for healing inflammation according to the results of pro-inflammatory protein release and the value of transepithelial electrical resistance (TEER).

Malaekeh-Nikouei et al. (2013) explored the ability of saffron (*Crocus sativus*) as a natural substance for the treatment of cancer. Safranal is an aromatic compound in saffron with antitumor activities. In this research, safranal was applied in its bare form and as the guest molecule in nanoliposomes. Subsequently, the antitumor together with their apoptosis role was investigated. The antitumor impacts of Bare (0.01–3 mM) and nanoliposomal safranal (0.04–0.32 mM) were tested on cultured HeLa, MCF7 plus L929 cell lines. In order to evaluate apoptosis, the cells were stained with propidium iodide, moreover the MTT (3-(4,5-dimethylthiazol-2-yl)-2-yl)-2,5-diphenyl tetrazolium) test was carried out to obtain the cell-toxicity properties. As a result, entrapped safranal revealed higher cytotoxic effect on HeLa and MCF7 cell lines compared to safranal alone.

Hamzeh and Sunahara (2013) studied the cytotoxicity effects of TiO_2 nanoparticles coated with polyacrylate. They suggested that micro and nanosized TiO_2 nanoparticles differed significantly in regard to toxicological properties as well as coated and uncoated nanoparticles.

Dhapte et al. (2014) fabricated hydrophobic silica nanoparticulate-PMMA (SNP@PMMA) and hydrophilic silica nanoparticulate-PVA (SNP@PVA) films. Regarding the two tested films during in vitro antimicrobial study, SNP@PVA failed to exhibit any considerable biocidal behavior against both Gram types of bacteria. Hydrophilic nature of PVA caused SNP@PVA film to retain more water and moisture, which would incite microbial growth along with physicochemical instability issues. On the contrary, SNP@PMMA exhibited noteworthy biocidal effect against Gram-positive and negative bacteria, as indicated by areas of inhibition in microbial assays. As a rule, bactericidal mechanism of polymeric composites like PMMA is derived from their lethal interaction on cell wall as well as cytoplasmic membranes.

Feng et al. (2014) tested the efficacy of chitosan together with o-car-boxymethyl chitosan nanoparticles as possible carriers for delivering doxorubicin hydrochloride via the oral route, which was considered as an efficient delivery system.

Zhang et al. (2012) fabricated nanoparticles of soy protein isolate (SPI) through the modified cold gelation technique in the size range of 28–179 nm. The MTT test applied to evaluate the toxicity of SPI nanoparticles in the cells of Caco-2 after 24 and 48 h. These nanoparticles exhibited no toxicity as the cellular viability was approximately 100%, besides SPI nanoparticles were accumulated in cytoplasm as they entered Caco-2 cells.

15.2.3.2 In vivo studies

Lately, in vivo studies have been done mostly in rodents and for this reason, clinical trials should be implemented to assure the negative or positive effects of the nanomaterials in human body (Hunter et al., 2012). Here some of the in vivo researches are discussed to understand the effects of nanomaterials present in the food network on the biological system.

The action of SiO_2 nanoparticles (NPs) with diameters of about 30 nm on the biological system of mice was described by So et al. (2008). According to their results, after feeding mice with the total dose of 140 g SiO_2/kg bw during 10 weeks, higher ALT (alanine-aminotransferase) enzyme values were observed in their blood. Furthermore, micro-scale SiO_2 (30 μm) led to fatty liver symptoms. It is interesting that the amount of both micro and nanosized in the liver were approximately equal, indicating that they both have difficulty getting through the intestinal barriers.

Yang et al. (2013) analyzed the toxicity of graphene oxide (GO) nanocomponents and PEGylated GO derivatives, via the oral and intraperitoneal (i.p.) injection delivery routes in mouse models. PEGylated GO derivatives delivered via the oral route, indicated low tissue uptake, therefore it can be inferred that their intestinal absorption is restricted. On the other hand, after the i.p injection, PEGylated GO derivatives accumulated in the reticuloendothelial (RES) system, especially in spleen and liver. Moreover, there was no noticeable toxicity after the serum biochemistry, blood panel test and histological measurements. According to this research, nanomaterial's size, surface coatings, and delivery routes greatly impact the toxicology of nanostructures.

Liu et al. (2014) assessed the biodistribution of quercetin entrapped in cationic nanostructured lipid carriers (QR-CNLC) in healthy male mice following oral administration. They reported that QR-CNLC exhibited higher AUC (area under tissue concentration-time curve) plus greater C_{max} in organs, such as kidney, liver, and lung than the control group. Besides, quercetin was aggregated to a remarkable amount in lung; liver and kidney as their relative intake rate were 1.57, 1.51, and 1.68, respectively.

Chou et al. (2016) fabricated nanofibers loaded with lidocaine as a biodegradable membrane which was then placed on the femoral fracture zone of three groups of rabbits (A, B, and C). Group A involved rabbits with femoral fraction which went through fixation, group B was somehow similar to group A and the fabricated electrospun were implanted inside their body and group C was the anesthetized control category. Eventually, faster recovery level was observed for rabbits of group B which explains the pain relief efficacy of lidocaine electrospun.

Vecchione et al. (2016) investigated the toxicity and delivery of curcumin by a food-based nanoemulsion. The antiinflammatory properties of curcumin was tested in IL-1β, TNF-α, IL-6, and IL-8 LPS rats as they were fed 0.8 mg curcumin per kg of rat. The authors suggested that with the promising antiinflammatory features of curcumin they can be applied in clinical trials for healing inflammatory bowel disease and the chemoprevention effect in colorectal cancer.

Bodaghi et al. (2013) studied the photoinduced biocidal effects of TiO_2 nanocomposite films by performing in vivo tests. In this test, TiO_2 nanocomposite was utilized as direct contact material for fresh pears which were activated upon the irradiation of UV rays. In conclusion, the number of mesophilic bacteria plus yeast cells decreased notably when the photocatalyst films were fabricated by extrusion and where accompanied by UV illumination.

15.3 Regulatory principles legislated by various organizations and countries

Although there are still not enough regulations about the application of food nanomaterials and nanocapsules, some countries and organizations have published legislations, standards, and reports on possible risks of nanomaterials. Regarding the Nanomaterials being applied in Food Contact Materials (FCM), they may migrate into food products, however this sort of exposure has not yet been examined thoroughly. According to this interpretation, human is exposed to the nanopackaging used for food; furthermore the safe levels of exposure are determined via animal exposure trials (Cushen et al., 2012). In this section the regulatory principles of different countries and organizations are reviewed, especially the European Union (EU) and US food and Drug Administration (FDA). Furthermore, the legislations determined by authorities in different countries are summarized in Table 15.2.

15.3.1 European countries

15.3.1.1 The European Union (EU)

Nanofood products are governed under the Regulation No. 178/2002 provided by EU (Regulation Council, 2002). The provision of food data for the customers includes the principles for labeling within the EU market and some nanospecific information. Meanwhile, (European Parliament, 2011) notifies the consumers of the nanomaterials present in the food network (preamble, 25).

The Food Labeling Regulation is responsible for providing the labels of nanobased foods. According to Article 18 of Regulation 178/2002, foods containing nanomaterials should include the "nano" word on the label similar to the principle applied in

Table 15.2 Current regulations for nanofoods in different countries

Country	Authority	Key regulation	References (URL)
USA	Food and Drug Administration (FDA)	Final Guidance for Industry—Use of Nano Materials in Food Products	http://www.fda.gov/ForConsumers/ConsumerUpdates/ucm258462.htm
Canada	Health Canada	Food and Drugs Act	http://laws-lois.justice.gc.ca/eng/acts/F-27/
Switzerland	Swiss Secretariat for Economic Affairs (SECO)	2011/696/EU	http://www.bag.admin.ch/nanotechnologie/12171/12176/index.html?lang=en
United Kingdom	European Parliament and of the Council of 27	Regulation (EC) No 258/97	https://www.food.gov.uk/science/novel/nano
Turkey	Turkish Food Safety Association		http://www.ggd.org.tr/en.php
Russia	The Federal Service for protecting customer Rights and Human Well-being in the Ministry of Health and Social Development (Rospotrebnadzor)	Sanitary Rules and Regulations (SanPiN)	http://www.rospotrebnadzor.ru/endeyatelnost/bilateral.php
Australia and New Zealand	Food Standards Australia New Zealand (FSANZ)	Australia New Zealand Food Standards Code	http://www.foodstandards.gov.au/consumer/foodtech/nanotech/Pages/default.aspx
Iran	The Iran Nanotechnology Initiative Council (INIC)		http://nanohealth.ir/pages/static_page.php?id=14&site=1&lang=2
India	Food Safety and Standards Authority of India (FSSAI)	Food Safety and Standards Act	http://www.fssai.gov.in/
China	Ministry of Health Ministry of Agriculture	Food Safety Law of China	www.gov.cn/english/2005-10/09/content_75326.htm http://english.agri.gov.cn/
Malaysia	Ministry of Science Technology and Innovation	Food regulations 1985 The Food Act 1983	http://www.mosti.gov.my/index.php?option=com_content&view=frontpage&Itemid=27&lang=en

Japan	Ministry of Health, Labor and Welfare	Food Sanitation Law	http://www.mhlw.go.jp/english/policy/health-medical/food/index.html http://www.jetro.go.jp/en/reports/regulations/
Thailand	Food and Drug administration of the Ministry of Public Health	Food Act B.E.2522	http://eng.moph.go.th/index.php/safety
Korea	Ministry of Food and Drug Safety (MFDS)	Food Sanitation Act	http://www.kfda.go.kr/eng/index.do?nMenuCode=61
South Africa		Foodstuff, Cosmetics and Disinfectant Amendment Act 2007	http://www.sani.org.za/
Brazil	National Agency of SanitarySurveillance (ANVISA) Ministry of Agriculture, Livestock, and Food Supply (MAPA) Ministry of Health (MS)		http://portal.anvisa.gov.br/wps/portal/anvisa-ingles http://www.agricultura.gov.br/

Source: Adapted with permission from Amenta, V., Aschberger, K., Arena, M., Bouwmeester, H., Moniz, F.B., Brandhoff, P., Gottardo, S., Marvin, H.J., Mech, A., Pesudo, L.Q., 2015. Regulatory aspects of nanotechnology in the agri/feed/food sector in EU and non-EU countries. Regul. Toxicol. Pharmacol. 73(1), 463–476.

cosmetic products. Regarding Article 8(1), the food business operator undertakes the safety testing and labeling task of marketed food if it is not provided in the Union (Law, 2002).

Directive 2001/95/EC established in December 2001 accentuates that merely safe products can be put into markets. The General Product Safety Directive also deals with safety issues that are not supported by the other EU provisions as it considers the safety of manufactured nanoproducts (Floyd et al., 2006). The chemical regulation of EU recognized as REACH, monitors all the substances manufactured and sold in the EU in amounts of higher than 1000 kg annually. Nevertheless, REACH does not cover nanomaterials used in food but the environmental effects of food contact materials (Friege, 2005).

The first regulatory description of nanomaterials was defined in the cosmetics sector (European commission regulation No. 1223/2009: OJ L 342, of 22.12.2009, p. 59). According to Article 2(1) (k), nanomaterials are biopersistent or nonsoluble substances with at least one dimension within their structure varying between 1 and 100 nm in size. Moreover, under Article 2(3) of this regulation a definition is predicted for the nanomaterials on the condition that it will be agreed internationally in the future. The definition is as follows:

"In regard to the recent scientific developments and various definitions defined by organizations in the area of nanotechnology, the commission should update case (k) of paragraph 1 due to the scientific and technical developments plus the points which are approved internationally."

Another vertical regulation in relation to novel foods and components is the European Commission Regulation 258/97, which enables a compulsive approved premarket system for novel food products (Regulation, 1997). According to Article 1 of this principle, a novel edible material should not have been consumed before May 1997 and be related to one of the defined divisions. Divisions relating to nanotechnologies are found in Article 1(c): "Foods and food components modified deliberately by novel components" and 1(f): "Foods or food materials processed in a way that considerable structure changes are observable in their properties, which influence the nutritional value, ingestion and amount of unwanted materials."

Therefore, nanofood products won't be incorporated within the novel food group unless its nutritional value, ingestion and amount of unwanted materials are changed in comparison to its macroscale structure. Later, a new draft replaced the former regulation with a detailed definition of nanomaterials. The definition provided in Article 3(2)(c) is described further:

Materials produced deliberately with dimensions in the range of 100 nm or less comprising functional parts at the surface or inside. Moreover, agglomerates, aggregates, and structures representing the features of nanoscale materials are considered as nanomaterials even if they are above 100 nm in size. Indicator of nanoscale materials are:

1. High specific surface area,
2. special physicochemical characteristics, which render them distinctive from the nonnanoform substance.

The above description of nanomaterials was ultimately approved by the Food Information Regulation (FIR), furthermore in July 2011 an agreement was reached within the EU parliament so as to employ the nanomaterials definition in the labeling of nanofood products.

European Commission Regulation No. 1333/2008 entered EU in regard to the regulation of food additives and established a well-organized system along with regulations of food enzymes and food flavorings (European Commission, 2008). According to the Article 12 of new food additive principle 1333/2008, a food additive is considered novel when there is an alteration in the manufacturing methods, the raw substances or a modification in material size (e.g., via nanotechnology). The food additive produced via novel methods and materials shall be classified as a different additive and to be included in a new production list, thus a modification in the characteristics seems necessary prior to the market entry of the food additive. Therefore, a risk assessment should be considered for approved new additives available in the market as they have undergone a considerable change. A considerable change may involve a change in the manufacturing technique, a change in initial materials and change in regard to the particle size related to nanotechnology (European Commission, 2008).

According to Article 26, consumers or manufacturers of a food additive should notify the commission about new data that is related to the safety issues of the novel food additive.

Generally, the safety concepts of feed are considered similar to the food in EU. The guidance of ESFA applied for the possible risks of nanotechnology in food/feed (Barlow et al., 2009) and the risk evaluation of food/feed nanomaterials (EFSA Scientific Committee, 2011), both underline the equal application of the guidance in food and feed nanomaterials. Moreover, the guidance (EFSA Scientific Committee, 2011) of ENM (engineered nanomaterials) consider all interested classes, such as consumers and risk assessors. Finally, the ENM guidance in food is employed as similar to the food guidance on account of the similarity of the impact on both animals and humans.

The ESFA panel for additives and substances utilized in animal feed (FEEDAP) has entered the concept of size within its guideline information. Therefore, a chapter should be provided on the particle size regarding the employment of new feed additives.

Regarding the packaging of foods, the European Commission has prepared a regulation (No. 1935/2004) in regard to the surfaces that in contact with foodstuff (the "framework regulation"). Article 3 of this arrangement insists that all the FCMs should be prepared in compliance with the good manufacturing practice (GMP) in order to restrict the migration of FCM materials into the food, which may threat human health and deteriorate the sensory properties of food. In order to ensure the safety of FCMs considering Article 3 of the regulation, it should be adjusted by a useful listing, which involves potential restrictions. To put it briefly, in all conditions FCMs must be safe for the intended use, particularly the nanomaterials used in their structures (Commission, 2004).

As approved in May 2011, plastics and nanomaterials to be used in FCMs can be utilized in regard to Annex I of EU Regulation No. 10/2011. Furthermore, Article 9(2) suggests that nanomaterials have the potential to be implemented only if they exist in Annex I. In another EU regulation (Article 6 of European Commission Regulation

No. 450/2009) intelligent packaging materials for foods are defined as "materials engineered on purpose (nanosized), which render different physicochemical properties compared to their macroscale," besides they would be applied only if they are specifically authorized (Communities, 2009; European Commission, 2011).

The principle related to the quality of drinking water is also another interesting area in which nanomaterials can be exerted. Several directives are prepared by World Health Organization (WHO) for determining the quality of drinking water (World Health Organization, 2003) plus WHO, FAO, and UNEP directives for the safety of wastewater recovery in farming and cultivation (Carr et al., 2004). Besides, these guidelines are reinforced by the groundwater recharge guidelines.

Regulations concerning nanoparticles in food products for different countries in EU are to some extent similar and are described for Switzerland and UK in the following section.

In Switzerland, the safety of nanomaterials is determined via the existing principles related to the traditional chemicals. According to Swiss Ordinance in 2010 some characteristics regarding the nanomaterials used in pesticides should be reported which are the composition, size, shape, aggregation status, surface area, and functionalization (Schweizerischer-Bundesrat, 2010). The Federal Office of Public Health (FOPH) did not deliver any approval requests for nanomaterials used in nourishment until 2013 (FAO/WHO, 2013) and the same trend was true for utilization of nanobased pesticides (Bucheli et al., 2013). As for the risk assessment and chemical regulation of nanomaterials, a program was initiated in 2008. In 2008, the Federal Office for the Environment (FOEFN) and FOPH provided a precautionary report for unnatural nanomaterials (Piegorsch and Schuler, 2008). This precautionary report is aimed to measure the risks for consumers, employees, and the environment. Moreover, it is based on a set of factors connected with the nanomaterials including reactivity, stability, release rate, particle's concentration, and their size (Höck et al., 2008).

Swiss Secretariat for Economic Affairs (SECO, 2012), has prepared a guideline for the safety of synthetic nanomaterials. According to this institute, nanomaterials are defined as follows:

- A material comprising over 1% nanoparticles in its particle size distribution, which can even exist in the form of agglomerates and aggregates.
- Fullerenes, carbon nanotubes and graphene flakes are defined as nanomaterials even if they are under 1nm in size.
- If the present amount of particles according to their size is unknown, any material with the mean particle size less than 500 nm belongs to the nanomaterials.

The responsibility and regulating food safety in the UK is undertaken by the UK food standards (FSA). If a firm desires to market foods containing nanomaterials, the products should be evaluated using the guidance prepared by the European food safety authority (ESFA) in 2011. Some characteristics, such as composition, nutritional value, metabolism, toxicological, chemical, microbiological, and manufacturing process to investigate the safety of nanomaterials. Furthermore, The UK FSA collaborates with the advisory committee on novel foods and processes (ACNFP). It is notable that this committee is consisted of numerous food specialists on the basis of food safety.

This group holds discussion sessions regarding nanotechnology and food sector. Stakeholders from academia, industry, consumers, and regulating departments are members of this group. This group is focusing on exchanging ideas and studying nanotechnology-based foods with food contact materials in the UK market. Concerning the record of UK House of Lords, there were nearly no nanobased food products available in the market of UK and EU (Couch et al., 2016). Recently, the foods standard agency (FSA) accepted the usage of nano silica dioxide (an anticaking substance), nanosilver as an antimicrobial agent, and titanium nitride to empower the food packaging industry (Couch et al., 2016).

15.3.1.2 Non-EU countries

In Turkey, a regional or national policy manages nanotechnology (OECD, 2013). Today, there are 32 research centers active in the area of nanotechnology in Turkey. Moreover, Turkey takes part in the ESFA Nano Network with other European candidates as an observer (EFSA, 2012). The Ministry of Environment and Urban Planning prepared a CLP regulation in Turkish language, which will enter the force in June 2016 (Intertek).

The Russian nanofood safety legislations are originated from documents like Sanitary rules and Regulations (SanPiN), technical regulations and state standards (GOST) (Office of Agricultural Affairs of the USDA, 2009). The principles in the regard to industrial nanotechnology has entered the Russian corporation of Nanotechnologies, being established in 2007 plus some national programs, such as Nanotechnology Industry Development Program. According to FAO/WHO, 17 documents were published in such programs regarding the safety of nanomaterials within the food industry. Furthermore, the sanitary regulations for nanomaterials, such as silver, titanium oxide etc. and their impact on human body were established (FAO/WHO, 2013).

15.3.2 North America

The US FDA is responsible for the development of the emerging nanofood products and pursuing the related legislations in partnership with the National Nanotechnology Initiative (NNI). Nevertheless not a comprehensive and accurate principle has been established for the nanomaterials within food industry. Accordingly, the US FDA has organized a safety guidance with regulatory texts for the manufacturers entitled "Draft Guidance for Industry" concerning the novel technologies within the food sector (FDA, 2011). In this draft guidance, nanomaterials are defined as:

- Materials or products between the size of 1–100 nm in one or more external dimensions (from 1 to 100 nm).
- Materials or products that are akin to nanomaterials according to their physicochemical and biological characteristics even though they are not nanosized.

In addition, the draft has introduced some responsibilities for the manufacturers, which are listed below:

- Monitor the changes being exerted to the food materials, such as physicochemical properties and impurities.

- Test the nanofood products concerning the safety challenges following their modifications.
- Submit a regulatory assessment to US FDA.
- Consider regulatory policies for the consumption of the novel nourishment.

Regarding the following notes, the US FDA insists that the current legislations can be exerted in assuring the safety of nanomaterials, furthermore the organization accentuates that the whole processed nanofood products should be approved in accordance with the principles present in their guidance.

The regulatory corporation for foods in Canada (Health Canada), has introduced nanomaterials policies, also considering their aim of use and scope (Neethirajan and Jayas, 2011). This definition is explained below:

- Manufactured nanomaterials size varies from 1 to 100 nm.
- Agglomerates and aggregates displaying one or more nanoscale properties are also known as nanomaterials, although their size is not within 1–100 nm.

With the rapid development in the field of nanoscience and technology, the policies and regulations will also be updated and novel international standard will enter this promising area.

Furthermore, the policy statement will admire manufacturers to prepare safety information for their products. The information is as follows:

- Manufacturing steps,
- physicochemical features
- toxicity, ADME profile, and environmental information
- analyzing the potential risks and suggest strategies to overcome these concerns.

At the moment, there is not a legislation regarding the use of nanofood in Canada (Neethirajan and Jayas, 2011). For this reason, potential benefits and risks should be considered in these products prior to entering the market.

15.3.3 South America

Recently, Brazil has been so successful in developing the nanotechnology field among the other south American countries (Eguchi et al., 2013; Silva, 2016). In 2011, Brazilian nanotechnology experts gathered in São Paulo in order to legislate rules, guidelines and standards for nanotechnology applications (NIA, 2011; World Health Organization, 2013). A proposal was prepared in 2013 considering a new label for food, pharmaceutical and cosmeceutical products manufactured through nanotechnology, however this proposal was denied by the Brazilian congress as it had the possibility to harm the companies, which have greatly invested in the area of nanotechnology and even result in the bankruptcy of the corporations Nevertheless, some investigators believe that regulations will develop the novel field of nanotechnology and even open new horizons for further research with the aim of providing public interest. The latest proposal accentuates the labeling of products prepared by employing nanotechnology, however the other principles are still intact. Unfortunately, to date, there is still no change in the proposal although the new version of the proposal was supposed to be established soon (Almeida, 2013).

15.3.4 Africa

In the year 2002, the South African Nanotechnology Initiative was found. Afterwards, a strategy was planned for the next 10 years regarding the development of nanotechnology (Cele and Coville, 2009). However, there is no certain principle defined for nanomaterials and therefore food nanotechnology is put under the other food-related legislations (Foodstuff, Cosmetics and Disinfectant Amendment Act 2007).

15.3.5 Australia and New Zealand

The regulatory organization of food products in Australia and New Zealand known as the Food Standards Australia New Zealand (FSANZ) strongly believes that functional nanofood should first undergo the safety and toxicity assays before the wholesale. The authorities' viewpoint is optimistic through current regulatory tools for nanofoods. In effect, there is not a specific regulatory issue or legislation for nanomaterials applied in the food sector.

Department of health and aging (DoHA), located in Australia (DoHA, 2010a) has introduced a definition for industrial nanomaterials, which is described below:

- Nanomaterials size ranges between 1 and 100 nm, which are manufactured or engineered for a particular purpose.
- Aggregation and agglomeration are structures related to nanoingredients.
- Nanomaterial is defined as a material containing 10% or more particles possessing the features mentioned above.

DoHA proposed legislation that novel industrial nanomaterials should be tested for their safety before entering the market. Conversely, nanofood and food additives are not involved in industrial chemical groups, therefore this legislation in not executed in the food industry (DoHA, 2010b).

15.3.6 Asia

15.3.6.1 Iran

The Iran Nanotechnology Initiative Council (INIC) initiated its work in 2003 as it determined the commonplace principles for the improvement of nanotechnology market. The Food and Drug Organization (FDO) determines whether a food product can enter the market or not. This organization established a nanotechnology committee with the aim of developing certain guidelines for the evaluation and the acceptance of food, beverages, pharmaceutical, nutraceutical, cosmetics, and medical equipment.

Moreover, a nano symbol is designed for industrial standard certification[a] referring to products in food and beverages, cosmetics, pharmacy, medical equipment, and dietary supplements. According to FDO, right now there are 12 products available based on nanotechnology.[b]

[a]http://www.nano.ir/index.php?ctrl=news&actn=news_view&id=42433&lang=2
[b]http://www.nanotechia.org/news/news-articles/iran-introduces-certification-nano-foods-and-health-products

15.3.6.2 *China*

The food Safety constitution handles the safety issues of food products in China, however it has not characterized any specifications for nanomaterials. Two institutes legislate and handle the national standards in the field of nanotechnology, the National Centre for Nanoscience and Technology (NCNST) and the Commission on Nanotechnology Standardization.

Among the several standards related to nanotechnology, one standard has defined nanomaterials (GB/T 19619-2004), which is explained below (ISO, 2013):

- Three- dimensional materials that has at least one nanoscaled dimension (1–100 nm) or a substance, which comprise nanostructures units and has specific properties.

By now, Chinese regulatory authorities have not approved the nanomaterials to be applied in food products (FAO/WHO, 2013).

15.3.6.3 *India*

Food Safety and Standards case governs the food safety regulations in India. Two programs were launched in the field of nanotechnology by the Indian government, first the Nano Science and Technology Initiative (NSTI) in 2001 and second the Nano Mission in 2007.[c] Investigations are still being undertaken under the running programs; moreover some centers have initiated to address the potential risks due of nanofoods. Nevertheless, standardization in this field has not progressed much and is still at the basic levels. According to reports, India does not have a legislation for the risk and hazards caused by nanoparticles (Sharma and Chugh, 2009) and therefore miss the expertise to manage the risks of nanotechnology (Barpujari, 2011).

15.3.6.4 *Japan*

Food sanitation law regulates the safety of food products in Japan. Currently, there is no specific principle for nanomaterials in Japan, however several studies are being performed in the area of nanotechnology. The current position of food nanotechnology in Japan is available in a document entitled "Basic Survey Report on Safety Assessment Information on the Use of Nanotechnology in the Food Sector" (FSCJ, 2010).

15.3.6.5 *Malaysia*

A National Nanotechnology committee focusing on the safe use of these promising materials is governed by the National Nanotechnology panel centering on the toxicity and environmental impacts of nanomaterials. Food nanotechnology regulations are derived from Food act 1983 plus the revisions of Food Regulations 1985, which also involved among other specifications (NanoMalaysia, 2013).

[c]http://nanomission.gov.in/

15.3.6.6 *Thailand*

Monitoring the safety and controlling the food products are carried out by the Food and Drug Administration of the organization of Public Health in Thailand. Recently, 10 programs with national priority are running involving the industrial standards for products made via nanotechnology, which are identified by National Nanotechnology Centre (NANOTEC) in Thailand. These standards are known as NANOMARKS and Food Quality in order to yield high-quality food products. NanoQ label is considered for the nanobased products in Thailand,[d,e] which is applied to textile, plastics and other industries. However, nanolabels are not yet used for foods and beverages.

15.3.6.7 *Korea*

Up to now, Korea has manufactured various products by employing nanotechnology. The Korea ministry of food and drug safety (MFDS), has brought the scientific knowledge and data together to include the definitions of nanomaterials and nanoingredients in the food industry (MFDS, 2012a).

Nanomaterials are referred to agents within the size range of 1–100 nm, which can be observed using electron microscopy.

Nanoingredients are attributed to:

- Manufactured materials ranging between 1 and 100 nm in size;
- Manufactured materials having 50% or more particles according to size distribution that varies between 1 and 100 nm;
- Manufacturing materials representing nanoscale characteristics or reactions, if the average size doesn't exceed 1000 nm considering minimum one dimension;
- Aggregation and agglomeration is a possible phenomenon in these materials.

Korea has not defined a specific code for utilizing nanomaterials in food/beverages. MFDS states that current regulatory systems support the nanofood products. Nevertheless, a safety guideline regarding nanomaterials in foods was designed by Korea MFDS in 2012 (MFDS, 2012b). The guideline establishes a practical tool for the safety measurement via nanoparticles and nanomaterials. The guideline involves; physicochemical properties, surface potential, surface area, toxicological tests comprising in vivo plus in vitro assays seems necessary to determine the nanoparticles possible toxicity. Furthermore, this guideline offers detailed properties for nanoingredients in food, such as content existing structure or condition along with their intended application.

Table 15.3 shows the list of organizations and initiatives active in the developing of legislations and regulations for applying food nanotechnology around the world.

[d]http://www.nanotec.or.th/en/?p=4014
[e]http://www.nanotechia.org/news/news-articles/first-label-nanomaterial-product-awarded-thailand

Table 15.3 **Global organizations striving for the growth of legislations in food nanotechnology**

Number	Acronym	Full name
International corporations		
1	ASTM International	American Society for Testing and Materials
2	GNN	Global Nanotechnology Network
3	IANT	International Association of Nanotechnology
4	IFIC	International Food Information Council
5	ILSI	International Life Sciences Institute
6	ISO	International Organization for Standardization
7	IRGC	international Risk Governance Council
8	IUFoST	International Union of Food Science and Technology
9	UN	United Nations
10	FAO	Food and Agriculture Organization
11	WHO	World Health Organization
North America		
1	AFMNet	Advanced Foods and Materials Network
2	ANSI	American National Standards Institute
3	CIFAR	Canadian Institute for Advanced Research
4	CNS	Centers for Nanotechnology in Society. Council of Canadian Academies
5	DOE	Department of Energy
6	DTSC	Department of Toxic Substances Control
7	EPA	Environmental Protection Agency
8	FDA	Food and Drug Administration
9	NTIG	NanoTechnology Interest Group
10	HC	Health Canada
11	IFT	Institute of Food Technologists
12	ISN	Institute for Soldier Nanotechnologies
13	ICON	International Council on Nanotechnology
14		NanoCenter at the University of South Carolina
15	NSTI	Nano Science and Technology Institute
16	NCI	National Cancer Institute- Alliance for Nanotechnology in Cancer
17	NCEM	National Center for Electron Microscopy
18	PEN	Program of Excellence in Nanotechnology
19	NIH	National Institutes of Health
20	DHHS	Department of Health and Human Services
21	CDC-NIOSH	National Institute for Occupational Safety and Health of the Centers for Disease Control and Prevention

Table 15.3 Global organizations striving for the growth of legislations in food nanotechnology (*cont.*)

North America		
22	NIST	National Institute of Standards and Technology
23	CNST	Center for Nanoscale Science and Technology
24	NNI	National Nanotechnology Initiative
25	NSF	National Science Foundation
26	NRL	Naval Research Laboratory, Institute for Nanoscience
27	OSTP & NSET	Office of Science and Technology Policy with the Nanoscale Science, Engineering and Technology
28	ONAMI	Oregon Nanoscience and Microtechnologies Institute
29	SOT	Society of Toxicology
30	USDA	United States Department of Agriculture
31	NIFA	National Institute of Food and Agriculture
32	PEN	Woodrow Wilson International Center for Scholars Project on Emerging Nanotechnologies
Europe		
1		AlphaGalileo
2		Austrian NANO Initiative and Forum
3	BBSRC	Biotechnology and Biological Sciences Research Council
4	BSI	British Standards Institution
5	Defra	Department for Environment, Food and Rural Affairs
6	NRCG	Nanotechnology Research Coordination Group
7	Cefic	European Chemical Industry Council
8	DG SANCO	Directorate General for Health and Consumer Affairs
9	SCENIHR	Scientific Committee on Emerging and Newly Identified Hazards and Risks
10	EFSA	European Food Safety Authority
11	ESF	European Science Foundation
12	FOEN	Federal Office for the Environment
13		FinNano (Academy of Finland)
14	Fera	Food and Environment Research Agency
15	FSAI	Food Safety Authority of Ireland
16	FSA	Food Standards Agency
17	HSE	Health and Safety Executive
18	HPA	Health Protection Agency

(*Continued*)

Table 15.3 **Global organizations striving for the growth of legislations in food nanotechnology** (*cont.*)

Europe		
19	IFR	Institute of Food Research
20	IFST	Institute of Food Science and Technology
21	IoN	Institute of Nanotechnology
22	IOM	Institute of Occupational Medicine, SAFENANO Initiative
23	SnIRC	Safety of Nanomaterials Interdisciplinary Research Centre
24	INA	Irish Nanotechnology Association
25		NANO: The Magazine for Small Science
26		NanoHand
27		Nanologue
28		Nanojury 2005
29	AIRI	Nanotec IT
30	ITA	NanoTrust by Institute of Technology Assessment of the Austrian Academy of Sciences
31	RMNT	Réseau Micro Nano Technologie /R3N/ ANR-PNANO
32		Royal Society and the Royal Academy of Engineering (Joint Study in Nanotechnology)
33		Technology Centre AS CR
Asia		
1	ANF	Asia Nano Forum
2		Nano Technology Research Association and the NanoKorea International Event
3		NanoChina
4		Nanonet by the Japanese National Institute for Materials Science
5	AIST	National Institute of Advanced Industrial Science of Technology
Australia and New Zeeland		
1	ARCNN	Australian Research Council Nanotechnology Network
2		Department of Innovation, Industry, Science and Research
3	FSANZ	Food Standards Australia New Zealand
4		MacDiarmid Institute for Advanced Materials and Nanotechnology
Africa		
2	SANi	South African Nanotechnology Initiative (SANi)

15.4 Panorama and challenges for the future

Nanotechnology has recently found its place in food and beverage products, followed by electronics and computers class. In spite of being a promising technology, some companies are somehow reluctant to apply nanotechnology and the term nano in their food and beverage products. The major reasons for this reluctance are attributed to consumer's advocacy parties and government agencies monitoring the safety of food products, which will impose the fear of loss of reputation and being rejected. As a result, this sense of reluctance may affect the public in a way they observe food-nanoscience in future. The most important fact, which causes disconformity between the commercialization of food nanotechnology and the rapid advance in nanoscience is the effect of nanomaterials on public health and environment (Hobson, 2009).

Several investigations have been performed in order to determine how people view the concept of food nanotechnology (Macoubrie, 2004). According to a survey, the majority of Americans have no idea about the nanotechnology and its application in the food sector, which has not changed much in the past few years (Satterfield et al., 2009; Scheufele et al., 2009). Nonetheless, the public conception of nanofood varies between neutral and optimistic in a conservative trend, besides the advantages seems to cover the concerns and possible risks (Currall et al., 2006; Kyle and Dodds, 2009; Lee et al., 2005; Satterfield et al., 2009). Myriad factors are responsible for affecting the perception of food nanoscience including trust, emotion, knowledge and awareness of nanotechnology, governance philosophy, religious beliefs, age, gender, etc. Recently, a metaanalysis has reviewed these factors and their impact on judgments of nanoscience (Satterfield et al., 2009). Mass media is one the crucial factors, which forms Consumer attitudes and points of view towards this novel technology. For instance, mass media has created mistrust among consumers in relation to the government regulatory organizations in the UK and their series of food crises in late decades (Maurer and Sobal, 1995; Pidgeon et al., 2003). Moreover, it has greatly declined the admissibility of genetically modified food products (GMOs) among the European consumers. Therefore, European citizens are less likely to consume nanofood products than Americans (Gaskell et al., 2005). Compared with USA, studies have shown that in Europe normally the negative aspects of nanotechnology are introduced via the mass media (Gaskell et al., 2005; Stephens, 2005).

The attitudes of the public towards novel technologies are greatly influenced by the trust of consumers in corporations and individuals supporting public interests. An introduction to a single product or application based on nanotechnology via the media can trigger a series of reactions that not only affects the marketability of the related products but also lower the government authority to regulate these achievements. Regarding the aforementioned information, the future of nanofoods strongly depends on the consumer's willingness to approve this novel technology as it is susceptible to fear or negative emotions (Stationaryoffice, 2010a,b). Finally, the key issues, which may cause optimistic views on the consumption of nanofoods should be carefully studied to establish successful marketing and to manufacture beneficial and healthy products.

Here some topics are discussed in relation to the factors developing marketing and commercialization of food nanotechnology plus the potential advantages and concerns of this technology in near future.

15.4.1 Enhancing marketing and positive attitudes toward nanoengineered materials in food industry

Today, nanoencapsulation is finding its way into the food and beverage industry. Flavorings that can remain active for longer times in mouth and at the market, providing a combination of tastes and flavorings that are designed to be released at the right time to cause the maximum desirability to the consumers are all achieved via the encapsulation technique in the food industry, especially nanoencapsulation so as to enhance the reactions by many folds.

There is a direct relationship between customer's attitude toward the merchandise and their willingness to purchase it (Rodríguez-Entrena et al., 2013; Verdurme and Viaene, 2003). Nanofood is a novel product in the modern world, hence a better consumer acceptance is supposed to be achieved through informing people about the possible benefits of these promising product.

Studies considering customer's willingness to buy nanofood products have not been done in depth, as the actual products were not presented to consumers. In New Zealand, the concept of producing low-fat lamb via nanotechnology was supported by the citizens and eagerness to purchase the functional product (Cook and Fairweather, 2007). Nevertheless, presenting the nanofood and packaging products in Switzerland, people were satisfied with nanobased packaging rather food prepared via nanotechnology (Siegrist et al., 2007).

Here several factors, influencing the market and acceptance of nanofood products are explored.

15.4.1.1 Trust of consumers toward nanofood products

In order to get information about a special subject or product, it is normally preferred to adopt the ideas of people who are familiar with the particular issue or product. When people appraise or systematically approves a certain product or topic, others will be also persuaded that this is the right product to purchase although the information might not be completely true and the product might have some deficiencies (Lindzey et al., 1998). Proved honesty without the aim of jobbery are also crucial in the persuasion of customers (Frewer et al., 1996), moreover expertise should also be considered in order to convince the public. It is also important for the message contained in the information to be the same as the values approved by the public. For instance, the sort of information, which is meant to encourage environment protection, is more trustworthy compared to a source, like a private firm, which is mainly based on their own interests (Van Kleef et al., 2007).

Personal experience and most importantly habits play a key role in people's shopping attitudes. Therefore, at first, people would probably reject novel functional food products, although they have no idea about it. Regarding nanotechnology as an advent technology, persuasive information has its own place as people's choice is less likely to be affected by contradictory discussions, however persuasive information has a considerable effect on the attitudes (Eiser et al., 2002; Frewer et al., 1998). To sum up, probably the food and beverage companies will soon inform the public about beneficial impacts of the functional foods and their impact on the body and the environment (e.g. nanopackaging materials) by providing logical and scientific reasons to convince the consumers and to gain their trust.

15.4.1.2 Emotions and their effect on selecting nanofood products

Emotion is an important factor in which the benefits are inversely proportional to the potential risks (Alhakami and Slovic, 1994). For instance, when a person grasps the perfect quality or emotions of a particular product or event, the risks will be ignored to a high extent as the benefits outnumber the disadvantages (Alhakami and Slovic, 1994). Accordingly, an individual is eager to move toward a more positive emotion regarding the activity or product on the condition that the provided information is also positive and vice versa (Slovic et al., 2004). Moreover, the impact of inducing certain emotions related to risk may cause simplicity for information processing. Associating the syndromes of an illness with a certain behavior or event has been employed as an incentive to consumers so that their attitudes will be changed, nevertheless applying fear as a motivator might have equivocal results (Ruiter et al., 2001). Above all, the motivational influence of a particular emotion should be aligned with the persuasive message, otherwise the emotions may lead to unexpected results (Meijnders et al., 2001). For instance, self-protection may rise as a response to fear regardless of the message definition. Consequently, when the public has a sense of fear toward nanotechnology, this sense will affect the possible benefits of this technology too. Recently, the psychology and communication science scholars are studying the difference between negative and positive emotions, while some argue that emotions may represent specific functions (Lerner and Keltner, 2000). As an example, the emotion of anger and anger triggers a position in which an individual will face and fight a threatening situation, thus short-term goals are preferred over the long ones (Oatley and Johnson-Laird, 1987). However, it is believed that in certain conditions the self-protection due to aggression may bring about long-term advantages, for example, by standing up and showing resistance against bullying and aggressions. In this way, the long-term benefit of declining the possibility of threat will dominate the short-term negative effect of getting hurt (Evans, 2002). According to (Nauta et al., 2008) when discussions about the prevention of food-related illnesses were coupled with photos arousing the sense of disgust, consumers were more likely to change their food preparation procedures so as to have a better lifestyle. However, when the same message was transferred by photos inducing aggression, the message was no longer effective and even equal to the messages without any emotional provocations (Nauta et al., 2008). Finally, it is currently vague, whether the unnaturalness of the novel technologies applied to food materials would evoke emotions that cause consumers to reject these functional products or not.

15.4.1.3 Provision of fair information

Persuasive strategy is an effective trend leading to the attitude change in consumers. Furthermore, consumers trust can be gained by supporting information about the pros and cons of nanofood products. Due to ethical reasons, it is necessary to provide balanced information to the consumers about the possible benefits and drawbacks of modern technologies compared to the persuasive one (Frewer and Fischer, 2010). Different points of views are supported via the presentation of fair information, nevertheless complex messages may influence consumers in a way that they may express different attitudes. Previously, presentation of complex messages was avoided and didn't reflect the truth about the situations. Meanwhile, when an individual has already defined his attitude, therefore cannot be adapted to the other attitudes easily and messages perceived from others might only lead to greater distrust (Frewer et al., 2003). At last, in this era of communications and robust tools like the internet, normally people do not count on the notions of experts anymore, especially those they do not approve (Frewer and Salter, 2002).

15.4.1.4 Power of attitude versus uncertainty

As an individual makes a strong decision and sticks to it, then the attitude is less susceptible to be affected by persuasive information (Pomerantz et al., 1995). Of course, after continuous personation endeavors, the strength of attitude may slightly decline but the main concept of the attitude will still the same (Clarkson et al., 2008). As an example, people's attitude towards the employment of nanotechnology within the food industry is very weak, in other words, public do not have a general background for nanofood products, thus their attitudes are more prone to be affected by persuasive information regarding the pros and cons of this novel technology, nevertheless for the genetically modified food products the opposite is true as negative characteristics are attributed to them and people are aware of it (van Dijk et al., 2012). On the other hand, when consumers hear both advantages and disadvantages of food nanotechnology they will somehow get hesitant. Usually, neutral attitudes may trigger the sense of doubt, whereas extra positive or negative attitudes with regard to the technology may cause low indecision. Balanced information can for sure reduce the amount of doubt and build vigorous attitudes due to the provision of firm attitudes. Considering the food nanotechnology, consumers will immediately be influenced by the information given, therefore once the attitudes are formed according to the distributed information it will be hard to influence the formed notions by different information. All in all, the pros and cons perceived by the consumers plus the attitudes formed by persuasive information are assumed to be of great importance in the commercialization of nanotechnology applied in the food sector.

Nanobiotechnology is also another novel field within the food industry which is under improvement, lately with the presence of genetically modified foods this technology is compared to the other novel technologies used in the food industry. The attitudes of consumers towards nanofood are strongly affected by the way it is introduced to the public according to the type of communication and advertising strategies. Conducting studies in order to determine the psychological mechanisms (e.g., emotion, type of information processing), which highlight the attitude of individuals towards the emerging nanofood products seems to be inevitable. Moreover, this point is now obvious that the acceptability of certain food products in addition to both their risks and benefits have a direct effect on the consumer's marketing decision.

Implicit memory is another factor, which determines the customer's attitude change due to balanced information (Spence and Townsend, 2006; van Dijk et al., 2012). It is probable that the activation of attitude via providing relevant information may be more effective than establishing risk and safety information (Spence and Townsend, 2006). However, different attitudes may rise from various hazard types and safety issues, thus it is suggested that nanotechnology in agro-food sector must be further studied (Fischer et al., 2007; Poortinga and Pidgeon, 2005). Regarding trust and information, every customer is different in attitude due to their perceptions; therefore the psychology of each person shouldn't be neglected. All in all, according to psychological and demographic parameters, consumers view the emerging technologies differently and therefore targeted information needs to be prepared so as to support the whole consumers (Kornelis et al., 2007).

15.4.1.5 Predicting the market for food nanotechnology in near future

Nanotechnology is considered as a new concept in the food sector but there are some factors limiting the application of this discipline in food products, such as lack of definite regulations for food nanotechnology, potential health risks and the labeling of products (Coles and Frewer, 2013). However, it is globally believed that nanotechnology has interesting applications and benefits in food industry. Today, there is great interest of research in this field, thus companies, government, and academic foundations are investing heavily in this area (Gruère, 2012; Momin et al., 2013).

According to the rapid growth in the global population, it is assumed that the overall population will exceed 9.3 billion by 2050 and by 2100 it will be up to 10.1 billion, which represents a great demand for more food supplies (Parry and Hawkesford, 2010). The consumption trend has changed the food supply system in a way that demands for some products like meat and cereals are considerably increased. It is predicted that by 2050 the entire production of cereals and meat will reach 3 billion tons and 455 million tons, respectively (Tilman et al., 2011).

Nanotechnology within the food industry is able to maintain mass-produced food in order to decrease poverty and develop the food security (Sastry et al., 2011). It can also elevate the rate of exports via elaborating the processing of food commodities, such as sugar, tea, coffee, etc. leading to greater profits (Bradley et al., 2011).

Another market in the area of food nanotechnology is nanopackaging, which ensures the shelf life of the final product and simplifies the transportation in long distances. The basic aim in nanopackaging is to decline the number of preservatives and antibiotics plus the materials used in packaging. Moreover, environmental pollution will also be reduced with the aid of nanopackaging materials.

The market for Nutraceuticals and nanobased food products are developing as they provide health benefits for the consumers. According to research and market reports in 2013, the revenue earned from these functional products was approximately \$175 billion (Handford et al., 2014). It is predicted that Universal market of nutraceuticals will expand from \$23.8 billion in 2013 to \$33.6 billion in the year 2018 (Wire, 2014). USA, the greatest market for nutraceutical food products is expected to dedicate 21% growth in 2015 and exceed \$8.62 billion in the market (Handford et al., 2014).

Tager (2015) suggested that regulatory principles and safety assays should be carried out before the food product enters the market, otherwise the latter regulations and proceedings will be inconspicuous and ineffective. Moreover, when the market is established the industry resist against this process since a great loss of profit will take place as a result of the prohibitions by further regulations, for instance tobacco is one of these markets which still exists all over the world. Regarding the impacts of a novel technology on public health, premarket safety assays seem inevitable. However, the premarket testing does not equal a very strict and standard that totally rejects the food product, actually it reflects the importance of the produced food on public health. Therefore, with a former knowledge and existing data, appropriate measurements can be taken in order to establish safe food nanotechnology market.

15.4.2 Health promotion and manufacturing high-quality foods through the administration of nanomaterials

Nanobased food may bring about benefits for the consumers. As an example, the fat content is reduced in nanotextured food products, such as mayonnaise, whipped cream, or ice cream, which is low-fat but demonstrates creamy features. Thus, consumers can not only eat healthier food but also perceive the great taste of these products (Thangavel and Thiruvengadam, 2014). Moreover, they cause palatability in the food products in spite of the alteration made in the food structure. In this procedure, nanostructured food products display improved taste, texture and consistency (Huang et al., 2011; Tsukamoto et al., 2010). By implementing nanotechnology it is possible to improve the public health in a way that it enables disease treatment, design efficient drug delivery systems and nano disinfectants in packaging materials.

For instance, chitosan (Chien et al., 2007; De Campos et al., 2004; Georgantelis et al., 2007; Gray et al., 2004; Jo et al., 2001; Roy et al., 1999), ε-polylysine (Geornaras and Sofos, 2005; Njongmeta et al., 2011; Shih et al., 2006; Shima et al., 1984; Hamburg, 2012), titanium dioxide (Do Kim et al., 2006; Krishna et al., 2005) can be exerted as antimicrobial nanomaterials, moreover nanoscale antioxidant materials are good choices to treat diseases related to the oxidative stress phenomenon. The majority of studies to develop the targeting ability of therapeutic antioxidants are centered on cationic peptides (Hoshino et al., 2004). Nevertheless, due to the recent advances in food nanotechnology the procedures normally applied in the pharmaceutical industry can be utilized in the food sector to increase the public health via functional foods and beverages. As a final point, this trend should be assessed through in vivo assays and regulatory framework for these nanosized bioactives should be prepared.

15.4.3 Environmental concerns

Nanomaterials used in agrifood and food packaging materials may affect the environment (Chaudhry et al., 2008). Yet, few investigations have analyzed the possible risks of nanomaterials on people and particularly the environment, moreover it is not clear how to regulate these potential risks (Bai and Liu, 2015; Chaudhry et al., 2008; Garner and Keller, 2014; Nowack et al., 2012; Silva, 2016).

Regarding the need for evaluating the risk of nanomaterials and their effect on the environment some agencies are established for monitoring the novel substances including nanomaterials. For instance, Toxic substance control Act (TSCA) is a corporation responsible for establishing rules in regard to all the substances that are produced and utilized in USA (TSCA, 2010). Furthermore, the Environmental Protection Agency (EPA) regulates and manages the production and utilization of chemical and novel products and determines the necessity of legislating regulations. EPA has provided a list of chemical materials, which are produced, processed or traded commercially in the USA (TSCA, 2011a). A material, which does not exist within that list is considered as a new chemical, nevertheless not a single new chemical can be mass produced or utilized unless it is approved by the EPA through a procedure termed as "premanufacture notification" (PMN). Thus, if a new chemical substance is applied to a great amount under the significant New Use Rule (SNUR) (TSCA, 2011b), its PMN practice should also be documented.

The following cases do not need PMNs:

- Side products
- Impurities
- Materials produced merely for export
- Materials produced for test marketing
- Research and development investigations
- Low-volume manufacturing in which the amount of produced materials are 10,000 kg or less annually, moreover it includes strictly controlled exposures during manufacturing, transportation, and utilization.

Upon the advent of a new chemical material into the TSCA list, it will be called an "existing chemical." According to section 6 of TSCA (Griffin, 1996), if a chemical material exposes possible injury or risk during its production, processing utilization and disposal, the production process and distribution of the chemical substance will be characterized as harmful and will be prohibited by EPA.

Nanomaterials are also regulated by the conventional TSCA list updated with some additional information starting from 2008. Carbon nanotubes (CNTs) were characterized as novel chemical structures by the EPA in 2008 since their molecular features differ from the macrostructures, such as graphite and other carbon allotrope (Gulliford, 2008). Moreover, in November 2008 a significant new use rule (SNUR) was defined by the EPA in regard to the silica nanoparticles and siloxane-modified alumina nanomaterials. In brief, TSCA has carefully provided requirements for the mass producers to manufacture specific products, furthermore SNUR is assumed to become a trend in near future.

Below a recent project centered on the effects of nanomaterials on the environment is discussed.

15.4.3.1 *Project DaNa$^{2.0}$*

During the last 10 years, many researches have been conducted in regard to the possible hazards of nanomaterials on the ecosystem and environment (Keshwani et al., 2015; Kunhikrishnan et al., 2015; Warheit and Donner, 2016). However, these studies represent different results and therefore both the knowledge and risk evaluation of these tiny particles are somehow retarded. As a solution, Project DaNa$^{2.0}$ has been developed with the aim of improving trusted insight in nanomaterials. The experts in DaNa$^{2.0}$ team established a knowledge-based webpage (www.nanopartikel.info/en) to provide reliable information regarding the utilized nanomaterials and their effect on public and environment in a way to be understandable for everyone. Moreover, this data can also be employed to determine the aims and prioritization of future researches. Fig. 15.3 illustrates the procedure in which you can get the information related to a particular nanomaterial with its application and its effect on human and the environment plus other safety issues available at http://www.nanopartikel.info/en/nanoinfo.

Some of the applications of nanoproducts for the protection of ecosystem are briefly discussed here, which include nanopesticides, water remediation and exerting zeolite nanomaterials for the fertilization and treatment of water.

Due to the high reactivity of nanomaterials, they are capable of disintegrating the pollutants present in wastewater; moreover they are more economical compared to the traditional pump-and-treat techniques. The nanomaterials normally used for this purpose are based on iron like, zero-valent iron nanoparticles (Torrey et al., 2015) and nanocomposite Carbo-Iron (Chorawalaa and Mehta, 2015). Furthermore, nanomaterials can be employed to treat the waste water of the plants but in this process, the nanomaterials are isolated from the treated water before discharging into the environment (Hildebrand et al., 2010).

Zeolite nanostructures consist of channels and pores in which nutrients and minerals are accumulated. These structures can be utilized as fertilizers since their nutrients are released slowly, which inhibits extra fertilization. The most important benefit of these nanostructures is their natural occurrence, which reduce the toxicity of the environment and their ability to absorb cadmium considered as a hazard for fish population (Ghiasi et al., 2011).

Applying nanomaterials for encapsulating pesticides results in controlled release so that their hazardous effects are decreased to a considerable amount as they are released on target species. Today, there is not much information about the nanopesticides effect on the environment, thus their risks and benefits should be considered carefully in future studies (Chhipa and Joshi, 2016; Pandey et al., 2016).

Other running projects regarding the toxicological and potential risks of nanomaterials on the environment include CaNTser, DENANA, InhalT-90, NanoBEL, NanoBioDetect, nanoColt, nanoGRAVUR, Nanomobil, Nanosuppe, NanoUmwelt and ProCycle, which are being conducted in Germany.[f]

[f]http://www.nanopartikel.info/en/projects/current-projects

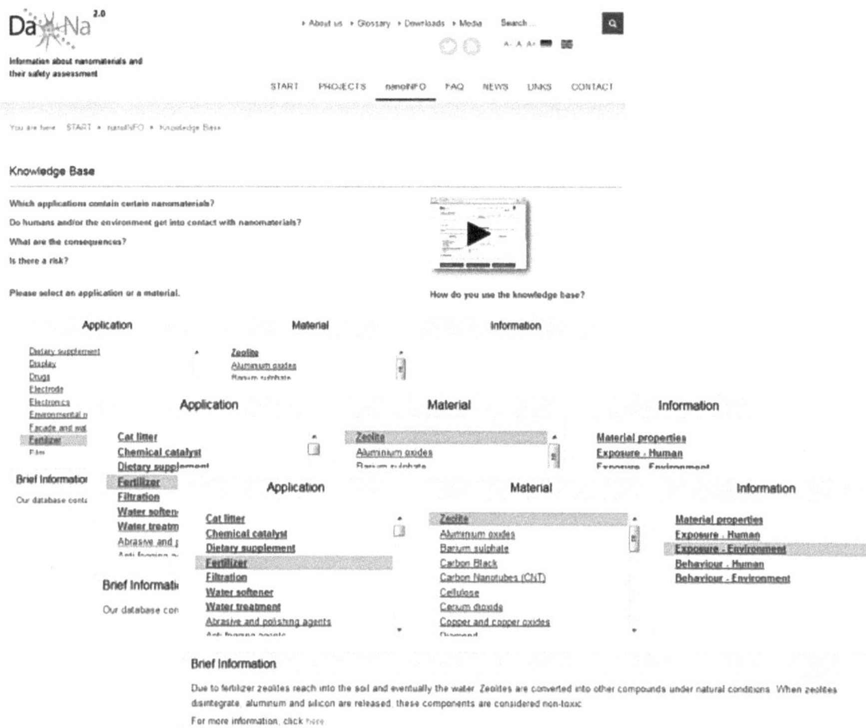

Figure 15.3 An example of obtaining information about a nanomaterial, its application and further information as well as its possible effect on man and the environment. The recent project and information is available at http://www.nanopartikel.info/en/nanoinfo. *Source:* Adapted with permission from Kühnel, D., Marquardt, C., Nau, K., Krug, H., Paul, F., Steinbach, C., 2016. Environmental benefits and concerns on safety: communicating latest results on nanotechnology safety research—the project DaNa2. 0. Environ. Sci. Pollut. Res. 1-6.

15.5 Conclusions

Since 2000 it was predicted that the revenue of nanobased products will reach $1 trillion and this promising technology will lead to entrepreneurship through the employment of workers (Corbett et al., 2000). Currently, well-known food firms, such as Kraft, Unilever, Nestle, and Heinz are investing greatly in this novel technology suggesting that food nanotechnology is a robust candidate to revolutionize the food industry and it is in the stage of infancy.

Nanotechnology is expected to influence the food industry significantly as in the near future it will be implemented to promote health via functional food products, formulate exclusive textures and flavors for food materials, and manufacture food contact materials with high self-sealing properties, present efficient processing technologies and so on.

However, lack of enough information and studies in regard to safety issues of nanobased food products requires case by case studies to determine the potential risks of the nanofood products and packaging materials. Second, the regulations legislated by different organizations and ministries should be developed to a great extent to fill the gaps in food nanotechnology, for instance labeling, risk-benefit evaluation, and ethical issues must be addressed. Furthermore, there is not sufficient scientific information concerning the impacts of nanotechnology on the environment and ecosystem, which seems beneficial to be invested on as the possible hazards may threaten not only the wildlife but also the human. Finally, we need to know the impact of nanosized food and packaging materials on human's general health and immune system via performing animal studies and clinical trials.

References

Aillon, K.L., Xie, Y., El-Gendy, N., Berkland, C.J., Forrest, M.L., 2009. Effects of nanomaterial physicochemical properties on in vivo toxicity. Adv. Drug Deliv. Rev. 61 (6), 457–466.

Akbulut, O., Mace, C.R., Martinez, R.V., Kumar, A.A., Nie, Z., Patton, M.R., Whitesides, G.M., 2012. Separation of nanoparticles in aqueous multiphase systems through centrifugation. Nano Lett. 12 (8), 4060–4064.

Alhakami, A.S., Slovic, P., 1994. A psychological study of the inverse relationship between perceived risk and perceived benefit. Risk Anal. 14 (6), 1085–1096.

Almeida, 2013. Brazil struggles to regulate emerging nanotechnology. Science and technology for development news and analysis website. Available from: http://www.scidev.net/global/technology/feature/brazil-struggles-to-regulate-emerging-nanotechnology.html#.

Anandharamakrishnan, C., 2014. Techniques for Nanoencapsulation of Food Ingredients. Springer, New York.

Arora, S., Rajwade, J.M., Paknikar, K.M., 2012. Nanotoxicology and in vitro studies: the need of the hour. Toxicol. Appl. Pharmacol. 258 (2), 151–165.

Bai, H., Liu, X., 2015. Food nanotechnology and nano food safety, 2015 IEEE Nanotechnology Materials and Devices Conference (NMDC). IEEE, pp. 1–4.

Baltes, M., 2008. Nanotechnology and food safety. Journal de pharmacie de Belgique 63 (1), 7–14.

Barlow, S., Chesson, A., Collins, J., Flynn, A., Hardy, A., Jany, K., Knaap, A., Kuiper, H., Larsen, J., Neindre, P., 2009. The potential risks arising from nanoscience and nanotechnologies on food and feed safety. EFSA J. Eur. Comm. 958, 1–39.

Barpujari, I., 2011. Attenuating risks through regulation: issues for nanotechnology in India. J. Biomed. Nanotechnol. 7 (1), 85–86.

Blasco, C., Picó, Y., 2011. Determining nanomaterials in food. TrAC Trends Anal. Chem. 30 (1), 84–99.

Bodaghi, H., Mostofi, Y., Oromiehie, A., Zamani, Z., Ghanbarzadeh, B., Costa, C., Conte, A., Del Nobile, M.A., 2013. Evaluation of the photocatalytic antimicrobial effects of a TiO_2 nanocomposite food packaging film by in vitro and in vivo tests. LWT-Food Sci. Technol. 50 (2), 702–706.

Bradley, E.L., Castle, L., Chaudhry, Q., 2011. Applications of nanomaterials in food packaging with a consideration of opportunities for developing countries. Trends Food Sci. Technol. 22 (11), 604–610.

Bucheli, T.D., K.K., Gogos, A, 2013. Overview of agricultural applications of 872 nanotechnology. Nanotechnology for the agricultural sector: from research to the 873 field. Workshop organized by JRC IPTS Agrilife Unit, Seville.

Carr, R., Blumenthal, U., Mara, D., 2004. Health guidelines for the use of wastewater in agriculture: developing realistic guidelines. CABI Publishing, UK, pp. 41–58.
Cele, L.M., Coville, N.J., 2009. Nanoscience and Nanotechnology in South Africa.
Chang, C., 2010. The immune effects of naturally occurring and synthetic nanoparticles. J. Autoimmun. 34 (3), J234–J246.
Chau, C.-F., Wu, S.-H., Yen, G.-C., 2007. The development of regulations for food nanotechnology. Trends Food Sci. Technol. 18 (5), 269–280.
Chaudhry, Q., Scotter, M., Blackburn, J., Ross, B., Boxall, A., Castle, L., Aitken, R., Watkins, R., 2008. Applications and implications of nanotechnologies for the food sector. Food Addit. Contam. 25 (3), 241–258.
Chen, G., Shim, S.M., Wen, J., 2014. Potential toxicity of food ingredients loaded in nano-and microparticles. Wiley Blackwell, Chichester, UK, pp. 363–381.
Chhipa, H., Joshi, P., 2016. Nanofertilisers, nanopesticides and nanosensors in agriculture, Nanoscience in Food and Agriculture 1. New Delhi, Springer, 247–282.
Chien, P.-J., Sheu, F., Yang, F.-H., 2007. Effects of edible chitosan coating on quality and shelf life of sliced mango fruit. J. Food Eng. 78 (1), 225–229.
Chorawalaa, K.K., Mehta, M.J., 2015. Applications of nanotechnology in wastewater treatment. Int. J. Innov. Emerg. Res. Eng. 2 (1), 21–26.
Chou, Y.-C., Cheng, Y.-S., Hsu, Y.-H., Yu, Y.-H., Liu, S.-J., 2016. Biodegradable nanofiber-membrane for sustainable release of lidocaine at the femoral fracture site as a periosteal block: in vitro and in vivo studies in a rabbit model. Colloids Surf. B Biointerfaces 140, 332–341.
Clarkson, J.J., Tormala, Z.L., Rucker, D.D., 2008. A new look at the consequences of attitude certainty: the amplification hypothesis. J. Pers. Soc. Psychol. 95 (4), 810.
Coles, D., Frewer, L., 2013. Nanotechnology applied to European food production–a review of ethical and regulatory issues. Trends Food Sci. Technol. 34 (1), 32–43.
Commission, E., 2004. Regulation (EC) No. 1935/2004 of the European Parliament and of the Council of 27 October 2004, on materials and articles intended to come into contact with food and repealing Directives 80/509/EEC and 89/109/EEC. European Commission. Off. J. Eur. Commun. Legis. 338, 4–17.
Communities, C.o.t.E., 2009. No. 450/2009 of 29 May on active and intelligent materials and articles intended to come into contact with food. Off. J. Eur. Union, 3–11.
Cook, A.J., Fairweather, J.R., 2007. Intentions of New Zealanders to purchase lamb or beef made using nanotechnology. Br. Food J. 109 (9), 675–688.
Corbett, J., McKeown, P., Peggs, G., Whatmore, R., 2000. Nanotechnology: international developments and emerging products. CIRP Annals-Manufacturing Technol. 49 (2), 523–545.
Couch, L.M., Wien, M., Brown, J.L., Davidson, P., 2016. Food nanotechnology: proposed uses, safety concerns and regulations. Agro Food Ind. Hi Tech. 27, 1.
Currall, S.C., King, E.B., Lane, N., Madera, J., Turner, S., 2006. What drives public acceptance of nanotechnology? Nat. Nanotechnol. 1 (3), 153–155.
Cushen, M., Kerry, J., Morris, M., Cruz-Romero, M., Cummins, E., 2012. Nanotechnologies in the food industry–recent developments, risks and regulation. Trends Food Sci. Technol. 24 (1), 30–46.
Dandekar, P.P., Jain, R., Patil, S., Dhumal, R., Tiwari, D., Sharma, S., Vanage, G., Patravale, V., 2010. Curcumin-loaded hydrogel nanoparticles: application in anti-malarial therapy and toxicological evaluation. J. Pharm. Sci. 99 (12), 4992–5010.
De Campos, A.M., Diebold, Y., Carvalho, E.L., Sánchez, A., Alonso, M.J., 2004. Chitosan nanoparticles as new ocular drug delivery systems: in vitro stability, in vivo fate, and cellular toxicity. Pharm. Res. 21 (5), 803–810.

Dehalu, V., Weigel, S., Rebe, S., Grombe, R., Löbenberg, R., Delahaut, P., 2012. Production and characterization of antibodies against crosslinked gelatin nanoparticles and first steps toward developing an ELISA screening kit. Anal. Bioanal. Chem. 403 (10), 2851–2857.

Dell'Orco, D., Lundqvist, M., Cedervall, T., Linse, S., 2012. Delivery success rate of engineered nanoparticles in the presence of the protein corona: a systems-level screening. Nanomed. Nanotechnol. Biol. Med. 8 (8), 1271–1281.

des Rieux, A., Fievez, V., Garinot, M., Schneider, Y.-J., Préat, V., 2006. Nanoparticles as potential oral delivery systems of proteins and vaccines: a mechanistic approach. J. Control. Release 116 (1), 1–27.

Dhapte, V., Kadam, S., Pokharkar, V., Khanna, P.K., Dhapte, V., 2014. Versatile SiO_2 nanoparticles@ polymer composites with pragmatic properties. ISRN Inorg. Chem. 2014, 8.

Diaz, B., Sánchez-Espinel, C., Arruebo, M., Faro, J., de Miguel, E., Magadán, S., Yagüe, C., Fernández-Pacheco, R., Ibarra, M.R., Santamaria, J., 2008. Assessing methods for blood cell cytotoxic responses to inorganic nanoparticles and nanoparticle aggregates. Small 4 (11), 2025–2034.

Dingman, J., REHS, D., 2008. Nanotechnology: its impact on food safety. J. Environ. Health 70 (6), 47–50.

Do Kim, K., Han, D.N., Lee, J.B., Kim, H.T., 2006. Formation and characterization of Ag-deposited TiO_2 nanoparticles by chemical reduction method. Scripta materialia 54 (2), 143–146.

DoHA, 2010a. Adjustments to NICNAS new chemicals processes for industrial nanomaterials (NICNAS Information Sheet).

DoHA, 2010b. Official website on nanomaterials Available from: http://www.nicnas.gov.au/Current_Issues/Nanomaterials.asp

Dufort, S., Sancey, L., Coll, J.-L., 2012. Physico-chemical parameters that govern nanoparticles fate also dictate rules for their molecular evolution. Adv. Drug Deliv. Rev. 64 (2), 179–189.

EFSA, 2012. Annual report of the EFSA Scientific Network of Risk Assessment of Nanotechnologies in Food and Feed for 2012. Available from: http://www.efsa.europa.eu/en/search/doc/362e.pdf

EFSA Scientific Committee, 2011. Guidance on the risk assessment of the application of nanoscience and nanotechnologies in the food and feed chain. EFSA J. 9 (5), 2140.

Eguchi, E., Cecato, U., Silva, S., 2013. The way of nanotechnology in Brazil animal production. PUBVET 7 (8).

Eiser, J.R., Miles, S., Frewer, L.J., 2002. Trust, perceived risk, and attitudes toward food technologies1. J. Appl. Soc. Psychol. 32 (11), 2423–2433.

Elsabahy, M., Wooley, K.L., 2013. Cytokines as biomarkers of nanoparticle immunotoxicity. Chem. Soc. Rev. 42 (12), 5552–5576.

European Commission, 2008. Proposal for a Regulation of the European Parliament and of the Council on novel foods and amending Regulation (EC) No. XXX/XXX [common procedure]. COM(2007) 872 Final, 14 January, European Commission, Brussels.

European Commission, 2011. Commission Regulation (EU) No 10/2011 of 14 January 2011 on plastic materials and articles intended to come into contact with food. Off. J. Eur. Comm. 50, 1–89.

European Parliament, C.o.t.E.U., 2011. Regulation (EU) No 1169/2011 of the European Parliament and of the Council of 25 October 2011 on the Provision of Food Information to Consumers, Amending Regulations (EC) No 1924/2006 and (EC) No 1925/2006 of the European Parliament and of the Council, and Repealing Commission Directive 87/250/EEC, Council Directive 90/496/EEC, Commission Directive 1999/10/EC, Directive 2000/13/EC of the European Parliament and of the Council, Commission Directives

2002/67/EC and 2008/5/EC and Commission Regulation (EC) No 608/2004 Text with EEA Relevance. European Parliament, Council of the European Union Brussels, Belgium.

Evans, D., 2002. The search hypothesis of emotion. Br. J. Philos. Sci. 53 (4), 497–509.

FAO/WHO, 2013. State of the art on the initiatives and activities relevant to risk assessment and risk management of nanotechnologies in the food and agriculture sectors, FAO/WHO. Available from: http://www.fao.org/docrep/018/i3281e/i3281e.pdf

FDA, U., 2011. Draft guidance for industry: assessing the effects of significant manufacturing process chances, including emerging technologies, on the safety and regulatory status of food ingredients and food contact substances, including food ingredients that are color additives.

Feng, C., Sun, G., Wang, Z., Cheng, X., Park, H., Cha, D., Kong, M., Chen, X., 2014. Transport mechanism of doxorubicin loaded chitosan based nanogels across intestinal epithelium. Eur. J. Pharm. Biopharm. 87 (1), 197–207.

Fischer, H.C., Chan, W.C., 2007. Nanotoxicity: the growing need for in vivo study. Curr. Opin. Biotechnol. 18 (6), 565–571.

Fischer, A.R., De Jong, A.E., Van Asselt, E.D., De Jonge, R., Frewer, L.J., Nauta, M.J., 2007. Food safety in the domestic environment: an interdisciplinary investigation of microbial hazards during food preparation. Risk Anal. 27 (4), 1065–1082.

Floyd, P., Nwaogu, T., Salado, R., George, C., Vernon, J., 2006. Establishing a comparative inventory of approaches and methods used by enforcement authorities for the assessment of the safety of consumer products covered by the Directive 2001/95/EC on general product safety and identification of best practices. Final Report Prepared for DG/SANCO, European Commission by Risk & Policy Analysts Limited, United Kingdom.

Frewer, L., Fischer, A., 2010. The Evolution of Food Technology, Novel Foods, and the Psychology of Novel Food 'Acceptance'. Nanotechnologies in Food. RSC Publishing, Cambridge, 18–35.

Frewer, L., Salter, B., 2002. Public attitudes, scientific advice and the politics of regulatory policy: the case of BSE. Sci. Public Policy 29 (2), 137–145.

Frewer, L.J., Howard, C., Hedderley, D., Shepherd, R., 1996. What determines trust in information about food-related risks? Underlying psychological constructs. Risk Anal. 16 (4), 473–486.

Frewer, L.J., Howard, C., Shepherd, R., 1998. The influence of initial attitudes on responses to communication about genetic engineering in food production. Agric. Human Values 15 (1), 15–30.

Frewer, L.J., Scholderer, J., Bredahl, L., 2003. Communicating about the risks and benefits of genetically modified foods: The mediating role of trust. Risk Anal. 23 (6), 1117–1133.

Friege, H., 2005. REACH—registration, evaluation and authorisation of chemicals. Umweltwissenschaften und Schadstoff-Forschung 17 (3), 184–1184.

FSCJ, 2010. A Basic Survey Report on Safety Assessment Information on the Use of Nanotechnology in the Food Sector, Conducted by the Toray Research Certer as 'A comprehensive study for ensuring food safety under the Cabinet Office Food Safety Commission (FY 2009)' for the Food Safety Commission of Japan. Available from: http://www.fsc.go.jp/english/survey_programs/use_nanotechnology.pdf

Garner, K.L., Keller, A.A., 2014. Emerging patterns for engineered nanomaterials in the environment: a review of fate and toxicity studies. J. Nanoparticle Res. 16 (8), 1–28.

Gaskell, G., Ten Eyck, T., Jackson, J., Veltri, G., 2005. Imagining nanotechnology: cultural support for technological innovation in Europe and the United States. Public Underst. Sci. 14 (1), 81–90.

Georgantelis, D., Blekas, G., Katikou, P., Ambrosiadis, I., Fletouris, D.J., 2007. Effect of rosemary extract, chitosan and α-tocopherol on lipid oxidation and colour stability during frozen storage of beef burgers. Meat Sci. 75 (2), 256–264.

Geornaras, I., Sofos, J.N., 2005. Activity of ε–polylysine against *Escherichia coli* O157: H7, *Salmonella Typhimurium, and Listeria monocytogenes*. J. Food Sci. 70 (9), M404–M408.

Ghiasi, F., Mirzargar, S., Badakhshan, H., Amoli, J., 2011. Influence of Iranian natural zeolite on accumulation of cadmium in cyprinus carpio tissues following exposure to low concentration of cadmium. Asian J. Anim. Veter. Adv. 6 (6), 636–641.

Giese, C., Marx, U., 2014. Human immunity in vitro—solving immunogenicity and more. Adv. Drug Deliv. Rev. 69, 103–122.

Gray, H.C., Hutcheson, P.S., Slavin, R.G., 2004. Is glucosamine safe in patients with seafood allergy? J. Allergy Clin. Immunol. 114 (2), 459–460.

Griffin, G.L., 1996. The TSCA Compliance Handbook. John Wiley & Sons, New York.

Gruère, G.P., 2012. Implications of nanotechnology growth in food and agriculture in OECD countries. Food Policy 37 (2), 191–198.

Gulliford, J.B., 2008. Toxic Substances Control Act Inventory Status of Carbon Nanotubes. Environmental Protection Agency, Washington, DC.

Hamburg, M.A., 2012. FDA's approach to regulation of products of nanotechnology. Science 336 (6079), 299–300.

Hamzeh, M., Sunahara, G.I., 2013. In vitro cytotoxicity and genotoxicity studies of titanium dioxide (TiO_2) nanoparticles in Chinese hamster lung fibroblast cells. Toxicol. In Vitro 27 (2), 864–873.

Handford, C.E., Dean, M., Henchion, M., Spence, M., Elliott, C.T., Campbell, K., 2014. Implications of nanotechnology for the agri-food industry: opportunities, benefits and risks. Trends Food Sci. Technol. 40 (2), 226–241.

Hildebrand, H., Kühnel, D., Potthoff, A., Mackenzie, K., Springer, A., Schirmer, K., 2010. Evaluating the cytotoxicity of palladium/magnetite nano-catalysts intended for wastewater treatment. Environ. Pollut. 158 (1), 65–73.

Hobson, D.W., 2009. Commercialization of nanotechnology. Wiley Interdiscip. Rev. Nanomed. Nanobiotechnol. 1 (2), 189–202.

Höck, J., Hofmann, H., Krug, H., Lorenz, C., Limbach, L., Nowack, B., Riediker, M., Schirmer, K., Som, C., Stark, W., 2008. Precautionary matrix for synthetic nanomaterials. Federal Office for Public Health and Federal Office for the Environment, Berne 2008. There is no corresponding record for this reference.

Hoshino, A., Fujioka, K., Oku, T., Nakamura, S., Suga, M., Yamaguchi, Y., Suzuki, K., Yasuhara, M., Yamamoto, K., 2004. Quantum dots targeted to the assigned organelle in living cells. Microbiol. Immunol. 48 (12), 985–994.

Huang, Y., Chen, S., Bing, X., Gao, C., Wang, T., Yuan, B., 2011. Nanosilver migrated into food-simulating solutions from commercially available food fresh containers. Packag. Technol. Sci. 24 (5), 291–297.

Hunter, A.C., Elsom, J., Wibroe, P.P., Moghimi, S.M., 2012. Polymeric particulate technologies for oral drug delivery and targeting: a pathophysiological perspective. Nanomed. Nanotechnol. Biol. Med. 8, S5–S20.

ISO, 2013. External Liaiason report Asia Nano Forum to ISO/TC 229.

Jo, C., Lee, J., Lee, K., Byun, M., 2001. Quality properties of pork sausage prepared with water-soluble chitosan oligomer. Meat Sci. 59 (4), 369–375.

Karakoti, A., Hench, L., Seal, S., 2006. The potential toxicity of nanomaterials—the role of surfaces. Jom 58 (7), 77–82.

Katouzian, I., Jafari, S.M., 2016. Nano-encapsulation as a promising approach for targeted delivery and controlled release of vitamins. Trends Food Sci. Technol. 53, 34–48.
Keshwani, A., Malhotra, B., Kharkwal, H., 2015. Advancements of nanotechnology in food packaging. World J. Pharm. Pharm. Sci. 4 (4), 1054–1057.
Kim, H.K., Lee, J.G., Lee, S.Y., 2014. Current regulation of nanomaterials used as food ingredients. John Wiley and Sons, pp. 368–394.
Kornelis, M., De Jonge, J., Frewer, L., Dagevos, H., 2007. Consumer selection of food-safety information sources. Risk Anal. 27 (2), 327–335.
Kowalczyk, B., Lagzi, I., Grzybowski, B.A., 2011. Nanoseparations: strategies for size and/or shape-selective purification of nanoparticles. Curr. Opin. Colloid Interface Sci. 16 (2), 135–148.
Krishna, V., Pumprueg, S., Lee, S.-H., Zhao, J., Sigmund, W., Koopman, B., Moudgil, B., 2005. Photocatalytic disinfection with titanium dioxide coated multi-wall carbon nanotubes. Process Saf. Environ. Prot. 83 (4), 393–397.
Kunhikrishnan, A., Shon, H.K., Bolan, N.S., El Saliby, I., Vigneswaran, S., 2015. Sources, distribution, environmental fate, and ecological effects of nanomaterials in wastewater streams. Crit. Rev. Environ. Sci. Technol. 45 (4), 277–318.
Kyle, R., Dodds, S., 2009. Avoiding empty rhetoric: Engaging publics in debates about nanotechnologies. Sci. Eng. Ethics 15 (1), 81–96.
Lai, D.Y., 2012. Toward toxicity testing of nanomaterials in the 21st century: a paradigm for moving forward. Wiley Interdiscip. Rev. Nanomed. Nanobiotechnol. 4 (1), 1–15.
Law, G.F., 2002. Regulation (EC) No. 178/2002 of the European Parliament and of the Council of 28 January 2002, laying down the general principles and requirements of food law, establishing the European Food Safety Authority, and laying down procedures in matters of food safety. OJ L 31(1.2).
Lee, C.-J., Scheufele, D.A., Lewenstein, B.V., 2005. Public attitudes toward emerging technologies examining the interactive effects of cognitions and affect on public attitudes toward nanotechnology. Sci. Commun. 27 (2), 240–267.
Lee, J., Ko, S., Kim, H., Kwon, H., 2011. Integrity and cell-monolayer permeability of chitosan nanoparticles in simulated gastrointestinal fluids. Food Sci. Biotechnol. 20 (4), 1033–1042.
Leonard, F., Ali, H., Collnot, E.-M., Crielaard, B.J., Lammers, T., Storm, G., Lehr, C.-M., 2012. Screening of budesonide nanoformulations for treatment of inflammatory bowel disease in an inflamed 3D cell-culture model. ALTEX 29 (3), 275.
Lerner, J.S., Keltner, D., 2000. Beyond valence: toward a model of emotion-specific influences on judgement and choice. Cognit. Emotion 14 (4), 473–493.
Lindzey, G., Gilbert, D., Fiske, S.T., 1998. The Handbook of Social Psychology. Oxford University Press, Oxford.
Linsinger, T., Chaudhry, Q., Dehalu, V., Delahaut, P., Dudkiewicz, A., Grombe, R., Von der Kammer, F., Larsen, E.H., Legros, S., Löschner, K., 2013. Validation of methods for the detection and quantification of engineered nanoparticles in food. Food Chem. 138 (2), 1959–1966.
Liu, L., Tang, Y., Gao, C., Li, Y., Chen, S., Xiong, T., Li, J., Du, M., Gong, Z., Chen, H., 2014. Characterization and biodistribution in vivo of quercetin-loaded cationic nanostructured lipid carriers. Colloids Surf. B Biointerfaces 115, 125–131.
Loh, J.W., Saunders, M., Lim, L.-Y., 2012. Cytotoxicity of monodispersed chitosan nanoparticles against the Caco-2 cells. Toxicol. Appl. Pharmacol. 262 (3), 273–282.
Lozano, T., Rey, M., Rojas, E., Moya, S., Fleddermann, J., Estrela-Lopis, I., Donath, E., Wang, B., Mao, Z., Gao, C., 2011. Cytotoxicity effects of metal oxide nanoparticles in human tumor cell lines. J. Phys. Conf. Ser. 304 (1), 012046.

Macoubrie, J., 2004. Public perceptions about nanotechnology: risks, benefits and trust. J. Nanopart. Res. 6 (4), 395–405.
Mahmoudi, M., Meng, J., Xue, X., Liang, X.J., Rahman, M., Pfeiffer, C., Hartmann, R., Gil, P.R., Pelaz, B., Parak, W.J., 2014. Interaction of stable colloidal nanoparticles with cellular membranes. Biotechnol. Adv. 32 (4), 679–692.
Malaekeh-Nikouei, B., Mousavi, S.H., Shahsavand, S., Mehri, S., Nassirli, H., Moallem, S.A., 2013. Assessment of cytotoxic properties of safranal and nanoliposomal safranal in various cancer cell lines. Phytother. Res. 27 (12), 1868–1873.
Manke, A., Wang, L., Rojanasakul, Y., 2013. Mechanisms of nanoparticle-induced oxidative stress and toxicity. BioMed Res. Int. 2013, 15.
Maurer, D., Sobal, J. (Eds.), 1995. Eating Agendas: Food and Nutrition as Social Problems. Hawthorne, Aldine de Gruyter.
Meijnders, A.L., Midden, C.J., Wilke, H.A., 2001. Role of negative emotion in communication about CO_2 risks. Risk Anal. 21 (5), 955–1955.
MFDS, K., 2012a. Definition of nanotechnology: applied foods, [In Korean]. Available from: www.mfds.go.kr
MFDS, K., 2012b. Industry's voluntary safety assessment guideline on nanotechnology: applied foods. [In Korean]. Available from: www.mfds.go.kr
Miller, T., Rachel, R., Besheer, A., Uezguen, S., Weigandt, M., Goepferich, A., 2012. Comparative investigations on in vitro serum stability of polymeric micelle formulations. Pharm. Res. 29 (2), 448–459.
Momin, J.K., Joshi, B., 2015. Nanotechnology in Foods, Nanotechnologies in Food and Agriculture. Springer, pp. 3–24.
Momin, J.K., Jayakumar, C., Prajapati, J.B., 2013. Potential of nanotechnology in functional foods. Emirates J. Food. Agric. 25 (1), 10.
Moros, M., Hernáez, B., Garet, E., Dias, J.T., Sáez, B., Grazú, V., González-Fernández, Á., Alonso, C., de la Fuente, J.s.M., 2012. Monosaccharides versus PEG-functionalized NPs: influence in the cellular uptake. ACS Nano 6 (2), 1565–1577.
Mwilu, S.K., El Badawy, A.M., Bradham, K., Nelson, C., Thomas, D., Scheckel, K.G., Tolaymat, T., Ma, L., Rogers, K.R., 2013. Changes in silver nanoparticles exposed to human synthetic stomach fluid: effects of particle size and surface chemistry. Sci. Total Environ. 447, 90–98.
NanoMalaysia, 2013. Towards a Malaysian Regulatory Framework, Asia Nano Forum, Nanotech 2013, Japan 31st January, (power point presentation). Available from: http://www.google.com/url?sa=t&rct=j&q=&esrc=s&frm=1&source=web&cd=1&ved=0CB8QFjAA&url=http%3A%2F%2Firannano.org%2Ffilereader.php%3Fp1%3Dmain_47e24b47020343c728d44a00681a79b5.pdf%26p2%3Dstatic_page%26p3%3D22%26p4%3D1&ei=QbAFVNT8PMriatPlgsAF&usg=AFQjCNFA69SOMsrseyu1NemcbZYc8Jxe1Q&sig2=9UloR7UCc4DmKnrRw_hDUw&bvm=bv.74115972,d.d2s
Nauta, M.J., Fischer, A.R., Van Asselt, E.D., De Jong, A.E., Frewer, L.J., De Jonge, R., 2008. Food safety in the domestic environment: the effect of consumer risk information on human disease risks. Risk Anal. 28 (1), 179–192.
Neethirajan, S., Jayas, D.S., 2011. Nanotechnology for the food and bioprocessing industries. Food Bioprocess Technol. 4 (1), 39–47.
NIA, 2011. Opportunities & risks in nanomedicine how to build an infrastructure in support of the next generation of health care. 5° ENIFarMed; São Paulo; 29–31 August 2011. Brussels, Belgium:Nanotechnology Industries Association.
Njongmeta, N., Benli, H., Dunkley, K., Dunkley, C., Miller, D., Anderson, R., O'BRYAN, C., Keeton, J., Nisbet, D., Crandall, P., 2011. Application of acidic calcium sulfate and ε-polylysine to pre-rigor beef rounds for reduction of pathogens. J. Food Saf. 31 (3), 395–400.

Nowack, B., Ranville, J.F., Diamond, S., Gallego-Urrea, J.A., Metcalfe, C., Rose, J., Horne, N., Koelmans, A.A., Klaine, S.J., 2012. Potential scenarios for nanomaterial release and subsequent alteration in the environment. Environ. Toxicol. Chem. 31 (1), 50–59.
Oatley, K., Johnson-Laird, P.N., 1987. Towards a cognitive theory of emotions. Cognit. Emotion 1 (1), 29–50.
Oberdörster, G., Maynard, A., Donaldson, K., Castranova, V., Fitzpatrick, J., Ausman, K., Carter, J., Karn, B., Kreyling, W., Lai, D., 2005. Principles for characterizing the potential human health effects from exposure to nanomaterials: elements of a screening strategy. Part. Fibre Toxicol. 2 (1), 1.
OECD, 2013. Regulatory frameworks for nanotechnology in foods and medical products. Summary results of a survey activity, Paris. Available from: http://search.oecd.org/officialdocuments/publicdisplaydocumentpdf/?cote=DSTI/STP/NANO(2012)22/FINAL&docLanguage=En
Office of Agricultural Affairs of the USDA, 2009. Russian Federation—Food and Agricultural Import Regulations and Standards—FAIRS Country Report, Moscow. Available from: http://gain.fas.usda.gov/Recent%20GAIN%20Publications/Food%20and%20Agricultural%20Import%20Regulations%20and%20Standards%20-%20Narrative_Moscow_Russian%20Federation_11.08.2009.pdf
Pandey, S., Giri, K., Kumar, R., Mishra, G., Rishi, R.R., 2016. Nanopesticides: Opportunities in Crop Protection and Associated Environmental Risks. Proceedings of the National Academy of Sciences, India Section B: Biological Sciences, Springer, pp. 1–22.
Parry, M.A., Hawkesford, M.J., 2010. Food security: increasing yield and improving resource use efficiency. Proc. Nutr. Soc. 69 (04), 592–600.
Peng, Q., Zhang, S., Yang, Q., Zhang, T., Wei, X.-Q., Jiang, L., Zhang, C.-L., Chen, Q.-M., Zhang, Z.-R., Lin, Y.-F., 2013. Preformed albumin corona, a protective coating for nanoparticles based drug delivery system. Biomaterials 34 (33), 8521–8530.
Peng, Q., Wei, X.-Q., Shao, X.-R., Zhang, T., Zhang, S., Fu, N., Cai, X.-X., Zhang, Z.-R., Lin, Y.-F., 2014. Nanocomplex based on biocompatible phospholipids and albumin for long-circulation applications. ACS Appl. Mater. Interfaces 6 (16), 13730–13737.
Peters, R., ten Dam, G., Bouwmeester, H., Helsper, H., Allmaier, G., vd Kammer, F., Ramsch, R., Solans, C., Tomaniová, M., Hajslova, J., 2011. Identification and characterization of organic nanoparticles in food. Trends Anal. Chem. 30 (1), 100–112.
Pidgeon, N., Kasperson, R.E., Slovic, P., 2003. The Social Amplification of Risk. Cambridge University Press, Cambridge.
Piegorsch, W.W., Schuler, E., 2008. Communicating the risks, and the benefits, of nanotechnology. Int. J. Risk Assess. Manag. 10 (1–2), 57–69.
Pinheiro, A.C., Lad, M., Silva, H.D., Coimbra, M.A., Boland, M., Vicente, A.A., 2013. Unravelling the behaviour of curcumin nanoemulsions during in vitro digestion: effect of the surface charge. Soft Matter 9 (11), 3147–3154.
Pomerantz, E.M., Chaiken, S., Tordesillas, R.S., 1995. Attitude strength and resistance processes. J. Pers. Social Psychol. 69 (3), 408.
Poortinga, W., Pidgeon, N.F., 2005. Trust in risk regulation: cause or consequence of the acceptability of GM food? Risk Anal. 25 (1), 199–209.
Pornwilard, M., Somchue, W., Shiowatana, J., Siripinyanond, A., 2014. Flow field-flow fractionation for particle size characterization of selenium nanoparticles incubated in gastrointestinal conditions. Food Res. Int. 57, 203–209.
Regulation, H.A.T., 1997. Regulation (EC) No 258/97 of the European Parliament and of the Council of 27 January 1997 concerning novel foods and novel food ingredients. Off. J. L 43 (14/02), 0001–0006.

Regulation Council, 2002. No 178/2002 of the European Parliament and of the Council of 28 January 2002: laying down the general principles and requirements of food law, establishing the European Food Safety Authority and laying down procedures in matters of food safety. Off. J. Eur. Commun., 1–24.

Risk Assessment Section, 2010. C.f.F.S., Food and Environmental Hygiene Department. Nanotechnology and Food Safety, 41, Hong Kong.

Rodríguez-Entrena, M., Salazar-Ordóñez, M., Sayadi, S., 2013. Applying partial least squares to model genetically modified food purchase intentions in southern Spain consumers. Food Policy 40, 44–53.

Roy, K., Mao, H.-Q., Huang, S.-K., Leong, K.W., 1999. Oral gene delivery with chitosan–DNA nanoparticles generates immunologic protection in a murine model of peanut allergy. Nature Med. 5 (4), 387–391.

Ruh, H., Kühl, B., Brenner-Weiss, G., Hopf, C., Diabaté, S., Weiss, C., 2012. Identification of serum proteins bound to industrial nanomaterials. Toxicol. Lett. 208 (1), 41–50.

Ruiter, R.A., Abraham, C., Kok, G., 2001. Scary warnings and rational precautions: a review of the psychology of fear appeals. Psychol. Health 16 (6), 613–630.

Sahoo, S., Parveen, S., Panda, J., 2007. The present and future of nanotechnology in human health care. Nanomed. Nanotechnol. Biol. Med. 3 (1), 20–31.

Sastry, R.K., Rashmi, H., Rao, N., 2011. Nanotechnology for enhancing food security in India. Food Policy 36 (3), 391–400.

Satterfield, T., Kandlikar, M., Beaudrie, C.E., Conti, J., Harthorn, B.H., 2009. Anticipating the perceived risk of nanotechnologies. Nature Nanotechnol. 4 (11), 752–758.

Sayes, C.M., Warheit, D.B., 2009. Characterization of nanomaterials for toxicity assessment. Wiley Interdiscip. Rev. Nanomed. Nanobiotechnol. 1 (6), 660–670.

Scheufele, D.A., Corley, E.A., Shih, T.-j., Dalrymple, K.E., Ho, S.S., 2009. Religious beliefs and public attitudes toward nanotechnology in Europe and the United States. Nature Nanotechnol. 4 (2), 91–94.

Schweizerischer-Bundesrat, 2010. Verordnung über das Inverkehrbringen von 1161 Pflanzenschutzmitteln (Pflanzenschutzmittelverordnung, PSMV). 916.161. F. O. 1162 o. P. Health.

SECO, S.S.f.E.A., 2012. Safety data Sheet (SDS): Guidelines for Synthetic Nanomaterials.

Semete, B., Booysen, L., Kalombo, L., Venter, J.D., Katata, L., Ramalapa, B., Verschoor, J.A., Swai, H., 2010. In vivo uptake and acute immune response to orally administered chitosan and PEG coated PLGA nanoparticles. Toxicol. Appl. Pharmacol. 249 (2), 158–165.

Sharifi, S., Behzadi, S., Laurent, S., Forrest, M.L., Stroeve, P., Mahmoudi, M., 2012. Toxicity of nanomaterials. Chem. Soc. Rev. 41 (6), 2323–2343.

Sharma, K., Chugh, A., 2009. Legal aspects of nanobiotechnology inventions: an indian perspective. J. Law Tech. Soc. 6 (2), 433–448.

Shih, L., Shen, M.-H., Van, Y.-T., 2006. Microbial synthesis of poly (ε-lysine) and its various applications. Bioresour. Technol. 97 (9), 1148–1159.

Shima, S., Matsuoka, H., Iwamoto, T., Sakai, H., 1984. Antimicrobial action of, EPSILON-poly-L-lysine. J. Antibiot. 37 (11), 1449–1455.

Siegrist, M., Cousin, M.-E., Kastenholz, H., Wiek, A., 2007. Public acceptance of nanotechnology foods and food packaging: the influence of affect and trust. Appetite 49 (2), 459–466.

Silva, T., 2016. Challenges and opportunities of nanotechnology and other technological advances for the health and environment, Third ISA Forum of Sociology, July 10–14, 2016. Isaconf.

Simón-Vázquez, R., Lozano-Fernández, T., Peleteiro-Olmedo, M., González-Fernández, Á., 2014. Conformational changes in human plasma proteins induced by metal oxide nanoparticles. Colloids Surf. B Biointerfaces 113, 198–206.

Slovic, P., Finucane, M.L., Peters, E., MacGregor, D.G., 2004. Risk as analysis and risk as feelings: some thoughts about affect, reason, risk, and rationality. Risk Anal. 24 (2), 311–322.
So, S.J., Jang, I.S., Han, C.S., 2008. Effect of micro/nano silica particle feeding for mice. J. Nanosci. Nanotechnol. 8 (10), 5367–5371.
Spence, A., Townsend, E., 2006. Implicit attitudes towards genetically modified (GM) foods: a comparison of context-free and context-dependent evaluations. Appetite 46 (1), 67–74.
Stationaryoffice, 2010a. Government Response to the Lord's Science and Technology Select Committee Report into Nanotechnologies and Food. Available from: http://go.nature.com/KApSYN
StationaryOffice, 2010b. House of Lords Science and Technology Committee Nanotechnologies and Food Volume 1. Available from: http://go.nature.com/cHRZfV
Stephens, L.F., 2005. News narratives about nano S&T in major US and non-US newspapers. Sci. Commun. 27 (2), 175–199.
Tager, J., 2015. Pre-market testing of nanomaterials in food is both practical and necessary. Nature Nanotechnol. 10 (4), 285–286.
Thangavel, G., Thiruvengadam, S., 2014. Nanotechnology in food industry–a review. Int. J. Chem. Tech. Res. 16 (9), 4096–4101.
Tilman, D., Balzer, C., Hill, J., Befort, B.L., 2011. Global food demand and the sustainable intensification of agriculture. Proc. Natl. Acad. Sci. 108 (50), 20260–20264.
Torrey, J.D., Killgore, J.P., Bedford, N.M., Greenlee, L.F., 2015. Oxidation behavior of zero-valent iron nanoparticles in mixed matrix water purification membranes. Environ. Sci. Water Res. Technol. 1 (2), 146–152.
TSCA, 2010. Statute, Regulations & Enforcement. Available from: www.epa.gov/compliance/civil/tsca/tscaenfstatreq.html
TSCA, 2011a. Chemical Substance Inventory. Available from: www.epa.gov/opptintr/existingchemicals/pubs/tscainventory/index.html
TSCA, 2011b. Section 5 Significant New Use Rules. Available from: www.epa.gov/oppt/existingchemicals/pubs/sect5a2.html
Tsukamoto, K., Wakayama, J., Sugiyama, S., 2010. Nanobiotechnology approach for food and food related fields, Poster presented at the International Conference on Food Applications of Nanoscale Science (ICOFANS), Tokyo, Japan, pp. 9–11.
van Dijk, H., Fischer, A.R., de Jonge, J., Rowe, G., Frewer, L.J., 2012. The impact of balanced risk–benefit information and initial attitudes on post-information attitudes1. J. Appl. Soc. Psychol. 42 (8), 1958–1983.
Van Kleef, E., Houghton, J., Krystallis, A., Pfenning, U., Rowe, G., Van Dijk, H., Van der Lans, I., Frewer, L.J., 2007. Consumer evaluations of food risk management quality in Europe. Risk Anal. 27 (6), 1565–1580.
Vecchione, R., Quagliariello, V., Calabria, D., Calcagno, V., De Luca, E., Iaffaioli, R.V., Netti, P.A., 2016. Curcumin bioavailability from oil in water nano-emulsions: In vitro and in vivo study on the dimensional, compositional and interactional dependence. J. Control. Release 233, 88–100.
Verdurme, A., Viaene, J., 2003. Consumer attitudes towards GM food: Literature review and recommendations for effective communication. J. Int. Food Agribus. Mark. 13 (2-3), 77–98.
Verma, A., Stellacci, F., 2010. Effect of surface properties on nanoparticle–cell interactions. Small 6 (1), 12–21.
Wang, Y., Ma, Y., Zheng, Y., Song, J., Yang, X., Bi, C., Zhang, D., Zhang, Q., 2013. In vitro and in vivo anticancer activity of a novel puerarin nanosuspension against colon cancer, with high efficacy and low toxicity. Int. J. Pharm. 441 (1), 728–735.

Warheit, D.B., Donner, E.M., 2016. How meaningful are risk determinations in the absence of a complete dataset? Making the case for publishing standardized test guideline and 'no effect'studies for evaluating the safety of nanoparticulates versus spurious 'high effect'results from single investigative studies. Sci. Technol. Adv. Mater. 16 (3), 034603.

Wei, X., Lee, Y.K., Huh, K.M., Kim, S., Park, K., 2009. Safety and efficacy of nano/micro materials. In: Webster, T.J. (Ed.), Safety of Nanoparticles: From Manufacturing to Medical Applications. Springer, New York, NY, pp. 63–88.

Wire, B., 2014. Research and Markets: Global Functional Food and Nutraceuticals Market 2013-2018-Analysis of the $175 Billion Industry, publicado on-line em14 de abril de 2014.

Wolfram, J., Yang, Y., Shen, J., Moten, A., Chen, C., Shen, H., Ferrari, M., Zhao, Y., 2014. The nano-plasma interface: implications of the protein corona. Colloids Surf. B Biointerfaces 124, 17–24.

World Health Organization, 2003. Guidelines for Safe Recreational Water Environment, vol. 1, Coastal and Fresh Waters, World Health Organization, Geneva.

World Health Organization, 2013. State of the art on the initiatives and activities relevant to risk assessment and risk management of nanotechnologies in the food and agriculture sectors: FAO/WHO technical paper report. Rome, Italy: FAO/WHO. p. 48.

Yaktine, A., Pray, L., 2009. Nanotechnology in Food Products: Workshop Summary. National Academies Press, Washington, DC, USA.

Yan, L., Gu, Z., Zhao, Y., 2013. Chemical mechanisms of the toxicological properties of nanomaterials: generation of intracellular reactive oxygen species. Chem. Asian J. 8 (10), 2342–2353.

Yang, K., Gong, H., Shi, X., Wan, J., Zhang, Y., Liu, Z., 2013. In vivo biodistribution and toxicology of functionalized nano-graphene oxide in mice after oral and intraperitoneal administration. Biomaterials 34 (11), 2787–2795.

Zattoni, A., Roda, B., Borghi, F., Marassi, V., Reschiglian, P., 2014. Flow field-flow fractionation for the analysis of nanoparticles used in drug delivery. J. Pharm. Biomed. Anal. 87, 53–61.

Zhang, J., Liang, L., Tian, Z., Chen, L., Subirade, M., 2012. Preparation and in vitro evaluation of calcium-induced soy protein isolate nanoparticles and their formation mechanism study. Food Chem. 133 (2), 390–399.

Zolnik, B.S., González-Fernández, Á., Sadrieh, N., Dobrovolskaia, M.A., 2010. Minireview: nanoparticles and the immune system. Endocrinology 151 (2), 458–465.

Index

E

MIX
Papier | Fördert gute Waldnutzung
FSC® C083411

Kolibri 360 GmbH
Ferdinand-Jühlke-Straße 7
99095 Erfurt, Deutschland
produktsicherheit@kolibri360.de